STUDY GUIDE
FOR
CAMPBELL BIOLOGY

STUDY GUIDE
FOR
CAMPBELL BIOLOGY

JANE B. REECE • LISA A. URRY • MICHAEL L. CAIN
STEVEN A. WASSERMAN • PETER V. MINORSKY
ROBERT B. JACKSON

MARTHA R. TAYLOR
Cornell University

NINTH EDITION

Benjamin Cummings

Boston   Columbus   Indianapolis   New York   San Francisco   Upper Saddle River
Amsterdam   Cape Town   Dubai   London   Madrid   Milan   Munich   Paris   Montréal   Toronto
Delhi   Mexico City   São Paulo   Sydney   Hong Kong   Seoul   Singapore   Taipei   Tokyo

Vice President/Editor-in-Chief: Beth Wilbur
Acquisitions Editor: Josh Frost
Senior Editorial Manager: Ginnie Simione Jutson
Senior Supplements Project Editor: Susan Berge
Executive Marketing Manager: Lauren Harp
Managing Editor, Production: Michael Early
Production Project Manager: Jane Brundage
Photo Editor: Donna Kalal
Image Permissions Coordinator: Elaine Soares
Manufacturing Buyer: Michael Penne
Production Management and Composition: S4Carlisle, Lori Bradshaw
Cover Production: Seventeenth Street Studios
Text and Cover Printer: Bind-Rite, Robbinsville

Cover Photo Credit: "Succulent I" ©2005 Amy Lamb, www.amylamb.com

ISBN-13: 978-0-321-62992-0
ISBN-10: 0-321-62992-2

Benjamin Cummings
is an imprint of

www.pearsonhighered.com   2 3 4 5 6 7 8 9 10—BRR—14 13 12 11

Dedication

To my dad, Kenneth E. Taylor, who still shares his love of learning with me.

Contents

Credits

Figures IQ 3.1, IQ 5.3, IQ 5.8, IQ 6.4, SYK 6.3, IQ 7.1, IQ 7.5, SYK 7.2, IQ 8.5, IQ 8.6, IQ 9.5, IQ 9.7, IQ 9.8, IQ 10.1, IQ 10.2, IQ 10.5, IQ 10.8, IQ 12.3, IQ 13.3, IQ 16.2, IQ 16.5, IQ 17.6, IQ 18.4, IQ 19.1, IQ 20.3, SYK 25.2, IQ 26.1, SYK 26.2, IQ 28.3, SYK 28.1, IQ 29.5, SYK 29.2, IQ 30.5, SYK 30.3, IQ 31.4, IQ 33.5, SYK 33.1, IQ 34.1, IQ 34.3, IQ 34.8, IQ 35.1, IQ 35.4, IQ 35.5, SYK 35.2, IQ 36.3, IQ 36.6, IQ 37.5, IQ 38.1, IQ 38.4, IQ 40.2, IQ 40.3, IQ 40.4, SYK 41.1, IQ 42.3, SYK 42.1, IQ 43.5, IQ 44.5, IQ 45.1, IQ 46.3, IQ 46.5, IQ 47.2, IQ 47.3, IQ 47.4, IQ 47.6, IQ 48.1, IQ 48.3, IQ 48.4, IQ 49.3, IQ 50.3, IQ 50.5, IQ 50.8, IQ 52.2, IQ 52.4, IQ 53.2, IQ 53.4, IQ 54.9 : Adapted from Reece, J. B., et al., *Campbell Biology*, 9th ed. (San Francisco, CA: Benjamin Cummings, 2011) © 2011 Pearson Education Inc., publishing as Benjamin Cummings.

Photos [12.03, page 87] USDA/ARS/Agricultural Research Service; [27.02, page 199 - top] Biological Photo Service; [27.03, page 199 - bottom] Susan M. Barns, Ph.D.; [35.02, page 255 - top] Ed Reschke; [35.02, page 255 - bottom] Natalie Bronstein; [40.02, page 293 - top] Nina Zanetti; [40.02, page 293 - middle] Nina Zanetti; [40.02, page 293 - bottom] Dr. Gopal Murti/SPL/Photo Researchers; [40.03a, page 293] Manfred Kage/Peter Arnold/Photolibrary; [40.03b, page 293] Carolina Biological Supply Company/Alamy.

Another name for the *Student Study Guide* for *Campbell Biology*, Ninth Edition, could be "A Student Structuring Guide." This guide is designed to help you structure and organize your developing knowledge of biology and create your own personal understanding of the topics covered in the text. Take a few minutes to learn about some strategies of successful students, as well as some strategies for using this study guide.

Strategies of Successful Students

What does it take to be a successful student? How can you learn more efficiently and earn better grades? And how can you enjoy your biology course more?

Every successful student tackles learning in his or her own unique way. But there are commonalities that can be shared and results from educational research that can inform you as you develop your own approach. Here is a short list of suggestions that may help.

Make it interesting. Biology is fascinating and so much easier to learn when you love it. Try to bring a sense of wonder to your classes and your studying. Appreciate how much has been discovered and how much we have yet to learn about the living world.

Focus in lecture. What does it mean to focus? Obviously sleeping or texting are out. But so is taking down every word your professor says without thinking. If a lecture outline is provided, take notes on it. Or try taking notes on one page of a notebook and then on the facing page rework, elaborate, or at least make legible what you wrote down in class. Doing this shortly after class helps you review the material while it is fresh and develop an organized set of notes to use in studying for exams.

Make reading active. Research has shown that simply rereading material does not improve test scores, but interacting with the reading assignment does. How do you interact with the printed word? Ask yourself questions and look for answers. Reword every concept heading as a question before you start reading. Make

questions out of subheadings. If you are trying to answer a question, you are looking for meaning. Your brain is not just decoding words, it is working toward understanding those words.

Use your resources. What does your course provide—review sessions, office hours, tutors, study groups? Science is a social endeavor; interact with people as well as the material. Use the resources provided by your course and by your textbook: Web sites, practice exams, tutorials. When possible, prepare for exams by practicing with questions from previous exams. Each professor has his/her own approach to writing questions. One way to improve your grade is to study how your professor asks questions.

Learn from your mistakes. Carefully review your exams—the questions you missed and even those you got right. If possible, go over the exam with an instructor. Review both *what* you missed and *why* you missed it, so you can learn *how* to choose the correct answer on the next exam. Work to improve your test-taking skills. (See the section on tips for taking tests.)

Try something new. If what you are doing isn't working, don't just keep trying to do it harder. Experiment with new strategies. Are you a visual learner? Try drawing diagrams. If you are an auditory learner, listen to lecture or study tapes as you commute or exercise. Do you need to interact physically with ideas? Build models and be active in your learning.

Strategies for Using This Study Guide

This study guide is not a replacement for your textbook or biology class. However, it should help support and even streamline your learning process. Some students read the text chapters before lecture and then read the study guide chapters after class to reinforce and review. Others reverse that order, skimming the study guide before lecture and then carefully dissecting the text after the professor's presentation. Use this study guide to prepare for tests. With the textbook open so you can refer to essential diagrams, read the relevant study guide chapters for a quick review.

Each study guide chapter has the following seven sections:

- The listing of the **Key Concepts** from the text keeps you focused on the major themes of the chapter.
- The **Framework** identifies the overall picture; it provides a conceptual framework into which the chapter's information fits.
- The **Chapter Review** is a condensation of the textbook chapter, including a brief summary for each concept heading. All the bold terms in the text are also shown in boldface in this section. Interspersed in this review section are **Interactive Questions** that help you stop and actively review the material just covered.
- The **Word Roots** section presents the meanings of key biological prefixes, suffixes, and word roots. Examples of terms from the chapter are then defined. Breaking a complicated term into identifiable components will help you to recognize and learn many new biological terms.
- The **Structure Your Knowledge** section directs you to organize and relate the main concepts of the chapter. It helps you to piece together the key ideas into a bigger picture.
- In the **Test Your Knowledge** section, you are provided with objective questions to test your understanding. The multiple choice questions ask you to choose the best answer. Some answers may be partly correct; almost all choices have been written to test your ability to discriminate among alternatives or to point out common misconceptions.
- Suggested answers to the Interactive Questions, the Structure Your Knowledge section, and the Test Your Knowledge questions are provided in the **Answer Section** at the end of the book.

Using Concept Maps. What are the concept maps that appear throughout this study guide? A **concept map** is a diagram that organizes and relates ideas. The *structure* of a concept map is a hierarchically organized cluster of concepts, enclosed in boxes and connected with labeled lines that explicitly state the relationships among concepts. The *function* of a concept map is to help you structure your understanding of a topic and create meaning. The *value* of a concept map is in the thinking and organizing required to create it.

Developing a concept map requires you to evaluate the relative importance of a group of concepts (Which are most inclusive and important? Which are subordinate to other concepts?), arrange the concepts in meaningful clusters, and draw and label the connections between them.

This book uses concept maps in several ways. A map of a chapter may be presented in the Framework section to show the organization of the key concepts in that chapter. An Interactive Question may provide a skeleton map, with some concepts provided and some empty boxes for you to label. This technique is intended to help you become more familiar with concept maps and to illustrate one possible approach to organizing the concepts of a particular section.

You will also be asked to develop your own concept maps on certain subsets of ideas. In these cases, the Answer Section will present a suggested map. A concept map is an individual picture of your understanding at the time you make the map. As your understanding of an area develops, your map will evolve—sometimes becoming more complex and interrelated, sometimes becoming simplified and more streamlined. Do not look to the Answer Section for the "right" concept map. After you have organized your own thoughts, look at the answer map to make sure you have included the key concepts (although you may have added more), to check that the connections you have made are reasonable, and perhaps to see another way to organize the information.

Tips for Taking Multiple Choice Tests. Interact with each question. Read the stem of the question carefully, underlining or using a highlighter to identify the key concept. Read each answer slowly. Cross out the ones you know are wrong. Circle the key idea that you think identifies the correct answer. Now read the question and the answer you chose together, making sure your choice really does answer what is asked. If you aren't sure about a question, try rereading the question and each choice individually. Don't do all of your thinking in your head. Write in the margins and blank spaces of your test. Draw yourself diagrams and pictures. Write down what you do know, and it may jog your memory. If you still are not sure, mark the question and come back to it later. As you work through related questions, you may find information that helps you figure out that question. And remember, there is no substitute for good preparation, proper rest and nutrition, and a positive attitude.

Biology is a fascinating, broad, and exciting subject. *Campbell Biology*, Ninth Edition, is filled with information organized in a manner that will help you build a conceptual framework of the major themes of biology. This *Student Study Guide* is intended to help you learn and recall information and, most importantly, to encourage and guide you as you develop your own understanding of and appreciation for biology.

Martha R. Taylor
Cornell University

Chapter 1

Introduction: Themes in the Study of Life

Key Concepts

1.1 The themes of this book make connections across different areas of biology

1.2 The Core Theme: Evolution accounts for the unity and diversity of life

1.3 In studying nature, scientists make observations and then form and test hypotheses

1.4 Science benefits from a cooperative approach and diverse viewpoints

Framework

This chapter outlines the broad scope of biology, describes themes that unify the study of life, and examines the scientific construction of biological knowledge. A course in biology is neither a vocabulary course nor a classification exercise for the diverse forms of life. Biology is a collection of facts and concepts structured within theories and organizing principles. Recognizing the common themes within biology will help you structure your knowledge of the fascinating and challenging study of life.

Chapter Review

Biology is the scientific study of life, with **evolution,** the process of change that has shaped life from its origin on Earth to today's diversity, as its organizing principle. The properties and processes of life include highly ordered structure, evolutionary adaptation, response to the environment, regulation, energy processing, reproduction, and growth and development.

1.1 The themes of this book make connections across different areas of biology

Theme: New Properties Emerge at Each Level in the Biological Hierarchy The scale of biology extends from the biosphere to molecules.

INTERACTIVE QUESTION 1.1

Write a brief description of each of the following levels of biological organization.

a. biosphere

b. ecosystem

c. community

d. population

e. organism

f. organs and organ systems

g. tissues

h. cells

i. organelles

j. molecules

Interactions among components at each level of biological organization lead to the emergence of novel properties at the next level. These **emergent properties** result from the structural arrangement and interaction of parts.

Biology combines the powerful and pragmatic strategy of reductionism, which breaks down complex systems into simpler components, with the study of the holistic, highly complex interactions in higher levels of life.

Many researchers are seeking to understand the emergent properties of life by looking at the functional integration of a system's parts. **Systems biology** studies the interactions of the parts of a system and models the system's dynamic behavior.

INTERACTIVE QUESTION   1.2

Give examples of how systems biology may affect medical practice or environmental policy making.

Theme: Organisms Interact with Other Organisms and the Physical Environment   Both organisms and the environment are affected by interactions between them. These interactions result in the cycling of chemical nutrients between organisms and the environment. The continual burning of fossil fuels, which releases increasing amounts of CO_2 to the atmosphere, is causing global warming and contributing to **global climate change.** The resulting increase in temperature affects organisms' habitats and interactions.

Theme: Life Requires Energy Transfer and Transformation   Living organisms require energy. Producers transform light energy to the chemical energy in sugar, which powers the cellular activities of plants. Consumers eat plants and other organisms, using the chemical energy in their foods to power their movement, growth, and other activities. In each energy transformation, some energy is converted to thermal energy, which is dissipated to the surroundings as heat.

INTERACTIVE QUESTION   1.3

Compare the movement of chemical nutrients and energy in an ecosystem.

Theme: Structure and Function Are Correlated at All Levels of Biological Organization   The form of a biological structure is usually well matched to its function. Form fits function at all of life's structural levels.

Theme: Cells Are an Organism's Basic Units of Structure and Function   The cell is the lowest structural level capable of performing all the activities of life. Every cell uses DNA as its genetic information and is enclosed by a membrane. The simpler and smaller **prokaryotic cell,** unique to bacteria and archaea, lacks both a nucleus to enclose its DNA and cytoplasmic organelles. The **eukaryotic cell**—with a nucleus containing DNA, and numerous membrane-bound organelles—is typical of all other living organisms.

Theme: The Continuity of Life Is Based on Heritable Information in the Form of DNA   The heritable information of a cell is coded in **DNA,** deoxyribonucleic acid, the substance of genes. **Genes** are the units of inheritance that transmit information from parents to offspring. Genes are located on chromosomes, long DNA molecules that replicate before cell division and provide identical copies to daughter cells.

The biological instructions for the development and functioning of organisms are coded in the arrangement of the four kinds of nucleotides on the two strands of a DNA double helix. Most genes program the cell's production of proteins, and almost all cellular structures and actions involve one or more proteins.

Gene expression is the process by which a gene's information is converted into a protein or an RNA product. All forms of life use essentially the same genetic code of nucleotides.

INTERACTIVE QUESTION   1.4

Describe the pathway from DNA nucleotides to proteins.

All the genetic instructions an organism inherits make up its **genome.** One set of human chromosomes contains about 3 billion nucleotide pairs, and codes for the production of about 75,000 proteins and a large number of non-protein-coding RNA molecules.

With the sequencing of the human genome and the genomes of many other organisms now complete, current research focuses on the coordination of the proteins coded for by the DNA sequences on a cellular and organismal level. Using a systems approach called **genomics,** scientists are analyzing whole sets of genes of a species and comparing genomes between species.

Three research developments contribute to genomics: "high-throughput" technology that can produce enormous amounts of data, such as the automatic DNA-sequencing machines; **bioinformatics,** which provides the computational tools to process and integrate data from large data sets; and interdisciplinary research teams with specialists from many diverse scientific fields.

Theme: Feedback Mechanisms Regulate Biological Systems Protein enzymes catalyze a cell's chemical reactions. Many biological systems self-regulate by a mechanism called feedback. In **negative feedback,** an end product slows down a process, often by inhibiting an enzyme early in a chemical pathway. In **positive feedback,** less common in biological processes, an end product speeds up its own production. Regulatory mechanisms operate at all levels of the biological hierarchy.

Evolution, the Overarching Theme of Biology Evolution explains how diverse organisms of the past and the present are related through common ancestry, and it presents the mechanism through which organisms come to fit their environments.

1.2 The Core Theme: Evolution accounts for the unity and diversity of life

Classifying the Diversity of Life Of an estimated total of 10–100 million species, only about 1.8 million species have been identified and named. Taxonomy is the branch of biology that names organisms and groups species into ever broader categories, from genera to family, order, class, phylum, kingdom, and domain.

The number of kingdoms is an ongoing debate, but all of life is now grouped into three domains. The prokaryotes are divided into domains **Archaea** and **Bacteria.** All eukaryotes are placed in domain **Eukarya.** Within the Eukarya, the traditional kingdom Protista is being split to better reflect evolutionary relationships.

Within this diversity, living forms share a universal genetic language of DNA and similarities in cell structure.

INTERACTIVE QUESTION 1.5

What is a commonly used criterion for placing plants, fungi, and animals into separate kingdoms?

Charles Darwin and the Theory of Natural Selection In *The Origin of Species,* published in 1859, Charles Darwin presented his case for "descent with modification," the idea that present forms have diverged from a succession of ancestral forms. Darwin proposed the theory of **natural selection** as the mechanism of evolution by drawing an inference from three observations: Individuals vary in many heritable traits, the overproduction of offspring sets up a competition for survival, and species are generally matched to their environments. From this, Darwin inferred that individuals with traits best suited for an environment leave more offspring than do less-fit individuals. This natural selection, or unequal reproductive success within a population, results in the gradual accumulation of favorable adaptations to an environment.

The Tree of Life The underlying unity seen in the structures of related species, both living and in the fossil record, reflects the inheritance of that structure from a common ancestor. The diversity of species results from natural selection acting over millions of generations in different environments. The tree-like diagrams of evolutionary relationships reflect the branching genealogy extending from ancestral species. Similar species share a common ancestor at a more recent branch point on the tree of life. Distantly related species share a more ancient common ancestor.

INTERACTIVE QUESTION 1.6

Describe in your own words Darwin's theory of natural selection as the mechanism of evolutionary adaptation and the origin of new species.

1.3 In studying nature, scientists make observations, and then form and test hypotheses

Science is an approach to understanding the natural world that involves **inquiry,** the search for information by asking questions and endeavoring to answer them.

Making Observations Careful and verifiable observation and analysis of data are the basis of scientific inquiry. Observations involve our senses and tools that extend our senses; **data,** both *quantitative* and *qualitative*, are recorded observations. Using **inductive reasoning,** a generalized conclusion can often be drawn from collections of observations.

Forming and Testing Hypotheses Observations and inductions lead to the search for natural causes and explanations. A **hypothesis** is a tentative answer to a question or an explanation of observations, and it leads to predictions that can be tested. **Deductive reasoning** uses "if . . . then" logic to proceed from the general to the specific—from a general hypothesis to specific predictions of results if the general premise is correct.

A hypothesis must be *testable* and *falsifiable*—that is, there must be some observation or experiment that can reveal if the hypothesis is actually not true. The ideal is to frame two or more alternative hypotheses and design experiments to test each candidate explanation. A hypothesis cannot be *proven*; the more attempts to falsify it that fail, however, the more a hypothesis gains credibility.

Science seeks natural causes for natural phenomena; it does not address questions of the supernatural.

The Flexibility of the Scientific Method   The *scientific method*, as outlined by a structured series of steps, is rarely adhered to rigidly in scientific inquiry. Scientists often backtrack to make more observations, or make progress in answering a question only after other research provides a new context.

A Case Study in Scientific Inquiry: Investigating Mimicry in Snake Populations   The scarlet king snake mimics the ringed coloration of the venomous coral snake. D. and K. Pfennig and W. Harcombe tested the hypothesis that mimicry is an adaptation that reduces a harmless animal's risk of being eaten. To test the prediction that predators will attack king snakes less frequently in areas where predators have adapted to the warning coloration of coral snakes, they placed equal numbers of plain brown and ringed-colored artificial king snakes in regions with and without coral snakes. Compared to the brown snakes, the ringed snakes were attacked less frequently only in field sites within the range of the venomous coral snakes.

This experimental design illustrates a **controlled experiment** in which subjects are divided into an *experimental group* and a *control group.* Both groups are alike except for the one variable that the experiment is trying to test. Another characteristic of science is that observations and experimental results must be repeatable.

INTERACTIVE QUESTION   1.7

a. How did predators "learn" to avoid coral snakes?

b. Why were the results of the mimicry study presented as the percent of attacks on king snakes in each area rather than as the total number of attacks?

Theories in Science   A **theory** is broader in scope than a hypothesis, generates many specific hypotheses, and is supported by a large body of evidence. Still, a theory can be modified or even rejected when results and new evidence no longer support it.

1.4   Science benefits from a cooperative approach and diverse viewpoints

Building on the Work of Others   Most scientists work in teams and share their results with a broader research community in seminars, publications, and websites. Scientists often attempt to confirm the observations and experimental results of other colleagues. Scientists often share data on **model organisms,** which are easy to grow in the lab and are useful for answering particular questions that may have broad applications. Biological questions can be approached from different angles and levels of biological organization.

Science, Technology, and Society   The political and cultural environment influences the ways in which scientists approach their work. But science is still distinguished by adherence to the criteria of verifiable observations and hypotheses that are testable and falsifiable.

Science and technology are interdependent: The information generated by science is applied by **technology** in the development of goods and services, and technological advances are used to extend scientific knowledge. The uses of scientific knowledge and technologies are influenced by and in turn influence politics, economics, and cultural values.

The Value of Diverse Viewpoints in Science   Women and many racial and ethnic groups have been underrepresented in scientific professions. A diversity of backgrounds and viewpoints is important to the progress of science.

INTERACTIVE QUESTION   1.8

a. Compare hypotheses and theories.

b. Compare science and technology.

Word Roots

bio- = life (*biology:* the scientific study of life; *bioinformatics:* the use of computers, software, and mathematical models to process and integrate biological information from large data sets)

-ell = small (*organelle:* a small membrane-enclosed structure with a specialized function, found in eukaryotic cells) eu- = true (*eukaryotic cell:* a type of cell with a membrane-enclosed nucleus and organelles)

pro- = before; **karyo-** = nucleus (*prokaryotic cell:* a type of cell lacking a membrane-enclosed nucleus and organelles)

Structure Your Knowledge

1. Briefly describe in your own words each of the eight unifying themes of biology presented in this chapter:
   a. emergent properties
   b. interaction with the biotic and physical environment
   c. energy transfer and transformation
   d. structure and function
   e. cells
   f. heritable information
   g. feedback mechanisms of regulation
   h. evolution

Test Your Knowledge

MULTIPLE CHOICE: *Choose the one best answer.*

1. The core idea that makes sense of the unity and the diversity of life is
   a. the scientific method.
   b. inductive reasoning.
   c. deductive reasoning.
   d. evolution.
   e. systems biology.

2. In an experiment similar to the mimicry experiment performed by the Pfennigs, a researcher found that more total predator attacks occurred on model king snakes in areas with coral snakes than in areas outside the range of coral snakes. From this the researcher concluded that
   a. the mimicry hypothesis is false.
   b. the predators in the areas with coral snakes were hungrier than the predators in other areas.
   c. king snakes do not resemble coral snakes enough to protect them from attack.
   d. the data that should be compared to draw a conclusion must include a control—a comparison with the number of attacks on model brown snakes.
   e. more data must be collected before a conclusion can be drawn.

3. Why can a hypothesis never be "proven" to be true?
   a. One can never collect enough data to be 100% sure.
   b. There may always be alternative untested hypotheses that might account for the results.
   c. Science is limited by our senses.
   d. Experimental error is involved in every research project.
   e. Science "evolves"; hypotheses and even theories are always changing.

4. Which of the following statements is an example of positive feedback regulation?
   a. The hormones insulin and glucagon regulate blood-sugar levels.
   b. In the birth of a baby, uterine contractions stimulate release of chemicals that stimulate more uterine contractions.
   c. A rise in temperature when you exercise stimulates sweating and increased blood flow to the skin.
   d. When cells have sufficient energy available, the pathways that break down sugars are turned off.
   e. A rise in CO_2 in the atmosphere correlates with increasing global temperature.

5. Which of the following areas is mismatched with its description?
   a. model organisms—using type organisms to characterize each domain and kingdom
   b. scientific inquiry—generating hypotheses; formulating predictions; conducting experiments or making observations
   c. genomics—studying whole sets of genes of a species and between species
   d. taxonomy—identifying and naming organisms, and placing them in hierarchical categories
   e. technology—inventing practical uses of scientific knowledge

6. In a pond sample, you find a unicellular organism that has numerous chloroplasts and a whiplike flagellum. In which of the following groups do you think it should be classified?
   a. plant
   b. animal
   c. domain Archaea
   d. one of the proposed kingdoms of protists
   e. You cannot tell unless you see if it has a nucleus.

7. What is DNA?
   a. the substance of heredity
   b. a double helix made of four types of nucleotides
   c. a code for protein synthesis
   d. a component of chromosomes
   e. all of the above

8. Which of the following sequences correctly lists life's hierarchical levels from lowest to highest?

   a. organ, tissue, organ system, organism, population
   b. organism, community, population, ecosystem, biosphere
   c. molecule, organelle, cell, tissue, organ, organism
   d. tissue, cell, organ, organism, community
   e. Both b and c are correct sequences.

9. Which of the following themes of biology is most related to the goals and practices of systems biology?

   a. Evolution accounts for the unity and diversity of life.
   b. Cells are an organism's basic units of structure and function.
   c. The continuity of life is based on heritable information in the form of DNA.
   d. Life requires energy transfer and transformation.
   e. New properties emerge at each level in the biological hierarchy.

The Chemistry of Life

Chapter 2

The Chemical Context of Life

Key Concepts

2.1 Matter consists of chemical elements in pure form and in combinations called compounds

2.2 An element's properties depend on the structure of its atoms

2.3 The formation and function of molecules depend on chemical bonding between atoms

2.4 Chemical reactions make and break chemical bonds

Framework

This chapter considers the basic principles of chemistry that explain the behavior of atoms and molecules. You will learn how the subatomic particles—protons, neutrons, and electrons—are organized in atoms, how atoms are connected by covalent bonds, and how ions are attracted to each other in ionic bonds. Weak chemical bonds help to create the shapes and functions of molecules. Emergent properties are associated with each new level of structural organization in the hierarchy from atoms to life.

Chapter Review

2.1 Matter consists of chemical elements in pure form and in combinations called compounds

Elements and Compounds **Matter** is anything that takes up space and has mass. (Although sometimes used interchangeably, mass is the amount of matter in an object, whereas weight reflects gravity's pull on that mass.) **Elements** are substances that cannot be chemically broken down to other types of matter. A **compound** is made up of two or more elements combined in a fixed ratio. The characteristics of a compound differ from those of its constituent elements, an example of emergent properties in higher levels of organization.

The Elements of Life Carbon (C), oxygen (O), hydrogen (H), and nitrogen (N) make up 96% of living matter. The seven elements listed in Interactive Question 2.1 make up most of the remaining 4%. Some elements, like iron (Fe) and iodine (I), may be required in very minute quantities and are called **trace elements.**

Fill in the names beside the symbols of the following elements commonly found in living matter.

Ca   Na

P   Cl

K   Mg

S

Evolution of Tolerance to Toxic Elements   Serpentine soil contains toxic elements and low levels of essential elements. Serpentine plant communities exhibit evolutionary adaptations that allow such plants to grow in toxic soils.

2.2   An element's properties depend on the structure of its atoms

An **atom** is the smallest unit of matter retaining the properties of a particular element.

Subatomic Particles   Three stable *subatomic particles* are important to our understanding of atoms. Uncharged **neutrons** and positively charged **protons** are packed tightly together to form the **atomic nucleus** of an atom. Negatively charged **electrons** form a cloud around the nucleus.

Protons and neutrons have a similar mass of about 1.7×10^{-24} g, or close to 1 dalton each. A **dalton** is the measurement unit for atomic mass. Electrons have negligible mass.

Atomic Number and Atomic Mass   Each element has a characteristic **atomic number,** or number of protons in the nucleus of each of its atoms. Unless otherwise indicated, an atom has a neutral electrical charge, and thus the number of protons is equal to the number of electrons. A subscript to the left of the symbol for an element indicates its atomic number; a superscript indicates its mass number. The **mass number** is equal to the number of protons and neutrons in the nucleus and approximates the mass of an atom of that element in daltons. The term **atomic mass** refers to the total mass of an atom.

The difference between the mass number and the atomic number of an atom is equal to the number of _____. An atom of phosphorus, $^{31}_{15}P$, contains _____ protons, _____ electrons, and _____ neutrons. The atomic mass of phosphorus is approximately _____.

Isotopes   Although the number of protons is constant, the number of neutrons can vary among the atoms of an element, creating different **isotopes** that have slightly different masses but the same chemical behavior. Some isotopes are unstable; the nuclei of **radioactive isotopes** spontaneously decay, giving off particles and energy. Radioactive isotopes are important tools in biological research and medicine. Too great an exposure to radiation from decaying isotopes poses a significant health hazard.

The Energy Levels of Electrons   **Energy** is defined as the capacity to cause change. **Potential energy** is energy stored in matter as a consequence of its position or structure. The potential energy of electrons increases as their distance from the positively charged nucleus increases. Electrons can be located in different **electron shells** surrounding the nucleus.

To move to a shell farther from the nucleus, an electron must _____ energy; an electron _____ energy when it moves to a closer shell.

Electron Distribution and Chemical Properties   The chemical behavior of an atom is a function of the distribution of its electrons—in particular, the number of **valence electrons** in its outermost electron shell, or **valence shell.** A valence shell of eight electrons is complete, resulting in an unreactive or inert atom. (The first shell, however, can hold only two electrons.) Atoms with incomplete valence shells are chemically reactive. The elements in each row, or period, of the *periodic table of the elements* have the same number of electron shells and are arranged in order of increasing number of electrons.

Draw an electron shell diagram for the following atoms.
a. $_6C$ **c.** $_8O$

b. $_7N$ **d.** $_{12}Mg$

Electron Orbitals   An **orbital** is the three-dimensional space or volume within which an electron is most likely to be found. No more than two electrons can occupy the same orbital. The first electron shell can contain two electrons in a single spherical orbital, called the 1s orbital. The second electron shell can hold a maximum of eight electrons in its four orbitals, which are a 2s spherical orbital and three dumbbell-shaped p orbitals located along the x, y, and z axes.

INTERACTIVE QUESTION 2.5

Look again at the electron shell diagram you drew for carbon (a) in Interactive Question 2.4. Did you show the outer shell electrons unpaired? Why?

INTERACTIVE QUESTION 2.6

Fill in the blanks in the following concept map to help you review the atomic structure of atoms.

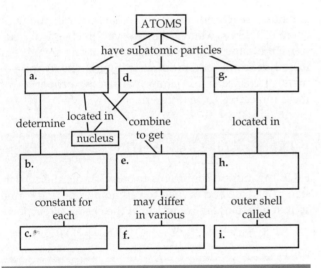

2.3 The formation and function of molecules depend on chemical bonding between atoms

Atoms with incomplete valence shells can either share electrons with or completely transfer electrons to or from other atoms such that each atom is able to complete its valence shell. These interactions usually result in attractions called **chemical bonds,** which hold the atoms close together.

Covalent Bonds   When two atoms share a pair of valence electrons, a **covalent bond** is formed. A **molecule** consists of two or more atoms held together by covalent bonds. A *structural formula*, such as H—H, indicates both the atoms and bonds in a molecule. The dash indicates a **single bond.** A *molecular formula*, such as O_2,

indicates only the kinds and numbers of atoms. In an oxygen molecule, two pairs of valence electrons are shared between oxygen atoms, forming a double covalent bond, or simply a **double bond** (O=O). The **valence,** or bonding capacity, of an atom usually equals the number of unpaired electrons in its valence shell.

INTERACTIVE QUESTION 2.7

a. What are the valences of the four most common elements of living matter?

b. Draw the *Lewis dot structures* for the following molecules.

H_2 O_2 N_2 CH_4

Electronegativity is the attraction of a particular atom for shared electrons. If the atoms in a molecule have similar electronegativities, the electrons remain equally shared, and the bond is said to be a **nonpolar covalent bond.** If one element is more electronegative, it pulls the shared electrons closer to itself, creating a **polar covalent bond.** This unequal sharing of electrons results in a slight negative charge ($\delta-$) associated with the more electronegative atom and a slight positive charge ($\delta+$) associated with the atom from which the electrons are pulled.

INTERACTIVE QUESTION 2.8

Explain whether the following molecules contain nonpolar or polar covalent bonds. (Hint: N and O both have high electronegativities.)

a. nitrogen molecule N≡N   **c.** methane H—C—H (with H above and H below)

b. ammonia H—N—H (with H below)   **d.** formaldehyde (H, H) C=O

Ionic Bonds   If two atoms are very different in their attraction for valence electrons, the more electronegative atom may completely transfer an electron from the other atom, resulting in the formation of charged atoms called **ions.** The atom that lost the electron is a positively charged **cation.** The negatively charged atom that gained the electron is called an **anion.** An **ionic bond** may hold these ions together because of the attraction of their opposite charges.

Ionic compounds, called **salts,** often exist as three-dimensional crystalline lattice arrangements held together by electrical attractions. The number of ions present in a salt crystal is not fixed, but the atoms are present in specific ratios. Salts have strong ionic bonds when dry, but the crystal dissolves in water.

Ion also refers to entire covalent molecules that are electrically charged.

INTERACTIVE QUESTION   2.9

Calcium ($_{20}$Ca) and chlorine ($_{17}$Cl) can combine to form the salt calcium chloride. Based on the number of electrons in their valence shells and their bonding capacities, what would the formula for this salt be? **a.** _____ Which atom becomes the cation? **b.** _____

Weak Chemical Bonds   Ionic bonds and other weak bonds may form temporary interactions between molecules. Weak bonds within many large molecules help to create those molecules' three-dimensional functional shapes.

A hydrogen atom that is covalently bonded to an electronegative atom has a partial positive charge and can be attracted to another nearby electronegative atom. This attraction is called a **hydrogen bond.**

All atoms and molecules are attracted to each other when in close contact by **van der Waals interactions.** Momentary uneven electron distributions produce changing positive and negative regions that create these weak attractions.

INTERACTIVE QUESTION   2.10

Sketch a water molecule, showing oxygen's electron shells and the covalently shared electrons. Indicate the areas with slight negative and positive charges that enable a water molecule to form hydrogen bonds with other polar molecules. Then draw a second water molecule and indicate a hydrogen bond between the two.

Molecular Shape and Function   A molecule's characteristic size and shape affect how it interacts with other molecules. When atoms form covalent bonds, their *s* and three *p* orbitals hybridize to form four teardrop-shaped orbitals in a tetrahedral arrangement. These hybrid orbitals dictate the specific shapes of different molecules.

INTERACTIVE QUESTION   2.11

Look at your diagram of a water molecule in Interactive Question 2.10. Why is it roughly V shaped?

2.4   Chemical reactions make and break chemical bonds

Chemical reactions involve the making or breaking of chemical bonds. Matter is conserved in chemical reactions; the same number and kinds of atoms are present in both **reactants** and **products,** although the rearrangement of electrons and atoms causes the properties of these molecules to be different.

INTERACTIVE QUESTION   2.12

Fill in the missing coeffcients for respiration, the conversion of glucose and oxygen to carbon dioxide and water, so that all atoms are conserved in the chemical reaction.

$$C_6H_{12}O_6 + \_\_O_2 \rightarrow \_\_CO_2 + \_\_H_2O$$

Chemical reactions are reversible—the products of the forward reaction can become reactants in the reverse reaction. Increasing the concentrations of reactants can speed up the rate of a reaction. **Chemical equilibrium** is reached when the forward and reverse reactions proceed at the same rate, and the relative concentrations of reactants and products no longer change.

Word Roots

an- = not (*anion:* a negatively charged ion)

co- = together; **-valent** = strength (*covalent bond:* a strong bond between atoms that share one or more pairs of valence electrons)

electro- = electricity (*electronegativity:* the attraction of a given atom for the electrons of a covalent bond)

iso- = equal (*isotope:* one of several forms of an element, each with the same number of protons but a different number of neutrons)

neutr- = neither (*neutron:* a subatomic particle with a neutral electrical charge)

pro- = before (*proton:* a subatomic particle with a single positive electrical charge)

Structure Your Knowledge

Take the time to write out or discuss your answers to the following questions. Then refer to the suggested answers at the end of the book.

1. Fill in the following chart concerning the major subatomic particles of an atom.

Particle	Charge	Mass	Location

2. Atoms can have various numbers associated with them.
   a. Define the following and show where each of them is placed relative to the symbol of an element such as C (use the most common isotope, C-12): atomic number, mass number, atomic mass.
   b. Define valence.
   c. Which of these four numbers is most related to the chemical behavior of an atom? Explain.

3. Arrange the two types of covalent bonds and ionic bonds in an order that reflects the degree of electron sharing.

Test Your Knowledge

MULTIPLE CHOICE: *Choose the one best answer.*

1. Each element has its own characteristic atom in which
   a. the atomic mass is constant.
   b. the atomic number is constant.
   c. the mass number is constant.
   d. Two of the above are correct.
   e. All of the above are correct.

2. Which of the following is *not* a trace element in the human body?
   a. iodine
   b. zinc
   c. iron
   d. calcium
   e. fluorine

3. A sodium ion (Na^+) contains 10 electrons, 11 protons, and 12 neutrons. What is the atomic number of sodium?
   a. 10   c. 12   e. 33
   b. 11   d. 23

4. Radioactive isotopes can be used in studies of metabolic pathways because
   a. their half-life allows a researcher to time an experiment.
   b. they are more reactive.
   c. the cell does not recognize the extra protons in the nucleus, so isotopes are readily used in metabolism.
   d. their location or quantity can be experimentally determined because of their radioactivity.
   e. their extra neutrons produce different colors that can be traced through the body.

5. Which of the following atomic numbers would describe the element that is least reactive?
   a. 1   c. 12   e. 18
   b. 8   d. 16

Use this information to answer questions 6 through 11.

The six elements most common in living organisms are
$${}^{12}_{6}C \quad {}^{16}_{8}O \quad {}^{1}_{1}H \quad {}^{14}_{7}N \quad {}^{32}_{16}S \quad {}^{31}_{15}P$$

6. How many electrons does phosphorus have in its valence shell?
   a. 3   c. 7   e. 16
   b. 5   d. 15

7. What is the atomic mass of phosphorus?
   a. 15   c. 31   e. 62
   b. 16   d. 46

8. A radioactive isotope of carbon has the mass number 14. How many neutrons does this isotope have?
   a. 2   c. 8   e. 14
   b. 6   d. 12

9. How many covalent bonds is a sulfur atom most likely to form?
   a. 1   c. 3   e. 5
   b. 2   d. 4

10. Based on electron configuration, which of the following elements would have chemical behavior most like that of oxygen?
    a. C   c. N   e. S
    b. H   d. P

11. How many of the elements listed on the previous page are found next to each other (side by side) on the periodic table?
    a. one group of two
    b. two groups of two
    c. one group of two and one group of three
    d. one group of three
    e. all of them

12. Which of the following describes what happens as a chlorophyll pigment absorbs energy from sunlight?
    a. An electron moves to a higher electron shell and the electron's potential energy increases.
    b. An electron moves to a higher electron shell and its potential energy decreases.
    c. An electron drops to a lower electron shell and releases its energy as heat.
    d. An electron drops to a lower electron shell and its potential energy increases.
    e. An electron of sunlight is transferred to chlorophyll, producing a chlorophyll ion with higher potential energy.

13. How are the electrons of an oxygen atom arranged?
    a. eight in the second electron shell, creating an inert element
    b. two in the first electron shell and six in the second, creating a valence of six
    c. two in the $1s$ orbital and two each in the three $2p$ orbitals, creating a valence of two
    d. two in the $1s$ orbital, one each in the $2s$ and three $2p$ orbitals, and two in the $3s$ orbital, creating a valence of two
    e. two in the $1s$ orbital, two in both the $2s$ and $2px$ orbitals, and one each in the $2py$ and $2pz$ orbitals, creating a valence of two

14. A covalent bond between two atoms is likely to be nonpolar if
    a. one of the atoms is much more electronegative than the other.
    b. the two atoms are about equally electronegative.
    c. the two atoms are of the same element.
    d. one atom is an anion and the other is a cation.
    e. Both b and c are correct.

15. A triple covalent bond would
    a. be very polar.
    b. involve the bonding of three atoms.
    c. involve the bonding of six atoms.
    d. produce a triangularly shaped molecule.
    e. involve the sharing of six electrons.

16. A cation
    a. has gained an electron.
    b. can easily form hydrogen bonds.
    c. is more likely to form in an atom with seven electrons in its valence shell.
    d. has a positive charge.
    e. Both c and d are correct.

For questions 17–19, choose from the following answers to identify the types of bonds in this diagram of a water molecule interacting with an ammonia molecule.

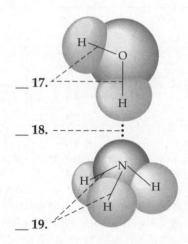

____ 17.

____ 18.

____ 19.

    a. nonpolar covalent bond
    b. polar covalent bond
    c. ionic bond
    d. hydrogen bond
    e. cannot determine without more information

20. In what type of bond would you expect potassium ($^{39}_{19}K$) to participate?
    a. ionic; it would lose one electron and carry a positive charge
    b. ionic; it would gain one electron and carry a negative charge
    c. covalent; it would share one electron and make one covalent bond
    d. covalent; it would share two electrons and form two bonds
    e. none; potassium is an inert element

21. Which of the following may form between any closely aligned molecules?
    a. nonpolar covalent bonds
    b. polar covalent bonds
    c. ionic bonds
    d. hydrogen bonds
    e. van der Waals interactions

22. What is the molecular shape of methane (CH_4)?
    a. planar or flat, with the four H around the carbon
    b. pentagonal, or a flat five-sided arrangement
    c. tetrahedral, due to the hybridization of the *s* and three *p* orbitals of carbon
    d. circular, with the four H attached in a ring around the carbon
    e. linear, since all the bonds are nonpolar covalent bonds

23. The ability of morphine to mimic the effects of the body's endorphins is due to
    a. a chemical equilibrium developing between morphine and endorphins.
    b. the one-way conversion of morphine into endorphin.
    c. molecular shape similarities that allow morphine to bind to endorphin receptors.
    d. the similarities between morphine and heroin.
    e. hydrogen bonding and other weak bonds forming between morphine and endorphins.

24. Which of the following molecules would you predict is capable of forming hydrogen bonds?
    a. CH_4
    b. CH_4O
    c. NaCl
    d. H_2
    e. a, b, and d can form hydrogen bonds.

25. Chlorine has an atomic number of 17 and a mass number of 35. How many electrons would a chloride ion have?
    a. 16   c. 18   e. 34
    b. 17   d. 33

26. Taking into account the bonding capacities or valences of carbon (C) and oxygen (O), how many hydrogen (H) must be added to complete the following structural diagram of this molecule?

    $$O \diagdown \atop O \diagup {}^{\diagdown}_{\diagup} C-C-C-C=C-C-C$$

    a. 9   c. 11   e. 13
    b. 10   d. 12

27. What is the difference between a molecule and a compound?
    a. There is no difference; the terms are interchangeable.
    b. Molecules contain atoms of a single element, whereas compounds contain two or more elements.
    c. A molecule consists of two or more covalently bonded atoms; a compound contains two or more atoms held by ionic bonds.
    d. A compound consists of two or more elements in a fixed ratio; a molecule has two or more covalently bonded atoms of the same or different elements.
    e. Compounds always consist of molecules, but molecules are not always compounds.

28. In a reaction in chemical equilibrium,
    a. the forward and reverse reactions are occurring at the same rate.
    b. the reactants and products are in equal concentration.
    c. the forward reaction has gone further than the reverse reaction.
    d. there are equal numbers of atoms on both sides of the equation.
    e. a, b, and d are correct.

29. What would be the probable effect of adding more product to a reaction that is in equilibrium?
    a. There would be no change because the reaction is in equilibrium.
    b. The reaction would stop because excess product is present.
    c. The reaction would slow down but still continue.
    d. The forward reaction would increase and more product would be formed.
    e. The reverse reaction would increase and more reactants would be formed.

30. What coefficients must be placed in the blanks to balance the following chemical reaction?

    $$C_5H_{12} + \underline{\quad}O_2 \rightarrow \underline{\quad}CO_2 + \underline{\quad}H_2O$$

    a. 5; 5; 5
    b. 6; 5; 6
    c. 6; 6; 6
    d. 8; 4; 6
    e. 8; 5; 6

Chapter 3

Water and Life

Key Concepts

3.1 Polar covalent bonds in water molecules result in hydrogen bonding

3.2 Four emergent properties of water contribute to Earth's suitability for life

3.3 Acidic and basic conditions affect living organisms

Framework

Water makes up 70% to 95% of the cell content of living organisms and covers 75% of Earth's surface. Its unique properties make the planet's environment suitable for life and the internal environments of organisms suitable for the chemical and physical processes of life.

Hydrogen bonding between polar water molecules creates a cohesive liquid with a high specific heat and high heat of vaporization, both of which help to regulate environmental temperature. Ice floats and protects oceans and lakes from freezing. The polarity of water makes it a versatile solvent. The [H⁺] in a solution is expressed as pH. Buffers regulate an organism's pH.

Chapter Review

3.1 Polar covalent bonds in water molecules result in hydrogen bonding

A water molecule consists of two hydrogen atoms, each bonded to a more electronegative oxygen atom by a **polar covalent bond.** This **polar molecule** has a shape like a wide V with a slight positive charge on each hydrogen atom ($\delta+$) and a slight negative charge ($\delta-$) associated with the oxygen. Hydrogen bonds, electrical attractions between the hydrogen atom of one water molecule and the oxygen atom of a nearby water molecule, create a structural organization that leads to the emergent properties of water.

INTERACTIVE QUESTION 3.1

Draw the four water molecules that can hydrogen-bond to this water molecule. Show the bonds and the slight negative and positive charges that account for the formation of these hydrogen bonds.

3.2 Four emergent properties of water contribute to Earth's suitability for life

Cohesion of Water Molecules Liquid water is unusually cohesive due to the constant forming and reforming of hydrogen bonds that hold the molecules close together. This **cohesion** creates a more structurally organized liquid and helps water to be pulled upward in plants. The **adhesion** of water molecules to the walls of plant vessels also contributes to water transport. Hydrogen bonding between water molecules produces a high **surface tension** at the interface between water and air.

Moderation of Temperature by Water In a body of matter, **heat** is a measure of the total quantity of **kinetic energy,** the energy associated with the movement of atoms and molecules. **Temperature** measures the average kinetic energy of the molecules in a substance.

Temperature is measured using a **Celsius scale.** Water at sea level freezes at 0°C and boils at 100°C. A **calorie (cal)** is the amount of heat energy it takes to raise 1 g of water 1°C. A **kilocalorie (kcal)** is 1,000 calories, the amount of heat required or released to change the temperature of 1 kg of water by 1°C. A **joule (J)** equals 0.239 cal; a calorie is 4.184 J.

Specific heat is the amount of heat absorbed or lost when 1 g of a substance changes its temperature by 1°C. Water's specific heat of 1 cal/g·°C is unusually high compared with other common substances. Why does water absorb or release a relatively large quantity of heat as its temperature changes? Heat must be absorbed to break hydrogen bonds before water molecules can move faster and the temperature can rise, and conversely, heat is released when hydrogen bonds form as the temperature of water drops. The high proportion of water in the environment and within organisms keeps temperature fluctuations within limits that permit life.

Vaporization or evaporation occurs when molecules of a liquid with sufficient kinetic energy overcome their attraction to other molecules and escape into the air as gas. The **heat of vaporization** is the quantity of heat that must be absorbed for 1 g of a liquid to be converted to a gas. Water has a high heat of vaporization (580 cal/g at 25°C) because a large amount of heat is needed to break the hydrogen bonds holding water molecules together. Water helps moderate Earth's climate as solar heat absorbed by tropical seas is dissipated during evaporation, and heat is released as moist tropical air moving poleward condenses to form rain.

As a liquid vaporizes, the surface left behind loses the kinetic energy of the escaping molecules and cools down. **Evaporative cooling** helps to protect terrestrial organisms from overheating and contributes to the stability of temperatures in lakes and ponds.

Floating of Ice on Liquid Water As water cools below 4°C, it expands. By 0°C, each water molecule is hydrogen-bonded to four other molecules, creating a crystalline lattice that spaces the molecules apart. Ice is thus less dense than liquid water and so it floats.

Water: The Solvent of Life A **solution** is a liquid homogeneous mixture of two or more substances; the dissolving agent is called the **solvent,** and the substance that is dissolved is the **solute.** Water is the solvent in an **aqueous solution.** The positive and negative regions of water molecules are attracted to oppositely charged ions or partially charged regions of polar molecules. Thus, solute molecules become surrounded by water molecules (a **hydration shell**) and dissolve into solution.

Ionic and polar substances are **hydrophilic;** they have an affinity for water due to electrical attractions and hydrogen bonding. Large hydrophilic substances may not dissolve but become suspended in an aqueous solution, forming a mixture called a **colloid.** Nonpolar and nonionic substances are **hydrophobic;** they will not easily mix with or dissolve in water.

INTERACTIVE QUESTION 3.2

The following concept map is one way to show how the breaking and forming of hydrogen bonds are related to temperature moderation. Fill in the blanks and compare your choice of concepts to those given in the answer section. Or, even better, create your own map to help you understand how water stabilizes temperature.

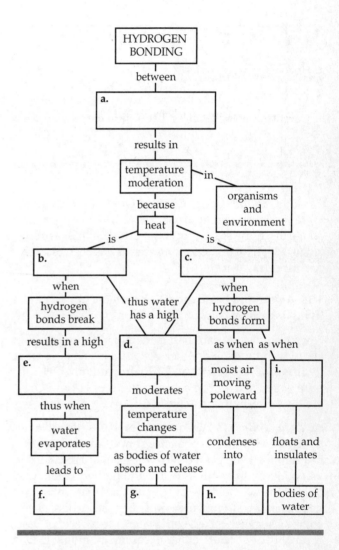

INTERACTIVE QUESTION 3.3

Indicate whether the following substances are hydrophilic or hydrophobic. Do they contain ionic bonds, polar covalent bonds, or nonpolar covalent bonds?

a. olive oil   **c.** salt

b. sugar   **d.** candle wax

Most of the chemical reactions of life take place in water. A **mole (mol)** is the amount of a substance that has a mass in grams numerically equivalent to its **molecular mass** (the sum of the mass of all atoms in the molecule) in daltons. A mole of any substance has exactly the same number of molecules—6.02×10^{23}, called Avogadro's number. The **molarity** of a solution (abbreviated *M*) refers to the number of moles of a solute dissolved in 1 liter of solution.

INTERACTIVE QUESTION   3.4

a. How many grams of lactic acid ($C_3H_6O_3$) are in 1 liter of a 0.5 *M* solution of lactic acid? (^{12}C, ^{1}H, ^{16}O)

b. How many molecules of lactic acid are in the solution in a?

Possible Evolution of Life on Other Planets with Water   The emergent properties of water support life on Earth. In the search for extraterrestrial life, astrobiologists look for evidence of water, as has been recently confirmed on Mars.

3.3   Acidic and basic conditions affect living organisms

A water molecule can dissociate into a **hydrogen ion,** H^+ (which binds to another water molecule to form a **hydronium ion,** H_3O^+) and a **hydroxide ion,** OH^-. In pure water at 25°C, the concentrations of H^+ and OH^- are the same; both are equal to 10^{-7} *M*.

Acids and Bases   When acids or bases dissolve in water, the H^+ and OH^- balance shifts. An **acid** adds H^+ to a solution, whereas a **base** reduces H^+ in a solution by accepting hydrogen ions or by adding hydroxide ions (which then combine with H^+ and thus remove hydrogen ions). A strong acid or strong base dissociates completely when mixed with water. A weak acid or base reversibly dissociates, either releasing or binding H^+.

The pH Scale   In an aqueous solution, the *product* of the $[H^+]$ and $[OH^-]$ is constant at 10^{-14}. Brackets, [ ], indicate molar concentration. If the $[H^+]$ is higher, then the $[OH^-]$ is lower, because the excess hydrogen ions combine with the hydroxide ions in solution and form water. Likewise, an increase in $[OH^-]$ causes an equivalent decrease in $[H^+]$.

The **pH** of a solution is defined as the negative log (base 10) of the $[H^+]$: pH $= -\log [H^+]$. For a neutral aqueous solution, $[H^+]$ is 10^{-7} *M*, and the pH equals 7. As the $[H^+]$ increases in an acidic solution, the pH value decreases. The difference between each unit of

the pH scale represents a tenfold difference in the concentration of $[H^+]$ and $[OH^-]$.

Most cells have an internal pH close to 7. **Buffers** within the cell maintain a stable pH by accepting excess H^+ or donating H^+ when H^+ concentration decreases. Weak acid-base pairs that reversibly bind hydrogen ions are typical of most buffering systems.

INTERACTIVE QUESTION   3.5

Complete the following table to review your understanding of pH.

$[H^+]$	$[OH^-]$	pH	Acidic, Basic, or Neutral?
	$[10^{-11}]$	3	
10^{-8}			
	$[10^{-7}]$		
		1	

INTERACTIVE QUESTION   3.6

The carbonic acid/bicarbonate system is an important biological buffer. Label the molecules and ions in this equation, and indicate which is the H^+ donor and which is the acceptor.

$$H_2CO_3 \rightleftharpoons HCO_3^- + H^+$$

In which direction will this reaction proceed

a. when the pH of a solution begins to fall?

b. when the pH rises above normal level?

Acidification: A Threat to Water Quality   The increasing release of CO_2 to the atmosphere is linked to fossil fuel combustion. The oceans absorb about 25% of this CO_2, which lowers the pH of seawater. The resulting **ocean acidification** decreases the concentration of carbonate (CO_3^{2-}), an important ion needed for coral reef calcification.

Acid precipitation—rain, snow, or fog with a pH lower than 5.2 (normal is 5.6)—is due to the reaction of water in the atmosphere with the sulfur oxides and nitrogen oxides released by the combustion of fossil fuels. Aquatic life and terrestrial plants are damaged by acid precipitation.

INTERACTIVE QUESTION 3.7

a. Use a formula to explain why increasing $[CO_2]$ in water leads to a lower pH.

b. Use a formula to explain how a lower pH would affect the $[CO_3^{2-}]$ in the ocean.

c. Assuming a fairly constant $[Ca^{2+}]$ in the ocean, how would a change in $[CO_3^{2-}]$ affect the calcification rate—the production of calcium carbonate ($CaCO_3$)—by the coral in a reef ecosystem?

Word Roots

kilo- = a thousand (*kilocalorie:* a thousand calories)

hydro- = water; **-philos** = loving; **-phobos** = fearing (*hydrophilic:* having an affinity for water; *hydrophobic:* having no affinity for water)

Structure Your Knowledge

1. Fill in the following table, which summarizes the emergent properties of water that contribute to the fitness of the environment for life.

2. To become proficient in the use of the concepts relating to pH, develop a concept map to organize your understanding of the following terms: pH, $[H^+]$, $[OH^-]$, acidic, basic, neutral, buffer, 0–14, acid-base pair. Remember to label connecting lines and add additional concepts as you need them.

A suggested concept map is given in the answer section, but remember that your concept map should represent your own understanding. The value of this exercise is in organizing these concepts for yourself.

Test Your Knowledge

MULTIPLE CHOICE: *Choose the one best answer.*

1. Each water molecule is capable of forming
   a. one hydrogen bond.
   b. three hydrogen bonds.
   c. four hydrogen bonds.
   d. two covalent bonds and two hydrogen bonds.
   e. four polar covalent bonds.

2. The polar covalent bonds of water molecules
   a. promote the formation of hydrogen bonds.
   b. help water to dissolve nonpolar solutes.
   c. lower the heat of vaporization and lead to evaporative cooling.
   d. create a crystalline structure in liquid water.
   e. do all of the above.

3. What accounts for the movement of water up the vessels of a tall tree?
   a. cohesion
   b. hydrogen bonding
   c. adhesion
   d. hydrophilic cell walls
   e. all of the above

Property	Explanation of Property	Example of Benefit to Life
a.	Hydrogen bonds hold water molecules together and adhere them to hydrophilic surface.	b.
High specific heat	c.	Temperature changes in environment and organisms are moderated.
d.	Hydrogen bonds must be broken for water to evaporate.	e.
f.	Water molecules with high kinetic energy evaporate; remaining molecules are cooler.	g.
Less dense as a solid	h.	i.
j.	k.	Most chemical reactions in life involve solutes dissolved in water.

4. Climates tend to be moderate near large bodies of water because
   a. a large amount of solar heat is absorbed during the gradual rise in temperature of the water.
   b. water releases heat to the environment as it cools.
   c. the high specific heat of water helps to moderate air temperatures.
   d. a great deal of heat is absorbed and released as hydrogen bonds break or form.
   e. all of the above are true.

5. Temperature is a measure of
   a. specific heat.
   b. average kinetic energy of molecules.
   c. total kinetic energy of molecules.
   d. Celsius degrees.
   e. joules.

6. You have three flasks containing 100 mL of different liquids. Each is warmed with 100 calories of heat. The temperature of the liquid in flask 1 rises 1°C; in flask 2 it rises 1.5°C; and in flask 3 it rises 2°C. Which of these liquids has the highest specific heat?
   a. the liquid in flask 1
   b. the liquid in flask 2
   c. the liquid in flask 3
   d. You cannot tell unless you know what liquid is in each flask.
   e. This type of experiment does not relate to the specific heat of a substance.

7. A burn from steam at 100°C is more severe than a burn from boiling water because
   a. the steam is hotter than boiling water.
   b. steam releases a great deal of heat as it condenses on the skin.
   c. steam has a higher heat of vaporization than does water.
   d. a person is more likely to come into contact with steam than with boiling water.
   e. steam stays on the skin longer than does boiling water.

8. Evaporative cooling is a result of
   a. a low heat of vaporization.
   b. the release of heat during the breaking of hydrogen bonds when water molecules escape.
   c. the absorption of heat as hydrogen bonds break.
   d. the reduction in the average kinetic energy of a liquid after energetic water molecules enter the gaseous state.
   e. both c and d.

9. Ice floats because
   a. air is trapped in the crystalline lattice.
   b. the formation of hydrogen bonds releases heat; warmer objects float.
   c. it has a smaller surface area than liquid water.

d. it insulates bodies of water so they do not freeze from the bottom up.
e. hydrogen bonding spaces the molecules farther apart, creating a less dense structure.

10. Why is water such an excellent solvent?
   a. As a polar molecule, it can surround and dissolve ionic and polar molecules.
   b. It forms ionic bonds with ions, hydrogen bonds with polar molecules, and hydrophobic interactions with nonpolar molecules.
   c. It forms hydrogen bonds with itself.
   d. It has a high specific heat and a high heat of vaporization.
   e. It is liquid and has a high surface tension.

11. Which of the following, when mixed with water, would form a colloid?
   a. a large hydrophobic protein
   b. a large hydrophilic protein
   c. sugar
   d. cotton
   e. NaCl

12. A hydration shell is likely to form around
   a. an ion.
   b. a fat.
   c. a sugar.
   d. both a and c.
   e. both b and c.

13. Which of the following are least soluble in water?
   a. polar molecules
   b. nonpolar molecules
   c. ionic compounds
   d. hydrophilic molecules
   e. anions

14. The molarity of a solution is equal to
   a. Avogadro's number of molecules in 1 liter of solvent.
   b. the number of moles of a solute in 1 liter of solution.
   c. the molecular mass of a solute in 1 liter of solution.
   d. the number of solute molecules in 1 liter of solvent.
   e. 342 g if the solute is sucrose.

15. Which of the following substances would you add to enough water to yield 1 liter of solution in order to make a 0.1 M solution of glucose ($C_6H_{12}O_6$)? The mass numbers for these elements are approximately $C = 12$, $O = 16$, and $H = 1$.
   a. 6 g C, 12 g H, and 6 g O
   b. 72 g C, 12 g H, and 96 g O
   c. 18 g of glucose
   d. 29 g of glucose
   e. 180 g of glucose

16. How many molecules of glucose would be in 1 liter of the 0.1 *M* solution made in question 15?
    a. 0.1
    b. 6
    c. 60
    d. 6×10^{23}
    e. 6×10^{22}

17. Adding a base to a solution would
    a. raise the pH.
    b. lower the pH.
    c. decrease [H$^+$].
    d. do both a and c.
    e. do both b and c.

18. Some archaea are able to live in lakes with pH values of 11. How does pH 11 compare with the pH 7 typical of your body cells?
    a. It is four times more acidic than pH 7.
    b. It is four times more basic than pH 7.
    c. It is a thousand times more acidic than pH 7.
    d. It is a thousand times more basic than pH 7.
    e. It is ten thousand times more basic than pH 7.

19. A buffer
    a. releases excess OH$^-$.
    b. releases excess H$^+$.
    c. is often a weak acid-base pair.
    d. always maintains a neutral pH.
    e. Both c and d are correct.

Use the following pH values to answer questions 20–22: cola–2; orange juice–3; beer–4; coffee–5; human blood–7.4.

20. Which of these liquids has the *highest* molar concentration of OH$^-$?
    a. cola
    b. orange juice
    c. beer
    d. coffee
    e. human blood

21. Comparing the [H$^+$] of orange juice and coffee, the [H$^+$] of
    a. orange juice is 10 times higher.
    b. orange juice is 100 times higher.
    c. orange juice is 1,000 times higher.
    d. coffee is 2 times higher.
    e. coffee is 100 times higher.

22. What is the concentration of hydroxide ions in 1 liter of beer?
    a. 1.0×10^{-4}
    b. 1.0×10^{-10}
    c. 6.02×10^{23}
    d. 6.02×10^{19}
    e. 6.02×10^{-13}

23. Which would be the best method for reducing acid precipitation?
    a. Raise the height of smokestacks so that exhaust enters the upper atmosphere.
    b. Add buffers and bases to bodies of water whose pH has dropped.
    c. Use coal-burning generators rather than nuclear power to produce electricity.
    d. Tighten emission control standards for factories and automobiles.
    e. Reduce the concentration of heavy metals in industrial exhaust.

24. In the past century, the average temperature of the oceans has increased by 0.74°C. Is this evidence of global warming?
    a. No, the rise in temperature is too small to be significant.
    b. No, global warming affects air temperature, not water temperature.
    c. No, the change of average temperature does not reflect the quantity of heat in the oceans.
    d. Yes, because of the high specific heat of water and the huge volume of water in the oceans, a small rise in temperature would reflect a large amount of heat absorbed by the oceans.
    e. Yes, the decreased rate of calcification is directly related to this temperature increase.

Chapter 4

Carbon and the Molecular Diversity of Life

Key Concepts

4.1 Organic chemistry is the study of carbon compounds

4.2 Carbon atoms can form diverse molecules by bonding to four other atoms

4.3 A few chemical groups are key to the functioning of biological molecules

Framework

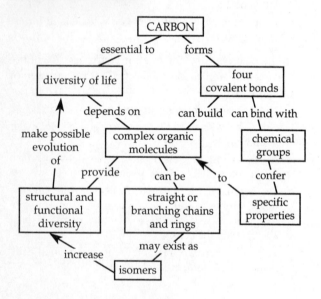

Chapter Review

4.1 Organic chemistry is the study of carbon compounds

Organic compounds are those containing carbon and usually hydrogen. Early chemists could not synthesize the complex molecules found in living organisms and therefore attributed their formation to a life force independent of physical and chemical laws, a belief known as *vitalism*.

Organic Molecules and the Origin of Life on Earth Stanley Miller's 1953 experiment showed that organic molecules could form under conditions believed to simulate those on early Earth. *Mechanism*, the philosophy underlying modern **organic chemistry**, holds that physical and chemical explanations are sufficient to account for all natural phenomena.

INTERACTIVE QUESTION   4.1

How did Miller's classic experiment apply mechanism to the origin of life?

4.2   Carbon atoms can form diverse molecules by bonding to four other atoms

The Formation of Bonds with Carbon How many covalent bonds must carbon (with an atomic number of 6) form to complete its valence shell? Carbon's valence of four is at the center of its ability to form large and complex molecules. When carbon forms four single covalent bonds, its hybrid orbitals create a tetrahedral shape. When two carbons are joined by a double bond, the other atoms bonded to the carbons are in the same plane, forming a flat molecule.

Molecular Diversity Arising from Carbon Skeleton Variation Carbon skeletons can vary in length, branching, placement of double bonds, and the presence of rings. **Hydrocarbons** consist of only carbon and hydrogen. Hydrocarbon chains are hydrophobic due to their nonpolar C—H bonds, and they release energy when broken down.

Isomers are compounds with the same molecular formula but different structural arrangements and thus different properties. **Structural isomers** differ in the covalent arrangement of atoms and often in the location of double bonds. *Cis-trans* isomers (formerly called geometric isomers) have the same sequence of covalently bonded atoms overall but differ in structure due to the inflexibility of double bonds. A *cis* isomer has the same atoms attached to the carbons on the same side of the double bond; a *trans* isomer has these atoms on opposite sides of the double bond. **Enantiomers** are left- and right-handed versions of a molecule and can differ greatly in their biological activity. An *asymmetric carbon* is one that is covalently bonded to four different kinds of atoms or groups of atoms, whose arrangement can result in mirror images.

INTERACTIVE QUESTION 4.2

Identify the structural isomers, *cis-trans* isomers, and enantiomers from the following compounds. Which of the *cis-trans* isomers is the *cis* isomer?

of organic compounds in water. A seventh group, the nonpolar methyl group, alters molecular shape and may serve as a signal on organic molecules.

The **hydroxyl group** consists of an oxygen and hydrogen (—OH). Organic molecules with hydroxyl groups are called alcohols, and their names often end in -*ol*.

A **carbonyl group** consists of a carbon double-bonded to an oxygen ($\rangle$CO). If the carbonyl group is at the end of the carbon skeleton, the compound is called an aldehyde. Otherwise, the compound is called a ketone.

A **carboxyl group** consists of a carbon double-bonded to an oxygen and also attached to a hydroxyl group (—COOH). Compounds with a carboxyl group are called carboxylic acids or organic acids because they tend to release H^+, becoming a carboxylate ion (—COO$^-$).

An **amino group** consists of a nitrogen atom bonded to two hydrogens (—NH$_2$). Compounds with an amino group, called amines, can act as bases, picking up a hydrogen ion and becoming —NH$_3^+$.

The **sulfhydryl group** consists of a sulfur atom bonded to a hydrogen (—SH). Thiols are compounds containing sulfhydryl groups.

A **phosphate group** is bonded to the carbon skeleton by an oxygen attached to a phosphorus atom that is bonded to three other oxygen atoms (—OPO$_3^{2-}$). This anion contributes negative charge to organic phosphates.

A **methyl group** is a carbon bonded to three hydrogens (—CH$_3$). Methylated compounds may have their function modified due to the addition of the methyl group.

ATP: An Important Source of Energy for Cellular Processes **Adenosine triphosphate,** or **ATP,** consists of the organic molecule adenosine to which three phosphate groups are attached. When ATP reacts with water, the third phosphate is split off and energy is released.

The Chemical Elements of Life: A Review Carbon, oxygen, hydrogen, nitrogen, and smaller quantities of sulfur and phosphorus, all capable of forming strong covalent bonds, are combined into the complex organic molecules of living matter.

4.3 A few chemical groups are key to the functioning of biological molecules

The Chemical Groups Most Important in the Processes of Life The properties of organic molecules are largely determined by characteristic chemical groups attached to the carbon skeleton. The following six **functional groups** may participate in chemical reactions. These hydrophilic groups increase the solubility

Word Roots

carb- = coal (*carboxyl group:* a chemical group present in organic acids, consisting of a carbon atom double-bonded to an oxygen atom and also bonded to a hydroxyl group)

enanti- = opposite (*enantiomer:* molecules that are mirror images of each other due to the presence of an asymmetric carbon)

hydro- = water (*hydrocarbon:* an organic molecule consisting only of carbon and hydrogen)

iso- = equal (*isomer:* one of several organic compounds with the same molecular formula but different structures and therefore different properties)

sulf- = sulfur (*sulfhydryl group:* a chemical group that consists of a sulfur atom bonded to a hydrogen atom)

Structure Your Knowledge

1. Construct a concept map that illustrates your understanding of the characteristics and significance of the three types of isomers. *A suggested map is in the answer section. Comparing and discussing your map with that of a study partner would be most helpful.*

2. Fill in the following table on the important chemical groups of organic compounds.

Chemical Group	Molecular Formula	Names and Characteristics of Compounds Containing Group
	—OH	
		Aldehyde or ketone; polar group
Carboxyl		
	—NH₂	
		Thiols; cross-links stabilize protein structure
Phosphate		
	—CH₃	

Test Your Knowledge

MULTIPLE CHOICE: *Choose the one best answer.*

1. Which of the following did the results of Stanley Miller's experiments support?
   a. abiotic synthesis of organic compounds
   b. possible first stage in the origin of life
   c. mechanism
   d. vitalism
   e. Answers a, b, and c are correct.

2. Carbon's valence of four most directly results from
   a. its tetrahedral shape.
   b. its very slight electronegativity.
   c. its four electrons in the valence shell that can form four covalent bonds.
   d. its ability to form single, double, and triple bonds.
   e. its ability to form chains and rings of carbon atoms.

3. Hydrocarbons are not soluble in water because
   a. they are hydrophilic.
   b. the C—H bond is nonpolar.
   c. they do not ionize.
   d. they store energy in the many C—H bonds along the carbon backbone.
   e. they are lighter than water.

4. Which of the following is *not* true of an asymmetric carbon atom?
   a. It is attached to four different atoms or groups.
   b. It results in right- and left-handed versions of a molecule.
   c. It is found in all enantiomers.
   d. Its configuration is in the shape of a tetrahedron.
   e. It is found in all *cis-trans* isomers.

5. Which statement is *not* true about structural isomers?
   a. They have different chemical properties.
   b. They have the same molecular formula.
   c. Their atoms and bonds are arranged in different sequences.
   d. They are a result of restricted movement around a carbon double bond.
   e. Their possible numbers increase as carbon skeletons increase in size.

6. The fats stored in your body consist mostly of
   a. methyl groups.
   b. alcohols.
   c. carboxylic acids.
   d. hydrocarbons.
   e. organic phosphates.

7. The following ribose molecule contains how many asymmetric carbons?

   a. 1   c. 3   e. 5
   b. 2   d. 4

$$H-C(=O)$$
$$H-C-OH$$
$$H-C-OH$$
$$H-C-OH$$
$$H-C-OH$$
$$H$$

8. Which of the following is mismatched with its description?

   a. phosphate—forms bonds that stabilize protein structure

   b. hydroxyl and carbonyl—components of sugars

   c. ATP—source of energy for cellular processes

   d. methyl—its addition changes the shape and function of molecules

   e. amino and carboxyl—components of amino acids

9. The chemical group that can cause an organic molecule to act as a base is

   a. —COOH.   **c.** —SH.   **e.** —CH$_3$.

   b. —OH.   **d.** —NH$_2$.

10. The chemical group that confers acidic properties to organic molecules is

   a. —COOH.   **c.** —SH.   **e.** —CH$_3$.

   b. —OH.   **d.** —NH$_2$.

MATCHING: *Match the formulas (**a–f**) to the terms at the right. Choices may be used more than once, and terms may have more than one correct answer.*

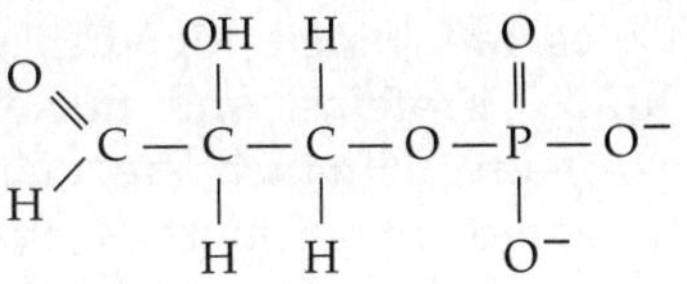

a.

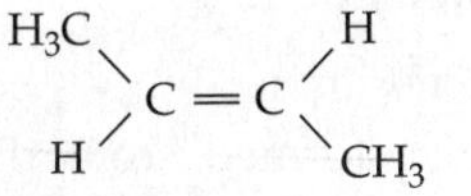

b.

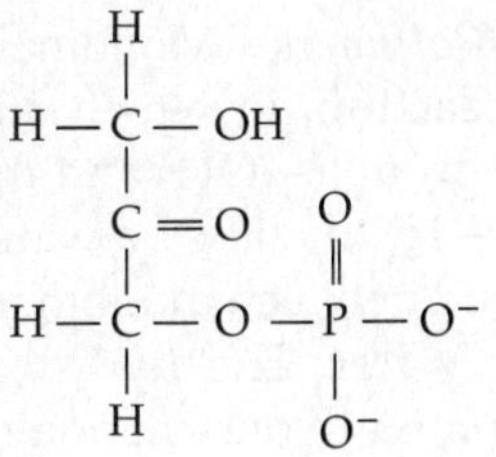

c.

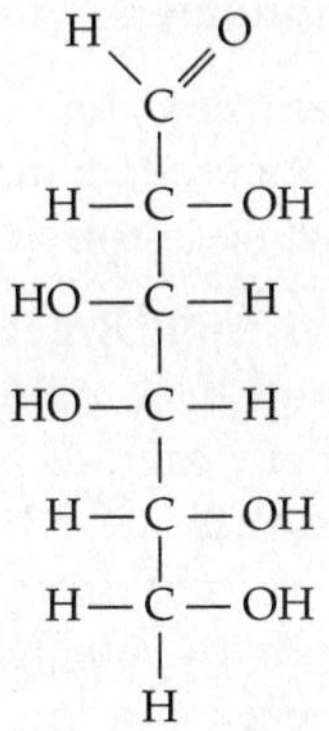

d.

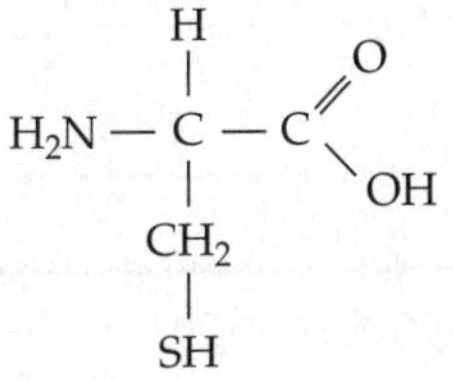

e.

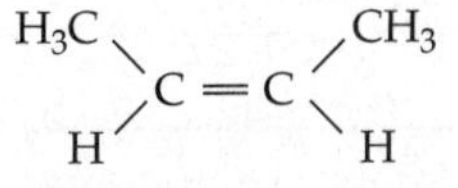

f.

_____ **1.** structural isomers

_____ **2.** *cis-trans* isomers

_____ **3.** can have enantiomers

_____ **4.** carboxylic acid

_____ **5.** can make cross-link in protein

_____ **6.** hydrophilic

_____ **7.** hydrocarbon

_____ **8.** amino acid

_____ **9.** organic phosphate

_____ **10.** aldehyde

_____ **11.** amine

_____ **12.** ketone

Chapter 5

The Structure and Function of Large Biological Molecules

Key Concepts

5.1 Macromolecules are polymers, built from monomers

5.2 Carbohydrates serve as fuel and building materials

5.3 Lipids are a diverse group of hydrophobic molecules

5.4 Proteins include a diversity of structures, resulting in a wide range of functions

5.5 Nucleic acids store, transmit, and help express hereditary information

Framework

A small number of monomers or subunits joined into unique sequences and forming three-dimensional structures create a huge variety of large molecules with diverse functions. The following table briefly summarizes the major characteristics of the four classes of biological molecules.

Chapter Review

Smaller organic molecules are joined together to form carbohydrates, lipids, proteins, and nucleic acids. These molecules, many of which are giant **macromolecules,** have emergent properties that arise from their complex and unique structures.

5.1 Macromolecules are polymers, built from monomers

Polymers are chainlike molecules formed from the linking together of many similar or identical small molecules, called **monomers.**

Synthesis and Breakdown of Polymers Monomers are joined by a **dehydration reaction,** in which one monomer provides a hydroxyl group (—OH) and the other contributes a hydrogen (—H) to release a water molecule. In **hydrolysis,** the bond between monomers is broken by the addition of water. The hydroxyl group of a water molecule is joined to one monomer while the hydrogen is bonded with the other. **Enzymes** catalyze both dehydration reactions and hydrolysis.

Class	Monomers or Components	Functions
Carbohydrates	Monosaccharides	Energy source, raw materials, energy storage, structural compounds
Lipids	Glycerol and fatty acids → fats; phospholipids; steroids	Energy storage, membrane components, hormones
Proteins	Amino acids	Enzymes, transport, movement, receptors, defense, structure
Nucleic acids	Nucleotides	Heredity, various functions in gene expression

Diversity of Polymers Polymers are constructed from about 40 to 50 common monomers and a few rarer molecules. The seemingly endless variety of macromolecules arises from the essentially infinite number of possibilities in the sequencing and arrangement of these basic building blocks.

INTERACTIVE QUESTION 5.1

Monomers are linked into polymers by _____ _____, which involve the _____ of a water molecule.

Polymers are broken down to monomers by _____, which involves the _____ of a water molecule.

5.2 Carbohydrates serve as fuel and building materials

Carbohydrates include sugars and their polymers.

Sugars **Monosaccharides** have the general formula of $(CH_2O)_n$. The number (n) of these units forming a sugar varies from three to seven, with hexoses ($C_6H_{12}O_6$), trioses, and pentoses being most common. Sugar molecules may also vary due to the spatial arrangement of parts around asymmetric carbons.

INTERACTIVE QUESTION 5.2

Fill in the blanks to review monosaccharides.

You can recognize a monosaccharide by its multiple (a) _____ groups and its one (b) _____ group, whose location determines whether the sugar is an (c) _____ or a (d) _____. In aqueous solutions, most five- and six-carbon sugars form (e) _____.

Glucose is broken down to yield energy in cellular respiration. Monosaccharides also serve as the raw materials for the synthesis of other organic molecules. Two monosaccharides are joined by a **glycosidic linkage** to form a **disaccharide.**

Polysaccharides **Polysaccharides** are storage or structural macromolecules. **Starch,** a storage molecule in plants, is a polymer made of glucose molecules joined by 1–4 glycosidic linkages that give starch a helical shape. Animals use **glycogen,** a highly branched polymer of glucose, as their energy storage molecule.

Cellulose, the major component of plant cell walls, is the most abundant organic compound on Earth. It differs from starch by the configuration of the ring form of glucose (beta instead of alpha) and the resulting geometry of the glycosidic bonds. In a plant cell wall, hydrogen bonds between hydroxyl groups hold parallel cellulose molecules together to form strong microfibrils.

Enzymes that digest the α linkages of starch are unable to hydrolyze the β linkages of cellulose. Only a few organisms (some prokaryotes and fungi) have enzymes that can digest cellulose.

Chitin is a structural polysaccharide formed from glucose monomers with a nitrogen-containing group. Chitin is found in the exoskeleton of arthropods and the cell walls of many fungi.

INTERACTIVE QUESTION 5.3

Number the carbons in the following glucose molecules (each unlabeled corner of the ring represents a carbon). Circle the atoms that will be removed by a dehydration reaction. Then draw the resulting maltose molecule with its 1–4 glycosidic linkage.

Glucose   Glucose

Maltose

INTERACTIVE QUESTION   5.4

Fill in the following concept map that summarizes this section on carbohydrates.

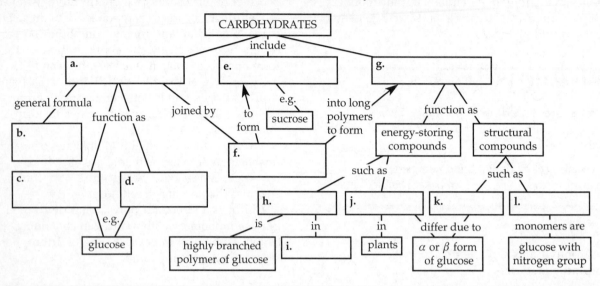

5.3   Lipids are a diverse group of hydrophobic molecules

Fats, phospholipids, and steroids are part of a diverse assemblage of biological molecules that are grouped together as **lipids** based on their hydrophobic behavior. Lipids do not form polymers.

Fats   Fats are composed of **fatty acids** attached to the three-carbon alcohol, glycerol. A fatty acid consists of a long hydrocarbon chain with a carboxyl group at one end. The nonpolar hydrocarbons make a fat hydrophobic.

A **triacylglycerol,** or fat, consists of three fatty acid molecules, each linked to glycerol by an ester linkage, a bond that forms between a hydroxyl and a carboxyl group. Triglyceride is another name for fats.

Fatty acids with double bonds in their carbon chain are called **unsaturated fatty acids.** The *cis* double bonds create a kink in the hydrocarbon chain and prevent fat molecules with unsaturated fatty acids from packing closely together and becoming solidified at room temperature. The fats of plants and fish are generally unsaturated and are called oils. **Saturated fatty acids** have no double bonds in their carbon chains. Most animal fats are saturated and solid at room temperature. Diets rich in saturated fats and in **trans fats** made in the process of hydrogenating vegetable oils have been linked to cardiovascular disease.

Fats are excellent energy storage molecules, containing twice the energy reserves of carbohydrates such as starch. Adipose tissue, made of fat storage cells, also cushions organs and insulates the body.

Phospholipids   **Phospholipids** consist of a glycerol linked to two fatty acids and a negatively charged phosphate group, to which other small molecules are attached. The phosphate head of this molecule is hydrophilic and water soluble, whereas the two fatty acid chains are hydrophobic. The unique structure of phospholipids makes them ideal constituents of cell membranes.

INTERACTIVE QUESTION   5.5

Sketch a section of a phospholipid bilayer of a membrane, and label the hydrophilic head and hydrophobic tails of one of the phospholipids.

Steroids   **Steroids** are a class of lipids distinguished by four connected carbon rings with various chemical groups attached. **Cholesterol** is a common component of animal cell membranes and a precursor for other steroids, including many hormones.

INTERACTIVE QUESTION 5.6

Fill in this concept map to help you organize your understanding of lipids.

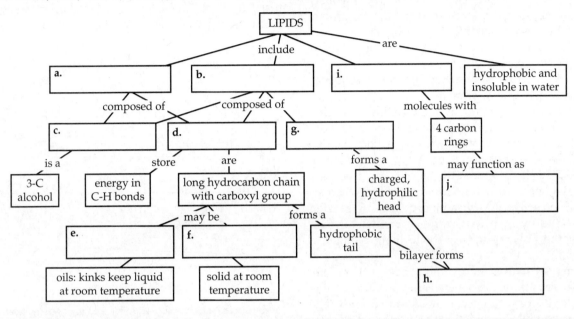

5.4 Proteins include a diversity of structures, resulting in a wide range of functions

Proteins are central to almost every function of life. Most enzymes, which function as **catalysts** that selectively speed up the chemical reactions of a cell, are proteins.

Polypeptides A **polypeptide** is a polymer of amino acids. A **protein** is a functional molecule that consists of one or more polypeptides, each folded into a specific three-dimensional shape.

Amino acids are composed of an asymmetric carbon (called the *alpha [α] carbon*) bonded to a hydrogen atom, a carboxyl group, an amino group, and a variable side chain called the R group. At the pH in a cell, the amino and carboxyl groups are usually ionized. The R group confers the unique physical and chemical properties of each amino acid. Side chains may be either nonpolar and hydrophobic, or polar or charged (acidic or basic) and thus hydrophilic.

A **peptide bond** links the carboxyl group of one amino acid with the amino group of another.

INTERACTIVE QUESTION 5.7

a. Draw the amino acids alanine (R group: —CH_3) and serine (R group: —CH_2OH) and then show how a dehydration reaction will form a peptide bond between them.

b. Which of these amino acids has a polar R group? a nonpolar R group?

c. What does the following molecule segment represent? (Note the N—C—C—N—C—C sequence.)

Protein Structure and Function A protein has a unique three-dimensional shape, or structure, created by the twisting or folding of one or more polypeptide chains. Protein structure usually arises spontaneously as the protein is synthesized in the cell. The unique structure of a protein enables it to recognize and bind to other molecules. *Globular proteins* are roughly spherical; *fibrous proteins* are long fibers.

Primary structure is the genetically coded sequence of amino acids within a protein.

Secondary structure involves the coiling or folding of the polypeptide backbone, stabilized by hydrogen bonds between the oxygen (with a partial negative charge) of one peptide bond and the partially positive hydrogen attached to the nitrogen of another peptide bond. An **α helix** is a coil produced by hydrogen bonding between every fourth amino acid. A **β pleated sheet** is held by repeated hydrogen bonds along regions of the polypeptide backbone lying parallel to each other.

Tertiary structure, the overall shape of a protein, results from interactions between the various side chains (R groups) of the constituent amino acids. The following chemical interactions help produce the stable and unique shape of a protein: **hydrophobic interactions** between nonpolar side groups clumped in the center of the molecule due to their repulsion by water, van der Waals interactions among those non-polar side chains, hydrogen bonds between polar side chains, and ionic bonds between negatively and positively charged side chains. Strong covalent bonds, called **disulfide bridges,** may occur between the sulfhydryl side groups of cysteine monomers that have been brought close together by the folding of the polypeptide.

Quaternary structure occurs in proteins that are composed of more than one polypeptide. The individual polypeptide subunits are held together in a precise structural arrangement to form a functional protein.

INTERACTIVE QUESTION   5.8

In the following diagram of a portion of a protein, label the types of interactions that are shown. What level of structure are these interactions producing?

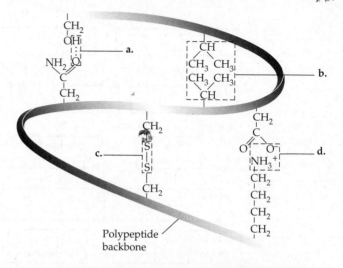

Polypeptide backbone

In the inherited blood disorder **sickle-cell disease,** a change in one amino acid affects the structure of a hemoglobin molecule, causing red blood cells to deform into a sickle shape that clogs tiny blood vessels.

The bonds and interactions that maintain the three-dimensional shape of a protein may be disrupted by changes in pH, salt concentration, or temperature, causing a protein to unravel. **Denaturation** also occurs if a protein is transferred to an organic solvent; in that case, its hydrophilic regions cluster on the inside while its hydrophobic regions are on the outside interacting with the nonpolar solvent.

The primary structure (sequence of amino acids) determines where the interactions and bonds that maintain a protein's shape can form. Within a cell, **chaperonins,** or chaperone proteins, assist newly made polypeptides during the folding process, perhaps by providing a sheltered environment.

Using the techniques of **X-ray crystallography,** nuclear magnetic resonance (NMR) spectroscopy, and bioinformatics, biochemists have identified the structure of thousands of proteins. These structures can then be related to the specific functions of different regions of a protein.

INTERACTIVE QUESTION 5.9

Now that you have gained experience with concept maps, create your own map to review what you have learned about proteins. Try to include the concepts of structure and function, and look for cross-links on your map. You may want to include the functions of proteins. *One version of a protein concept map is included in the answer section, but remember that the real value is in the thinking process you must go through to create your own map.*

5.5 Nucleic acids store, transmit, and help express hereditary information

Genes are the units of inheritance that determine the primary structure of proteins. **Nucleic acids** are the polymers that carry and transmit this information.

The Roles of Nucleic Acids DNA, **deoxyribonucleic acid,** is the genetic material that is inherited from one generation to the next and is replicated whenever a cell divides so that all cells of an organism contain identical DNA. The instructions coded in DNA are transcribed to **RNA, ribonucleic acid,** which directs the synthesis of proteins, the ultimate enactors of the genetic program. In a eukaryotic cell, DNA resides in the nucleus. *Messenger RNA (mRNA)* carries the instructions for protein synthesis to ribosomes located in the cytoplasm. Recent research has revealed other important functions of RNA.

The Components of Nucleic Acids **Polynucleotides** are polymers of **nucleotides**—monomers that consist of a pentose (five-carbon sugar) covalently bonded to a phosphate group and a nitrogenous (nitrogen-containing) base. A nucleotide may contain more than one phosphate group; without the phosphate group it is called a *nucleoside.*

There are two families of nitrogenous bases. **Pyrimidines,** including cytosine (C), thymine (T), and uracil (U), are characterized by six-membered rings of carbon and nitrogen atoms. **Purines,** adenine (A) and guanine (G), add a five-membered ring to the pyrimidine ring. Thymine is only in DNA; uracil is only in RNA. In DNA, the sugar is **deoxyribose;** in RNA, it is **ribose.**

Nucleotide Polymers Nucleotides are linked together into a polynucleotide by phosphodiester linkages, which join the sugar of one nucleotide with the phosphate of the next. The polymer has two distinct ends: a 5′ end with a phosphate attached to the 5′ carbon of a sugar, and a 3′ end with a hydroxyl group on the 3′ carbon of a sugar. The nitrogenous bases extend from this backbone of repeating sugar-phosphate units. The unique sequence of bases in a gene codes for the specific amino acid sequence of a protein.

INTERACTIVE QUESTION 5.10

a. Label the three parts of this nucleotide. Indicate with an arrow where the phosphate group of the next nucleotide would attach to build a polynucleotide. Number the carbons of the pentose.

b. Is this a purine or a pyrimidine? How do you know?

c. Is this a DNA nucleotide or an RNA nucleotide? How do you know?

The Structure of DNA and RNA Molecules DNA molecules consist of two polynucleotides (strands) spiraling in a **double helix.** The two sugar-phosphate backbones run in opposite 5′ to 3′ directions, an arrangement called **antiparallel.** In 1953, Watson and Crick first proposed the double helix structure, with the nitrogenous bases pairing and hydrogen-bonding together in the inside of the molecule. Adenine pairs only with thymine; guanine always pairs with cytosine. Thus, the sequences of nitrogenous bases on the two strands of DNA are *complementary.* Because of this specific base-pairing property, DNA can replicate itself and precisely copy the genes of inheritance.

RNA molecules are usually single polynucleotides, although base-pairing within or between RNA molecules is common. For example, the functional shape of *transfer RNA (tRNA)*, an RNA involved in protein synthesis, involves four regions of complementary base-pairing.

DNA and Proteins as Tape Measures of Evolution Genes form the hereditary link between generations. Closely related members of the same species share many common DNA sequences and proteins. More closely related species have a larger proportion of their DNA and proteins in common. This "molecular genealogy" provides evidence of evolutionary relationships.

INTERACTIVE QUESTION   5.11

Take the time to create a concept map that summarizes what you have just reviewed about nucleic acids. Compare your map with that of a study partner or explain it to a friend. Refer to Figures 5.26 and 5.27 in your textbook to help you visualize polynucleotides and the three-dimensional structures of DNA and RNA.

The Theme of Emergent Properties in the Chemistry of Life: A Review   At each stage in the hierarchy from atoms through large biological molecules, we have seen that novel properties arise with increasing structural organization.

Word Roots

di- = two; **-sacchar** = sugar (*disaccharide:* two monosaccharides [simple sugars] joined through a dehydration reaction)

glyco- = sweet (*glycogen:* an extensively branched glucose polysaccharide that stores energy in animals)

hydro- = water; **-lyse** = break (*hydrolysis:* a chemical reaction that breaks bonds between two molecules by the addition of water)

macro- = large (*macromolecule:* a giant molecule formed by the joining of smaller molecules, such as a polysaccharide, a protein, or a nucleic acid)

poly- = many; **meros-** = part (*polymer:* a long molecule consisting of many similar or identical monomers)

tri- = three (*triacylglycerol:* three fatty acids linked to one glycerol molecule; also called a fat or triglyceride)

Structure Your Knowledge

1. Describe the four structural levels in the functional shape of a protein.

2. Identify the type of monomer or group shown by the formulas a–g. Then match the chemical formulas with their description. Answers may be used more than once.

_____ 1. molecules that would combine to form a fat

_____ 2. molecule that would be attached to other monomers by a peptide bond

_____ 3. molecules or groups that would combine to form a nucleotide

_____ 4. molecules that are carbohydrates

_____ 5. molecule that is a purine

_____ 6. monomer of a protein

_____ 7. groups that would be joined by phosphodiester bonds

_____ 8. most nonpolar (hydrophobic) molecule

a. _____

b. _____

c. _____

d. _____

e. _____

f. _____

g. _____

Test Your Knowledge

MATCHING: *Match the molecule with its class of molecule.*

_____ **1.** glycogen
_____ **2.** cholesterol
_____ **3.** RNA
_____ **4.** collagen
_____ **5.** hemoglobin
_____ **6.** a gene
_____ **7.** triacylglycerol
_____ **8.** enzyme
_____ **9.** cellulose
_____ **10.** chitin

A. carbohydrate
B. lipid
C. protein
D. nucleic acid

MULTIPLE CHOICE: *Choose the one best answer.*

1. Polymerization (the formation of polymers) is a process that
  a. creates bonds between amino acids in the formation of a polypeptide.
  b. involves the removal of a water molecule.
  c. links the sugar of one nucleotide with the phosphate of the next.
  d. involves a dehydration reaction.
  e. may involve all of the above.

2. Which of the following statements is *not* true of a pentose?
  a. It can be found in nucleic acids.
  b. It can occur in a ring structure.
  c. It has the formula $C_5H_{12}O_5$.
  d. It has one carbonyl and four hydroxyl groups.
  e. It may be an aldose or a ketose.

3. Disaccharides can differ from each other in all of the following ways *except*
  a. in the number of their monosaccharides.
  b. in their structural formulas.
  c. in the monomers involved.
  d. in the location of their glycosidic linkage.
  e. They can differ in all of these ways.

4. Which of the following statements is *not* true of cellulose?
  a. It is the most abundant organic compound on Earth.
  b. It differs from starch because of the configuration of glucose and the geometry of the glycosidic linkage.
  c. It may be hydrogen-bonded to neighboring cellulose molecules to form microfibrils.

  d. Few organisms have enzymes that hydrolyze its glycosidic linkages.
  e. Its monomers are glucose with nitrogen-containing appendages.

5. Plants store most of their energy for later use as
  a. unsaturated fats.
  b. glycogen.
  c. starch.
  d. sucrose.
  e. cellulose.

6. Sucrose is made from joining a glucose molecule and a fructose molecule in a dehydration reaction. What is the molecular formula for this disaccharide?
  a. $C_6H_{12}O_6$
  b. $C_{10}H_{20}O_{10}$
  c. $C_{12}H_{22}O_{11}$
  d. $C_{12}H_{24}O_{12}$
  e. $C_{12}H_{24}O_{13}$

7. A cow can derive nutrients from cellulose because
  a. it can produce the enzymes that break the β linkages between glucose molecules.
  b. it chews and rechews its cud so that cellulose fibers are finally broken down.
  c. its rumen contains prokaryotes that can hydrolyze the bonds of cellulose.
  d. its intestinal tract contains termites, which harbor microbes that hydrolyze cellulose.
  e. it can convert cellulose to starch and then hydrolyze starch to glucose.

8. Which of the following substances is the major component of the cell membrane of a fungus?
  a. cellulose
  b. chitin
  c. cholesterol
  d. phospholipids
  e. unsaturated fatty acids

9. A fatty acid that has the formula $C_{16}H_{32}O_2$ is
  a. saturated.
  b. unsaturated.
  c. branched.
  d. hydrophilic.
  e. part of a steroid molecule.

10. Three molecules of the fatty acid in question 9 are joined to a molecule of glycerol ($C_3H_8O_3$). The resulting molecule has the formula
  a. $C_{48}H_{96}O_6$.
  b. $C_{48}H_{98}O_9$.
  c. $C_{51}H_{102}O_8$.
  d. $C_{51}H_{98}O_6$.
  e. $C_{51}H_{104}O_9$.

11. What are trans fats?
  a. hydrogenated fish oils that have been identified with health risks
  b. fats made from cholesterol that are components of plaques in the walls of blood vessels
  c. fats that are derived from animal sources and are associated with cardiovascular disease
  d. fats that contain *trans* double bonds and may contribute to atherosclerosis
  e. polyunsaturated fats produced by removing H from fatty acids and forming *cis* double bonds

12. Which of the following molecules is the most hydrophobic?
    a. cholesterol
    b. nucleotide
    c. chitin
    d. phospholipid
    e. glucose

13. Which of the following molecules provides the most energy (kcal/g) when eaten and digested?
    a. glucose
    b. starch
    c. glycogen
    d. fat
    e. protein

14. Which of the following is *not* one of the many functions performed by proteins?
    a. acting as signals and receptors
    b. acting as an enzymatic catalyst for metabolic reactions
    c. providing protection against disease
    d. serving as contractile components of muscle
    e. forming primary structural component of membranes

15. What happens when a protein denatures?
    a. Its primary structure is disrupted.
    b. Its secondary and tertiary structures are disrupted.
    c. It always flips inside out.
    d. It hydrolyzes into component amino acids.
    e. Its hydrogen bonds, ionic bonds, hydrophobic interactions, disulfide bridges, and peptide bonds are disrupted.

16. The α helix of proteins is
    a. part of the protein's tertiary structure and is stabilized by disulfide bridges.
    b. a double helix.
    c. stabilized by hydrogen bonds and is commonly found in fibrous proteins.
    d. found in some regions of globular proteins and is stabilized by hydrophobic interactions.
    e. a complementary sequence to messenger RNA.

17. β pleated sheets are characterized by
    a. disulfide bridges between cysteine amino acids.
    b. parallel regions of the polypeptide chain held together by hydrophobic interactions.
    c. folds stabilized by hydrogen bonds between segments of the polypeptide backbone.
    d. membrane sheets composed of phospholipids.
    e. hydrogen bonds between adjacent cellulose molecules.

18. What is the best description of the following molecule?

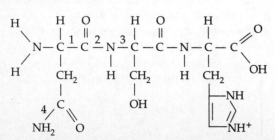

19. Which number(s) in the molecule in question 18 refer(s) to a peptide bond?
    a. 1   c. 3   e. both 2 and 4
    b. 2   d. 4

    a. chitin   d. nucleotide
    b. amino acid   e. protein
    c. tripeptide

20. What *determines* the sequence of the amino acids in a particular protein?
    a. its primary structure
    b. the sequence of nucleotides in RNA, which was determined by the sequence of nucleotides in the gene for that protein
    c. the sequence of nucleotides in DNA, which was determined by the sequence of nucleotides in RNA
    d. the sequence of RNA nucleotides making up the ribosome
    e. the three-dimensional shape of the protein

21. Both hydrophobic and hydrophilic interactions are important for which of the following types of molecules?
    a. proteins   d. polynucleotides
    b. saturated fats   e. all of the above
    c. glycogen and cellulose

22. How are nucleotide monomers connected to form a polynucleotide?
    a. by hydrogen bonds between complementary nitrogenous base pairs
    b. by ionic attractions between phosphate groups
    c. by disulfide bridges between cysteines
    d. by covalent bonds between the sugar of one nucleotide and the phosphate of the next
    e. by ester linkages between the carboxyl group of one nucleotide and the hydroxyl group on the ribose of the next

23. If the nucleotide sequence of one strand of a DNA helix is 5'GCCTAA3', what would be the 3'–5' sequence on the complementary strand?
    a. GCCTAA   d. ATTCGG
    b. CGGAUU   e. TAAGCC
    c. CGGATT

24. Monkeys and humans share many of the same DNA sequences and have similar proteins, indicating that
    a. the two groups belong to the same species.
    b. the two groups share a relatively recent common ancestor.
    c. humans evolved from monkeys.
    d. monkeys evolved from humans.
    e. the two groups evolved about the same time.

UNIT 2 The Cell

Chapter 6

A Tour of the Cell

Key Concepts

6.1 Biologists use microscopes and the tools of biochemistry to study cells

6.2 Eukaryotic cells have internal membranes that compartmentalize their functions

6.3 The eukaryotic cell's genetic instructions are housed in the nucleus and carried out by the ribosomes

6.4 The endomembrane system regulates protein traffic and performs metabolic functions in the cell

6.5 Mitochondria and chloroplasts change energy from one form to another

6.6 The cytoskeleton is a network of fibers that organizes structures and activities in the cell

6.7 Extracellular components and connections between cells help coordinate cellular activities

Framework

The cell is the fundamental unit of life. The complexities in the processes of life are reflected in the complexities of the cell. It is easy to become overwhelmed by the number of new vocabulary terms for this array of cell structures. The diagram on the following page provides an organizational framework for the wealth of detail found in "a tour of the cell."

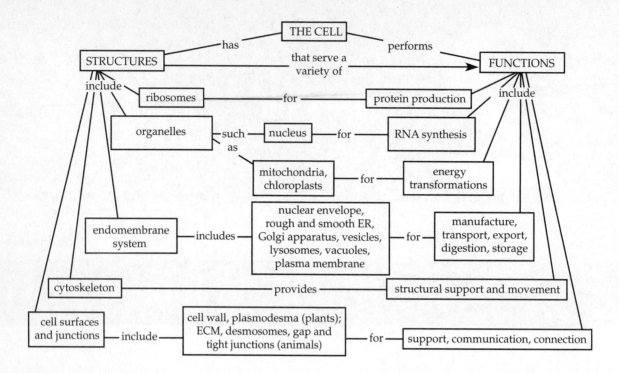

Chapter Review

The cell is the basic structural and functional unit of all organisms. In the hierarchy of biological organization, the capacity for life emerges from the structural order of the cell. All cells are related through common descent, but evolution has shaped diverse adaptations.

6.1 Biologists use microscopes and the tools of biochemistry to study cells

Microscopy Cells were first described by Robert Hooke in 1665. The glass lenses of a **light microscope (LM)** refract (bend) the visible light passing through a specimen such that the projected image is magnified. *Magnification* is the ratio of the size of an image to the real size of the object. *Resolution* is a measure of the clarity of an image and is determined by the minimum distance two points must be separated to be seen as separate. The resolving power of the light microscope is limited, so that details finer than 0.2 μm (micrometers = 10^{-3} mm) cannot be distinguished. Staining specimens and using techniques such as fluorescence, phase-contrast, and confocal microscopy improve visibility by increasing *contrast* between structures.

Many cellular structures, including most membrane-enclosed **organelles,** cannot be resolved by the light microscope. Their structures were first revealed when the **electron microscope (EM)** was developed in the 1950s. The electron microscope focuses a beam of electrons through a specimen or onto its surface. The short wavelength of the electron beam allows a resolution of about 2 nm (nanometers = 10^{-3} μm), about a hundred times greater than that of the light microscope.

In a **scanning electron microscope (SEM),** an electron beam scans the surface of a specimen that is usually coated with a thin gold film. The beam excites electrons from the specimen, which are detected and translated into an image on a video screen. This image appears three-dimensional.

In a **transmission electron microscope (TEM),** a beam of electrons is passed through a thin section of a specimen stained with atoms of heavy metals. Electromagnets, acting as lenses, then focus the image onto a screen or film.

New techniques such as super-resolution microscopy are providing high-resolution images. Modern cell biology integrates *cytology* with *biochemistry* to understand relationships between cellular structure and function.

INTERACTIVE QUESTION 6.1

a. Define cytology.

b. What do cell biologists use a TEM to study?

c. What does an SEM show best?

d. What advantages does light microscopy have over both TEM and SEM?

Cell Fractionation **Cell fractionation** is a technique that separates organelles and other subcellular structures of a cell so that they can be identified and their functions studied. Cells are broken apart and the homogenate is separated into component fractions by centrifugation at increasing speeds.

6.2 Eukaryotic cells have internal membranes that compartmentalize their functions

Comparing Prokaryotic and Eukaryotic Cells All cells are bounded by a *plasma membrane,* which encloses a semifluid medium called **cytosol.** All cells contain *chromosomes* and *ribosomes.*

Only members of the domains Bacteria and Archaea have **prokaryotic cells,** which are cells with no nucleus or membrane-enclosed organelles. The DNA of prokaryotic cells is concentrated in a region called the **nucleoid. Eukaryotic cells** have a true *nucleus* enclosed in a double membrane and numerous organelles suspended in cytosol. **Cytoplasm** refers to the region between the nucleus and the plasma membrane, and also to the interior of a prokaryotic cell.

Most bacterial cells range from 1 to 10 μm in diameter, whereas eukaryotic cells are 10 times larger, ranging from 10 to 100 μm. The small size of cells is influenced by geometry. Area is proportional to the square of linear dimension, while volume is proportional to its cube. The **plasma membrane** surrounding every cell must provide sufficient surface area for exchange of oxygen, nutrients, and wastes relative to the volume of the cell.

A Panoramic View of the Eukaryotic Cell Membranes compartmentalize the eukaryotic cell, providing local environments for specific metabolic functions and participating in metabolism through membrane-bound enzymes. All types of eukaryotic cells share many common structures, although there are also important differences between them.

INTERACTIVE QUESTION 6.2

a. Describe the molecular structure of the plasma membrane.

b. If a eukaryotic cell has a diameter that is 10 times that of a bacterial cell, proportionally how much more surface area would the eukaryotic cell have?

c. Proportionally how much more volume would it have?

6.3 The eukaryotic cell's genetic instructions are housed in the nucleus and carried out by the ribosomes

The Nucleus: Information Central The **nucleus** is surrounded by the **nuclear envelope,** a *double* membrane perforated by pores. A protein *pore complex* lining each pore regulates the movement of materials between the nucleus and the cytoplasm. The inner membrane is lined by the **nuclear lamina,** a layer of protein filaments that helps to maintain the shape of the nucleus. A *nuclear matrix* of fibers extends throughout the nucleus.

Most of the cell's DNA is located in the nucleus, where it is organized into units called **chromosomes,** which are made up of **chromatin,** a complex of DNA and proteins. Each eukaryotic species has a characteristic chromosomal number. Individual chromosomes are visible only when condensed in a dividing cell.

The **nucleolus,** a dense structure visible in the nondividing nucleus, synthesizes *ribosomal RNA (rRNA)* and combines it with protein to assemble ribosomal subunits, which then pass through nuclear pores to the cytoplasm.

Ribosomes: Protein Factories **Ribosomes** are composed of protein and ribosomal RNA. Most of the proteins produced on *free ribosomes* are used within the cytosol. *Bound ribosomes,* attached to the endoplasmic reticulum or nuclear envelope, usually make proteins that will be included within membranes, packaged into organelles, or exported from the cell.

INTERACTIVE QUESTION 6.3

How does the nucleus control protein synthesis in the cytoplasm?

6.4 The endomembrane system regulates protein traffic and performs metabolic functions in the cell

The **endomembrane system** of a cell consists of the nuclear envelope, endoplasmic reticulum, Golgi apparatus, lysosomes, vesicles, vacuoles, and the plasma membrane. Although not identical, all these membranes are related either through direct contact or by the transfer of membrane segments by membrane-bound sacs called **vesicles.**

The Endoplasmic Reticulum: Biosynthetic Factory The **endoplasmic reticulum (ER)** is a membranous system that is continuous with the nuclear envelope and encloses a network of interconnected tubules or compartments called cisternae. Ribosomes are attached to the cytoplasmic surface of **rough ER; smooth ER** lacks ribosomes.

Smooth ER serves diverse functions in different cells: Its enzymes are involved in phospholipid and steroid (including sex hormone) synthesis, carbohydrate metabolism, and detoxification of drugs and poisons. Barbiturates, alcohol, and other drugs increase a liver cell's production of smooth ER, thus leading to an increased tolerance (and thus reduced effectiveness) for these and other drugs. Smooth ER also functions in storage and release of calcium ions during muscle contraction.

Proteins intended for secretion are manufactured by membrane-bound ribosomes and then threaded into the lumen of the rough ER. Many are covalently bonded to small carbohydrates to form **glycoproteins.** Secretory proteins are transported from the rough ER in membrane-bound **transport vesicles.**

Rough ER also manufactures membranes. Enzymes built into the membrane assemble phospholipids, and membrane proteins formed by bound ribosomes are inserted into the ER membrane. Transport vesicles transfer ER membrane to other parts of the endomembrane system.

The Golgi Apparatus: Shipping and Receiving Center

The **Golgi apparatus** consists of a stack of flattened sacs. Vesicles that bud from the ER join to the *cis* face of a Golgi stack, adding to it their contents and membrane.

According to the *cisternal maturation model,* Golgi products are processed and tagged as the cisternae themselves progress from the *cis* to the *trans* face. Glycoproteins often have their attached carbohydrates modified. The Golgi apparatus of plant cells manufactures some polysaccharides, such as pectins. Golgi products are sorted into vesicles, which pinch off from the *trans* face of the Golgi apparatus. These vesicles may have surface molecules that help direct them to the plasma membrane or to other organelles.

Lysosomes: Digestive Compartments

In animal cells, **lysosomes** are membrane-enclosed sacs containing hydrolytic enzymes that digest macromolecules. Lysosomes provide an acidic pH for these enzymes.

In some protists, lysosomes fuse with *food vacuoles* to digest material ingested by **phagocytosis.** Macrophages, a type of white blood cell, use lysosomes to destroy ingested bacteria. Lysosomes also recycle a cell's own macromolecules by fusing with vesicles enclosing damaged organelles or small bits of cytosol, a process known as *autophagy.*

Vacuoles: Diverse Maintenance Compartments

Vacuoles are large vesicles. **Food vacuoles** are formed as a result of phagocytosis. **Contractile vacuoles** pump excess water out of freshwater protists. Vacuoles in plant cells may store organic compounds and inorganic ions for the cell, or contain dangerous metabolic byproducts and poisonous or unpalatable compounds, which may protect the plant from predators. A large

---

INTERACTIVE QUESTION   6.4

Name the components of the endomembrane system shown in this diagram, and list the functions of each of these membranes.

a.

b.

c.

d.

e.

f.

g.

central vacuole is found in mature plant cells and encloses a solution called cell sap. A plant cell increases in size with a minimal addition of new cytoplasm as its vacuole absorbs water and expands.

The Endomembrane System: A Review As membranes move from the ER to the Golgi apparatus and then to other organelles, their compositions, functions, and contents are modified.

6.5 Mitochondria and chloroplasts change energy from one form to another

Cellular respiration, the metabolic processing of fuels to produce ATP, occurs within the **mitochondria** of eukaryotic cells. Photosynthesis occurs in the **chloroplasts** of plants and photosynthetic protists, which produce sugars from carbon dioxide and water by absorbing solar energy.

The Evolutionary Origins of Mitochondria and Chloroplasts According to the **endosymbiont theory,** both mitochondria and chloroplasts originated as prokaryotic cells engulfed by an ancestral eukaryotic cell. The double membranes surrounding mitochondria and chloroplasts appear to have been part of the prokaryotic *endosymbiont.* These organelles grow and reproduce independently within the cell. They contain a small amount of DNA and ribosomes (both similar to prokaryotic structures), and they synthesize some of their proteins.

Mitochondria: Chemical Energy Conversion Two membranes, each a phospholipid bilayer with unique embedded proteins, enclose a mitochondrion. A narrow intermembrane space exists between the smooth outer membrane and the convoluted inner membrane. The folds of the inner membrane, called **cristae,** create a large surface area and enclose the **mitochondrial matrix.** Many respiratory enzymes, mitochondrial DNA, and ribosomes are housed in this matrix. Other respiratory enzymes and proteins are built into the inner membrane.

Chloroplasts: Capture of Light Energy Two membranes separated by a thin intermembrane space enclose a chloroplast. Inside the inner membrane is a membranous system of connected flattened sacs called **thylakoids,** inside of which is the thylakoid space. Photosynthetic enzymes are embedded in the thylakoids, which may be stacked together to form structures called **grana.** Chloroplast DNA, ribosomes, and many enzymes are found in the **stroma,** the fluid surrounding the thylakoids.

Plastids are plant organelles that also include amyloplasts, which store starch, and chromoplasts, which contain pigments.

INTERACTIVE QUESTION 6.5

Sketch a mitochondrion and a chloroplast, and label their membranes and compartments.

Peroxisomes: Oxidation **Peroxisomes** are oxidative organelles filled with enzymes that function in a variety of metabolic pathways, such as breaking down fatty acids for energy or detoxifying alcohol and other poisons. An enzyme that converts hydrogen peroxide (H_2O_2), a toxic by-product of these pathways, to water is also packaged into peroxisomes. *Glyoxysomes*, found in plant seeds, contain enzymes that convert fatty acids to sugars for emerging seedlings.

INTERACTIVE QUESTION 6.6

Why are peroxisomes not considered part of the endomembrane system?

6.6 The cytoskeleton is a network of fibers that organizes structures and activities in the cell

Roles of the Cytoskeleton: Support and Motility The **cytoskeleton** is a network of protein fibers that give mechanical support and function in cell motility (of both internal structures and the cell as a whole). The cytoskeleton interacts with special proteins called **motor proteins** to produce cellular movements.

Components of the Cytoskeleton The three main types of fibers involved in the cytoskeleton are *microtubules, microfilaments,* and *intermediate filaments.*

All eukaryotic cells have **microtubules,** which are hollow rods constructed of columns of globular proteins called tubulins. Microtubules change length through the addition or subtraction of tubulin dimers. In addition to providing the supporting framework of the cell, microtubules separate chromosomes during cell division and serve as tracks along which organelles move with the aid of motor molecules.

In animal cells, microtubules grow out from a region near the nucleus called a **centrosome.** A pair of **centrioles,** each composed of nine sets of triplet microtubules arranged in a ring, is associated with the centrosome and replicates before cell division. Fungi and plant cells lack centrosomes with centrioles and evidently have some other microtubule-organizing center.

Cilia and **flagella** are locomotor extensions of some eukaryotic cells. Cilia are numerous and short; flagella occur one or two to a cell and are longer. Many protists use cilia or flagella to move through aqueous media. Cilia or flagella attached to stationary cells of a tissue move fluid past the cell. A signal-receiving cilium, such as the *primary cilium* found on vertebrate animal cells, transmits environmental signals to a cell's interior.

Both cilia and flagella are composed of two single microtubules surrounded by a ring of nine doublets of microtubules (a nearly universal "9 + 2" arrangement), all of which are enclosed in an extension of the plasma membrane. A **basal body,** with a "9 + 0" pattern of microtubule triplets, anchors a cilium or flagellum in the cell. ATP drives the sliding of the microtubule doublets past each other as the two "feet" of large motor proteins called **dyneins** alternately attach to adjacent doublets, pull down, release, and reattach. In conjunction with anchoring cross-linking proteins and radial spokes, this action causes the bending of the flagellum or cilium.

Microfilaments, probably present in all eukaryotic cells, are solid rods consisting of a twisted double chain of molecules of the globular protein **actin.** Thus, they are also called actin filaments. *Cortical microfilaments* form a network just inside the plasma membrane that creates a semisolid or gel consistency in this cytoplasmic layer called the **cortex.** Microfilaments form the supportive core of small cytoplasmic extensions called microvilli.

In muscle cells, thousands of actin filaments interdigitate with thicker filaments made of the motor protein **myosin.** The sliding of actin and myosin filaments past each other causes the contraction of muscles.

Actin and myosin also interact in localized contractions such as cleavage furrows in animal cell division and amoeboid movements. Actin subunits reversibly assemble into microfilament networks, driving the conversion of cytoplasm from sol to gel during the formation of cellular extensions called **pseudopodia.** Actin filaments interacting with myosin may propel cytoplasm forward into pseudopodia. **Cytoplasmic streaming** in plant cells appears to involve both actin–myosin interactions and sol–gel conversions.

Intermediate filaments are intermediate in size between microtubules and microfilaments and are more diverse in their composition. Intermediate filaments appear to be important in maintaining cell shape. The nucleus is securely held in a web of intermediate filaments, and the nuclear lamina lining the inside of the nuclear envelope is composed of intermediate filaments.

6.7   Extracellular components and connections between cells help coordinate cellular activities

Cell Walls of Plants   Plant **cell walls** are composed of microfibrils of cellulose embedded in a matrix of polysaccharides and protein.

INTERACTIVE QUESTION   6.7

Fill in the following table to organize what you have learned about the components of the cytoskeleton. You may wish to refer to the textbook for additional details.

Cytoskeleton Component	Structure and Monomers	Functions
Microtubules	a.	b.
Microfilaments (actin filaments)	c.	d.
Intermediate filaments	e.	f.

The **primary cell wall** secreted by a young plant cell is relatively thin and flexible. Microtubules in the cell cortex appear to guide the path of cellulose synthase, determining the pattern of cellulose fibril deposition and thus the direction of cell expansion. Adjacent cells are glued together by the **middle lamella,** a thin layer of polysaccharides (called pectins). When they stop growing, some cells secrete a thicker and stronger **secondary cell wall** between the plasma membrane and the primary cell wall.

INTERACTIVE QUESTION 6.8

Sketch two adjacent plant cells, and show the location of the primary and secondary cell walls and the middle lamella.

The Extracellular Matrix (ECM) of Animal Cells Animal cells secrete an **extracellular matrix (ECM)** composed primarily of glycoproteins and other carbohydrate-containing molecules. **Collagen** forms strong fibers that are embedded in a network of proteoglycan complexes. These large complexes form when multiple **proteoglycans,** each consisting of a small core protein with many carbohydrate chains, attach to a long polysaccharide. Cells may be attached to the ECM by **fibronectins** and other glycoproteins that bind to **integrins,** proteins that span the plasma membrane and bind, via other proteins, to microfilaments of the cytoskeleton. Thus, information about changes outside the cell can be communicated through a mechanical signaling pathway involving fibronectins, integrins, and the microfilaments of the cytoskeleton. Signals from the ECM appear to influence the activity of genes in the nucleus.

Cell Junctions **Plasmodesmata** are channels in plant cell walls through which the plasma membranes of bordering cells connect, thus linking most cells of a plant into a living continuum. Water, small solutes, and even some proteins and RNA molecules can move through these channels.

There are three main types of cell junctions between animal cells. At **tight junctions,** proteins hold adjacent cell membranes tightly together, creating an impermeable seal across a layer of epithelial cells.

Desmosomes (also called *anchoring junctions*) are reinforced by intermediate filaments and rivet cells into strong sheets. **Gap junctions** (also called *communicating junctions*) are cytoplasmic connections that allow for the exchange of ions and small molecules between cells through protein-lined pores.

INTERACTIVE QUESTION 6.9

a. Return to your sketch of plant cells in Interactive Question 6.8 and draw in a plasmodesma.

b. List some of the functions of the extracellular matrix of animal cells.

The Cell: A Living Unit Greater Than the Sum of Its Parts The compartmentalization and the many specialized organelles typical of cells exemplify the principle that structure correlates with function. The intricate functioning of a living cell emerges from the complex interactions of its multiple parts.

Word Roots

centro- = the center; **-soma** = a body (*centrosome:* a structure in the cytoplasm of animal cells that functions as a microtubule-organizing center)

chloro- = green (*chloroplast:* the site of photosynthesis in plants and photosynthetic protists)

cili- = hair (*cilium:* a short cellular appendage containing microtubules)

cyto- = cell (*cytosol:* the semifluid portion of the cytoplasm)

-ell = small (*organelle:* a membrane-enclosed structure with a specialized function, suspended in the cytosol of eukaryotic cells)

endo- = inner (*endomembrane system:* the system of membranes inside and surrounding a eukaryotic cell that includes the plasma membrane, nuclear envelope, endoplasmic reticulum, Golgi apparatus, lysosomes, vesicles, and vacuoles)

eu- = true (*eukaryotic cell:* a cell with a membrane-enclosed nucleus and organelles)

extra- = outside (*extracellular matrix:* the meshwork of glycoproteins, polysaccharides, and proteoglycans surrounding animal cells)

flagell- = whip (*flagellum:* a long cellular appendage specialized for locomotion)

glyco- = sweet (*glycoprotein:* a protein covalently bonded to one or more carbohydrates)

lamin- = sheet/layer (*nuclear lamina:* a netlike array of protein filaments lining the inner surface of the nuclear envelope)

lyso- = loosen (*lysosome:* a membrane-enclosed sac of hydrolytic enzymes that digest macromolecules in animal cells)

micro- = small; **-tubul** = a little pipe (*microtubule:* a hollow rod of tubulin proteins that makes up part of the cytoskeleton in all eukaryotic cells)

nucle- = nucleus; **-oid** = like (*nucleoid:* a non-membrane-bound region in prokaryotic cells where the DNA is concentrated)

phago- = to eat; **-kytos** = vessel (*phagocytosis:* a type of endocytosis in which large particulate substances are taken up by a cell)

plasm- = molded; **-desma** = a band or bond (*plasmodesma:* an open channel in a plant cell wall, connecting the cytoplasm of adjacent cells)

pro- = before; **karyo-** = nucleus (*prokaryotic cell:* a cell lacking a membrane-enclosed nucleus and organelles)

pseudo- = false; **-pod** = foot (*pseudopodium:* a cellular extension of amoeboid cells used in moving and feeding)

thylaco- = sac or pouch (*thylakoid:* a flattened membranous sac inside a chloroplast)

vacu- = empty (*vacuole:* a relatively large vesicle with a specialized function)

Structure Your Knowledge

1. In the following table, write the organelles or structures that are associated with each of the listed functions in animal cells.

Functions	Associated Organelles and Structures
Cell division	a.
Information storage and transferal	b.
Energy conversions	c.
Manufacture of proteins, membranes, and other products	d.
Lipid synthesis, drug detoxification	e.
Digestion, recycling	f.
Oxidation, conversion of H_2O_2 to water	g.
Structural integrity	h.
Movement	i.
Exchange with the environment	j.
Cell-to-cell connections	k.

2. Write the functions of each of the following plant cell structures.

Plant Cell Structures	Functions
Cell wall	**a.**
Central vacuole	**b.**
Chloroplast	**c.**
Amyloplast	**d.**
Plasmodesmata	**e.**

3. Label the indicated structures in the following diagram of an animal cell.

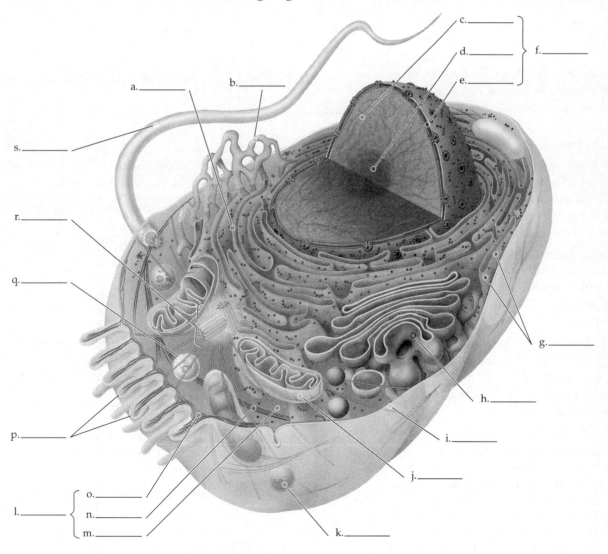

4. Create a diagram or flowchart in the following space to trace the development of a secretory product (such as a digestive enzyme) from the DNA code to its export from the cell.

Test Your Knowledge

MULTIPLE CHOICE: *Choose the one best answer.*

1. Resolution of a microscope refers to
   a. the distance between two separate points.
   b. the sharpness or clarity of an image.
   c. the amount of magnification of an image.
   d. the depth of focus on a specimen's surface.
   e. the wavelength of light.

2. The detailed structure of a chloroplast can be seen with the best resolution by using
   a. transmission electron microscopy.
   b. scanning electron microscopy.
   c. phase-contrast light microscopy.
   d. deconvolution microscopy.
   e. confocal fluorescence microscopy.

3. In an animal cell fractionation procedure, one of the last pellets from the series of centrifugations would most likely contain
   a. ribosomes.   d. lysosomes.
   b. peroxisomes.   e. nuclei.
   c. mitochondria.

4. Which of the following is/are *not* found in a prokaryotic cell?
   a. ribosomes   d. a and c
   b. plasma membrane   e. a, b, and c
   c. mitochondria

5. The cells of an ant and an elephant are, on average, the same size; an elephant just has more cells. What is the main advantage of small cell size?
   a. Small cells are easier to organize into tissues and organs.
   b. A small cell has a larger plasma membrane surface area than does a large cell, facilitating the exchange of sufficient materials with its environment.

c. The cytoskeleton of a large cell would have to be so large that cells would be too heavy.
   d. Small cells require less oxygen and nutrients than do large cells.
   e. A small cell has a larger surface area relative to cytoplasmic volume, facilitating sufficient exchange of materials.

6. Proteins that function within the cytosol are generally synthesized
   a. by ribosomes bound to rough ER.
   b. by free ribosomes.
   c. by the nucleolus.
   d. within the Golgi apparatus.
   e. by mitochondria and chloroplasts.

7. The pores in the nuclear envelope provide for the movement of
   a. proteins into the nucleus.
   b. ribosomal subunits out of the nucleus.
   c. mRNA out of the nucleus.
   d. signal molecules into the nucleus.
   e. all of the above.

8. A fluorescent green tag is attached to a protein being synthesized and inserted into an area of rough ER membrane. Which of the following is a possible route for that protein to follow?
   a. rough ER → Golgi → lysosome → nuclear membrane → plasma membrane
   b. rough ER → transport vesicle → Golgi → vesicle → plasma membrane → food vacuole
   c. rough ER → nuclear envelope → Golgi → smooth ER → lysosome
   d. rough ER → transport vesicle → Golgi → smooth ER → plasma membrane
   e. rough ER → transport vesicle → Golgi → vesicle → extracellular matrix

9. A growing plant cell elongates primarily by
   a. increasing the number of vacuoles.
   b. synthesizing more cytoplasm.
   c. taking up water into its central vacuole.
   d. synthesizing cellulose fibrils in an orientation perpendicular to the axis of cell elongation.
   e. producing a secondary cell wall.

10. Which of the following is *not* a similarity among nuclei, chloroplasts, and mitochondria?
    a. They contain DNA.
    b. They are bounded by two (or more) phospholipid bilayer membranes.
    c. They can divide to reproduce themselves.
    d. They are derived from the endoplasmic reticulum system.
    e. Their membranes are associated with specific proteins.

11. The contractile elements of muscle cells are
    a. intermediate filaments.
    b. centrioles.
    c. microtubules.
    d. microfilaments (actin filaments).
    e. fibronectins.

12. Microtubules are components of all of the following *except*
    a. centrioles.
    b. the spindle apparatus for separating chromosomes in cell division.
    c. tracks along which organelles can move using motor proteins.
    d. flagella and cilia.
    e. the cleavage furrow that pinches apart cells in animal cell division.

13. Which of the following characteristics are shared by cristae of mitochondria, thylakoids, and microvilli?
    a. have a large surface area of membrane that enhances their particular function
    b. are bounded by a double membrane
    c. have a shape reinforced by microfilaments
    d. are not derived from the endoplasmic reticulum system
    e. arose by endosymbiosis of an aerobic bacterium, a photosynthetic bacterium, and a motile bacterium, respectively

14. The innermost portion of the cell wall of a plant cell specialized for support is the
    a. primary cell wall.
    b. secondary cell wall.
    c. middle lamella.
    d. plasma membrane.
    e. cytoskeleton.

15. Plasmodesmata in plant cells are similar in function to
    a. desmosomes.
    b. tight junctions.
    c. gap junctions.
    d. the extracellular matrix.
    e. integrins.

16. Which of the following structures is/are involved in receiving external information and relaying it into a cell?
    a. a primary cilium
    b. fibronectins
    c. integrins
    d. microfilaments of the cytoskeleton
    e. all of the above

17. Which of the following is *incorrectly* paired with its function?
    a. peroxisome—contains enzymes that break down H_2O_2
    b. nucleolus—produces ribosomal RNA, assembles ribosome subunits
    c. Golgi apparatus—processes, tags, and ships cellular products
    d. lysosome—food sac formed by phagocytosis
    e. ECM (extracellular matrix)—supports and anchors cells, communicates information with inside of cell

Answer questions 18–22 using the following five choices.
   a. muscle cells in the thigh muscle of a runner
   b. pancreatic cells that manufactures digestive enzymes
   c. macrophages that engulf bacteria
   d. epithelial cells lining digestive tract
   e. ovarian cells that produce estrogen (a steroid hormone)

18. In which cells would you expect to find the most tight junctions?

19. In which cells would you expect to find the most lysosomes?

20. In which cells would you expect to find the most smooth endoplasmic reticulum?

21. In which cells would you expect to find the most bound ribosomes?

22. In which cells would you expect to find the most mitochondria?

Chapter 7

Membrane Structure and Function

Key Concepts

7.1 Cellular membranes are fluid mosaics of lipids and proteins

7.2 Membrane structure results in selective permeability

7.3 Passive transport is diffusion of a substance across a membrane with no energy investment

7.4 Active transport uses energy to move solutes against their gradients

7.5 Bulk transport across the plasma membrane occurs by exocytosis and endocytosis

Framework

This chapter presents the fluid mosaic model of membrane structure, relating the molecular structure of biological membranes to their function of regulating the passage of substances into and out of the cell.

Chapter Review

The plasma membrane is the boundary of life; this **selectively permeable** membrane allows some materials to cross it more easily than others and enables the cell to maintain a unique internal environment.

7.1 Cellular membranes are fluid mosaics of lipids and proteins

According to the **fluid mosaic model**, biological membranes consist of various proteins that are attached to or embedded in a bilayer of **amphipathic** phospholipids (having both hydrophilic and hydrophobic regions).

Membrane Models: Scientific Inquiry Early chemical analysis of membranes revealed a lipid and protein composition. Scientists suggested that cell membranes must be a phospholipid bilayer, with the hydrophobic hydrocarbon tails in the center and the hydrophilic heads facing the aqueous solution on both sides of the membrane. In 1935 H. Davson and J. Danielli proposed a sandwich model in which a bilayer of phospholipids is covered with a coat of hydrophilic proteins.

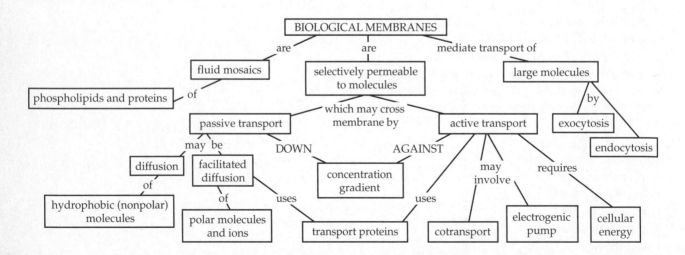

In 1972 S. J. Singer and G. Nicolson proposed the fluid mosaic model in which amphipathic membrane proteins are embedded in the phospholipid bilayer with their hydrophilic regions extending into the aqueous surroundings. The phospholipid bilayer is envisioned as fluid, with a mosaic of proteins floating in it.

The fluid mosaic model is supported by evidence from freeze-fracture electron microscopy. This technique involves freezing a specimen, fracturing it with a cold knife, and examining the exposed interior of the membrane with an electron microscope.

This membrane model is currently the most useful model for organizing existing knowledge and extending research on membrane structure. Some of this research now indicates that both proteins and lipids may be localized in specialized regions.

INTERACTIVE QUESTION 7.1

Label the components in the following diagram of the fluid mosaic model of membrane structure. Indicate whether regions are hydrophobic or hydrophilic.

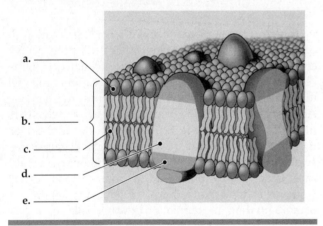

a. _____

b. _____

c. _____

d. _____

e. _____

The Fluidity of Membranes   Membranes are held together primarily by weak hydrophobic interactions that allow the lipids and some of the proteins to drift laterally. Some membrane proteins seem to be held by attachments to the cytoskeleton or ECM; others appear to be directed in their movements.

Phospholipids with unsaturated hydrocarbon tails maintain membrane fluidity at lower temperatures. The steroid cholesterol, common in plasma membranes of animals, restricts movement of phospholipids and thus reduces fluidity at warmer temperatures. Cholesterol also prevents the close packing of lipids and thus enhances fluidity at lower temperatures.

Evolution of Differences in Membrane Lipid Composition   Variations in membrane lipid composition and the ability to change that composition in response to changing temperatures are evolutionary adaptations.

INTERACTIVE QUESTION 7.2

a. Cite some experimental evidence that indicates that membrane proteins drift.

b. How might the membrane composition differ for two subspecies of fish if one lives in cold, mountain lakes and the other in warm valley ponds?

Membrane Proteins and Their Functions   Each membrane has its own unique set of proteins, which determine most of the specific functions of that membrane. **Integral proteins** often extend through the membrane (are *transmembrane* proteins), with two hydrophilic ends. The hydrophobic midsection usually consists of one or more α helical stretches of nonpolar amino acids. **Peripheral proteins** are attached to the surface of the membrane. Attachments of membrane proteins to the cytoskeleton and to fibers of the extracellular matrix provide support for the plasma membrane.

INTERACTIVE QUESTION 7.3

List the six major kinds of functions that membrane proteins may perform.

The Role of Membrane Carbohydrates in Cell–Cell Recognition   The ability of a cell to distinguish other cells is based on recognition of membrane carbohydrates. The **glycolipids** and **glycoproteins** attached to the outside of plasma membranes vary from species to species, from individual to individual, and even among cell types.

Synthesis and Sidedness of Membranes   Membranes have distinct inner and outer faces, related to the composition of the lipid layers and the directional orientation of their proteins. Carbohydrates attached to membrane proteins as they are synthesized in the ER are modified in the Golgi apparatus. Carbohydrates may also be attached to lipids in the Golgi apparatus. When transport vesicles fuse with the plasma membrane, these interior glycoproteins and glycolipids become located on the extracellular face of the membrane.

7.2   Membrane structure results in selective permeability

The plasma membrane permits a regular exchange of nutrients, waste products, oxygen, and inorganic ions. Biological membranes are selectively permeable; the ease and rate at which small molecules pass through them differ.

The Permeability of the Lipid Bilayer   Hydrophobic, nonpolar molecules, such as hydrocarbons, CO_2, and O_2, can dissolve in and cross a membrane.

INTERACTIVE QUESTION   7.4

What types of molecules have difficulty crossing the plasma membrane? Why?

Transport Proteins   Ions and polar molecules may move across the plasma membrane with the aid of **transport proteins.** Hydrophilic passageways through a membrane are provided for specific molecules or ions by *channel proteins,* such as **aquaporins,** which facilitate water passage. *Carrier proteins* may physically bind and transport a specific molecule.

7.3   Passive transport is diffusion of a substance across a membrane with no energy investment

Diffusion is the movement of a substance down its **concentration gradient** due to random thermal motion. The diffusion of one solute is unaffected by the concentration gradients of other solutes. The cell does not expend energy when substances diffuse across membranes down their concentration gradient; therefore, the process is called **passive transport.**

Effects of Osmosis on Water Balance   **Osmosis** is the diffusion of free water across a selectively permeable membrane. Water diffuses down its own concentration gradient, which is affected by the solute concentration.

Clustering of water molecules around solute particles lowers the proportion of free water that is available to cross the membrane.

INTERACTIVE QUESTION   7.5

A solution of 1 *M* glucose is separated by a selectively permeable membrane from a solution of 0.2 *M* fructose and 0.7 *M* sucrose. The membrane is not permeable to the sugar molecules. Indicate which side initially has more free water molecules, and which side has fewer. Show the direction of osmosis.

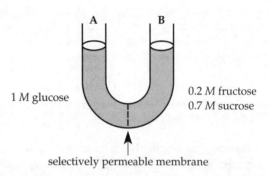

1 *M* glucose

0.2 *M* fructose
0.7 *M* sucrose

selectively permeable membrane

Tonicity, the tendency of a surrounding solution to cause a cell to gain or lose water, is affected by the relative concentrations of those solutes in the solution and in the cell that cannot cross the membrane. An animal cell will neither gain nor lose water in an **isotonic** environment. An animal cell placed in a **hypertonic** solution (which has more nonpenetrating solutes) will lose water and shrivel. If placed in a **hypotonic** solution, the cell will gain water, swell, and possibly lyse (burst). Cells without rigid walls must either live in an isotonic environment, such as salt water or isotonic body fluids, or have adaptations for **osmoregulation**, the regulation of water balance.

INTERACTIVE QUESTION   7.6

What osmotic problems do freshwater protists face, and what adaptations help them to osmoregulate?

The cell walls of plants, fungi, prokaryotes, and some protists play a role in water balance in hypotonic environments. Water moving into the cell causes the cell to swell against its cell wall, creating *turgor pressure.* **Turgid** cells provide mechanical support for nonwoody plants. Plant cells in an isotonic surrounding are

flaccid. In a hypertonic medium, a plant cell undergoes **plasmolysis**—that is, the plasma membrane pulls away from the cell wall as water exits and the cell shrivels.

INTERACTIVE QUESTION 7.7

a. The ideal osmotic environment for animal cells is
_____.

b. The ideal osmotic environment for plant cells is
_____.

Facilitated Diffusion: Passive Transport Aided by Proteins **Facilitated diffusion** involves the diffusion of polar molecules and ions across a membrane with the aid of transport proteins, either channel proteins or carrier proteins. Many **ion channels** are **gated channels,** which open or close in response to electrical or chemical stimuli. The binding of a solute to a carrier protein may cause a change in the protein's shape that translocates the binding site and the attached solute across the membrane.

INTERACTIVE QUESTION 7.8

Why is facilitated diffusion considered a form of passive transport?

7.4 Active transport uses energy to move solutes against their gradients

The Need for Energy in Active Transport **Active transport,** which requires the expenditure of energy to transport a solute against its concentration gradient, is essential if a cell is to maintain internal concentrations of small molecules that differ from their concentrations outside the cell. The terminal phosphate group of ATP may be transferred to a carrier protein, inducing it to change its shape and translocate the bound solute across the membrane. The **sodium-potassium pump** works this way to exchange N^+ and K^+ across animal cell membranes, creating a higher concentration of potassium ions and a lower concentration of sodium ions within the cell.

How Ion Pumps Maintain Membrane Potential Cells have a **membrane potential,** a voltage across the plasma membrane due to the unequal distribution of ions on either side. This electrical potential energy results from the separation of opposite charges: The cytoplasm of a cell is negatively charged relative to the extracellular fluid. The membrane potential favors the diffusion of cations into the cell and anions out of the cell. Both the membrane potential and the concentration gradient affect the diffusion of an ion; thus, an ion diffuses down its **electrochemical gradient.**

Electrogenic pumps are membrane proteins that generate voltage across a membrane by the active transport of ions. A **proton pump** that transports H^+ out of the cell generates voltage across membranes in plants, fungi, and bacteria.

INTERACTIVE QUESTION 7.9

The sodium-potassium pump, the major electrogenic pump in animal cells, exchanges sodium ions for potassium ions, both of which are cations. How does this exchange generate a membrane potential?

Cotransport: Coupled Transport by a Membrane Protein **Cotransport** is a mechanism through which the active transport of a solute is indirectly driven by an ATP-powered pump that transports another substance against its gradient. As that actively transported substance diffuses back down its concentration gradient through a cotransporter, the solute is carried against its concentration gradient across the membrane.

7.5 Bulk transport across the plasma membrane occurs by exocytosis and endocytosis

Bulk transport, like active transport, requires energy to transport larger biological molecules, packaged in vesicles, across the membrane.

Exocytosis In **exocytosis,** the cell secretes large molecules by the fusion of vesicles with the plasma membrane.

Endocytosis In **endocytosis,** a region of the plasma membrane sinks inward and pinches off to form a vesicle containing material that had been outside the cell. **Phagocytosis** is a form of endocytosis in which pseudopodia wrap around a food particle or another large particle, creating a vacuole that then fuses with a lysosome containing hydrolytic enzymes. In **pinocytosis,** droplets of extracellular fluid are taken into the cell in small vesicles. **Receptor-mediated endocytosis** enables a cell to acquire specific substances from extracellular fluid. **Ligands,** molecules that bind specifically to receptor sites, attach to receptor proteins usually clustered in coated pits on the cell surface and are carried into the cell when a vesicle forms.

INTERACTIVE QUESTION   7.10

a. How is cholesterol transported into human cells?

b. Explain why cholesterol accumulates in the blood of individuals with the disease familial hypercholesteremia.

Word Roots

amphi- = dual (*amphipathic molecule:* a molecule that has both hydrophobic and hydrophilic regions)

aqua- = water; **-pori** = a small opening (*aquaporin:* a channel protein in the plasma membrane that specifically facilitates the diffusion of water across the membrane)

co- = together; **trans-** = across (*cotransport:* the coupling of the "downhill" diffusion of one substance to the "uphill" transport of another against its concentration gradient)

electro- = electricity; **-genic** = producing (*electrogenic pump:* an ion transport protein that generates voltage across a membrane)

endo- = inner; **cyto-** = cell (*endocytosis:* cellular uptake of matter via formation of new vesicles from the plasma membrane)

exo- = outer (*exocytosis:* cellular secretion by the fusion of vesicles with the plasma membrane)

hyper- = exceeding; **-tonus** = tension (*hypertonic:* referring to a surrounding solution that will cause a cell to lose water)

hypo- = lower (*hypotonic:* referring to a surrounding solution that will cause a cell to take up water)

iso- = same (*isotonic:* referring to a surrounding solution that causes no net movement of water into or out of a cell)

phago- = eat (*phagocytosis:* "cell eating"; endocytosis in which large particulate substances are taken up by a cell)

pino- = drink (*pinocytosis:* "cell drinking"; endocytosis in which the cell ingests extracellular fluid)

plasm- = molded; **-lyso** = loosen (*plasmolysis:* a phenomenon in walled cells in which the cytoplasm shrivels and the plasma membrane pulls away from the cell wall when the cell loses water to a hypertonic environment)

Structure Your Knowledge

1. Create a concept map to illustrate your understanding of osmosis. This exercise will help you to practice using the words *hypotonic, isotonic,* and *hypertonic,* and to focus on the effect of these osmotic environments on plant and animal cells. Explain your map to a friend.

2. The following diagram illustrates passive and active transport across a plasma membrane. Use it to answer questions a–d.

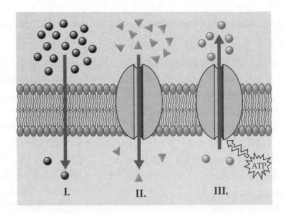

a. Which section represents facilitated diffusion? How can you tell? Does the cell expend energy in this transport? Why or why not? What types of solute molecules may be moved by this type of transport?

b. Which section shows active transport? List two ways that you can tell.

c. Which section shows diffusion? What types of solute molecules may be moved by this type of transport?

d. Which of these sections are considered passive transport?

Test Your Knowledge

MULTIPLE CHOICE: *Choose the one best answer.*

1. If a single layer of phospholipids coats the water in a beaker, which parts of the molecules face the air?
   a. the phosphate groups
   b. the hydrocarbon tails
   c. both heads and tails because the molecules are amphipathic and lie sideways
   d. the glycolipid regions
   e. No parts of the molecules face the air, because the phospholipids dissolve in the water and do not form a layer.

2. Support for the fluid mosaic model of membrane structure comes from
   a. freeze-fracture preparation and electron microscopy.
   b. the movement of proteins in hybrid cells.
   c. the amphipathic nature of many membrane proteins.
   d. a and c.
   e. a, b, and c.

3. Glycoproteins and glycolipids are important for
   a. facilitated diffusion.
   b. active transport.
   c. cell–cell recognition.
   d. intercellular joining.
   e. signal-transduction pathways.

4. Which of the following is the most probable description of an integral, transmembrane protein?
   a. amphipathic with a hydrophilic head and a hydrophobic tail region
   b. a globular protein with hydrophobic amino acids in the interior and hydrophilic amino acids arranged around the outside
   c. a fibrous protein coated with hydrophobic fatty acids
   d. a glycolipid attached to the portion of the protein facing the exterior of the cell and cytoskeletal elements attached to the interior portion
   e. a middle region composed of α helical stretches of hydrophobic amino acids, with hydrophilic regions at both ends of the protein

5. A cell is manufacturing receptor proteins for cholesterol. How would those proteins be oriented before they reach the plasma membrane?
   a. facing inside the ER lumen but outside the transport vesicle membrane
   b. facing inside the ER lumen and inside the transport vesicle
   c. attached outside the ER and outside the transport vesicle
   d. attached outside the ER but facing inside the transport vesicle
   e. embedded in the hydrophobic center of both the ER and transport vesicle membranes

6. The fluidity of membranes in a plant in cold weather may be maintained by increasing the
   a. proportion of peripheral proteins.
   b. action of an H^+ pump.
   c. concentration of cholesterol in the membrane.
   d. number of phospholipids with unsaturated hydrocarbon tails.
   e. number of phospholipids with saturated hydrocarbon tails.

Use the following U-tube setup to answer questions 7 through 9.

The solutions in the two arms of this U-tube are separated by a membrane that is permeable to water and glucose but not to sucrose. Side A is filled with a solution of 2.0 *M* sucrose and 1.0 *M* glucose. Side B is filled with 1.0 *M* sucrose and 2.0 *M* glucose.

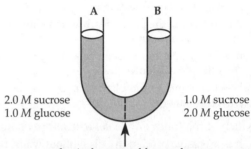

2.0 *M* sucrose
1.0 *M* glucose

1.0 *M* sucrose
2.0 *M* glucose

selectively permeable membrane

7. *Initially*, the solution in side A, with respect to that in side B,
   a. has a lower solute concentration.
   b. has a higher solute concentration.
   c. has an equal solute concentration.
   d. is lower in the tube.
   e. is higher in the tube.

8. During the period *before* equilibrium is reached, which molecule(s) will show net movement through the membrane?
   a. water
   b. glucose
   c. sucrose
   d. water and sucrose
   e. water and glucose

9. *After* the system reaches equilibrium, what changes can be observed?
   a. The water level is higher in side A than in side B.
   b. The water level is higher in side B than in side A.
   c. The molarity of glucose is higher in side A than in side B.
   d. The molarity of sucrose has increased in side A.
   e. Both a and c have occurred.

10. An animal cell placed in a hypotonic environment will
    a. become flaccid.
    b. become turgid.
    c. burst (lyse).
    d. plasmolyze.
    e. shrivel.

11. You observe plant cells under a microscope as they are placed in an unknown solution. First the cells plasmolyze; after a minute, the plasmolysis reverses and the cells appear normal. What would you conclude about the unknown solution?

    a. It is hypertonic to the plant cells, and its solute cannot cross the plant cell membranes.

    b. It is hypotonic to the plant cells, and its solute cannot cross the plant cell membranes.

    c. It is isotonic to the plant cells, but its solute can cross the plant cell membranes.

    d. It is hypertonic to the plant cells, but its solute can cross the plant cell membranes.

    e. It is hypotonic to the plant cells, but its solute can cross the plant cell membranes.

12. Which of the following is *not* true about osmosis?

    a. It is a passive process in cells without walls, but an active one in cells with walls.

    b. Water moves into a cell from a hypotonic environment.

    c. Solute molecules bind to water and decrease the free water available to move.

    d. It can occur more rapidly through channel proteins known as aquaporins.

    e. There is no net osmosis when cells are in isotonic solutions.

13. Which of the following is *not* true of carrier molecules involved in facilitated diffusion?

    a. They increase the speed of transport across a membrane.

    b. They can concentrate solute molecules on one side of the membrane.

    c. They may have specific binding sites for the molecules they transport.

    d. They may undergo a change in shape upon binding of solute.

    e. They do not require an energy investment from the cell to operate.

14. Facilitated diffusion of ions across a cellular membrane requires _____; and the ions move _____.

    a. energy and channel proteins; against their electrochemical gradient

    b. energy and channel proteins; against their concentration gradient

    c. cotransport proteins; against their electrochemical gradient

    d. channel proteins; down their electrochemical gradient

    e. channel proteins; down their concentration gradient

15. The membrane potential of a cell favors

    a. the movement of cations into the cell.

    b. the movement of anions into the cell.

    c. the action of an electrogenic pump.

    d. the movement of sodium out of the cell.

    e. both b and d.

16. Which of the following describes cotransport?

    a. active transport of two solutes through a cotransport protein

    b. passive transport of two solutes through a cotransport protein

    c. ion diffusion against the electrochemical gradient created by an electrogenic pump

    d. a pump such as the sodium-potassium pump that moves ions in two different directions

    e. transport of one solute against its concentration gradient in tandem with another that is diffusing down its concentration gradient

17. The proton pump in plant cells is the functional equivalent of an animal cell's

    a. cotransport mechanism.

    b. sodium-potassium pump.

    c. contractile vacuole for osmoregulation.

    d. receptor-mediated endocytosis of cholesterol.

    e. ATP pump.

18. Which of the following is an example of active transport?

    a. the transport of glucose when the carrier protein GLUT-1 binds glucose, changes shape, and moves glucose across a membrane

    b. the flow of K^+ through an open ion channel out of a cell

    c. the movement of water into a plant cell

    d. the movement of LDL particles into a cell

    e. the movement of O_2 into a cell

19. Pinocytosis involves

    a. the fusion of a newly formed food vacuole with a lysosome.

    b. receptor-mediated endocytosis that involves binding of a ligand.

    c. the pinching in of the plasma membrane around small droplets of external fluid.

    d. the secretion of cell fluid.

    e. the accumulation of specific molecules in a cell.

20. Exocytosis may involve all of the following *except*

    a. ligands and coated pits.

    b. the fusion of a vesicle with the plasma membrane.

    c. a mechanism to export some carbohydrates during the formation of plant cell walls.

    d. a mechanism to rejuvenate the plasma membrane.

    e. a means of exporting large molecules.

Chapter 8

An Introduction to Metabolism

Key Concepts

8.1 An organism's metabolism transforms matter and energy, subject to the laws of thermodynamics

8.2 The free-energy change of a reaction tells us whether or not the reaction occurs spontaneously

8.3 ATP powers cellular work by coupling exergonic reactions to endergonic reactions

8.4 Enzymes speed up metabolic reactions by lowering energy barriers

8.5 Regulation of enzyme activity helps control metabolism

Framework

This chapter considers metabolism, the totality of the chemical reactions that take place in living organisms. Two key topics are emphasized: the energy transformations that underlie all chemical reactions and the role of enzymes in the "cold chemistry" of the cell. Enzymes are macromolecules that act as catalysts, lowering the activation energy of a reaction and thereby greatly speeding metabolic processes. Intricate control and feedback mechanisms produce the metabolic integration necessary for life.

The reactions in a cell either release energy or consume energy. The following illustration summarizes some of the components of the energy changes within a cell.

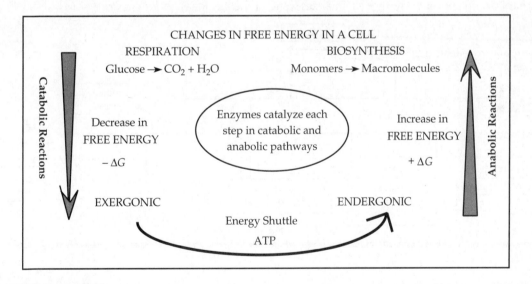

Chapter Review

8.1 An organism's metabolism transforms matter and energy, subject to the laws of thermodynamics

Organization of the Chemistry of Life into Metabolic Pathways **Metabolism** includes all the chemical reactions in an organism. These reactions are ordered into **metabolic pathways,** a sequence of steps, each controlled by an enzyme, that converts specific molecules to products.

Catabolic pathways release the energy stored in complex molecules through the breakdown of these molecules into simpler compounds. **Anabolic pathways,**

sometimes called biosynthetic pathways, require energy to combine simpler molecules into more complicated ones. The energy released from catabolic pathways drives the anabolic pathways in a cell. The study of energy transformations in organisms, called **bioenergetics,** is central to understanding metabolism.

Forms of Energy   Energy can be defined as the capacity to cause change. Some forms of energy can do work, such as moving matter against an opposing force. **Kinetic energy** is the energy of motion, of matter that is moving. This matter does its work by transferring its motion to other matter. The kinetic energy of randomly moving atoms or molecules is **heat,** or **thermal energy. Potential energy** is the capacity of matter to cause change as a consequence of its location or arrangement. **Chemical energy** is a form of potential energy that is available for release in chemical reactions.

Energy can be converted from one form to another. Plants convert light energy to the chemical energy in sugar, and cells release this potential energy to drive cellular processes.

The Laws of Energy Transformation   Thermodynamics is the study of energy transformations in a collection of

matter. In an *open system,* energy and matter may be exchanged between the *system* and its *surroundings.*

The **first law of thermodynamics** states that energy can be neither created nor destroyed. According to this *principle of conservation of energy,* energy can be transferred and transformed from one kind to another, but the total energy of the universe is constant.

The **second law of thermodynamics** states that every energy transformation or transfer results in increasing disorder within the universe. **Entropy** is used as a measure of disorder or randomness. In every energy transfer or transformation, some of the energy is converted to heat, a disordered form of energy.

A **spontaneous process** is one that is "energetically favorable," that occurs without an input of energy. For a process to occur spontaneously, it must result in an increase in entropy. A nonspontaneous process will occur only if energy is added to the system.

A cell may become more ordered, but it does so with an attendant increase in the entropy of its surroundings. For example, an animal takes in and uses highly ordered organic molecules as a source of the matter and energy needed to create and maintain its own organized structure, but it returns heat and the simple molecules of carbon dioxide and water to the environment.

INTERACTIVE QUESTION   8.1

Complete the following concept map that summarizes some of the key ideas about energy.

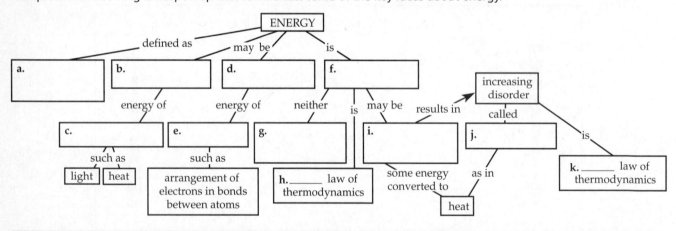

8.2   The free-energy change of a reaction tells us whether or not the reaction occurs spontaneously

Free-Energy Change, ΔG   The portion of a system's energy available to perform work when the system's temperature and pressure are uniform is defined as **free energy.** The change in free energy during a chemical reaction is represented by $\Delta G = \Delta H - T\Delta S$.

Enthalpy (H) is the total energy of a system; entropy (S) is a measure of its disorder; and T is absolute temperature (°Kelvin, or °C + 273). The free energy of the system must decrease ($-\Delta G$) for a reaction to be spontaneous: The system must lose energy (H must decrease), become more disordered (TS must increase), or both. Processes with a positive or zero ΔG are never spontaneous.

Free Energy, Stability, and Equilibrium   When ΔG is negative, the final state has less free energy than the initial state; thus, the final state is less likely to change and is more stable. A system rich in free energy has a tendency to change spontaneously to a more stable state. This change may be harnessed to perform work. A state of maximum stability is called *equilibrium.* At equilibrium in a chemical reaction, the forward and backward reactions occur at the same rate, and the relative concentrations of products and reactants stay the same. Moving toward equilibrium is spontaneous; the ΔG of the reaction is negative. Once at equilibrium, a system is at its minimum of free energy; it will not spontaneously change, and it can do no work.

INTERACTIVE QUESTION 8.2

Complete the following table to indicate how the free energy of a system (or a reaction) relates to the system's stability, tendency for spontaneous change, equilibrium, and capacity to do work.

	System with High Free Energy	System with Low Free Energy
Stability		
Spontaneous		
Equilibrium		
Work capacity		

Free Energy and Metabolism   An **exergonic reaction** (−ΔG) proceeds with a net release of free energy and is spontaneous. The magnitude of ΔG indicates the maximum amount of work an exergonic reaction can do. **Endergonic reactions** (+ΔG) are nonspontaneous; they must absorb free energy from the surroundings. The energy released by an exergonic reaction (−ΔG) is equal to the energy required by the reverse reaction (+ΔG).

Metabolic disequilibrium is essential to life. Metabolic reactions are reversible and could reach equilibrium if the cell did not maintain a steady supply of reactants and siphon off the products (as reactants for other processes or as waste products to be expelled).

INTERACTIVE QUESTION 8.3

Develop a concept map on free energy and ΔG. The value in this exercise is for you to wrestle with and organize these concepts for yourself. Do not turn to the suggested concept map until you have worked on your own understanding. Remember that the concept map in the answer section is only one way of structuring these ideas—have confidence in your own organizational scheme.

8.3   ATP powers cellular work by coupling exergonic reactions to endergonic reactions

Central to a cell's bioenergetics is **energy coupling,** the use of exergonic processes to drive endergonic ones. A cell usually uses ATP as the immediate source of energy for its *chemical, transport,* and *mechanical work.*

The Structure and Hydrolysis of ATP   ATP **(adenosine triphosphate)** consists of the nitrogenous base adenine bonded to the sugar ribose, which is connected to a chain of three phosphate groups. ATP can be hydrolyzed to ADP (adenosine diphosphate) and an inorganic phosphate molecule (P_i), releasing 7.3 kcal (30.5 kJ) of energy per mole of ATP when measured under standard conditions. The ΔG of the reaction in the cell is estimated to be closer to −13 kcal/mol.

INTERACTIVE QUESTION 8.4

Label the three components (a through c) of the following ATP molecule.

d. Indicate which bond is likely to break. By what chemical mechanism is the bond broken?

e. Explain why this reaction releases so much energy.

How the Hydrolysis of ATP Performs Work   In coupling exergonic and endergonic reactions in a cell, the free energy released from the hydrolysis of ATP is used to transfer the phosphate group to a reactant molecule, producing a **phosphorylated intermediate** that is more reactive (less stable). The hydrolysis of ATP also forms the basis for almost all transport and mechanical work in a cell by changing the shapes and binding affinities of proteins.

The Regeneration of ATP   A cell regenerates ATP at a phenomenal rate. The formation of ATP from ADP and $\textcircled{P}_i$ is endergonic, with a ΔG of $+7.3$ kcal/mol (standard conditions). Cellular respiration (the catabolic processing of glucose and other organic molecules) provides the energy for the regeneration of ATP. Plants can also produce ATP using light energy.

8.4 Enzymes speed up metabolic reactions by lowering energy barriers

Enzymes are biological **catalysts**—agents that speed the rate of a reaction but are unchanged by the reaction.

The Activation Energy Barrier   Chemical reactions involve both the breaking and forming of chemical bonds. Energy must be absorbed to contort molecules to an unstable state in which bonds can break. Energy is released when new bonds form and molecules return to stable, lower energy states. **Activation energy,** or the *free energy of activation* (E_A), is the energy that must be absorbed by reactants to reach the unstable *transition state*, in which bonds are likely to break, and from which the reaction can proceed.

How Enzymes Lower the E_A Barrier   The E_A barrier is essential to life because it prevents the energy-rich macromolecules of the cell from decomposing spontaneously. For metabolism to proceed in a cell, however, E_A must be reached. Enzymes are able to lower E_A so that specific reactions can proceed at cellular temperatures. Enzymes do not change the ΔG for a reaction.

INTERACTIVE QUESTION   8.5

In the following graph of an exergonic reaction with and without an enzyme catalyst, label parts a through e.

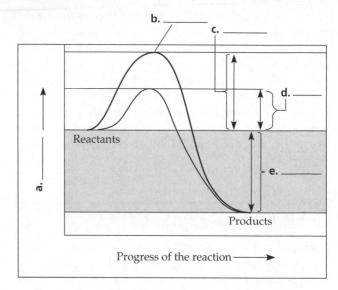

Substrate Specificity of Enzymes   Protein enzymes are macromolecules with characteristic three-dimensional shapes that result in their specificity for a particular **substrate.** The substrate attaches at the enzyme's **active site,** a pocket or groove found on the surface of the enzyme that has a shape compatible to that of the substrate. The substrate is temporarily bound to its enzyme, forming an **enzyme-substrate complex.** Interactions between substrate and active site cause the enzyme to change shape slightly, creating an **induced fit** that enhances the ability of the enzyme to catalyze the chemical reaction.

Catalysis in the Enzyme's Active Site   The substrate is often held in the active site by hydrogen bonds or ionic bonds. The side chains (R groups) of some of the surrounding amino acids in the active site facilitate the conversion of substrate to product. The product then leaves the active site, and the catalytic cycle repeats, often at astonishing speed.

Whether an enzyme catalyzes the forward or backward reaction is influenced by the relative concentrations of reactants and products and the ΔG of the reactions. Enzymes catalyze reactions moving toward equilibrium.

Enzymes may catalyze reactions involving the joining of two reactants by properly orienting the substrates closely together. An induced fit can stretch critical bonds in the substrate molecule and make them easier to break. An active site may provide a microenvironment, such as a lower pH, that is necessary for a particular reaction. Enzymes may also actually participate in a reaction by forming brief covalent bonds with the substrate.

The rate of a reaction will increase with increasing substrate concentration up to the point at which all enzyme molecules are *saturated* with substrate molecules and working at full speed. At that point, only adding more enzyme molecules will increase the rate of the reaction.

INTERACTIVE QUESTION 8.6

In the following diagram of a catalytic cycle, sketch two appropriate substrate molecules and two products, and identify the key steps of the cycle.

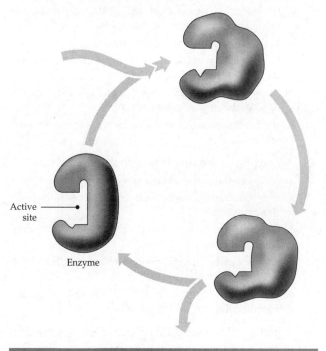

Active
site

Enzyme

Effects of Local Conditions on Enzyme Activity The speed of an enzyme-catalyzed reaction may increase with rising temperature up to the point at which increased thermal agitation begins to disrupt the weak bonds and interactions that stabilize protein shape. Each enzyme has *optimal conditions* that include a temperature and pH that favor its most active shape.

Cofactors are small molecules that bind either permanently or reversibly with enzymes and are necessary for enzyme function. They may be inorganic, such as various metal ions, or organic molecules called **coenzymes.** Most vitamins are coenzymes or precursors of coenzymes.

Enzyme inhibitors disrupt the action of enzymes, either reversibly by binding with the enzyme with weak bonds, or irreversibly by attaching with covalent bonds. **Competitive inhibitors** compete with the substrate for the active site of the enzyme. Increasing the concentration of substrate molecules may overcome this type of inhibition. **Noncompetitive inhibitors** bind to a part of the enzyme separate from the active site and impede enzyme action by changing the shape of the enzyme.

INTERACTIVE QUESTION 8.7

Return to the diagram in Interactive Question 8.6. Draw a competitive inhibitor and a noncompetitive inhibitor, and indicate where each would bind to the enzyme molecule.

The Evolution of Enzymes Mutations in genes may alter the amino acid sequence and thus change the substrate specificity or activity of an enzyme. If this novel function is beneficial, natural selection would be expected to preserve the mutated gene in the population. Researchers tested this hypothesis with the *E. coli* gene for β-galactosidase. After seven rounds of mutation and selection, the "evolved" enzyme digested a new substrate much better than did the original enzyme.

8.5 Regulation of enzyme activity helps control metabolism

Allosteric Regulation of Enzymes In **allosteric regulation,** molecules may inhibit or activate enzyme activity when they bind to a site separate from the active site. Enzymes made of two or more polypeptides, each with its own active site, may have regulatory sites (sometimes called allosteric sites) located where subunits join. The entire unit may oscillate between two forms. The binding of an *activator* stabilizes the catalytically active shape, whereas an *inhibitor* reinforces the inactive form of the enzyme. Allosteric enzymes may be critical regulators of both catabolic and anabolic pathways.

Through a phenomenon called **cooperativity,** the binding of a substrate molecule to one subunit changes the shape of all subunits such that their active sites are stabilized in the active form.

Researchers search for allosteric regulators that may function as drugs that can regulate enzyme activity. Metabolic pathways are commonly regulated by **feedback inhibition,** in which the end product acts as an allosteric inhibitor of an enzyme early in the pathway.

INTERACTIVE QUESTION   8.8

Both ATP and ADP serve as regulators of enzyme activity. In catabolic pathways, which of these two molecules would you predict acts as an inhibitor?

Which would act as an activator?

Specific Localization of Enzymes Within the Cell
Enzymes for several steps of a metabolic pathway may be associated in a multienzyme complex, facilitating the sequence of reactions. Specialized cellular compartments may contain high concentrations of the enzymes and substrates needed for a particular pathway. Enzymes are often incorporated into the membranes of cellular compartments. The complex internal structures of the cell facilitate metabolic order.

Word Roots

allo- = different (*allosteric regulation:* the binding of a regulatory molecule at one site that affects the function of the protein at a different site)

ana- = up (*anabolic pathway:* a metabolic pathway that consumes energy to synthesize a complex molecule from simpler compounds)

bio- = life (*bioenergetics:* the overall flow and transformation of energy in an organism)

cata- = down (*catabolic pathway:* a metabolic pathway that releases energy by breaking down complex molecules into simpler compounds)

endo- = within (*endergonic reaction:* a nonspontaneous chemical reaction in which free energy is absorbed from the surroundings)

ex- = out (*exergonic reaction:* a spontaneous reaction, in which there is a net release of free energy)

kinet- = movement (*kinetic energy:* the energy of motion)

therm- = heat (*thermodynamics:* the study of the energy transformations that occur in a collection of matter)

Structure Your Knowledge

This chapter introduced many new and complex concepts. See if you can step back from the details and answer the following general questions.

1. What is the relationship between the concept of free energy and metabolism?

2. What role do enzymes play in metabolism?

Test Your Knowledge

MULTIPLE CHOICE: *Choose the one best answer.*

1. Catabolic and anabolic pathways are often coupled in a cell because
   a. the intermediates of a catabolic pathway are used in the anabolic pathway.
   b. both pathways use the same enzymes.
   c. the free energy released from one pathway is used to drive the other pathway.
   d. the activation energy of the catabolic pathway can be used in the anabolic pathway.
   e. their enzymes are controlled by the same activators and inhibitors.

2. According to the first law of thermodynamics,
   a. for every action there is an equal and opposite reaction.
   b. every energy transfer results in an increase in disorder or entropy.
   c. the total amount of energy in the universe is conserved or constant.
   d. energy can be transferred or transformed, but disorder always increases.
   e. potential energy is converted to kinetic energy, and kinetic energy is converted to heat.

3. When a cell breaks down glucose, only about 34% of the energy is captured in ATP molecules. The remaining 66% of the energy is
   a. used to increase the order necessary for life to exist.
   b. lost as heat, in accordance with the second law of thermodynamics.
   c. used to increase the entropy of the system by converting kinetic energy into potential energy.
   d. stored in starch or glycogen for later use by the cell.
   e. released when the ATP molecules are hydrolyzed.

4. When glucose and O_2 are converted to CO_2 and H_2O, changes in total energy, entropy, and free energy are correctly represented as
   a. $-\Delta H, -\Delta S, -\Delta G$.
   b. $-\Delta H, +\Delta S, -\Delta G$.
   c. $-\Delta H, +\Delta S, +\Delta G$.

    d. $+\Delta H$, $+\Delta S$, $+\Delta G$.

    e. $+\Delta H$, $-\Delta S$, $+\Delta G$.

5. When amino acids join to form a protein, which of the following energy and entropy changes apply?

    a. $-\Delta H$, $+\Delta S$, $+\Delta G$

    b. $-\Delta H$, $-\Delta S$, $+\Delta G$

    c. $+\Delta H$, $+\Delta S$, $+\Delta G$

    d. $+\Delta H$, $-\Delta S$, $+\Delta G$

    e. $+\Delta H$, $+\Delta S$, $-\Delta G$

6. A negative ΔG means that

    a. the quantity G of energy is available to do work.

    b. the reaction is spontaneous.

    c. the reactants have more free energy than the products.

    d. the reaction is exergonic.

    e. all of the above are true.

7. One way in which a cell maintains metabolic disequilibrium is to

    a. siphon products of a reaction off to the next step in a metabolic pathway.

    b. provide a constant supply of enzymes for critical reactions.

    c. use feedback inhibition to turn off pathways.

    d. use allosteric enzymes that can bind to activators or inhibitors.

    e. use the energy from anabolic pathways to drive catabolic pathways.

8. At equilibrium,

    a. no enzymes are functioning.

    b. free energy is decreasing.

    c. the forward and backward reactions have stopped.

    d. the products and reactants have equal values of H.

    e. ΔG is 0.

9. An endergonic reaction could be described as one that

    a. proceeds spontaneously with the addition of activation energy.

    b. produces products with more free energy than the reactants.

    c. is not able to be catalyzed by enzymes.

    d. releases energy.

    e. produces ATP for energy coupling.

10. The formation of ATP from ADP and inorganic phosphate

    a. is an exergonic process.

    b. transfers the phosphate to an intermediate that becomes more reactive.

    c. produces an unstable energy compound that can drive cellular work.

    d. has a ΔG of -7.3 kcal/mol under standard conditions.

    e. involves the hydrolysis of a phosphate bond.

11. What is meant by an induced fit?

    a. The binding of the substrate is an energy-requiring process.

    b. A competitive inhibitor can outcompete the substrate for the active site.

    c. The binding of the substrate changes the shape of the active site, which can stress or bend substrate bonds.

    d. The active site creates a microenvironment ideal for the reaction.

    e. The binding of an activator to an allosteric site induces a more active form of the subunits of an enzyme.

12. In an experiment, changing the pH from 7 to 6 resulted in an increase in product formation. From this we could conclude that

    a. the enzyme became saturated at pH 6.

    b. the enzyme's optimal pH is 6.

    c. this enzyme works best at a neutral pH.

    d. the temperature must have increased when the pH was changed to 6.

    e. the enzyme was in a more active shape at pH 6.

13. When substance A was added to an enzyme reaction, product formation decreased. The addition of more substrate did not increase product formation. From this we conclude that substance A could be

    a. product molecules.

    b. a cofactor.

    c. an allosteric enzyme.

    d. a competitive inhibitor.

    e. a noncompetitive inhibitor.

14. Which of the following characteristics is most directly responsible for the specificity of a protein enzyme?

    a. its primary structure

    b. its secondary and tertiary structures

    c. the shape and characteristics of its allosteric site

    d. its cofactors

    e. the R groups of the amino acids in its active site

15. An enzyme raises which of the following parameters?

    a. ΔG

    b. ΔH

    c. the free energy of activation

    d. the speed of a reaction

    e. the equilibrium of a reaction

16. Zinc, an essential trace element, may be found bound to the active site of some enzymes. Such zinc ions most likely function as

    a. a coenzyme derived from a vitamin.

    b. a cofactor necessary for catalysis.

    c. a substrate of the enzyme.

    d. a competitive inhibitor of the enzyme.

    e. an allosteric activator of the enzyme.

Use the following diagram to answer questions 17–19.

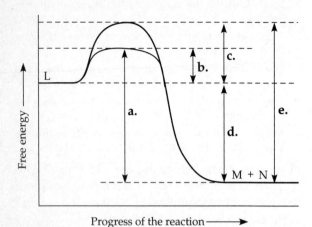

17. Which line in the diagram indicates the ΔG of the enzyme-catalyzed reaction L → M + N?

18. Which line in the diagram indicates the activation energy of the noncatalyzed reaction?

19. Which of the following terms best describes this reaction?
   a. nonspontaneous
   b. $-\Delta G$
   c. endergonic
   d. coupled reaction
   e. anabolic reaction

20. In cooperativity,
   a. a cellular organelle contains all the enzymes needed for a metabolic pathway.
   b. a product of a pathway serves as a competitive inhibitor of an enzyme early in the pathway.
   c. a molecule bound to the active site of one subunit of an enzyme affects the active site of other subunits.
   d. the allosteric site is filled with an activator molecule.
   e. the product of one reaction serves as the substrate for the next reaction in intricately ordered metabolic pathways.

21. In the metabolic pathway A → B → C → D → E, what effect would molecule E likely have on the enzyme that catalyzes A → B?
   a. allosteric inhibitor
   b. allosteric activator
   c. competitive inhibitor
   d. feedback activator
   e. coenzyme

FILL IN THE BLANKS

_____ **1.** the totality of an organism's chemical processes

_____ **2.** pathways that use energy to synthesize complex molecules

_____ **3.** the energy resulting from location or structure

_____ **4.** the most random form of energy

_____ **5.** term for the measure of disorder or randomness

_____ **6.** the energy that must be absorbed by molecules to reach the transition state

_____ **7.** inhibitors that decrease an enzyme's activity by binding to the active site

_____ **8.** organic molecules that bind to enzymes and are necessary for their functioning

_____ **9.** regulatory device in which the product of a pathway binds to an enzyme early in the pathway

_____ **10.** more reactive molecules created by the transfer of a phosphate group from ATP

Chapter 9

Cellular Respiration and Fermentation

Key Concepts

9.1 Catabolic pathways yield energy by oxidizing organic fuels

9.2 Glycolysis harvests chemical energy by oxidizing glucose to pyruvate

9.3 After pyruvate is oxidized, the citric acid cycle completes the energy-yielding oxidation of organic molecules

9.4 During oxidative phosphorylation, chemiosmosis couples electron transport to ATP synthesis

9.5 Fermentation and anaerobic respiration enable cells to produce ATP without the use of oxygen

9.6 Glycolysis and the citric acid cycle connect to many other metabolic pathways

Framework

The catabolic pathways of glycolysis and cellular respiration release the chemical energy in glucose and other fuels and store it in ATP. Glycolysis, occurring in the cytosol, produces ATP, pyruvate, and NADH; the latter two may then enter the mitochondria for respiration. A mitochondrion consists of a matrix in which the enzymes of the citric acid cycle are localized and a highly folded inner membrane in which enzymes and the molecules of the electron transport chain are embedded. The redox reactions of the electron transport chain pump H^+ into the intermembrane space between the two membranes. ATP is produced by oxidative phosphorylation, using a chemiosmotic mechanism in which a proton-motive force drives protons through ATP synthases located in the membrane. Anaerobic respiration uses a final electron acceptor other than oxygen. Fermentation degrades glucose to pyruvate, producing 2 ATP and recycling NADH to NAD^+.

Chapter Review

9.1 Catabolic pathways yield energy by oxidizing organic fuels

Catabolic Pathways and Production of ATP **Fermentation** occurs without oxygen and partially degrades glucose to release energy. **Aerobic respiration** uses oxygen in the breakdown of glucose (or of other energy-rich organic compounds) to yield carbon dioxide and water and release energy as ATP and heat. The *anaerobic respiration* of some prokaryotes does not use oxygen as a reactant but is a similar process. Aerobic respiration is often referred to as **cellular respiration.** This exergonic process has a free energy change of −686 kcal/mol of glucose.

INTERACTIVE QUESTION   9.1

Fill in the following summary equation for cellular respiration.

_____ + 6 O_2 → _____ + 6 H_2O + _____

Redox Reactions: Oxidation and Reduction Oxidation–reduction reactions or **redox reactions** involve the partial or complete transfer of one or more electrons from one reactant to another. **Oxidation** is the loss of electrons from one substance; **reduction** is the addition of electrons to another substance. The substance that loses electrons becomes oxidized and acts as a **reducing agent** (electron donor). By gaining electrons, a substance acts as an **oxidizing agent** (electron acceptor) and becomes reduced.

INTERACTIVE QUESTION 9.2

Fill in the appropriate terms in the following equation.

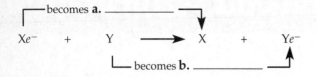

Xe⁻ is the reducing agent; it **c.** _____ electrons.
Y is the **d.** _____ ; it **e.** _____ electrons.

Oxygen strongly attracts electrons and is one of the most powerful oxidizing agents. As electrons shift toward a more electronegative atom, they give up potential energy. Thus, chemical energy is released in a redox reaction that relocates electrons closer to oxygen.

Organic molecules with an abundance of hydrogen are rich in "hilltop" electrons, which release their potential energy when they "fall" closer to oxygen.

INTERACTIVE QUESTION 9.3

a. In the conversion of glucose and O_2 to CO_2 and H_2O, which molecule becomes reduced?

b. Which molecule becomes oxidized?

c. What happens to the energy that is released in this redox reaction?

At certain steps in the oxidation of glucose, two hydrogen atoms are removed by enzymes called dehydrogenases, and the two electrons and one proton are passed to a coenzyme, **NAD⁺** (nicotinamide adenine dinucleotide), reducing it to NADH.

Energy from respiration is slowly released in a series of redox reactions as electrons are passed from NADH down an **electron transport chain,** a group of carrier molecules located in the inner mitochondrial membrane (or in the plasma membrane of aerobic prokaryotes), to a stable location close to a highly electronegative oxygen atom, forming water.

INTERACTIVE QUESTION 9.4

a. NAD⁺ is called an _____.

b. Its reduced form is _____.

The Stages of Cellular Respiration: A Preview **Glycolysis,** which occurs in the cytosol, breaks glucose into two molecules of pyruvate. Within the mitochondrial matrix or in the cytosol of prokaryotes, pyruvate is oxidized to acetyl CoA. The **citric acid cycle** then oxidizes acetyl CoA to CO_2. In some steps, electrons are transferred to NAD⁺. NADH passes electrons to the electron transport chain, from which they combine with H⁺ and oxygen to form water. The energy released in this chain of redox reactions is used to synthesize ATP by **oxidative phosphorylation,** a process that includes electron transport and chemiosmosis.

Up to 32 molecules of ATP may be generated for each glucose molecule oxidized to CO_2. About 10% of this ATP is produced by **substrate-level phosphorylation,** in which an enzyme transfers a phosphate group from a substrate molecule to ADP.

INTERACTIVE QUESTION 9.5

Fill in the three stages of cellular respiration (a–c). Indicate whether ATP is produced by substrate-level or oxidative phosphorylation (d–f). Label the arrows indicating electrons carried by NADH.

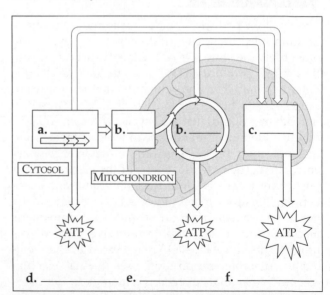

9.2 Glycolysis harvests chemical energy by oxidizing glucose to pyruvate

Glycolysis, a ten-step process occurring in the cytosol, has an energy-investment phase and an energy-payoff phase. Two molecules of ATP are consumed as glucose is split into two three-carbon sugars (glyceraldehyde-3-phosphate). The conversion of these molecules to pyruvate produces 2 NADH and 4 ATP by substrate-level phosphorylation. For each molecule of glucose, glycolysis yields a net gain of 2 ATP and 2 NADH.

Enzymes catalyze each step in glycolysis. Kinases transfer phosphate groups; a dehydrogenase oxidizes glyceraldehyde-3-phosphate and reduces NAD^+; and other enzymes cleave the six-carbon sugar and rearrange atoms in substrate molecules.

INTERACTIVE QUESTION 9.6

Fill in the blanks in the following summary diagram of glycolysis.

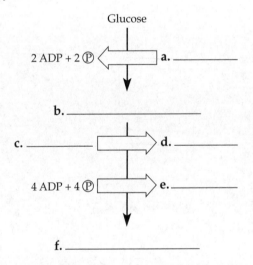

9.3 After pyruvate is oxidized, the citric acid cycle completes the energy-yielding oxidation of organic molecules

Oxidation of Pyruvate to Acetyl CoA Pyruvate is actively transported into the mitochondrion. In a series of steps within a multienzyme complex, a carboxyl group is removed from the three-carbon pyruvate and released as CO_2; the remaining two-carbon group is oxidized to form acetate, with the accompanying reduction of NAD^+ to NADH; and coenzyme A is attached by its sulfur atom to the acetate, forming **acetyl CoA.**

The Citric Acid Cycle In the citric acid cycle, the acetyl group of acetyl CoA is added to oxaloacetate to form citrate, which is progressively decomposed back to oxaloacetate. For each turn of the citric acid cycle, two carbons enter from acetyl CoA; two carbons exit completely oxidized as CO_2; three NADH and one $FADH_2$ are formed; and one ATP (GTP in most animal cells) is made by substrate-level phosphorylation. It takes two turns of the citric acid cycle to oxidize a single glucose molecule.

INTERACTIVE QUESTION 9.7

Fill in the blanks in the following diagram of the citric acid cycle. Gray balls represent carbon atoms.

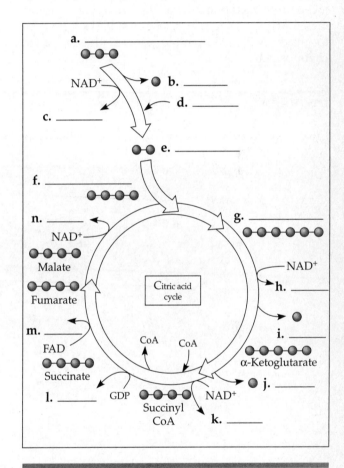

9.4 During oxidative phosphorylation, chemiosmosis couples electron transport to ATP synthesis

The Pathway of Electron Transport Thousands of electron transport chains are embedded in the cristae (infoldings) of the inner mitochondrial membrane (or in the plasma membrane of prokaryotes). The components of the electron transport chain are organized into four complexes, and most are proteins to which nonprotein *prosthetic groups* are tightly bound.

The electron carriers shift between reduced and oxidized states as they accept and donate electrons. The transfer of electrons proceeds from NADH to a flavoprotein to an iron-sulfur protein (Complex I) to a mobile hydrophobic molecule called ubiquinone (Q or CoQ). Next, electrons are passed down a series of molecules called **cytochromes,** which are proteins with an iron-containing heme group. The last cytochrome, cyt a_3, passes electrons to oxygen, which picks up two H^+, forming water.

$FADH_2$ adds its electrons to the chain at a lower energy level (at Complex II); thus, less energy is provided for ATP synthesis by $FADH_2$ as compared to NADH.

Chemiosmosis: The Energy-Coupling Mechanism **ATP synthase,** a protein complex embedded in the inner mitochondrial membrane or in the prokaryotic plasma membrane, uses the energy of a proton (H^+) gradient to make ATP, an example of the process called **chemiosmosis.**

The flow of H^+ down their gradient through the rotor and stator part of the ATP synthase complex causes the rotor and attached rod to rotate, activating catalytic sites in the knob portion, where ADP and inorganic phosphate join to make ATP.

The electron transport chain creates the proton gradient. When some members of the chain pass electrons, they also accept and release protons, which are deposited in the intermembrane space at three sites. The potential energy of the proton gradient is referred to as the **proton-motive force.**

In mitochondria, exergonic redox reactions produce the proton gradient that drives the production of ATP. Chloroplasts use light energy to create the proton-motive force used to make ATP by chemiosmosis. Prokaryotes use proton gradients generated across the plasma membrane to transport molecules, make ATP, and rotate flagella.

INTERACTIVE QUESTION   9.8

Label the following diagram of oxidative phosphorylation in a mitochondrial membrane.

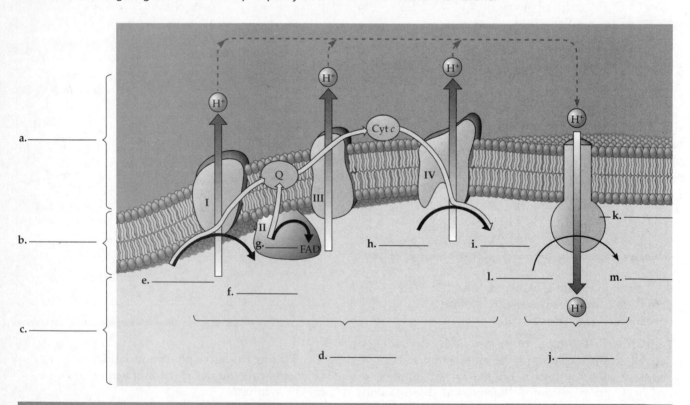

An Accounting of ATP Production by Cellular Respiration   About 30 to 32 ATPs may be produced per glucose molecule oxidized. These numbers are only estimates for three reasons: Experimental data indicate the production of 2.5 ATPs/NADH and of 1.5 ATPs/$FADH_2$; the electrons from NADH produced by glycolysis may be passed across the mitochondrial membrane to NAD^+ or FAD, depending on the type of shuttle used in the cell; and the proton-motive force generated by the electron transport chain is also used to power other work in the mitochondrion.

The efficiency of energy conversion in respiration is about 34%. The remaining 66% is lost as heat.

INTERACTIVE QUESTION 9.9

Fill in the following tally for the maximum ATP yield from the oxidation of one molecule of glucose to six molecules of carbon dioxide.

Process	# ATP
Initial phosphorylation of glucose:	**a.** _____
Substrate-level phosphorylation: in glycolysis	**b.** _____
in **c.** _____	2
Oxidative phosphorylation:*	**d.** _____
Maximum Total	**e.** _____

*2.5 ATP for each of the **f.** _____ NADH from pyruvate → acetyl CoA and the **g.** _____ NADH from citric acid cycle; 1.5 ATP for each of the **h.** _____ $FADH_2$ from citric acid cycle; 2.5 or 1.5 ATP for each of the **i.** _____ NADH from glycolysis, depending on which shuttle passes electrons across the membrane.

9.5 Fermentation and anaerobic respiration enable cells to produce ATP without the use of oxygen

Organisms that generate ATP through anaerobic respiration have an electron transport chain that does not use oxygen as the final electron acceptor. Some bacteria use sulfate ions to accept electrons, generating H_2S instead of H_2O. Fermentation generates ATP without using either oxygen or an electron transport chain. In fermentation, oxidation of glucose in glycolysis produces a net of 2 ATP by substrate-level phosphorylation, and NADH is recycled to NAD^+ by the transfer of electrons to pyruvate or derivatives of pyruvate.

Types of Fermentation In **alcohol fermentation,** pyruvate is converted to acetaldehyde, and CO_2 is released. Acetaldehyde is then reduced by NADH to form ethanol (ethyl alcohol), and NAD^+ is regenerated. In **lactic acid fermentation,** pyruvate is reduced directly by NADH to lactate, recycling NAD^+. Muscle cells make ATP by lactic acid fermentation when energy demand is high and O_2 supply is low.

Comparing Fermentation with Anaerobic and Aerobic Respiration Both fermentation and aerobic and anaerobic respiration use glycolysis, with NAD^+ as the oxidizing agent to convert glucose and other organic fuels to pyruvate. To oxidize NADH back to NAD^+, fermentation uses pyruvate or acetaldehyde as the final electron acceptor. Respiration uses oxygen or another electronegative molecule (in anaerobic respiration) as the final electron acceptor after electrons are passed down an electron transport chain.

Facultative anaerobes, such as yeasts and many bacteria, can make ATP by fermentation or respiration, depending upon the availability of oxygen. **Obligate anaerobes** make ATP by fermentation or anaerobic respiration and are poisoned by oxygen.

INTERACTIVE QUESTION 9.10

How much more ATP can be generated by respiration than by fermentation? Explain why.

The Evolutionary Significance of Glycolysis Glycolysis is common to fermentation and respiration. This most widespread of all processes probably evolved in ancient prokaryotes. Its cytosolic location is also evidence of its antiquity.

9.6 Glycolysis and the citric acid cycle connect to many other metabolic pathways

The Versatility of Catabolism Fats, proteins, and carbohydrates can all be used in cellular respiration. Proteins are digested into amino acids, which are then *deaminated* (the amino group is removed) and can enter glycolysis or the citric acid cycle at several points. The digestion of fats yields glycerol, which is fed into glycolysis, and fatty acids, which are broken down by **beta oxidation** to two-carbon fragments that enter the citric acid cycle as acetyl CoA.

Biosynthesis (Anabolic Pathways) The organic molecules of food also provide carbon skeletons for biosynthesis. Some monomers, such as amino acids, can be directly incorporated into the cell's macromolecules. Intermediates of glycolysis and of the citric acid cycle serve as precursors for anabolic pathways. Carbohydrates, fats, and proteins can be interconverted to provide for a cell's needs.

Regulation of Cellular Respiration via Feedback Mechanisms   Through feedback inhibition, the end product of an anabolic pathway inhibits an enzyme early in the pathway, thus preventing a cell from producing an excess of a particular substance.

The supply of ATP in the cell regulates respiration. The allosteric enzyme that catalyzes the third step of glycolysis, phosphofructokinase, is inhibited by ATP and activated by AMP (derived from ADP). Phosphofructokinase is also inhibited by citrate released from the mitochondria, thereby synchronizing the rates of glycolysis and the citric acid cycle. Other enzymes located at key intersections help to maintain metabolic balance.

Word Roots

aero- = air (*aerobic respiration:* a catabolic pathway using oxygen as the final electron acceptor in an electron transport chain and ultimately producing ATP)

an- = not (*anaerobic:* refers to a chemical reaction that does not use oxygen)

chemi- = chemical (*chemiosmosis:* an energy-coupling mechanism that uses energy stored in the form of a H^+ gradient across a membrane to drive cellular work, such as the synthesis of ATP)

glyco- = sweet; **-lysis** = split (*glycolysis:* a series of reactions that splits glucose into pyruvates)

Structure Your Knowledge

1. This chapter describes how catabolic pathways release chemical energy and store it in ATP. One of the best ways to learn the three main components of cellular respiration is to teach them to someone. Find two study partners and have each person learn and explain the important concepts of glycolysis, pyruvate oxidation and the citric acid cycle, or oxidative phosphorylation. Use diagrams to illustrate the process you are explaining.

2. Create a concept map to organize your understanding of oxidative phosphorylation.

3. Fill in the following table to summarize the major aspects of glycolysis, the citric acid cycle, oxidative phosphorylation, and fermentation. Base inputs and outputs on one glucose molecule.

Process	Main Function	Inputs	Outputs
Glycolysis			
Pyruvate to acetyl CoA			
Citric acid cycle			
Oxidative phosphorylation			
Fermentation			

Test Your Knowledge

MULTIPLE CHOICE: *Choose the one best answer.*

1. When electrons move closer to a more electronegative atom,
   a. energy is released.
   b. energy is consumed.
   c. a proton gradient is established.
   d. water is produced.
   e. ATP is synthesized.

2. In the reaction $C_6H_{12}O_6 + 6\,O_2 \rightarrow 6\,CO_2 + 6\,H_2O$,
   a. glucose becomes reduced.
   b. oxygen becomes reduced.
   c. oxygen becomes oxidized.
   d. water is a reducing agent.
   e. oxygen is a reducing agent.

3. Some prokaryotes use anaerobic respiration, a process that
   a. does not involve an electron transport chain.
   b. produces ATP solely by substrate-level phosphorylation.
   c. uses a substance other than oxygen as the final electron acceptor.
   d. does not rely on chemiosmosis for the production of ATP.
   e. Both a and b are correct.

4. Which of the following reactions is *incorrectly* paired with its location?
   a. ATP synthesis—inner membrane of the mitochondrion, mitochondrial matrix, and cytosol
   b. fermentation—cell cytosol
   c. glycolysis—cell cytosol
   d. substrate-level phosphorylation—cytosol and mitochondrial matrix
   e. citric acid cycle—cristae of mitochondrion

5. Which of the following enzymes uses NAD^+ as a coenzyme?
   a. phosphofructokinase
   b. phosphoglucoisomerase
   c. triose phosphate dehydrogenase
   d. hexokinase
   e. phosphoglyceromutase

6. Which of the following compounds produces the most ATP when oxidized?
   a. acetyl CoA
   b. glucose
   c. pyruvate
   d. fructose-1,6-bisphosphate
   e. glyceraldehyde-3-phosphate

7. When pyruvate is converted to acetyl CoA,
   a. CO_2 and ATP are released.
   b. a multienzyme complex removes a carboxyl group, transfers electrons to NAD^+, and attaches a coenzyme.
   c. one turn of the citric acid cycle is completed.
   d. NAD^+ is regenerated so that glycolysis can continue to produce ATP by substrate-level phosphorylation.
   e. phosphofructokinase is activated and glycolysis continues.

8. How many molecules of CO_2 are generated for each molecule of acetyl CoA introduced into the citric acid cycle?
   a. 1   c. 3   e. 6
   b. 2   d. 4

9. Which of the following statements correctly describes the role of oxygen in cellular respiration?
   a. It is reduced in glycolysis as glucose is oxidized.
   b. It combines with H^+ diffusing through ATP synthase to produce H_2O.
   c. It provides the activation energy needed for oxidation to occur.
   d. It is the final electron acceptor for the electron transport chain.
   e. It combines with the carbon removed during the citric acid cycle to form CO_2.

10. In the chemiosmotic mechanism,
    a. ATP production is linked to the proton gradient established by the electron transport chain.
    b. the difference in pH between the intermembrane space and the cytosol drives the formation of ATP.
    c. the flow of H^+ through ATP synthases rotates a rotor and rod, driving the hydrolysis of ADP.
    d. the energy released by the reduction and subsequent oxidation of electron carriers transfers a phosphate to ADP.
    e. the production of water in the mitochondrial matrix by the reduction of oxygen leads to a net flow of water out of a mitochondrion.

11. When glucose is oxidized to CO_2 and water, approximately 66% of its energy is transformed to
    a. heat.
    b. ATP.
    c. a proton-motive force.
    d. molecular movement.
    e. NADH.

12. Which of the following statements is *incorrect* concerning oxidative phosphorylation?
    a. It produces about 2.5 ATP for every NADH that is oxidized.
    b. It involves the redox reactions of the electron transport chain.
    c. It involves an ATP synthase located in the inner mitochondrial membrane.
    d. It uses oxygen as the final electron donor.
    e. It is an example of chemiosmosis.

13. Which of the following statements correctly describes a metabolic effect of cyanide, a poison that blocks the passage of electrons along the electron transport chain?
    a. The pH of the intermembrane space becomes much lower than normal.
    b. Electrons are passed directly to oxygen, causing cells to explode.
    c. Alcohol would build up in the cells.
    d. NADH supplies would be exhausted, and ATP synthesis would cease.
    e. No proton gradient would be produced, and ATP synthesis would cease.

14. Substrate-level phosphorylation
    a. involves the shifting of a phosphate group from ATP to a substrate.
    b. can use NADH or $FADH_2$.
    c. takes place only in the cytosol.
    d. accounts for 10% of the ATP formed by fermentation.
    e. is the energy source for facultative anaerobes under anaerobic conditions.

15. Fermentation produces less ATP than cellular respiration because
    a. NAD^+ is regenerated by alcohol or lactate production, without the electrons of NADH passing through the electron transport chain.
    b. pyruvate still contains most of the "hilltop" electrons that were present in glucose.
    c. its starting reactant is pyruvate and not glucose.
    d. a and b are correct.
    e. a, b, and c are correct.

16. Which of the following conversions represents a reduction reaction?
    a. pyruvate to acetyl CoA and CO_2
    b. $C_6H_{12}O_6$ to CO_2
    c. NADH to NAD^+
    d. glucose to pyruvate
    e. acetaldehyde (C_2H_4O) to ethanol (C_2H_6O)

17. Muscle cells in oxygen deprivation gain which of the following from the reduction of pyruvate?
    a. ATP
    b. ATP and NAD^+
    c. CO_2 and NAD^+
    d. ATP, alcohol, and NAD^+
    e. ATP and CO_2

18. Glucose made from six radioactively labeled carbon atoms is fed to yeast cells in the absence of oxygen. How many molecules of radioactive alcohol (C_2H_5OH) are formed from each molecule of glucose?
    a. 0   c. 2   e. 6
    b. 1   d. 3

19. Glycolysis is considered one of the first metabolic pathways to have evolved because
    a. it relies on fermentation, which is characteristic of archaea and bacteria.
    b. it is found only in prokaryotes, whereas eukaryotes use mitochondria to produce ATP.
    c. it produces much less ATP than does the electron transport chain and chemiosmosis.
    d. it produces ATP only by substrate phosphorylation and does not involve redox reactions.
    e. it is nearly universal, is located in the cytosol, and does not involve O_2.

20. Which of the following substances produces the most ATP per gram?
    a. glucose, because it is the starting place for glycolysis
    b. glycogen or starch, because they are polymers of glucose
    c. fats, because they are highly reduced compounds
    d. proteins, because of the energy stored in their tertiary structure
    e. amino acids, because they can be fed directly into the citric acid cycle

21. Fats and proteins can be used as fuel in the cell because they
    a. can be converted to glucose by enzymes.
    b. can be converted to intermediates of glycolysis or the citric acid cycle.
    c. can pass through the mitochondrial membrane to enter the citric acid cycle.
    d. contain phosphate groups.
    e. contain more energy than glucose.

22. Which of the following statements is *false* concerning the enzyme phosphofructokinase?
    a. It is an allosteric enzyme.
    b. It is inhibited by citrate.
    c. It is the pacemaker of glycolysis and respiration.
    d. It is inhibited by AMP.
    e. It is an early enzyme in the glycolytic pathway.

23. Which of the following statements describes a possible function of brown fat, which has uncoupler proteins that, when activated, make the inner mitochondrial membrane leaky to H^+?
    a. It produces more ATP than does regular fat and is found in the flight muscles of ducks and geese, providing more energy for long-distance migrations.
    b. It lowers the pH of the intermembrane space, which results in the production of more ATP per gram than is produced by the oxidation of glucose or regular fat tissue.
    c. Because it dissipates the proton gradient, it generates heat through cellular respiration without producing ATP, thereby raising the body temperature of hibernating mammals or newborn infants.
    d. Its main function is insulation in the endothermic animals in which is it common.
    e. Both a and b are correct.

Chapter 10

Photosynthesis

Key Concepts

10.1 Photosynthesis converts light energy to the chemical energy of food

10.2 The light reactions convert solar energy to the chemical energy of ATP and NADPH

10.3 The Calvin cycle uses the chemical energy of ATP and NADPH to reduce CO_2 to sugar

10.4 Alternative mechanisms of carbon fixation have evolved in hot, arid climates

Framework

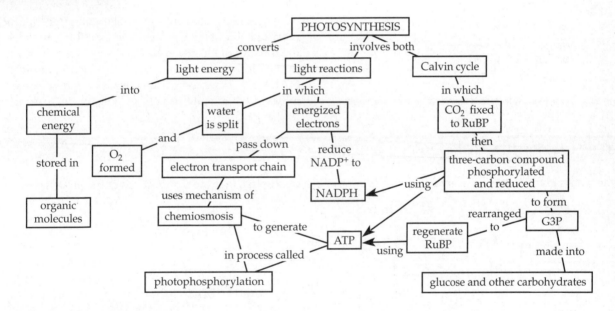

Chapter Review

In **photosynthesis,** the light energy of the sun is converted into chemical energy stored in organic molecules. **Autotrophs** "feed themselves" in the sense that they make their own organic molecules from inorganic raw materials. Autotrophs are the *producers* of the biosphere. Plants, algae, some other protists, and some prokaryotes are photoautotrophs.

Heterotrophs are *consumers.* They may eat plants or animals or decompose organic litter, but almost all ultimately depend on photoautotrophs for food and O_2.

10.1 Photosynthesis converts light energy to the chemical energy of food

The first photosynthetic organisms may have been bacteria with clusters of photosynthetic enzymes and other molecules embedded in infoldings of the plasma membrane.

Chloroplasts: The Sites of Photosynthesis in Plants Chloroplasts are located mainly in the **mesophyll** tissue of the leaf. CO_2 enters and O_2 exits the leaf through **stomata.** Veins carry water from the roots to leaves and distribute sugar to nonphotosynthetic tissue.

A chloroplast consists of a double membrane surrounding a dense fluid called the **stroma** and an elaborate membrane system called **thylakoids,** which encloses the *thylakoid space.* Thylakoid sacs may be stacked to form *grana.* **Chlorophyll,** the green pigment that absorbs the light energy that drives photosynthesis, is embedded in the thylakoid membrane.

INTERACTIVE QUESTION   10.1

Label the indicated parts in the following diagram of a chloroplast.

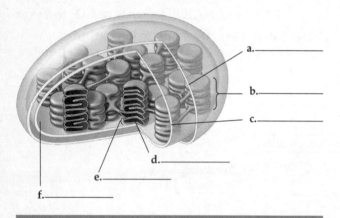

Tracking Atoms through Photosynthesis: Scientific Inquiry Considering only the net consumption of water and considering glucose as the product (even though the direct product is a three-carbon sugar), the equation for photosynthesis is the reverse of respiration:

$$6\ CO_2 + 6\ H_2O + \text{Light energy} \rightarrow C_6H_{12}O_6 + 6\ O_2$$

The simplest equation is $CO_2 + H_2O \rightarrow [CH_2O] + O_2$.

Using evidence from bacteria that utilize hydrogen sulfide (H_2S) for photosynthesis, C. B. van Niel hypothesized that photosynthetic organisms need a hydrogen source and that plants use H_2O as their hydrogen (and electron) source and release O_2. This hypothesis was later confirmed by experiments that used a heavy isotope of oxygen (^{18}O). Labeled O_2 was produced only when water, but not carbon dioxide, contained ^{18}O.

Photosynthesis, like respiration, is a redox process, but it differs in the direction of electron flow. The electrons increase their potential energy when they travel from water to reduce CO_2 into sugar, and light provides the energy for this endergonic process.

The Two Stages of Photosynthesis: A Preview Solar energy is converted into chemical energy in the **light reactions.** Light energy absorbed by chlorophyll drives the transfer of electrons and hydrogen ions from water to the electron acceptor **NADP$^+$**, which is reduced to NADPH and temporarily stores electrons. Oxygen is released when water is split. ATP is formed during the

INTERACTIVE QUESTION   10.2

Fill in the blanks in the following overview of photosynthesis in a chloroplast. Indicate the locations of the processes c and h.

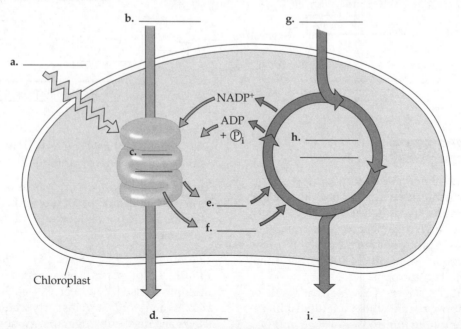

light reactions, using chemiosmosis in a process called **photophosphorylation.**

In the **Calvin cycle,** CO_2 is incorporated into existing organic compounds by **carbon fixation,** and these compounds are then reduced to form carbohydrate. NADPH and ATP from the light reactions supply the reducing power and chemical energy needed for the Calvin cycle.

10.2   The light reactions convert solar energy to the chemical energy of ATP and NADPH

The Nature of Sunlight   Electromagnetic energy, also called electromagnetic radiation, travels as rhythmic wave disturbances of electric and magnetic fields. The distance between the crests of electromagnetic waves, called their **wavelength,** ranges across the **electromagnetic spectrum,** from short gamma waves to long radio waves. The small band of radiation from about 380 nm to 750 nm is called **visible light** and is the radiation that drives photosynthesis.

Light also behaves as if it consists of discrete particles called **photons,** which have a fixed quantity of energy. The amount of energy in a photon is inversely related to its wavelength.

Photosynthetic Pigments: The Light Receptors *Pigments* are substances that absorb light. A **spectrophotometer** measures the amount of light of different wavelengths absorbed by a pigment. The **absorption spectrum** of **chlorophyll *a*,** the pigment that participates directly in the light reactions, shows that it absorbs violet-blue and red light best. Accessory pigments such as **chlorophyll *b*** and some **carotenoids** absorb light of different wavelengths and broaden the spectrum of colors useful in photosynthesis. Some carotenoids function in *photoprotection* by absorbing excessive light energy that might damage chlorophyll or interact with oxygen to form reactive molecules. (These pigments act as antioxidants.)

Excitation of Chlorophyll by Light   When a pigment molecule absorbs energy from a photon, one of the molecule's electrons is elevated to an orbital where it has more potential energy. Only photons whose energy is equal to the difference between the ground state and the excited state for that molecule are absorbed.

The excited state is unstable. Energy is released as heat as the electron drops back to its ground-state orbital. Isolated chlorophyll molecules also emit photons of light, called fluorescence, as their electrons return to the ground state.

INTERACTIVE QUESTION   10.3

An **action spectrum** shows the relative rates of photosynthesis under different wavelengths of light. On the following graph, label the line that represents the absorption spectrum for chlorophyll *a* and the line for the action spectrum for photosynthesis. Why are these lines different?

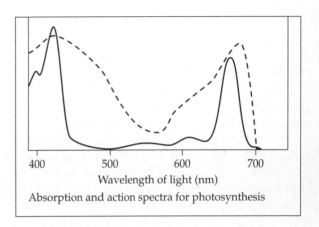

Absorption and action spectra for photosynthesis

A Photosystem: A Reaction-Center Complex Associated with Light-Harvesting Complexes Embedded in thylakoid membranes are numerous **photosystems,** each composed of several **light-harvesting complexes** and a **reaction-center complex.** A reaction-center complex contains two special chlorophyll *a* molecules and a **primary electron acceptor.** When a pigment molecule in a light-harvesting complex absorbs a photon, the energy is passed from pigment to pigment until it reaches the reaction center. In a redox reaction, an excited electron of a reaction-center chlorophyll *a* is captured by the primary electron acceptor before it can return to the ground state.

There are two types of photosystems in the thylakoid membrane. The chlorophyll *a* molecule at the reaction center of **photosystem II (PSII)** is called P680, after the wavelength of light (680 nm) it absorbs best. At the reaction center of **photosystem I (PSI)** is a chlorophyll *a* molecule called P700.

INTERACTIVE QUESTION   10.4

Describe the components of a photosystem.

Linear Electron Flow   Through a sequence called **linear electron flow,** electrons pass from water to $NADP^+$ through the two photosystems. A pigment molecule absorbs a photon of light, and the energy is relayed through other pigment molecules of the light-harvesting complex to the P680 pair of chlorophyll *a* molecules in the PS II reaction-center complex. An excited electron of P680 is trapped by the primary electron acceptor. $P680^+$ is a strong oxidizing agent, and its electron hole is filled when an enzyme removes electrons from water, splitting it into two electrons, two H^+, and an oxygen atom that immediately combines with another oxygen atom to form O_2.

The primary electron acceptor passes the photoexcited electron to an electron transport chain made up of plastoquinone (Pq), a cytochrome complex, and plastocyanin (Pc). The energy released as electrons "fall" through the electron transport chain is used to pump protons into the thylakoid space, contributing to the proton gradient used for the synthesis of ATP.

At the bottom of the electron transport chain, the electron passes to $P700^+$ in photosystem I. It replaces the photoexcited electron that was captured by its primary electron acceptor when PSI absorbed a photon. This primary electron acceptor passes the electron down a second electron transport chain through ferredoxin (Fd), from which the enzyme $NADP^+$ reductase transfers electrons to reduce $NADP^+$ to NADPH.

INTERACTIVE QUESTION   10.5

Fill in the steps of electron flow in the following diagram. Circle the important products that will provide chemical energy and reducing power to the Calvin cycle.

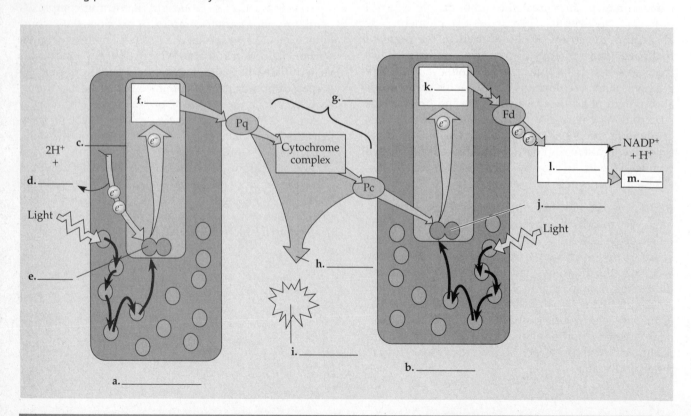

Cyclic Electron Flow In **cyclic electron flow**, electrons excited from P700 in PSI are passed from Fd to the cytochrome complex and back to P700. Several groups of photosynthetic bacteria have only photosystem I, and photosynthesis may have evolved in a form similar to cyclic electron flow. Cyclic electron flow may be photoprotective in eukaryotic photosynthesizers.

INTERACTIVE QUESTION 10.6

a. On the diagram in Interactive Question 10.5, sketch the path that electrons from P700 take during cyclic electron flow.

b. Why is neither O_2 nor NADPH generated by cyclic electron flow?

c. How, then, is ATP produced by cyclic electron flow?

A Comparison of Chemiosmosis in Chloroplasts and Mitochondria Chemiosmosis in mitochondria and in chloroplasts is very similar. Electron transport chains built into a membrane pump protons across the membrane as electrons are passed down the chain in a series of redox reactions. In respiration, however, organic molecules provide the electrons, and chemical energy is transferred to ATP; in chloroplasts, by contrast, water provides the electrons, and light energy is transformed to the chemical energy of ATP.

In chloroplasts, the electron transport chain pumps protons from the stroma into the thylakoid space. As H^+ diffuses back through ATP synthase, ATP is formed on the stroma side, where it is available to the Calvin cycle.

INTERACTIVE QUESTION 10.7

a. In the light, the proton gradient across the thylakoid membrane is as great as 3 pH units. On which side is the pH lowest?

b. What three factors contribute to the formation of this large difference in H^+ concentration between the thylakoid space and the stroma?

10.3 The Calvin cycle uses the chemical energy of ATP and NADPH to reduce CO_2 to sugar

It takes three turns of the Calvin cycle to fix three molecules of CO_2 and produce one molecule of the three-carbon sugar **glyceraldehyde-3-phosphate (G3P).** The cycle can be divided into three phases:

1. *Carbon fixation:* CO_2 is added to a five-carbon sugar, ribulose bisphosphate (RuBP), in a reaction catalyzed by the enzyme RuBP carboxylase (**rubisco**). The resulting unstable six-carbon intermediate splits into two molecules of 3-phosphoglycerate.

2. *Reduction:* Each molecule of 3-phosphoglycerate is then phosphorylated by ATP to form 1,3-bisphosphoglycerate. Two electrons from NADPH reduce this compound to G3P. The cycle must turn three times to create a net gain of one molecule of G3P.

3. *Regeneration of CO_2 acceptor* (**RuBP**): The rearrangement of five molecules of G3P into three molecules of RuBP requires three more ATP.

INTERACTIVE QUESTION   10.8

Label the three phases (a, b, and c) and key molecules in the following diagram of the Calvin cycle. How many ATP and NADPH are needed to synthesize one G3P molecule?

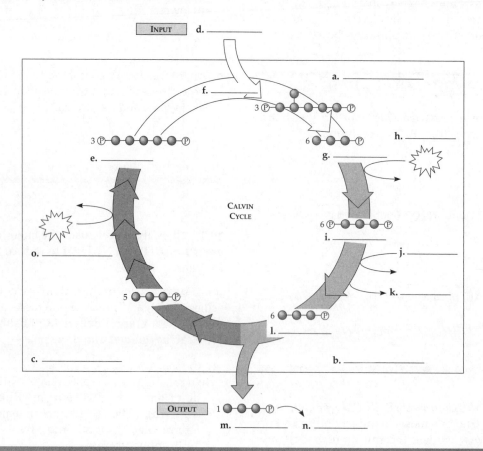

10.4   Alternative mechanisms of carbon fixation have evolved in hot, arid climates

Photorespiration: An Evolutionary Relic?   In most plants, the first product of carbon fixation is a three-carbon compound, 3-phosphoglycerate, formed in the Calvin cycle. When these **C₃ plants** close their stomata on hot, dry days to limit water loss, CO_2 concentration in the leaf air spaces falls, slowing the Calvin cycle. As more O_2 than CO_2 accumulates, rubisco adds O_2 to RuBP in place of CO_2. The product splits, and a two-carbon compound exits the chloroplast and is broken down to release CO_2. This seemingly wasteful process is called **photorespiration**.

INTERACTIVE QUESTION   10.9

What are two possible explanations for the existence of photorespiration, a process that can result in the loss of as much as 50% of the carbon fixed in the Calvin cycle?

C₄ Plants   In **C₄ plants,** CO_2 is first added to the three-carbon compound PEP, with the aid of an enzyme (**PEP carboxylase**) that has a high affinity for CO_2. The resulting four-carbon compound, formed in the meso-phyll cells of the leaf, is transported to **bundle-sheath cells** tightly packed around the veins of the leaf. The compound is then broken down to release CO_2, which rubisco then fixes in the Calvin cycle.

Increasing atmospheric CO_2 concentrations and the resulting global climate change may affect the distribution of C_3 and C_4 plants and change the structure of various plant communities.

INTERACTIVE QUESTION   10.10

a. Where does the Calvin cycle take place in C_4 plants?

b. How can C_4 plants successfully utilize the Calvin cycle in hot, dry conditions when C_3 plants would be undergoing photorespiration?

c. Why does C_4 photosynthesis requires more ATP than does C_3 photosynthesis?

CAM Plants   Many desert succulent plants close their stomata during the day, helping to prevent water loss. At night, they open their stomata and take up CO_2, incorporating it into a variety of organic acids in a mode of carbon fixation called **crassulacean acid metabolism (CAM)**. During daylight, **CAM plants** break these compounds down and release CO_2, allowing the Calvin cycle to proceed. Unlike the C_4 pathway, the CAM pathway does not structurally separate carbon fixation from the Calvin cycle; instead, the two processes are separated in time.

The Importance of Photosynthesis: A Review   About 50% of the organic material produced by photosynthesis is used as fuel for cellular respiration in the mitochondria of plant cells; the rest is used as carbon skeletons for the synthesis of organic molecules (proteins, lipids, and a great deal of cellulose), stored as starch, or lost through photorespiration. Each year about 160 billion metric tons of carbohydrate are produced by photosynthesis.

Word Roots

auto- = self; **-troph** = food (*autotroph:* an organism that obtains organic food molecules not by eating other organisms, but by using energy from the sun or from the oxidation of inorganic substances to make organic molecules)

chloro- = green; **-phyll** = leaf (*chlorophyll:* a green, photosynthetic pigment located in membranes within chloroplasts)

electro- = electricity; **magnet-** = magnetic (*electromagnetic spectrum:* the entire spectrum of radiation)

hetero- = other (*heterotroph:* an organism that obtains organic food molecules by eating other organisms or substances derived from them)

meso- = middle (*mesophyll:* leaf cells specialized for photosynthesis, located between the upper and lower epidermis)

photo- = light (*photosystem:* a light-capturing unit located in the thylakoid membrane, consisting of a reaction-center complex surrounded by numerous light-harvesting complexes)

Structure Your Knowledge

1. You have already filled in the blanks in several diagrams of photosynthesis. To really understand this process, however, try to create your own representation. In a diagrammatic form, trace the flow of electrons and the production of ATP and NADPH in the light reactions. Then outline the three major stages in the production of G3P. Indicate where these reactions occur in the chloroplast. You can compare your diagram to the sketch in the answers section. Perhaps your study group can collaboratively create a clear, concise summary of this chapter.

2. Create a concept map to confirm your understanding of the chemiosmotic synthesis of ATP in photophosphorylation.

Test Your Knowledge

MULTIPLE CHOICE: *Choose the one best answer.*

1. Which of the following processes or structures is mismatched with its location?
   a. light reactions—grana
   b. electron transport chain—thylakoid membrane
   c. Calvin cycle—stroma
   d. ATP synthase—double membrane surrounding chloroplast
   e. splitting of water—thylakoid space

2. Photosynthesis is a redox process in which
   a. CO_2 is reduced and water is oxidized.
   b. $NADP^+$ is reduced and RuBP is oxidized.
   c. CO_2, $NADP^+$, and water are reduced.
   d. O_2 acts as an oxidizing agent and water acts as a reducing agent.
   e. G3P is reduced and the electron transport chain is oxidized.

3. Which of the following statements is *false*?
   a. When isolated chlorophyll molecules absorb photons, their electrons fall back to ground state, giving off heat and light.
   b. Accessory pigments, cyclic electron flow, and photorespiration may all contribute to photoprotection, protecting plants from the detrimental effects of intense light.
   c. In the cyclic electron flow of purple sulfur bacteria, the electron transport chain pumps H^+ across a membrane, creating a proton-motive force used in ATP synthesis.
   d. In both photosynthetic prokaryotes and eukaryotes, ATP synthases catalyze the production of ATP within the cytosol of the cell.
   e. In sulfur bacteria, H_2S is the hydrogen (and thus electron) source for photosynthesis.

4. A spectrophotometer can be used to measure
   a. the absorption spectrum of a substance.
   b. the action spectrum of a reaction.
   c. the amount of energy in a photon.
   d. the wavelength of visible light.
   e. the efficiency of photosynthesis.

5. Accessory pigments within chloroplasts are responsible for
   a. driving the splitting of water molecules.
   b. absorbing photons of different wavelengths of light and passing that energy to P680 or P700.
   c. providing electrons to the reaction-center chlorophyll after photoexcited electrons pass to $NADP^+$.
   d. pumping H^+ across the thylakoid membrane to create a proton-motive force.
   e. anchoring chlorophyll *a* within the reaction center.

6. The following diagram is an absorption spectrum for an unknown pigment molecule. What color would this pigment appear to you?

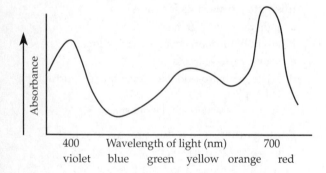

   a. violet
   b. blue
   c. green
   d. yellow
   e. red

7. Linear electron flow along with chemiosmosis in the chloroplast results in the formation of
   a. ATP only.
   b. ATP and NADPH.
   c. ATP and G3P.
   d. ATP and O_2.
   e. ATP, NADPH, and O_2.

8. The chlorophyll known as $P680^+$ has its electron "holes" filled by electrons from
   a. photosystem I.
   b. photosystem II.
   c. water.
   d. NADPH.
   e. accessory pigments.

9. Which of the following substances is/are the final electron acceptors for the electron transport chains in the light reactions of photosynthesis and in cellular respiration?
   a. O_2 in both
   b. CO_2 in both
   c. H_2O in the light reactions, and O_2 in respiration
   d. P700 and NAD^+ in the light reactions, and NAD^+ or FAD in respiration
   e. $NADP^+$ in the light reactions, and O_2 in respiration

10. In the chemiosmotic synthesis of ATP in a chloroplast, H^+ diffuses through the ATP synthase
    a. from the stroma into the thylakoid space.
    b. from the thylakoid space into the stroma.
    c. from the intermembrane space into the matrix.
    d. from the cytoplasm into the intermembrane space.
    e. from the matrix into the stroma.

11. Which of the following parts of an illuminated plant cell would you expect to have the lowest pH?
    a. nucleus
    b. cytosol
    c. chloroplast
    d. stroma of chloroplast
    e. thylakoid space

12. A difference between electron transport in photosynthesis and respiration is that in photosynthesis,
    a. NADPH rather than NADH passes electrons to the electron transport chain.
    b. ATP synthase releases ATP into the stroma rather than into the cytosol.
    c. light provides the energy to push electrons to the top of the electron chain, rather than energy from the oxidation of food molecules.
    d. an H^+ concentration gradient rather than a proton-motive force drives the phosphorylation of ATP.
    e. Both a and c are correct.

13. How does cyclic electron flow differ from linear electron flow?
    a. No NADPH is produced by cyclic electron flow.
    b. No O_2 is produced by cyclic electron flow.
    c. The cytochrome complex in the electron transport chain is not involved in cyclic electron flow.
    d. Both a and b are correct.
    e. a, b, and c are correct.

14. Chloroplasts can make carbohydrate in the dark if provided with
    a. ATP, NADPH, and CO_2.
    b. an artificially induced proton gradient.
    c. organic acids or four-carbon compounds.
    d. a source of hydrogen.
    e. photons and CO_2.

15. How many turns of the Calvin cycle does it take to produce one molecule of glucose?
    a. 1   c. 3   e. 12
    b. 2   d. 6

16. Six molecules of G3P formed from the fixation of $3CO_2$ in the Calvin cycle are used to produce
    a. three molecules of glucose.
    b. three molecules of RuBP and one G3P.
    c. one molecule of glucose and four molecules of 3-phosphoglycerate.
    d. one G3P and three four-carbon intermediates.
    e. none of the above, because three molecules of G3P result from three turns of the Calvin cycle.

17. Both NADPH and ATP from the light reactions are needed
    a. in the carbon fixation stage to provide energy and reducing power to rubisco.
    b. to regenerate three RuBP from five G3P.
    c. to combine two molecules of G3P to produce glucose.
    d. to reduce 3-phosphoglycerate to G3P.
    e. to reduce the H^+ concentration in the stroma and contribute to the proton-motive force.

18. Rubisco
    a. reduces CO_2 to G3P.
    b. regenerates RuBP with the aid of ATP.
    c. combines electrons and H^+ to reduce $NADP^+$ to NADPH.
    d. adds CO_2 to RuBP in the carbon fixation stage.
    e. transfers electrons from NADPH to 1,3-bisphosphoglycerate to produce G3P.

19. In C_4 plants,
    a. initial carbon fixation takes place in the mesophyll cells.
    b. photorespiration requires more energy than it does in C_3 plants.
    c. the Calvin cycle, which takes place in the bundle-sheath cells, uses PEP carboxylase instead of rubisco because of its greater affinity for CO_2.
    d. a and b are correct.
    e. a and c are correct.

20. CAM plants avoid photorespiration by
    a. keeping their stomata closed during the day.
    b. performing the Calvin cycle at night.
    c. fixing CO_2 into four-carbon compounds in the mesophyll, which then release CO_2 in the bundle-sheath cells.
    d. storing water in their succulent stems and leaves.
    e. fixing CO_2 into organic acids during the night, which then provide CO_2 during the day.

21. In green plants, most of the ATP for synthesis of proteins, cytoplasmic streaming, and other cellular activities comes directly from
    a. photosystem I.
    b. photosystem II.
    c. the Calvin cycle.
    d. oxidative phosphorylation.
    e. photophosphorylation.

For each of the events listed in questions 22 through 27, indicate whether the event occurs during
    a. respiration
    b. photosynthesis
    c. both respiration and photosynthesis
    d. neither respiration nor photosynthesis

22. Chemiosmotic synthesis of ATP

23. Reduction of oxygen

24. Reduction of CO_2

25. Reduction of NAD^+

26. Oxidation of $NADP^+$

27. Oxidative phosphorylation

Chapter 11

Cell Communication

Key Concepts

11.1 External signals are converted to responses within the cell

11.2 Reception: A signaling molecule binds to a receptor protein, causing it to change shape

11.3 Transduction: Cascades of molecular interactions relay signals from receptors to target molecules in the cell

11.4 Response: Cell signaling leads to regulation of transcription or cytoplasmic activities

11.5 Apoptosis integrates multiple cell-signaling pathways

Framework

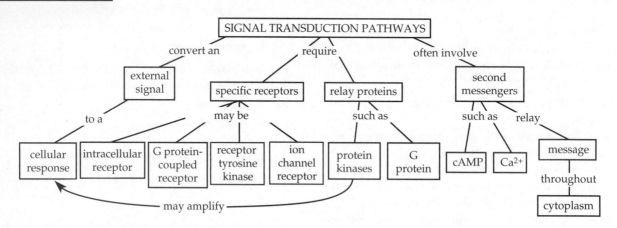

Chapter Review

Cell-to-cell communication is critical to the development and functioning of multicellular organisms and also occurs between unicellular organisms. The similarity of mechanisms of cellular interaction provides evidence for the evolutionary relatedness of all life.

11.1 External signals are converted to responses within the cell

Evolution of Cell Signaling Each of the two mating types of yeast (**a** and **α**) secretes chemical factors that bind to receptors on the other mating type, initiating fusion (mating) of the cells. The series of steps involved in the conversion of a signal on a cell's surface to a cellular response is called a **signal transduction pathway.** Similarities among these pathways in prokaryotes, yeast, plants, and animals suggest an early evolution of cell-signaling mechanisms.

The concentration of signaling molecules allows some bacteria to sense their local density, a process called *quorum sensing.* Aggregations of bacterial cells called *biofilms* may form in response to signaling within a population.

Local and Long-Distance Signaling   Chemical signals may be communicated between cells through direct cytoplasmic connections (gap junctions or plasmodesmata) or through contact of membrane-bound surface molecules (cell–cell recognition in animal cells).

In *paracrine signaling* in animals, a signaling cell releases messenger molecules into the extracellular fluid, and these **local regulators** influence nearby cells. *Growth factors* are one class of local regulators. In another type of local signaling called *synaptic signaling,* a nerve cell releases neurotransmitter molecules, which diffuse across the narrow synapse to its target cell.

Hormones are chemical signals that travel to more distant cells. In hormonal or *endocrine signaling* in animals, the circulatory system transports hormones throughout the body to reach and bind to target cells that have appropriate receptors.

Transmission of electrical and chemical signals within the nervous system is also a type of long-distance signaling.

INTERACTIVE QUESTION   11.1

How do plant hormones travel between secreting cells and target cells?

The Three Stages of Cell Signaling: A Preview   E. W. Sutherland's studies of epinephrine's effect on the hydrolysis of glycogen in liver cells established that cell signaling involves three stages: *reception* of a chemical signal by binding to a receptor protein either inside the cell or on its surface; *transduction* of the signal, often by a signal transduction pathway—a sequence of changes in relay molecules; and the specific *response* of the cell.

11.2   Reception: A signaling molecule binds to a receptor protein, causing it to change shape

A signaling molecule acts as a **ligand,** which specifically binds to a receptor protein and usually induces a change in the receptor protein's shape or the aggregation of receptors.

Receptors in the Plasma Membrane   There are three major types of transmembrane receptors that bind with water-soluble signaling molecules and transmit information into the cell. Malfunctions of these receptors are associated with many human diseases.

The various receptors that work with the aid of a G protein, called **G protein-coupled receptors** (GPCRs), are structurally similar, with seven helices of a single polypeptide spanning the plasma membrane. Binding of the appropriate extracellular signaling molecule to a G protein-coupled receptor activates the receptor, which then binds to and activates a specific **G protein** located on the cytoplasmic side of the membrane. This activation occurs when a GTP nucleotide replaces the GDP bound to the G protein. The G protein then activates a membrane-bound enzyme, after which it hydrolyzes its GTP and becomes inactive again. The activated enzyme triggers the next step in the pathway to the cell's response.

G protein-coupled receptor systems are involved in the function of many hormones and neurotransmitters and in embryological development and sensory reception. Many bacteria produce toxins that interfere with G-protein function; up to 60% of all medicines influence G-protein pathways.

INTERACTIVE QUESTION   11.2

Explain why G protein-coupled receptor pathways shut down rapidly in the absence of a signal molecule.

Receptor tyrosine kinases (RTKs) are membrane receptors with enzymatic activity that can trigger several pathways at once. Part of the receptor protein extending into the cytoplasm is tyrosine kinase, an enzyme that transfers phosphate groups from ATP to the amino acid tyrosine of a protein. Many receptor tyrosine kinases exist as a single transmembrane monomer with a ligand binding site, a transmembrane α helix, and a cytoplasmic tail with a series of tyrosine amino acids. Ligand binding causes two receptor monomers to form a dimer, which causes the tyrosine kinases to phosphorylate the tyrosines on each other's cytoplasmic tails. Different relay proteins then bind to specific phosphorylated tyrosines and become activated, triggering many different transduction pathways in response to one type of signal.

INTERACTIVE QUESTION   11.3

Label the parts in the following diagram of an activated receptor tyrosine kinase dimer.

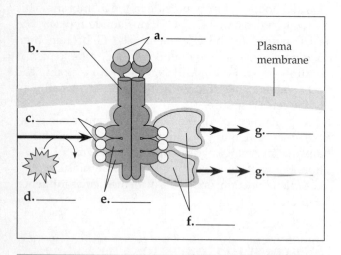

The binding of a signaling molecule to a **ligand-gated ion channel** opens or closes a "gate," thereby allowing or blocking the flow of specific ions through the receptor channel. The resulting change in ion concentration inside the cell triggers a cellular response. Neurotransmitters often bind to ligand-gated ion channels in the transmission of neural signals. Some *voltage-gated ion channels* respond to electrical signals.

A human G protein-coupled receptor has recently been crystallized and its structure determined. Understanding the structure and function of GPCRs and TRKs may facilitate the development of treatments for diseases such as asthma, heart disease, and cancers associated with malfunctioning receptors.

Intracellular Receptors   Hydrophobic chemical messengers and small signaling molecules such as the gas nitric oxide may cross a cell's plasma membrane and bind to receptors in the cytoplasm or nucleus of target cells. Steroid hormones activate receptors in target cells that function as *transcription factors* that regulate gene expression.

11.3   Transduction: Cascades of molecular interactions relay signals from receptors to target molecules in the cell

Multistep pathways enable a small number of extracellular signals to be amplified to produce a large cellular response and also provide opportunities for regulation and coordination.

Signal Transduction Pathways   The relay molecules in a signal transduction pathway are usually proteins, which interact as they pass the message from the extracellular signaling molecule to the protein that produces the cellular response.

Protein Phosphorylation and Dephosphorylation   **Protein kinases** are enzymes that transfer phosphate groups from ATP to proteins, often to the amino acids serine or threonine. Relay molecules in signal transduction pathways are often protein kinases, which are sequentially phosphorylated. Phosphorylation produces a shape change that usually activates each enzyme. Hundreds of different kinds of protein kinases regulate the activity of a cell's proteins.

INTERACTIVE QUESTION   11.4

a. What does a protein kinase do?

b. What does a **protein phosphatase** do?

c. What is a "phosphorylation cascade?"

Small Molecules and Ions as Second Messengers   Small, water-soluble molecules or ions often function as **second messengers,** which rapidly relay the signal from the membrane-receptor-bound "first messenger" into a cell's interior.

Binding of an extracellular signal to a G protein-coupled receptor activates a G protein that may activate **adenylyl cyclase,** a membrane protein that converts ATP to cyclic adenosine monophosphate (**cyclic AMP** or **cAMP**). The cAMP often activates *protein kinase A,* which phosphorylates other proteins. Phosphodiesterase, a cytoplasmic enzyme, converts cAMP to inactive AMP, thereby removing the second messenger.

Some signal molecules may activate an inhibitory G protein that inhibits adenylyl cyclase.

INTERACTIVE QUESTION 11.5

Label the components in the following diagram depicting the steps in a signal transduction pathway that uses cAMP as a second messenger.

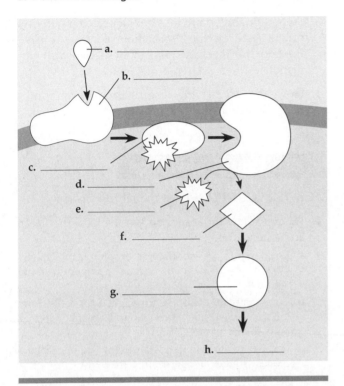

a. _____

b. _____

c. _____

d. _____

e. _____

f. _____

g. _____

h. _____

Calcium ions are widely used as a second messenger in many G protein-coupled and receptor tyrosine kinase pathways. Cytosolic concentration of Ca^{2+} is kept low by active transport of calcium out of the cell and into the endoplasmic reticulum. The release of Ca^{2+} from the ER involves another second messenger, **inositol trisphosphate (IP$_3$)**. In response to the reception of a signal, the enzyme phospholipase C cleaves a membrane phospholipid into two second messengers, IP$_3$ and **diacylglycerol (DAG)**. IP$_3$ binds to and opens IP$_3$-gated calcium channels in the ER. The calcium ions then activate cellular responses.

INTERACTIVE QUESTION 11.6

Fill in the blanks to review a G protein-coupled pathway that uses Ca^{2+} as a second messenger.

A a._____ binds to a G protein-coupled receptor. An activated b._____ activates the enzyme phospholipase C, which cleaves a c._____ into DAG and d._____, which binds to and opens a ligand-gated channel, releasing e._____ from the f._____.

11.4 Response: Cell signaling leads to regulation of transcription or cytoplasmic activities

Nuclear and Cytoplasmic Responses Signal transduction pathways may lead to the activation of transcription factors, which regulate the expression of specific genes. Signaling pathways may also activate existing cytoplasmic enzymes, open or close protein channels in membranes, or influence overall cell activity by orienting the growth of the cytoskeleton.

Fine-Tuning of the Response A signal transduction pathway amplifies a signal in an enzyme cascade, because each successive enzyme in the pathway can process multiple molecules, which then activate the next step.

As a result of their particular set of receptor proteins, relay proteins, and response proteins, different cells can respond to different signals or can exhibit different responses to the same molecular signal. Pathways may branch to produce multiple responses, or two pathways may interact ("cross-talk") to mediate a single response.

Scaffolding proteins are large relay proteins to which other relay proteins attach, increasing the efficiency of signal transduction in a pathway. Scaffolding proteins that permanently attach networks of signaling-pathway proteins at synapses have been identified in brain cells.

INTERACTIVE QUESTION 11.7

How does each of the following *inactivation* mechanisms discontinue a cell's response to a signal and maintain the cell's ability to respond to fresh signals?

a. reversible binding of signaling molecules

b. GTPase activity of G protein

c. phosphodiesterase

d. protein phosphatases

11.5 Apoptosis integrates multiple cell-signaling pathways

In the best understood type of "programmed cell death," called **apoptosis**, cellular components are chopped up and packaged into vesicles that are released as "blebs" and then engulfed by scavenger cells.

Apoptosis in the Soil Worm *Caenorhabditis elegans*

Apoptosis occurs numerous times during normal development in *C. elegans*. The *ced-9* gene produces a protein that inhibits the activity of Ced-3 and Ced-4, protein products of genes *ced-3* and *ced-4*. When a cell receives a death signal on its membrane receptor, Ced-9 protein becomes inactivated, and the apoptosis pathway activates proteases and nucleases that hydrolyze the cell's proteins and DNA. Ced-3 is the main *caspase* (protease) of apoptosis in this nematode.

Apoptotic Pathways and the Signals That Trigger Them

One type of apoptotic pathway in mammals involves proteins, such as cytochrome *c*, that are released through pores formed by apoptotic proteins in mitochondrial membranes. In other cases, binding of a death-signaling ligand to a cell-surface receptor leads to the activation of caspases that carry out apoptosis. Other signals can come from the nucleus when the DNA has suffered irreparable damage, or from the endoplasmic reticulum in response to extensive protein misfolding.

Programmed apoptosis is part of normal development. Faulty cell suicide programs have been implicated in some neurological diseases and in cancer.

INTERACTIVE QUESTION 11.8

a. What can one conclude from the fact that the mitochondrial apoptosis proteins in mammals are similar to the Ced-3, Ced-4, and Ced-9 proteins of nematodes?

b. Give some examples of programmed cell death in humans.

Word Roots

liga- = bound or tied (*ligand:* a molecule that binds specifically to another, usually larger molecule)

trans- = across (*signal transduction pathway:* a series of steps linking a mechanical, chemical, or electrical stimulus to a specific cellular response)

-yl = substance or matter (*adenylyl cyclase:* an enzyme that converts ATP to cAMP in response to an extracellular signal)

Structure Your Knowledge

1. Why is cell signaling such an important aspect of a cell's life?

2. Briefly describe the three stages of cell signaling.

3. Some signaling pathways alter a protein's activity; others result in the production of new proteins. Explain the mechanisms for these two different responses.

4. How does an enzyme cascade produce an amplified response to a signal molecule?

Test Your Knowledge

MULTIPLE CHOICE: *Choose the one best answer.*

1. What is a key difference between a local regulator and a hormone?
   a. Local regulators are small, hydrophobic molecules; hormones are either larger polypeptides or steroids.
   b. Local regulators diffuse to neighboring cells; hormones usually travel throughout the plant or animal body to distant target cells.
   c. Local regulators initiate short-term responses; hormones trigger longer-lasting responses to environmental stimuli.
   d. The signal transduction pathways of local regulators do not involve second messengers; pathways triggered by hormones do involve second messengers.
   e. Local regulators often open ligand-gated channels and affect ion concentrations in a cell; hormones bind with intracellular receptors and affect gene expression.

2. Which of the following substances is used in the type of local signaling called paracrine signaling in animals?
   a. the neurotransmitter acetylcholine
   b. the hormone epinephrine
   c. the neurotransmitter norepinephrine
   d. a local regulator such as a growth factor
   e. both a and c

3. A signaling molecule that binds to a plasma-membrane protein receptor functions as a
   a. ligand.
   b. second messenger.
   c. protein phosphatase.
   d. protein kinase.
   e. receptor protein.

4. A G protein is
   a. a specific type of membrane-receptor protein.
   b. a protein on the cytoplasmic side of a membrane that becomes activated by a receptor protein.
   c. a membrane-bound enzyme that converts ATP to cAMP.
   d. a membrane-bound protein that cleaves phospholipids to produce second messengers.
   e. a guanine nucleotide that converts between GDP and GTP to activate and inactivate relay proteins.

5. How do receptor tyrosine kinases transduce a signal?
   a. They transport the signaling molecule into the cell, where it binds to and activates a transcription factor. The transcription factor then alters gene expression.
   b. Signaling molecule binding causes a shape change that activates membrane-bound tyrosine kinase relay proteins that then phosphorylate serine and threonine amino acids.
   c. Their activated tyrosine kinases convert ATP to cAMP; cAMP then acts as a second messenger to activate other protein kinases.
   d. When activated, they cleave a membrane phospholipid into two second-messenger molecules, one of which opens Ca^{2+} ion channels on the endoplasmic reticulum.
   e. They form a dimer; they phosphorylate each other's tyrosines; specific proteins bind to and are activated by these phosphorylated tyrosines.

6. Many human diseases (including bacterial infections) and the medicines used to treat them produce their effects by influencing which of the following?
   a. cAMP concentrations in the cell
   b. Ca^{2+} concentrations in the cell
   c. G protein-coupled receptor pathways
   d. gene expression
   e. ligand-gated ion channels

7. Which of the following compounds can activate a protein by transferring a phosphate group to it?
   a. G protein
   b. phosphodiesterase
   c. protein phosphatase
   d. protein kinase
   e. both a and c

8. Many signal transduction pathways use second messengers to
   a. transport a signaling molecule through the hydrophobic center of the plasma membrane.
   b. relay a signal from the outside to the inside of the cell.
   c. relay the message from the inside of the membrane throughout the cytoplasm.
   d. amplify the message by phosphorylating cascades of proteins.
   e. dampen the message once the signaling molecule has left the receptor.

9. A function of the second messenger IP_3 is to
   a. bind to and activate protein kinase A.
   b. activate transcription factors.
   c. activate other membrane-bound relay molecules.
   d. convert ATP to cAMP.
   e. bind to and open ligand-gated calcium channels on the ER.

10. Signal amplification is most often achieved by
    a. an enzyme cascade involving multiple protein kinases.
    b. the binding of multiple signaling molecules.
    c. branching pathways that produce multiple cellular responses.
    d. the activation of transcription factors that affect gene expression.
    e. the action of adenylyl cyclase in converting ATP to ADP.

11. From studying the effects of epinephrine on liver cells, Sutherland concluded that
    a. there is a one-to-ten thousand correlation between the number of epinephrine molecules bound to receptors and the number of glucose molecules released from glycogen.
    b. epinephrine enters liver cells and binds to receptors that function as transcription factors to turn on the gene for glycogen phosphorylase.
    c. there is a "second messenger" that transmits the signal of epinephrine binding on the plasma membrane to the enzymes involved in glycogen breakdown inside the cell.
    d. the signal transduction pathway through which epinephrine signals glycogen breakdown involves receptor tyrosine kinases and Ca^{2+} that activate glycogen phosphorylase.
    e. epinephrine functions as a ligand to open ion channels in the plasma membrane that allow Ca^{2+} to enter and initiate a response.

12. Which of the following characteristics is a similarity between G protein-coupled receptors and receptor tyrosine kinases?
    a. signaling molecule-binding sites specific for steroid hormones
    b. formation of a dimer following the binding of a signaling molecule
    c. activation that results from the binding of GTP
    d. phosphorylation of specific amino acids in direct response to ligand binding
    e. α-helix regions of the receptor that span the plasma membrane

13. Which of the following molecules is *incorrectly* matched with its description?

    a. scaffolding protein—large relay protein that may bind with several other relay proteins to increase the efficiency of a signaling pathway

    b. protein phosphatase—enzyme that transfers a phosphate group from ATP to a protein, causing a shape change that usually activates that protein

    c. adenylyl cyclase—enzyme attached to plasma membrane that converts ATP to cAMP in response to an extracellular signal

    d. phospholipase C—enzyme that may be activated by a G protein or receptor tyrosine kinase and cleaves a plasma-membrane phospholipid into the second messengers IP_3 and DAG

    e. G protein—relay protein attached to the inside of plasma membrane that, when activated by an activated G protein-coupled receptor, binds GTP and then usually activates another membrane-attached protein

14. When epinephrine binds to cardiac (heart) muscle cells, it speeds their contraction. When it binds to muscle cells of the small intestine, it inhibits their contraction. How can the same hormone have different effects on muscle cells?

    a. Cardiac cells have more receptors for epinephrine than do intestinal muscle cells.

    b. Epinephrine circulates to the heart first and is in higher concentration around cardiac cells.

    c. The two types of muscle cells have different signal transduction pathways for epinephrine and thus have different cellular responses.

    d. Cardiac muscle is stronger than intestinal muscle and thus has a stronger response to epinephrine.

    e. Epinephrine binds to G protein-coupled receptors in cardiac cells, and these receptors always increase a response to the signal. Epinephrine binds to receptor tyrosine kinases in intestinal muscle cells, and these receptors always inhibit a response to the signal.

15. The gene *ced-9* codes for a protein that inactivates the proteins of suicide genes found in *C. elegans*. For development to proceed normally, *ced-9*

    a. should be expressed in all cells, but its protein product must remain inactive.

    b. should be expressed in all cells, but its product will be inactivated when cells programmed to die receive the proper signal.

    c. should not be expressed except when a cell receives a signal to die.

    d. should not be expressed in those cells that must die if proper development is to occur.

    e. should code for a transcription factor that activates the enhancers of other suicide genes.

Chapter 12

The Cell Cycle

Key Concepts

12.1 Most cell division results in genetically identical daughter cells

12.2 The mitotic phase alternates with interphase in the cell cycle

12.3 The eukaryotic cell cycle is regulated by a molecular control system

Framework

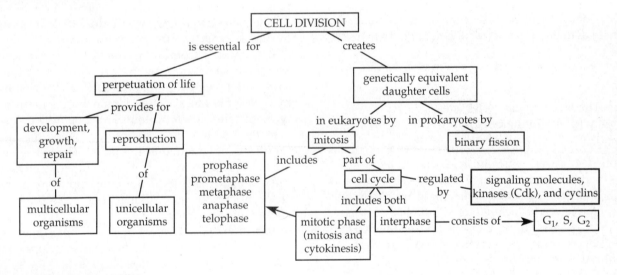

Chapter Review

Cell division, the reproduction of cells, creates duplicate offspring in unicellular organisms and provides for growth, development, and repair in multicellular organisms. The **cell cycle** extends from the creation of a new cell by the division of its parent cell to its own division into two daughter cells.

12.1 Most cell division results in genetically identical daughter cells

The process of recreating a structure as intricate as a cell necessitates the exact duplication and equal division of the DNA containing the cell's genetic program.

Cellular Organization of the Genetic Material A cell's complete complement of DNA is called its **genome,** which is organized into **chromosomes.** Each is a very long DNA molecule with associated proteins that help structure the chromosome and control the activity of genes, the units of inheritance. This DNA–protein complex is called **chromatin.**

Each diploid eukaryotic species has a characteristic number of chromosomes in each **somatic cell;** reproductive cells, or **gametes** (egg and sperm), have half that number of chromosomes.

Distribution of Chromosomes During Eukaryotic Cell Division Prior to cell division, a cell replicates its DNA. Duplicated chromosomes consist of two identical **sister chromatids,** attached by proteins called *cohesins* in *sister chromatid cohesion.* Each sister chromatid has a **centromere,** a region where proteins bind to specific centromeric DNA sequences and hold the chromatids closely together. Regions on either side of the centromere are called the chromatid *arms.*

The two sister chromatids separate during **mitosis** (the division of the genetic material in the nucleus), and then the cytoplasm divides during **cytokinesis,** producing two separate, genetically equivalent cells.

A type of cell division called *meiosis* produces daughter cells that have half the number of chromosomes as the parent cell. When a sperm fertilizes an egg (both of which are formed by meiosis in animals), the chromosome number is restored, and the somatic cells of the new offspring again have two sets of chromosomes.

INTERACTIVE QUESTION   12.1

a. How many chromosomes are in your somatic cells?

b. How many chromosomes are in your gametes?

c. How many chromatids are in one of your somatic cells that has duplicated its chromosomes prior to mitosis?

12.2   The mitotic phase alternates with interphase in the cell cycle

Phases of the Cell Cycle   The cell cycle consists of the **mitotic (M) phase,** which includes mitosis and cytokinesis, and **interphase,** during which the cell grows and duplicates its chromosomes. Interphase, usually lasting 90% of the cell cycle, includes the **G_1 phase,** the **S phase,** and the **G_2 phase.** Mitosis is conventionally divided into five stages: **prophase, prometaphase, metaphase, anaphase,** and **telophase.**

INTERACTIVE QUESTION   12.2

a. How are the three subphases of interphase alike?

b. What key event happens during the S phase?

The Mitotic Spindle: A Closer Look   The **mitotic spindle** consists of fibers made of microtubules and associated proteins. The assembly of the spindle begins in the **centrosome,** or *microtubule-organizing center.* A pair of centrioles is centered in each centrosome of an animal cell, but centrioles are not required for cell division. The single centrosome duplicates during interphase. As spindle microtubules grow out from them, the two centrosomes move to opposite poles of the cell. Radial arrays of shorter microtubules, called **asters,** extend from the centrosomes.

During prophase, the nucleoli disappear and the chromatin fibers coil and fold into visible chromosomes, consisting of sister chromatids joined at their centromeres and by sister chromatid cohesion. During prometaphase, some of the spindle microtubules attach to each chromatid's **kinetochore,** a structure of protein associated with DNA located at the centromere region. Alternate tugging on the chromosome by opposite kinetochore microtubules moves the chromosome to the midline of the cell. At metaphase, the chromosomes are aligned at the **metaphase plate,** a plane across the midline of the spindle. Nonkinetochore microtubules (or "polar" microtubules) extend out from each centrosome and overlap at the midline. Aster microtubules contact the plasma membrane.

The cohesins joining sister chromatids are cleaved by the enzyme *separase* to begin anaphase, and the now separate chromosomes move toward the poles. Motor proteins appear to "walk" a chromosome along the kinetochore microtubules as these shorten by depolymerizing at their kinetochore end. In some types of cells, motor proteins at the spindle poles appear to "reel in" the chromosomes while the microtubules depolymerize at the poles. The extension of the spindle poles away from each other as an animal cell elongates is due to the overlapping nonkinetochore microtubules walking past each other, also using motor proteins.

In telophase, equivalent sets of chromosomes are at the two poles of the cell. Nuclear envelopes form, nucleoli reappear, cytokinesis begins, and the spindle microtubules depolymerize.

Cytokinesis: A Closer Look   **Cleavage** is the process that separates the two daughter cells in animals. A **cleavage furrow** forms, as a ring of actin microfilaments interacting with myosin proteins begins to contract on the cytoplasmic side of the membrane. The cleavage furrow deepens until the dividing cell is pinched in two.

In plant cells, a **cell plate** forms from the fusion of membrane vesicles derived from the Golgi apparatus. The membrane of the enlarging cell plate joins with the plasma membrane, separating the two daughter cells. A new cell wall develops between the cells from the contents of the cell plate.

INTERACTIVE QUESTION 12.3

The following diagrams will depict the five stages of mitosis in an animal cell, after you draw in the missing chromosomes and kinetochore microtubules in stages c, d, and e. For simplicity, this cell has only four chromosomes, as you can see in stages b and f. Identify the stages and label the indicated structures.

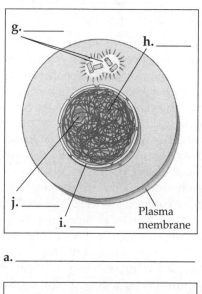

g. _____
h. _____
j. _____
i. _____
Plasma membrane

a. _____

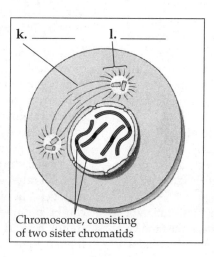

k. _____
l. _____
Chromosome, consisting of two sister chromatids

b. _____

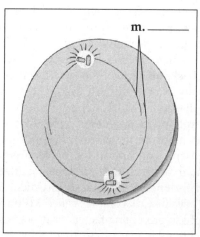

m. _____

c. _____

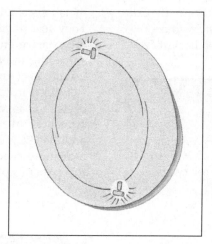

n. _____
o. _____

d. _____

e. _____

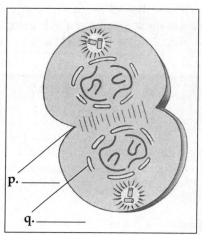

p. _____
q. _____

f. _____

Binary Fission of Bacteria   Single-celled eukaryotes reproduce asexually by a "dividing in half" process known as **binary fission,** which includes mitosis. The binary fission of prokaryotes does not include mitosis. The *bacterial chromosome* is a single circular DNA molecule with associated proteins. It begins to replicate at the **origin of replication,** and one of these duplicated origins moves to the opposite pole of the cell. Replication is completed as the cell doubles in size, and the plasma membrane grows inward to divide the two identical daughter cells. The mechanism of chromosome movement is not fully understood but involves some proteins that resemble eukaryotic actin and tubulin.

The Evolution of Mitosis   Evidence for the evolution of mitosis from prokaryotic cell division includes (1) the relatedness of several proteins involved in both types of division and (2) some possible intermediate stages seen in some eukaryotes, in which chromosomal separation takes place within an intact nuclear envelope.

12.3 The eukaryotic cell cycle is regulated by a molecular control system

Normal cell growth, development, and maintenance depend on proper control of the timing and rate of cell division.

Evidence for Cytoplasmic Signals   Experiments that fuse two cells at different phases of the cell cycle indicate that cytoplasmic signaling molecules drive the cell cycle.

The Cell Cycle Control System   A **cell cycle control system,** consisting of a set of molecules that function cyclically, coordinates the events of the cell cycle.

Important internal and external signals are monitored to determine whether the cell cycle will proceed past the three main **checkpoints** in the G_1, G_2, and M phases. If a mammalian cell does not receive a go-ahead signal at the G_1 checkpoint, called the "restriction point," the cell will usually exit the cell cycle to a nondividing state called the G_0 **phase.**

Protein kinases are enzymes that activate or inactivate other proteins by phosphorylating them. The changing concentrations of **cyclins,** regulatory proteins that attach to these kinases, affect the activity of **cyclin-dependent kinases,** or **Cdks.**

A cyclin–Cdk complex called **MPF,** for "maturation or M-phase-promoting factor," triggers passage past the G_2 checkpoint into M phase. In addition to phosphorylating proteins and other kinases that initiate mitotic events, MPF activates a protein breakdown process that destroys its cyclin and thus MPF activity during anaphase. The Cdk portion of the complex remains to associate with new cyclin synthesized during S and G_2 phases of the next cycle.

INTERACTIVE QUESTION   12.4

a. What is MPF?

b. Describe the relative concentrations of MPF and its constituent molecules throughout the cell cycle:

MPF

Cdk

cyclin

Other Cdk proteins and cyclins appear to control the movement of a cell past the G_1 checkpoint.

An internal signal is required to move past the M phase checkpoint into anaphase. Only after all chromosomes are attached at their kinetochores to spindle microtubules is the enzyme separase activated and the cohesins holding sister chromatids together cleaved.

Growing animal cells in cell culture has allowed researchers to identify chemical and physical factors that affect cell division. Certain nutrients and regulatory proteins called **growth factors** have been found to be essential for cells to divide in culture. Mammalian fibroblasts have receptors on their plasma membranes for *platelet-derived growth factor (PDGF)*, which is released from blood platelets at the site of an injury. Binding of PDGF initiates a signal-transduction pathway that enables cells to pass the G_1 checkpoint.

Density-dependent inhibition of cell division appears to involve the binding of cell-surface proteins of adjacent cells, which sends a growth-inhibiting signal to both cells. Most animal cells also show **anchorage dependence** and must be attached to a substratum in order to divide.

Loss of Cell Cycle Controls in Cancer Cells   When grown in cell culture, cancer cells do not exhibit density-dependent inhibition or anchorage dependence, do not depend on growth factors to divide, and may continue to divide indefinitely instead of stopping after the typical 20 to 50 divisions of normal mammalian cells in culture.

When a normal cell is **transformed** or converted to a cancer cell, the body's immune system may destroy it. If it proliferates, a mass of abnormal cells develops within a tissue. A **benign tumor** remains at its original site and can be removed by surgery. The cells of a **malignant tumor** have significant genetic changes, which enable them to invade and disrupt the functions of one or more organs. An individual with a malignant tumor is said to have cancer. Malignant tumor cells may have abnormal metabolism and unusual numbers of chromosomes. They lose their attachments to other cells and may **metastasize,** entering the blood and lymph systems and spreading to other sites. Radiation and chemicals are used to treat known or suspected metastatic tumors.

Defects in cell signaling pathways that affect the cell cycle are involved in the development of cancer. Advances in molecular techniques to analyze DNA and the levels of specific cancer-associated proteins increasingly enable physicians to match chemotherapy to a patient's particular type of tumor.

Word Roots

ana- = up, throughout, again (*anaphase:* the mitotic stage in which the chromatids of each chromosome have separated and the daughter chromosomes are moving to opposite ends of the cell)

bi- = two (*binary fission:* a method of asexual reproduction by "division in half." In prokaryotes, binary fission does not involve mitosis)

centro- = the center; **-mere** = a part (*centromere:* the region of a chromosome where proteins bind to specific sequences of DNA and hold the two sister chromatids most closely together)

chroma- = colored (*chromatin:* the complex of DNA and proteins that makes up a eukaryotic chromosome)

cyclo- = a circle (*cyclin:* a protein whose concentration fluctuates cyclically and which plays an important regulatory role in the cell cycle) cyto- = cell; -kinet = move (*cytokinesis:* division of the cytoplasm to form two separate daughter cells)

gamet- = a wife or husband (*gamete:* a haploid egg or sperm cell)

gen- = produce (*genome:* the genetic material of an organism)

inter- = between (*interphase:* the period in the cell cycle when the cell is not dividing)

mal- = bad or evil (*malignant tumor:* a cancerous tumor containing cells that have significant genetic and cellular changes and are capable of invading and surviving in new sites)

meta- = between (*metaphase:* the mitotic stage in which the chromosomes are attached to microtubules at their kinetochores and are aligned at the metaphase plate)

mito- = a thread (*mitosis:* a process of nuclear division that allocates replicated chromosomes equally to each of the daughter nuclei)

pro- = before (*prophase:* the first mitotic stage in which the chromatin condenses and the spindle begins to form)

-soma- = body (*centrosome:* a structure in animal cells that functions as a microtubule-organizing center and is important during cell division)

telos- = an end (*telophase:* the final stage of mitosis in which daughter nuclei are forming and cytokinesis has typically begun)

trans- = across; -form shape (*transformation:* the conversion of a normal animal cell to a cancerous cell)

Structure Your Knowledge

1. In the following photomicrograph of onion root tip cells, identify the cell cycle stage for the indicated cells.

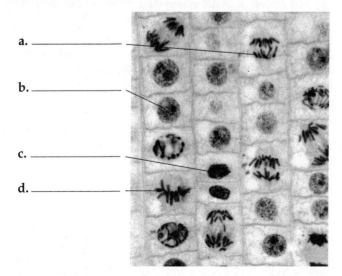

a. _____

b. _____

c. _____

d. _____

2. Describe one chromosome as it proceeds through an entire cell cycle, starting with G_1 of interphase and ending with telophase of mitosis.

3. Draw a sketch of one half of a mitotic spindle. Identify and list the functions of the components.

Test Your Knowledge

FILL IN THE BLANK: *Identify the appropriate phase of the cell cycle.*

_____ 1. most cells that will no longer divide are in this phase

_____ 2. sister chromatids separate and chromosomes move apart

_____ 3. mitotic spindle begins to form

_____ 4. cell plate forms or cleavage furrow pinches cells apart

_____ 5. chromosomes duplicate

_____ 6. chromosomes line up at central plane

_____ 7. nuclear membranes form around separated chromosomes

_____ 8. chromosomes become visible

_____ 9. kinetochore–microtubule interactions move chromosomes to midline

_____ 10. restriction point occurs in this phase

MULTIPLE CHOICE: *Choose the one best answer.*

1. One of the differences between prokaryotic binary fission and eukaryotic mitosis is that
   a. cytokinesis does not occur in prokaryotic cells.
   b. mitosis involves microtubules composed of tubulin and actin microfilaments, but no such molecules are involved in prokaryotic binary fission.
   c. mitosis occurs within the nucleus but binary fission does not.
   d. the duplicated single chromosome does not separate along a mitotic spindle in prokaryotic cells.
   e. the chromosome number is reduced by half in eukaryotic cells but not in prokaryotic cells.

2. A plant cell has 12 chromosomes at the end of mitosis. How many *chromosomes* would it have in the G_2 phase of its next cell cycle?
   a. 6   c. 12   e. 48
   b. 9   d. 24

3. How many *chromatids* would this plant cell have in the G_2 phase of its cell cycle?
   - **a.** 6
   - **b.** 9
   - **c.** 12
   - **d.** 24
   - **e.** 48

4. The longest part of the cell cycle is
   - **a.** prophase.
   - **b.** G_1 phase.
   - **c.** G_2 phase.
   - **d.** mitosis.
   - **e.** interphase.

5. In animal cells, cytokinesis involves
   - **a.** the separation of sister chromatids.
   - **b.** the contraction of a ring of actin microfilaments.
   - **c.** depolymerization of kinetochore microtubules.
   - **d.** a protein kinase that phosphorylates other enzymes.
   - **e.** sliding of nonkinetochore microtubules past each other.

6. Humans have 46 chromosomes. That number of chromosomes will be found
   - **a.** in cells in anaphase.
   - **b.** in egg and sperm cells.
   - **c.** in somatic cells.
   - **d.** in all the cells of the body.
   - **e.** only in cells in G_1 of interphase.

7. Sister chromatids
   - **a.** have one-half the amount of genetic material as does the original chromosome.
   - **b.** start to move along kinetochore microtubules to opposite poles of a cell during telophase.
   - **c.** each have their own kinetochore.
   - **d.** are formed during S phase but do not join by sister chromatid cohesion until prophase.
   - **e.** slide past each other as nonkinetochore microtubules elongate.

8. A cell in which of the following phases has the *least* amount of DNA?
   - **a.** G_0
   - **b.** G_2
   - **c.** prophase
   - **d.** metaphase
   - **e.** anaphase

9. Which of the following is *not* true of a cell plate?
   - **a.** It forms at the site of the metaphase plate.
   - **b.** It results from the fusion of microtubules.
   - **c.** It fuses with the plasma membrane.
   - **d.** A cell wall is laid down between its membranes.
   - **e.** It forms during telophase in plant cells.

10. A cell that passes the restriction point in G_1 will most likely
    - **a.** undergo chromosome duplication.
    - **b.** have just completed cytokinesis.
    - **c.** continue to divide only if it is a cancer cell.
    - **d.** show a drop in MPF concentration.
    - **e.** move into the G_0 phase.

11. The rhythmic changes in cyclin concentration in a cell cycle are related to
    - **a.** its increased production once the restriction point is passed.
    - **b.** the cascade of increased production once its enzyme is phosphorylated by MPF.
    - **c.** its degradation, which is initiated by active MPF.
    - **d.** the correlation of its production with the production of Cdk.
    - **e.** the binding of the growth factor PDGF.

12. Which of the following would *not* be exhibited by cancer cells?
    - **a.** changing levels of MPF concentration
    - **b.** passage through the restriction point
    - **c.** density-dependent inhibition
    - **d.** metastasis
    - **e.** G_1 phase of the cell cycle

13. What initiates the separation of sister chromatids?
    - **a.** the drop in MPF concentration
    - **b.** a rapid rise in Cdk concentration
    - **c.** movement past the G_2 checkpoint
    - **d.** a signal pathway initiated by the binding of a growth factor
    - **e.** activation of a regulatory protein complex after all kinetochores are attached to microtubules

14. Knowledge of the cell cycle control system will be most beneficial to understanding
    - **a.** human reproduction.
    - **b.** plant genetics.
    - **c.** prokaryotic growth and development.
    - **d.** cancer.
    - **e.** cardiovascular disease.

UNIT 3

Genetics

Chapter 13

Meiosis and Sexual Life Cycles

Key Concepts

13.1 Offspring acquire genes from parents by inheriting chromosomes

13.2 Fertilization and meiosis alternate in sexual life cycles

13.3 Meiosis reduces the number of chromosome sets from diploid to haploid

13.4 Genetic variation produced in sexual life cycles contributes to evolution

Framework

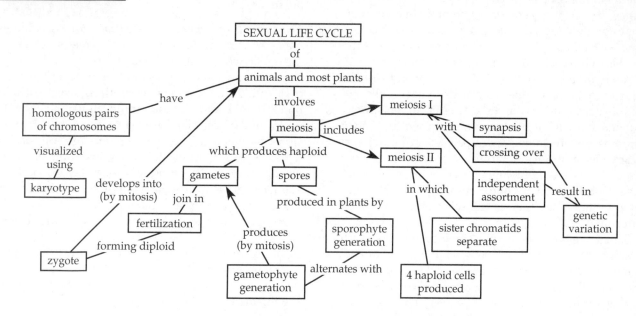

Chapter Review

Genetics is the scientific study of the transmission of traits from parents to offspring (**heredity**) and the **variation** between and within generations.

13.1 Offspring acquire genes from parents by inheriting chromosomes

Inheritance of Genes   The inheritance of traits from parents by offspring involves the transmission of discrete units of information coded in segments of DNA known as **genes.** Most genes contain instructions for synthesizing enzymes and other proteins that then guide the development of inherited traits.

When **gametes** (sperm and eggs) fuse in fertilization, genes from both parents are transmitted to offspring. The DNA of a eukaryotic cell is packaged into a species-specific number of chromosomes present in all **somatic cells.** A gene's **locus** is its location on a chromosome.

Comparison of Asexual and Sexual Reproduction   In **asexual reproduction,** a single parent passes copies of all its genes to its offspring. A **clone** is a group of genetically identical offspring. In **sexual reproduction,** an individual receives a unique combination of genes inherited from two parents.

13.2 Fertilization and meiosis alternate in sexual life cycles

An organism's **life cycle** is the sequence of stages from conception to production of its own offspring.

Sets of Chromosomes in Human Cells   Two chromosomes of each type, known as **homologous chromosomes** or homologs, are present in each somatic cell. A gene controlling a particular character is found at the same locus on each chromosome of a homologous pair.

A **karyotype** is an ordered display of an individual's condensed chromosomes. Isolated somatic cells are stimulated to undergo mitosis, arrested in metaphase, and stained. Digital photographs are used to arrange chromosomes into homologous pairs by size and shape.

Sex chromosomes determine the sex of a person: Females have two homologous X chromosomes; males have nonhomologous X and Y chromosomes. Chromosomes other than the sex chromosomes are called **autosomes.**

Somatic cells contain a set of chromosomes from each parent. These are **diploid cells,** each with a diploid number of chromosomes, abbreviated $2n$. Gametes are **haploid cells** and contain a single set of chromosomes. The haploid number (n) of chromosomes for humans is 23.

INTERACTIVE QUESTION   13.1

a. How many chromosomes are there in the somatic cells of an animal in which $2n = 14$? _____
How many chromosomes are in its gametes? _____

b. If $n = 14$, how many chromosomes are there in diploid somatic cells?_____
How many sets of homologous chromosomes are in the gametes? _____

c. If $2n = 28$, how many chromatids are there in a cell in which DNA replication has occurred prior to cell division? _____ What is the difference between sister chromatids and nonsister chromatids?

Behavior of Chromosome Sets in the Human Life Cycle   **Fertilization,** or fusion of sperm and egg, produces a **zygote** containing both a paternal and a maternal set of chromosomes. The diploid zygote then divides by mitosis to produce the somatic cells of the body, all of which contain the diploid number ($2n$) of chromosomes.

Meiosis is a special type of cell division that halves the chromosome number and provides a haploid set of chromosomes to each gamete. Gametes are produced by meiosis from specialized *germ cells* in the gonads. An alternation between diploid and haploid numbers of chromosomes, involving the processes of fertilization and meiosis, is characteristic of sexually reproducing organisms.

The Variety of Sexual Life Cycles   In most animals, meiosis occurs in the formation of gametes, which are the only haploid cells in the life cycle.

Plants and some species of algae have a type of life cycle called **alternation of generations,** which includes both diploid and haploid multicellular stages. The multicellular diploid *sporophyte* produces haploid *spores* by meiosis. These spores undergo mitosis and develop into a multicellular haploid organism, the *gametophyte,* which produces gametes by mitosis. Gametes fuse to form a diploid zygote, which develops into the next sporophyte generation.

In most fungi and some protists, the only diploid stage is the zygote. Meiosis occurs after the gametes fuse, producing haploid cells that divide by mitosis to create a unicellular or multicellular haploid organism. Gametes are produced by mitosis in these organisms.

INTERACTIVE QUESTION 13.2

Complete the following diagrams of sexual life cycles by filling in the type of cell division, event, or cells, and indicating whether cell l. is *n* or *2n*.

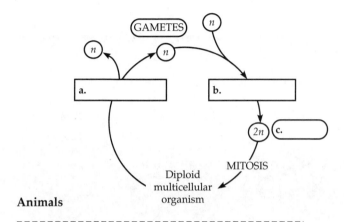

Animals

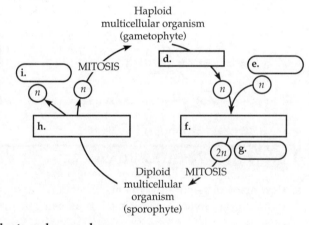

Plants and some algae

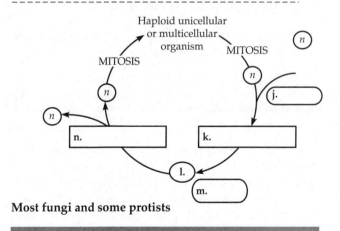

Most fungi and some protists

13.3 Meiosis reduces the number of chromosome sets from diploid to haploid

In meiosis, chromosome replication is followed by two consecutive cell divisions—**meiosis I** and **meiosis II**—producing four haploid daughter cells, each with one set of chromosomes.

The Stages of Meiosis In interphase, each chromosome replicates, producing two genetically identical sister chromatids that remain attached at the centromere and along their length by *sister chromatid cohesion.* **Synapsis** occurs during prophase I, when homologous chromosomes pair up and are held together along their lengths by the *synaptonemal complex.* Genetic material is exchanged by **crossing over** between nonsister chromatids. After synapsis ends, crossovers remain visible as X-shaped regions called **chiasmata,** where sister chromatid cohesion continues to hold the original (but now crossed-over) sister chromatids together.

In metaphase I, homologous pairs line up at the metaphase plate with their kinetochores attached to spindle fibers from opposite poles. Each pair separates in anaphase I, with one homolog moving toward each pole. In telophase I, a haploid set of chromosomes, each composed of two sister chromatids (usually containing some regions of nonsister chromatid DNA), reaches each pole. Cytokinesis usually occurs during telophase I. There is no replication of genetic material prior to the second division of meiosis.

Meiosis II looks like a regular mitotic division, in which chromosomes line up individually on the metaphase plate. Sister chromatid cohesion at the centromere breaks down, and sister chromatids separate and move apart in anaphase II. These sister chromatids, however, are not genetically identical due to crossing over in prophase I. At the end of telophase II, there are four haploid daughter cells.

A Comparison of Mitosis and Meiosis Mitosis produces daughter cells that are genetically identical to the parent cell. Meiosis produces haploid cells that differ genetically from their parent cell and from each other.

The three unique events that produce this result occur during meiosis I. In prophase I, homologous chromosomes pair and are held together (synapsis), and genetic material is exchanged by crossing over between nonsister chromatids. In metaphase I, chromosomes line up in homologous pairs, not as individuals, at the metaphase plate. During anaphase I, the homologous pairs separate and one homolog (with sister chromatids still attached at the centromere) goes to each pole.

Homologs are held together by sister chromatid cohesion at regions of crossing over. In anaphase I, enzymes cleave the protein complexes called *cohesins* along the arms of sister chromatids, and the homologs can separate. The cohesins at the centromere hold the sister chromatids together until anaphase II.

Meiosis I is called a *reductional division* because it reduces the chromosome sets from two (diploid) to one (haploid). The sister chromatids of each homolog do not separate until meiosis II, sometimes called the *equational division*.

INTERACTIVE QUESTION   13.3

The following diagrams represent some of the stages of meiosis (not in the right order). Label these stages.

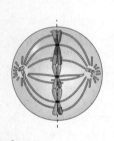     

a. _____   b. _____   c. _____   d. _____   e. _____   f. _____

Now place these stages in the correct sequence.

_____ _____ _____ _____ _____ _____

13.4   Genetic variation produced in sexual life cycles contributes to evolution

Origins of Genetic Variation Among Offspring   Mutations that result in different *alleles* are the original source of genetic variation. In sexual reproduction, three mechanisms produce the genetic variation evident in each generation: independent assortment of chromosomes, crossing over, and random fertilization.

Each homologous pair lines up independently at the metaphase plate; the orientation of the original maternal and paternal chromosomes is random. Due to this *independent assortment* of chromosomes, the number of possible combinations of maternal and paternal chromosomes in gametes is 2^n, where n is the haploid number for that species.

In prophase I, homologous segments of nonsister chromatids are exchanged by crossing over, forming **recombinant chromosomes** with new genetic combinations of maternal and paternal alleles on the same chromosome. These no-longer-equivalent sister chromatids assort independently during meiosis II.

The random nature of fertilization adds to the genetic variability established in meiosis.

INTERACTIVE QUESTION   13.4

a. How many different assortments of maternal and paternal chromosomes are possible in a human gamete?

b. In a human zygote, how many diploid combinations of parental chromosomes are possible?

c. Why does the answer to question b still underestimate the possible genetic variations in a zygote?

Evolutionary Significance of Genetic Variation within Populations   In Darwin's theory of evolution by natural selection, the genetic variation present in a population results in adaptation as the individuals with the variations best suited to the environment produce the most offspring. The processes of sexual reproduction and mutation are the sources of this variation. Although sexual reproduction is more energetically expensive than asexual reproduction and may disrupt successful combinations of alleles in stable environments, it is widespread among all organisms and almost universal among animals. Genetic variation appears to be evolutionarily advantageous.

Word Roots

a- = not or without (*asexual:* generation of offspring from a single parent, occurring without the fusion of gametes)

-apsis = juncture (*synapsis:* the pairing and physical connection of duplicated homologous chromosomes during prophase I of meiosis)

auto- = self (*autosome:* a chromosome that is not directly involved in determining sex)

chiasm- = marked crosswise (*chiasma:* the X-shaped microscopically visible region where crossing over has occurred between homologous nonsister chromatids)

di- = two (*diploid:* a cell containing two sets of chromosomes [2*n*], one set inherited from each parent)

fertil- = fruitful (*fertilization:* the union of haploid gametes to produce a diploid zygote)

haplo- = single (*haploid:* a cell containing only one set of chromosomes [*n*])

homo- = like (*homologous chromosomes:* a pair of like chromosomes that possess genes for the same characters at corresponding loci)

karyo- = nucleus (*karyotype:* a display of the chromosome pairs of a cell)

meio- = less (*meiosis:* a modified type of cell division that yields daughter cells with half the number of chromosome sets as the original cell)

soma- = body (*somatic cell:* any cell in a multicellular organism except a gamete or its precursors)

Structure Your Knowledge

1. Label the following diagram to review the terms that describe duplicated chromosomes in a diploid cell.

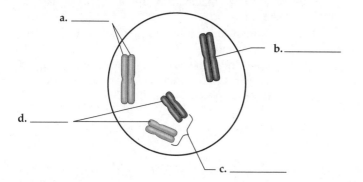

2. Describe the key events of the following stages of meiosis.

a. Interphase
b. Prophase I
c. Metaphase I
d. Anaphase I
e. Metaphase II
f. Anaphase II

3. Create a concept map to help you organize your understanding of the similarities and differences between mitosis and meiosis. *Compare your map with those of some classmates to see different ways of organizing the material.*

Test Your Knowledge

MULTIPLE CHOICE: *Choose the one best answer.*

1. While of the following is *not* an accurate statement?
   a. Gametes transmit genes from two parents to offspring.
   b. You have 46 chromosomes in your somatic cells and 23 chromosomes in your gametes.
   c. Offspring differ from their parents because mutations occur during sexual reproduction.
   d. Each gene has a specific locus on a chromosome.
   e. A clone of genetically identical offspring is produced in asexual reproduction.

2. Asexual reproduction of a diploid organism
   a. is impossible.
   b. involves meiosis.
   c. involves spores produced by mitosis.
   d. shows variation among sibling offspring.
   e. produces identical offspring.

3. Homologous chromosomes
   a. have identical DNA sequences in their genes.
   b. have genes for the same characters at the same loci.
   c. are found in gametes.
   d. separate in meiosis II.
   e. have all of the characteristics in a through d.

4. What are autosomes?
   a. sex chromosomes
   b. chromosomes that occur singly
   c. chromosomal abnormalities that result in genetic defects
   d. chromosomes found in mitochondria and chloroplasts
   e. none of the above

5. What is a karyotype?
   a. a genotype of an individual
   b. a pictorial display of an individual's chromosomes
   c. a blood type determination of an individual
   d. a unique combination of chromosomes found in a gamete
   e. a species-specific diploid number of chromosomes

6. The DNA content of a diploid cell is measured in the G_1 phase. After meiosis I, the DNA content of one of the two cells produced is
   a. equal to that of the G_1 cell.
   b. twice that of the G_1 cell.
   c. one-half that of the G_1 cell.
   d. one-fourth that of the G_1 cell.
   e. impossible to estimate due to independent assortment of homologous chromosomes.

7. In most fungi and some protists,
   a. the zygote is the only haploid stage.
   b. gametes are formed by meiosis.
   c. the multicellular or unicellular organism is haploid.
   d. the gametophyte generation produces gametes by mitosis.
   e. reproduction is exclusively asexual.

8. In the alternation of generations found in plants,
   a. the sporophyte generation produces spores by mitosis.
   b. the gametophyte generation produces gametes by mitosis.
   c. the zygote develops into a sporophyte generation by meiosis.
   d. spores develop into the haploid sporophyte generation.
   e. the gametophyte generation produces spores by meiosis.

9. A synaptonemal complex would be found during
   a. early prophase of mitosis.
   b. fertilization or syngamy of gametes.
   c. metaphase I of meiosis.
   d. early prophase of meiosis I.
   e. both a and d.

10. During meiosis I,
    a. homologous chromosomes separate.
    b. the chromosome number is reduced in half.
    c. crossing over between nonsister chromatids occurs.
    d. paternal and maternal chromosomes assort randomly.
    e. all of the above occur.

11. Compared with one of the four cells produced by meiosis, a cell in G_2 before meiosis has
    a. twice as much DNA and twice as many chromosomes.
    b. four times as much DNA and twice as many chromosomes.
    c. four times as much DNA and four times as many chromosomes.
    d. half as much DNA but the same number of chromosomes.
    e. half as much DNA and half as many chromosomes.

12. Meiosis II is similar to mitosis because
    a. sister chromatids separate.
    b. homologous chromosomes separate.
    c. DNA replication precedes the division.
    d. they both take the same amount of time.
    e. haploid cells are almost always produced.

13. How many *chromatids* are present in metaphase II in a cell undergoing meiosis from an organism in which $2n = 24$?
    a. 12   c. 36   e. 96
    b. 24   d. 48

14. Which of the following would *not* be considered a haploid cell?
    a. a daughter cell after meiosis II
    b. a gamete
    c. a daughter cell after mitosis in the gametophyte generation of a plant
    d. a cell in prophase I
    e. a cell in prophase II

15. Which of the following is *not* true of homologous chromosomes?
    a. They behave independently in mitosis.
    b. They synapse during the S phase of meiosis.
    c. They travel together to the metaphase plate in meiosis I.
    d. They are held together during synapsis by a synaptonemal complex.
    e. Crossing over between nonsister chromatids of homologous chromosomes is indicated by the presence of chiasmata.

16. Which of the following statements is *not* true?
    a. The restoration of the diploid chromosome number after halving in meiosis is due to fertilization.
    b. Sister chromatid cohesion at areas where crossing over has occurred holds homologous chromosomes together until anaphase I.
    c. Recombinant chromosomes are produced when both maternal and paternal chromosomes independently assort into the same gamete.
    d. In mitosis, separation of sister chromatids in anaphase results in two identical cells.
    e. Separation of sister chromatids in anaphase II does not usually result in two identical cells.

17. In a species with a diploid number of 6, how many different combinations of maternal and paternal chromosomes would be possible in *gametes*?
    a. 6   c. 12   e. 128
    b. 8   d. 64

18. In a sexually reproducing species with a diploid number of 8, how many different diploid combinations of chromosomes would be possible in the *offspring?*
    a. 8   c. 64   e. 512
    b. 16   d. 256

19. The calculation of offspring in question 18 includes only variation resulting from
    a. crossing over.
    b. random fertilization.
    c. independent assortment of chromosomes.
    d. a, b, and c.
    e. b and c.

20. Which of the following is *least* likely to be a source of genetic variation in sexually reproducing organisms?
    a. crossing over
    b. replication of DNA during S phase before meiosis I

c. independent assortment of chromosomes
d. random fertilization of gametes
e. All of the above contribute equally to the genetic variation produced in sexual life cycles.

21. Which of the following statements describes why or how recombinant chromosomes add to genetic variability?
    a. They are formed as a result of random fertilization when two sets of chromosomes combine in a zygote.
    b. They are the result of mutations that change alleles.
    c. They randomly orient during metaphase II, and the nonequivalent sister chromatids separate in anaphase II.
    d. Genetic material from two parents is combined on the same chromosome.
    e. Both c and d are true.

22. What is the evolutionary significance of bdelloid rotifers, which have not engaged in sex in 40 million years?
    a. They show that asexual reproduction in more advantageous because it ensures that successful combinations of alleles are passed to offspring.
    b. Although their 400 species originated through sexual reproduction, each species has successfully cloned itself since it arose.
    c. They exchange genetic material through horizontal gene transfer while in a state of suspended animation during dry periods, supporting the idea that genetic variation is evolutionarily advantageous.
    d. Their long history shows that sexual reproduction and genetic variation are not required for evolutionary success.
    e. Both a and d are correct.

Chapter 14

Mendel and the Gene Idea

Key Concepts

14.1 Mendel used the scientific approach to identify two laws of inheritance

14.2 The laws of probability govern Mendelian inheritance

14.3 Inheritance patterns are often more complex than predicted by simple Mendelian genetics

14.4 Many human traits follow Mendelian patterns of inheritance

Framework

Through his work with garden peas in the 1860s, Gregor Mendel developed the fundamental principles of inheritance and the laws of segregation and independent assortment. This chapter describes the basic monohybrid and dihybrid crosses that Mendel performed to establish that inheritance involves particulate genetic factors (genes) that segregate independently in the formation of gametes and recombine to form offspring. The laws of probability can be applied to predict the outcome of genetic crosses.

The phenotypic expression of genotype may be affected by such factors as incomplete dominance, codominance, multiple alleles, pleiotropy, epistasis, and polygenic inheritance, as well as the environment. Genetic screening and counseling for inherited disorders use new technologies and Mendelian principles to analyze human pedigrees.

Chapter Review

14.1 Mendel used the scientific approach to identify two laws of inheritance

Mendel's Experimental, Quantitative Approach Mendel worked with garden peas, a good choice of study organism because they are available in many varieties, their fertilization is easily controlled, and the characteristics of their many offspring can be quantified.

Mendel studied seven **characters,** or heritable features, that occurred in alternative forms called **traits.** He used **true-breeding** varieties of pea plants, which means that self-fertilizing parents always produce offspring with the parental form of the character. To follow the transmission of these well-defined traits, Mendel performed **hybridizations** in which he mated (*crossed* by cross-pollinating) contrasting true-breeding varieties, and then allowed the next generation to self-pollinate. The true-breeding parental plants are the **P generation** (parental); the offspring of the first cross are the **F$_1$ generation** (first filial); and the next generation, from the self- or cross-pollination of the F$_1$, is known as the **F$_2$ generation.**

The Law of Segregation Mendel found that the F$_1$ offspring did not show a blending of the parental traits. Instead, only one of the parental traits of the character was found in the hybrid offspring. In the F$_2$ generation, however, the missing parental trait reappeared in the ratio of 3:1—three offspring with the *dominant* trait shown by the F$_1$ to one offspring with the reappearing *recessive* trait.

Mendel's explanation for this phenomenon contains four parts: (1) alternate forms of genes, called **alleles,** account for variations in characters; (2) an organism has two copies of a gene for each trait, one inherited from each parent; (3) when two different alleles occur together, one of them, called the **dominant allele,** determines the organism's appearance while the other, the **recessive allele,** has no observable effect on appearance; and (4) allele pairs separate (segregate) during the formation of gametes (Mendel's **law of segregation**), so an egg or sperm carries only one allele for each inherited character. This explanation is consistent with the behavior of chromosomes during meiosis.

A **Punnett square** can be used to predict the results of simple genetic crosses. Dominant alleles are often symbolized by a capital letter, recessive alleles by a small letter.

INTERACTIVE QUESTION   14.1

Fill in the following diagram of a cross of round-seeded and wrinkled-seeded pea plants. The round allele (*R*) is dominant; the wrinkled allele (*r*) is recessive.

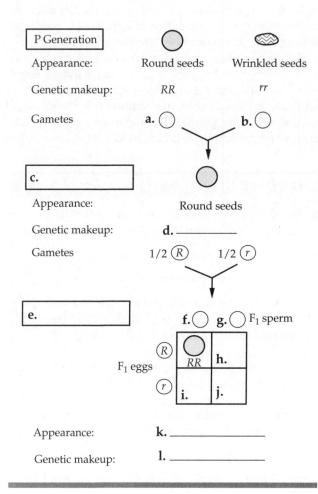

An organism that has a pair of identical alleles is said to be **homozygous** for that gene. If the organism has two different alleles, it is said to be **heterozygous.** Homozygotes are true-breeding; heterozygotes are not, since they produce gametes with one or the other allele that can combine to produce offspring that are dominant homozygotes, heterozygotes, and recessive homozygotes. **Phenotype** is an organism's observable traits; **genotype** is its genetic makeup.

In a **testcross,** an organism expressing the dominant phenotype is crossed with a recessive homozygote to determine the genotype of the phenotypically dominant organism.

INTERACTIVE QUESTION   14.2

A tall pea plant is crossed with a recessive dwarf pea plant. What will the phenotypic and genotypic ratio of offspring be if

a. the tall plant was *TT*?

b. the tall plant was *Tt*?

The Law of Independent Assortment   In a **monohybrid cross,** the inheritance of a single character is followed through the crossing of **monohybrids,** F_1 offspring that are heterozygous for that character. Mendel used **dihybrid crosses** between F_1 **dihybrids,** which are heterozygous for two characters, to determine whether the two characters from the parent plants were transmitted together or independently.

If the two pairs of alleles segregate independently, then gametes from an F_1 hybrid generation (*AaBb*) should contain four combinations of alleles in equal quantities (*AB, Ab, aB, ab*). The random fertilization of these four types of gametes would result in 16 (4 × 4) gamete combinations that produce four phenotypic categories in a ratio of 9:3:3:1 (nine offspring showing both dominant traits, three showing one dominant trait and one recessive, three showing the opposite dominant and recessive traits, and one showing both recessive traits). Mendel obtained these ratios in the F_2 progeny of dihybrid crosses, providing evidence for the **law of independent assortment.** This principle states that each pair of alleles segregates independently from other pairs in the formation of gametes. This law applies to genes located on different chromosomes or far apart on the same chromosome.

INTERACTIVE QUESTION   14.3

A true-breeding tall, purple-flowered pea plant (*TTPP*) is crossed with a true-breeding dwarf, white-flowered plant (*ttpp*).

a. What is the phenotype of the F_1 generation?

b. What is the genotype of the F_1 generation?

c. What four types of gametes are formed by F_1 plants?

_____ _____ _____ _____

d. Fill in the following Punnett square to show the offspring of the F_2 generation. Shade each phenotype a different color so you can see the ratio of offspring.

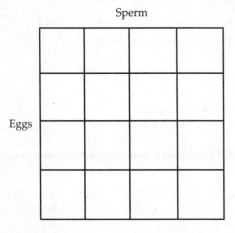

Sperm

Eggs

e. List the phenotypes and ratios found in the F_2 generation.

_____   _____

_____   _____

f. What is the ratio of tall plants to dwarf plants? _____

Of purple-flowered plants to white-flowered plants?

_____

(Note that the alleles for each individual character segregate as in a monohybrid cross.)

14.2 The laws of probability govern Mendelian inheritance

The probability scale goes from 0 to 1; the probabilities of all possible outcomes must add up to 1. The probability of an event occurring is the number of times that event could occur over all the possible events; for example, the probability of drawing an ace of any suit from a deck of cards is $^4/_{52}$. The outcome of independent events is not affected by previous or simultaneous trials.

The Multiplication and Addition Rules Applied to Monohybrid Crosses The **multiplication rule** states that the probability that a certain combination of independent events will occur together is equal to the product of the separate probabilities of the independent events. The probability of a particular genotype being formed by fertilization is equal to the product of the probabilities of forming each type of gamete needed to produce that genotype.

If a genotype can be formed in more than one way, then, according to the **addition rule,** its probability is equal to the sum of the separate probabilities of the different, mutually exclusive ways it can occur. For example, a heterozygote offspring can occur if the egg contains the dominant allele and the sperm the recessive ($\frac{1}{2} \times \frac{1}{2} = \frac{1}{4}$ probability) or vice versa ($\frac{1}{4}$). Therefore, a heterozygote offspring would be the predicted result from a monohybrid cross half of the time ($\frac{1}{4} + \frac{1}{4} = \frac{1}{2}$).

Solving Complex Genetics Problems with the Rules of Probability The probability of a particular genotype arising from a cross can be determined by considering each gene involved as a separate monohybrid cross and then multiplying the probabilities of all the independent events involved in the final genotype. When more than one outcome is involved, the addition rule is also used.

The larger the sample size, the more closely the results will conform to statistical predictions.

INTERACTIVE QUESTION   14.4

a. In the following cross, what is the probability of obtaining offspring that show all three dominant traits?

AaBbcc × *AabbCC*

probability of offspring that are *A_B_C_* = _____

(_ indicates that the second allele can be either dominant or recessive without affecting the phenotype determined by the first dominant allele.)

b. What is the probability that the offspring of this cross will show at least two dominant traits? _____

c. What is the probability of obtaining offspring that show only one dominant trait? _____

14.3 Inheritance patterns are often more complex than predicted by simple Mendelian genetics

Extending Mendelian Genetics for a Single Gene In the case of an allele showing **complete dominance,** the phenotype of the heterozygote is indistinguishable from that of the dominant homozygote. Intermediate phenotypes are characteristic of alleles showing **incomplete dominance.** The F_1 hybrids have a phenotype intermediate between that of the parents. The F_2 offspring show a 1:2:1 phenotypic and genotypic ratio. Alleles that exhibit **codominance** will both affect the phenotype in separate, distinguishable ways, as in the case of the MN blood group.

Tay-Sachs disease is a lethal disorder in which brain cells are unable to metabolize certain lipids that then accumulate and damage the brain. In a heterozygote, the Tay-Sachs allele is recessive at the organismal level; at the biochemical level, the enzyme activity level is intermediate between that of both homozygotes; and at the molecular level, the alleles are codominant in that each produces its enzyme product, either normal or dysfunctional.

Whether or not an allele is dominant or recessive has no relation to how common it is in a population.

Most genes exist in more than two allelic forms. The gene that determines human blood groups has three alleles. The alleles I^A and I^B are codominant with each other; each codes for an enzyme that attaches a carbohydrate to the surface of red blood cells. The allele *i* results in no attached carbohydrate.

INTERACTIVE QUESTION 14.5

List all the possible genotypes for the following blood groups.

a. A

b. B

c. AB

d. O

Pleiotropy is the ability of a single gene to produce multiple phenotypic effects. Certain hereditary diseases with complex sets of symptoms are caused by pleiotropic alleles.

Extending Mendelian Genetics for Two or More Genes In **epistasis,** the expression of a gene at one locus may affect the expression of another gene. F_2 ratios that differ from the typical 9:3:3:1 often indicate epistasis.

INTERACTIVE QUESTION 14.6

Suppose that a dominant allele *M* is necessary for the production of the black pigment melanin; mm individuals are white. A dominant allele *B* results in the deposition of a lot of pigment in an animal's hair, producing a black color. The genotype *bb* produces gray hair, assuming the presence of at least one *M allele*. Two black animals heterozygous for both genes are bred. Fill in the following table for the offspring of this cross.

Phenotype	Genotype	Ratio
Black		
	M_bb	

Quantitative characters, such as height or skin color, vary along a continuum in a population. Such phenotypic gradations are usually due to **polygenic inheritance,** in which two or more genes have an additive effect on one character. Each dominant allele contributes one "unit" to the phenotype. A polygenic character may result in a normal distribution (forming a bell-shaped curve) of the character within a population.

INTERACTIVE QUESTION 14.7

The height of spike weed is a result of polygenic inheritance involving three genes, each of which can contribute an additional 5 cm to the base height of the plant, which is 10 cm. The tallest plant (*AABBCC*) can reach a height of 40 cm.

a. If a tall plant (*AABBCC*) is crossed with a base-height plant (*aabbcc*), what is the height of the F_1 plants?

b. How many different phenotypes will there be in the F_2?

Nature and Nurture: The Environmental Impact on Phenotype   The phenotype of an individual is the result of complex interactions between its genotype and the environment. Genotypes have a phenotypic range called a **norm of reaction** within which the environment influences phenotypic expression. Polygenic characters are often **multifactorial,** meaning that a combination of genetic and environmental factors affects phenotype.

Integrating a Mendelian View of Heredity and Variation   The phenotypic expression of most genes is influenced by other genes and by the environment. The theory of particulate inheritance and Mendel's principles of segregation and independent assortment, however, still form the basis for modern genetics.

14.4   Many human traits follow Mendelian patterns of inheritance

Pedigree Analysis   A family **pedigree** is a family tree with the history of a particular trait shown across the generations. By convention, circles represent females, squares are used for males, and solid symbols indicate individuals that express the trait in question. Parents are joined by a horizontal line, and offspring are listed below parents from left to right in order of birth. The genotype of individuals in the pedigree can often be deduced by following the patterns of inheritance.

Recessively Inherited Disorders   Only homozygous recessive individuals express the phenotype for the thousands of genetic disorders that are inherited as simple recessive traits. **Carriers** of the disorder are heterozygotes, who are usually phenotypically normal but may transmit the recessive allele to their offspring.

Consider the following pedigree for the trait albinism (lack of skin pigmentation) in three generations of a family. (Solid symbols represent individuals who are albinos.) From your knowledge of Mendelian inheritance, answer the questions that follow.

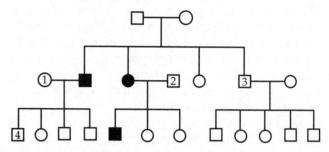

a. Is this trait caused by a dominant or recessive allele? How can you tell?

b. Determine the genotypes of the parents in the first generation. (Let *A* and *a* represent the alleles.) Genotype of father _____; genotype of mother _____.

c. Determine the probable genotypes of the mates of the albino offspring in the second generation and the grandson # 4 in the third generation. Genotypes: mate # 1 _____; mate # 2 _____; grandson # 4 _____.

d. Can you determine the genotype of son # 3 in the second generation? Why or why not?

The likelihood of two individuals carrying the same rare deleterious allele increases when they have common ancestors. Consanguineous matings, between siblings or close relatives, are indicated on pedigrees by double lines.

Cystic fibrosis, the most common lethal genetic disease in the United States, occurs more frequently in people of European descent than in other groups. The recessive allele results in defective chloride channels in certain cell membranes. Accumulating extracellular chloride leads to the buildup of thickened mucus in various organs and a predisposition to bacterial infections.

Sickle-cell disease is the most common inherited disease among African Americans. Due to a single amino acid substitution in the hemoglobin protein, red blood cells deform into a sickle shape when blood-oxygen concentration is low, triggering blood clumping and other pleiotropic effects. Heterozygous individuals are said to have sickle-cell trait but are usually healthy.

INTERACTIVE QUESTION 14.9

a. What is the probability that a mating between two carriers will produce an offspring with a recessively inherited disorder? _____

b. What is the probability that a phenotypically normal child produced by a mating of two heterozygotes will be a carrier? _____

Dominantly Inherited Disorders   A few human disorders are due to dominant genes. In *achondroplasia*, dwarfism results from a single copy of a mutant allele.

Dominant lethal alleles are rarer than recessive lethals because the harmful allele cannot be masked in the heterozygote. A late-acting lethal dominant allele can be passed on if the symptoms do not develop until after reproductive age. Molecular geneticists have developed a method to detect the lethal gene for **Huntington's disease,** a degenerative disease of the nervous system that does not develop until later in life.

Multifactorial Disorders   Many diseases have genetic (usually polygenic) and environmental components. These multifactorial disorders include heart disease, diabetes, and cancer.

Genetic Testing and Counseling   The probability of a child having a genetic defect may be determined by considering the family history of the disease.

INTERACTIVE QUESTION 14.10

If two prospective parents both have siblings who have a recessive genetic disorder, what is the chance that they would have a child who inherits the disorder? If this couple has two children without the disorder, what is the chance that a third child would inherit the disorder? Explain both answers.

DNA-based genetic tests to identify carriers have been developed for an increasing number of heritable disorders. Insurance companies and employers are now legally prohibited from making decisions based on results of genetic testing.

Amniocentesis is a procedure that extracts a small amount of amniotic fluid from the sac surrounding the fetus. The fluid is analyzed biochemically, and fetal cells are cultured for several weeks and then tested for certain genetic disorders and karyotyped to check for chromosomal defects.

Chorionic villus sampling (CVS) is a technique in which a small amount of fetal tissue is suctioned from the placenta. These rapidly growing cells can be karyotyped immediately, and this procedure can be performed earlier in pregnancy. A new technique isolates fetal cells from maternal blood, which can then be cultured and tested.

Ultrasound is a simple, noninvasive procedure that can reveal major abnormalities. *Fetoscopy,* the insertion of a needle-thin viewing scope and light into the uterus, allows the fetus to be checked for anatomical problems.

Some genetic disorders can be detected at birth. Routine screening tests are used to detect phenylketonuria (PKU) in newborns. If detected, dietary adjustments can lead to normal development.

Word Roots

-centesis = a puncture (*amniocentesis:* a technique of prenatal diagnosis in which amniotic fluid is obtained by aspiration from a needle inserted into the uterus; the fluid and fetal cells are analyzed to detect certain genetic defects in the fetus)

co- = together (*codominance:* situation in which the phenotypes of both alleles are exhibited in the heterozygote)

di- = two (*dihybrid cross:* a cross between two organisms that are each heterozygous for both of the characters being followed)

epi- = beside; **-stasis** = standing (*epistasis:* a type of gene interaction in which the phenotypic expression of one gene alters that of another independently inherited gene)

geno- = offspring (*genotype:* the genetic makeup of an organism)

hetero- = different (*heterozygous:* having two different alleles for a given gene)

homo- = alike (*homozygous:* having two identical alleles for a given gene)

mono- = one (*monohybrid cross:* a cross between two organisms that are heterozygous for the character

being followed or the self-pollination of a heterozygous plant)

pedi- = a child (*pedigree:* a diagram of a family tree
showing the occurrence of heritable characters in
parents and offspring over multiple generations)

pheno- = appear (*phenotype:* the physical and physiological traits of an organism, determined by its genetic makeup)

pleio- = more (*pleiotropy:* the ability of a single gene to
have multiple effects)

poly- = many; **gene-** = produce (*polygenic:* an additive
effect of two or more gene loci on a single phenotypic character)

Structure Your Knowledge

1. Relate Mendel's two laws of inheritance to the behavior of chromosomes in meiosis that you studied
   in Chapter 13.

2. How many different types of gametes can be
   formed by the following genotypes? What generalized formula describes this number?
   $Aa =$ $AaBb =$
   $AaBbCc =$ $AABbCc =$

3. How can you tell whether alleles are completely
   dominant/recessive, incompletely dominant, or
   codominant?

Genetics Problems

*One of the best ways to learn genetics is to work problems.
You can't memorize genetic knowledge; you have to practice
using it. Work through the following problems, methodically
setting down the information you are given and what you are
to determine. Write the symbols used for the alleles and genotypes, and the phenotypes resulting from those genotypes.
Avoid using Punnett squares; while useful for learning basic
concepts, they are much too laborious. Instead, break complex crosses into their monohybrid components and rely on
the rules of multiplication and addition. Always look at the
answers you get to see if they make logical sense. And remember that study groups are great for going over your
problems and helping each other "see the light."*

1. Summer squash are either white or yellow. To get
   white squash, at least one of the parental plants must
   be white. The allele for which color is dominant?

2. For the following crosses, determine the probability of obtaining an offspring with the indicated
   genotype.

Cross	Offspring	Probability
$AAbb \times AaBb$	$AAbb$	**a.**
$AaBB \times AaBb$	$aaBB$	**b.**
$AABbcc \times aabbCC$	$AaBbCc$	**c.**
$AaBbCc \times AaBbcc$	$aabbcc$	**d.**

3. True-breeding tall red-flowered plants are crossed
   with dwarf white-flowered plants. The resulting
   F_1 generation consists of all tall pink-flowered
   plants. Assuming that height and flower color are
   each determined by a single gene locus on different
   chromosomes, predict the results of an F_1 cross of
   dihybrid plants. Choose appropriate symbols for
   the alleles of the genes for height and flower color.
   List the phenotypes and predicted ratios for the
   F_2 generation.

4. Blood typing has been used as evidence in paternity cases, when the blood types of the mother and
   child may indicate that a man alleged to be the father could not possibly have fathered the child. For
   the following combinations of mother and child,
   indicate the blood group(s) of men who would be
   exonerated (could *not* be the father of that child).

Blood Group of Mother	Blood Group of Child	Man Exonerated If He Belongs to Blood Group(s)
AB	A	**a.**
O	B	**b.**
A	AB	**c.**
O	O	**d.**
B	A	**e.**

5. In rabbits, CC is normal, Cc results in rabbits with
   deformed legs, and cc is lethal. For a gene for coat
   color, the genotype BB produces black, Bb brown,
   and bb white. Give the phenotypic ratio of offspring from a cross of a deformed-leg, brown rabbit with a deformed-leg, white rabbit.

6. Polydactyly (extra fingers and toes) results from
   a dominant gene. A father is polydactyl, the
   mother has the normal phenotype, and they have
   had one normal child. What is the genotype of
   the father? Of the mother? What is the probability that a second child will have the normal number of digits?

7. In dogs, black (*B*) is dominant to chestnut (*b*), and solid color (*S*) is dominant to spotted (*s*). What are the genotypes of the parents in a mating that produced ⅜ black solid, ⅜ black spotted, ⅛ chestnut solid, and ⅛ chestnut spotted puppies? (*Hint:* First determine what genotypes the offspring must have before you deal with the fractions.)

8. When hairless hamsters are mated with normal-haired hamsters, about one-half the offspring are hairless and one-half are normal. When hairless hamsters are crossed with each other, the ratio of normal-haired to hairless is 1:2. How do you account for the results of the first cross? How would you explain the unusual ratio obtained in the second cross?

9. Two pigs whose tails are exactly 25 cm in length are bred over 10 years and they produce 96 piglets with the following tail lengths: 6 piglets at 15 cm, 25 at 20 cm, 37 at 25 cm, 23 at 30 cm, and 5 at 35 cm.
   a. How many pairs of genes are regulating the tail length character? *Hint:* Count the number of phenotypic classes, or determine the sum of the ratios of the classes. In a monohybrid cross, the F_2 ratios add up to 4 (3:1 or 1:2:1). In a dihybrid cross, the F_2 ratios add up to 16 (9:3:3:1 or some variation if the genes are epistatic or quantitative).
   b. What offspring phenotypes would you expect from a mating between a pig with a 15-cm-long tail and a pig with a 30-cm-long tail?

10. Fur color in rabbits is determined by a single gene locus, for which there are four alleles. Four phenotypes are possible: black, Chinchilla (gray color caused by white hairs with black tips), Himalayan (white with black patches on extremities), and white. The black allele (*C*) is dominant over all other alleles, the Chinchilla allele (C^{ch}) is dominant over Himalayan (C^h), and the white allele (*c*) is recessive to all others.
    a. A black rabbit is crossed with a Himalayan, and the F_1 consists of a ratio of 2 black to 2 Chinchilla. Explain what you can and cannot determine about the genotypes of the parents.
    b. A second cross was done between a black rabbit and a Chinchilla. The F_1 contained a ratio of 2 black to 1 Chinchilla to 1 Himalayan. Explain what you can and cannot determine about the genotypes of the parents.

11. Suppose that in hamsters, a dominant gene *B* determines black coat color and *bb* produces brown. A separate gene *E/e*, however, shows epistasis over the *B/b* gene. The dominant *E* allele prevents the deposition of pigment in the hair, resulting in a golden coat color. An *ee* genotype is necessary for a black or brown hamster. A pet owner wants to know the genotypes of her three hamsters, so she breeds them and makes note of the offspring of the litters. Determine the genotypes of the three hamsters.
    a. golden female (hamster 1) × golden male (hamster 2) offspring: 7 golden, 1 black, 1 brown
    b. black female (hamster 3) × golden male (hamster 2) offspring: 8 golden, 5 black, 2 brown

12. The ability to taste phenylthiocarbamide (PTC) is controlled in humans by a single dominant allele (*T*). A nontasting woman married a tasting man, and they had three children: two tasting boys and a nontasting girl. All the grandparents could taste. Create a pedigree for this family for this trait. Solid symbols should signify nontasters (*tt*). Whenever possible, indicate whether tasters are *TT* or *Tt*.

13. Two true-breeding varieties of garden peas are crossed. One parent had red, axial flowers, and the other had white, terminal flowers. All F_1 individuals had red, terminal flowers. If 100 F_2 offspring were counted, how many of them would you expect to have red, axial flowers?

14. You cross true-breeding red-flowered plants with true-breeding white-flowered plants, and the F_1s are all red-flowered plants. The F_2s, however, occur in a ratio of 9 red:6 pale purple:1 white. How many genes are involved in the inheritance of this color character? Explain why the F_1s are all red and how the 9:6:1 ratio of phenotypes in the F_2 occurred.

Test Your Knowledge

MULTIPLE CHOICE: *Choose the one best answer.*

1. According to Mendel's law of segregation,
   a. there is a 50% probability that a gamete will get a dominant allele.
   b. allele pairs segregate independently of other pairs of alleles in gamete formation.
   c. allele pairs separate in gamete formation.
   d. the laws of probability determine gamete formation.
   e. there is a 3:1 ratio in the F_2 generation.

2. The F_2 generation
   a. has a phenotypic ratio of 3:1.
   b. is the result of the self-fertilization or crossing of F_1 individuals.
   c. can be used to determine the genotype of individuals with the dominant phenotype.
   d. has a phenotypic ratio that equals its genotypic ratio.
   e. has 16 different genotypic possibilities.

3. A 1:1 phenotypic ratio in a testcross indicates that
  a. the alleles are dominant.
  b. one parent must have been homozygous dominant.
  c. the parent with a dominant phenotype was a heterozygote.
  d. the alleles segregated independently.
  e. the alleles are codominant.

4. Which phase of meiosis is most directly related to the law of independent assortment?
  a. prophase I   **d.** metaphase II
  b. prophase II   **e.** anaphase II
  c. metaphase I

5. After obtaining two heads from two tosses of a coin, the probability of obtaining a head on the next toss is
  a. 1/2.   **c.** 1/8.   **e.** 3/4.
  b. 1/4.   **d.** 3/8.

6. The probability of tossing three coins simultaneously and obtaining three heads is
  a. 1/2.   **c.** 1/8.   **e.** 3/4.
  b. 1/4.   **d.** 3/8.

7. The probability of tossing three coins simultaneously and obtaining two heads and one tail is
  a. 1/2.   **c.** 1/8.   **e.** 3/4.
  b. 1/4.   **d.** 3/8.

8. In guinea pigs, the brown coat color allele (B) is dominant over red (b), and the solid color allele (S) is dominant over spotted (s). The F_1 offspring of a cross between true-breeding brown, solid-colored guinea pigs and red, spotted pigs are crossed. What proportion of their offspring (F_2) would be expected to be red and solid colored?
  a. 1/9   **c.** 3/16   **e.** 3/4
  b. 1/16   **d.** 9/16

9. The base height of the dingdong plant is 10 cm. Four genes contribute to the height of the plant, and each dominant allele contributes 3 cm to height. If you cross a 10-cm plant (quadruply homozygous recessive) with a 34-cm plant, how many phenotypic classes will there be in the F_2?
  a. 4   **c.** 8   **e.** 64
  b. 5   **d.** 9

10. You think that two alleles for coat color in mice show incomplete dominance. What is the best and simplest cross to perform in order to test your hypothesis?
  a. a testcross of a homozygous recessive mouse with a mouse of unknown genotype
  b. a cross of F_1 mice to look for a 1:2:1 ratio in the offspring
  c. a reciprocal cross in which the sex of the mice of each coat color is reversed

  d. a cross of two true-breeding mice of different colors to look for an intermediate phenotype in the F_1
  e. a cross of F_1 mice to look for a 9:7 ratio in the offspring

11. A mother with type B blood has two children, one with type A blood and one with type O blood. Her husband has type O blood. Which of the following could you conclude from this information?
  a. The husband could not have fathered either child.
  b. The husband could have fathered both children.
  c. The husband must be the father of the child with type O blood and could be the father of the type A child.
  d. Neither the mother nor the husband could be the biological parent of the type A child.
  e. The husband could be the father of the child with type O blood, but not the type A child.

12. A dominant allele P causes the production of purple pigment; pp individuals are white. A dominant allele C is also required for color production; cc individuals are white. What proportion of offspring will be purple from a $ppCc \times PpCc$ cross?
  a. 1/8   **c.** 1/2   **e.** None
  b. 3/8   **d.** 3/4

13. Carriers of a genetic disorder
  a. are indicated by solid symbols on a family pedigree.
  b. are heterozygous at the genetic locus for the disorder.
  c. will produce children with the disease.
  d. are often the result of consanguineous matings.
  e. have a homozygous recessive genotype but do not phenotypically exhibit the disease.

14. Which of the following human diseases is inherited as a simple recessive trait?
  a. cystic fibrosis
  b. cancer
  c. Huntington's disease
  d. achondroplasia
  e. cardiovascular disease

15. A multifactorial disorder
  a. can usually be traced to consanguineous matings.
  b. is caused by recessively inherited lethal genes.
  c. has both genetic and environmental causes.
  d. has a collection of symptoms traceable to an epistatic gene.
  e. is usually associated with quantitative traits.

16. Which of the following procedures would be best for identifying a chromosomal abnormality in a fetus?
  a. chorionic villus sampling and karyotyping
  b. biochemical tests
  c. ultrasound
  d. fetoscopy
  e. analysis of fetal DNA

Chapter 15

The Chromosomal Basis of Inheritance

Key Concepts

15.1 Mendelian inheritance has its physical basis in the behavior of chromosomes

15.2 Sex-linked genes exhibit unique patterns of inheritance

15.3 Linked genes tend to be inherited together because they are located near each other on the same chromosome

15.4 Alterations of chromosome number or structure cause some genetic disorders

15.5 Some inheritance patterns are exceptions to standard Mendelian inheritance

Framework

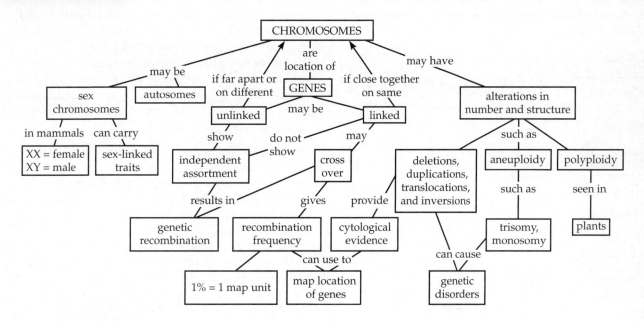

Chapter Review

15.1 Mendelian inheritance has its physical basis in the behavior of chromosomes

Mendel's laws, combined with cytological evidence of the process of meiosis, led to the **chromosome theory of inheritance:** Genes occupy specific positions (loci) on chromosomes, and it is the chromosomes that undergo segregation and independent assortment during meiosis in gamete formation.

Morgan's Experimental Evidence: Scientific Inquiry T. H. Morgan worked with fruit flies, *Drosophila melanogaster,* which are prolific and rapid breeders. Fruit flies have only four pairs of chromosomes; the sex chromosomes occur as XX in female flies and as XY in male flies.

The normal phenotype found most commonly in nature for a character is called the **wild type,** whereas alternative traits, assumed to have arisen as mutations, are called *mutant phenotypes*. A mutant allele is designated with a small letter, the wild-type allele with a superscript +.

Morgan discovered a mutant white-eyed male fly, which he then mated with a wild-type red-eyed female. The F_1 offspring were all red-eyed. In the F_2, however, all female flies were red-eyed, whereas half of the males were red-eyed and half were white-eyed. Morgan deduced that the gene for eye color was located on the X chromosome. Males have only one X, so their phenotype is determined by the eye-color allele they inherit from their mother. This association of a specific gene with a chromosome provided evidence for the chromosome theory of inheritance.

INTERACTIVE QUESTION   15.1

Complete the following summary of Morgan's crosses involving the mutant white-eyed fly by filling in the Punnett square and indicating the phenotypes and genotypes of the F_2 generation. (X^w stands for the mutant recessive white allele; X^{w+} for the wild-type red allele.) The Y reminds you that this eye color gene is not present on the Y chromosome.

P Generation		
Phenotype	red-eyed (wild-type) female ×	white-eyed male
Genotype	$X^{w+}X^{w+}$	X^wY

F_1 Generation		
Phenotype	red-eyed female ×	red-eyed male
Genotype	$X^{w+}X^w$	$X^{w+}Y$

F_2 Generation

Sperm

Eggs

Phenotypes

_____   _____   _____   _____

Genotypes

_____   _____   _____

15.2   Sex-linked genes exhibit unique patterns of inheritance

The Chromosomal Basis of Sex   Sex is a phenotypic character usually determined by sex chromosomes. In humans and other mammals, females, who are XX, produce eggs that each contain an X chromosome. Males, who are XY, produce two kinds of sperm, each with either an X or a Y chromosome.

Whether the gonads of an embryo develop into testes or ovaries depends on the presence or absence of the gene *SRY*, found on the Y chromosome, whose protein product regulates many other genes. Genes located on either sex chromosome are called **sex-linked genes.** The relatively few genes located on the Y chromosome are called *Y-linked genes*. About half of these genes are expressed only in the testis. The term **X-linked genes** refers to the approximately 1,100 genes located on the X chromosome.

Other sex-determination systems include X-0 (in grasshoppers and some other insects), Z-W (in birds and some fishes and insects), and haplo-diploid (in bees and ants).

Inheritance of X-Linked Genes   X-linked genes may code for characters that are not related to sex. Males inherit X-linked alleles from their mothers; daughters inherit X-linked alleles from both parents. Recessive X-linked traits are seen more often in males, because they are *hemizygous* for X-linked genes.

Duchenne muscular dystrophy is an X-linked disorder resulting from the lack of a key muscle protein. **Hemophilia** is an X-linked trait characterized by excessive bleeding due to the absence of one or more blood-clotting proteins.

X Inactivation in Female Mammals   Only one of the X chromosomes is fully active in most mammalian female somatic cells. The other X chromosome is contracted into a **Barr body** located inside the nuclear membrane. M. Lyon demonstrated that the selection of which X chromosome is inactivated is a random event occurring independently in embryonic cells. As a result of X inactivation, both males and females have an equal dosage of most X-linked genes. A gene called *XIST* becomes active on the X chromosome that forms the Barr body. Its RNA products appear to trigger DNA methylation and X inactivation.

INTERACTIVE QUESTION 15.2

Two normal color-sighted individuals have two children and seven grandchildren. Fill in the probable genotype of each of the numbered individuals in the following pedigree. *Squares are males, circles are females, and solid symbols represent color blindness. Use the superscript N for the normal allele, and n for the recessive allele for color blindness.*

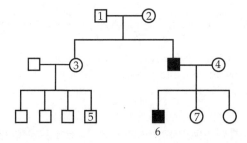

Genotypes:

1. _____   4. _____   7. _____

2. _____   5. _____

3. _____   6. _____

15.3 Linked genes tend to be inherited together because they are located near each other on the same chromosome

Genes located near each other on the same chromosome tend to be inherited together and are called **linked genes.**

How Linkage Affects Inheritance   Morgan performed a testcross of F_1 dihybrid wild-type flies with flies that were homozygous recessive for black bodies and vestigial wings. He found that the offspring were not in the predicted 1:1:1:1 phenotypic ratio. Rather, most of the offspring were the same phenotypes as the P generation parents—either wild type (gray, normal wings) or double mutant (black, vestigial). Morgan deduced that these traits were inherited together because their genes were located on the same chromosome.

Genetic recombination results in offspring with combinations of traits that differ from those of either parent.

Genetic Recombination and Linkage   In a cross between a dihybrid heterozygote and a doubly recessive homozygote, one-half of the offspring will be **parental types** and have phenotypes like one or the other of the parents, and one-half of the offspring, called **recombinant types (recombinants),** will have combinations of the two traits that are unlike those of the parents. This 50% frequency of recombination is observed when two genes are located on different chromosomes, and it results from the random alignment of homologous chromosomes at metaphase I and the resulting independent assortment of the two unlinked genes.

INTERACTIVE QUESTION 15.3

In a testcross between a heterozygote tall, purple-flowered pea plant and a dwarf, white-flowered plant,

a. what are the phenotypes of offspring that are parental types?

b. what are the phenotypes of offspring that are recombinants?

Linked genes do not assort independently, and one would not expect to see recombination of parental traits in the offspring. Recombination of linked genes does occur, however, due to **crossing over,** the reciprocal trade between nonsister (a maternal and a paternal) chromatids of paired homologous chromosomes during prophase of meiosis I.

The combination of recombinant chromosomes, independent assortment, and random fertilization generates new assortments of alleles, providing an abundance of genetic variation on which natural selection can work.

INTERACTIVE QUESTION 15.4

With unlinked genes, an equal number of parental and recombinant offspring are produced. With linked genes, are more or fewer parentals than recombinants produced? Explain your answer.

Mapping the Distance Between Genes Using Recombination Data: Scientific Inquiry   The percentage of recombinant offspring is called the *recombination frequency.* A **genetic map** is an ordered list of genes on a chromosome. A. H. Sturtevant suggested that recombination frequencies reflect the relative distance between genes; for genes that are farther apart, the probability that crossing over will occur between them is greater than for genes that are closer together. Sturtevant used recombination data to create a **linkage map,** in which one **map unit** is defined as equal to a 1% recombination frequency.

The sequence of genes on a chromosome can be determined by finding the recombination frequency between different pairs of genes. Linkage cannot be determined if genes are so far apart that crossovers between them are almost certain. They would then have the 50% recombination frequency typical of unlinked genes. Such genes are *physically linked* but *genetically unlinked.* Distant genes on the same chromosome may be mapped by adding the recombination frequencies between them and intermediate genes.

Sturtevant and his colleagues found that the genes for the various known mutations of *Drosophila* clustered into four groups of linked genes, providing additional evidence that genes are located on chromosomes—in this case, on the four *Drosophila* chromosomes.

The frequency of crossing over may vary along the length of a chromosome, and a linkage map provides the sequence but not the exact location of genes on chromosomes. **Cytogenetic maps** locate gene loci in reference to visible chromosomal features.

INTERACTIVE QUESTION 15.5

The following recombination frequencies have been determined for several gene pairs. Create a linkage map for these genes, showing the map unit distance between loci.

j, k: 12%   *j, m:* 9%   *k, l:* 6%   *l, m:* 15%

15.4 Alterations of chromosome number or structure cause some genetic disorders

Abnormal Chromosome Number   **Nondisjunction** occurs when a pair of homologous chromosomes does not separate properly in meiosis I, or when sister chromatids do not separate in meiosis II. As a result, a gamete may receive either two copies or no copies of that chromosome. A zygote formed with one of these aberrant gametes has a chromosomal alteration known as **aneuploidy,** a nontypical number of a particular chromosome. The zygote will be either **trisomic** for that chromosome (chromosome number is $2n + 1$) or

monosomic ($2n - 1$). Aneuploid organisms usually have a set of symptoms caused by the abnormal dosage of genes. A mitotic nondisjunction early in embryonic development is also likely to be harmful.

Polyploidy is a chromosomal alteration in which an organism has more than two complete chromosomal sets, as in *triploidy* ($3n$) or *tetraploidy* ($4n$). Polyploidy is common in the plant kingdom and has played an important role in the evolution of plants.

INTERACTIVE QUESTION 15.6

a. What is the difference between an organism with a trisomy and a triploid organism?

b. Which of these two organisms is likely to exhibit the more deleterious effects as a result of its chromosomal anomaly?

Alterations of Chromosome Structure   Chromosome breakage can result in chromosome fragments that are lost, called **deletion;** that join to a sister chromatid (or a nonsister chromatid), called **duplication;** that rejoin the original chromosome in the reverse orientation, called **inversion;** or that join a nonhomologous chromosome, called **translocation.** An unequal crossover can result in a deletion and duplication in nonsister chromatids, caused by unequal exchange between chromatids.

INTERACTIVE QUESTION 15.7

Two nonhomologous chromosomes have gene orders, respectively, of *A-B-C-D-E-F-G-H-I-J* and *M-N-O-P-Q-R-S-T.* What types of chromosome alterations would have occurred if daughter cells were found to have a gene sequence of *A-B-C-O-P-Q-G-J-I-H* on the first chromosome?

Human Disorders Due to Chromosomal Alterations The frequency of aneuploid zygotes may be fairly high in humans, but development is usually so disrupted that embryos spontaneously abort. Some genetic disorders, expressed as *syndromes* of characteristic traits, are the result of aneuploidy.

Down syndrome, caused by trisomy of chromosome 21, results in characteristic facial features, short stature, correctable heart defects, and developmental delays. The incidence of Down syndrome increases for older mothers.

XXY males exhibit *Klinefelter syndrome,* a condition in which the individual has abnormally small testes, is sterile, and may have subnormal intelligence. Males with an extra Y chromosome do not exhibit any well-defined syndrome.

Trisomy X results in females who are healthy and distinguishable only by karyotype, although they are slightly taller than average and at risk of learning disabilities. Monosomy X individuals (X0) exhibit *Turner syndrome* and are phenotypically female, sterile individuals with short stature and usually normal intelligence.

Structural alterations of chromosomes, such as deletions or translocations, may be associated with specific human disorders, such as *cri du chat* syndrome and *chronic myelogenous leukemia* (CML). The reciprocal translocation involved in CML produces a short, recognizable *Philadelphia chromosome.*

INTERACTIVE QUESTION 15.8

Why do most sex chromosome aneuploidies have less deleterious effects than do autosomal aneuploidies?

15.5 Some inheritance patterns are exceptions to standard Mendelian inheritance

Genomic Imprinting A few dozen traits in mammals depend on which parent supplied the alleles for the trait. **Genomic imprinting,** which occurs during gamete formation, determines whether or not an allele will be expressed in an offspring. When the next generation makes gametes, old maternal or paternal imprints are removed, and alleles are again imprinted according to the sex of the parent. Most of the mammalian genes subject to imprinting identified so far are involved in embryonic development. The addition of methyl groups to cytosine nucleotides may inactivate (or activate) the imprinted gene, assuring that the developing embryo has only one active copy.

Inheritance of Organelle Genes Exceptions to Mendelian inheritance occur for *extranuclear* (or *cytoplasmic*) *genes.* Such genes are located on small circles of DNA in mitochondria and plant plastids and are transmitted to offspring in the cytoplasm of the egg. Some rare human disorders are caused by mitochondrial mutations, and

maternally inherited mitochondrial defects may contribute to some cases of diabetes, heart disease, and Alzheimer's disease.

Word Roots

aneu- = without (*aneuploidy:* a chromosomal aberration in which one or more chromosomes are present in extra copies or are deficient in number)

cyto- = cell (*cytogenetic map:* a map of a chromosome that locates genes with respect to distinguishable chromosomal features)

hemo- = blood (*hemophilia:* a human genetic disease that is caused by an X-linked recessive allele and characterized by excessive bleeding following injury)

mono- = one (*monosomic:* referring to a cell that has only one copy of a particular chromosome instead of the normal two)

non- = not; **dis-** = separate (*nondisjunction:* an error in meiosis or mitosis in which members of a pair of homologous chromosomes or a pair of sister chromatids fail to separate properly)

poly- = many (*polyploidy:* a chromosomal alteration in which the organism possesses more than two complete chromosome sets)

re- = again; **com-** = together; **bin-** = two at a time (*recombinant:* an offspring whose phenotype differs from that of the true-breeding P generation parents; also refers to the phenotype itself)

trans- = across (*translocation:* an aberration in chromosome structure resulting from attachment of a chromosomal fragment to a nonhomologous chromosome)

tri- = three; **soma-** = body (*trisomic:* referring to a diploid cell that has three copies of a particular chromosome)

Structure Your Knowledge

1. Mendel's law of independent assortment applies to genes that are on different chromosomes. However, at least two of the genes Mendel studied were actually located on the same chromosome. Explain why genes located more than 50 map units apart behave as though they are not linked. How can one determine whether these genes are linked, and what the relative distance is between them?

2. You have found a new mutant phenotype in fruit flies that you suspect is recessive and X-linked.

What is the single, best cross you could make to confirm your predictions?

3. Various human disorders or syndromes are related to chromosomal abnormalities. What explanation can you give for the adverse phenotypic effects associated with these chromosomal alterations?

Genetics Problems

Again, one of the best ways to learn genetics is to do problems.

1. *The following pedigree traces the inheritance of a genetic trait.*

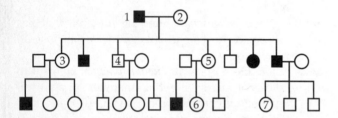

   a. What type of inheritance does this trait show?
   b. Chose an appropriate allele labeling system, and give the predicted genotype for the following individuals:

   1. _____   5. _____
   2. _____   6. _____
   3. _____   7. _____
   4. _____

   c. What is the probability that a child of individual #6 and a phenotypically normal male will have this trait?

2. Use the following recombination frequencies to determine the order of the genes on the chromosome.
   a, c: 10%   a, d: 30%   b, c: 24%   b, d: 16%

3. In guinea pigs, black (*B*) is dominant to brown (*b*), and solid color (*S*) is dominant to spotted (*s*). A heterozygous black, solid-colored pig is mated with a brown, spotted pig. The offspring from several litters are as follows: black solid: 16; black spotted: 5; brown solid: 5; and brown spotted: 14. Are these genes linked or unlinked? If they are linked, how many map units are they apart?

4. A woman is a carrier for an X-linked lethal allele that causes an embryo with the allele to spontaneously abort. She has nine children. How many of these children do you expect to be boys?

5. A dominant sex-linked allele *B* produces white bars on black chickens, as seen in the Barred Plymouth Rock breed. A clutch of chicks has equal numbers of black and barred chicks. (Remember that sex is determined by the Z-W system in birds: ZZ are males, ZW are females.)
   a. If only the females are found to be black, what were the genotypes of the parents?
   b. If males and females are evenly represented in the black and barred chicks, what were the genotypes of the parents?

Test Your Knowledge

MULTIPLE CHOICE: *Choose the one best answer.*

1. Which of the following statements is part of the chromosome theory of inheritance?
   a. Genes are located on chromosomes.
   b. Homologous chromosomes and their associated genes undergo segregation during meiosis.
   c. Chromosomes and their associated genes undergo independent assortment in gamete formation.
   d. Mendel's laws of inheritance relate to the behavior of chromosomes in meiosis.
   e. all of the above

2. A wild type is
   a. the phenotype found most commonly in nature.
   b. the dominant allele.
   c. designated by a small letter if it is recessive or a capital letter if it is dominant.
   d. a trait found on the X chromosome.
   e. your basic party animal.

3. Sex-linked traits
   a. are carried on an autosome but expressed only in males.
   b. are coded for by genes located on a sex chromosome.
   c. are found in only one sex, depending on the sex-determination system of the species.
   d. are always inherited from the mother in mammals and fruit flies.
   e. depend on whether the gene was inherited from the mother or the father.

4. In which of the following structures would you expect to find a Barr body?
   a. an egg
   b. a sperm
   c. a liver cell of a man
   d. a liver cell of a woman
   e. a mitochondrion

5. A cross of a wild-type red-eyed female *Drosophila* with a violet-eyed male produces all red-eyed offspring. If the gene is X-linked, which of the following should the reciprocal cross (violet-eyed female × red-eyed male) produce? (Assume that the red allele is dominant to the violet allele.)
   a. all violet-eyed flies
   b. 3 red-eyed flies to 1 violet-eyed fly
   c. a 1:1 ratio of red and violet eyes in both males and females
   d. red-eyed females and violet-eyed males
   e. all red-eyed flies

6. Linkage and cytogenetic maps for the same chromosome
   a. are both based on mutant phenotypes and recombination data.
   b. may have different orders of genes.
   c. have both the same order of genes and intergenic distances.
   d. have the same order of genes but different intergenic distances.
   e. are created using chromosomal abnormalities.

7. A 1:1:1:1 ratio of offspring from a dihybrid test-cross indicates that
   a. the genes are linked.
   b. the dominant organism was homozygous.
   c. crossing over has occurred.
   d. the genes are 25 map units apart.
   e. the genes are not linked or are more than 50 map units apart.

8. Genes *A* and *B* are linked and 12 map units apart. A heterozygous individual, whose parents were *AAbb* and *aaBB*, would be expected to produce gametes in which of the following frequencies?
   a. 44% *AB*   6% *Ab*   6% *aB*   44% *ab*
   b. 6% *AB*   44% *Ab*   44% *aB*   6% *ab*
   c. 12% *AB*   38% *Ab*   38% *Ab*   12% *ab*
   d. 6% *AB*   6% *Ab*   44% *aB*   44% *ab*
   e. 38% *AB*   12% *Ab*   12% *aB*   38% *ab*

9. Two true-breeding *Drosophila* are crossed: a normal-winged, red-eyed female and a miniature-winged, vermilion-eyed male. The F_1s all have normal wings and red eyes. When F_1 offspring are crossed with miniature-winged, vermilion-eyed flies, the following offspring resulted:
   233 normal wing, red eye
   247 miniature wing, vermilion eye
   7 normal wing, vermilion eye
   13 miniature wing, red eye

From these results, you could conclude that the alleles for miniature wings and vermilion eyes are
   a. both X-linked and dominant.
   b. located on autosomes and dominant.
   c. recessive, and that these genes are located 4 map units apart.
   d. recessive, and that these genes are located 20 map units apart.
   e. recessive, and that the deviation from the expected 9:3:3:1 ratio is due to epistasis.

10. A female tortoiseshell cat is heterozygous for the gene that determines black or orange coat color, which is located on the X chromosome. A male tortoiseshell cat
   a. cannot occur.
   b. is hemizygous at this locus.
   c. must have resulted from a nondisjunction and has a Barr body in each of his cells.
   d. must have two alleles for coat color in each of his cells, one from his father and one from his mother.
   e. would be hermaphroditic.

11. When we say that a few of the genes for Mendel's pea characters were physically linked but genetically unlinked, we mean that
   a. the genes are on the same chromosome, but they are more than 50 map units apart.
   b. the genes assort independently even though the chromosomes they are on travel to the metaphase plate together.
   c. their alleles segregate in anaphase I, and each gamete receives a single allele for all of these genes.
   d. dihybrid crosses with these genes produce more than 50% recombinant offspring even though they are on the same chromosome.
   e. Mendel could not determine that the genes were on the same chromosome because he did not perform crosses with these gene pairs.

12. Which of the following chromosomal alterations does not alter genic balance but may affect gene expression and thus phenotype?
   a. deletion   d. nonidentical duplication
   b. inversion   e. nondisjunction
   c. duplication

13. Genomic imprinting
   a. explains cases in which the phenotypic effect of an allele depends on the sex of the parent from whom that allele is inherited.
   b. may involve the silencing of an allele by methylation such that offspring inherit only one active copy of a gene.
   c. occurs more often in females because of the larger maternal contribution of cytoplasm.
   d. is more likely to occur in the offspring of older mothers.
   e. involves both a and b.

14. Which of the following statements is *not* true about genetic recombination?
    a. Recombination of linked genes occurs by crossing over.
    b. Recombination of unlinked genes occurs by independent assortment of chromosomes.
    c. Genetic recombination results in offspring with combinations of traits that differ from the phenotypes of both parents.
    d. Recombinant offspring outnumber parental type offspring when two genes are 50 map units apart on a chromosome.
    e. The number of recombinant offspring is proportional to the distance between two gene loci on a chromosome.

15. Tests of social behavior given to Turner syndrome volunteers (who are X0) found a correlation between scores on "behavioral inhibition" tasks and the source of the lone X chromosome. These test results appear to be an example of
    a. genomic imprinting of a gene on the X chromosome.
    b. Y-linked inheritance.
    c. meiotic nondisjunction.
    d. chromosomal reciprocal translocation.
    e. nonreciprocal crossover.

16. Suppose that alleles for an X-linked character for wing shape in flies show incomplete dominance. The X^+ allele codes for pointed wings, the X^r for round wings, and X^+X^r individuals have oval wings. In a cross between an oval-winged female and a round-winged male, the following offspring were observed: oval-winged females, round-winged females, pointed-winged males, and round-winged males. A rare pointed-winged female was noted. Cytological study revealed that she had two X chromosomes. Which of the following events could account for this unusual offspring?
    a. a crossover between the two X chromosomes
    b. a crossover between the X and Y chromosomes
    c. a nondisjunction in meiosis II between two X^+ chromatids
    d. a nondisjunction between the X and Y chromosomes, producing some sperm with no sex chromosome
    e. Both c and d together could produce an X^+X^+ female when an XX egg was fertilized by a sperm in which there was no sex chromosome.

17. Some girls who fail to undergo puberty are found to have Swyer syndrome, a condition in which they are externally female but have an XY genotype. Which of the following statements may explain the origin of this syndrome?
    a. A mutation in the *XIST* gene, which codes for RNA molecules that coat the X chromosome and initiate X-inactivation, must have occurred.
    b. A nondisjunction in the egg from the mother resulted in both sex chromosomes coming from the father. Genomic imprinting of the father's X chromosome then caused the development of a female phenotype.
    c. These individuals are actually XXY; the second X is not seen because it is condensed into a Barr body. They have small testes and are sterile but otherwise appear female.
    d. A mutation or deletion of the *SRY* gene on the Y chromosome prevented development of testes and production of the male sex hormones required for a male phenotype.
    e. A translocation of part of an X chromosome to the Y chromosome resulted in a double dose of female-determining genes.

18. A mutation in a mitochondrial gene has been linked to a rare muscle-wasting disease in humans. This disease is
    a. found more often in males than in females.
    b. found more often in females than in males.
    c. inherited in a simple Mendelian fashion.
    d. caused by a translocation of a nuclear gene to the mitochondria.
    e. inherited from the mother and is equally likely to occur in male offspring as in female offspring.

19. The genetic event that results in Turner syndrome (X0) is probably
    a. nondisjunction.
    b. deletion.
    c. genomic imprinting.
    d. monoploidy.
    e. independent assortment.

20. A cross between a wild-type mouse and a dwarf mouse homozygous for a recessive mutation in the *Igf2* gene produces heterozygous offspring that are normal size if the dwarf parent was the mother, but dwarf if the dwarf parent was the father. Which of the following genetic events explains these results?
    a. sex-linked inheritance
    b. monosomy
    c. genomic imprinting
    d. inheritance of mitochondrial genes
    e. a mutant *XIST* allele

Chapter 16

The Molecular Basis of Inheritance

Key Concepts

16.1 **DNA is the genetic material**

16.2 **Many proteins work together in DNA replication and repair**

16.3 **A chromosome consists of a DNA molecule packed together with proteins**

Framework

This chapter outlines the key evidence that was gathered to establish DNA as the molecule of inheritance. Watson and Crick's double helix, with its rungs of specifically paired nitrogenous bases and twisting phosphate-sugar side ropes, provided the three-dimensional model that explained DNA's ability to encode a great variety of information and produce exact copies of itself through semiconservative replication. The replication of DNA is an extremely fast and accurate process involving many enzymes and proteins. Eukaryotic chromosomes are complexes of DNA and protein that exhibit varying levels of packing during the cell cycle.

Chapter Review

Deoxyribonucleic acid, or DNA, is the genetic material that is transmitted from one generation to the next and encodes the blueprints that direct and control the biochemical, anatomical, physiological, and behavioral traits of organisms. DNA is precisely copied in the process of **DNA replication.**

16.1 DNA is the genetic material

The Search for the Genetic Material: Scientific Inquiry Chromosomes were shown to carry hereditary information and to consist of proteins and DNA.

Until the 1940s, most scientists believed that proteins were the genetic material. The role of DNA was first established through work with microorganisms—bacteria and viruses.

In 1928, F. Griffith was working with two strains of *Streptococcus pneumoniae.* When he mixed the remains of heat-killed pathogenic bacteria with harmless bacteria, some bacteria were changed into disease-causing bacteria. These bacteria incorporated external genetic material in a process called **transformation,** which results in a change in genotype and phenotype.

O. Avery worked to identify the transforming agent by testing the DNA, RNA, and proteins from heat-killed pathogenic cells. In 1944, Avery, McCarty, and MacLeod announced that DNA was the molecule that transformed bacteria.

Viruses consist of little more than DNA (or sometimes RNA) contained in a protein coat. They reproduce by infecting a cell and commandeering that cell's metabolic machinery. **Bacteriophages,** or **phages,** are viruses that infect bacteria. In 1952, A. Hershey and M. Chase showed that DNA was the genetic material of a phage known as T2 that infects the bacterium *E. coli.*

In 1950, E. Chargaff noted that the percentages of nitrogenous bases in DNA were species specific. Chargaff also determined that the number of adenines and thymines was approximately equal, and the number of guanines and cytosines was also equal. The variation in base composition among species and the A=T and G=C properties of DNA became known as *Chargaff's rules.*

INTERACTIVE QUESTION 16.1

Hershey and Chase devised an experiment using radioactive isotopes to determine whether it was a phage's DNA or protein that entered the bacteria and served as the genetic material of T2 phage.

a. How did they label phage protein?

b. How did they label phage DNA?

Separate samples of *E. coli* were infected with the differently labeled T2 cells, then blended and centrifuged to isolate the bacterial cells from the lighter viral particles.

c. Where was the radioactivity found in the samples with labeled phage protein?

d. Where was the radioactivity found in the samples with labeled phage DNA?

e. What did Hershey and Chase conclude from these results?

Building a Structural Model of DNA: Scientific Inquiry   By the early 1950s, the arrangement of covalent bonds in a nucleic acid polymer was established, but the three-dimensional structure of DNA was yet to be determined.

In X-ray crystallography, an X-ray beam passed through a substance produces an X-ray diffraction photo with a pattern of spots that a crystallographer interprets as information about three-dimensional molecular structure. J. Watson saw an X-ray photo produced by R. Franklin that indicated the helical shape of DNA. The width of the helix suggested that it consisted of two strands—thus the term **double helix.**

Watson and F. Crick constructed wire models of a double helix that had the paired nitrogenous bases on the inside of the helix and two sugar-phosphate chains running in opposite directions (**antiparallel**) on the outside. The helix makes one full turn every 3.4 nm; ten layers of nucleotide pairs, stacked 0.34 nm apart, are present in each turn of the helix.

To produce the molecule's uniform 2-nm width, a purine base must pair with a pyrimidine base. The side groups of the bases permit two hydrogen bonds to form between adenine and thymine, and three hydrogen bonds between guanine and cytosine. This complementary pairing explains Chargaff's rules. Van der Waals attractions between the closely stacked bases help hold the molecule together.

In 1953, Watson and Crick published a paper in *Nature* reporting the double helix as the molecular model for DNA.

INTERACTIVE QUESTION   16.2

Review the structure of DNA by labeling the following diagrams.

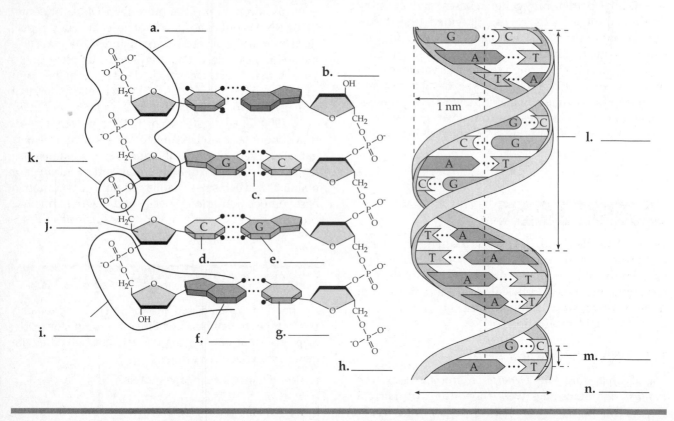

16.2 Many proteins work together in DNA replication and repair

The Basic Principle: Base Pairing to a Template Strand Watson and Crick noted that the base-pairing rule of DNA sets up a mechanism for its replication. Each side of the double helix is an exact complement to the other. When the two sides separate, each strand serves as a template for rebuilding a double-stranded molecule identical to the "parental" molecule.

The **semiconservative model** of DNA replication predicts that the two daughter DNA molecules each have one parental strand and one newly formed strand. In contrast, a conservative model predicts that the parental double helix reforms and the duplicated molecule is totally new, whereas a dispersive model predicts that all four strands of the two DNA molecules are a mixture of parental and new DNA.

M. Meselson and F. Stahl tested these models by growing *E. coli* in a medium with ^{15}N, a heavy isotope that the bacteria incorporated into their nitrogenous bases. Cells with labeled DNA were transferred to a medium with a lighter isotope, ^{14}N. Samples were removed after one and two generations of bacterial growth, and DNA was extracted and centrifuged. The locations of the density bands in the centrifuge tubes confirmed the semiconservative model of DNA replication.

INTERACTIVE QUESTION 16.3

Using different colors for heavy (parental) and light (new) strands of DNA, sketch the DNA molecules formed in two replication cycles after *E. coli* were moved from medium containing ^{15}N to medium containing ^{14}N. Show the resulting density bands in the centrifuge tubes.

DNA molecules   Density bands

Parental DNA
(grown on ^{15}N
medium)

First generation
(grown on light
^{14}N medium)

Second generation
(grown on light
^{14}N medium)

DNA Replication: A Closer Look Replication of most bacterial chromosomes begins at a single **origin of replication,** where proteins that initiate replication bind to a specific sequence of nucleotides and separate the two strands to form a replication "bubble." Replication proceeds in both directions in the two Y-shaped **replication forks.** Eukaryotic chromosomes have many origins of replication.

An enzyme called **helicase** unwinds the helix and separates the parental strands at replication forks. **Single-strand binding proteins** keep the separated strands apart while they serve as templates. **Topoisomerase** helps relieve the strain from the tighter twisting of DNA strands in front of helicase. An enzyme called **primase** joins about 5 to 10 RNA nucleotides base-paired to the parental strand to form the **primer** needed to start the new DNA strand.

Enzymes called **DNA polymerases** connect nucleotides to the growing end of a new DNA strand. DNA polymerase III and I are involved in replication in *E. coli*; at least 11 different DNA polymerases have been discovered so far in eukaryotes. A nucleoside triphosphate lines up with its complementary base on the template strand; it loses two phosphate groups, and the hydrolysis of this pyrophosphate to two inorganic phosphates ($\textcircled{P}_i$) provides the energy for polymerization.

The two strands of a DNA molecule are antiparallel. The deoxyribose sugar of each nucleotide is connected to its own phosphate group at its 5′ carbon and connects to the phosphate group of the adjacent nucleotide by its 3′ carbon. Thus, a strand of DNA has polarity, with a 5′ end where the first nucleotide's phosphate group is exposed, and a 3′ end where the nucleotide at the other end has a hydroxyl group attached to its 3′ carbon.

INTERACTIVE QUESTION 16.4

Look back to Interactive Question 16.2 and label the 5′ and 3′ ends of the left strand of the DNA molecule.

DNA polymerases add nucleotides only to the free 3′ end of a primer or growing DNA strand; DNA is replicated in a 5′ → 3′ direction. The **leading strand** is the new continuous strand being formed along one template strand as DNA polymerase III (DNA pol III) remains in the progressing replication fork.

The **lagging strand** is created as a series of short segments, called **Okazaki fragments,** which are elongated in the 5′ → 3′ direction away from the replication fork. Each fragment on the lagging strand requires a primer.

A continuous strand of DNA is produced after both DNA polmerase I (DNA pol I) replaces the RNA primer with DNA nucleotides and an enzyme called **DNA ligase** joins the sugar-phosphate backbones of the fragments.

The various proteins that function in replication form a large DNA replication complex. In eukaryotic cells, many such complexes may anchor to the nuclear matrix, and the DNA polymerase molecules may pull the parental DNA strands through them.

INTERACTIVE QUESTION   16.5

In this diagram of bacterial DNA replication, label the following items: leading and lagging strands, Okazaki fragment, DNA pol III, DNA pol I, DNA ligase, helicase, single-strand binding proteins, primase, RNA primer, and 5' and 3' ends of parental DNA.

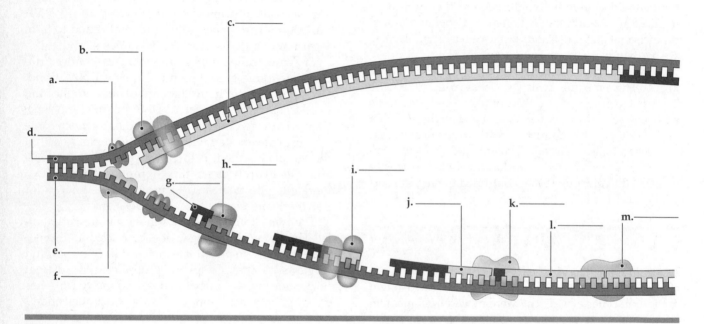

Proofreading and Repairing DNA   Initial pairing errors in nucleotide placement may occur as often as 1 per 100,000 base pairs. The amazing accuracy of DNA replication (one error in ten billion nucleotides) is achieved because DNA polymerases check each newly added nucleotide against its template and remove incorrect nucleotides. Other enzymes also fix incorrectly paired nucleotides, a process called **mismatch repair.**

A large number of different DNA repair enzymes monitor and repair damaged DNA. In **nucleotide excision repair,** the damaged strand is cut out by a **nuclease** and the gap is correctly filled through the action of a DNA polymerase and ligase. In skin cells, nucleotide excision repair frequently corrects *thymine dimers* caused by ultraviolet rays in sunlight.

Evolutionary Significance of Altered DNA Nucleotides   *Mutations* result when uncorrected mismatched nucleotides are replicated and the change is passed on to daughter cells (or to offspring, if the mutation is in a gamete). Although usually harmful, these permanent genetic changes provide the variation on which natural selection operates.

Replicating the Ends of DNA Molecules   Because DNA polymerase cannot attach nucleotides to the 5' end of a growing DNA strand, repeated replications cause a progressive shortening of linear DNA molecules. Multiple repetitions of a short nucleotide sequence at the ends of chromosomes, called **telomeres,** protect an organism's genes from being shortened during successive DNA replications.

The shortening of telomeres may limit cell division. In eukaryotic germ cells, however, the enzyme **telomerase** lengthens telomeres. Some somatic cancer cells and "immortal" strains of cultured cells produce telomerase and are thus capable of unregulated cell division.

INTERACTIVE QUESTION 16.6

Draw the last Okazaki fragment being formed on the lagging strand of a linear DNA molecule. Indicate how this results in a shortening of the end of the DNA molecule.

INTERACTIVE QUESTION 16.7

List the multiple levels of packing in a metaphase chromosome in order of increasing complexity.

16.3   A chromosome consists of a DNA molecule packed together with proteins

A bacterial cell's double-stranded, circular DNA molecule (associated with a small amount of protein) is supercoiled and tightly packed into a region of the cell called the **nucleoid.**

In eukaryotes, each chromosome consists of a single, extremely long DNA double helix precisely associated with a large amount of protein, forming a complex called **chromatin.** This DNA fits into the nucleus through a multilevel system of coiling and folding.

Histones are small, positively charged proteins that bind tightly to the negatively charged DNA. Unfolded chromatin appears as a string of beads, each bead a **nucleosome** consisting of the DNA helix wound around a protein core of four pairs of different histones. The "string" between beads is called *linker DNA.* The amino end (*histone tail*) of each of the eight histones extends outward. Also called the *10-nm fiber* because of its diameter, nucleosomes are the basic unit of DNA packing. A fifth histone, H1, attaches to the DNA near the bead.

The *30-nm fiber* is a tightly coiled fiber of nucleosomes organized with the aid of histone H1, histone tails, and linker DNA. The *looped domain* is a loop of the 30-nm chromatin fiber attached to a protein scaffold, forming a *300-nm fiber.* In a metaphase chromosome, looped domains coil and fold, further compacting the chromatin into chromatids that are 700 nm wide.

Some DNA packing is evident in interphase chromosomes, and their looped domains appear to be attached to specific locations of the nuclear lamina inside the nuclear envelope. A technique for "painting" individual chromosomes different colors has shown that each chromosome occupies a discrete area in the interphase nucleus.

Certain regions of chromatin, called **heterochromatin,** are in a highly condensed state during interphase. The more open form of interphase chromatin, called **euchromatin,** is available for the transcription of genes.

Word Roots

helic- = a spiral (*helicase:* an enzyme that untwists the double helix of DNA at the replication forks and separates the two strands)

liga- = bound or tied (*DNA ligase:* a linking enzyme essential for DNA replication that catalyzes the bonding of the 3′ end of one DNA fragment to the 5′ end of another DNA fragment)

-phage = to eat (*bacteriophage:* a virus that infects bacteria)

telos- = an end (*telomere:* the repetitive DNA at the end of a eukaryotic chromosome that protects the organism's genes from being eroded during successive rounds of replication)

trans- = across (*transformation:* a change in genotype and phenotype due to the assimilation of external DNA by a cell)

Structure Your Knowledge

1. Summarize the evidence and techniques Watson and Crick used to deduce the double-helix structure of DNA.

2. Review your understanding of DNA replication by describing the key enzymes and proteins (in the order of their functioning) that direct replication.

Test Your Knowledge

MULTIPLE CHOICE: *Choose the one best answer.*

1. One of the reasons most scientists believed proteins were the carriers of genetic information was that
   a. proteins are more heat stable than nucleic acids.
   b. the protein content of duplicating cells always doubles prior to division.
   c. proteins are much more complex and heterogeneous than nucleic acids.
   d. early experimental evidence pointed to proteins as the hereditary material.
   e. proteins are found in DNA.

2. Transformation involves
   a. the uptake of external genetic material, often from one bacterial strain to another.
   b. the creation of a strand of RNA from a DNA molecule.
   c. the infection of bacterial cells by phage.
   d. the type of semiconservative replication shown by DNA.
   e. the replication of DNA along the lagging strand.

3. The DNA of an organism has thymine as 20% of its bases. What percentage of its bases would be guanine?
   a. 20%   c. 40%   e. 80%
   b. 30%   d. 60%

4. In his work with pneumonia-causing bacteria, Griffith found that
   a. DNA was the transforming agent.
   b. the pathogenic and harmless strains mated.
   c. heat-killed harmless cells could cause pneumonia when mixed with heat-killed pathogenic cells.
   d. a substance was transferred to harmless cells to transform them into pathogenic cells.
   e. a T2 phage transformed harmless cells to pathogenic cells.

5. T2 phage is grown in *E. coli* with radioactive phosphorus and then allowed to infect other *E. coli*. The culture is blended to separate the viral coats from the bacterial cells and then centrifuged. Which of the following statements best describes the expected results of such an experiment?
   a. Both viral and bacterial DNA molecules are labeled; radioactivity is found in the liquid above the pellet.
   b. Viral DNA is labeled; radioactivity is found in the pellet.
   c. Viral proteins are labeled; radioactivity is found in the liquid but not in the pellet.
   d. Both viral and bacterial proteins are labeled; radioactivity is present in both the liquid and the pellet.
   e. The virus destroyed the bacteria; no pellet is formed.

6. Watson and Crick concluded that each base could not pair with itself because
   a. there would not be room for the helix to make a full turn every 3.4 nm.
   b. the uniform width of 2 nm would not permit two purines or two pyrimidines to pair together.
   c. the bases could not be stacked 0.34 nm apart.
   d. identical bases could not hydrogen-bond together.
   e. they would be on antiparallel strands.

7. In their classic experiment, Meselson and Stahl
   a. provided evidence for the semiconservative model of DNA replication.

   b. were able to separate phage protein coats from *E. coli* by using a blender.
   c. found that DNA labeled with ^{15}N was of intermediate density.
   d. grew *E. coli* on labeled phosphorus and sulfur.
   e. found that DNA composition was species specific.

8. The joining of nucleotides in the polymerization of DNA requires energy from
   a. DNA polymerase.
   b. the hydrolysis of the terminal phosphate group of ATP.
   c. RNA nucleotides.
   d. the phosphate groups of the sugar-phosphate backbone.
   e. the hydrolysis of the pyrophosphates removed from nucleoside triphosphates.

9. The continuous elongation of a new DNA strand along one of the template strands of DNA
   a. requires the action of DNA ligase as well as polymerase.
   b. occurs because DNA ligase can only elongate in the $5' \rightarrow 3'$ direction.
   c. occurs on the leading strand.
   d. occurs on the lagging strand.
   e. a, b, and c are correct.

10. Which of the following statements about DNA polymerase is *incorrect*?
    a. It forms the bonds between complementary base pairs.
    b. It is able to proofread and correct errors in base pairing.
    c. It is unable to initiate synthesis; it requires an RNA primer.
    d. It only works in the $5' \rightarrow 3'$ direction.
    e. It is found in eukaryotes and prokaryotes.

11. Thymine dimers—covalent links between adjacent thymine bases in DNA—may be induced by UV light. When these dimers occur, they are repaired by
    a. excision enzymes (nucleases).
    b. DNA polymerase.
    c. ligase.
    d. primase.
    e. a, b, and c are all needed.

12. How does DNA synthesis along the lagging strand differ from that on the leading strand?
    a. Nucleotides are added to the 5′ end instead of the 3′ end.
    b. Ligase is the enzyme that polymerizes DNA on the lagging strand.
    c. An RNA primer is needed on the lagging strand but not on the leading strand.
    d. Okazaki fragments, which each grow $5' \rightarrow 3'$, must be joined along the lagging strand.
    e. Helicase synthesizes Okazaki fragments, which are then joined by ligase.

13. Which of the following enzymes or proteins is paired with an incorrect or inaccurate function?
    a. Helicase—unwinds and separates parental double helix
    b. Telomerase—adds telomere repetitions to ends of chromosomes
    c. Single-strand binding protein—holds strands of unwound DNA apart and straight
    d. Nuclease—cuts out (excises) damaged DNA strands
    e. Primase—forms DNA primer to start replication

Use the following diagram to answer Questions 14 through 17.

a. O⁻ | O=P—O—CH₂ | O⁻ — A ... (diagram)

14. Which letter indicates the 5′ end of this single DNA strand?
    a.   b.   c.   d.   e.

15. At which letter would the next nucleotide be added?
    a.   b.   c.   d.   e.

16. Which letter indicates a phosphodiester bond formed by DNA polymerase?
    a.   b.   c.   d.   e.

17. The base sequence of the DNA strand made from this template would be (from top to bottom)
    a. A T C.   c. T A C.   e. A T G.
    b. C G A.   d. U A C.

18. Which of the following statements about telomeres is correct?
    a. They are ever-shortening tips of chromosomes that may signal cells to stop dividing at maturity.
    b. They are highly repetitive sequences at the tips of chromosomes that protect the lagging strand during replication.
    c. They are repetitive sequences of nucleotides at the centromere region of a chromosome.
    d. They are enzymes, in germ cells that allow these cells to undergo repeated divisions.
    e. Both a and b are correct.

19. You are trying to test your hypothesis that DNA replication is *conservative*—that is, that the parental strands separate, newly made complementary strands join together to make a new DNA molecule, and the parental strands then rejoin. You take a sample of *E. coli* grown in a medium containing only heavy nitrogen (^{15}N) and transfer it to a medium containing light nitrogen (^{14}N). After allowing time for only one DNA replication, you centrifuge a sample and compare the density band(s) formed to the bands formed from bacteria grown on either normal ^{14}N or ^{15}N medium. Which band location would support your hypothesis of *conservative* DNA replication?

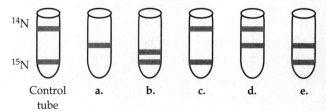

Bands after one replication on ^{14}N

20. Given the experimental procedure explained in question 19, which centrifuge tube (obtained after one DNA replication) would represent the band distribution indicating that DNA replication is *semiconservative*?

21. If the following structures were put in order from smallest to largest, which structure would be in the middle of that size range?
    a. looped domain
    b. histone
    c. nucleosome
    d. 30-nm fiber
    e. metaphase chromosome

22. Biologists have learned from the technique of "painting chromosomes" with different colored molecular tags that
    a. DNA replication proceeds at different rates on different chromosomes.
    b. the two homologs of a pair of chromosomes differ enough that they stain different colors.
    c. chromosome packing occurs only as the cell prepares for mitosis or meiosis.
    d. heterochromatin is concentrated at the tips and centromeres of chromosomes.
    e. in the interphase nucleus, each chromosome appears to occupy a specific area.

Chapter 17

From Gene to Protein

Key Concepts

17.1 Genes specify proteins via transcription and translation

17.2 Transcription is the DNA-directed synthesis of RNA: *a closer look*

17.3 Eukaryotic cells modify RNA after transcription

17.4 Translation is the RNA-directed synthesis of a polypeptide: *a closer look*

17.5 Mutations of one or a few nucleotides can affect protein structure and function

17.6 While gene expression differs among the domains of life, the concept of a gene is universal

Framework

This chapter examines the pathway from DNA to RNA to proteins. The instructions of DNA are transcribed to a sequence of codons in mRNA. In eukaryotes, mRNA is processed before it leaves the nucleus. Complexed with ribosomes, mRNA is translated as a sequence of amino acids in a polypeptide as tRNAs match their anticodons to the mRNA codons. Mutations, which alter the nucleotide pairs in DNA, may alter the protein product. A gene may be defined as a DNA sequence whose final product is either a polypeptide or an RNA molecule.

Chapter Review

The DNA-directed synthesis of proteins (or sometimes just RNA) is called **gene expression.**

17.1 Genes specify proteins via transcription and translation

Evidence from the Study of Metabolic Defects In 1909, A. Garrod first suggested that genes determine phenotype through the action of enzymes, reasoning that inherited diseases were caused by an inability to make certain enzymes.

G. Beadle and E. Tatum worked with mutants of a bread mold, *Neurospora crassa.* By transferring samples of nutritional mutants growing on *complete growth medium* to various combinations of *minimal medium* and one added nutrient, they were able to identify the specific metabolic defect for each mutant. By supplementing different precursors of the arginine pathway, A. Srb and N. Horowitz identified three classes of *Neurospora* mutants. They reasoned that the arginine pathway was blocked at a different enzymatic step in each class of mutants. Such research supported Beadle and Tatum's *one gene–one enzyme hypothesis.*

Biologists revised this hypothesis to one gene–one protein. Because many proteins consist of more than one polypeptide chain, the axiom changed to *one gene–one polypeptide.* (Exceptions include genes that code for more than one protein due to alternative splicing and genes that code for RNA molecules.)

Basic Principles of Transcription and Translation RNA is the link between a gene and the protein for which it codes. RNA differs from DNA in three ways: The sugar component of its nucleotides is ribose, rather than deoxyribose; uracil (U) replaces thymine as one of its nitrogenous bases; and RNA is usually single stranded.

Transcription is the transfer of information from DNA to **messenger RNA (mRNA)** or another type of RNA. **Translation** transfers information from mRNA to a polypeptide, changing from the language of nucleotides to that of amino acids. **Ribosomes** are the sites of translation—the sites of the synthesis of a polypeptide.

In prokaryotes, which lack a nucleus, transcription and translation can occur simultaneously. In eukaryotes, the mRNA, called *pre-mRNA,* is processed before it exits the nucleus and enters the cytoplasm, where translation occurs. The initial RNA transcript of any gene is called the **primary transcript.**

INTERACTIVE QUESTION 17.1

Fill in the following sequence in the flow of genetic information, often called the *central dogma.* Above each arrow, write the name of the process involved.

_____ → _____ → _____

The Genetic Code   A sequence of three nucleotides provides 4^3 (64) possible unique sequences of nucleotides, more than enough to code for the 20 amino acids. The translation of nucleotides into amino acids uses a **triplet code** to specify each amino acid.

The nucleotide base triplets along the **template strand** of a gene are transcribed into mRNA **codons.** The same strand of a DNA molecule can be the template strand for one gene and the nontemplate strand for another. The mRNA is complementary to the DNA template because its nucleotides follow the same base-pairing rules, with the exception that uracil substitutes for thymine in RNA. The term *codon* can also refer to the DNA nucleotide triplets on the *nontemplate* strand.

During translation, the sequence of codons, read in the $5' \rightarrow 3'$ direction, determines the sequence of amino acids in the polypeptide.

In the early 1960s, M. Nirenberg added artificial "poly U" mRNA to a test tube containing all the biochemical ingredients necessary for protein synthesis and obtained a polypeptide containing a single amino acid. By the mid-1960s, molecular biologists had deciphered all 64 codons. Three codons function as stop signals, or termination codons. The codon AUG both codes for methionine and functions as an initiation codon, a start signal for translation.

The code is often redundant, meaning that more than one codon may specify a single amino acid. The code is never ambiguous: No codon specifies two different amino acids.

The nucleotide sequence on mRNA is read in the correct **reading frame,** starting at a start codon and reading each triplet sequentially.

INTERACTIVE QUESTION 17.2

Practice using the codon table in your textbook. Determine the amino acid sequence for a polypeptide coded for by the following mRNA transcript (written $5' \rightarrow 3'$):

AUGCCUGACUUUAAGUAG

The genetic code of codons and their corresponding amino acids is almost universal. A bacterial cell can translate the genetic messages of human cells. The near universality of a common genetic language provides compelling evidence of the antiquity of the code and the evolutionary connection of all living organisms.

17.2   Transcription is the DNA-directed synthesis of RNA: *a closer look*

Molecular Components of Transcription   The **promoter** is the DNA sequence where **RNA polymerase** attaches and initiates transcription. RNA polymerase joins RNA nucleotides that are complementary to the DNA template strand in a $5' \rightarrow 3'$ direction. In bacteria, the **terminator** is the sequence that signals the end of transcription. A **transcription unit** is the sequence of DNA that is transcribed into one RNA molecule.

Bacteria have one type of RNA polymerase. Eukaryotes have three types; the one that synthesizes mRNA is called RNA polymerase II.

Synthesis of an RNA Transcript   The specific binding of RNA polymerase to the promoter determines where transcription starts and which DNA strand is used as the template. The promoter includes the **start point** and recognition sequences, such as the **TATA box** common in eukaryotes, upstream from the start point. In eukaryotes, **transcription factors** must first recognize and bind to the promoter before RNA polymerase II can attach, at which point the assembly is called the **transcription initiation complex.**

RNA polymerase untwists the double helix, exposing DNA nucleotides for base pairing with RNA nucleotides, and joins the nucleotides to the 3' end of the growing polymer. The new RNA peels away from the DNA template, and the DNA double helix re-forms. Several molecules of RNA polymerase may be transcribing simultaneously along a gene, enabling a cell to produce large quantities of mRNA.

In prokaryotes, transcription ends after RNA polymerase transcribes the terminator sequence. In eukaryotes, polymerase continues past a polyadenylation signal sequence (coding for AAUAAA), and proteins cut loose the pre-mRNA.

INTERACTIVE QUESTION　17.3

Describe the key steps of transcription in eukaryotes:

a.

b.

c.

17.3　Eukaryotic cells modify RNA after transcription

In eukaryotes, pre-mRNA is modified by **RNA processing** before it leaves the nucleus.

Alteration of mRNA Ends　A modified guanine nucleotide is attached to the 5' end of a pre-mRNA, and a string of adenine nucleotides, called a **poly-A tail,** is added to the 3' end. The **5' cap** and poly-A tail may facilitate transport of mRNA from the nucleus, aid ribosome attachment, and protect the ends of mRNA from hydrolytic enzymes. The cap and tail are attached to the untranslated regions (UTRs) at the 5' and 3' ends.

Split Genes and RNA Splicing　Long segments of noncoding nucleotide sequences, known as **introns** or intervening sequences, occur within the boundaries of eukaryotic genes. The remaining coding regions are called **exons,** since they are expressed in protein synthesis (with the exception of the 5' and 3' UTRs, which are not translated). A primary transcript is made of the gene—but introns are removed and exons joined together before the mRNA leaves the nucleus—in a process called **RNA splicing.**

The signals for RNA splicing are sets of a few nucleotides at either end of each intron. *Small nuclear ribonucleoproteins (snRNPs),* composed of proteins and *small nuclear RNA (snRNA),* are components of a large molecular complex called a **spliceosome.** The spliceosome snips an intron out of the RNA transcript and connects the adjoining exons. In addition to splice-site recognition and spliceosome assembly, snRNA may catalyze RNA splicing.

RNA molecules that act as enzymes are called **ribozymes.** In some cases of RNA splicing, intron RNA catalyzes its own removal. The ability to function as enzymes relates to three properties of RNA: It is single-stranded and can base-pair with itself, forming a specific three-dimensional structure; some of its bases contain functional groups that can participate in catalysis; and it

can hydrogen-bond with other nucleic acid molecules, allowing it to precisely locate splicing regions.

Some introns are involved in regulating gene activity, and splicing is necessary for the export of mRNA from the nucleus. **Alternative RNA splicing** allows different polypeptides to be produced from a single gene. Exons may code for polypeptide **domains,** functional segments of a protein, such as binding and active sites. Introns may facilitate recombination of exons between different alleles or even between different genes. Such *exon shuffling* can result in novel proteins.

INTERACTIVE QUESTION　17.4

How does the mRNA that leaves the nucleus differ from pre-mRNA?

17.4　Translation is the RNA-directed synthesis of a polypeptide: *a closer look*

Molecular Components of Translation　**Transfer RNA (tRNA)** molecules carry specific amino acids to ribosomes, where they are added to a growing polypeptide. Each tRNA carries a specific amino acid and has a nucleotide triplet, called an **anticodon,** that base-pairs with a complementary codon on mRNA, thus assuring that amino acids are arranged in the sequence prescribed by the transcription from DNA.

Transfer RNA is transcribed in the nucleus of a eukaryote and moves into the cytoplasm, where it can be used repeatedly. These single-stranded, short RNA molecules fold into a three-dimensional, roughly L-shaped structure due to hydrogen bonding between complementary nucleotide base sequences. The anticodon is at one end of the L; the 3' end is the attachment site for its specific amino acid.

Each amino acid has a specific **aminoacyl-tRNA synthetase** that attaches it to its appropriate tRNA molecules to create an aminoacyl tRNA, or charged tRNA. The hydrolysis of ATP drives this process.

Sixty-one codons for amino acids can be read from mRNA, but there are only about 45 different tRNA molecules. A phenomenon known as **wobble** enables the third nucleotide of some tRNA anticodons to pair with more than one kind of nucleotide in the codon. Thus, one tRNA can recognize more than one mRNA codon, all of which code for the same amino acid carried by that tRNA.

INTERACTIVE QUESTION 17.5

Using some of the codons and the amino acids you identified in Interactive Question 17.2, fill in the following table.

DNA Triplet $3' \rightarrow 5'$	mRNA Codon $5' \rightarrow 3'$	Anticodon $3' \rightarrow 5'$	Amino Acid
$3' \rightarrow 5'$	$5' \rightarrow 3'$	$3' \rightarrow 5'$	methionine
		GCA	
TTC			
	UAG		

Ribosomes facilitate the specific pairing of tRNA anticodons with mRNA codons during protein synthesis. Ribosomes consist of a large and a small subunit, each composed of proteins and a form of RNA called **ribosomal RNA (rRNA).** Subunits are constructed in the nucleolus in eukaryotes. Prokaryotic ribosomes are smaller and differ enough in molecular composition that some antibiotics can inhibit them without affecting eukaryotic ribosomes.

A large and a small subunit join to form a ribosome when they attach to an mRNA molecule. Ribosomes have a binding site for mRNA and three tRNA binding sites: a **P site** (peptidyl tRNA-binding site) that holds the tRNA carrying the growing polypeptide chain, an **A site** (aminoacyl tRNA-binding site) that holds the tRNA carrying the next amino acid, and an **E site** (exit site) from which discharged tRNAs leave the ribosome. A ribosome also has an *exit tunnel* through which the growing polypeptide passes out of the ribosome.

Building a Polypeptide   Each of the three stages of protein synthesis—initiation, elongation, and termination—requires the aid of protein "factors." The first two stages also require energy provided by the hydrolysis of GTP (guanosine triphosphate).

The initiation stage begins as the small subunit of the ribosome binds to an mRNA and an initiator tRNA carrying methionine, which attaches to the start codon AUG on the mRNA. With the aid of proteins called *initiation factors* and the hydrolysis of GTP, the large subunit of the ribosome attaches to the small one, forming a *translation initiation complex.* The initiator tRNA fits into the P site.

The addition of amino acids in the elongation stage involves several proteins called *elongation factors* and occurs in a three-step cycle.

In codon recognition, an aminoacyl-tRNA base-pairs with the mRNA codon in the A binding site. This step requires energy from the hydrolysis of GTP.

In the peptide bond formation step, an RNA molecule of the large subunit catalyzes the formation of a peptide bond between the carboxyl end of the polypeptide held in the P site and the amino group of the new amino acid in the A site. The polypeptide is now held by the tRNA in the A site.

In translocation, the tRNA carrying the growing polypeptide is translocated to the P site, a process requiring energy from the hydrolysis of GTP. The empty tRNA from the P site is moved to the E site and released. The next mRNA codon moves into the A site as the mRNA moves through the ribosome.

Termination occurs when a stop codon—UAA, UAG, or UGA—reaches the A site of the ribosome. A *release factor* binds to the stop codon and hydrolyzes the bond between the polypeptide and the tRNA in the P site. The completed polypeptide leaves through the exit tunnel of the large subunit. With the hydrolysis of GTP and the aid of other protein factors, the two ribosomal subunits and other components dissociate. Review translation by completing Interactive Question 17.6.

An mRNA may be translated simultaneously by several ribosomes in strings called **polyribosomes** (or **polysomes**).

INTERACTIVE QUESTION 17.6

In the following diagrams of polypeptide synthesis, name the stages (1–4), identify the components (a–l), and then briefly describe what happens in each stage. (This diagram does not include the initiation stage.)

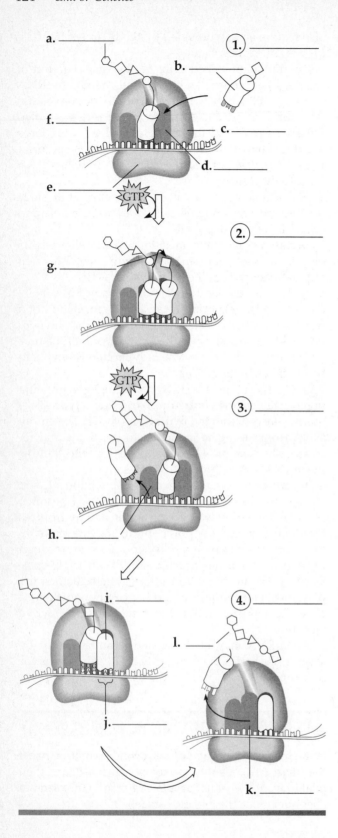

a. _____
b. _____
①. _____
f. _____
c. _____
d. _____
e. _____
GTP
②. _____
g. _____
GTP
③. _____
h. _____
i. _____
④. _____
l. _____
j. _____
k. _____

Completing and Targeting the Functional Protein
During and following translation, a polypeptide folds spontaneously into its secondary and tertiary structures. Chaperone proteins often facilitate the correct folding.

The protein may need to undergo *post-translational modifications:* Amino acids may be chemically modified; one or more amino acids at the beginning of the chain may be removed; segments of the polypeptide may be excised; or several polypeptides may associate into a protein with a quaternary structure.

All ribosomes are identical, whether they are free ribosomes that synthesize cytosolic proteins or ER-bound ribosomes that make membrane and secretory proteins. Polypeptide synthesis begins in the cytoplasm. If a protein is destined for the endomembrane system or for secretion, its polypeptide chain will begin with a **signal peptide.** This short sequence of amino acids is recognized by a protein-RNA complex called a **signal-recognition particle (SRP),** which attaches the ribosome to a receptor protein that is part of a translocation complex on the ER membrane. As the growing polypeptide threads into the ER, the signal peptide is usually removed.

Other signal peptides direct some proteins made in the cytosol to specific sites such as mitochondria, chloroplasts, or the interior of the nucleus.

INTERACTIVE QUESTION   17.7

What determines if a ribosome becomes bound to the ER?

17.5   Mutations of one or a few nucleotides can affect protein structure and function

Mutations, changes in the genetic information of a cell (or virus), may be either large scale (involving long segments of a chromosome) or small scale (such as **point mutations,** which affect just one nucleotide pair). If the mutation occurs in a cell that gives rise to a gamete, it may be passed on to offspring.

Types of Small-Scale Mutations   A **nucleotide-pair substitution** replaces one nucleotide and its complementary partner with another pair of nucleotides. Due to the redundancy of the genetic code, some substitutions do not change the amino acid translation and are called **silent mutations.** A **missense mutation,** which results in the insertion of a different amino acid, may not alter the character of the protein if the new amino acid has similar properties or is not located in a region crucial to that protein's function.

A nucleotide-pair substitution that results in a different amino acid in a critical portion of a protein, such as the active site of an enzyme, may significantly impair protein function. Occasionally such a change proves beneficial.

Nonsense mutations occur when a point mutation changes an amino acid codon into a stop codon, prematurely halting the translation of the polypeptide chain and usually creating a nonfunctional protein.

Nucleotide-pair **insertions** or **deletions** that are not in multiples of three alter the reading frame. All nucleotides downstream from the mutation will be improperly grouped into codons, creating extensive missense mutations and usually ending in nonsense. These **frameshift mutations** almost always produce nonfunctional proteins.

Mutagens *Spontaneous mutations* include nucleotide-pair substitutions, insertions, deletions, and longer mutations that occur during DNA replication, repair, or recombination. Physical agents (such as X-rays and UV light) and a variety of chemical agents can cause mutations and are called **mutagens.** Chemical mutagens include nucleotide analogs that substitute for normal nucleotides and then pair incorrectly in DNA synthesis, chemicals that insert into and distort the double helix, and other agents that chemically alter nucleotides and change their pairing. Tests can measure the mutagenic effects of chemicals and thus their potential carcinogenic risk.

INTERACTIVE QUESTION 17.8

Define the following terms, and explain what type of small-scale mutation could cause each of these types of mutations.

a. silent mutation

b. missense mutation

c. nonsense mutation

d. frameshift mutation

17.6 While gene expression differs among the domains of life, the concept of a gene is universal

Comparing Gene Expression in Bacteria, Archaea, and Eukarya The RNA polymerase of archaea resembles that of eukaryotes; both depend on transcription factors. Termination of transcription in archaea may resemble that of eukaryotes. Archaeal ribosomes are the same size as bacterial ribosomes, but their sensitivity to chemical inhibitors is more similar to that of eukaryotic ribosomes. The initiation of translation is most similar between archaea and bacteria.

In bacteria and probably archaea, transcription and translation occur almost simultaneously, whereas in eukaryotes, transcription is physically separated from translation by the nuclear envelope, allowing extensive RNA processing to occur before mRNA leaves the nucleus.

What Is a Gene? Revisiting the Question Our definition of a gene has evolved from Mendel's heritable factors, to Morgan's loci along chromosomes, to one gene–one polypeptide. Research continually refines our understanding of the structural and functional aspects of genes, which now include introns, promoters, and other regulatory regions. Currently, the best working definition of a gene is that it is a region of DNA whose final product is either a polypeptide or an RNA molecule.

Word Roots

anti- = opposite (*anticodon:* a nucleotide triplet on one end of a tRNA molecule that base-pairs with a particular complementary codon on an mRNA molecule)

exo- = out, outside, without (*exon:* a sequence within a primary transcript that remains in the RNA after RNA processing; also the region of DNA from which this sequence was transcribed)

intro- = within (*intron:* a noncoding, intervening sequence within a primary transcript that is removed; also the region of DNA from which this sequence was transcribed)

muta- = change; **-gen** = producing (*mutagen:* a chemical or physical agent that interacts with DNA and can cause a mutation)

poly- = many (*polyribosome:* a group of ribosomes attached to, and translating, the same mRNA molecule)

trans- = across; **-script** = write (*transcription:* the synthesis of RNA using a DNA template)

Structure Your Knowledge

1. You have been introduced to several types of RNA in this chapter. List four of these types and their functions. What explains the functional versatility of RNA molecules?

2. Make sure you understand and can explain the processes of transcription and translation. To help you review these processes, fill in the following table describing various aspects of eukaryotic gene expression (either by yourself or in a study group).

	Transcription	Translation
Template		
Location		
Molecules involved		
Enzymes involved		
Control—start and stop		
Product		
Product processing		
Energy source		

3. What is the genetic code? Explain redundancy and the wobble phenomenon. What is the significance of the fact that the genetic code is nearly universal?

4. Prepare a concept map showing the types and consequences of small-scale mutations.

Test Your Knowledge

MULTIPLE CHOICE: *Choose the one best answer.*

1. In their study of *Neurospora*, Srb and Horowitz were able to identify three classes of mutants that needed arginine added to minimal media in order to grow. The production of arginine includes the following steps: precursor → ornithine → citrulline → arginine. What nutrient(s) have to be supplied to the mutants with a defective enzyme for the ornithine → citrulline step in order for them to grow?
   a. the precursor
   b. ornithine
   c. citrulline
   d. either ornithine or citrulline
   e. the precursor, ornithine, and citrulline

2. Transcription involves the transfer of information from
   a. DNA to RNA.
   b. RNA to DNA.
   c. mRNA to an amino acid sequence.
   d. DNA to an amino acid sequence.
   e. the nucleus to the cytoplasm.

3. If the 5' → 3' nucleotide sequence on the nontemplate DNA strand is CAT, what is the corresponding codon on mRNA?
   a. UAC   d. GTA
   b. CAU   e. CAT
   c. GUA

4. A bacterial gene 600 nucleotides long can code at most for a polypeptide of how many amino acids?
   a. 100   d. 600
   b. 200   e. 1,800
   c. 300

5. RNA polymerase
   a. is the protein responsible for the production of ribonucleotides.
   b. is the enzyme that creates hydrogen bonds between nucleotides on the DNA template strand and their complementary RNA nucleotides.
   c. is the enzyme that transcribes exons but does not transcribe introns.
   d. is a ribozyme composed of snRNPs.
   e. moves along the template strand of DNA, elongating an RNA molecule in a 5' → 3' direction.

6. How is the template strand for a particular gene determined?
   a. It is the DNA strand that runs from the 5' → 3' direction.
   b. It is the DNA strand that runs from the 3' → 5' direction.
   c. It is established by the promoter.
   d. It doesn't matter which strand is the template because they are complementary and will produce the same mRNA.
   e. It is signaled by a polyadenylation signal sequence.

7. Which enzyme synthesizes tRNA?
   a. DNA polymerase
   b. RNA polymerase
   c. reverse transcriptase
   d. aminoacyl-tRNA synthetase
   e. ribosomal RNA

8. Which of the following is *not* involved in the formation of a eukaryotic transcription initiation complex?
   a. TATA box
   b. transcription factors
   c. snRNA
   d. RNA polymerase II
   e. promoter

9. Which of the following is *true* of RNA processing?
   a. Exons are excised before the mRNA is translated.
   b. The RNA transcript that leaves the nucleus may be much longer than the original transcript.
   c. Assemblies of protein and snRNPs, called spliceosomes, may catalyze splicing.
   d. Large quantities of rRNA are assembled into ribosomes.
   e. Signal peptides are added to the 5′ end of the transcript.

10. All of the following are transcribed from DNA *except*
    a. exons.
    b. introns.
    c. tRNA.
    d. 3′ and 5′ UTRs.
    e. promoter.

11. What might introns have to do with the evolution of new proteins?
    a. The excised introns are transcribed and translated as new proteins by themselves.
    b. Introns are more likely to accumulate mutations than exons, and these mutations then result in the production of novel proteins.
    c. Introns that are self-excising may also function as hydrolytic enzymes for other processes.
    d. Introns provide more area where crossing over may occur (without interfering with the coding sequences) and thus increase the probability of exon shuffling between alleles.
    e. Introns often correspond to domains in proteins that fold independently and have specific functions. Switching domains between nonallelic genes could produce novel proteins.

12. A ribozyme is
    a. an exception to the one gene–one RNA molecule axiom.
    b. an enzyme that adds the 5′ cap and poly-A tail to mRNA.
    c. an example of rearrangement of protein domains caused by RNA splicing.
    d. an RNA molecule that functions as an enzyme.
    e. an enzyme that produces both small and large ribosomal subunits.

13. Which of the following would *not* be found in a bacterial cell?
    a. mRNA
    b. rRNA
    c. snRNA
    d. RNA polymerase
    e. simultaneous transcription and translation

14. Which of the following is transcribed and then translated to form a protein product?
    a. a gene for tRNA
    b. an intron
    c. a gene for a transcription factor
    d. 5′ and 3′ UTRs
    e. a gene for rRNA

15. Transfer RNA
    a. translocates a growing polypeptide destined for export to the endoplasmic reticulum.
    b. binds to its specific amino acid in the active site of an aminoacyl-tRNA synthetase.
    c. has catalytic activity and is thus a ribozyme.
    d. is translated from mRNA.
    e. is produced in the nucleolus.

16. Place the following events in the synthesis of a polypeptide in the proper order.
    1. A peptide bond forms.
    2. An aminoacyl tRNA matches its anticodon to the codon in the A site.
    3. A tRNA translocates from the A site to the P site, and an unattached tRNA exits from the E site.
    4. The large subunit attaches to the small subunit, with the initiator tRNA in the P site.
    5. A small subunit binds to an mRNA and an initiator tRNA.
       a. 4-5-3-2-1
       b. 4-5-2-1-3
       c. 5-4-3-2-1
       d. 5-4-1-2-3
       e. 5-4-2-1-3

17. Translocation in the process of translation involves
    a. the hydrolysis of GTP.
    b. movement of the tRNA in the A site to the P site.
    c. movement along the mRNA a distance of one triplet.
    d. the release of the unattached tRNA from the E site.
    e. all of the above.

18. Which of the following type of molecule catalyzes the formation of a peptide bond?
    a. RNA polymerase
    b. rRNA
    c. mRNA
    d. aminoacyl-tRNA synthetase
    e. proteinase

19. Which of the following is *not* true of an anticodon?

a. It consists of three nucleotides.

b. It lines up in the 5′ → 3′ direction along the 5′ → 3′ mRNA strand.

c. It extends from one loop of a tRNA molecule.

d. It may pair with more than one codon.

e. Its base uracil base-pairs with adenine.

20. Changes in a polypeptide following translation may involve

a. the addition of sugars or lipids to certain amino acids.

b. the enzymatic addition of amino acids at the beginning of the chain.

c. the removal of poly-A from the end of the chain.

d. the addition of a 5′ cap of a modified guanosine.

e. all of the above.

21. Several proteins may be produced at the same time from a single mRNA by

a. the action of several ribosomes in a string, called a polyribosome.

b. several RNA polymerase molecules working sequentially.

c. signal peptides that associate ribosomes with rough ER.

d. the action of several promoter regions.

e. the involvement of multiple spliceosomes.

22. A signal peptide

a. is most likely to be found on cytosolic proteins produced by bacterial cells.

b. directs an mRNA molecule into the lumen of the ER.

c. is a sign to bind the small ribosomal unit at the initiation codon.

d. would be the first 20 or so amino acids of a protein destined for a membrane location or for secretion from the cell.

e. is part of the UTR following the 5′ cap.

23. A nucleotide deletion early in the coding sequence of a gene would most likely result in

a. a nonsense mutation.

b. a frameshift mutation.

c. multiple missense mutations.

d. a nonfunctional protein.

e. all of the above.

24. The type of mutation responsible for sickle cell anemia is

a. a silent mutation.

b. a nucleotide-pair insertion.

c. a point mutation.

d. a nucleotide-pair substitution.

e. Both c and d describe the type of mutation.

25. Which of the following statements best characterizes gene expression in archaea?

a. Their molecular machinery and processes are exactly like those in bacteria.

b. Their molecular machinery and processes are exactly like those in eukaryotes.

c. Their processes of transcription and translation differ substantially from both bacteria and eukaryotes.

d. Their mechanisms of gene expression are most similar to eukaryotes, although they share some similarities with bacteria.

e. Their processes of transcription and translation have not yet been studied.

Use the following diagram of coupled transcription and translation in bacteria to answer questions 26–30.

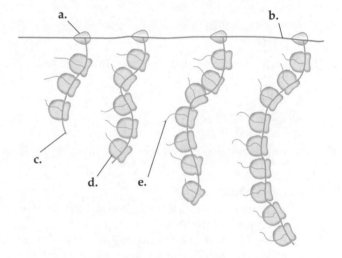

26. Which letter refers to an mRNA molecule?

27. Which letter refers to a forming polypeptide?

28. Which letter refers to RNA polymerase?

29. Which letter refers to a ribosome?

30. Which letters indicate structures or molecules containing nucleotides?

a. a and b

b. a, b, and d

c. b, c, and d

d. b, d, and e

e. c and d

Chapter 18

Regulation of Gene Expression

Key Concepts

18.1 Bacteria often respond to environmental change by regulating transcription

18.2 Eukaryotic gene expression is regulated at many stages

18.3 Noncoding RNAs play multiple roles in controlling gene expression

18.4 A program of differential gene expression leads to the different cell types in a multicellular organism

18.5 Cancer results from genetic changes that affect cell cycle control

Framework

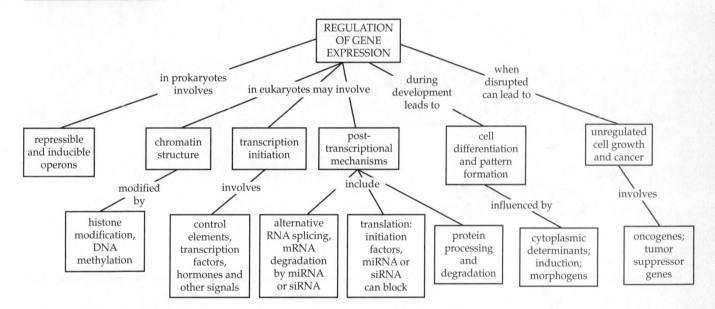

Chapter Review

18.1 Bacteria often respond to environmental change by regulating transcription

Bacteria can respond to short-term environmental fluctuations by regulating enzyme activity through *feedback inhibition*. Regulation of gene transcription controls enzyme production as metabolic needs change. F. Jacob and J. Monod described the *operon model* for gene regulation in 1961.

Operons: The Basic Concept   In bacteria, genes for the different enzymes of a single metabolic pathway may be grouped together into one transcription unit and served by a single promoter. Thus, these genes are *coordinately controlled*. An **operator** is a segment of DNA within or near the promoter that controls the access of RNA polymerase to the genes. An **operon** is the DNA segment that includes the clustered genes, the promoter, and the operator.

A **repressor** is a protein that binds to a specific operator, blocking attachment of RNA polymerase and thus turning an operon "off." **Regulatory genes** code for repressor proteins. These allosteric proteins, which may assume active or inactive shapes, are usually produced at a slow but continuous rate. The activity of the repressor protein may be determined by the presence or absence of a **corepressor.**

In the *trp* operon, tryptophan is the corepressor that binds to the *trp* repressor, changing it into its active shape, which has a high affinity for the operator and switches off the *trp* operon. Should the tryptophan concentration of the cell fall, repressor proteins are no longer bound with tryptophan; the operator is no longer repressed; RNA polymerase attaches to the promoter; and mRNA for the enzymes needed for tryptophan synthesis is produced.

Repressible and Inducible Operons: Two Types of Negative Gene Regulation The transcription of a *repressible operon,* such as the *trp* operon, is inhibited when a specific small molecule binds to and activates a repressor. The transcription of an *inducible operon* is stimulated when a specific small molecule binds to and inactivates a repressor.

The *lac* operon, controlling lactose metabolism in *E. coli,* is an inducible operon that contains three genes. The *lac* repressor protein, coded for by the regulatory gene *lacI,* is innately active, binding to the *lac* operator and switching off the operon. Allolactose, an isomer of lactose, acts as an **inducer,** a small molecule that binds to and inactivates the repressor protein, so that the operon can be transcribed.

INTERACTIVE QUESTION 18.1

In the following diagram of the *lac* operon, an operon for inducible enzymes, identify components a through i.

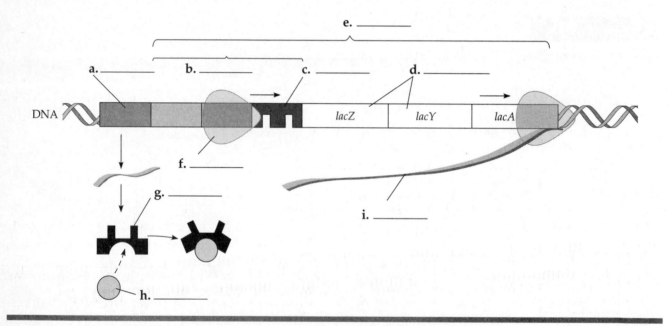

INTERACTIVE QUESTION 18.2

a. *Repressible enzymes* usually function in _____ pathways. The pathway's product serves as a _____ to activate the repressor and turn off enzyme synthesis, thus preventing overproduction of the product of the pathway. Genes for repressible enzymes are usually switched _____ and the repressor is synthesized in an _____ form.

b. *Inducible enzymes* usually function in _____ pathways. Nutrient molecules serve as _____ to stimulate production of the enzymes necessary for their breakdown. Genes for inducible enzymes are usually switched _____ and the repressor is synthesized in an _____ form.

Positive Gene Regulation   E. coli cells preferentially use glucose. Should glucose levels fall, transcription of operons for other catabolic pathways can be increased through the action of the allosteric regulatory protein *catabolite activator protein (CAP)*, which is an **activator.** When glucose is scarce, **cyclic AMP (cAMP)** accumulates in the cell and binds with CAP, changing it to its active shape. The active CAP attaches above the promoter region and stimulates transcription by facilitating the binding of RNA polymerase.

The regulation of the *lac* operon includes both negative control by the repressor protein that is inactivated by the presence of lactose, and positive control by CAP when complexed with cAMP.

18.2 Eukaryotic gene expression is regulated at many stages

All cells control the expression of their genes in response to their external and internal environments. Eukaryotic cells in multicellular organisms also regulate gene expression to produce specialized cells.

Differential Gene Expression   Differences in cells with the same genome are the result of **differential gene expression.** In eukaryotes (as in prokaryotes), gene expression is most commonly regulated at transcription, but it can be regulated at various stages.

Regulation of Chromatin Structure   DNA packing and the location of a gene's promoter (relative to nucleosomes and to the attachment to the chromosome scaffold) may help control which genes are available for transcription. Genes within highly condensed heterochromatin are usually not expressed.

Histone acetylation, the attachment of acetyl groups (—$COCH_3$) to lysines in histone tails, interrupts the binding to neighboring nucleosomes, giving transcription proteins greater access to genes. Some enzymes that acetylate or deacetylate histones are associated with or are part of transcription factors.

The addition of methyl groups (—CH_3) to histone tails leads to condensation of the chromatin. Adding a phosphate group to an amino acid next to a methylated amino acid undoes that effect. According to the *histone code hypothesis,* the specific patterns of modifications to histone tails determine chromatin shape and influence transcription.

In a process called **DNA methylation,** methyl groups are added to DNA bases (usually cytosine). Genes are generally more heavily methylated in cells in which they are not expressed. DNA methylation may reinforce gene regulatory decisions of early

development because methylation records are correctly passed on during subsequent DNA replications. Such methylation patterns account for *genomic imprinting* in mammals, in which either a maternal or a paternal allele of certain genes is permanently regulated.

Epigenic inheritance is the transmission of traits by mechanisms that do not involve changes in DNA sequences, such as chromatin modifications that affect gene expression in future generations of cells. In some cases, it appears that DNA methylating proteins and histone-modifying enzymes act in concert to repress gene expression.

INTERACTIVE QUESTION   18.3

a. Give an example of highly methylated and inactive DNA that is common in mammalian cells.

b. Would histone tail deacetylation in a nucleosome increase or decrease the transcription of a gene?

Regulation of Transcription Initiation   A typical eukaryotic gene consists of both a promoter sequence, to which a *transcription initiation complex* that includes RNA polymerase II attaches, and a sequence of introns interspersed among the coding exons. After transcription, RNA processing of the pre-mRNA removes the introns and adds a 5' cap and a poly-A tail at the 3' end. **Control elements** are noncoding sequences that help regulate transcription by serving as binding sites for transcription factors.

General transcription factors bind to RNA polymerase and to each other to initiate the transcription of all protein-coding genes. The binding of *specific transcription factors* to control elements, however, greatly increases or decreases transcription rates of particular genes.

Proximal control elements are located close to the promoter. **Enhancers** are groups of *distal control elements* located far upstream or downstream from a gene or within an intron. According to a current model, a protein-mediated bend in the DNA brings activators bound to enhancers into contact with *mediator proteins,* which interact with proteins at the promoter. These protein-protein interactions assemble the initiation complex. Hundreds of transcription activators have been identified. Most have a DNA-binding domain and one or more activation domains that bind other regulatory proteins.

Some specific transcription factors act as repressors in various ways, such as by blocking the binding of activators or by binding to specific control elements in an enhancer. Some activators and repressors influence chromatin structure by recruiting proteins that either acetylate or deacetylate histones. Most eukaryotic gene repression may occur by this *silencing* at the level of chromatin modification.

Only a dozen or so unique nucleotide sequences have been found in control elements, but each enhancer contains about ten different control elements. The *combination* of these elements ensures that a specific collection of activator proteins must be present to activate transcription. Thus, the specific activators and repressors made in a cell determine which genes are expressed.

Prokaryotic genes are often organized into operons and are transcribed together. Most eukaryotic genes coding for enzymes in the same metabolic pathway are scattered on the chromosomes. The integrated control of these genes may involve a specific collection of control elements that are associated with each related gene and require the same activators.

For example, steroid hormones may activate multiple genes when they bind to specific receptor proteins within a cell that function as transcription activators for several genes. The same chemical signal will activate genes with the same control elements and thus coordinate gene expression.

Although each chromosome occupies a distinct area in the interphase nucleus, loops of chromatin from various chromosomes may extend into RNA polymerase-rich areas called *transcription factories*.

Mechanisms of Post-Transcriptional Regulation Gene expression, measured by the amount of functional protein that is made, can be regulated at any post-transcriptional step. At the level of RNA processing, **alternative RNA splicing** may produce different mRNA molecules when regulatory proteins control which segments of the primary transcript are chosen as introns and which are chosen as exons. The majority of human genes that have multiple exons probably undergo alternative splicing.

In contrast to prokaryotic mRNA, which is degraded after a few minutes, eukaryotic mRNA can last hours or even weeks. The untranslated region (UTR) at the 3'

INTERACTIVE QUESTION   18.4

Label the components of the following diagram that illustrates how enhancers, mediator proteins, and transcription factors facilitate the formation of a transcription initiation complex.

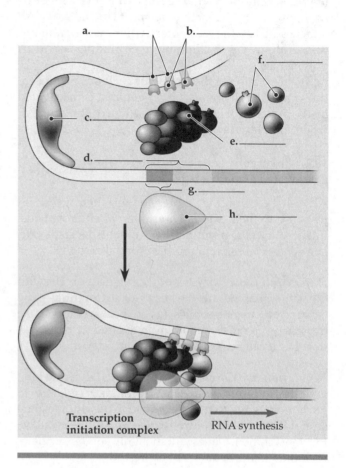

end of mRNA may contain nucleotide sequences that affect how long an mRNA lasts.

The translation of eukaryotic mRNA can be delayed by regulatory proteins that bind to the 5' or 3' UTR and prevent ribosome attachment. A great deal of mRNA is synthesized and stored in egg cells; translation may begin when enzymes add more A nucleotides to poly-A tails or when translation initiation factors are activated following fertilization.

Following translation, polypeptides are often cleaved or chemical groups added to yield an active protein. Some proteins must be transported to target locations.

Selective degradation of proteins may serve as a control mechanism in a cell. Molecules of the protein ubiquitin are added to mark proteins for destruction. **Proteasomes** recognize and degrade the marked proteins.

INTERACTIVE QUESTION 18.5

a. The untranslated regions (UTR) at both the 5′ and 3′ ends of an mRNA may contribute to regulation of gene expression. Describe their effects.

b. Describe the process of protein degradation in a cell.

18.3 Noncoding RNAs play multiple roles in controlling gene expression

Only 1.5% of the human genome is protein-coding DNA; an additional small percentage codes for genes for ribosomal and transfer RNA (rRNA and tRNA). Mounting evidence indicates that a large amount of the genome may be transcribed into non-protein-coding RNAs (*noncoding RNAs* or *ncRNAs*). A large population of small RNA molecules may be involved in regulating gene expression.

Effects on mRNAs by MicroRNAs and Small Interfering RNAs Long RNA transcripts fold on themselves forming hairpin structures. These hairpins are cut off, and an enzyme called Dicer trims the ends. One strand of the fragment becomes a **microRNA (miRNA).** These small, single-stranded RNA molecules associate with a complex of proteins and then base-pair with complementary sequences on target mRNA. This miRNA-protein complex either blocks translation or degrades the mRNA. An estimated one-half of all human genes may be regulated by miRNAs.

This inhibition of gene expression by RNA molecules was first observed experimentally and called **RNA interference (RNAi).** Injection of double-stranded RNA into a cell turned off genes with matching sequences. It is now known that cells cut these longer double-stranded RNA molecules into many short RNA molecules called **small interfering RNAs (siRNAs).** siRNAs appear to inhibit the expression of genes by the same mechanism as miRNAs. Many species apparently produce double-stranded RNA precursors that are cut into small RNAs such as siRNAs.

Chromatin Remodeling and Effects on Transcription by ncRNAs In some yeasts, siRNAs are involved in the formation of heterochromatin. Evidence indicates that an RNA transcript produced from DNA in the centromere region is copied into double-stranded RNA and associated with proteins. This complex then targets the centromere region, where it initiates chromatin remodeling into heterochromatin.

Newly discovered *piwi-associated RNAs (piRNAs)* also induce heterochromatin formation, thus blocking transcription of some transposons. They also appear to help re-establish proper methylation patterns in the genome during gamete formation. Some cases in which miRNAs and piRNAs block (or activate) specific gene expression have been reported.

The Evolutionary Significance of Small ncRNAs The evolution of morphological complexity may be linked to the extra levels of gene regulation that are provided by small ncRNAs. Evidence indicates that siRNAs evolved first, followed by miRNAS, and then piRNAs, which are found only in animals.

INTERACTIVE QUESTION 18.6

a. Describe the sequence from the formation of a primary miRNA transcript to the blocking of transcription of an mRNA.

b. What is the difference between microRNAs and small interfering RNAs?

c. Do small RNAs regulate gene expression by affecting translation or transcription?

18.4 A program of differential gene expression leads to the different cell types in a multicellular organism

A Genetic Program for Embryonic Development The three key processes of embryonic development are (1) cell division, the production of large numbers of cells; (2) cell **differentiation,** the formation of cells specialized in structure and function; and (3) **morphogenesis,** the physical processes that produce body shape. All three processes are based on differential gene expression resulting from differences in gene regulation in cells.

Cytoplasmic Determinants and Inductive Signals   The cytoplasm of an unfertilized egg cell contains maternal mRNA, proteins, and other substances that are unevenly distributed, and the first few mitotic divisions separate these components and expose the nuclei in these new cells to different environments. These maternal components of the egg that influence early development by regulating gene expression are called **cytoplasmic determinants.**

The other important source of developmental control is signals from other embryonic cells in the form of contact with cell-surface molecules or secreted molecules. Change in the gene expression of target cells resulting from signals from other cells is called **induction.**

Sequential Regulation of Gene Expression During Cellular Differentiation   A cell's developmental history leads to its eventual differentiation as a cell with a specific structure and function. The term **determination** is used to describe the condition in which a cell is irreversibly committed to its fate. When a cell becomes differentiated, it expresses genes for *tissue-specific proteins,* and that expression is usually controlled at the level of transcription.

The embryonic precursor cells from which muscle cells arise have the potential to develop into a number of different cell types. Once they become committed to becoming muscle cells, they are called *myoblasts.* Researchers have identified several "master regulatory genes" that cause myoblast determination. One of these is *myoD,* which codes for a transcription factor named MyoD that binds to specific control elements and initiates transcription of other muscle-specific transcription factors. These secondary transcription factors then activate muscle-protein genes. MyoD also turns on genes that block the cell cycle and stop cell division, and it activates its own transcription, thereby maintaining the cell's differentiated state.

INTERACTIVE QUESTION   18.7

a. What is the difference between determination and differentiation?

b. The MyoD protein is able to transform some, but not all, differentiated cells into muscle cells. Why doesn't it work on all kinds of cells?

Pattern Formation: Setting Up the Body Plan   **Pattern formation** is the ordering of cells and tissues into their characteristic structures and locations. An animal's three major body axes are laid out early in development. Cells and their progeny develop in response to molecular cues called **positional information** that tell a cell where it is located relative to the body axes and neighboring cells.

The genetic study of *Drosophila* development has led to the discovery of some common developmental principles. A fruit fly's body consists of a series of segments grouped into the head, thorax, and abdomen. The anterior-posterior and dorsal-ventral axes are determined by positional information provided by cytoplasmic determinants localized in the unfertilized egg. Each egg cell is surrounded by nurse cells and follicle cells that supply nutrients and other molecules needed for development. Eggs are laid following fertilization, and a wormlike larva develops. Following three larval stages, the larva becomes a pupa. Metamorphosis produces an adult fly.

In the 1940s, E. B. Lewis studied developmental mutants and was able to map certain mutations that control pattern formation to specific genes, called **homeotic genes.** In the 1970s, C. Nüsslein-Volhard and E. Wieschaus undertook a search for the genes that control segment formation. They studied mutations that were **embryonic lethals,** which prevented the development of viable larvae. They exposed flies to a chemical mutagen and then performed many thousands of crosses to detect recessive mutations that caused the death of embryos or resulted in larvae with abnormal segmentation. They identified 120 genes involved in pattern formation leading to normal segmentation.

Maternal effect genes are genes of the mother that code for proteins or mRNA that are deposited in the unfertilized egg. These genes are also called **egg-polarity genes** because they determine the anterior-posterior and dorsal-ventral axes of the egg and consequently of the embryo.

One egg-polarity gene is *bicoid.* The product of the *bicoid* gene is concentrated at one end of the embryo and responsible for determining its anterior end. Offspring of a mother defective for this gene have two tail regions and lack the front half of the body. Researchers located bicoid mRNA concentrated in the most anterior end of egg cells. Following fertilization, the mRNA is translated into Bicoid protein, which diffuses posteriorly, forming a gradient in the early embryo. Gradients of such substances, which are called **morphogens,** establish an embryo's axes or other features—an example of the *morphogen gradient hypothesis.* Proteins whose gradients determine the posterior end and establish the dorsal-ventral axis have also been identified. Positional information then establishes the proper number of segments and finally triggers the formation of each segment's characteristic structures.

INTERACTIVE QUESTION 18.8

What type of evidence established that Bicoid protein is a morphogen that determines the anterior end of a fruit fly?

18.5 Cancer results from genetic changes that affect cell cycle control

Types of Genes Associated with Cancer Chemical carcinogens, X-rays, or certain viruses most often cause the changes in the genes that regulate cell growth and division that lead to cancer. **Oncogenes,** or cancer-causing genes, were first found in certain types of viruses. Similar genes were later recognized in the genomes of humans and other animals. Cellular **proto-oncogenes,** which code for proteins that stimulate cell growth and division, may become oncogenes by several mechanisms, resulting in an overproduction of growth-stimulating proteins.

Mutations in **tumor-suppressor genes** can contribute to the onset of cancer when they result in a decrease in the activity of proteins that prevent uncontrolled cell growth.

INTERACTIVE QUESTION 18.9

a. List three genetic changes that can convert a proto-oncogene into an oncogene.

b. List three possible functions of tumor-suppressor proteins.

Interference with Normal Cell-Signaling Pathways In about 30% of human cancers, the *ras* proto-oncogene is mutated. The *ras* **gene** codes for a G protein that connects a growth-factor receptor on the plasma membrane to a cascade of protein kinases that leads to the production of a cell cycle stimulating protein. A mutation may create a hyperactive version of the Ras protein that relays a signal without the binding of a growth factor.

The *p53* **gene** is mutated in about 50% of human cancers. It codes for a tumor-suppressor protein that is a specific transcription factor for several genes. It often activates the *p21* gene, whose product binds to cyclin-dependent kinases, halting the cell cycle and allowing time for the cell to repair damaged DNA. It also activates several miRNAS that inhibit the cell cycle. The p53 protein can also activate genes involved in DNA repair. Should DNA damage be irreparable, p53 activates "suicide genes" that initiate apoptosis.

The Multistep Model of Cancer Development More than one mutation appears to be needed to produce a cancerous cell. Mutation of a single proto-oncogene can stimulate cell division, but usually both alleles for several tumor-suppressor genes must be defective to allow uncontrolled cell growth.

Inherited Predisposition and Other Factors Contributing to Cancer A genetic predisposition to certain cancers may involve the inheritance of an oncogene or a recessive mutant allele for a tumor-suppressor gene. Approximately 15% of colorectal cancers involve inherited mutations, often in the tumor-suppressor gene *APC*, which regulates cell migration and adhesion.

About 5–10% of breast cancer cases are linked to an inherited mutant allele for either *BRCA1* or *BRCA2*, both of which appear to be tumor-suppressor genes.

The ultraviolet radiation in sunlight and the chemicals in cigarette smoke may contribute to cancer through their DNA-damaging effects.

Tumor viruses appear to be involved in about 15% of human cancers. Viruses can interfere with gene regulation when the insertion of their genetic material into a cell's DNA introduces an oncogene or affects a proto-oncogene or tumor-suppressor gene. Viral proteins may also inactivate p53 or other tumor-suppressor proteins.

Word Roots

morph- = form; **-gen** = produce (*morphogen:* a substance that provides positional information in the form of a concentration gradient along an embryonic axis)

proto- = first, original; **onco-** = tumor (*proto-oncogene:* a normal cellular gene that has the potential to become an oncogene, which is involved in triggering molecular events that lead to cancer)

Structure Your Knowledge

1. Complete the following concept map to help you review the mechanisms by which bacteria regulate gene expression in response to varying metabolic needs.

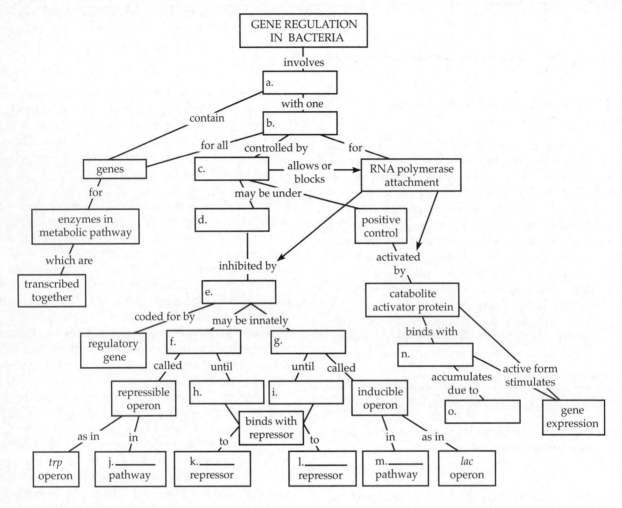

2. Fill in the following table to help you organize the major mechanisms that can regulate the expression of eukaryotic genes.

Level of Control	Examples	
Chromatin modification	**a.**	
Transcriptional regulation	**b.**	
RNA processing	**c.**	
RNA degradation	**d.**	
Translational regulation	**e.**	
Protein processing; degradation	**f.**	

3. How might the mechanism of transcriptional regulation differ for cytoplasmic determinants and for the cell-cell signaling involved in induction?

Test Your Knowledge

MULTIPLE CHOICE: *Choose the one best answer.*

1. Inducible enzymes
   a. are usually involved in anabolic pathways.
   b. are produced when a small molecule inactivates the repressor protein.
   c. are produced when an activator molecule enhances the attachment of RNA polymerase to the operator.
   d. are regulated by inherently inactive repressor molecules.
   e. are regulated almost entirely by feedback inhibition.

2. In *E. coli,* tryptophan switches off the *trp* operon by
   a. inactivating the repressor protein.
   b. inactivating the gene for the first enzyme in the pathway (feedback inhibition).
   c. binding to the repressor and increasing the latter's affinity for the operator.
   d. binding to the operator.
   e. binding to the promoter.

3. In the control of gene expression in bacteria, a regulatory gene
   a. has its own promoter.
   b. is transcribed continuously.
   c. is not contained in the operon it controls.
   d. codes for repressor proteins.
   e. is or does all of the above.

4. A mutation that renders nonfunctional the product of a regulatory gene for a repressible operon would result in
   a. continuous transcription of the genes of the operon.
   b. complete blocking of the attachment of RNA polymerase to the promoter.
   c. irreversible binding of the repressor to the operator.
   d. no difference in transcription rate when an activator protein was present.
   e. negative control of transcription.

5. The control of gene expression is more complex in eukaryotic cells because
   a. DNA is associated with protein.
   b. gene expression differentiates specialized cells.
   c. the chromosomes are linear and more numerous.
   d. operons are controlled by more than one promoter region.
   e. inhibitory or activating molecules may help regulate transcription.

6. DNA methylation of cytosine residues
   a. initiates the acetylation of histones.
   b. may be a mechanism of epigenic inheritance when methylation patterns are repeated in daughter cells.
   c. occurs in the promoter region and enhances binding of RNA polymerase.
   d. is a signal for proteasomes to degrade a protein.
   e. may be related to the transformation of proto-oncogenes to oncogenes.

7. Which of the following is *not* true of enhancers?
   a. They may be located thousands of nucleotides upstream from the genes they affect.
   b. When bound with activators, they interact with the promoter region and other transcription factors to produce an initiation complex.
   c. They may complex with steroid-activated receptor proteins, which selectively activate specific genes.
   d. They may coordinate the transcription of enzymes involved in the same metabolic pathway when they contain the same combination of control elements.
   e. Each gene may have several enhancers, and each enhancer may be associated with and regulate several genes.

8. Which of the following is *not* an example of the control of gene expression after transcription?
   a. mRNA stored in the cytoplasm needing activation of translation initiation factors
   b. the length of time mRNA lasts before it is degraded
   c. rRNA genes amplified in multiple copies in the genome
   d. alternative RNA splicing before mRNA leaves the nucleus
   e. splicing or modification of a polypeptide

9. A eukaryotic gene typically has all of the following associated with it *except*
   a. a promoter.
   b. an operator.
   c. enhancers.
   d. introns and exons.
   e. control elements.

10. Which of the following would you expect to find as part of a receptor protein that binds with a steroid hormone?
    a. a TATA box
    b. a domain that binds to DNA and protein-binding domains
    c. an activated operator region that allows attachment of RNA polymerase
    d. an enhancer sequence located at some distance upstream or downstream from the promoter
    e. transmembrane domains that facilitate the protein's localization in a plasma membrane

11. Proteasomes are
    a. complexes of proteins that excise introns.
    b. single-stranded RNA molecules complexed with proteins that block translation of or degrade mRNA.
    c. small, positively charged proteins that form the core of nucleosomes.
    d. enormous protein complexes that degrade proteins marked with ubiquitin.
    e. complexes of transcription factors whose protein-protein interactions are required for enhancing gene transcription.

12. Which of the following statements explains why a larger portion of the DNA in a eukaryotic cell is transcribed than would be predicted by the number of proteins made by the cell?
    a. Multiple enhancer regions are being transcribed to amplify the transcription of protein-coding genes.
    b. Much of this non-protein-coding RNA functions to regulate the translation or degradation of mRNAs.
    c. Many of these transcriptions produce double-stranded siRNAs that regulate the transcription of other genes.
    d. The additional DNA that is transcribed represents introns that are excised from the primary transcript in the production of mRNA.
    e. These transcriptions are of noncoding "junk" DNA that is a remnant of mutated protein-coding segments, and the transcripts are degraded by nuclear enzymes.

13. Which of the following is *not* descriptive of small ncRNAs?
    a. They are thought to have facilitated the evolution of morphological complexity.
    b. The inhibition of gene expression by RNA was first observed experimentally and called RNA interference (RNAi).
    c. The newly discovered piRNAs induce heterochromatin formation, and evidence indicates that they are the most recent ncRNA to have evolved.
    d. The regulatory functions of ncRNAs include effects on both transcription and translation.
    e. They are often found in "transcription factories" within an interphase nucleus, where they loosen loops of chromatin and enhance transcription of genes on multiple chromosomes.

14. Which of the following descriptions is *not* part of the process by which miRNA regulates gene expression?
    a. A long miRNA transcript folds on itself, forming loops called hairpins.
    b. An enzyme cuts the hairpins, and dicer trims the ends.

c. One strand is degraded, and the remaining RNA strand associates with proteins.
    d. If the miRNA and an mRNA molecule are complementary all along their length, the miRNA is degraded and translation proceeds.
    e. If the match between the miRNA and an mRNA is less complete, then translation is blocked.

15. Cytoplasmic determinants are
    a. unevenly distributed cytoplasmic components of an unfertilized egg.
    b. often involved in transcriptional regulation.
    c. usually separated in the first few mitotic divisions following fertilization.
    d. maternal contributions that help to direct the initial stages of development.
    e. all of the above.

16. Pattern formation in animals is based on
    a. positional information a cell receives from gradients of morphogens.
    b. the induction of cells by the nurse cells in the mother's ovary.
    c. the packing of chromatin in the nucleus.
    d. the differentiation of cells that then migrate together to form tissues and organs.
    e. the first few mitotic divisions.

17. What would be the fate of a *Drosophila* larva that inherits two copies of a mutant *bicoid* gene (one mutant allele from each heterozygous parent)?
    a. It develops two heads, one at each end of the larva.
    b. It develops two tails, one at each end of the larva.
    c. It develops normally but, if female, produces mutant larvae that have two tail regions.
    d. It develops into an adult with legs growing out of its head.
    e. It receives no *bicoid* mRNA from the nurse cells of its mother.

18. In the following hypothetical embryo, a high concentration of a morphogen called morpho is needed to activate gene *P*; gene *Q* is active at or above medium concentrations of morpho; and gene *R* is expressed so long as any quantity of morpho is present. A different morphogen, called phogen, activates gene *S* and inactivates gene *Q* when at medium to high concentrations. If morpho and phogen are diffusing from their sites of production at the opposite ends of the embryo, which genes will be expressed in region 2 of this embryo? (Assume a gradient of morphogen concentrations in the three

regions, from high at the source, to medium in the middle, and to low at the opposite end.)

Morpho produced here

Phogen produced here

a. genes *P, Q, R,* and *S*
b. genes *P, Q,* and *R*
c. genes *Q* and *R*
d. genes *R* and *S*
e. gene *R*

19. How is the coordinated transcription of genes involved in the same pathway regulated?
    a. The genes are transcribed in one transcription unit, although each gene has its own promoter.
    b. The genes are located in the same region of the chromosome, and enzymes acetylate the entire region so that transcription may begin.
    c. All the genes respond to the same general transcription factors, although they may respond to different specific transcription factors.
    d. Asteroid hormone selectively binds to the promoters for all the genes.
    e. The genes have the same combination of control elements in their enhancers.

20. Which of the following might a proto-oncogene code for?
    a. DNA polymerase
    b. RNA polymerase
    c. receptor protein for growth factors
    d. an enhancer
    e. transcription factors that inhibit cell division genes

21. A gene can develop into an oncogene when
    a. it is present in more copies than normal.
    b. it undergoes a translocation that removes it from its normal control region.
    c. a mutation results in a more active or resistant protein.
    d. a mutation in a control element increases expression.
    e. any of the above occur.

22. A tumor-suppressor gene could cause the onset of cancer if
    a. both alleles have mutations that decrease the activity of the gene product.
    b. only one allele has a mutation that alters the gene product.
    c. it is inherited from a parent in mutated form.
    d. a proto-oncogene has also become an oncogene.
    e. both a and d have occurred.

23. Apoptosis is
    a. a cell suicide program that may be initiated by p53 protein in response to DNA damage.
    b. metastasis, or the spread of cancer cells to a new location in the body.
    c. the transformation of a normal cell to a cancer cell.
    d. the mutation of a G protein into a hyperactive form.
    e. the transformation of a proto-oncogene to an oncogene by a point mutation.

24. Which of the following would most likely account for a family history of colorectal cancer?
    a. a diet that is low in fats and high in fiber
    b. inheritance of one mutated *APC* allele that regulates cell adhesion and migration
    c. a family history of breast cancer
    d. inheritance of the *ras* oncogene, which locks the G protein in an active configuration
    e. inheritance of a proto-oncogene

Chapter 19

Viruses

Key Concepts

19.1 A virus consists of a nucleic acid surrounded by a protein coat

19.2 Viruses replicate only in host cells

19.3 Viruses, viroids, and prions are formidable pathogens in animals and plants

Framework

A virus is an infectious particle consisting of a genome of single-stranded or double-stranded DNA or RNA enclosed in a protein capsid, and sometimes within a membrane envelope derived from the host. Viruses replicate using the metabolic machinery of their bacterial, animal, or plant host. Viral infections may destroy the host cell and cause diseases within the host organism. Viruses may have evolved from plasmids or transposons.

Chapter Review

19.1 A virus consists of a nucleic acid surrounded by a protein coat

The Discovery of Viruses: Scientific Inquiry The search for the cause of tobacco mosaic disease led to the discovery of viruses. The infectious agent could neither be filtered from infected sap nor cultivated on nutrient media, but it was able to replicate within plants. In 1935, W. Stanley crystallized the infectious particle, now known as tobacco mosaic virus (TMV). Since that time, many viruses have been seen with the electron microscope.

Structure of Viruses A **virus** is an infectious particle consisting of genes inside a protein coat. Viral genomes may be single-stranded or double-stranded DNA or RNA. Viral genes are usually contained on a single linear or circular nucleic acid molecule.

The **capsid**, or protein shell, is built from a large number of often identical protein subunits (*capsomeres*) and may be rod-shaped (*helical viruses*), polyhedral (*icosahedral viruses*), or more complex in shape. **Viral envelopes,** which are derived from membranes of the host cell but also include viral proteins and glycoproteins, may cloak the capsids of viruses that infect animals. Some viruses also contain a few viral enzymes.

Complex capsids are found among **bacteriophages,** or **phages,** viruses that infect bacteria. Of the phages that infect the bacterium *E. coli*, T2, T4, and T6 have similar capsid structures consisting of a polyhedral head and a protein tail piece with tail fibers for attaching to a bacterium.

19.2 Viruses replicate only in host cells

Viruses are obligate intracellular parasites that lack metabolic enzymes and other equipment needed to replicate. Each virus type has a limited **host range** due to proteins on the outside of the virus that recognize only specific receptor molecules on the host cell surface.

General Features of Viral Replicative Cycles Once the viral genome enters the host cell, the cell's enzymes, nucleotides, amino acids, ribosomes, ATP, and other resources are used to replicate the viral genome and produce capsid proteins. Many DNA viruses use host DNA polymerases to copy their genome, whereas RNA viruses use virus-encoded RNA polymerases for replicating their RNA genome.

After replication, viral nucleic acid and capsid proteins spontaneously self-assemble to form new viruses within the host cell. Hundreds or thousands of newly formed virus particles are released, often destroying the host cell in the process.

Replicative Cycles of Phages A **lytic cycle** culminates in lysis of the host cell and release of newly produced phages. **Virulent phages** replicate only by a lytic cycle.

The T4 phage uses its tail fibers to stick to a receptor site on the surface of an *E. coli* cell. The sheath of the tail contracts and thrusts its viral DNA into the cell, leaving the empty capsid behind. The *E. coli* cell begins to transcribe and translate phage genes, one of which codes for an enzyme that chops up host cell DNA. Nucleotides from the degraded bacterial DNA are used to produce viral DNA. Capsid proteins are assembled into phage tails, tail fibers, and heads. The viral components assemble into phage particles, which are released after the manufacture of an enzyme that damages the bacterial cell wall.

Mutations that change their receptor sites and **restriction enzymes** that chop up viral DNA once it enters the cell help to defend bacteria against viral infection.

In a **lysogenic cycle,** a virus replicates its genome without killing its host. **Temperate phages** can replicate by lytic and lysogenic cycles.

When the phage lambda (λ) injects its DNA into an *E. coli* cell, it can begin a lytic cycle, or its DNA may be incorporated into the host cell's chromosome and begin a lysogenic cycle as a **prophage.** Most of the genes of the inserted phage genome are repressed by a protein coded for by a prophage gene. Reproduction of the host cell replicates the phage DNA along with the bacterial DNA. The prophage may exit the bacterial chromosome, usually in response to environmental stimuli, and start a lytic cycle.

Several disease-causing bacteria would be harmless except for the expression of prophage genes that code for toxins.

INTERACTIVE QUESTION   19.1

In the following diagram of lytic and lysogenic cycles, describe steps 1–8 and label structures a–e.

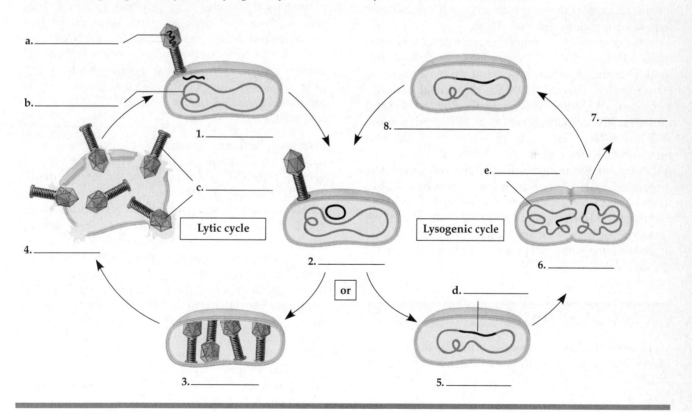

a._____

b._____

1._____

c._____

Lytic cycle

4._____

3._____

2._____

or

5._____

d._____

Lysogenic cycle

6._____

e._____

8._____

7._____

Replicative Cycles of Animal Viruses Animal viruses are classified based on their type of genome. A viral envelope surrounds the capsid of almost all animal viruses that have RNA genomes. Glycoproteins extending from the viral membrane attach to receptor sites on a host cell plasma membrane. Either the two membranes fuse, or endocytosis transports the capsid into the cell. The viral genome replicates and directs the synthesis of viral proteins. Glycoproteins are produced and embedded in the ER membrane, and then transported to the plasma membrane. New viruses bud off within an envelope that is derived from the host's plasma membrane and bears viral glycoproteins.

Herpesviruses replicate within the host cell nucleus and are temporarily cloaked in host cell nuclear membrane. The herpesvirus' double-stranded DNA can remain latent as a minichromosome in the host cell nucleus until it initiates herpes infections in times of stress.

The single-stranded RNA of animal class IV viruses can serve directly as mRNA. The RNA genome of class V viruses must first be transcribed into a strand of complementary RNA (using a viral enzyme packaged inside the capsid) that then serves as mRNA and as a template for making genome RNA.

In the complicated replicative cycle of **retroviruses** (class VI), the viral RNA genome is transcribed into double-stranded DNA by a viral enzyme, **reverse transcriptase.** This viral DNA is then integrated into a chromosome, where it is transcribed by the host cell into viral RNA, which acts both as new viral genome and as mRNA for viral proteins. **HIV (human immunodeficiency virus)** is a retrovirus that causes **AIDS (acquired immunodeficiency syndrome).** The integrated viral DNA remains as a **provirus** within the host cell DNA. New viruses, assembled with two copies of the RNA genome and two molecules of reverse transcriptase within a capsid, bud off covered in host cell plasma membrane studded with viral glycoproteins.

INTERACTIVE QUESTION   19.2

Summarize the flow of genetic information during replication of a retrovirus. Indicate the enzymes that catalyze this flow.

_____ → _____ → _____

Enzymes:

Evolution of Viruses Viruses may have evolved from fragments of cellular nucleic acids that moved from one cell to another and eventually evolved special packaging. Sources of viral genomes may have been *plasmids,* self-replicating circles of DNA found in bacteria and yeast, and *transposons,* segments of DNA that can change locations within a cell's genome. Thus, viruses, plasmids, and transposons are all *mobile genetic elements.*

The genomes of viruses are often more similar to those of their host cells than to the genomes of viruses infecting other hosts. Recent sequencing of viral genomes, however, has revealed genetic similarities among viruses that otherwise seem distantly related.

19.3   Viruses, viroids, and prions are formidable pathogens in animals and plants

Viral Diseases in Animals The symptoms of a viral infection may be caused by viral-programmed toxins produced by infected cells, cells killed or damaged by the virus, or the body's defense mechanisms fighting the infection.

Vaccines are harmless variants or derivatives of pathogens that induce the immune system to react against the actual disease agent. Vaccinations have greatly reduced the incidence of many viral diseases.

Unlike bacteria, viruses use the host's cellular machinery to replicate, and few drugs have been found to treat or cure viral infections. Some antiviral drugs resemble nucleosides and interfere with viral nucleic acid synthesis.

Emerging Viruses Examples of *emerging viruses* include HIV, Ebola virus, and West Nile virus. A general outbreak of a disease is called an **epidemic.** The flu epidemic of 2009 spread rapidly, becoming a global epidemic or **pandemic.** The sudden emergence of viral diseases may be linked to the mutation of an existing virus (more common in RNA viruses, which have higher mutation rates), to the dissemination of an existing virus to a more widespread population (as in HIV), or to the spread from one host species to another (as in the 2009 flu pandemic, which likely passed to humans from pigs). Many new human diseases are thought to originate by the third mechanism.

Influenza types B and C infect only humans. If different strains of virus undergo genetic recombination within an animal's cells and accumulate mutations that allow the virus to infect human cells, the recombinant virus may be highly pathogenic. The H1N1 virus (named for the form of the viral surface proteins hemagglutinin and neuraminidase) caused both the 1918 and 2009 flu pandemics. The H5N1 virus is carried by wild and domestic birds. The human mortality rate for "avian flu" infections is greater than 50%, but thus far the virus is not easily transmitted from person to person.

Viral Diseases in Plants   Most plant viruses are RNA viruses. Plant viral diseases may spread through *vertical transmission* from a parent plant or through *horizontal transmission* from an external source. Plant injuries increase susceptibility to viral infections, and insects can act as carriers of viruses.

INTERACTIVE QUESTION   19.3

How does a virus spread throughout a plant? Are there cures for viral plant diseases?

Viroids and Prions: The Simplest Infectious Agents
Viroids are very small infectious molecules of circular RNA that can replicate in host plant cells, causing abnormal development and stunted growth.

Prions are protein infectious agents that may be linked to several degenerative brain diseases, such as mad cow disease and Creutzfeldt-Jacob disease in humans. Prions, which are misfolded forms of a protein, apparently cause disease by converting normal cellular proteins into the misfolded prion version.

Word Roots

capsa- = a box (*capsid:* the protein shell that encloses a viral genome)

lyto- = loosen (*lytic cycle:* a type of phage replicative cycle resulting in the release of new phages by lysis (and death) of the host cell)

-phage = to eat (*bacteriophage:* a virus that infects bacteria)

pro- = before (*provirus:* a viral genome that is permanently inserted into a host genome)

retro- = backward (*retrovirus:* an RNA virus that replicates by transcribing its RNA into DNA and then inserting the DNA into a cellular chromosome)

virul- = poisonous (*virulent phage:* a phage that replicates only by a lytic cycle)

Structure Your Knowledge

1. Create a concept map that describes the lytic and lysogenic cycles of a phage.

2. Complete the following concept map to help organize your understanding of viruses.

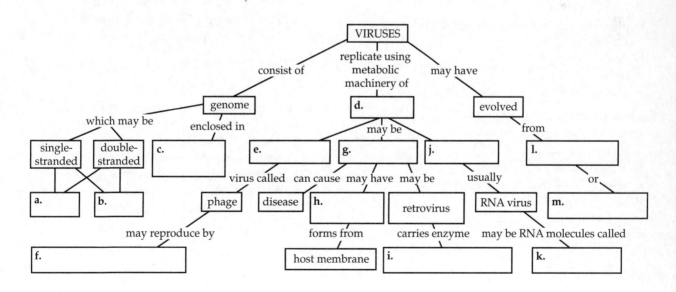

Test Your Knowledge

MULTIPLE CHOICE: *Choose the one best answer.*

1. The study of viruses has provided information on all of the following topics *except*
   a. the molecular biology of all organisms.
   b. the sexual replicative cycles of viruses.
   c. new techniques for manipulating genes.
   d. the causes of diseases.
   e. the role of mutation in the relationship between host and virus.

2. Beijerinck concluded that the cause of tobacco mosaic disease was not a filterable toxin because
   a. the infectious agent could not be cultivated on nutrient media.
   b. a plant sprayed with filtered sap would develop the disease.
   c. the infectious agent could be crystallized.
   d. the infectious agent replicated and could be passed on from a plant infected with filtered sap.
   e. the filtered sap was infectious even though microbes could not be found in it.

3. Viral genomes may be any of the following *except*
   a. single-stranded DNA.
   b. double-stranded RNA.
   c. misfolded infectious proteins.
   d. a linear single-stranded RNA molecule.
   e. a circular double-stranded DNA molecule.

4. The reverse transcriptase carried by retroviruses
   a. uses viral RNA as a template for making complementary RNA strands.
   b. protects viral DNA from degradation by restriction enzymes.
   c. destroys the host cell DNA.
   d. translates RNA into proteins.
   e. uses viral RNA as a template for DNA synthesis.

5. Virus particles are formed from capsid proteins and nucleic acid molecules
   a. by spontaneous self-assembly.
   b. at the direction of viral enzymes.
   c. using host cell enzymes.
   d. using ATP stored in the tail piece.
   e. by both b and d.

6. A virus has a base ratio of $(A + G)/(U + C) = 1$. What type of virus is this?
   a. a single-stranded DNA virus
   b. a single-stranded RNA virus
   c. a double-stranded DNA virus
   d. a double-stranded RNA virus
   e. a retrovirus

7. Vertical transmission of a plant virus involves
   a. movement of viral particles through plasmodesmata.
   b. inheritance of an infection from a parent.
   c. a bacteriophage transmitting viral particles.
   d. insects carrying viral particles between plants.
   e. the transfer of filtered sap.

8. Bacteria defend against viral infection
   a. with antibiotics they produce.
   b. with restriction enzymes that chop up foreign DNA.
   c. through thetransfer of R plasmids.
   d. with reverse transcriptase.
   e. through the incorporation of viral DNA into the bacterial chromosome.

9. Drugs that are effective in treating viral infections
   a. induce the body to produce antibodies.
   b. inhibit the action of viral ribosomes.
   c. interfere with the synthesis of viral nucleic acid.
   d. change the cell-recognition sites on the host cell.
   e. are vaccines that stimulate the immune system to create immunity.

10. Which of the following is true of prions?
    a. They are emerging viruses.
    b. They are fast-acting infectious agents.
    c. They probably evolved from transposons.
    d. They are infectious proteins that may convert brain proteins into misfolded forms.
    e. They may be transferred between animals by sexual contact.

11. An RNA viral genome may be replicated by
    a. DNA polymerase from the host.
    b. RNA polymerase coded by viral genes and carried in the viral capsid.
    c. reverse transcriptase that synthesizes RNA.
    d. RNA polymerase from the host.
    e. restriction enzymes from the host.

Chapter 20

Biotechnology

Framework

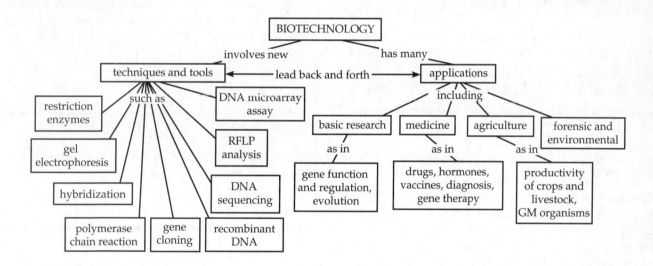

Chapter Review

DNA technology began with techniques for making **recombinant DNA,** which combines nucleotide sequences from different organisms or species into the same DNA molecule *in vitro.* **Genetic engineering,** the direct manipulation of genetic material for practical purposes, has begun a revolution in **biotechnology.** This use of living organisms or their components to manufacture desirable products dates back centuries, but advances in DNA technology have resulted in hundreds of new products and applications.

20.1 DNA cloning yields multiple copies of a gene or other DNA segment

DNA Cloning and Its Applications: A Preview   One of the approaches used to clone pieces of DNA makes use of the **plasmids** of bacterial cells. Recombinant DNA may be made by inserting foreign DNA into plasmids. These plasmids are put back into bacterial cells, where they will replicate as the *recombinant bacteria* reproduce to form clones of identical cells. Such **gene cloning** *amplifies* or provides multiple copies of the gene and may also be used to produce protein coded for by the foreign DNA.

Using Restriction Enzymes to Make Recombinant DNA **Restriction enzymes** protect bacteria from phages or other organisms by cutting up foreign DNA. Most restriction enzymes recognize short nucleotide sequences, called **restriction sites,** and cut at specific points within them. The cell protects its own DNA from restriction by methylating nucleotide bases within its own restriction sequences.

A restriction site is usually symmetrical; the same sequence, often of four to eight nucleotides, runs in opposite directions on the two strands. The most useful restriction enzymes cut the sugar-phosphate backbone in a staggered way, leaving **sticky ends** of short single-stranded sequences on both sides of the resulting **restriction fragment.**

DNA from different sources can be combined in the laboratory when the various DNA molecules are cut by the same restriction enzyme, and the complementary bases on the resulting sticky ends of the restriction fragments form hydrogen-bonded base pairs. **DNA ligase** is used to seal the strands together.

INTERACTIVE QUESTION   20.1

Which of the following DNA sequences would most likely function as a restriction site for a restriction enzyme? Why?

··CAGCAG··   ··GTGCTG··   ··GAATTC··
··GTCGTC··   ··CACGAC··   ··CTTAAG··

Cloning a Eukaryotic Gene in a Bacterial Plasmid **Cloning vectors** are DNA molecules that can move foreign DNA into a cell and can replicate there. Recombinant plasmids returned to bacterial cells will replicate the foreign DNA as the bacteria reproduce.

The plasmid method of gene cloning involves treating plasmids containing an antibiotic-resistant gene with a restriction enzyme that cuts the DNA ring at a single restriction site and disrupts a gene whose activity is easily determined, such as *lacZ*, the gene for β-galactosidase. The clipped plasmids are mixed with DNA containing the gene of interest. This DNA has also been treated with the same restriction enzyme, yielding many different fragments. The sticky ends form hydrogen bonds with each other, and DNA ligase seals the recombinant molecules.

The plasmids are introduced by transformation into bacterial cells that have a mutation in their *lacZ* gene and thus are unable to produce β-galactosidase.

Bacteria are plated onto a medium containing ampicillin and X-gal, a compound that is cleaved by β-galactosidase and yields a blue product. Colonies that are able to grow on the medium (because they contain a plasmid with the *amp^R* gene) and are not blue (because foreign DNA inserted in the middle of the β-galactosidase gene) are carrying a recombinant plasmid.

An organism's DNA can be cut into thousands of pieces with restriction enzymes and inserted into plasmids. The collection of the thousands of clones of bacteria containing recombinant plasmids derived from this "shotgun" approach is called a **genomic library.**

Bacteriophages have also been used as vectors for creating genomic libraries. DNA fragments are spliced into phage DNA and used to infect bacteria. The production of new phages clones the foreign DNA.

Bacterial artificial chromosomes (BACs) are widely used as cloning vectors. These large plasmids can carry longer inserts than do standard plasmid vectors. Both plasmid and BAC genomic libraries are usually stored in multiwelled plastic plates.

A partial genomic library can be produced using the mRNA molecules isolated from a cell. Using reverse transcriptase from retroviruses, mRNA is used as a template to produce **complementary DNA (cDNA).** Restriction sites are added to the ends and the cDNA is inserted into vector DNA, which is used to create a **cDNA library.**

INTERACTIVE QUESTION   20.2

Compare genomic and cDNA libraries with regard to their advantages and disadvantages.

The colonies with recombinant plasmids may be tested to find the ones that contain the gene of interest by **nucleic acid hybridization,** using a **nucleic acid probe** that has complementary sequences to segments of the gene. Cells from each clone are applied to a nylon membrane to create an *arrayed library.* The membrane is treated to break open the cells and denature the DNA. The probe can then hybridize to the single-stranded DNA and be located by its radioactively labeled molecules or fluorescent tags. Once the desired clone is identified, it can be grown in liquid culture and the gene of interest isolated in large quantities.

INTERACTIVE QUESTION 20.3

The following schematic diagram depicts the steps in plasmid cloning of a human gene. Identify components a–j. Briefly describe the five steps of the process. How are bacterial clones that have picked up the recombinant plasmid containing the human gene of interest identified?

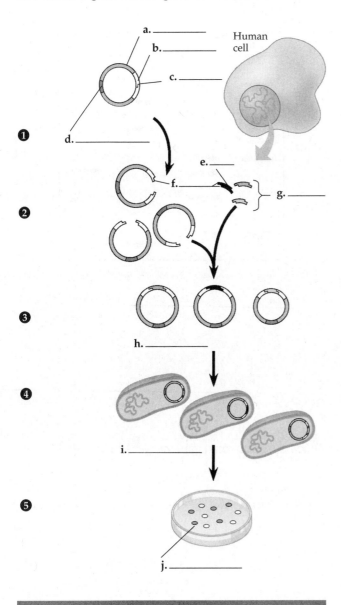

Expressing Cloned Eukaryotic Genes Differences between prokaryotic and eukaryotic mechanisms for gene expression can be overcome by using an **expression vector,** a cloning vector that has an active prokaryotic promoter just upstream from the eukaryotic

gene insertion site. Using a cDNA gene removes the problem of long introns, especially given that bacteria lack RNA-splicing machinery.

Yeast cells are eukaryotic hosts used to clone and express eukaryotic genes; they are easy to grow and have plasmids that serve as vectors. Plant and animal cells in culture can serve as hosts and may be necessary when a protein must be modified following translation.

More efficient means for introducing DNA into cells include **electroporation,** in which an electric pulse briefly opens holes in the plasma membrane through which DNA can enter; injection into cells using microscopically thin needles; and the use of *Agrobacterium* to introduce DNA into plant cells.

The ability of species to express genes transferred from very different species illustrates the shared evolutionary history of all life. For example, the vertebrate *Pax-6* gene, which triggers development of the single lens vertebrate eye, will lead to the formation of a compound eye when introduced into a fly embryo. Similarly, the fly *Pax-6* gene, when transferred into a frog embryo, will produce a vertebrate eye.

Amplifying DNA in Vitro: The Polymerase Chain Reaction (PCR) The **polymerase chain reaction (PCR)** can produce billions of copies of a section of DNA in only a few hours. DNA containing the region of interest is incubated with the four nucleotides, a special heat-resistant type of DNA polymerase, and specially synthesized primers that bind upstream from the target sequence on both DNA strands. The solution is heated to separate the DNA strands, then cooled so the primers can anneal (hydrogen-bond) to complementary sequences. DNA polymerase then adds nucleotides to the 3′ ends of the primers. The solution is heated again and the process repeated. The desired DNA segment need not be purified from the starting material, and very small samples can be used.

INTERACTIVE QUESTION 20.4

a. Why is PCR often used prior to cloning a gene in cells?

b. Why even bother cloning genes in cells, given that PCR produces so many copies so fast?

20.2   DNA technology allows us to study the sequence, expression, and function of a gene

Gel Electrophoresis and Southern Blotting   Many of the methods for analyzing and comparing DNA make use of **gel electrophoresis,** a technique that separates nucleic acids and proteins on the basis of their size and electrical charge. Due to the negative charge of their phosphate groups, DNA molecules migrate through the electric field produced in a thin slab of gel (often made of *agarose*) toward the positive electrode. Linear molecules of DNA move at a rate inversely proportional to their length, producing in the gel band patterns containing fragments of decreasing size.

Cutting a DNA molecule with a particular restriction enzyme and separating the resulting restriction fragments by gel electrophoresis produces a characteristic pattern of bands, a process called *restriction fragment analysis.* Pure samples of such bands can be recovered from the gel and retain their biological activity.

Two DNA samples, such as alleles of a gene, will produce different patterns of bands when differences in their DNA sequences add or delete restriction sites for a specific restriction enzyme. Such a sequence change is called a **restriction fragment length polymorphism (RFLP),** and it results in different sets of fragments. Using nucleic acid hybridization with a probe allows two or more unpurified samples of DNA (such as the entire genome) to be compared for the presence and band location of a particular DNA sequence. In a technique known as **Southern blotting,** DNA is treated with a restriction enzyme; the resulting fragments are separated on a gel and then transferred to nitrocellulose paper by blotting; labeled probes of single-stranded DNA are added and hybridize with complementary DNA sequences; and the restriction fragment bands of interest may be identified by autoradiography.

A bloody crime has occurred. Police have collected several samples: blood from the victim and two suspects, and blood found at the scene. Briefly list the steps laboratory personnel performed to produce the following autoradiograph (Southern blotting).

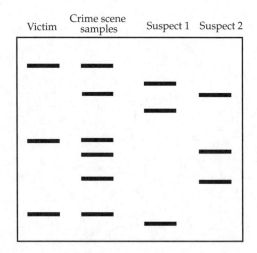

a.

b.

c.

d.

e.

Which suspect would you charge with the crime?

DNA Sequencing In the *dideoxyribonucleotide chain termination method* of DNA sequencing, developed by F. Sanger, a sample of a denatured DNA fragment is incubated with a primer, DNA polymerase, the four deoxyribonucleotides, and the four modified nucleotides (dideoxyribonucleotides, each labeled with a different fluorescent tag). These ddNTPs randomly block further synthesis when they are incorporated into a growing DNA strand. The sets of strands of varying lengths are separated through a polyacrylamide gel in a capillary tube. A fluorescence detector reads the sequence of their colors, which corresponds to the nucleotide sequence complementary to the DNA fragment.

"Next generation sequencing" techniques use a single template strand and allow *sequencing by synthesis* of a complementary strand, one nucleotide at a time. "Third-generation sequencing" uses technical modifications to speed up and lessen the expense of DNA sequencing.

Analyzing Genes Expression Labeled nucleic acid probes that hybridize with mRNAs can identify the location and timing of the expression of a gene in an organism. **Northern blotting,** which separates samples of mRNA from different developmental stages, identifies when a gene is expressed.

With **reverse transcriptase-polymerase chain reaction (RT-PCR),** reverse transcriptase makes cDNA from mRNAs isolated from different samples. The cDNA is amplified by PCR using primers specific for that gene, and gel electrophoresis produces bands of amplified DNA only in samples in which the gene has been transcribed to mRNA.

In *in situ* **hybridization,** fluorescently labeled probes can identify the location of specific mRNAs in an intact organism.

In order to study patterns of gene expression, researchers isolate the mRNA made in different cells, create a cDNA library using reverse transcriptase, and then use the cDNA as probes to explore collections of genomic DNA. Using this cDNA in **DNA microarray assays,** scientists can test all the genes expressed in a tissue for hybridization with short, single-stranded DNA fragments from thousands of genes arrayed on a grid (called a *DNA chip*). Gene expression in different tissues and at different stages of development can be compared.

Determining Gene Function The function of unknown genes can be studied using *in vitro* **mutagenesis,** in which changes are made to a cloned gene, the gene is returned to the cell, and it "knocks out" the normal gene. Changes in physiology or developmental patterns that result from the missing gene product are monitored. A new way to stop the expression of selected genes in cells is called **RNA interference (RNAi).** Synthetic double-stranded RNA molecules that match a gene sequence trigger the breakdown or block the translation of that gene's mRNA.

In **genome-wide association studies,** researchers test for *genetic markers*—sequence variations or *polymorphisms*—that may be associated with certain conditions or diseases. **Single nucleotide polymorphisms (SNPs)** are single base-pair variations that occur in at least 1% of the human population. These genetic markers can now be detected by microarray analysis or PCR. When an SNP is identified in individuals with a given disease, it is assumed to be close to or within a gene that contributes to the disease, and the region of DNA can be sequenced and studied.

INTERACTIVE QUESTION   20.6

a. What are some of the benefits of determining the nucleotide sequence of a gene of unknown function?

b. Describe a recent research method you could use to identify the tissues in which a particular gene is expressed.

c. Describe a research method that enables you to compare both which genes are expressed and what their relative rates of expression are in several tissues.

d. In one study, RNAi was used to prevent the expression, one gene at a time, of 86% of the genes in early nematode embryos. What did this allow researchers to do?

20.3   Cloning organisms may lead to production of stem cells for research and other applications

Organismal cloning involves producing genetically identical individuals (clones) from a single cell of a multicellular organism.

Cloning Plants: Single Cell Cultures   F. C. Steward demonstrated *genomic equivalence* in plants by growing new carrot plants from differentiated root cells. Most plant cells remain **totipotent,** retaining the ability to give rise to a complete new organism.

Cloning Animals: Nuclear Transplantation   Early evidence of genomic equivalence in animals was provided by the work of Briggs, King, and Gurdon, who transplanted nuclei from embryonic and tadpole cells into enucleated frog egg cells, a method called *nuclear transplantation.* The ability of the transplanted nucleus to direct normal development was inversely related to its developmental age.

In 1997, Scottish researchers reported cloning an adult sheep by transplanting a nucleus from a fully differentiated mammary cell into an unfertilized enucleated egg cell, and then implanting the resulting early embryo into a surrogate mother. The mammary cell was induced to dedifferentiate by culturing it in a nutrient-poor medium.

The *reproductive cloning* of numerous mammals has shown that cloned animals do not always look and behave identically. Environmental influences and random events play a role in development.

INTERACTIVE QUESTION   20.7

Although numerous mammals have now been cloned successfully, most cloned embryos fail to develop normally, and many cloned animals have various defects. What is a likely cause of these developmental failures?

Stem Cells of Animals   **Stem cells** are relatively unspecialized cells that continue to reproduce themselves and can, under proper conditions, differentiate into one or more types of cells. *Embryonic stem (ES) cells* taken from early embryos can be cultured indefinitely and can differentiate into all cell types. Thus, ES cells are **pluripotent.** *Adult stem cells* have been isolated from various tissues and grown in culture. Such cells are capable of producing multiple (but not all) types of cells. Both types of stem cells can be induced to differentiate into specialized cells.

Stem cell research has the potential to provide cells to repair organs that are damaged or diseased. *Therapeutic cloning* of embryonic stem cells, although different from reproductive cloning of humans, still raises ethical issues. The transformation of adult stem cells into *induced pluripotent stem cells* may provide sources of model cells for studying diseases and potential treatments, and may someday provide a patient's own iPS cells for regenerative treatments.

20.4   The practical applications of DNA technology affect our lives in many ways

Medical Applications   DNA technology is identifying genes responsible for genetic diseases; it is hoped this will lead to new ways to diagnose, treat, and even prevent those disorders. DNA microarray assays allow comparisons of gene expression in healthy and diseased tissues.

PCR, labeled DNA probes, and RT-PCR are being used to identify pathogens and to diagnose infectious diseases and genetic disorders. An allele for various diseases or increased risk for heart disease, Alzheimer's, and various cancers may be identified when closely linked with a known SNP.

Treatments for diseases have improved as it becomes increasingly possible to correlate an individual's genetic risk factors with treatment options, perhaps leading to a future of personalized medicine.

Gene therapy may provide the means for correcting genetic disorders in individuals by replacing or supplementing defective genes. New genes would be introduced into somatic cells of the affected tissue. For the correction to be permanent, the cells must be types that actively reproduce within the body, such as bone marrow cells. A few trials have used retroviral vectors to carry a normal allele into bone marrow cells to treat severe combined immunodeficiency (SCID), but results have been mixed.

INTERACTIVE QUESTION   20.8

What are some of the practical and ethical considerations in human gene therapy?

As the molecular bases for cancers and other diseases become known, researchers are developing drugs that can block the function of proteins responsible for such diseases. Many pharmaceutical proteins are produced using biotechnology. Engineering host cells so that they secrete a protein as it is made simplifies its purification.

Transgenic animals are produced by injecting a foreign gene into egg cells fertilized *in vitro*. The eggs are then transplanted into surrogate mothers. "Pharm" animals have been engineered to produce large quantities of a pharmaceutical protein, often by secretion in the animal's milk.

Forensic Evidence and Genetic Profiles In criminal cases, the **genetic profile** of a suspect can now be compared with that of crime-related DNA samples. Variations in the number of **short tandem repeats (STRs)** found at various loci are now commonly used in DNA analysis. Even if only very small samples are available, PCR can amplify specific STRs using different-colored fluorescent tags, which are then separated and their length (corresponding to the number of repeats) determined by electrophoresis. Forensic tests are able to provide a high statistical probability that matching genetic profiles come from the same individual.

Environmental Cleanup Genetically engineered microorganisms that are able to extract heavy metals (such as copper, lead, and nickel) may become important in mining and in cleaning up mining waste. Engineering organisms to degrade chlorinated hydrocarbons and other toxic compounds is an area of active research.

Agricultural Applications The production of transgenic animals is aimed at increasing productivity or beneficial traits.

The ability to regenerate plants from single cells growing in tissue culture has made plant cells easier to genetically manipulate than animal cells. The most commonly used vector is the **Ti plasmid** from the bacterium *Agrobacterium tumefaciens*, which integrates a segment of its DNA into plant chromosomes. Using DNA technology, foreign genes are inserted into Ti plasmids, and the recombinant plasmids are introduced into plant cells growing in culture. When these cells regenerate whole plants, the foreign gene is included in the plant genome.

Many crops have been engineered with bacterial genes for herbicide resistance. Crop plants are being engineered to be resistant to infectious pathogens and insects and to grow in soils having high salinity.

Safety and Ethical Questions Raised by DNA Technology Safety regulations have focused on potential hazards of engineered microbes. Most public concern, however, now centers on agriculture—on **genetically modified (GM) organisms,** which contain artificially acquired genes from either a different

variety or a different species. One environmental risk of transgenic plants is that their herbicide- or insect-resistant genes might pass to wild plants, creating "super weeds." Some fear that GM foodstuffs may be hazardous to human health. An international Biosafety Protocol requires exporters to label GM organisms in bulk food shipments so that importing countries can decide on potential environmental or health risks. In the United States, several federal agencies evaluate and regulate new products and procedures.

Ethical questions about human genetic information include who should have access to information about a person's genome and how that information should be used. Potential ethical, environmental, and health issues must be considered in the development of these powerful genetic techniques and remarkable products of biotechnology.

Word Roots

electro- = electricity (*electroporation:* a technique to introduce recombinant DNA into cells by applying a brief electrical pulse to a solution containing the cells. The pulse creates temporary holes in the cells' plasma membranes, through which DNA can enter.)

liga- = bound, tied (*DNA ligase:* a linking enzyme essential for DNA replication; it catalyzes the covalent bonding of the 3′ end of one DNA fragment to the 5′ end of another DNA fragment)

muta- = change; **-genesis** = origin, birth (*in vitro mutagenesis:* a technique used to discover the function of a gene by cloning it, introducing specific changes into the cloned gene's sequence, reinserting the mutated gene into a cell, and studying the phenotype of the mutant)

poly- = many; **morph-** = form (*single nucleotide polymorphism:* a single base-pair site in a genome where nucleotide variation is found in at least 1% of the population)

Structure Your Knowledge

1. Describe several examples of the many possible applications for DNA technology in agriculture and in medicine.
2. Fill in the following table on the basic tools of gene manipulation used in DNA technology.

Technique or Tool	Brief Description	Some Uses in DNA Technology
Restriction enzymes	a.	
Gel electrophoresis	b.	
cDNA	c.	
Nucleic acid probe	d.	
Southern blotting	e.	
DNA sequencing	f.	
PCR	g.	
DNA microarray assay	h.	
RT-PCR	i.	
in vitro mutagenesis	j.	
RNA interference (RNAi)	k.	

Test Your Knowledge

MULTIPLE CHOICE: *Choose the one best answer.*

1. The role of restriction enzymes in DNA technology is to
   a. provide a vector for the transfer of recombinant DNA.
   b. produce cDNA from mRNA.
   c. produce a cut (usually staggered) at specific restriction sites on DNA.
   d. reseal "sticky ends" after base pairing of complementary bases.
   e. denature DNA into single strands that can hybridize with complementary sequences.

2. Yeast has become important in genetic engineering because it
   a. has RNA splicing machinery.
   b. has plasmids that can be genetically engineered.
   c. enables the study of eukaryotic gene regulation and expression.
   d. grows readily and rapidly in the laboratory.
   e. does all of the above.

3. Which of the following DNA sequences would most likely be a restriction site?
   a. AACCGG
      TTGGCC
   b. GGTTGG
      CCAACC
   c. AAGG
      TTCC
   d. AATTCCGG
      TTAAGGCC
   e. GAATTC
      CTTAAG

4. A plasmid has two antibiotic resistance genes, one for ampicillin and one for tetracycline. It is treated with a restriction enzyme that cuts in the middle of the ampicillin gene. DNA fragments containing a human globin gene were cut with the same enzyme. The plasmids and fragments are mixed, treated with ligase, and used to transform bacterial cells. Clones that have taken up the recombinant DNA
   a. are blue and can grow on plates with both antibiotics.
   b. can grow on plates with ampicillin but not with tetracycline.
   c. can grow on plates with tetracycline but not with ampicillin.
   d. cannot grow with any antibiotics.
   e. can grow on plates with tetracycline and are blue.

5. If the first three nucleotides in a six-nucleotide restriction site are CTG, what would the next three nucleotides most likely be?
   a. AGG
   b. GTC
   c. CTG
   d. CAG
   e. GAC

6. The following segment of DNA has restriction sites I and II, which create restriction fragments a, b, and c. Which of the following gels produced by electrophoresis would represent the separation and identity of these fragments?

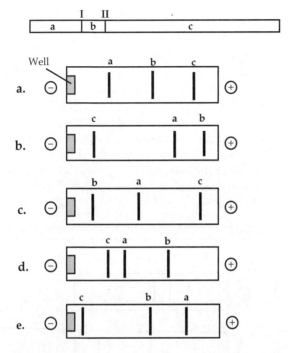

7. Which of the following processes or procedures does *not* involve any nucleic acid hybridization?
   a. separation of fragments by gel electrophoresis
   b. Southern blotting
   c. polymerase chain reaction
   d. DNA profiling
   e. DNA microarray assay

8. Which of the following statements is *not* true of restriction sites?
   a. Modification by methylation of bases within them prevents restriction of bacterial DNA.
   b. They are usually symmetrical sequences of four to eight nucleotides.
   c. They signal the attachment of RNA polymerase.
   d. Each is cut by a specific restriction enzyme.
   e. Cutting one in the middle of a functional and identifiable gene is used to screen clones that have taken up foreign DNA.

9. Which of the following statements describes a difficulty in getting prokaryotic cells to express eukaryotic genes?
   a. The signals that control gene expression are different, and prokaryotic promoter regions must be added to the vector.

   b. The genetic code differs because prokaryotes substitute the base uracil for thymine.
   c. Prokaryotic cells cannot transcribe introns because their genes do not have them.
   d. The ribosomes of prokaryotes are not large enough to handle long eukaryotic genes.
   e. The RNA splicing enzymes of bacteria work differently from those of eukaryotes.

10. Complementary DNA does not create as complete a library of genes as the shotgun approach because
    a. it has eliminated introns from the genes.
    b. a cell produces mRNA for only a small portion of its genes.
    c. the shotgun approach produces more restriction fragments.
    d. cDNA is not as easily integrated into plasmids.
    e. reverse transcriptase cannot transcribe introns.

Use the following choices to answer questions 11–14.
    a. restriction enzyme
    b. reverse transcriptase
    c. DNA ligase
    d. DNA polymerase
    e. RNA polymerase

11. Which is the last enzyme involved in making recombinant plasmids?

12. Which is the first enzyme used in the production of cDNA?

13. Which enzyme is used in the polymerase chain reaction?

14. Which is the first enzyme used in the production of RFLPs?

15. You are attempting to introduce a gene that imparts resistance to larval moths in bean plants. Which of the following vectors are you most likely to use?
    a. phage DNA
    b. *E. coli* plasmid
    c. Ti plasmid
    d. yeast plasmid
    e. bacterial artificial chromosome

16. STRs (short tandem repeats) are a valuable tool for
    a. DNA microarray assays.
    b. infecting plant cells with recombinant DNA.
    c. acting as probes in Southern blots.
    d. genetic profiling.
    e. PCR to produce multiple copies of a DNA segment.

17. You have affixed the chromosomes from a cell onto a microscope slide. Which of the following would *not* make a good radioactively labeled probe to help map a particular gene to one of those chromosomes? (Assume that the DNA of chromosomes and probes is single stranded.)
    a. cDNA made from the mRNA transcribed from the gene
    b. a portion of the amino acid sequence of that protein
    c. mRNA transcribed from the gene
    d. a piece of the restriction fragment on which the gene is located
    e. a sequence of nucleotides determined from a known sequence of amino acids in the protein product of the gene

18. Which of the following is *not* true of adult stem cells?
    a. They have been found not only in bone marrow, but also in other tissues, including the adult brain.
    b. They have been successfully grown in culture and made to differentiate into specialized cells.
    c. They are capable of developing into several (but not all) types of cells.
    d. These relatively unspecialized cells continually reproduce themselves in the body.
    e. They come from skin cells that have been induced to become pluripotent by the introduction of cloned "stem cell" master regulatory genes.

19. A "pharm" animal is
    a. a transgenic animal that produces large quantities of a pharmaceutical product.
    b. an animal used by the pharmaceutical industry to test new medical treatments.
    c. a cloned animal that was produced from an adult cell nucleus inserted into an egg.
    d. a transgenic animal that produces more meat.
    e. a genetically modified organism whose production is permitted in the United States but not in the European Union.

20. Petroleum-lysing bacteria are being engineered for the treatment of oil spills. Which of the following is the most realistic danger of these bacteria to the environment?
    a. mutations leading to the production of a strain pathogenic to humans
    b. extinction of natural microbes due to the competitive advantage of the "petro-bacterium"
    c. destruction of natural oil deposits
    d. poisoning of the food chain
    e. contamination of the water

21. Which of the following would be useful in signaling the presence of a disease-causing allele even if the gene has not yet been identified?
    a. RNA interference
    b. *in situ* hybridization
    c. short tandem repeats
    d. single nucleotide polymorphisms
    e. RT-PCR

22. The following restriction fragment contains a gene whose recessive allele is lethal. The normal allele has restriction sites for the restriction enzyme PSTI at sites I and II. The recessive allele lacks restriction site I. An individual whose sister had the lethal trait is being tested to determine if he is a carrier of that allele. Which of the following band patterns would be produced on a gel if he is a carrier (heterozygous for the gene)?

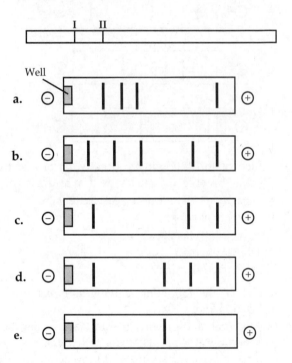

Chapter 21

Genomes and Their Evolution

Key Concepts

21.1 New approaches have accelerated the pace of genome sequencing

21.2 Scientists use bioinformatics to analyze genomes and their functions

21.3 Genomes vary in size, number of genes, and gene density

21.4 Multicellular eukaryotes have much noncoding DNA and many multigene families

21.5 Duplication, rearrangement, and mutation of DNA contribute to genome evolution

21.6 Comparing genome sequences provides clues to evolution and development

Framework

This chapter introduces genomics and bioinformatics, new approaches to analyzing and comparing the genomes of life's diverse organisms. Researchers in these fields address questions about genome organization, gene expression, growth and development, and evolution. The various types of noncoding DNA in the human genome are described. The chapter also covers the processes that contribute to genome evolution.

Chapter Review

The complete genome sequences of humans, chimpanzees, and numerous other organisms are enabling the study of whole sets of genes and their interactions, called **genomics.** The new field of **bioinformatics** is applying computational methods to the analysis of the ever-growing volume of biological data.

21.1 New approaches have accelerated the pace of genome sequencing

The international effort to sequence the human genome, called the **Human Genome Project,** was begun in 1990 and declared "virtually completed" in 2006.

Three-Stage Approach to Genome Sequencing Cytogenetic maps based on banding patterns and fluorescence *in situ* hybridization (FISH) provided the basis for mapping chromosomes. The first stage in mapping the human genome was to develop a **linkage map** of several thousand genetic markers, either genes or other identifiable sequences such as RFLPs or simple tandem repeats (STRs). Determining recombination frequencies using these known markers enabled researchers to order genes and locate other markers.

In a **physical map,** the actual distances between markers are determined. The DNA of each chromosome is cut into overlapping restriction fragments, and their order on the chromosome is determined. First large fragments are cut, cloned, and ordered. Then those fragments are cut, cloned, and their fragments ordered.

The nucleotide sequences of short fragments were determined by sequencing machines, using the dideoxy chain-termination method. The development of "high-throughput" sequencing machines greatly speeded the process.

Whole-Genome Shotgun Approach to Genome Sequencing J. C. Venter, founder of the company Celera Genomics, developed a *whole-genome shotgun approach* that relies on powerful computer programs to order the large number of sequenced, overlapping short fragments. The whole-genome shotgun approach is now widely used. Newer sequencing techniques, called *sequencing by synthesis,* have both increased the speed and decreased the cost of sequencing genomes. **Metagenomics** is an approach in which an environmental sample that includes DNA from many species is sequenced, and then computers sort the sequences into specific genomes.

INTERACTIVE QUESTION   21.1

Compare the strategies of the public consortium of the Human Genome Project and of the whole-genome shotgun approach for mapping a genome.

21.2   Scientists use bioinformatics to analyze genomes and their functions

Centralized Resources for Analyzing Genome Sequences
The National Center for Biotechnology Information (NCBI) and other genome centers maintain websites with databases of DNA sequences and protein sequences and structures, as well as software for analyzing and comparing sequences and structures. The ever-expanding NCBI database of sequences, called GenBank, makes the resources of bioinformatics available to researchers worldwide. A Protein Data Bank contains all the three-dimensional protein structures determined thus far.

Identifying Protein-Coding Genes and Understanding Their Functions   Now that genes can be studied directly, geneticists employ *reverse genetics* to determine phenotype from genotype. Through the process called **gene annotation,** sequences are analyzed to identify protein-coding genes and their functions. Computer software can scan DNA sequences for signs of genes, such as start and stop signals; RNA-splicing sites; and *expressed sequence tags,* or *ESTs* (sequences from known mRNAs and stored as cDNA in databases).

The sequences that might be genes are compared with sequences of known genes of other species to look for similarities that might indicate the gene's function. A combination of biochemical techniques (identifying three-dimensional structure and binding sites) and functional studies (blocking the gene with RNAi to determine the effect on phenotype) helps to reveal protein function.

Understanding Genes and Gene Expression at the Systems Level   A research project called ENCODE (Encyclopedia of DNA Elements) analyzed 1% of the human genome to identify protein-coding and noncoding RNA genes, regulatory sequences, and chromatin modifications. Researchers found that only 2% of the region they studied codes for proteins, but over 90% of the region was transcribed. This approach has now been extended to the entire human genome and to those of the nematode *C. elegans* and the fruit fly *D. melangaster.*

Proteomics is the identification and study of entire protein sets *(proteomes)* coded for by a genome. With compiled lists of DNA sequences and proteins now available, researchers are studying the functional integration of these components in biological systems. This systems biology approach seeks to define gene circuits and protein interaction networks, using computers, mathematics, and new software to process and integrate huge amounts of data. To map the protein interaction network in the yeast *S. cerevisiae,* for example, researchers created doubly mutant cells, knocking out one pair of genes at a time, in order to test for gene interactions.

The Cancer Genome Atlas is taking a systems biology approach to analyzing changes in genes and patterns of gene expression in cancer cells. The project catalogues the mutations found in several common cancers and identifies genes of known and unknown functions that may provide new targets for therapies. Silicon "chips" holding arrays of most known human genes are used to analyze gene expression in patients with various diseases. Such approaches may help to match treatments with a person's unique genetic makeup.

INTERACTIVE QUESTION   21.2

The software program BLAST, available on the NCBI website, allows a researcher to compare a DNA sequence to every sequence in Genbank to identify similar regions. What are some uses of this function?

21.3   Genomes vary in size, number of genes, and gene density

Genome Size   Of the 1,200 genomes that had been sequenced as of early 2010, most are of bacteria, whose genomes have between 1 and 6 million base pairs (Mb). Archaeal genomes appear to be of a similar size, whereas most animals and plants have genomes of at least 100 Mb. There does not appear to be a correlation between genome size and an organism's phenotype.

Number of Genes   Bacteria and archaea have from 1,500 to 7,500 genes; eukaryotes range from about 5,000 genes for unicellular fungi to 40,000 for some multicellular organisms. The human genome contains fewer than 21,000 genes. Alternative splicing of the ten or so exons in most human genes can yield many different proteins for each gene. Small RNAs (such as miRNAs) that regulate gene expression may contribute to greater organismal complexity.

Gene Density and Noncoding DNA Eukaryotes have fewer genes per million base pairs than bacteria or archaea. Mammals appear to have the lowest gene density. Most of the DNA of eukaryotic genomes is noncoding DNA that is located within and between genes and includes introns and complex regulatory sequences.

INTERACTIVE QUESTION 21.3

Refer to the organisms listed in Table 21.1 in your text to answer the following questions.

a. Which organism has the highest gene density? _____ the lowest gene density? _____

b. Which organism has the largest number of genes? _____ the smallest number? _____

c. Which organism has the largest haploid genome size? _____ the smallest genome size? _____

d. What is the estimated number of genes in the human genome? _____ Explain the fact that there are many more different polypeptides than genes.

21.4 Multicellular eukaryotes have much noncoding DNA and many multigene families

Noncoding DNA, which was previously referred to as "junk DNA," may turn out to have important functions. Almost 500 identical regions of noncoding DNA have been identified in humans, rats, and mice, a higher level of sequence conservation than for protein-coding regions in these species.

About 1.5% of the human genome consists of exons that code for proteins, rRNA, or tRNA. The rest includes gene-related regulatory sequences (5%), introns (20%), gene fragments and nonfunctional former genes called **pseudogenes,** and sequences present in many copies, called **repetitive DNA.** Much of this repetitive DNA (44% of the human genome) is either made up of or related to transposable elements. About 15% of the genome is unique noncoding DNA.

Transposable Elements and Related Sequences Stretches of DNA that can move about within a genome through a process called *transposition* are called *transposable genetic elements* or **transposable elements.**

Transposons move about a genome as a DNA intermediate, either by a "cut-and-paste" mechanism or a "copy-and-paste" mechanism. The enzyme *transposase,*

generally encoded by the transposon, is required for both mechanisms. **Retrotransposons,** which make up the majority of transposable elements, are first transcribed into an RNA intermediate. This RNA transcript is converted back to DNA by reverse transcriptase, which is coded for by the retrotransposon itself.

Transposable elements may be represented as multiple (although not identical) copies of transposons or as related sequences that have lost the ability to move. In humans, about 10% of the genome is made up of *Alu elements.* Many of these 300-nucleotide-long sequences are transcribed into RNA but are of unknown function.

About 17% of the human genome consists of *LINE-1,* or *L1,* retrotransposons. The introns of about 80% of analyzed human genes contain L1 sequences, suggesting that L1 may help regulate gene expression.

INTERACTIVE QUESTION 21.4

Why do retrotransposons always move by the "copy-and-paste" mechanism?

Other Repetitive DNA, Including Simple Sequence DNA About 14% of the human genome is repetitive DNA that appears to have arisen from mistakes in DNA replication. Scattered large-segment duplications account for 5–6% of the human genome. **Simple-sequence DNA,** by contrast, makes up 3% of the human genome and consists of multiple copies of tandemly repeated sequences. When the repeat consists of two to five nucleotides, the unit is called a **short tandem repeat,** or **STR.** The variation in repeat numbers between genomes is the basis of genetic profiles. Much of a genome's simple sequence DNA is located at centromeres, where it functions in cell division and chromatin organization, and at telomeres, which protect the tips of chromosomes.

Genes and Multigene Families Sequences coding for proteins, tRNAs, and rRNAs make up 1.5% of the human genome. Including introns and regulatory sequences, the amount of gene-related DNA is 25%. More than half of the coding DNA occurs in **multigene families,** collections of similar or identical genes.

With the exception of the genes for histone proteins, *identical* multigene families code for RNA products. The genes coding for the three largest rRNA molecules are arranged in a single transcription unit repeated in huge tandem arrays, enabling cells to produce the millions of ribosomes needed for protein synthesis.

Examples of multigene families of *nonidentical* genes are the two families of genes that code for globins, including the α and β polypeptide subunits of hemo-globin. Different versions of each globin subunit are clustered together on two different chromosomes and are expressed at the appropriate times during development.

Complete Interactive Question 21.5 to review the types of DNA found in the human genome.

INTERACTIVE QUESTION   21.5

For each of the following types of DNA sequences found in the human genome, write the letter of the correct description and the percentage of the genome (listed beneath the descriptions) in the blanks provided.

Types of DNA	Description	%
1. Exons or rRNA/ tRNA-coding	_____	_____
2. Introns	_____	_____
3. Regulatory sequences	_____	_____
4. Transposable elements and related sequences	_____	_____
5. *Alu* elements	_____	_____
6. L1 sequences	_____	_____
7. Unique noncoding DNA	_____	_____
8. Large-segment duplications	_____	_____
9. Simple sequence DNA	_____	_____

Descriptions

A. DNA in centromeres and telomeres, also STRs

B. multiple copies of mostly movable sequences

C. gene fragments and pseudogenes

D. protein- and RNA-coding sequences

E. family of short sequences related to transposable elements

F. multiple copies of large sequences

G. retrotransposons found in introns of most genes

H. enhancers, promoters, and other such sequences

I. noncoding sequences within genes

Choices of percentages: 1.5, 3, 5, 5–6, 10, 15, 17, 20, and 44 (These percentages do not add up to 100 because some of these types of DNA are subsets of other categories, and some types are not listed.)

21.5   Duplication, rearrangement, and mutation of DNA contribute to genome evolution

Duplication of Entire Chromosome Sets   Extra sets of chromosomes may arise by accidents in meiosis. The resulting extra genes might diverge through mutation, leading to genes with novel functions. Polyploidy is fairly common in plants.

Alterations of Chromosome Structure   Using genomic sequence information, researchers can compare the locations of DNA sequences on chromosomes among different species and reconstruct the evolutionary history of chromosomal rearrangements. Duplications and inversions of chromosomes are thought to contribute to speciation. The breakage points associated with these alterations appear to be recombination "hotspots." Such sites are associated with some human congenital diseases.

Duplication and Divergence of Gene-Sized Regions of DNA   Errors such as unequal crossing over during meiosis (as may occur between copies of a transposable element on misaligned nonsister chromatids) and slippage of template strands during DNA replication might lead to the duplication of individual genes.

The α-globin and β-globin gene families appear to have evolved from a common ancestral globin gene, which was duplicated and then diverged. Multiple duplications and mutations within each family have led to the current family of genes with related functions along with several intervening pseudogenes.

In other cases, mutation of a duplicated gene may lead to a protein product with a new function.

INTERACTIVE QUESTION   21.6

Lysozyme and α-lactalbumin have similar amino acid sequences but different functions. The genes for both proteins are found in mammals, but birds have only the gene for lysozyme. What does this observation suggest about the evolution of these genes?

Rearrangements of Parts of Genes: Exon Duplication and Exon Shuffling   Unequal crossing over can lead to a gene with a duplicated exon. Exons often code for domains of a protein, and their duplication could provide a protein with enhanced structure and function. Errors in meiotic recombination could also lead to *exon shuffling* within a gene or between nonallelic genes.

How Transposable Elements Contribute to Genome Evolution Recombination events can take place between homologous transposable element sequences that are scattered throughout the genome, causing chromosomal mutations that may occasionally be beneficial to the organism. Transposable elements that insert within a gene may disrupt its functioning; those that insert within regulatory sequences may increase or decrease gene expression. A transposable element can also move a copy of a gene or an exon to a new location. The increased genetic diversity provided by these mechanisms provides raw material for natural selection.

INTERACTIVE QUESTION   21.7

a. Explain two ways in which exon shuffling could occur.

b. What is a potential benefit of exon shuffling?

21.6   Comparing genome sequences provides clues to evolution and development

Comparing Genomes Comparisons of *highly conserved* genes illuminate the evolutionary relationships among species that are distantly related. Such analyses support the theory that bacteria, archaea, and eukaryotes represent the three domains of life, and also demonstrate the advantages of using model organisms to study both basic biological processes and human biology.

The similarity between genomes of two closely related species allows researchers to use one genome sequence as a framework for mapping the other genome. The identified small differences between genomes translate into the phenotypic divergence of the species. The human and chimpanzee genomes differ in single nucleotide substitutions by only 1.2%. Insertions or deletions of larger regions in the genome result in an additional 2.7% difference. There are more *Alu* elements in the human genome, and a third of the human duplications are not in the chimpanzee genome.

Comparisons of genetic changes since species diverged show that some genes are changing faster in humans than in the chimpanzee or mouse. Many of these more-quickly evolving genes code for transcription factors; one example is the *FOXP2* gene, which appears to function in vocalization in vertebrates and in speech and language in humans. Mutations in this gene cause verbal impairment in humans; it is expressed in the brains of songbirds during the period they are learning

their songs; and knock-out experiments with mice have shown that homozygous mutant mice had malformed brains and did not produce their normal vocalizations.

Comparisons of human genomes have revealed several million single nucleotide polymorphism (SNP) sites, as well as inversions, deletions, and duplications. A surprisingly high number of *copy-number variants* (*CNVs*) have been identified in which some individuals have one or multiple copies of a gene or genetic region rather than the normal two. These genetic markers will contribute to the study of human evolution.

Comparing Developmental Processes Biologists in the field of evolutionary developmental biology (**evo-devo**) compare developmental processes to understand how they have evolved and how minor changes in gene sequence or regulation may lead to diverse forms of life.

A sequence of 180 nucleotides called a **homeobox**, which codes for a *homeodomain*, has been found in *Drosophila* homeotic genes. The same or very similar homeobox nucleotide sequences have been identified in homeotic genes of many animals. Homeotic genes in the fruit fly and mouse are found in the same linear sequence on chromosomes. Related sequences are found in regulatory genes of yeast and plants. These similarities indicate that the homeobox sequence must have arisen early and been conserved through evolution as part of the genes involved in the regulation of gene expression and development.

Homeotic genes are often called *Hox* genes in animals. Proteins with homeodomains probably coordinate the transcription of groups of developmental genes.

Many other genes involved in development, such as those coding for components of signaling pathways, are highly conserved. The differing patterns of expression of these genes in different body areas may explain the development of animals with different body plans.

INTERACTIVE QUESTION   21.8

If all *Hox* genes contain the same or very similar homeobox, how can they control different developmental sequences?

The processes of development evolved independently in plants and animals due to their ancient divergence. Their rigid cell walls restrict movement, and morphogenesis in plants relies more on the orientation of cell divisions and cell enlargement. Some similarities persist from their common ancestral single-celled eukaryote. In both, development involves a cascade of transcriptional regulators, although the master control genes are different in the two groups. Although both *Hox* genes and *Mads-box* genes are found in animals and plants, the former act as master regulatory switches in animals, whereas the latter do so in plants.

Word Roots

pseudo- = false (*pseudogene:* a DNA segment that is very similar to a real gene but does not yield a functional product)

retro- = backward (*retrotransposon:* a transposable element that moves within a genome by means of an RNA intermediate, a transcript of the retrotransposon DNA)

Structure Your Knowledge

1. About 25% of the human genome relates to the production of proteins or RNA products (exons, introns, or regulatory sequences). Is the remaining 75% just "junk"? Describe the following types of noncoding DNA, including some of their possible functions.
   a. transposable elements
   b. *Alu* elements
   c. L1 sequences
   d. simple sequence DNA
   e. pseudogenes
2. Describe some of the processes that contribute to genome evolution.

Test Your Knowledge

MULTIPLE CHOICE: *Choose the one best answer.*

1. Why is the whole-genome shotgun approach now widely used to sequence genomes?
   a. It uses only one, very efficient restriction enzyme to create fragments.
   b. It avoids the steps of producing linkage and physical maps.
   c. Newer sequencing techniques, such as sequencing by synthesis, now rapidly clone fragments.
   d. It uses bioinformatics, whereas the three-stage approach to genome sequencing does not.
   e. All of the above contribute to its widespread use.

2. Metagenomics is a new approach that
   a. identifies proteomes and protein interaction networks.
   b. analyzes genomes for all functionally important elements.
   c. provides sequence data and software programs on Internet websites.
   d. sequences all the DNA in an environmental sample and uses computer software to assemble the sequences into specific genomes.
   e. applies genome-wide association studies to the identification of human genes of medical importance.

3. Computer software can identify putative genes in nucleotide sequences. For which of the following components is the software probably *not* scanning?
   a. promoters
   b. RNA-splicing sites
   c. STRs (simple tandem repeats)
   d. ESTs (expressed sequence tags)
   e. stop signals for transcription

4. Why is proteomics important in the systems biology approach?
   a. The interactions of networks of proteins are central to the functioning of cells and organisms.
   b. This bioinformatics field allows for the mathematical modeling of biological systems.
   c. Determining the proteins expressed in a cell identifies the genes more accurately than can be done through genomics.
   d. The three-dimensional structure of a protein can be used to predict its function.
   e. Comparing the proteins produced by a normal allele and the allele associated with a disease can facilitate improved treatments.

5. Bacterial genes have an average length of 1,000 base pairs; human genes average about 27,000 base pairs. Which of the following statements is the best explanation for that difference?
   a. Prokaryotes have smaller, but many more, individual genes.
   b. Prokaryotes are more ancient organisms; longer genes arose later in evolution.
   c. Prokaryotes are unicellular; humans have many types of differentiated cells.
   d. Prokaryotic genes do not have introns; human genes have multiple introns.
   e. Prokaryotic proteins are not as large and complex as human proteins.

6. Which of the following statements best explains the discovery that a complex human has roughly the same number of genes as the simple nematode *C. elegans*?

   a. The unusually long introns in human genes are involved in regulation of gene expression.

   b. More than one polypeptide can be produced from a human gene by alternative splicing.

   c. Human genes code for many more types of domains.

   d. The human genome has a high proportion of non-coding DNA.

   e. The large number of SNPs (single nucleotide polymorphisms) in the human genome provides a great deal of genetic variability.

7. Which of the following statements best describes what pseudogenes and introns have in common?

   a. They do not produce a functional product.

   b. They are DNA segments that lack a promoter but have other control regions.

   c. They are transcribed but their translation is blocked by miRNAs.

   d. They code for RNA products, not proteins.

   e. They appear to have arisen from retrotransposons.

8. Which of the following techniques can be used to determine the function of a newly identified gene?

   a. comparisons with genes of known functions that have similar sequences

   b. blockage of gene function with RNAi to see the effect on the phenotype

   c. searches for *Alu* sequences that may indicate alternate splicing sites

   d. both a and b

   e. a, b, and c

9. Which of the following statements is *not* descriptive of transposable elements?

   a. Barbara McClintock's work with maize provided the first evidence of such DNA segments.

   b. Transposable elements make up 85% of the corn genome.

   c. Retrotransposons called *LINE-1* are found within the introns of many human genes and may help regulate gene expression.

   d. Transposable elements often encode the enzymes, such as transposase or reverse transcriptase, necessary for their movement.

   e. Each transposable element is present as multiple identical copies, often clustered in the centromere or telomere regions of a chromosome.

10. The protein tissue plasminogen activator (TPA) has three types of domains. Which of the following statements explains why one of each of these types of domains is found in three different proteins (epidermal growth factor, fibronectin, and plasminogen).

    a. The genes for all four proteins are members of a multigene family involved in cell signaling.

    b. The gene for TPA was the first gene to evolve; the other three genes each lost two domains.

    c. The gene for TPA arose by exon shuffling involving the other three genes.

    d. The gene for TPA has many *Alu* elements that provide alternative splice sites to incorporate these exons.

    e. Several duplication events led to the evolution of the TPA gene.

11. Genes that are highly conserved are useful for

    a. identifying genes that led to new species.

    b. determining the function of newly discovered genes.

    c. establishing the sequence of divergence of closely related species.

    d. tracing the relationships of groups that diverged early in the evolution of life.

    e. both a and c.

12. A highly conserved nucleotide sequence that has been found in developmental regulatory genes in many diverse organisms is called

    a. a homeodomain.

    b. a homeobox.

    c. a retrotransposon.

    d. a homeotic gene.

    e. an L1 sequence.

13. Which of the following events has probably *not* contributed to genome evolution?

    a. duplications and inversions of chromosome sections

    b. exon duplication and shuffling

    c. recombination between transposable elements on different chromosomes

    d. divergence of duplicated genes through mutations

    e. All of the above have contributed to genome evolution.

14. Which of the following approaches would be most useful in tracing human evolution?

    a. evo-devo and the comparison of developmental genes in plants and animals

    b. metagenomics and proteomics

    c. systems biology and the use of "knock-out" experiments

    d. analysis of single nucleotide polymorphisms and copy-number variants across individuals from the same and different populations

    e. All of the above make important contributions to studying the evolution of human populations.

4 Mechanisms of Evolution

Chapter 22

Descent with Modification: A Darwinian View of Life

▮ Key Concepts

22.1 The Darwinian revolution challenged traditional views of a young Earth inhabited by unchanging species

22.2 Descent with modification by natural selection explains the adaptations of organisms and the unity and diversity of life

22.3 Evolution is supported by an overwhelming amount of scientific evidence

▮ Framework

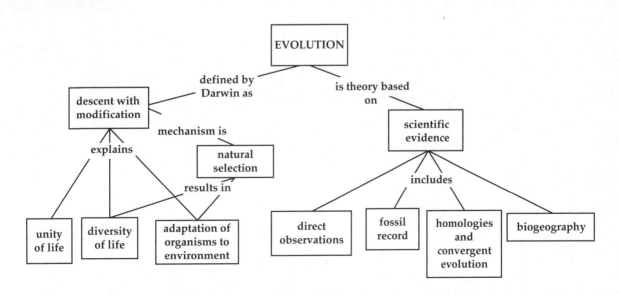

Chapter Review

Charles Darwin presented his scientific explanation for the adaptations of organisms to their environments, the diversity of life, and the unity of life in his book *On the Origin of Species*. **Evolution** may be defined in Darwin's terms as *descent with modification,* or more narrowly as the changes in a population's genetic composition over time.

Evolution can be viewed as both the *pattern* of evolutionary change observable in the natural world and the *process* or mechanisms underlying those changes.

22.1 The Darwinian revolution challenged traditional views of a young Earth inhabited by unchanging species

Scala Naturae and Classification of Species   The Greek philosopher Aristotle proposed that all forms of life were permanent and perfect and could be arranged on a "scale of nature" of increasing complexity.

The Old Testament account of creation asserts that species are perfect and fixed. In the 1700s, Linnaeus developed both a *binomial* system for naming organisms according to their genus and species, and a hierarchy of groups organizing his named species.

Ideas About Change over Time   **Fossils** are remnants or traces of past organisms, usually found in sedimentary rocks formed through the compression of layers of sand and mud into superimposed layers called **strata.** Cuvier developed **paleontology,** the study of fossils. Advocating **catastrophism,** he maintained that the extinctions and differences in the fossils found in different strata are the result of local sudden catastrophic events and are not indicative of evolution.

Hutton proposed that immense changes in Earth's geology are the cumulative result of slow but continuous processes. Lyell, a contemporary of Darwin, developed a theory of **uniformitarianism,** stating that the rates of geologic processes have remained the same throughout Earth's history and continue in the present.

Darwin took two ideas from the observations of Hutton and Lyell: Earth must be very old, and very slow processes could produce substantial changes in species.

Lamarck's Hypothesis of Evolution   Lamarck's hypothesis, published in 1809, explained the mechanism of evolution using two principles: the *use and disuse* of body parts, leading to their development or deterioration, and the *inheritance of acquired characteristics.* Although current genetic knowledge rejects his mechanism, Lamarck proposed the key idea that evolution is the best explanation for both the fossil record and the adaptation of organisms to the environment.

INTERACTIVE QUESTION   22.1

a. Write the capital letter representing the theory or philosophy, and the lowercase letter(s) representing its proponent(s), in the blanks preceding the following seven descriptions.

A. catastrophism

B. inheritance of acquired characteristics

C. gradual geologic changes

D. descent with modification

E. classification system

F. scale of nature

G. uniformitarianism

a. Aristotle

b. Cuvier

c. Darwin

d. Hutton

e. Lamarck

f. Linnaeus

g. Lyell

Theory   Proponent(s)

1. ——— ——— Describes the diversity of God's creations by naming and classifying species

2. ——— ——— The history of Earth is marked by sudden floods or droughts that resulted in extinctions

3. ——— ——— Proposed mechanism of evolution in which modifications due to use or disuse are passed on to offspring

4. ——— ——— Profound change is the cumulative product of slow but continuous processes

5. ——— ——— Earth contains fixed species on a continuum from simple to complex

6. ——— ——— All of life is related; present-day species differ from ancestral species

7. ——— ——— Geologic processes have constant rates throughout time

b. Now place the men listed in a through g in chronological order.

——— ——— ——— ——— ——— ——— ———

22.2 Descent with modification by natural selection explains the adaptations of organisms and the unity and diversity of life

Darwin's Research Darwin was 22 years old when he sailed from Great Britain on the HMS *Beagle*. He spent the voyage collecting thousands of specimens of the fauna and flora of South America, observing the various adaptations of organisms living in very diverse habitats, and making special note of the geographic distribution of the distinctly South American species. Darwin also read and was influenced by Lyell's *Principles of Geology.*

Adaptations are inherited characteristics that contribute to an organism's survival and reproduction in a specific environment. Darwin proposed that adaptations arise through **natural selection,** a process in which individuals with beneficial characteristics produce more offspring than others because of those characteristics. In 1844 he wrote an essay on the origin of species and natural selection but did not publish it. In 1858 Darwin received Wallace's manuscript describing a nearly identical theory. Darwin then published *On the Origin of Species by Means of Natural Selection* in 1859.

The Origin of Species Darwin's book developed two main points: Descent with modification is the basis of life's unity and diversity, and natural selection is the mechanism that matches organisms with their environment.

Darwin's concept of descent with modification explains that all organisms are related through descent from some unknown ancestor and develop increasing modifications as they adapt to various habitats. The history of life is analogous to a tree, with a common ancestor at the fork of each new branch and with present-day species at the tips of the youngest twigs. The taxonomy developed by Linnaeus provided a hierarchical organization of groups that suggested to Darwin this branching tree of life.

Using the example of **artificial selection** in the breeding of domesticated plants and animals, Darwin presented evidence that selection among the variations present in a population can lead to substantial changes.

Malthus' essay on human population growth supported Darwin's idea of the overproduction of offspring.

Remember these three points about evolution by natural selection: (1) Natural selection results in the evolution of populations, not individuals; (2) natural selection affects only those traits that are heritable and that differ in a population; and (3) natural selection depends on the specific environmental factors present in a region at a given time. If the environment changes, different adaptations will be favored.

INTERACTIVE QUESTION 22.2

List the two observations from which Darwin drew the two inferences that explain natural selection.

   Observation 1:

   Observation 2:

Inference 1: Individuals whose inherited characteristics give them a better chance of surviving and reproducing in a given environment are likely to leave more offspring.

Inference 2: Unequal survival and reproduction leads to an accumulation of favorable traits over generations.

22.3 Evolution is supported by an overwhelming amount of scientific evidence

Direct Observations of Evolutionary Change Studies have shown that the beak length of soapberry bugs correlates with the size of the fruits that house the seeds on which they feed. Populations of soapberry bugs have become adapted to introduced species, often in relatively short amounts of time.

The development of new antibiotics is usually followed rapidly by the evolution of resistance to them. Current strains of MRSA—methicillin-resistant *Staphylococus aureus*—are resistant to multiple antibiotics and can cause potentially lethal infections.

The evolution of drug-resistant bacteria and changes in soapberry bug beak length illustrate two facets of natural selection: It is an editing, not a creative, mechanism that selects for variations already present in a population. And it is regional and temporal, selecting for traits that are beneficial in the local environment at that current time.

INTERACTIVE QUESTION 22.3

a. Explain how the rapid evolution of drug resistance in bacteria is an example of natural selection.

b. How might multidrug-resistant strains of MRSA have evolved?

Homology   **Homology** is similarity resulting from common ancestry. The forelimbs of all mammals are **homologous structures,** containing the same skeletal elements regardless of function or external shape. Comparative anatomy and embryology illustrate that evolution is a remodeling process in which ancestral structures become modified for new functions.

Vestigial structures, which may be of little or no value to the organism, are historical remnants of ancestral structures.

Homologies can be seen on a molecular level. DNA, RNA, and an essentially universal genetic code, which have been passed along through all branches of evolution, are important evidence that all forms of life descended from the earliest organisms and are thus related.

Homology is evident on different hierarchical levels, reflecting evolutionary history in a nested pattern of descent. An **evolutionary tree** is a diagram of evolutionary relationships. Each branch point represents the common ancestor of all the species beyond that point. Evolutionary trees are hypotheses based on the best available data.

Distantly related organisms may appear similar as a result of **convergent evolution,** the independent evolution of similar characteristics. These **analogous** structures arise as a result of evolutionary adaptation to similar environments.

The Fossil Record   The fossil record documents that present and past organisms differ, and that many species have become extinct. Fossils also trace the evolution of new groups, as in the origin of whales from land mammals.

The major branches of evolutionary descent established with evidence from anatomy and molecular data have been tested and supported by the sequence of fossil forms found in the fossil record.

Biogeography   The geographic distribution of species, or **biogeography,** has been influenced by *continental drift,* the slow movements of Earth's continents. The single land mass called **Pangaea** formed 250 million years ago and began to break apart 200 million years ago. The gradual separation of continents helps to explain and predict where fossils of different groups and their descendants are found.

Islands often have **endemic** species, found nowhere else and usually closely related to species on the nearest island or mainland. Widely separated areas having similar environments are not likely to be populated by closely related species. Rather, each area is more likely to have species that are taxonomically related to those of their region, regardless of environment.

INTERACTIVE QUESTION   22.4

Complete the following concept map that summarizes the main sources of evidence for evolution.

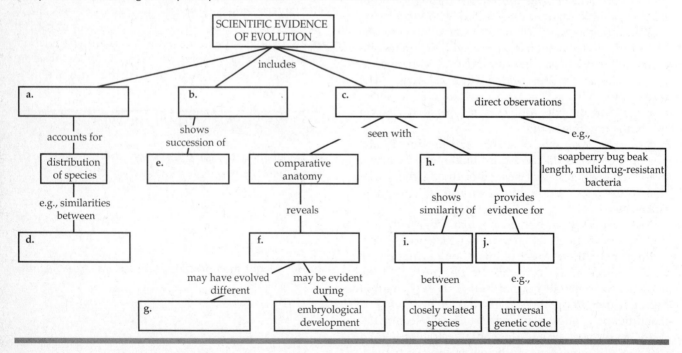

What Is Theoretical About the Darwinian View of Life? A scientific *theory* is a unifying concept with broad explanatory power and with predictions that have been and continue to be tested by experiments and observations.

Word Roots

bio- = life; **geo-** = the Earth (*biogeography:* the study of the past and present distribution of species)

end- = within (*endemic:* referring to a species that is confined to a specific geographic area)

homo- = like, resembling (*homology:* similarity in characteristics resulting from a shared ancestry)

paleo- = ancient (*paleontology:* the scientific study of fossils)

vestigi- = trace (*vestigial organ:* a feature of an organism that is an historical remnant of a structure that served a function in the organism's ancestors)

Structure Your Knowledge

1. Explain in your own words the main components of Darwin's theory of evolution.

Test Your Knowledge

MULTIPLE CHOICE: *Choose the one best answer.*

1. The best description of natural selection is
   a. the survival of the fittest.
   b. the struggle for existence.
   c. the reproductive success of the members of a population best adapted to the environment.
   d. the overproduction of offspring in environments with limited natural resources.
   e. a change in allele frequencies in a population.

2. To Cuvier, the differences in fossils from different strata were evidence for
   a. changes occurring as a result of cumulative but gradual processes.
   b. divine creation.
   c. evolution by natural selection.
   d. continental drift.
   e. local catastrophic events such as droughts or floods.

3. Darwin proposed that new species evolve from ancestral forms by
   a. the gradual accumulation of adaptations to changing or different environments.
   b. the inheritance of acquired adaptations to the environment.
   c. the struggle for limited resources.
   d. the accumulation of mutations.
   e. the excessive production of offspring.

4. All of the following influenced Darwin as he synthesized the theory of evolution by natural selection *except*
   a. the biogeographic distribution of species such as the mockingbirds on the Galápagos Islands.
   b. Lyell's book, *Principles of Geology*, on the gradualness of geologic changes.
   c. Linnaeus's hierarchical classification of species, which could be interpreted as evidence of evolutionary relationships.
   d. examples of artificial selection that produce rapid changes in domesticated species.
   e. Mendel's paper in which he described his "laws of inheritance."

5. The smallest unit that can evolve is
   a. an individual.
   b. a mating pair.
   c. a species.
   d. a population.
   e. a community.

6. Which of the following statements is *not* considered part of the process of natural selection?
   a. Many of the variations among individuals in a population are heritable.
   b. More offspring are produced than are able to survive and reproduce.
   c. Individuals with traits best adapted to the environment are likely to leave more offspring.
   d. Many adaptive traits may be acquired during an individual's lifetime, contributing to that individual's reproductive success.
   e. Unequal reproductive success leads to gradual change in a population.

7. What might you conclude from the observation that the bones in your arm and hand are similar to the bones that make up a bat's wing?
   a. The bones in the bat's wing are vestigial structures, no longer useful as "arm" bones.
   b. The bones in a bat's wing are homologous to your arm and hand bones.
   c. Bats and humans evolved in the same geographic area.
   d. Bats lost their opposable digits during the course of evolution.
   e. Our ancestors could fly.

8. The remnants of pelvic and leg bones in a snake
   a. are vestigial structures.
   b. show that lizards evolved from snakes.
   c. are homologous structures.
   d. provide evidence for inheritance of acquired characteristics.
   e. resulted from artificial selection.

9. The hypothesis that whales evolved from land-dwelling ancestors is supported by
   a. evidence from the biogeographic distribution of whales.
   b. molecular comparisons of whales, fish, and reptiles.
   c. historical accounts of walking whales.
   d. the ability of captive whales to be trained to walk.
   e. fossils of extinct whales that had increasingly reduced hind limbs.

10. Darwin's claim that all of life descended from a common ancestor may best be supported with evidence from
    a. the fossil record.
    b. comparative embryology.
    c. taxonomy.
    d. molecular biology.
    e. comparative anatomy.

11. When cytochrome *c* molecules are compared, yeasts and molds are found to differ by approximately 46 amino acids per 100 residues (amino acids in the protein); insects and vertebrates are found to differ by 29 amino acids per 100 residues. What can one conclude from these data?
    a. Very little, unless the DNA sequences for the cytochrome *c* genes are compared.
    b. Yeasts evolved from molds, but vertebrates did not evolve from insects.
    c. Insects and vertebrates diverged from a common ancestor more recently than did yeasts and molds.
    d. Yeasts and molds diverged from a common ancestor more recently than did insects and vertebrates.
    e. The evolution of cytochrome *c* occurred more rapidly in yeasts and molds than in insects and vertebrates.

Use the following evolutionary tree representing the relationships among a group of vertebrates to answer questions 12 and 13. The letters at each branch point indicate the common ancestor for groups beyond that point.

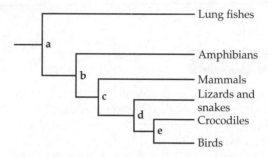

12. Which was the last common ancestor of crocodiles and the lineage of lizards and snakes?

13. Which of the following can be concluded from this evolutionary tree?
    a. Mammals are more closely related to amphibians than to birds.
    b. Mammals are more closely related to lizards and snakes than to birds.
    c. Birds are more closely related to lizards and snakes than to mammals.
    d. Birds and mammals are more closely related than are birds to lizards and snakes.
    e. Lungfishes are not related to any of these groups.

14. Which of the following is an example of convergent evolution?
    a. the evolution of multiple-drug resistance in MRSA
    b. similarities between the marsupial Tasmanian wolf and the eutherian North American wolf
    c. two very different plants that are found in different habitats but evolved from a fairly recent common ancestor
    d. the remodeling of the vertebrate forelimb in the evolution of a bird wing
    e. the many different bill sizes and shapes of finches on the Galápagos Islands

Chapter 23

The Evolution of Populations

Key Concepts

23.1 Genetic variation makes evolution possible

23.2 The Hardy-Weinberg equation can be used to test whether a population is evolving

23.3 Natural selection, genetic drift, and gene flow can alter allele frequencies in a population

23.4 Natural selection is the only mechanism that consistently causes adaptive evolution

Framework

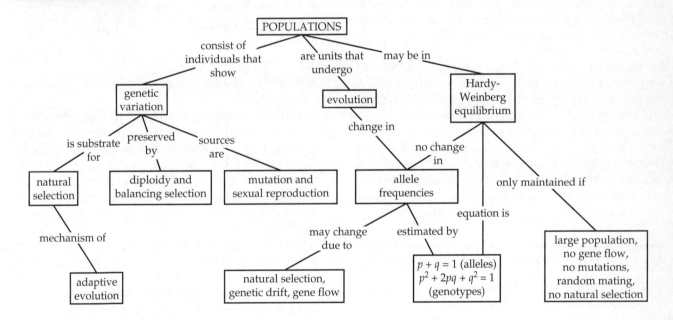

Chapter Review

Although individuals are selected for or against by natural selection, it is populations that evolve. **Microevolution** is defined as changes in the allele frequencies of a population from generation to generation.

23.1 Genetic variation makes evolution possible

Genetic Variation Genetic variation—slight differences in genes or other DNA sequences among individuals—is the raw material for natural selection. *Discrete characters* vary as distinct, either-or phenotypes and usually are determined by a single gene locus. *Quantitative characters*, those that vary along a continuum and are usually affected by two or more gene loci, represent most of the heritable variation within a population.

Measures of genetic variation address both *gene variability* and *nucleotide variability*. The **average heterozygosity** of a population is the average percent of loci that are heterozygous. Nucleotide variability measures the average percent of differences in nucleotide sites between individuals of a population.

Geographic variations are genetic differences in the gene pools of separate populations. A **cline**, or graded variation in a character along a geographic range, may parallel an environmental gradient.

Sources of Genetic Variation   New alleles can originate by *mutation,* a change in the sequence of nucleotides in DNA. Mutations that occur in somatic cells in animals cannot be passed on to the next generation. Point mutations that occur in noncoding DNA or do not change the amino acid sequence of a protein are often harmless. Mutations that do alter the phenotype are usually harmful. Rarely, however, a mutant allele may enhance an individual's reproductive success.

Chromosomal mutations are most often deleterious. Duplication of small DNA segments, introduced by transposable elements or errors in cell division, may provide extra loci that could eventually take on new functions by mutation.

Mutation rates in animals and plants average about one in every 100,000 genes per generation. Mutation produces genetic variation very rapidly in prokaryotes and viruses due to their short generation spans.

In a sexually reproducing population, most of the genetic variation comes from the reshuffling of alleles into new combinations in each individual. The processes of crossing over and independent assortment in meiosis and random combination of gametes in fertilization create unique genetic combinations in **offspring.**

INTERACTIVE QUESTION   23.1

a. What is a major source of genetic variation for prokaryotes and viruses?

b. What is the major source of genetic variation for plants and animals?

c. Explain why your answers to a and b are different.

23.2   The Hardy-Weinberg equation can be used to test whether a population is evolving

Gene Pools and Allele Frequencies   A **population** is a localized, interbreeding group of individuals of a species. The **gene pool** is the term for all the alleles at all the loci present in a population. If all individuals are homozygous for the same allele, the allele is said to be *fixed*. More often, two or more alleles are present in the gene pool in some relative proportion or frequency. In a case with two alleles at a particular gene locus, the letters p and q represent the frequencies of the two alleles within the population, and their combined frequencies must equal 1: $p + q = 1$.

INTERACTIVE QUESTION   23.2

In a population of 200 mice, 98 are homozygous dominant for brown fur (*BB*), 84 are heterozygous (*Bb*), and 18 are homozygous recessive for white fur (*bb*).

a. The genotype frequencies of this population are

_____*BB* _____*Bb* _____*bb*.

b. The allele frequencies of this population are

_____*B* allele _____*b* allele.

The Hardy-Weinberg Principle   If only Mendelian segregation and recombination of alleles in sexual reproduction are involved, the frequencies of alleles and genotypes for a particular locus in a population will remain constant from one generation to the next, as described by the **Hardy-Weinberg principle.** Such a nonevolving gene pool is said to be in *Hardy-Weinberg equilibrium*.

The allele frequency within a population determines the proportion of gametes that will contain that allele. The random combination of gametes will yield offspring with genotypes that reflect and reconstitute the allele frequencies of the previous generation.

With the equation for Hardy-Weinberg equilibrium, the frequencies of genotypes in the next generation can be calculated from the probability of each combination of alleles. According to the rule of multiplication, the probability that two gametes containing the same allele will come together is equal to ($p \times p$) or p^2, or to ($q \times q$) or q^2. A p allele and a q allele can combine in two different ways, depending on which parent contributes which allele; therefore, the frequency of a heterozygous offspring is equal to $2pq$. The sum of the frequencies of all possible genotypes in the population adds up to 1: $p^2 + 2pq + q^2 = 1$.

INTERACTIVE QUESTION 23.3

Use the allele frequencies you determined in Interactive Question 23.2 to predict the genotype frequencies of the next generation.

Frequencies of

B (p) = _____ b (q) = _____

$BB = p^2 =$ _____ $Bb = 2pq =$ _____ $bb = q^2 =$ _____

Hardy-Weinberg equilibrium is maintained only if all of the following five conditions are met: (1) no mutations, (2) random mating (because nonrandom mating changes genotype frequencies), (3) no natural selection (no unequal survival and reproductive success), (4) an extremely large population to offset chance fluctuations (called genetic drift), and (5) no gene flow or movement of alleles into or out of the population.

If the frequency of homozygous recessive individuals is known (q^2), then the frequency of q may be estimated as the square root of q^2 (assuming the population is in Hardy-Weinberg equilibrium for that gene).

INTERACTIVE QUESTION 23.4

Practice using the Hardy-Weinberg equation so that you can easily determine genotype frequencies from allele frequencies, and vice versa.

a. The allele frequencies in a population are $A = 0.6$ and $a = 0.4$. Predict the genotype frequencies for the next generation.

   AA _____ *Aa* _____ *aa* _____

b. What would the allele frequencies be for the generation you predicted in part a?

   A _____ *a* _____

c. Suppose you are able to determine the actual genotype frequencies in the population and find that these frequencies differ significantly from what you predicted in part a. What would such results indicate?

23.3 Natural selection, genetic drift, and gene flow can alter allele frequencies in a population

Natural Selection   Individuals with traits that are better suited to their environment tend to be more successful in producing viable, fertile offspring, and they pass their alleles to the next generation in disproportionate numbers, resulting in *adaptive evolution*.

Genetic Drift   Chance deviations from expected results are more likely to occur in a small sample. Chance fluctuations in a population's allele frequencies from one generation to the next are called **genetic drift**.

Genetic drift that occurs when only a few individuals colonize a new area is known as the **founder effect**. Allele frequencies in the small sample are unlikely to be representative of the parent population.

The **bottleneck effect** occurs when some disaster or other factor reduces population size dramatically, and the few surviving individuals are unlikely to represent the genetic makeup of the original population. Genetic drift will remain a factor until the population grows large enough for chance events to be less significant.

Genetic drift has a larger effect on small populations, causes gene frequencies to randomly fluctuate over time, can lead to the loss of alleles within populations, and can cause fixation of harmful alleles.

Gene Flow   **Gene flow**, the migration of individuals or the transfer of gametes between populations, may change allele frequencies. Differences in allele frequencies between populations tend to be reduced by gene flow. Gene flow may transfer alleles that either improve or reduce the ability of populations to adapt to local conditions.

INTERACTIVE QUESTION   23.5

Fill in the following concept map that summarizes three causes of microevolution. Even better, create your own concept map to help you review the ways in which a population's genetic composition may be altered.

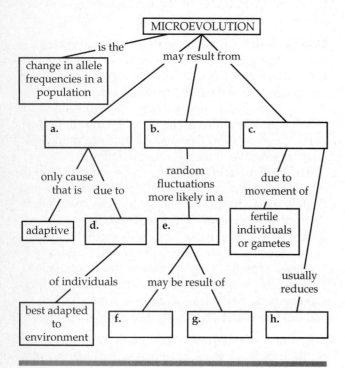

23.4   Natural selection is the only mechanism that consistently causes adaptive evolution

A Closer Look at Natural Selection   **Relative fitness** is a measure of an individual's contribution to the gene pool of the next generation, relative to the contributions of others.

The frequency distribution of a trait may be affected by three modes of selection. **Directional selection** occurs most frequently during periods of environmental change, when individuals on one end of a phenotypic range may be favored. **Disruptive selection** occurs when the environment favors individuals on both extremes of a phenotypic range. **Stabilizing selection** acts against extreme phenotypes and favors more intermediate forms, tending to reduce phenotypic variation.

The Key Role of Natural Selection in Adaptive Evolution   As natural selection increases the frequencies of alleles that enhance survival and reproduction, the match between organisms and their environment increases.

Sexual Selection   **Sexual selection** is the selection for characteristics that enhance an individual's chances of obtaining mates. Sexual selection can lead to **sexual dimorphism,** the distinction between males and females on the basis of secondary sexual characteristics. Sexual selection may involve **intrasexual selection,** in which individuals of the same sex compete for mates, or **intersexual selection,** in which individuals of one sex (usually female) discriminate in choosing a mate. Also called *mate choice,* intersexual selection may be based on showy traits that reflect the general health of the male, and thus the quality of his genes.

The Preservation of Genetic Variation   Some of the genetic variation seen in populations may be **neutral variations** that do not confer a selective advantage or disadvantage.

The diploidy of most eukaryotes maintains genetic variation by hiding recessive alleles in heterozygotes, enabling those alleles to persist in the population and be selected for, should the environment change.

Balancing selection can maintain stable frequencies of two of more phenotypes in a population. **Heterozygote advantage** tends to maintain two or more alleles at a locus when heterozygotes have survival and reproductive advantages. In **frequency-dependent selection,** a phenotype's reproductive success declines if it becomes too common in the population.

INTERACTIVE QUESTION   23.6

a. Why hasn't the highly deleterious sickle-cell allele been selected against and eliminated from the gene pool of the U.S. population?

b. Why is this allele at such a relatively high frequency in the gene pool of some African populations?

Why Natural Selection Cannot Fashion Perfect Organisms   Natural selection can act only on variations that are available; new alleles do not arise as they are needed. Each species has evolved from a long line of ancestral forms, many of whose structures have been co-opted for new situations. Adaptations are often compromises between the need to do different things, such as swim and walk. And finally, chance events affect a population's evolutionary history.

Word Roots

inter- = between (*intersexual selection:* selection that operates when individuals of one sex are choosy in selecting their mates from individuals of the other sex; also called mate choice)

intra- = within (*intrasexual selection:* selection involving direct competition among individuals of one sex for mates of the opposite sex)

micro- = small (*microevolution:* a change in the allele frequencies of a population over time)

Structure Your Knowledge

1. **a.** What is the Hardy-Weinberg principle?
   b. Define the variables of the equation for Hardy-Weinberg equilibrium. Make sure you can use this equation to determine allele frequencies and predict genotype frequencies.

2. It seems that natural selection would work toward genetic unity; the genotypes that are most fit produce the most offspring, increasing the frequency of adaptive alleles and eliminating less beneficial alleles from the population. Yet there remains a great deal of variability within populations of a species. Describe some of the factors that contribute to this genetic variability.

Test Your Knowledge

MULTIPLE CHOICE: *Choose the one best answer.*

1. Mutations are rarely a direct source for microevolution in eukaryotes because
   a. they are most often harmful and do not get passed on.
   b. they do not directly produce most of the genetic variation present in a diploid population.
   c. they occur very rarely.
   d. they are passed on only when they occur in gametes.
   e. all of the above are true.

2. The average heterozygosity of *Drosophila* is estimated to be about 14%, which means that
   a. 86% of fruit fly genes are identical.
   b. on average, 14% of a fruit fly's gene loci are heterozygous.
   c. 14% of nucleotide sites differ between individuals.
   d. nucleotide variability must be very great between individuals.
   e. the fruit fly population never experienced a bottleneck effect.

3. A scientist observes that the height of a certain species of asters decreases as the altitude on a mountainside increases. She gathers seeds from samples at various altitudes, plants them in a uniform environment, and measures the height of the new plants. All of her experimental asters grow to approximately the same height. From this she concludes that
   a. height is not a quantitative trait.
   b. the cline she observed was due to genetic variations.
   c. the differences in the parent plants' heights were due to directional selection.
   d. the height variation she initially observed was an example of nongenetic environmental influence.
   e. stabilizing selection was responsible for height differences in the parent plants.

4. Humans have an estimated 1,000 olfactory receptor genes. This is most likely a result of
   a. gene flow.
   b. gene duplication.
   c. frequency-dependent selection.
   d. neutral variation.
   e. disruptive selection.

5. Which of the following provides most of genetic variation found in plant and animal populations?
   a. mutations
   b. sexual reproduction
   c. sexual selection
   d. geographic variation
   e. recessive masking in heterozygotes

6. According to the Hardy-Weinberg principle,
   a. the allele frequencies of a population should remain constant from one generation to the next if the population is large and only sexual reproduction is involved.
   b. only natural selection, resulting in unequal reproductive success, will cause evolution.
   c. the square root of the frequency of individuals showing the dominant trait will equal the frequency of p.
   d. p and q can only be determined for a population that is not evolving.
   e. all of the above are correct.

7. If a population has the following genotype frequencies—$AA = 0.42$, $Aa = 0.46$, and $aa = 0.12$—what are the allele frequencies?
   a. $A = 0.42; a = 0.12$
   b. $A = 0.6; a = 0.4$
   c. $A = 0.65; a = 0.35$
   d. $A = 0.76; a = 0.24$
   e. $A = 0.88; a = 0.12$

8. In a population with two alleles, *B* and *b,* the allele frequency of *b* is 0.4. What would be the frequency of heterozygotes if the population is in Hardy-Weinberg equilibrium?
   a. 0.16
   b. 0.24
   c. 0.48
   d. 0.6
   e. You cannot tell from this information.

9. In a population that is in Hardy-Weinberg equilibrium for two alleles, *C* and *c,* 16% of the population show a recessive trait. Assuming *C* is dominant to *c,* what percent show the dominant trait?
   a. 36%
   b. 48%
   c. 60%
   d. 84%
   e. 96%

10. In a study of a population of field mice, you find that 48% of the mice have a coat color that indicates that they are heterozygous for a particular gene. What would be the frequency of the dominant allele in this population?
    a. 0.24
    b. 0.48
    c. 0.50
    d. 0.60
    e. You cannot estimate allele frequency from this information.

11. In a random sample of a population of shorthorn cattle, 73 animals were red ($C^R C^R$); 63 were roan, a mixture of red and white ($C^R C^r$); and 13 were white ($C^r C^r$). Estimate the allele frequencies of C^R and C^r, and explain whether or not the population is in Hardy-Weinberg equilibrium.
    a. $C^R = 0.64$, $C^r = 0.36$; because the population is large and a random sample was chosen, the population is in equilibrium.
    b. $C^R = 0.7$, $C^r = 0.3$; the genotype ratio is not what would be predicted from these frequencies, and the population is not in equilibrium.
    c. $C^R = 0.7$, $C^r = 0.3$; the genotype ratio is close to what would be predicted from these frequencies, and the population is in equilibrium.
    d. $C^R = 1.04$, $C^r = 0.44$; the allele frequencies add up to greater than 1, and the population is not in equilibrium.
    e. You cannot estimate allele frequency from this information.

12. Genetic drift is likely to be seen in a population
    a. that has a high migration rate.
    b. that has a low mutation rate.
    c. in which natural selection is occurring.
    d. that is very small.
    e. for which environmental conditions are changing.

13. Which of the following is a likely result of gene flow?
    a. a decrease in the adaptive evolution of neighboring populations that inhabit quite different environments.
    b. an increased migration of individuals to a favorable environment.
    c. a reduction of allele frequency differences between populations.
    d. Both a and c may result from gene flow.
    e. All three (a, b, and c) may result from gene flow.

14. Genetic analysis of a large population of mink inhabiting an island in Michigan revealed an unusual number of loci where one allele was fixed. Which of the following is the most probable explanation for this genetic homogeneity?
    a. The population exhibited nonrandom mating, producing a high proportion of homozygous genotypes.
    b. A very small number of mink may have colonized this island, and this founder effect and subsequent genetic drift could have fixed many alleles.
    c. The gene pool of this population never experienced gene flow.
    d. Natural selection has selected for and fixed the best-adapted alleles at these loci.
    e. The colonizing population may have had much more genetic diversity, but very recent genetic drift may have fixed these alleles by chance.

15. All of the following tend to maintain two or more alleles at a particular locus in a population *except*
    a. balancing selection.
    b. disruptive selection.
    c. heterozygote advantage.
    d. directional selection.
    e. frequency-dependent selection.

16. Sexual selection
    a. selects for traits that enhance an individual's chance of mating.
    b. increases the size of individuals.
    c. results in individuals better adapted to the environment.
    d. produces more offspring.
    e. selects for traits that increase fertility.

17. A plant population is found in an area that is becoming more arid. The average surface area of leaves has been decreasing over the generations. This trend is an example of
    a. a cline.
    b. directional selection.
    c. disruptive selection.
    d. gene flow.
    e. genetic drift.

18. Mice that are homozygous for a lethal recessive allele die shortly after birth. In a large breeding colony of mice, you find that a surprising 5% of all newborns die from this trait. In checking lab records, you discover that the same proportion of offspring have been dying from this trait in this colony for the past three years. (Mice breed several times a year and have large litters.) How might you explain the persistence of this lethal allele at such a high frequency?
    a. Homozygous recessive mice have a reproductive advantage.
    b. A large mutation rate keeps producing this lethal allele.
    c. There is some sort of heterozygote advantage and perhaps selection against the homozygous dominant trait.
    d. Genetic drift has kept the recessive allele at this high frequency in the population.
    e. Since this is a diploid species, the recessive allele cannot be selected against when it is in the heterozygote.

19. If an allele is recessive and lethal in homozygotes shortly after birth, then
    a. the allele is present in the population at a frequency of 0.001.
    b. the allele will be removed from the population by natural selection in approximately 1,000 years.
    c. the relative fitness of the homozygous recessive genotype is 0.
    d. the allele will most likely remain in the population at a low frequency because it cannot be selected against when in a heterozygote.
    e. Both c and d are correct.

20. Which of the following describes an organism's relative fitness?
    a. survival
    b. number of matings
    c. adaptation to the environment
    d. successful competition for resources
    e. relative number of viable offspring

21. Which of these types of selection is *mismatched* with its example?
    a. disruptive—a population of black-bellied seed-crackers consists of birds with either small bills (more effective at eating soft seeds) or large bills (able to crack hard seeds)
    b. intrasexual—elephant seal males are more than four times larger than females; males fight over areas of beach where females congregate during breeding season
    c. intersexual—female gray tree frogs choose mates that give long mating calls
    d. stabilizing—the frequencies of A, B, AB, and O blood groups remain constant in a population
    e. frequency-dependent—as fish of one coloration become more numerous, predators form a "search image" for that coloration and preferentially feed on them

22. In an area of erratic rainfall, a biologist found that in a certain wildflower population, homozygous plants with alleles for curled leaves reproduced better in dry years, whereas homozygous plants with alleles for flat leaves reproduced better in wet years. This situation would tend to
    a. result in genetic drift in the wildflower population.
    b. result in frequency-dependent selection against the most common leaf phenotype in these wildflowers.
    c. lead to balancing selection as a result of the heterozygote advantage of wildflowers with both alleles.
    d. preserve genetic variation in this wildflower population.
    e. result in disruptive selection during years of moderate rainfall.

Chapter 24

The Origin of Species

Key Concepts

24.1 The biological species concept emphasizes reproductive isolation

24.2 Speciation can take place with or without geographic separation

24.3 Hybrid zones reveal factors that cause reproductive isolation

24.4 Speciation can occur rapidly or slowly and can result from changes in few or many genes

Framework

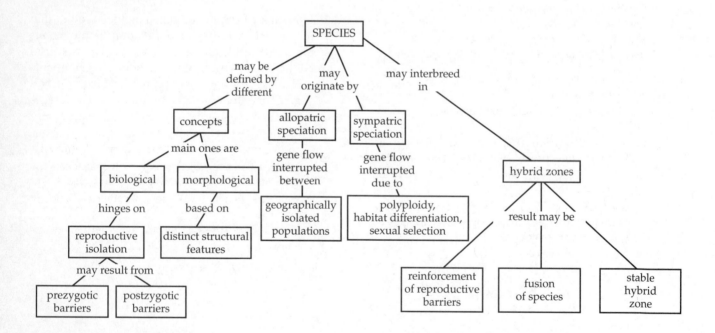

Chapter Review

Speciation, the origin of new species from an ancestral species, creates the diversity of life while maintaining the relatedness or unity of life. **Microevolution** explains evolutionary changes within the gene pool of a population. **Macroevolution** considers changes above the species level, such as the origin of new taxonomic groups.

24.1 The biological species concept emphasizes reproductive isolation

Different kinds of organisms are most often characterized by their physical form or morphology, although differences in physiology, biochemistry, and DNA sequences also support the existence of distinct species.

The Biological Species Concept According to the **biological species concept,** a **species** is a group of populations of individuals that have the potential to interbreed in nature and produce viable, fertile off-spring, but which do not successfully interbreed with members of other species. *Gene flow* between populations tends to hold a species together genetically, and thus morphologically.

The **reproductive isolation** that is necessary for the formation of a new species results from barriers that prevent individuals of different species from producing viable, fertile **hybrids. Prezygotic barriers** function before the formation of a zygote by preventing mating between members of different species or successful fertilization should gametes meet. Should a hybrid zygote form, **postzygotic barriers** prevent it from developing into a viable, fertile adult.

The biological species concept does not work for species that are asexual, and reproductive isolation cannot be determined for extinct species. Also, gene flow does occur between some species that remain morphologically and ecologically different.

Other Definitions of Species   Most species have been identified on the basis of physical characteristics, an approach called the **morphological species concept**. The **ecological species concept** defines species on the basis of their ecological niche, the role they play and the resources they use in the specific environments in which they are found. In the **phylogenetic species concept**, a species is an evolutionary lineage that represents one branch on the tree of life and has distinct morphology or molecular sequences.

INTERACTIVE QUESTION   24.1

For the following examples, indicate the type of reproductive isolation and whether it is a prezygotic barrier or a postzygotic barrier.

Type of Isolation	Pre- or Post-	Example
a.	**b.**	Two species of frogs mate in a laboratory setup and produce viable but sterile offspring.
c.	**d.**	Two species of sea urchins release gametes at the same time, but the sperm fail to fuse with eggs of a different species.
e.	**f.**	The genital openings of two species of land snails cannot line up because their shells spiral in opposite directions.
g.	**h.**	Two species of short-lived mayflies emerge during different weeks in spring.
i.	**j.**	Two species of salamanders mate and produce offspring, but the hybrid's offspring are sterile.
k.	**l.**	Two similar species of birds have different mating rituals.
m.	**n.**	Embryos of two species of mice bred in the lab usually abort.
o.	**p.**	Peepers breed in woodland ponds; leopard frogs breed in swamps.

INTERACTIVE QUESTION   24.2

Fill in the following table to review four of the approaches that biologists have proposed for conceptualizing a species.

Concept	Emphasis
biological	**a.**
b.	anatomical differences, most commonly used
c.	unique roles in specific environments
phylogenetic	**d.**

24.2   Speciation can take place with or without geographic separation

Allopatric ("Other Country") Speciation   **Allopatric speciation** occurs when geographic isolation interrupts gene flow between two subpopulations. Geographic separation alone is not a reproductive barrier in the biological sense. Intrinsic reproductive barriers may arise coincidentally as allopatric populations go down separate evolutionary paths due to mutation, genetic drift, and natural selection. Evidence of allopatric speciation has been found in both field and laboratory studies.

Sympatric ("Same Country") Speciation   In **sympatric speciation,** reproductive barriers prevent gene flow between populations that share the same area.

Mistakes during cell division may lead to **polyploidy,** the presence of extra sets of chromosomes.

An **autopolyploid** has more than two sets of chromosomes that have all come from the same species. Failures in cell division can produce tetraploids ($4n$), which can fertilize themselves or other tetraploids but cannot reproduce with diploids from the parent population, resulting in reproductive isolation in just one generation.

Polyploid species may also arise when two different species interbreed. The resulting hybrids may propagate asexually but are usually sterile due to difficulties in the meiotic production of gametes. Future cell division mistakes, however, can result in the production of a fertile **allopolyploid.**

Polyploid speciation has been frequent and important in plant evolution. Researchers now hybridize plants by inducing meiotic and mitotic errors to create new species.

INTERACTIVE QUESTION   24.3

a. A new plant species B forms by autopolyploidy from species A, which has a chromosome number of $2n = 10$. How many chromosomes would species B have?

b. If species A were to hybridize with species C ($2n = 14$) and produce a new allopolyploid species D, how many chromosomes would species D have?

Sympatric speciation in animals may involve isolation within the geographic range of the parent population based on different resource usage. Sexual selection may also lead to sympatric speciation. In a laboratory study of two sympatric species of cichlids, mate choice based on coloration was shown to be the reproductive barrier that normally separates the two species.

INTERACTIVE QUESTION   24.4

a. Differentiate between allopatric and sympatric speciation.

b. How might reproductive barriers arise in each type of speciation?

24.3   Hybrid zones reveal factors that cause reproductive isolation

Patterns Within Hybrid Zones   Areas where members of different species come into contact and interbreed are called **hybrid zones.** In the hybrid zone found between the yellow-bellied toad and the fire-bellied toad, the frequencies of alleles specific to one species range from 100% near that species' edge of the hybrid zone to 0% at the edge near the other species. What keeps these hybrids from introducing alleles of each other's species into the parent populations? Their poor survival and reproduction mean that these hybrids produce few viable offspring if they mate with the parent species.

Hybrid Zones over Time   In a process called **reinforcement,** natural selection may strengthen prezygotic barriers to reproduction when hybrids are less fit than members of the parent species. In other cases, weak reproductive barriers may allow sufficient gene flow that barriers break down even more, and the two hybridizing species fuse into a single species. Many hybrid zones appear to be stable, with hybrids continuing to be formed but with the two parent gene pools remaining separate.

INTERACTIVE QUESTION   24.5

Use the following diagrams to explain the three possible outcomes for a hybrid zone over time.

a.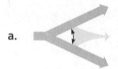

b.

c.

24.4 Speciation can occur rapidly or slowly and can result from changes in few or many genes

The Time Course of Speciation   In the fossil record, new forms often appear rather suddenly, persist unchanged for a long time, and then disappear. **Punctuated equilibria** is the term used to describe these long periods of stasis punctuated by episodes of relatively rapid speciation and change. For other species, the fossil record indicates a much more gradual divergence.

Various studies suggest that once divergence begins, the process of speciation may be relatively rapid. For example, the chromosomes of experimentally produced hybrid sunflowers came to resemble those of the wild sunflower hybrid species in only five generations of breeding in the lab.

INTERACTIVE QUESTION   24.6

If the process of speciation appears to occur relatively rapidly, why don't we see new species evolving all the time? And how does your answer relate to the concern about the current high rate of extinction of Earth's species?

Studying the Genetics of Speciation   Researchers have been able to identify the genes that play a key role in speciation in some organisms. For example, reproductive isolation in two species of monkey flowers has been linked to a few genes, including a flower color gene that affects pollinator preference.

From Speciation to Macroevolution   As species diverge and diversify, some differences may accumulate that lead to the formation of whole new groups of organisms, becoming part of the large-scale evolutionary changes of macroevolution.

Word Roots

auto- = self; **poly-** = many (*autopolyploid*: an individual that has more than two chromosome sets that are all derived from a single species)

macro- = large (*macroevolution*: evolutionary change above the species level)

post- = after (*postzygotic barrier*: a reproductive barrier that prevents hybrid zygotes produced by two different species from developing into viable, fertile adults)

sym- = together; **-patri** = father (*sympatric speciation*: the formation of new species in populations that live in the same geographic area)

Structure Your Knowledge

1. How are speciation and microevolution different?
2. In which type of speciation, sympatric or allopatric, would you expect reproductive barriers to arise more easily? Explain your answer.
3. What does the term *punctuated equilibria* describe?

Test Your Knowledge

MULTIPLE CHOICE: *Choose the one best answer.*

1. Which of the following is *not* a type of *intrinsic* reproductive isolation?
   a. mechanical isolation
   b. behavioral isolation
   c. geographic isolation
   d. gametic isolation
   e. temporal isolation

2. Which concept of species applies to the work of both Linnaeus, who first named organisms according to genus and species, and Darwin?
   a. biological
   b. ecological
   c. phylogenetic
   d. morphological
   e. both a and c

3. For which of the following is the biological species concept least appropriate?
   a. plants
   b. animals
   c. prokaryotes
   d. fossils
   e. both c and d

4. A horse ($2n = 64$) and a donkey ($2n = 62$) can mate and produce a mule. How many chromosomes would there be in a mule's cells?
   a. 31
   b. 62
   c. 63
   d. 64
   e. 126

5. Allopatric speciation is more likely to occur when an isolated population
   a. is large and thus has more genetic variation.
   b. is reintroduced to its original homeland.
   c. is small and exposed to different selection pressures in its new habitat.
   d. inhabits an island close to the mainland.
   e. All of the above contribute to allopatric speciation.

6. A tetraploid plant species (with four identical sets of chromosomes) is probably the result of
   a. allopolyploidy.
   b. autopolyploidy.
   c. hybridization and nondisjunction.
   d. allopatric speciation.
   e. a and c.

7. A botanist identifies a new species of plant that has 32 chromosomes. It grows in the same habitat with three similar species: species A ($2n = 14$), species B ($2n = 16$), and species C ($2n = 18$). Which of the following is a possible speciation mechanism for the new species?
   a. allopatric divergence by development of a reproductive isolating mechanism
   b. change in a key developmental gene that causes the plants to flower at different times
   c. autopolyploidy, perhaps due to a nondisjunction in the formation of gametes of species B
   d. allopolyploidy, a hybrid of species A and C
   e. Either answer c or d could account for the formation of this new plant species.

8. Which of the following would *not* contribute to allopatric speciation?
   a. geographic separation
   b. genetic drift
   c. gene flow
   d. different selection pressures
   e. founder effect

9. Morphological and genetic comparisons group 30 species of snapping shrimp (genus *Alpheus*) into 15 pairs of closely related species. What is the best explanation for the fact that one member of each pair lives on the Atlantic side of the Isthmus of Panama, whereas the other member of each pair lives on the Pacific side?
   a. Different predator pressures in the Atlantic and Pacific selected for differences that resulted in the reproductive isolation of these species.

   b. The pairs of species arose by allopatric speciation when the Isthmus of Panama separated their ancestral species.
   c. When these pairs meet in a hybrid zone, the reinforcement of reproductive barriers maintains each as a separate species.
   d. The 30 species of snapping shrimp evolved by sympatric speciation before the Isthmus of Panama formed.
   e. These 15 pairs of species illustrate the pattern of punctuated equilibrium, in which most speciation events take place in a short period of time.

10. When hybrids in a hybrid zone can breed with each other and with both parent species, and also have equal fitness to the parent species, one would predict that
   a. the hybrid zone would be stable.
   b. allopatric speciation would occur.
   c. reinforcement of reproductive barriers would keep the parent species separate.
   d. reproductive barriers would lessen and the two parent species would fuse.
   e. a new hybrid species would form by sympatric speciation.

11. Which of the following is the best description of punctuated equilibria?
   a. Long periods of stasis are punctuated by episodes of relatively rapid speciation and change.
   b. The equilibrium of separate species may be punctuated by gene flow within hybrid zones.
   c. Most rapid speciation events involve polyploidy in plants; speciation in animals is a much more gradual process.
   d. Rapid environmental changes produce rapid speciation events; gradual environmental changes result in gradual speciation events. Speciation does not occur if the environment remains constant.
   e. In the framework of geologic time periods, speciation events occur rapidly, and the equilibrium of species is punctuated by frequent extinctions.

12. This chapter introduced several research studies that illustrate various aspects of speciation. Which of the following is *not* an accurate description of a conclusion based on one of these studies?
   a. After several generations of being raised on different food sources, both "maltose flies" and "starch flies" tended to mate with like-adapted partners. These mating preferences appear to be the beginning of reproductive isolation by way of a behavioral prezygotic barrier between the two populations.

b. Two closely related species of monkey flowers have different pollinators (bumblebees and hummingbirds). Transferring an allele for flower color between the two species resulted in both types of pollinators visiting flowers with the transferred flower allele. Thus, this reproductive barrier based on pollinator choice may have been influenced by a change in a single gene locus.

c. In allopatric populations of the pied flycatcher and the collared flycatcher, males of the two species look very similar. In sympatric populations, the males have quite different colorations. Reinforcement of reproductive barriers is the best explanation for these observed differences.

d. Populations of mosquitofish inhabit ponds in the Bahamas. In ponds that contain predatory fishes, natural selection has favored a body shape that facilitates rapid bursts of speed. In mate choice experiments, female mosquitofish choose males with body shapes similar to their own. The reproductive isolation between populations is forming as a by-product of natural selection for predator avoidance.

e. In the many species of cichlids found in Lake Victoria, the reproductive barriers between species as a result of female mate choice based on male coloration appear to be breaking down because the introduced Nile perch is a strong predator of cichlids.

Chapter 25

The History of Life on Earth

Key Concepts

25.1 Conditions on early Earth made the origin of life possible

25.2 The fossil record documents the history of life

25.3 Key events in life's history include the origins of single-celled and multicelled organisms and the colonization of land

25.4 The rise and fall of dominant groups of organisms reflect differences in speciation and extinction rates

25.5 Major changes in body form can result from changes in the sequences and regulation of developmental genes

25.6 Evolution is not goal oriented

Framework

The history of life on Earth as chronicled in the fossil record illustrates **macroevolution,** evolutionary changes above the species level. Starting with hypotheses on how life may have originated, this chapter describes some of the major events in life's history, such as the rise of photosynthetic prokaryotes, the origin of eukaryotic cells and multicellularity, and the colonization of land. Divisions between eras in the geologic record are marked by major biological transitions linked with plate tectonics, mass extinctions, and adaptive radiations. The mechanisms underlying the major changes seen in the fossil record involve changes in and control of developmental genes.

Chapter Review

25.1 Conditions on early Earth made the origin of life possible

Scientists have proposed a four-stage hypothesis for the origin of life on early Earth.

Synthesis of Organic Compounds on Early Earth   Earth formed about 4.6 billion years ago and was bombarded by huge rocks and ice until about 3.9 billion years ago. As Earth cooled, water vapor condensed into oceans, and the atmosphere probably contained nitrogen and its oxides, carbon dioxide, methane, ammonia, and hydrogen sulfide.

In the 1920s, A. Oparin and J. Haldane independently hypothesized that conditions on early Earth—in particular the reducing atmosphere, lightning, and intense ultraviolet radiation—favored the synthesis of organic compounds from simpler molecules.

In 1953, S. Miller experimentally supported the Oparin-Haldane hypothesis using an apparatus that simulated the hypothetical conditions of early Earth. Numerous laboratory replications using various combinations of atmospheric gases have also produced organic compounds.

The openings of volcanoes and deep-sea vents may have provided the environments in which the first abiotic synthesis of organic molecules occurred. Meteorites that are carbonaceous chondrites have been found to contain carbon compounds, including many amino acids, lipids, sugars, and nitrogenous bases. Thus, meteorites may have been a second source of organic compounds.

Abiotic Synthesis of Macromolecules   Researchers have created polymers by dripping solutions of amino acids or RNA nucleotides onto hot sand, clay, or rock. Amino acid polymers may have served as weak catalysts on early Earth.

Protocells   Life requires accurate replication of genetic information and a metabolism to carry out this replication. Laboratory experiments have shown that membrane-bound *vesicles* can form spontaneously from mixtures of organic ingredients. Vesicle formation is speeded by the presence of *montmorillonite* clay, thought to have been common on early Earth. These vesicles may resemble early **protocells.** Such collections of abiotically produced organic molecules surrounded by a membrane can maintain a unique internal chemistry, undergo growth and reproduction, and perform some metabolic reactions.

Self-Replicating RNA and the Dawn of Natural Selection   RNA probably functioned as the first hereditary material. **Ribozymes** are enzyme-like RNA catalysts. Some laboratory-produced ribozymes can make complementary copies of their sequences. Vesicles with self-replicating, catalytic RNA could grow, split, and pass some of their RNA to daughter protocells. As the more successful offspring continued to reproduce and be acted on by natural selection, metabolic and hereditary improvements may have accumulated. DNA, a more stable genetic molecule, eventually replaced RNA as the carrier of genetic information. RNA may then have taken on its current role in the regulation and translation of genetic information.

INTERACTIVE QUESTION   25.1

Why do we say that, for life to have begun, the ability to carry out reproduction and metabolism must have evolved?

25.2   The fossil record documents the history of life

The Fossil Record   The fossil record, based on the sequence of fossils found in the *strata* of sedimentary rocks, provides evidence of great changes in the organisms that have lived on Earth. The record is incomplete, however, because (1) large numbers of species that lived probably left no fossils, (2) geologic processes destroy many fossils, and (3) only a fraction of existing fossils have been found.

How Rocks and Fossils Are Dated   The order in which fossils appear in the strata of sedimentary rocks indicates their relative age.

Radiometric dating is used to determine the actual ages of rocks and fossils. Each radioactive isotope has a fixed rate of decay—its **half-life**, which is the number of years it takes for 50% of the "parent" isotope to decay to its "daughter" isotope. During an organism's lifetime, it accumulates isotopes in proportions equal to their relative abundance in the environment. After

the organism dies, its radioactive isotopes decay at a fixed rate. Carbon-14, with its half-life of 5,730 years, can date fossils up to about 75,000 years old. Isotopes that decay more slowly can be used to date older fossils. The age of volcanic rock can be used to infer the age of fossils associated with such layers of rock.

INTERACTIVE QUESTION   25.2

A fossil has one-eighth of the atmospheric ratio of C-14 to C-12. Estimate the age of this fossil.

The Origin of New Groups of Organisms   The fossil record can shed light on the origin of new groups of organisms, as in the gradual evolution of mammals from a group of tetrapods called cynodonts.

25.3   Key events in life's history include the origins of single-celled and multicelled organisms and the colonization of land

Geologists have established a **geologic record** of Earth's history. The Archaean and the Proterozoic eons encompass the first 4 billion years. The Phanerozoic eon covers the last half billion years and is divided into three eras: the Paleozoic, Mesozoic, and Cenozoic. These eras are delineated by major extinction events.

The First Single-Celled Organisms   The oldest known fossils are 3.5-billion-year-old fossil **stromatolites,** which are rocklike layers of prokaryotes and sediment. It is possible that single-celled organisms originated as early as 3.9 billion years ago.

Banded iron formations in marine sediments and the rusting of iron in terrestrial rocks that began about 2.7 billion years ago provide evidence of oxygen accumulation from the photosynthesis of ancient cyanobacteria. The relatively rapid accumulation of atmospheric oxygen around 2.3 billion years ago caused the extinction of many prokaryotic groups and relegated others to anaerobic habitats.

Fossils that are generally accepted as eukaryotic date from 2.1 billion years ago. The cytoskeleton of eukaryotic cells allows them to change shape and engulf other cells. According to the **endosymbiont theory,** the mitochondria and plastids of eukaryotes originated from small prokaryotes that were undigested prey or internal parasites of larger cells.

Aerobic heterotrophic prokaryotes, living as endosymbionts within host cells, are the proposed ancestors of mitochondria. Plastids may have originated from photosynthetic endosymbionts. According to the hypothesis of **serial endosymbiosis,** mitochondria evolved first because they or their genetic remnants are present in all eukaryotes.

INTERACTIVE QUESTION　25.3

List some of the evidence supporting an endosymbiotic origin of mitochondria and plastids.

The Origin of Multicellularity　A great diversity of single-celled eukaryotes evolved from the structurally complex first eukaryotic cells. The first multicellular eukaryotes gave rise to algae, plants, fungi, and animals.

Molecular evidence suggests that the common multicellular ancestor arose 1.5 billion years ago. The oldest known fossils are small algae from about 1.2 billion years ago.

According to the "snowball Earth" hypothesis, severe ice ages that occurred from 750 to 580 million years ago accounted for the limited diversity and distribution of multicellular eukaryotes, with the first major diversification occurring after Earth thawed. Fossils of these larger and more diverse soft-bodied organisms, known as the Ediacaran biota, lived from 575 to 535 million years ago.

Fossils of sponges, cnidarians, and molluscs date from the late Proterozoic. The appearance of many of the major phyla of animals between 535 and 525 million years ago is referred to as the **Cambrian explosion.** But Chinese fossils dating back 575 million years as well as molecular (DNA) estimates indicate that many animal phyla began to diverge perhaps as early as 700 million to 1 billion years ago.

INTERACTIVE QUESTION　25.4

What lifestyle change may be related to the diversity of forms that evolved in the 10 million years known as the Cambrian explosion?

The Colonization of Land　Cyanobacteria coated damp terrestrial surfaces more than a billion years ago, but plants, fungi, and animals began to move onto land only about 500 million years ago. This colonization was associated with adaptations that helped prevent dehydration and permitted reproduction on land. Plants and fungi appear to have colonized the land together in symbiotic associations. Arthropods colonized land about 420 million years ago; the first tetrapod fossils appeared about 365 million years ago.

25.4 The rise and fall of groups of organisms reflect differences in speciation and extinction rates

Large-scale processes such as plate tectonics, mass extinctions, and adaptive radiations have influenced rates of speciation and extinction and thus the rise and fall of major groups.

Plate Tectonics　The theory of **plate tectonics** describes how the continents are part of great plates that float on Earth's molten mantle. Earthquakes occur and islands and mountains are formed in regions where these shifting plates abut. In the past 1.1 billion years, *continental drift* has brought most of the landmasses together to form a supercontinent three separate times.

The formation of the supercontinent **Pangaea** about 250 million years ago destroyed and altered habitats, causing many species to become extinct. The shifting of landmasses north or south greatly changes their climate. The separation and drifting apart of continents creates huge geographic isolation events, promoting allopatric speciation. Continental drift helps explain the past and current distributions of organisms.

INTERACTIVE QUESTION　25.5

Fossil evidence indicates that marsupials evolved in what is now Asia, yet their greatest diversity is found in Australia. How can you account for this biogeographic distribution?

Mass Extinctions　A species may become extinct due to a change in its physical or biological environment. **Mass extinctions** have occurred during periods of major environmental change. There have been five mass extinctions over the past 500 million years.

The Permian mass extinction, occurring at the boundary between the Paleozoic and Mesozoic eras 251 million years ago (mya), claimed about 96% of marine animal species and many terrestrial organisms. These extinctions may be related to massive volcanic eruptions in Siberia, which probably warmed the global climate and slowed the mixing of oceans, thus reducing the oxygen supply for marine organisms. This *ocean anoxia* would have favored anaerobic bacteria that emit H_2S. In addition to killing land organisms, this poisonous gas would contribute to the destruction of the protective ozone layer.

The Cretaceous mass extinction, which marks the boundary between the Mesozoic and Cenozoic eras about 65.5 mya, claimed more than one-half of the marine species and many families of land plants and animals, including all dinosaurs (except birds). The thin layer of clay rich in iridium (an element common in meteorites) may have accumulated from the fallout from a huge cloud of dust created when an asteroid or comet collided with Earth. This cloud would have blocked sunlight and severely affected weather.

The current rate of extinction may be 100 to 1,000 times the typical rate seen in the fossil record. A sixth mass extinction caused by human destruction of habitat may occur in the next few centuries or millennia.

The fossil record indicates it takes 5 to 10 million years for the diversity of life to recover following a mass extinction. Once an evolutionary lineage becomes extinct, it cannot reappear. Mass extinctions may alter the types of organisms making up an ecological community.

Adaptive Radiations **Adaptive radiations** are periods in which many new species evolve and fill various ecological niches. Adaptive radiations have occurred following mass extinctions, the evolution of major adaptations that allow organisms to exploit new ecological roles, and the colonization of new regions.

INTERACTIVE QUESTION 25.6

a. Mammals originated 180 million years ago but did not change much until their adaptive radiation 65.5 mya. Explain this observation.

b. Give some examples of adaptive radiations following major evolutionary innovations that opened new ecological roles.

c. What factors have contributed to the adaptive radiation of the thousands of endemic species of the Hawaiian Archipelago?

25.5 Major changes in body form can result from changes in the sequences and regulation of developmental genes

Effects of Developmental Genes The combination of evolutionary and developmental biology, called "evo-devo," explores how slight changes in developmental genes can result in major morphological differences between species.

Heterochrony is an evolutionary change in the rate or timing of development. A minor genetic alteration that affects the relative growth rates of body parts can produce a very differently proportioned adult form. **Paedomorphosis** is the retention in the adult of juvenile traits of an ancestral species and can result from changes in the timing of reproductive development.

Homeotic genes control the spatial arrangement of body parts. Mutations in homeotic genes called *Hox* genes, whose products provide positional information in animal embryos, can drastically alter body form. Changes in expression of *Mads-box* homeotic genes in different species of plants result in very different flowers.

The Evolution of Development Changes in the number and sequences of developmental genes as well as changes in gene regulation may produce new body forms.

INTERACTIVE QUESTION 25.7

a. Researchers identified the key difference in the *Hox* gene *Ubx* of *Artemia* (brine shrimp, a crustacean) and *Drosophila* (fruit fly, an insect) that is responsible for the suppression of legs in *Drosophila*. What was the significance of this study?

b. Marine populations of threespine stickleback fish have protective spines on their ventral surface; lake populations of these fish do not. How did researchers determine that the loss or reduction of these spines was caused by a difference in the regulation of a developmental gene *(Pitx1)* and not in the gene itself?

25.6 Evolution is not goal oriented

Evolutionary Novelties Often very complex organs, such as the eyes of vertebrates and some molluscs, have evolved gradually from simpler structures that served similar needs in ancestral species.

Evolutionary novelties may also evolve by the gradual modification of existing structures for new functions. *Exaptation* is the term for structures that evolved and functioned in one setting and were then co-opted for a new function.

Evolutionary Trends The fossil record documents that *Equus*, the modern horse, descended from its much smaller, browsing, multitoed ancestor, *Hyracotherium*, through a series of speciation episodes that produced many different species and diverging trends.

INTERACTIVE QUESTION   25.8

a. According to S. Stanley's model of *species selection,* "differential speciation success" plays a role in macroevolution similar to the role of differential reproductive success in microevolution. Explain.

b. Do evolutionary trends indicate that evolution is goal directed?

Word Roots

hetero = different (*heterochrony:* evolutionary changes in the timing or rate of development)

macro- = large (*macroevolution:* evolutionary change above the species level, such as the emergence of a new group of organisms through a series of speciation events)

paedo- = child (*paedomorphosis:* the retention in the adult organism of the juvenile features of its evolutionary ancestors)

proto- = first (*protocell:* an abiotic precursor of a living cell that had a membrane-like structure and maintained an internal chemistry different from its surroundings)

stromato- = something spread out; **-lite** = a stone (*stromatolite:* layered rock that results from the activities of prokaryotes that bind thin films of sediment together)

Structure Your Knowledge

1. List the sequence of four stages that could have led to the origin of the first living cells.
2. Label the eons, eras, and key events shown on the following clock analogy of Earth's history.

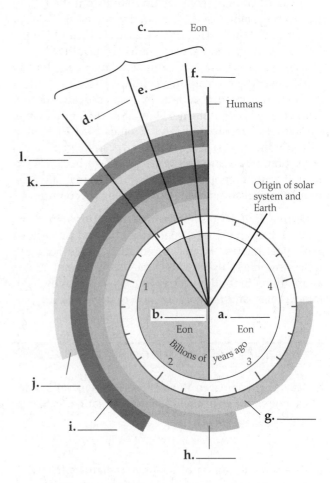

3. Describe three major processes that have influenced the changing diversity of life on Earth.
4. What genetic changes would most likely lead to the evolution of new morphological forms?

Test Your Knowledge

MULTIPLE CHOICE: *Choose the one best answer.*

1. The primitive atmosphere of Earth may have favored the abiotic synthesis of organic molecules because it
   a. was highly oxidative.
   b. was reducing (near openings of volcanoes) and had energy sources in the form of lightning and UV radiation.
   c. had a great deal of methane and organic fuels.

d. had plenty of water vapor, carbon, oxygen, and nitrogen, providing the C, H, O, and N needed for organic molecules.

e. consisted almost entirely of hydrogen gas, creating a very reducing environment.

2. Which of the following is a proposed hypothesis for the origin of genetic information?

   a. Early DNA molecules coded for RNA, which then catalyzed the production of proteins.

   b. Early polypeptides were the first "genes," followed by RNA and then DNA.

   c. Short RNA strands were capable of self-replication and evolved by natural selection.

   d. As protocells grew and split, they distributed copies of their molecules to their offspring.

   e. Early RNA molecules coded for the order of amino acids in a polypeptide.

3. All of the following types of evidence support the hypothesis that life on Earth formed spontaneously *except* for

   a. the discovery of ribozymes, showing that RNA molecules may be autocatalytic.

   b. laboratory studies involving the formation and reproduction of vesicles.

   c. the fossil record.

   d. the abiotic synthesis of polymers from monomers dripped onto hot rocks.

   e. the production of organic compounds in the Miller-Urey experimental apparatus.

4. The half-life of carbon-14 is 5,730 years. A fossil that is 22,920 years old would have what amount of the normal proportion of C-14 to C-12?

   a. ½   **c.** ⅙   **e.** ¹⁄₁₆

   b. ¼   **d.** ⅛

5. How many half-lives should have elapsed if a sample contains 12.5% of the normal quantity of a parent isotope?

   a. 1   **c.** 3   **e.** 5

   b. 2   **d.** 4

6. Which of the following characteristics would increase the likelihood that organisms would be well represented in the fossil record?

   a. abundant and widespread

   b. exist over a long time period

   c. have hard shells or skeletons

   d. live in regions where geology favors fossil formation

   e. All of the above contribute.

7. Life on Earth is thought to have begun

   a. 540 million years ago, at the beginning of the Paleozoic era.

   b. 635 million years ago, at the start of the Ediacaran.

   c. 2.7 billion years ago, when oxygen began to accumulate in the atmosphere.

   d. between 3.5 and 3.9 billion years ago, after Earth cooled and rocks began to form.

   e. 4.5 billion years ago, shortly after Earth first formed.

8. Stromatolites are

   a. early prokaryotic fossils found in sediments around hydrothermal vents.

   b. protocells that form when lipids assemble into a bilayer surrounding organic molecules.

   c. layers of rusted terrestrial rocks formed when O_2 produced by early photosynthetic prokaryotes entered the atmosphere.

   d. fossils appearing to be eukaryotes that are about twice the size of prokaryotes.

   e. layered communities of prokaryotes, fossils of which represent the oldest known prokaryotes.

9. Banded iron formations in marine sediments provide evidence of

   a. the first prokaryotes around 3.5 billion years ago.

   b. oxidized iron layers in terrestrial rocks.

   c. the accumulation of oxygen in the seas from the photosynthesis of ancient cyanobacteria.

   d. the evolution of photosynthetic archaea near deep-sea vents.

   e. the crashing of meteorites into Earth, possibly transporting abiotically produced organic molecules from space.

10. Diverse soft-bodied organisms, known as the Ediacaran biota, appear in the fossil record around 575 million years ago. This time period may correspond to

    a. the thawing of snowball Earth.

    b. the first appearance of multicellular eukaryotes in the fossil record.

    c. the Cambrian explosion of animal forms.

    d. the colonization of land by plants and fungi, providing the food source for animals to follow.

    e. the rapid increase in atmospheric oxygen that enabled the active lifestyle of animals.

11. According to the endosymbiont theory,

    a. the first primitive cells may have originated when a lipid bilayer spontaneously formed around abiotically produced polymers.

    b. plants were able to colonize land with fungi as symbionts within their roots.

    c. the mitochondria and plastids of eukaryotic cells were originally prokaryotic endosymbionts that became permanent parts of the host cell.

    d. the infoldings and specializations of the plasma membrane led to the evolution of the endomembrane system.

    e. the nuclear membrane evolved first, followed by plastids, and then mitochondria.

12. The Permian mass extinction between the Paleozoic and Mesozoic eras

  a. may have been caused by extreme volcanism, which produced multiple environmental effects.

  b. coincided with the breaking apart of Pangaea.

  c. made way for the adaptive radiation of mammals, birds, and pollinating insects.

  d. appears to have been caused by a large asteroid striking Earth.

  e. involved all of the above.

13. The more than 500 species of fruit fly on the various Hawaiian Islands, all apparently descended from a single ancestral species, are an excellent example of

  a. plate tectonics.

  b. adaptive radiation.

  c. biogeography.

  d. exaptation.

  e. an evolutionary trend.

14. Which of the following is *not* an accurate statement about ocean anoxia?

  a. Ocean anoxia is considered to be a major cause of the Permian mass extinction.

  b. About 50% of marine animal species were lost in the Permian mass extinction, but many were able to withstand the reduced O_2 concentration by gulping atmospheric oxygen.

  c. As global temperatures warm, reduced temperature differences between the poles and equator slow the mixing of ocean water, which causes a drop in oxygen concentrations.

  d. Anaerobic bacteria would thrive in such reduced oxygen environments and emit H_2S, a poisonous metabolic by-product.

  e. As H_2S bubbles into the atmosphere, it can kill land organisms and initiate chemical reactions that destroy the protective ozone layer.

15. The evolution of major changes in animal body form is most likely to involve

  a. duplication of *Hox* genes.

  b. changes in regulatory sequences that determine where developmental genes are expressed.

  c. mutations that alter the products of *Hox* genes.

  d. genetic changes that affect the growth rates of different body parts during development.

  e. All of the above can contribute to major morphological differences between species.

16. The evolution of the middle ear bones from bones of the jaw hinge of cynodonts is an example of

  a. heterochrony.

  b. exaptation.

  c. paedomorphosis.

  d. the gradual refinement of a structure with the same function.

  e. an evolutionary trend.

17. Pterosaurs are extinct flying reptiles whose membranous wings were supported by one greatly elongated finger. A bird's wing includes feathers, which are outgrowths of the skin along the whole length of the forelimb (arm). Four elongated fingers support most of the membrane that makes up a bat's wing. Which of the following phrases best describes the relationship of the "finger" wings of pterosaurs, the "arm" wings of birds, and the "hand" wings of bats?

  a. the gradual refinement of a structure that served the same function in all these species and their ancestors

  b. three different exaptations of the basic skeletal structure of the tetrapod forelimb to the function of flight

  c. changes in the regulation of the same developmental genes in all three groups

  d. the independent evolution of the skeletal support, membrane or feather coverings, and aerodynamic shape of wings in these three unrelated groups

  e. an evolutionary trend toward powered flight

18. What is meant by the concept of species selection?

  a. Reproductive isolating mechanisms are responsible for maintaining the integrity of individual species.

  b. Characteristics that increase the probability that a species will separate into two or more species will accumulate in the most successful species.

  c. Natural selection can act on the species level as well as on the population level.

  d. The species that last the longest and speciate the most often influence the direction of evolutionary trends.

  e. The colonization of new and diverse habitats can lead to the adaptive radiation of numerous species.

5

The Evolutionary History of Biological Diversity

Chapter 26

Phylogeny and the Tree of Life

▣ Key Concepts

26.1 Phylogenies show evolutionary relationships

26.2 Phylogenies are inferred from morphological and molecular data

26.3 Shared characters are used to construct phylogenetic trees

26.4 An organism's evolutionary history is documented in its genome

26.5 Molecular clocks help track evolutionary time

26.6 New information continues to revise our understanding of the tree of life

▣ Framework

A goal of systematics is to reconstruct the phylogenetic history of species. Cladistics uses the distribution of shared derived characters to identify monophyletic taxa, or clades. The best phylogenetic hypotheses are parsimonious trees based on morphological and molecular homologies as well as on fossil evidence. Current hypotheses for the basic tree of life suggest that eukaryotes originated from a fusion of a bacterium and archaean, and that horizontal gene transfers were common in the early history of life.

▣ Chapter Review

Phylogeny is the evolutionary history of a species or group. **Systematics** is the discipline focused on classifying organisms and determining evolutionary relationships.

26.1 Phylogenies show evolutionary relationships

Binomial Nomenclature **Taxonomy** is focused on naming and classifying species. As instituted by Linnaeus, each species is assigned a two-part Latinized name—a **binomial**—consisting of the name of the **genus** and the specific epithet, designating one species in that genus.

Hierarchical Classification In the Linnaean taxonomic system, genera are grouped into **families,** which are then placed into the broader categories of **orders, classes, phyla,** and **kingdoms,** and more recently into **domains.** A taxonomic unit at any level is called a **taxon.**

Linking Classification and Phylogeny A **phylogenetic tree** represents hypotheses about evolutionary relationships among groups. Its branching patterns may correspond to hierarchical Linnaean classification. The **PhyloCode** system of classification is based totally on evolutionary relationships, only naming groups that include a common ancestor and all descendants.

Each dichotomous **branch point** on a phylogenetic tree represents the divergence of two taxa from a common ancestor. **Sister taxa,** sharing an immediate common ancestor, are each other's closest relatives. Trees are **rooted,** with the common ancestor to all the taxa in the tree usually on the left. A **basal taxon** is one that diverged early in the evolutionary history of a group. A **polytomy** is a branch point involving more than two descendants, indicating that these evolutionary relationships are not yet clear.

What We Can and Cannot Learn from Phylogenetic Trees Phylogenetic trees show patterns of descent, which do not always correspond to morphological similarity. Branching sequences do not indicate when species evolved or the amount of genetic change in each lineage. Also, species located next to each other on the tree shared a common ancestor; one did not evolve from the other.

Applying Phylogenies The information contained in phylogenetic trees has practical applications in such areas as molecular forensics or crop improvement.

INTERACTIVE QUESTION   26.1

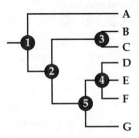

a. In this hypothetical phylogenetic tree, which number represents the common ancestor of all the taxa?

b. Which letter represents the basal taxon?

c. Which branch point is a polytomy?

d. Which taxa are sister taxa?

26.2   Phylogenies are inferred from morphological and molecular data

Morphological and Molecular Homologies Homologies are similarities due to shared ancestry. The more similar organisms' morphologies or DNA sequences are, the more likely those organisms are closely related.

Sorting Homology from Analogy **Analogy** is similarity due to convergent evolution, in which unrelated species develop similar features because natural selection has led to similar adaptations. Analogous structures are also called **homoplasies.** In general, when two complex structures share many similar features, the more likely it is that those structures were inherited from a common ancestor. And when genes from different organisms share many nucleotide sequences, the more likely the genes are to be homologous.

INTERACTIVE QUESTION   26.2

What two complications may make it difficult to determine phylogenetic relationships based on morphological similarities between species? Give examples.

Evaluating Molecular Homologies Molecular comparisons are complicated by insertion or deletion mutations that change the lengths of homologous regions of DNA. Computer programs can identify and align homologous DNA segments properly for nucleotide comparisons. Statistical tools help distinguish "distant" homologies in divergent sequences from coincidental molecular homoplasies.

Researchers in **molecular systematics** use DNA and other molecular comparisons to infer evolutionary relationships.

26.3   Shared characters are used to construct phylogenetic trees

Cladistics Classification based on common descent forms the basis of **cladistics.** A **clade** consists of an ancestral species and all of its descendant species. Such a clade is **monophyletic.** A **paraphyletic** group excludes some species that share a common ancestor with other species in the group, and a **polyphyletic** group includes several groups with different ancestors.

Shared ancestral characters are found in a particular clade but originated in an ancestor that is not part of that clade. **Shared derived characters** are unique to a particular clade.

To determine the branching sequence of a group of related species, the group is compared to an **outgroup,** a species or group of species that diverged before the group being studied (the **ingroup**). A comparison of the characters that are present in each taxon of the ingroup indicates the sequence in which shared derived characters evolved and determines the branch points used to produce a phylogenetic tree.

INTERACTIVE QUESTION 26.3

Place the taxa (outgroup, A, B, C, and D) on the following phylogenetic tree based on the presence or absence of the characters 1–4 as shown in the table. Indicate before each branch point the number for the shared derived character that evolved in the ancestor of the clade.

	Outgroup O	A	Taxa B	C	D
1	0	1	1	1	1
2	0	0	1	0	1
3	0	1	1	0	1
4	0	0	1	0	0

(left label: Characters)

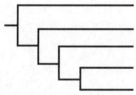

Phylogenetic Trees with Proportional Branch Lengths
The branching pattern of most phylogenetic trees is relative rather than absolute, indicating only the order in which members of each clade last shared a common ancestor. The branch length of some trees can be scaled to reflect rates of evolutionary change (for instance, number of changes in DNA sequence) or time (using the dates of branch points as indicated in the fossil record).

Maximum Parsimony and Maximum Likelihood
Systematists use morphological characters or molecular comparisons to choose among many possible phylogenetic trees using the principle of **maximum parsimony**—the approach that the smallest number of evolutionary changes is the simplest explanation and thus the best hypothesis to consider first. According to the principle of **maximum likelihood,** a tree can be found that represents the most likely sequence of events, given a certain set of assumptions (such as

equal rates of DNA changes). Computer programs search for the most parsimonious and most likely trees.

Phylogenetic Trees as Hypotheses A phylogenetic tree represents the best hypothesis of the relationships among a set of species; the more data that can be compared, the more reliable the tree becomes.

Phylogenetic hypotheses can be used to make and test predictions. *Phylogenetic bracketing* predicts that features shared by two closely related organisms will be present in their common ancestor and all its descendants. Fossil discoveries, for example, support the prediction that dinosaurs, as descendants of the ancestor of birds and crocodiles, built nests and brooded their eggs.

INTERACTIVE QUESTION 26.4

According to the principle of parsimony, the evolution of the four-chambered heart should place birds and mammals in the same clade. Why does the most accepted evolutionary tree show them as separate branches from the reptilian line?

26.4 An organism's evolutionary history is documented in its genome
Molecular systematics using nucleic acid or other molecular comparisons make it possible both to compare genetic divergence among groups within a species and to reconstruct phylogeny among very distantly related species. The ability to span this length of time depends on genes that evolve either very slowly or very rapidly.

Gene Duplications and Gene Families Gene duplications are important evolutionary mutations, increasing the number of genes in the genome. Repeated duplications can result in groups of related genes, or *gene families.*

Orthologous genes are homologous genes found in different species; such genes diverge after the two species separated. **Paralogous genes** result from gene duplication in the same genome, and as they diverge, they provide new opportunities for evolutionary change within a species.

Genome Evolution The existence of widespread orthologous genes indicates that all organisms share many biochemical and developmental pathways. The number of genes a species has does not necessarily correlate with morphological complexity.

INTERACTIVE QUESTION   26.5

a. Give an example of genes that would be compared to uncover phylogenetic relationships among the earliest branches on the tree of life.

b. Give an example of genes that evolve very rapidly and are used to discriminate among closely related species or even among populations of the same species.

26.5   Molecular clocks help track evolutionary time

Molecular Clocks   Some regions of DNA appear to evolve at constant rates, and comparisons of the number of nucleotide substitutions in orthologous genes can serve as **molecular clocks** to estimate the time since two species branched from their common ancestor. The number of differences in paralogous genes is proportional to the time since the ancestral gene was duplicated.

Some genes appear to have a reliable average rate of evolution. Graphs that plot nucleotide or amino acid differences against the times for known evolutionary branch points can be used to estimate phylogenetic branchings that are not evident from the fossil record.

According to the **neutral theory,** much of the evolutionary change in genes does not affect fitness. Harmful mutations are removed quickly from the gene pool, but neutral mutational changes that have little effect on fitness should occur at a constant rate.

INTERACTIVE QUESTION   26.6

Using the neutral theory of molecular evolution, explain why different genes might have a different molecular clock rate.

Natural selection, which favors some DNA changes over others, may disrupt the smooth running of the molecular clock. Evolutionists disagree about the extent of neutral mutations, and thus about whether molecular clocks are reliable for timing evolution. Many are skeptical when molecular clocks are used to date evolutionary divergences that occurred billions of years ago.

When molecular clocks are calibrated using many genes, fluctuations due to natural selection may average out. Molecular and fossil-based estimates of divergence times in vertebrate evolution agree closely.

Applying a Molecular Clock: The Origin of HIV   By comparing nucleotide sequences of HIV from samples taken at various times during the epidemic, researchers have observed a remarkably consistent rate of evolution. They estimate that the HIV-1 M strain first infected humans in the 1930s.

26.6   New information continues to revise our understanding of the tree of life

From Two Kingdoms to Three Domains   Historically, taxonomists divided the diversity of life into two kingdoms—plants and animals. In the five-kingdom system, the prokaryotes were set apart from the eukaryotes and placed in kingdom Monera. The Protista contained mostly unicellular eukaryotes while kingdoms Plantae, Fungi, and Animalia consisted of multicellular eukaryotes.

The current three-domain system creates a taxon above the kingdom level. The domains Bacteria and Archaea, although both consisting of single-celled prokaryotes, differ in many characteristics. The domain Eukarya includes all the eukaryotes: plants, fungi, animals, and many groups of mostly single-celled organisms.

A Simple Tree of All Life   The rRNA genes evolve so slowly that they have been used as the basis for constructing a basic tree of life. The tree's first major split represents the divergence of the bacteria from the other two domains. Analysis of other genes, however, suggests that eukaryotes share a more recent common ancestor with bacteria than with archaea.

Genome comparisons from the three domains indicate that **horizontal gene transfer,** perhaps through transposable elements and plasmid exchange, or even through fusions of different organisms, occurred during the early history of life.

Is the Tree of Life Really a Ring?   Some scientists suggest that the early history of life is best represented as a ring. If eukaryotes arose as a fusion between a bacterium and an archaean, then eukaryotes are most closely related to both.

INTERACTIVE QUESTION 26.7

This comprehensive tree of life represents current hypotheses on the origin of the three domains. Identify the domains a, b, and c. Number 1 represents the last common ancestor of all living things. Describe the two major episodes of horizontal gene transfer illustrated by numbers 2 and 3.

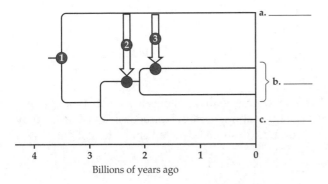

Word Roots

analog- = proportion (*analogy:* similarity due to convergence)

bi- = two; **nom-** = name (*binomial:* a two-part Latinized name of a species)

clad- = branch (*clade:* a group of species that includes an ancestral species and all its descendants)

homo- = like, resembling (*homology:* similarity in characteristics resulting from a shared ancestry)

mono- = one (*monophyletic:* pertaining to a group of taxa that consists of a common ancestor and all its descendants)

parsi- = few (*principle of parsimony:* the premise that a theory about nature should be the simplest explanation that is consistent with the facts)

phylo- = tribe; **-geny** = origin (*phylogeny:* the evolutionary history of a taxon)

Structure Your Knowledge

1. Draw a phylogenetic tree that best represents the relationships among taxa A–E described as follows: Taxon A is the basal taxon. B and C are sister taxa. They share a more recent common ancestor with D than with E.

2. Describe some of the tools that systematists use for constructing phylogenetic trees.

3. What is a molecular clock and how can it be used?

Test Your Knowledge

MULTIPLE CHOICE: *Choose the one best answer.*

1. Related families are grouped into the next-highest taxon, which is called a
   a. class.
   b. phylum.
   c. order.
   d. genus.
   e. kingdom.

2. Four of the following trees describe the same phylogenetic relationships among taxa A, B, C, D, E, and F. Which tree shows a different phylogeny?

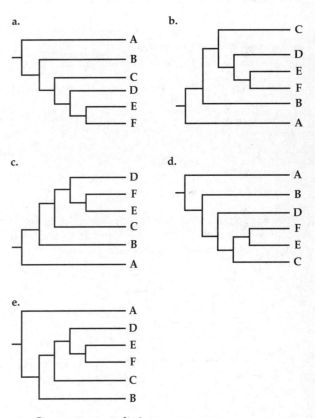

3. Convergent evolution may occur
   a. when ancestral structures are co-opted for new functions.
   b. when homologous structures are adapted for different functions.
   c. from adaptive radiation.
   d. when species are widely separated geographically.
   e. when species have similar ecological roles.

4. Which of the following provides the best example of analogous structures?
   a. forelimbs of bat and mole
   b. tree-like and shrub-like silversword plants of Hawaii
   c. four-chambered hearts of birds and bats
   d. skulls of apes and humans
   e. hindlegs of Australian and North American moles

5. Which of the following is a shared derived character for monotreme, marsupial, and eutherian mammals?
   a. parental care
   b. internal fertilization
   c. amnion
   d. production of milk for young
   e. complete embryonic development inside a uterus

6. Shared derived characters are
   a. those that characterize all the species on a branch of a phylogenetic tree.
   b. determined through a computer comparison of paralogous genes in a group of species.
   c. homologous structures that develop during adaptive radiation.
   d. characters found in the outgroup but not in the species under consideration.
   e. used to identify species but not higher taxa.

7. A taxon such as the class Reptilia—which does not include its relatives, the birds—is
   a. really an order.
   b. a clade.
   c. monophyletic.
   d. polyphyletic.
   e. paraphyletic.

The following phylogenetic tree includes the dates of divergence for taxa A through E as determined from the fossil record. Use this tree to answer questions 8–12.

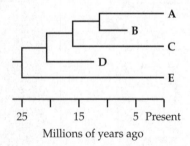

25   15   5 Present

Millions of years ago

8. Which letter represents the basal taxon?

9. Which letter(s) refer to extinct groups(s)?
   a. A   d. E and D
   b. B   e. B and D
   c. C

10. Which two extant taxa are most closely related?
    a. A and B
    b. A and C
    c. A and E
    d. C and D
    e. C and E

11. How many million years ago did taxa A and D last share a common ancestor?
    a. 25
    b. 20
    c. 15
    d. 10
    e. cannot tell

12. The technique that enables biologists to make predictions about certain characteristics of taxon B by comparing shared derived characters of taxa A and C is called
    a. cladistics.
    b. phylocode analysis.
    c. phylogenetic bracketing.
    d. maximum parsimony.
    e. maximum likelihood.

13. The greatest number of shared derived characters should be found in two organisms that were traditionally placed in the same
    a. order.
    b. domain.
    c. family.
    d. class.
    e. phylum.

14. Which of the following approaches would allow a biologist studying the evolution of four similar species of birds to choose the best phylogenetic tree from all possible phylogenies?
    a. Draw the simplest tree and choose that one.
    b. From a comparison of DNA sequences, determine the number of evolutionary events required for each tree and then choose the most parsimonious tree.
    c. Compare the entire genomes of each species; the two most similar genomes are the two species that are most closely related.
    d. Determine which species can interbreed; those that can interbreed evolved from a common ancestor most recently.
    e. Choose the tree that has the most evolutionary changes, as this would be the most likely explanation for how these very similar birds evolved into four distinct species.

15. A comparative study of which of the following would provide the best data on the early evolution of fungi and plants?
    a. mitochondrial DNA
    b. DNA from a chloroplast
    c. the amino acid sequence of chlorophyll
    d. DNA for ribosomal RNA
    e. the morphology of present-day specimens

16. A comparative study of which of the following would provide the best data on the ancestry of people from Germany, Italy, and Spain?
    a. mitochondrial DNA
    b. orthologous genes
    c. ribosomal RNA
    d. paralogous genes
    e. the fossil record

17. Analysis of which of the following data sets would produce the most reliable phylogenetic tree?
    a. DNA sequences of paralogous genes
    b. DNA sequences of orthologous genes
    c. morphology
    d. homoplasies
    e. Using all of these would produce the best supported phylogeny.

18. How could a molecular clock be useful for studying orthologous genes?
    a. It could help determine when the original gene duplication occurred.
    b. It could help date the divergence of different lineages that contain that homologous gene.
    c. Because orthologous genes do not have neutral mutations, the molecular clock could not be used to study them.
    d. It could date when the gene first arose.
    e. Because the evolution of orthologous genes does not occur at a consistent rate, a molecular clock would not be useful.

19. Which of the following segments of DNA would likely have the fastest molecular clock rate?
    a. noncoding DNA that has a regulatory function
    b. a pseudogene (gene that has lost sequences needed for expression)
    c. a gene for essential enzyme
    d. a gene for rRNA
    e. paralogous genes

20. Which of the following is the best description of our current hypothesis of the tree of life?
    a. The tree of life consists of three domains: Bacteria, Archaea, and Eukarya.
    b. The base of the tree of life is still uncertain because the molecular clock is not accurate for evolutionary events that occurred that long ago.
    c. The domain Archaea is known to be the first branch; domains Bacteria and Eukarya are more closely related to each other.
    d. There was substantial horizontal gene transfer and perhaps even fusion of different organisms during the early history of life.
    e. Both a and d represent our current hypothesis.

21. Which of the following makes establishing the phylogeny of the three domains of life most challenging?
    a. the lack of orthologous genes among the three domains
    b. the unreliability of the molecular clock for such early events
    c. the lack of an outgroup for cladistic analysis
    d. horizontal gene transfer
    e. the scarcity of fossils from that early time period

Chapter 27

Bacteria and Archaea

27.1 Structural and functional adaptations contribute to prokaryotic success

27.2 Rapid reproduction, mutation, and genetic recombination promote genetic diversity in prokaryotes

27.3 Diverse nutritional and metabolic adaptations have evolved in prokaryotes

27.4 Molecular systematics is illuminating prokaryotic phylogeny

27.5 Prokaryotes play crucial roles in the biosphere

27.6 Prokaryotes have both beneficial and harmful impacts on humans

Framework

This chapter presents the morphology, reproduction, phylogeny, and ecology of prokaryotes. These generally single-celled organisms greatly outnumber and outweigh all eukaryotes combined and flourish in all habitats, including those that are too harsh for any other forms of life. The collective impact of these microscopic organisms is huge.

Chapter Review

27.1 Structural and functional adaptations contribute to prokaryotic success

Prokaryotic cells are usually 0.5–5 μm in diameter. The three most common cell shapes are spheres (cocci), rods (bacilli), and spirals.

Cell-Surface Structures Prokaryotic cell walls maintain cell shape and protect the cell from bursting in a hypotonic surrounding. Most prokaryotes plasmolyze in a hypertonic environment. The cell walls of bacteria (but not those of archaea) contain **peptidoglycan,** a matrix composed of modified sugars cross-linked by short polypeptides.

The **Gram stain** is an important tool for identifying bacteria as **gram-positive** (bacteria with walls containing a thicker layer of peptidoglycan) or **gram-negative** (bacteria with more complex walls including an outer lipopolysaccharide membrane). Gram-negative bacteria are often more toxic and more resistant to antibiotics. Certain antibiotics, such as penicillin, inhibit the cross-linking of peptidoglycan in the bacterial cell wall.

Many prokaryotes secrete a dense, sticky **capsule** or a thinner *slime layer* outside the cell wall that serves as protection from a host's immune system and as glue for adhering to a substrate or other cells. Some prokaryotes may also attach by means of surface hair-like appendages called **fimbriae.** Fimbriae are shorter and more numerous than **pili,** sometimes called *sex pili,* which are specialized for the exchange of DNA between cells.

Motility Many prokaryotes exhibit **taxis,** an oriented movement in response to chemical, light, or other stimuli. Many prokaryotes are equipped with flagella, either scattered over the cell surface or concentrated at one or both ends of the cell. These flagella differ from eukaryotic flagella in their lack of a plasma membrane covering, and in their size, structure, and function. Though similar in size and rotational mechanism, the flagella of bacteria and archaea are composed of different proteins.

ATP-driven pumps and the diffusion of H^+ back into the cell power the motor (the rings embedded in the cell wall and plasma membrane), which turns a hook and rotates the filament of the flagellum. Of the 21 proteins shown to be universally required in bacterial flagella, 19 are homologous to cellular proteins with other functions. Thus, flagella may have evolved as a series of steps from simpler structures, as proteins took on new functions (*exaptation*).

Internal Organization and DNA Extensive internal compartmentalization is not found in prokaryotic cells, although some may have membranes, usually infoldings of the plasma membrane that function in respiration or photosynthesis.

The circular DNA chromosome, which has relatively little protein associated with it, is concentrated in a region called the **nucleoid.** Smaller rings of independently replicating DNA, called **plasmids,** may carry a few genes.

The ribosomes of prokaryotes are smaller than eukaryotic ribosomes and differ in their protein and RNA content. Some antibiotics work by binding to prokaryotic ribosomes and blocking protein synthesis.

Reproduction and Adaptation Prokaryotes reproduce rapidly by *binary fission* and have a huge reproductive potential. Their population growth, however, is regulated by the supply of nutrients, accumulation of toxic wastes, competition, or predation.

Some bacteria produce **endospores,** which are tough-walled, resistant, dormant cells formed in response to a lack of nutrients.

Due to the short generation times of prokaryotes, favorable mutations facilitate rapid adaptive evolution. In *E. coli* populations, scientists have documented rapid adaptation—occurring in 20,000 generations—to a challenging new, low-glucose environment.

INTERACTIVE QUESTION 27.1

Fill in the following table with brief descriptions of the characteristics of prokaryotic cells.

Property	Description
Cell shape	a.
Cell size	b.
Cell surface	c.
Motility	d.
Internal membranes	e.
Genome	f.
Reproduction and growth	g.

27.2 Rapid reproduction, mutation, and genetic recombination promote genetic diversity in prokaryotes

Rapid Reproduction and Mutation Mutations, although statistically rare, produce a great deal of genetic diversity because prokaryotes have such short generation times and large population sizes.

Genetic Recombination The mechanisms of *genetic recombination* in prokaryotes are different from the eukaryotic mechanisms of meiosis and fertilization. Three mechanisms can transfer DNA within species of prokaryotes and between different species (called *horizontal gene transfer*).

Prokaryotes can take up DNA from the environment in a process called **transformation.** The foreign DNA is integrated into the chromosome by an exchange of homologous DNA segments. Many bacteria have surface proteins specialized for uptake of DNA from closely related species.

Bacteriophages can transfer genes from one prokaryote to another by a process called **transduction.** A random piece of DNA may be accidentally packaged within a phage and introduced into another prokaryotic cell. Recombination may occur when the newly introduced DNA (from the donor cell) replaces the homologous region of the recipient cell's chromosome.

In **conjugation,** DNA is transferred while two prokaryotic cells are temporarily joined. In *E. coli,* a donor cell attaches to another cell by a sex pilus and transfers DNA to the recipient through a "mating bridge." The ability to form pili and donate DNA usually results from the presence of an **F factor,** a DNA segment that is either part of the chromosome or a plasmid.

Bacterial cells containing the F factor on the **F plasmid** are called F^+ cells. During conjugation, a single strand of the plasmid DNA is transferred through a mating bridge to the recipient cell. The single parental strands in both cells replicate, and the recipient cell is now also an F^+ cell.

Cells in which the F factor is inserted into the bacterial chromosome are called *Hfr cells,* for "high frequency of recombination." When an HFR cell undergoes conjugation, one parental strand of part of the F factor transfers an attached strand of the bacterial chromosome to the recipient cell. Movements disrupt the mating, usually resulting in a partial transfer of genes and F factor. Replication in both donor and recipient cells produces double-stranded DNA. Crossovers between the transferred DNA and the recipient cell's chromosome produce new genetic combinations in this recombinant F^- cell.

R plasmids carry genes that code for various mechanisms of resistance to antibiotics. Many R plasmids also have genes coding for pili and are transferred to nonresistant cells during conjugation, contributing to the medical problem of antibiotic-resistant pathogens.

INTERACTIVE QUESTION   27.2

Complete the following concept map that summarizes the genetic characteristics of prokaryotes.

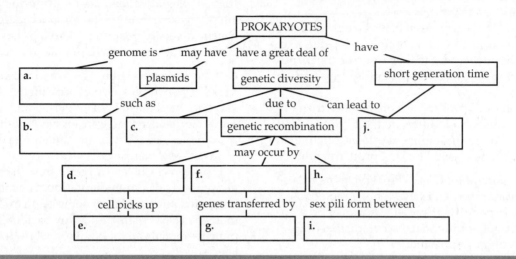

27.3 Diverse nutritional and metabolic adaptations have evolved in prokaryotes

Nutrition refers to how an organism obtains both the energy (*photo-* or *chemotroph*) and the carbon it needs for synthesizing organic compounds (*auto-* or *heterotroph*).

Thus, There are four major nutritional categories: (1) **photoautotrophs,** which use light energy and CO_2 (or HCO_3^-) to synthesize organic compounds; (2) **chemoautotrophs,** which obtain energy by oxidizing inorganic substances (such as H_2S, NH_3, or Fe^{2+}) and need only an inorganic carbon source; (3) **photoheterotrophs,** which use light energy but must obtain carbon in organic form; and (4) **chemoheterotrophs,** which use organic molecules as both an energy and a carbon source.

The Role of Oxygen in Metabolism   **Obligate aerobes** need O_2 for cellular respiration; **obligate anaerobes** are poisoned by O_2. Some obligate anaerobes use fermentation, others use **anaerobic respiration** to break down nutrients, with substances other than O_2 serving as the final electron acceptor. **Facultative anaerobes** can use O_2 but also can grow in anaerobic conditions using fermentation or anaerobic respiration.

Nitrogen Metabolism   Nitrogen is an essential component of proteins and nucleic acids. Some cyanobacteria and some methanogens obtain nitrogen through **nitrogen fixation,** the conversion of atmospheric N_2 to ammonia (NH_3).

Metabolic Cooperation   The cyanobacterium *Anabaena* forms filaments in which most cells carry out photosynthesis, while a few cells called **heterocytes** perform nitrogen fixation. Members of the filament share nutrients through intercellular connections. **Biofilms** are surface-coating colonies of one or more species. Biofilms are characterized by intercellular signaling, proteins that adhere cells to each other and to the substrate, and channels in the colony for movement of nutrients and wastes.

INTERACTIVE QUESTION   27.3

a. Identify the large structure inside the *Bacillus anthracis* cell in the following illustration. What is its significance? What mode of nutrition do you think this bacterium uses?

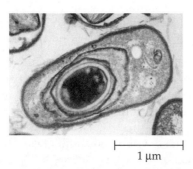

1 μm

b. The following micrograph shows the cyanobacterium *Anabaena*. What mode of nutrition does this bacterium use? What is the function of the spherical cell indicated by the arrow?

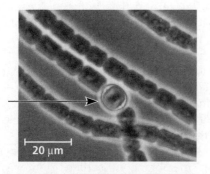

20 μm

27.4   Molecular systematics is illuminating prokaryotic phylogeny

While shape, staining characteristics, nutritional mode, and motility may be helpful characters for clinically identifying bacteria, they have not provided a phylogenetic classification.

Lessons from Molecular Systematics   Molecular comparisons of base sequences of SSU-rRNA indicate that domains Bacteria and Archaea diverged early in the history of life. Analyses of prokaryotic genomes have expanded with the field of *metagenomics*, which enables researchers to sample directly from the environment. This ongoing research has shown that the genetic diversity of prokaryotes is huge and that horizontal gene transfer has been significant in the evolution of prokaryotes.

Archaea   Archaea have some characteristics in common with bacteria, some in common with eukaryotes, and some unique to themselves. Many archaea are **extremophiles,** species that live in extreme habitats. **Extreme halophiles** (salt-lovers) live in extremely saline waters. **Extreme thermophiles** may be found in hot sulfur springs and near deep-sea hydrothermal vents.

Methanogens have a unique energy metabolism in which CO_2 is used to oxidize H_2, producing the waste product methane (CH_4). These strict anaerobes often live in swamps and marshes, are important decomposers in sewage treatment, and are gut inhabitants that contribute to the nutrition of cattle and other herbivores.

The methanogens and most extreme halophiles are placed in the clade Euryarchaeota. Most thermophiles fit into the Crenarchaeota. Not all archaea are extremeophiles; numerous archaea have been found in more moderate habitats.

Prokaryote phylogeny continues to change as new groups are discovered, including such clades as the Korarchaeota and Nanoarchaeota.

Bacteria   Bacteria show great diversity in their modes of nutrition and metabolism. Five major groups are described in the text.

Proteobacteria are a nutritionally diverse group of aerobic and anaerobic gram-negative bacteria. Its five subgroups are alpha proteobacteria, which include the nitrogen-fixing *Rhizobium* species and the *Agrobacterium* species used in the genetic engineering of plants; beta proteobacteria, which include the nitrogen-recycling soil bacterium *Nitrosomonas*; gamma proteobacteria, which include autotrophic sulfur bacteria and some serious pathogens and intestinal inhabitants such as *E. coli*; delta proteobacteria, including the colony-forming myxobacteria; and epsilon proteobacteria, the mostly pathogenic group that includes the stomach ulcer–causing *Helicobacter pylori*.

Chlamydias are obligate intracellular parasites of animals. One of these gram-negative species is the most common cause of blindness and nongonococcal urethritis, a sexually transmitted disease.

Spirochetes are helical heterotrophs that move in a corkscrew fashion. Spirochetes cause syphilis and Lyme disease.

Cyanobacteria perform plantlike, oxygen-generating photosynthesis and are important producers in freshwater and marine ecosystems. Some filamentous cyanobacteria have cells specialized for nitrogen fixation.

The gram-positive bacteria are a diverse group. The subgroup actinomycetes includes the species that cause tuberculosis and leprosy. Most actinomycetes are colonial soil bacteria, some of which are cultured to produce antibiotics. There are diverse solitary species of gram-positive bacteria, including those responsible for anthrax and botulism, and the species of *Staphylococcus* and *Streptococcus*. Mycoplasmas, the smallest of all cells, are the only bacteria that lack cell walls.

INTERACTIVE QUESTION 27.4

a. Evidence indicates that mitochondria evolved from an aerobic alpha proteobacterium through endosymbiosis. From which bacterial group did chloroplasts likely evolve?

b. What is genetic prospecting? How does it contribute to prokaryotic phylogeny?

27.5 Prokaryotes play crucial roles in the biosphere

Chemical Recycling As **decomposers,** prokaryotes return carbon, nitrogen, and other elements to the environment for assimilation into new living forms. Through photosynthesis and nitrogen fixation, prokaryotes convert inorganic compounds into forms that other organisms can use.

Ecological Interactions **Symbiosis** is an ecological relationship involving close contact between organisms of different species. The larger organism is called the **host,** and the smaller is known as the **symbiont.** In **mutualism,** both species benefit. In **commensalism,** one organism benefits while the other is neither harmed nor helped. In **parasitism,** a **parasite** eats portions of its host. **Pathogens** are parasites that cause disease.

INTERACTIVE QUESTION 27.5

Explain how cyanobacteria play an important role in chemical recycling.

27.6 Prokaryotes have both beneficial and harmful impacts on humans

Mutualistic Bacteria Many of the 500–1,000 species of bacteria living in the human intestines are mutualists that aid in food digestion and synthesize important nutrients that are used by the host.

Pathogenic Prokaryotes About one-half of all human diseases are caused by pathogenic bacteria. Pathogens most commonly cause disease by producing toxins. **Exotoxins** are secreted proteins that cause such diseases as botulism and cholera. **Endotoxins,** which are lipopolysaccharides released from the outer membrane

of gram-negative bacteria that have died, cause such diseases as typhoid fever and *Salmonella* food poisoning.

Improved sanitation and the development of antibiotics have decreased the incidence of bacterial disease in developed countries. The evolution of antibiotic-resistant strains of pathogenic bacteria, however, poses a serious health threat. Horizontal gene transfer also spreads genes connected with virulence, as in the emergence of the dangerous *E.coli* strain O157:H7. Some pathogenic prokaryotes are potential bioterrorism weapons.

Prokaryotes in Research and Technology **Bioremediation** is the use of organisms to remove environmental pollutants. Prokaryotes are used to treat sewage and clean up oil spills. They are used to produce biodegradable plastics. Prokaryotes have been engineered to make vitamins, antibiotics, ethanol, and other products.

Word Roots

an- = without, not; **aero-** = the air (*anaerobic respiration*: a catabolic pathway in which inorganic molecules other than oxygen accept electrons at the end of electron transport chains)

anti- = against; **-biot** = life (*antibiotic*: a chemical that kills bacteria or inhibits their growth)

chemo- = chemical; **hetero-** = different (*chemoheterotroph*: an organism that requires organic molecules for both energy and carbon)

endo- = inner, within (*endotoxin*: a toxic component of the outer membrane of certain gram-negative bacteria that is released only when the bacteria die)

exo- = outside (*exotoxin*: a toxic protein secreted by a prokaryote or other pathogen and that produces specific symptoms, even if the pathogen is no longer present)

-gen = produce (*methanogen*: organism that produces methane as a waste product of how it obtains energy; all known methanogens are in domain Archaea)

halo- = salt; **-philos** = loving (*extreme halophile*: an organism that lives in a highly saline environment)

mutu- = reciprocal (*mutualism*: an ecological interaction between two species in which both benefit)

-oid = like, form (*nucleoid*: a dense region of DNA in a prokaryotic cell)

photo- = light; **auto-** = self; **-troph** = food, nourish (*photoautotroph*: an organism that harnesses light energy to drive the synthesis of organic compounds from carbon dioxide)

sym- = with, together; **-bios** = life (*symbiosis*: an ecological relationship between organisms of two different species that live together in direct contact)

taxis = to arrange (*taxis*: an oriented movement toward or away from a stimulus)

thermo- = temperature (*extreme thermophile*: an organism that thrives in hot environments, often 60–80°C)

Structure Your Knowledge

1. One might think of prokaryotes as primitive, simple single cells. This chapter, however, provides many examples of how diverse, highly adaptable, metabolically versatile, cooperative, and communal they can be. List some examples of these prokaryotic characteristics.

2. Describe four positive ways in which prokaryotes have an impact on our lives and on the world around us.

Test Your Knowledge

MULTIPLE CHOICE: *Choose the one best answer.*

1. Many prokaryotes secrete a sticky capsule outside the cell wall that
   a. protects them from plasmolyzing in a hypertonic environment.
   b. serves as protection from host defenses and as glue for adherence.
   c. reacts with the Gram stain.
   d. is used for attaching cells during conjugation.
   e. is composed of peptidoglycan, modified sugars cross-linked by short polypeptides.

2. Some prokaryotes have specialized internal membranes that
   a. have attached ribosomes and function in protein synthesis.
   b. arose from endosymbiosis of smaller prokaryotes, creating mitochondria and chloroplasts.
   c. form from infoldings of the plasma membrane and may function in cellular respiration or photosynthesis.
   d. are produced by the endoplasmic reticulum but differ in composition from eukaryotic membranes.
   e. enclose the nucleoid and separate plasmids from the single prokaryotic chromosome.

3. Which of the following statements is *not* descriptive of conjugation between an Hfr bacterium and an F⁻ bacterium?
   a. The Hfr cell has an F plasmid integrated into its chromosome.
   b. The Hfr cell forms sex pili and transfers one strand of its chromosome into an F⁻ cell.

   c. Random movements often break the mating bridge before the entire single strand of its chromosome and F factor is transferred.
   d. Hfr conjugation is the major mechanism that transfers antibiotic resistance genes between bacteria.
   e. Crossing over between homologous genes creates a recombinant recipient cell.

4. Genetic variation in prokaryotes may be a result of
   a. horizontal gene transfer.
   b. mutation.
   c. conjugation.
   d. transformation and transduction.
   e. all of the above.

5. Chemoautotrophs
   a. are photosynthetic.
   b. use organic molecules for both energy and carbon sources.
   c. oxidize inorganic substances for energy and use CO_2 as a carbon source.
   d. use light to generate ATP but need organic molecules for a carbon source.
   e. use light energy to extract electrons from H_2S.

6. Facultative anaerobes
   a. can survive with or without oxygen.
   b. are poisoned by oxygen.
   c. can carry out anaerobic respiration or fermentation but not aerobic cellular respiration.
   d. are able to fix atmospheric nitrogen to make NH_3.
   e. include the methanogens, which oxidize H_2 and reduce CO_2 to CH_4.

7. Archaea
   a. may be more closely related to eukaryotes than to bacteria.
   b. have cell walls that lack peptidoglycan.
   c. are often found in harsh habitats, reminiscent of the environment of early Earth.
   d. include the methanogens, extreme halophiles, and extreme thermophiles, as well as groups found in more moderate habitats.
   e. All of the above are true.

8. The placement of prokaryotes into major clades within the domains Archaea and Bacteria has been based on
   a. fossil records of prokaryotes that have been recently discovered.
   b. molecular comparisons.
   c. Gram stain results and colony characteristics when grown on solid media.
   d. shape and motility.
   e. nutritional modes and ecology.

9. Which of the following groups is *incorrectly* described?

   a. Proteobacteria—diverse gram-negative bacteria, including pathogens such as *Salmonella* and *Helicobacter pylori* and beneficial species such as *Rhizobium*

   b. Chlamydias—intracellular parasites, including a species that causes blindness and nongonococcal urethritis

   c. Spirochetes—helical heterotrophs, including pathogens that cause syphilis and Lyme disease

   d. Cyanobacteria—photosynthetic filaments of cells; many species use the pigment bacteriorhodopsin and inhabit very saline waters

   e. Gram-positive bacteria—diverse group that includes actinomycetes, mycoplasmas, and pathogens that cause anthrax, botulism, and tuberculosis

10. Which of the following statements accurately describes a role of prokaryotes in chemical recycling?

    a. Chemoheterotrophic prokaryotes may function as decomposers to return chemical elements to an ecosystem.

    b. Photoheterotrophs convert atmospheric CO_2 into organic compounds.

    c. Chemoautotrophs such as the soil bacterium *Burkholderia glathei* increase the availability of nitrogen for growing seedlings.

    d. Photoautotrophic bacteria living on hydrothermal vents support communities of eukaryotic species.

    e. All of the above are correct.

11. Some cyanobacteria are capable of nitrogen fixation, a process that

    a. oxidizes nitrogen-containing compounds to produce ATP.

    b. allows these bacteria to live in anaerobic environments.

    c. removes soil nitrogen and returns N_2 gas to the atmosphere.

    d. converts N_2 to ammonia, making nitrogen available for incorporation into proteins and nucleic acids.

    e. is an essential part of the nitrogen cycle that extracts nitrogen trapped in mineral deposits and makes it available to organisms for their nitrogen metabolism.

12. O157:H7, a pathogenic strain of *E. coli*, has over 1,000 genes that are not present in the harmless strain K-12. Many of these genes are associated with phage DNA in the bacterial chromosome. These observations suggest that

    a. this virulent strain has many more R plasmids than does the harmless strain and is resistant to almost all antibiotics.

    b. at least some of these genes were acquired through phage-mediated horizontal gene transfer.

    c. these genes were not acquired through conjugation with K-12.

    d. most of these novel genes must be a result of mutations and short generation times.

    e. both b and c are correct.

FILL IN THE BLANKS

1. _____the name for spherical prokaryotes

2. _____region in which the prokaryotic chromosome is found

3. _____common laboratory technique that identifies two groups of bacteria

4. _____surface appendages of prokaryotes used for adherence to substrate

5. _____an oriented movement in response to light or chemical stimuli

6. _____resistant cell that can survive harsh conditions

7. _____surface-coating cooperative colonies of prokaryotes

8. _____proteins that are secreted by pathogens and are potent poisons

9. _____use of organisms to remove environmental pollutants

10. _____type of respiration that uses molecules other than O_2 as final electron acceptors

Chapter 28

Protists

Key Concepts

28.1 Most eukaryotes are single-celled organisms

28.2 Excavates include protists with modified mitochondria and protists with unique flagella

28.3 Chromalveolates may have originated by secondary endosymbiosis

28.4 Rhizarians are a diverse group of protists defined by DNA similarities

28.5 Red algae and green algae are the closest relatives of land plants

28.6 Unikonts include protists that are closely related to fungi and animals

28.7 Protists play key roles in ecological communities

Framework

This chapter surveys the incredible diversity of protists—eukaryotes that are not fungi, plants, or animals. The tree to the right represents a phylogenetic hypothesis of all extant eukaryote groups, organized into five larger "supergroups." Dotted lines indicate as-yet-uncertain evolutionary relationships.

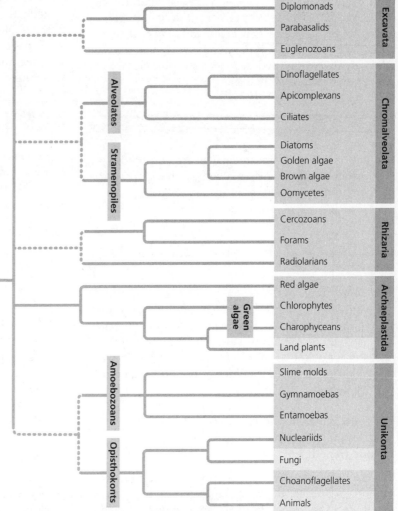

Chapter Review

The traditional kingdom Protista has been abandoned; many lineages of this polyphyletic group are recognized as their own kingdoms. The informal term **protist** is still used for this diverse assembly of eukaryotes.

28.1 Most eukaryotes are single-celled organisms

Structural and Functional Diversity in Protists Only three eukaryotic lineages are not protists. Most protists are unicellular, many with structurally complex cells. Nutritionally, protists can be photoautotrophs, heterotrophs, or **mixotrophs,** which are both photosynthetic and heterotrophic. These modes of nutrition are spread throughout the various protist lineages. Reproduction and life cycles are highly diverse.

Endosymbiosis in Eukaryotic Evolution Much evidence indicates that mitochondria evolved through **endosymbiosis,** originating from an endosymbiotic alpha proteobacteria within the earliest eukaryotes. Plastids evolved from a photosynthetic cyanobacterium that became an endosymbiont of a heterotrophic eukaryote. This lineage of cells eventually gave rise to two lineages of **algae:** red algae and green algae. In several instances of **secondary endosymbiosis,** a red or green alga was engulfed by a heterotrophic eukaryote, leading to new protist lineages.

INTERACTIVE QUESTION 28.1

The plastids of chlorarachniophytes are surrounded by four membranes and enclose a vestigial nucleus called a *nucleomorph*. What do these facts suggest?

Five Supergroups of Eukaryotes A current hypothesis of the phylogeny of protists includes five supergroups, shown on the phylogenetic tree on the previous page. The sequence of their divergence from a common ancestor is not known. The *amitochondriate protists* were once considered the oldest lineage, but new data refute that hypothesis.

28.2 Excavates include protists with modified mitochondria and protists with unique flagella

The proposed supergroup **Excavata** is based on morphological studies of the cytoskeleton and an "excavated" feeding groove found in some members. While evidence indicates that each of the three groups of excavates is monophyletic, the supergroup may not be.

Diplomonads and Parabasalids Most diplomonads and parabasalids are found in anaerobic environments. They have modified mitochondria and lack plastids.

The *mitosomes* of **diplomonads** lack functional electron transport chains. Diplomonads, such as the intestinal parasite *Giardia intestinalis*, have two nuclei and multiple flagella.

The reduced mitochondria of **parabasalids** are called *hydrogenosomes* and harvest energy anaerobically, releasing H_2. *Trichomonas vaginalis* is a sexually transmitted parabasalid parasite.

Euglenozoans The diverse clade of **euglenozoans** includes predatory heterotrophs, autotrophs, and pathogenic parasites, all of which have a spiral or crystalline rod inside their flagella.

A single large mitochondrion containing a mass of DNA called a kinetoplast is characteristic of the group called **kinetoplastids.** These protists include free-living heterotrophs as well as a number of parasites of plants, animals, and other protists.

The **euglenids** are characterized by one or two flagella that emerge from the anterior end. Many species of the photosynthetic *Euglena* switch to heterotrophy in the absence of sunlight. A pigmented eyespot allows light to strike a light detector from one direction, directing the organism toward sunlight.

INTERACTIVE QUESTION 28.2

Sleeping sickness is caused by the kinetoplastid *Trypanosoma* and transmitted by the African tsetse fly. Why is it so difficult for the human immune system to attack this parasite successfully?

28.3 Chromalveolates may have originated by secondary endosymbiosis

The large, diverse supergroup **Chromalveolata** is proposed based on some DNA sequence data and some evidence that these organisms originated by secondary endosymbiosis when their common ancestor engulfed a red alga. Future data may or may not support this supergroup.

Alveolates Membrane-bounded alveoli under the plasma membrane are characteristic of the **alveolates.** These sacs may function in stabilizing the cell surface or in osmoregulation. Molecular systematics supports the division of the monophyletic alveolates into three subgroups.

Dinoflagellates make up a large proportion of marine and freshwater *phytoplankton*. The beating of two flagella in grooves on the cell surface between the cellulose plates of the cell produces a characteristic spinning movement. *Blooms*, or population explosions, of dinoflagellates are responsible for the often-harmful "red tides." Many dinoflagellates are heterotrophic.

Nearly all **apicomplexans** are animal parasites, and most have complex life cycles that often include sexual and asexual stages and several host species. The parasites spread by tiny infectious cells called *sporozoites*. They have an apical complex of organelles specialized for invading host cells and a nonphotosynthetic plastid called an apicoplast.

The control of malaria, caused by *Plasmodium*, is complicated by multiple factors: development of insecticide resistance in *Anopheles* mosquitoes, which spread the disease; drug resistance in *Plasmodium;* the sequestering of the parasite within human liver and blood cells; and the ability of the parasite to change its surface proteins. The sequencing of the *Plasmodium* genome and the tracking of the expression of most genes may facilitate development of vaccines or therapeutic drugs.

Ciliates are characterized by the cilia they use to move and feed. Their numerous cilia may be widespread or clumped. Ciliates have two types of nuclei: Large macronuclei control everyday functions of the cell and contain multiple copies of the genome. Small micronuclei are exchanged in a process called **conjugation,** which results in genetic variation. Ciliates reproduce asexually by binary fission.

Stramenopiles The **stramenopiles** include several heterotrophic groups and various important groups of algae, all characterized by hairlike projections on a flagellum. In most cases, a shorter "smooth" flagellum is paired with the "hairy" one.

Diatoms are a major component of marine and freshwater phytoplankton; unique boxlike silica walls protect these unicellular protists. Massive quantities of fossilized diatom walls make up *diatomaceous earth*. Diatom blooms are part of a biological carbon "pump" that sequesters carbon in ocean deposits.

The color of **golden algae** results from yellow and brown carotenoids. Most golden algae are unicellular, biflagellated members of marine and freshwater phytoplankton.

The mostly marine **brown algae** are the largest and most complex algae. Many of these multicellular algae (and some large red and green algae) are called seaweeds. Some brown algae have tissues and organs that are analogous to those found in plants. They may have a plantlike **thallus,** or body, consisting of a rootlike **holdfast** and a stemlike **stipe,** which supports leaflike **blades.** Some brown algae have floats, which keep the photosynthesizing blades near the surface. The giant, fast-growing kelps live in deeper waters.

Some multicellular algae have an **alternation of generations.** The multicellular, haploid *gametophyte* produces gametes. After fertilization, the diploid zygote grows into the multicellular *sporophyte*, which then produces spores by meiosis. The gametophyte and sporophyte may be similar in appearance (**isomorphic**) or distinct (**heteromorphic**).

INTERACTIVE QUESTION   28.3

Label the indicated structures in the following diagram of a *Paramecium*. To which subgroup, clade, and supergroup does this organism belong?

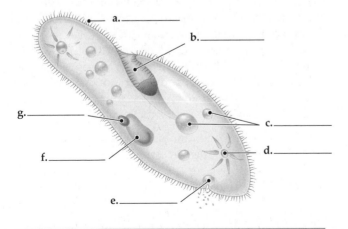

INTERACTIVE QUESTION   28.4

What adaptations help brown algal seaweeds inhabit intertidal zones?

Oomycetes include water molds, white rusts, and downy mildews. Unlike their plastid-bearing ancestors, oomycetes lack plastids. Although many resemble fungi in appearance, oomycetes have cell walls made of cellulose, and molecular systematics shows they are not closely related to fungi.

Water molds are important decomposers in aquatic ecosystems. Their life cycle includes both sexual and asexual reproduction.

White rusts and downy mildews can be destructive parasites of land plants. Genetic studies of *Phytophthora infestans*, the oomycete that causes potato late blight, indicate that in recent decades it has acquired genes that make it more aggressive and resistant to pesticides.

28.4   Rhizarians are a diverse group of protists defined by DNA similarities

Molecular systematics indicates that **Rhizaria** is a monophyletic yet diverse clade. Some studies suggest that Rhizaria should be included within the supergroup Chromalveolata.

Amoebas, originally defined on the basis of their cellular extensions called **pseudopodia,** are spread across many eukaryotic taxa. The rhizarian amoebas have threadlike pseudopodia.

Radiolarians   Slender pseudopodia help **radiolarians** phagocytize microscopic food organisms. These primarily marine organisms have delicate internal silica skeletons.

Forams   Foraminiferans, or **forams,** are known for their porous shells, called **tests,** made of organic material and calcium carbonate. Pseudopodia extending through the pores function in swimming, test formation, and feeding. Many forams obtain nourishment from symbiotic algae.

Cercozoans   This large group contains amoeboid and flagellated protists that feed with threadlike pseudopodia. Most **cercozoans** are parasites or predators, although one group is mixotrophic and at least one cercozoan is autotrophic.

INTERACTIVE QUESTION   28.5

a. How are foram fossils used?

b. The cercozoan *Paulinella chromatophora* has a unique photosynthetic structure called a chromatophore. What does genetic and morphological evidence suggest about the origin of the chromatophore?

28.5   Red algae and green algae are the closest relatives of land plants

Descendants of the heterotrophic protists that acquired a cyanobacterium endosymbiont evolved into red algae and green algae, probably more than a billion years ago. Land plants arose from the green algal lineage at least 475 million years ago. The supergroup **Archaeplastida**

is a monophyletic group that includes red algae, green algae, and land plants.

Red Algae   The accessory pigment phycoerythrin produces the color of **red algae** and allows them to absorb those wavelengths of light that penetrate into deep water. Most red algae species are marine and multicellular, the largest of which are called seaweeds. Alternation of generations is common, although life cycles are diverse and the algae have no flagellated stages.

Green Algae   The chloroplasts of **green algae** resemble those of land plants, and molecular and cytological evidence indicates that green algae and plants are closely related. Some systematists favor including green algae in an extended plant kingdom, Viridiplantae. The two main green algal groups are chlorophytes and charophytes. The latter group is very closely related to land plants.

Most chlorophytes live in fresh water. Unicellular forms may be phytoplankton, inhabitants of damp soil, or symbionts in other eukaryotes.

Size and complexity have increased in the green algae in three ways: the formation of colonies, cell division and differentiation to produce true multicellular forms, and repeated nuclear divisions to produce a multinucleated thallus.

The life cycles of most chlorophytes include sexual stages (with biflagellated gametes) and asexual stages. Some multicellular green algae have an alternation of generations.

28.6   Unikonts include protists that are closely related to fungi and animals

The proposed supergroup **Unikonta** includes animals, fungi, and some protists. Molecular systematics supports two major clades: the amoebozoans and the opisthokonts (animals, fungi, and closely related protists). Some studies support the close relationship of these two clades; other studies do not.

Research has yet to establish the root of the eukaryotic tree. Two Oxford researchers have proposed that the unikonts were the first eukaryotes to diverge. They base this hypothesis on the DHFR and TS genes, which in bacteria and the unikont groups remain separate (the ancestral character), and which are fused (the derived character) in members of the Excavata, Chromalveolata, Rhizaria, and Archaeplastida.

Amoebozoans   Slime molds, gymnamoebas, and entamoebas are grouped as **amoebozoans,** a clade that includes many species with lobe- or tube-shaped pseudopodia. The resemblance of the slime molds (or mycetozoans) to fungi is the result of convergent evolution.

Plasmodial slime molds engulf food particles by phagocytosis as they grow through leaf litter or rotting logs. This multinucleate mass, called a **plasmodium,** is the feeding stage. Under harsh conditions, fruiting bodies form that function in sexual reproduction.

During the feeding stage of the life cycle, **cellular slime molds** consist of haploid solitary amoeboid cells. As food is depleted, the individual cells congregate into a slug-like, motile mass. Some cells form a stalk that supports asexual fruiting bodies, which produce resistant spores.

The cellular slime mold *Dictyostelium discoideum* is an experimental model for studying the evolution of multicellularity. Research indicates that a recognition system involving cell surface proteins allows "noncheater" cells to aggregate differentially from "cheater" cells (mutant cells that always become spore cells).

Most amoebas of the group gymnamoebas are found free-living in freshwater, marine, or soil habitats, where they consume bacteria and other protists.

Amoebas of the genus *Entamoeba* are parasites of vertebrates and some invertebrates. *E. histolytica* causes amebic dysentery in humans.

Opisthokonts Of the protists in the diverse clade **opisthokonts,** nucleariids are more closely related to fungi, and choanoflagellates are more closely related to animals.

28.7 Protists play key roles in ecological communities

Symbiotic Protists Dinoflagellates are important photosynthetic symbionts in coral. Many termites harbor wood-digesting protists in their guts. Other protists are destructive parasites of plants and animals.

Photosynthetic Protists Many protists are **producers.** Most aquatic food webs are based on photosynthetic protists and prokaryotes.

INTERACTIVE QUESTION 28.7

Satellite data have shown a negative correlation, particularly in tropical and mid-latitude oceans, between increasing sea surface temperatures and the biomass of photosynthetic protists and prokaryotes. What may account for this decrease in the growth of producers? What effects might these changes have on marine ecosystems and the global carbon cycle?

INTERACTIVE QUESTION 28.6

Fill in the following table to compare three aspects of the life cycles of plasmodial slime molds and cellular slime molds.

Group	Ploidy of feeding stage	Body form of feeding stage	Response to lack of food
Plasmodial slime molds	a.	b.	c.
Cellular slime molds	d.	e.	f.

Word Roots

con- = with, together (*conjugation:* in ciliates, the transfer of micronuclei between two cells that are temporarily joined)

hetero- = different; **-morph** = form (*heteromorphic:* a condition in the life cycle of plants and certain algae in which the sporophyte and gametophyte generations differ in morphology)

iso- = same (*isomorphic:* alternating generations in which the sporophytes and gametophytes look alike, although they differ in chromosome number)

pseudo- = false; **-podium** = foot (*pseudopodium:* a cellular extension of amoeboid cells used in moving and feeding)

thallos- = sprout (*thallus:* a seaweed body that is plant-like but lacks true roots, stems, and leaves)

Structure Your Knowledge

1. Evidence indicates that all plastids (except for one recent example) evolved from a cyanobacterium that was engulfed by an ancestral heterotrophic eukaryote (primary endosymbiosis). Fill in the blanks in the following diagram, which depicts the diversification of this ancestral eukaryote into red algae and green algae and then to various protist groups.

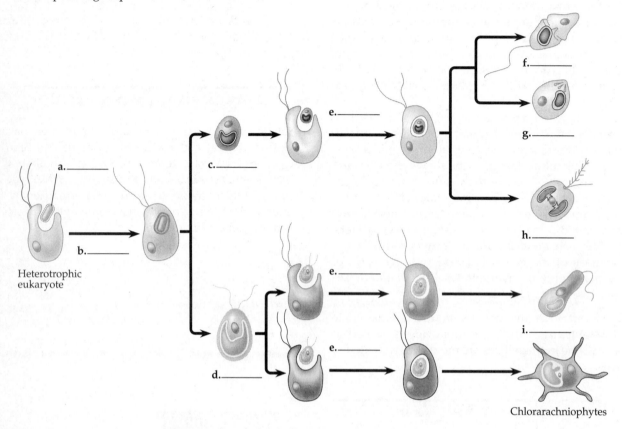

Heterotrophic eukaryote

Chlorarachniophytes

Test Your Knowledge

MULTIPLE CHOICE: *Choose the one best answer.*

1. Mixotrophs
   a. include the plasmodial slime molds.
   b. include the cellular slime molds.
   c. have gametophyte and sporophyte generations that differ in morphology.
   d. can be both heterotrophic and autotrophic.
   e. can be both parasitic and free-living.

2. According to the theory of secondary endosymbiosis,
   a. multicellularity evolved when primitive cells incorporated prokaryotic cells that then took on specialized functions.
   b. the symbiotic associations found in coral result from the incorporation of dinoflagellates into coral animals.
   c. cells that had obtained their plastids through endosymbiosis were engulfed and themselves became plastids in heterotrophic eukaryotic cells.

   d. the infoldings and specializations of the plasma membrane led to the evolution of the endomembrane system.
   e. mitochondria originated first from a heterotrophic prokaryotic endosymbiont; plastids originated from a secondary endosymbiosis of a photosynthetic prokaryote.

3. The diplomonads and parabasalids are unique among eukaryotes in that
   a. they lack double membranes around their nuclei, even though diplomonads have two nuclei.
   b. they lack ribosomes.
   c. they lack plastids, and all other protist lineages have at least some members that are autotrophic.
   d. they have modified mitochondria and obtain energy anaerobically.
   e. they were an early branch from bacteria rather than from archaea.

4. Genetic variation is generated in the ciliate *Paramecium* when
   a. a micronucleus replicates its genome many times and becomes a macronucleus.
   b. zoospores, which are gametes, fuse.
   c. micronuclei are exchanged in conjugation.
   d. mutations occur in the many copies of the genome contained in the macronucleus and are passed on to offspring.
   e. plasmids are exchanged in conjugation.

5. Which of the following is *not* true of seaweeds?
   a. They are found in the red, green, and brown algal groups.
   b. They have true roots that anchor them tightly to withstand the turbulence of waves.
   c. They may be a source of food and commercial products.
   d. Gel-forming polysaccharides in some of their cell walls protect them from abrasive wave action.
   e. They are multicellular photoautotrophs.

6. The Chlorophyta, or green algae, are believed to share a common ancestor with plants because
   a. they are the only multicellular algal protists.
   b. they do not have flagellated gametes.
   c. they are the only protists with plastids that evolved from cyanobacteria.
   d. their chloroplasts are similar in structure and pigment composition to those of plants.
   e. they are the only algae that exhibit alternation of generations.

7. Which protists are in the same eukaryotic supergroup as animals?
   a. dinoflagellates
   b. forams
   c. parabasalids
   d. stramenopiles
   e. amoebozoans

8. Which supergroup includes the land plants?
   a. Unikonta
   b. Archaeplastida
   c. Excavata
   d. Chromalveolata
   e. Rhizaria

9. Phytoplankton
   a. form the basis of most marine food chains.
   b. include multicellular green, red, and brown algae.
   c. are mutualistic symbionts that provide food for coral reef communities.
   d. are unicellular heterotrophs that float near the ocean surface.
   e. Both a and d are correct.

10. How do diatoms affect global CO_2 levels?
    a. As surface seawater warms, the photosynthetic output of diatoms increases, removing more CO_2.
    b. Blooms of diatoms kill fish and other consumers, increasing decomposition and raising CO_2 levels.
    c. CO_2 absorbed by diatoms is "pumped" to the ocean floor when they die, lowering CO_2 levels.
    d. Because their glass-like walls are made of silica and not carbon, diatoms do not affect CO_2 levels.
    e. As the basis of all marine food webs, diatoms perform 50% of the world's photosynthesis and lower CO_2.

MATCHING: *Match the protist clades (a–h) with their characteristics and examples (1–8).*

_____ 1. modified mitochondria; *Giardia* and *Trichomonas*

_____ 2. plant-type chloroplasts; *Chlamydomonas* and *Ulva*

_____ 3. lobe-shaped pseudopodia; free-living amoeba, parasites, slime molds

_____ 4. phycoerythrin, no flagellated stage; *Porphyra*

_____ 5. hairy and smooth flagella; water molds, diatoms, golden and brown algae

_____ 6. amoebas with threadlike pseudopodia

_____ 7. subsurface sacs; dinoflagellates, apicomplexans, ciliates

_____ 8. flagella with spiral or crystalline rod; *Trypanosoma* and *Euglena*

a. Alveolates
b. Amoebozoans
c. Diplomonads and parabasalids
d. Euglenozoans
e. Forams, radiolarians, cercozoans
f. Green algae
g. Red algae
h. Stramenopiles

Chapter 29

Plant Diversity I: How Plants Colonized Land

Key Concepts

29.1 **Land plants evolved from green algae**

29.2 **Mosses and other nonvascular plants have life cycles dominated by gametophytes**

29.3 **Ferns and other seedless vascular plants were the first plants to grow tall**

■ **Framework**

The evolution of land plants involved adaptations to terrestrial habitats. The kingdom Plantae includes multicellular, photoautotrophic eukaryotes that develop from embryos nourished by the parent plant. Plants exhibit an alternation of generations in which the diploid sporophyte is the more conspicuous stage in all groups except the bryophytes. This chapter describes the evolution, life cycles, and ecological importance of bryophytes and seedless vascular plants.

■ **Chapter Review**

For most of Earth's history, life was confined to aquatic environments. Plants began to move onto land about 500 million years ago and have now diversified into 290,000 known species. Plants enabled other forms of life to survive on land.

29.1 Land plants evolved from green algae

Morphological and Molecular Evidence The cells of land plants have chloroplasts containing chlorophylls *a* and *b* and cell walls of cellulose. These characteristics, however, are found in several algal groups and cannot be used to distinguish plants from algae. Only the

green algal charophytes and land plants share the following four traits: rings of cellulose-synthesizing proteins, which produce the cellulose microfibrils of cell walls; enzymes within peroxisomes that help minimize photorespiration losses; similarities in sperm cells (in those land plants that have flagellated sperm); and the formation of a **phragmoplast** in mitosis. This alignment of microtubules in the synthesis of a cell plate is found only in plants and certain charophytes, such as *Chara* and *Coleochaete.*

Results from the analysis of nuclear and chloroplast genes confirm that charophytes are the closest living relatives of land plants.

Adaptations Enabling the Move to Land The tough polymer **sporopollenin** protects charophyte zygotes during fluctuations in water levels. With the accumulation of such traits, ancient charophytes living along the edges of ponds and lakes may have given rise to the first plants to colonize land.

INTERACTIVE QUESTION   29.1

List some of the benefits and challenges of a terrestrial habitat for the first land plants.

Plant biologists debate three different options for the makeup of the plant kingdom: the traditional kingdom Plantae containing only embryophytes (plants with embryos); the kingdom Streptophyta, which includes the charophytes; or the kingdom Viridiplantae, which also includes the chlorophytes. This textbook uses the traditional embryophyte definition of the taxon Plantae.

Derived Traits of Plants The following four derived traits distinguish land plants: alternation of generations (along with a multicellular, dependent embryo), walled spores produced in sporangia, multicellular gametangia, and apical meristems.

Alternation of generations is found in all land plants. In this life cycle, the multicellular haploid **gametophyte** alternates with the multicellular diploid **sporophyte.** The sporophyte produces haploid **spores,** which are reproductive cells that develop directly into new organisms. Plant embryos, which are retained in the female gametophyte plant, have **placental transfer cells** that facilitate the transfer of nutrients from parental tissues. This derived trait is the basis for referring to land plants as **embryophytes.**

INTERACTIVE QUESTION 29.2

The gametophyte produces a. _____ by b. _____. Following fertilization, the c. _____ divides by d. _____ to develop into the e. _____. The sporophyte produces f. _____ by g. _____. Spores germinate and develop into the h. _____.

The tough polymer sporopollenin protects spores as they disperse on land. Spores are produced and protected within multicellular **sporangia** on the sporophyte plant. Diploid cells called **sporocytes,** or spore mother cells, undergo meiosis and produce haploid spores in sporangia.

In the gametophyte, gametes are produced within multicellular **gametangia.** Sperm are produced in **antheridia.** In many groups, sperm are flagellated and swim to the eggs. A single egg cell is produced and fertilized within an **archegonium,** where the zygote develops into an embryo.

Regions of cell division at the tips of roots and shoots, called **apical meristems,** produce linear growth and cells that differentiate into a protective epidermis and internal plant tissues. The elongation of roots and shoots provides increasing access to environmental resources.

The leaves and stems of many land plants are coated by a waxy **cuticle,** which protects from water loss and microbial attack. Early plant fossils show symbiotic associations between plants and fungi. Such *mycorrhizae* facilitate nutrient absorption by plants. Many plants produce *secondary compounds* through side branches of their main metabolic pathways. These compounds may deter herbivores, parasites, and pathogens or provide protection from UV radiation.

The Origin and Diversification of Plants Fossil spores embedded in plant tissue have been found that date back 475 million years.

Vascular plants have **vascular tissue,** a transport system composed of cells joined into tubes. Liverworts, mosses, and hornworts do not have an extensive transport system and are referred to as "nonvascular" plants and informally called **bryophytes.** Molecular and morphological evidence indicates that the bryophytes do not constitute a clade.

Vascular plants—about 93% of extant plant species— do form a clade and are divided into smaller clades. The **lycophytes** (club mosses and their relatives) and the **pterophytes** (ferns and their relatives) are informally called **seedless vascular plants.** The paraphyletic seedless vascular plants may be referred to as a **grade,** a group of organisms that shares a key adaptation.

A **seed** is an embryo enclosed with a supply of nutrients within a protective coat. The majority of living plants are seed plants, a clade that contains two groups. **Gymnosperms** are called "naked seed" plants because their seeds are not enclosed. **Angiosperms** are flowering plants, in which seeds develop inside chambers called ovaries in a flower. Taxonomists recognize ten phyla of extant plants.

INTERACTIVE QUESTION 29.3

Why are the seedless vascular plants considered a grade and not a clade?

29.2 Mosses and other nonvascular plants have life cycles dominated by gametophytes

Today there are three groups of nonvascular plants: **liverworts** (phylum Hepatophyta), **mosses** (phylum Bryophyta), and **hornworts** (phylum Anthocerophyta).

Bryophyte Gametophytes In the bryophytes, the gametophyte is the prevalent generation; sporophytes are present only part of the time. When moss spores germinate in a moist habitat, they grow into a mass of green filaments called a **protonema.** Meristem-containing "buds" grow into gamete-producing upright structures called **gametophores.** Some mosses have conducting tissues in their "stems." These tissues apparently evolved independently from the vascular tissue of vascular plants.

Bryophyte gametophytes are generally only a few cells thick and low growing. They are anchored by long, tubular cells or filaments of cells called **rhizoids,** which do not play a major role in water and mineral absorption.

Most mosses have separate male and female gametophytes. Gametangia are enclosed in jackets of protective tissue. Sperm are produced in antheridia, from which they swim to the single egg retained in each archegonium. Zygotes develop into embryos, which rely on nutrients transported by placental transfer cells to develop into sporophytes. Many bryophyte species also reproduce asexually, as in the *brood bodies* produced by some mosses.

Bryophyte Sporophytes   Although they are usually photosynthetic when young, bryophyte sporophytes remain attached to and dependent on the parental gametophytes. A sporophyte has a **foot** that obtains nutrients from the gametophyte, a **seta** (stalk) that conducts materials and may elongate for spore dispersal, and a sporangium or **capsule** in which millions of spores are produced by meiosis. The specialized "toothed" **peristome** gradually releases spores to be dispersed by wind currents.

The sporophytes of mosses and hornworts have **stomata,** pores through which gases are exchanged and water evaporates. There are three possible evolutionary origins of stomata: They were present in the ancestor of all three groups and later lost in the liverwort lineage; they were present in the ancestor of the branch leading to hornworts, mosses, and vascular plants; or they were present in the ancestor of mosses and vascular plants and arose independently in hornworts.

INTERACTIVE QUESTION   29.4

Review the life cycle of a typical moss plant by filling in the following blanks.

The dominant generation is the **a.** _____.

Female gametophytes produce eggs in **b.** _____.

Male gametophytes produce sperm in **c.** _____.

Sperm **d.** _____ through a film of water to fertilize the egg.

The zygote remains in the archegonium and develops into the **e.** _____, still attached to the female gametophyte.

Spores are formed by the process of **f.** _____ in the **g.** _____.

When shed, spores develop into the **h.** _____.

Ecological and Economic Importance of Mosses   Mosses are widely distributed and common in moist habitats. Some are adapted to very cold or dry habitats due to their ability to dehydrate and rehydrate without dying. Research has shown that mosses reduce the rate at which nitrogen is lost by water erosion in a sandy ecosystem. Some species harbor nitrogen-fixing cyanobacteria.

Sphagnum, or peat moss, is an abundant wetland moss that forms undecayed organic deposits known as **peat.** The decay-resistant compounds in peat bogs can preserve organisms for thousands of years. Peat is harvested for use as a fuel, a soil conditioner, and a packing material for plants. Peat serves as huge stores of organic carbon and helps to stabilize atmospheric CO_2 concentrations. Current overharvesting of *Sphagnum,* as well as its accelerated decomposition in situ due to the drying of some peatlands resulting from global climate change, would increase the release of stored CO_2 and contribute to global warming.

29.3   Ferns and other seedless vascular plants were the first plants to grow tall

Origins and Traits of Vascular Plants   Fossils of ancient relatives of modern vascular plants date back 425 million years. These tiny plants had branched, independent sporophytes and terminal sporangia. The branching made possible more complex bodies and multiple sporangia. They lacked roots and some other adaptations of vascular plants.

In modern vascular plants, the diploid sporophyte is the larger and more complex plant. Seedless vascular plants remain limited to damp habitats because of their swimming sperm and fragile gametophytes, as shown in the life cycle of a fern in Interactive Question 29.5.

INTERACTIVE QUESTION 29.5

In the following diagram of the life cycle of a fern, label the processes (in the boxes) and structures (on the lines). Indicate which portion of the life cycle is haploid and which is diploid.

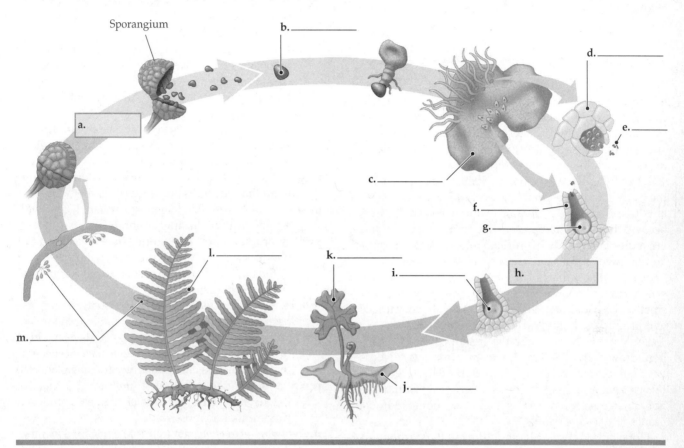

Sporangium

a.
b.
c.
d.
e.
f.
g.
h.
i.
j.
k.
l.
m.

Water and minerals are conducted up from roots in **xylem.** The xylem of most vascular plants includes tube-shaped cells called **tracheids** whose cell walls are strengthened by **lignin.** Vascular plants are also called tracheophytes. **Phloem** transports sugars and other organic nutrients through cells arranged into tubes.

Unlike the rhizoids of bryophytes, **roots** both anchor plants and absorb water and nutrients. Roots may have evolved from subterranean stems, either once in a common ancestor or independently in different lineages of vascular plants.

Leaves increase a plant's photosynthetic surface area. Lycophytes are the oldest lineage of extant vascular plants. The single-veined leaves of lycophytes, known as **microphylls,** are usually small and spine shaped. The branching vascular systems of **megaphylls,** typical of most other vascular plants, support larger leaves with greater photosynthetic capacity. Megaphylls appear in the fossil record 370 million years ago, 40 million years after the appearance of microphylls.

Microphylls may have originated from sporangia along the stem, each supported by a single vascular tissue strand. Megaphylls may have evolved from closely lying branches when one branch came to *overtop* lower branches, which then formed webbing that joined them together.

Sporophylls are modified leaves bearing sporangia. In ferns, clusters of sporangia called **sori** are produced on the undersides of sporophylls. Cones, or **strobili,** are formed from groups of sporophylls in many lycophytes and most gymnosperms.

Most seedless vascular plants are **homosporous,** producing only one kind of spore, which develops into bisexual gametophytes. **Heterosporous** species produce two kinds of spores: **megaspores** that develop into female gametophytes and **microspores** that develop into male gametophytes. All seed plants and some seedless vascular plants are heterosporous.

INTERACTIVE QUESTION   29.6

Describe the traits of vascular plants that enable them to grow tall. What are the adaptive advantages of increased height?

Classification of Seedless Vascular Plants   The two clades of extant seedless vascular plant are phylum Lycophyta and phylum Pterophyta. Some systematists still retain three separate phyla in the pterophyte clade: ferns, horsetails, and whisk ferns.

Lycophytes were a major part of the landscape during the Carboniferous period. One evolutionary line, the giant "tree lycophytes," became extinct when the climate became drier. The herbaceous lycophytes survived and are represented today by the club mosses, spike mosses, and quillworts. Many tropical species grow on trees as *epiphytes*—plants that anchor to other organisms but are not parasites. Sporophytes grow upright with many small leaves; horizontal stems grow along the soil surface and produce dichotomously branching roots.

Ferns were also found in the great forests of the Carboniferous period and are the most numerous seedless vascular plants in the modern flora. The large fern leaves (megaphylls), or fronds, often grow from horizontal stems. Most fern species are homosporous; gametophytes are small.

Horsetails grew as tall plants during the Carboniferous period. A few species of the genus *Equisetum* are the only representatives found today. These small, upright plants grow in damp locations and have green, jointed stems.

In spite of their dichotomous branching and lack of true roots, molecular evidence indicates that *Psilotum*, the whisk fern, is closely related to ferns.

The Significance of Seedless Vascular Plants   The seedless vascular plants of the Carboniferous forests decreased CO_2 levels in the atmosphere, perhaps contributing to the global cooling at the end of the Carboniferous period. Dead plant material did not completely decay in the stagnant swamp waters, and great accumulations of peat developed. When the sea later covered the swamps and marine sediments piled on top, heat and pressure converted the peat to coal.

Word Roots

-angio = vessel (*gametangium:* multicellular plant structure in which gametes are formed. Female gametangia are called archegonia; male gametangia are called antheridia)

bryo- = moss; **-phyte** = plant (*bryophyte:* an informal name for a moss, liverwort, or hornwort; a nonvascular plant that lives on land but lacks some of the terrestrial adaptations of vascular plants)

gymno- = naked; **-sperm** = seed (*gymnosperm:* a vascular plant that bears naked seeds—seeds not enclosed in specialized chambers)

hetero- = different; **-sporo** = a seed (*heterosporous:* referring to a plant species that has two kinds of spores: microspores, which develop into male gametophytes, and megaspores, which develop into female gametophytes)

homo- = like (*homosporous:* referring to a plant species that has a single kind of spore, which develops into a bisexual gametophyte)

mega- = large (*megaspores:* a spore from a heterosporous plant that develops into a female gametophyte)

micro- = small; **-phyll** = leaf (*microphyll:* in lycophytes. a small leaf with a single unbranched vein)

peri- = around; **-stoma** = mouth (*peristome:* a ring of interlocking, tooth-like structures on the upper part of the moss capsule, often specialized for gradual spore discharge)

-phore = bearer (*gametophore:* the mature gamete-producing structure of a moss gametophyte)

phragmo- = a partition; **-plast** = formed, molded (*phragmoplast:* an alignment of cytoskeletal elements that forms across the midline of a dividing plant cell)

proto- = first; **-nema** = thread (*protonema:* a mass of green, branched, one-cell-thick filaments produced by germinating moss spores)

pter- = fern (*pterophytes:* an informal name for a member of phylum Pterophyta, which includes ferns, horsetails, and whisk ferns)

rhizo- = root; **-oid** = like, form (*rhizoid:* a long tubular single cell or filament of cells that anchors bryophytes to the ground)

Structure Your Knowledge

1. The evolution of land plants involved adaptations to a terrestrial habitat. List some of these adaptations that evolved in the bryophytes and the seedless vascular plants.

2. The following diagram presents a widely held view of plant phylogeny. Fill in blanks a–e with the informal names for the major groups, and blanks f–k with the common names for the modern plant groups (list the three groups included in clade i). To what events do the numbers 1, 2, and 3 refer?

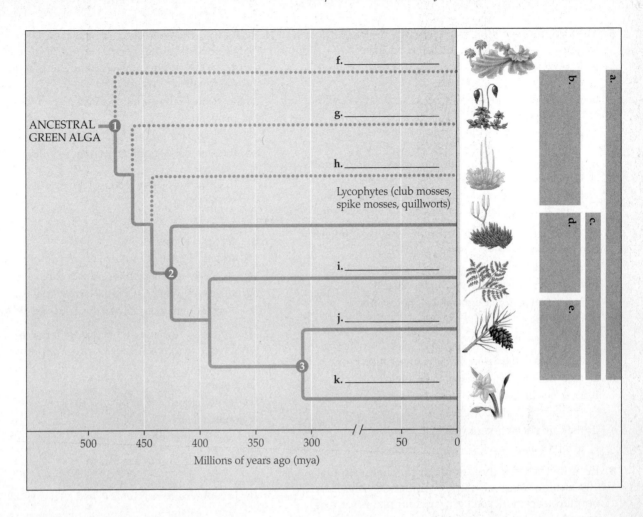

f._____

g._____

ANCESTRAL GREEN ALGA

h._____

Lycophytes (club mosses, spike mosses, quillworts)

i._____

j._____

k._____

500   450   400   350   300   50   0

Millions of years ago (mya)

Test Your Knowledge

MULTIPLE CHOICE: *Choose the one best answer.*

1. Which of the following are adaptations for terrestrial life seen in all plants?
   a. chlorophylls *a* and *b*.
   b. cell walls of cellulose and lignin.
   c. sporopollenin and an embryo protected and nourished by a gametophyte.
   d. vascular tissue.
   e. cuticle and stomata.

2. Plants are thought to be most closely related to charophytes based on
   a. peroxisome enzymes that minimize photorespiration.
   b. rosette cellulose-synthesizing proteins in their plasma membranes.
   c. phragmoplast formation of cell plates during mitosis.
   d. similarities in sperm cells (in those land plants with flagellated sperm).
   e. all of the above characteristics.

3. Which of the following clades has been proposed by plant biologists as the clade that defines the plant kingdom?
   a. kingdom Plantae, which includes only the embryophyte clade
   b. kingdom Streptophyta, which also includes the charophytes
   c. kingdom Viridiplantae, which also includes all green algae (chlorophytes)
   d. kingdom Vasculata, which includes only the vascular plants
   e. a, b, and c have all been proposed as the clade that defines the plant kingdom.

4. Which of the following groups is most likely the closest relative of seed plants?
   a. charophytes
   b. pterophytes
   c. hornworts
   d. lycophytes
   e. mosses

5. The evolution of sporopollenin was important to the movement of plants onto land because
   a. it provided a mechanism for the dispersal of spores and pollen.
   b. it provided the structural support necessary to withstand gravity.
   c. it enclosed developing gametes and embryos in maternal tissue and prevented desiccation.
   d. it provided a tough coating for spores so they could disperse on land.
   e. it initiated the alternation of generations that is characteristic of all plants.

6. Which of the following functions may secondary compounds serve?
   a. protect plants from ultraviolet radiation
   b. deter herbivores, parasites, and pathogens
   c. serve as accessory pigments for photosynthesis
   d. both a and b
   e. all of the above

7. Bryophytes differ from other plant groups because
   a. their gametophyte generation is dominant.
   b. they lack gametangia.
   c. they have flagellated sperm.
   d. they are not embryophytes.
   e. all of the above are true.

8. Bryophytes were the dominant plants in the first 100 million years of plant evolution. By the Carboniferous period, seedless vascular plants formed giant forests. Why were these plants able to outcompete bryophytes?
   a. Their protected embryos were better able to withstand dry conditions, providing a selective advantage in dominating terrestrial habitats.
   b. They did not require water for their sperm to swim to fertilize the eggs, allowing them to colonize dry habitats.
   c. They were diploid, so they could grow faster and taller than haploid bryophytes.
   d. As heterosporous plants, their spores had more genetic variation, enabling natural selection to favor taller growth.
   e. Their vascular tissue enabled them to grow tall, competing for light and more widely dispersing their spores.

9. Which of the following plant groups is *incorrectly* paired with its gametophyte generation?
   a. charophyte—no alternation of generations
   b. lycophyte—either green or nonphotosynthetic tiny plants
   c. moss—green matlike plant
   d. fern—frond growing from horizontal stem
   e. liverwort—flattened, low-growing green plant

10. Which of the following constitute a large store of organic carbon that could contribute to increasing CO_2 levels and global climate change should its decomposition rate increase?
    a. huge tropical swamps dominated by mosses and ferns
    b. peatlands
    c. coal deposits formed from Cambrian forests
    d. the abundant epiphytic lycophytes found in tropical regions
    e. large tracts of small seedless vascular plants found in boreal regions

11. Megaphylls
    a. are large leaves.
    b. are gametophyte plants that develop from megaspores.
    c. are leaves specialized for reproduction.
    d. are leaves with branching vascular systems.
    e. were a dominant group of the great coal forests.

12. If a plant's life cycle includes separate male and female gametophytes, the sporophyte plant must be
    a. heterosporous.
    b. homosporous.
    c. homologous.
    d. analogous.
    e. megasporous.

13. Xylem and phloem are found in
    a. all plants.
    b. bryophytes, ferns, conifers, and angiosperms.
    c. only the gametophytes of vascular plants.
    d. the vascular plants, which include lycophytes, ferns, conifers, and flowering plants.
    e. only the vascular plants with seeds.

14. If you could take a time machine back to the Carboniferous period, which of the following would you most likely encounter?
    a. creeping mats of low-growing bryophytes
    b. fields of tall grasses swaying in the wind
    c. swampy forests dominated by tree lycophytes, horsetails, and ferns
    d. huge forests of naked-seed trees filling the air with pollen
    e. the dominance of flowering plants

Chapter 30

Plant Diversity II: The Evolution of Seed Plants

Key Concepts

30.1 Seeds and pollen grains are key adaptations for life on land

30.2 Gymnosperms bear "naked" seeds, typically on cones

30.3 The reproductive adaptations of angiosperms include flowers and fruits

30.4 Human welfare depends greatly on seed plants

Framework

The reduction and protection of the gametophyte within the parent sporophyte, the protection and dispersal of embryos in seeds, and the use of pollen for the transfer of sperm are reproductive adaptations of seed plants that enhance their success on land.

Gymnosperms bear their seeds "naked" on modified sporophylls. Angiosperms are the most diverse and widespread of plants, owing much of their success to their efficient reproductive apparatus—the flower—and their seed dispersal mechanism—the fruit. Agriculture is based almost entirely on angiosperms.

Chapter Review

Seed plants originated about 360 million years ago. One of the key adaptations that enabled seed plants to become the dominant members of terrestrial ecosystems is the **seed**—a mobile embryo packaged with food within a protective coat. The cultivation of seed plants, which transformed human societies, began about 12,000 years ago.

30.1 Seeds and pollen grains are key adaptations for life on land

Advantages of Reduced Gametophytes The extremely reduced gametophytes of seed plants develop within the sporangium and thus are protected from drying out and UV radiation. They are also nourished by the sporophyte plant.

Heterospory: The Rule Among Seed Plants All seed plants are *heterosporous*, with *megaspores* that give rise to female gametophytes, and *microspores* that give rise to male gametophytes. A megaspongium produces a single functional megaspore; a microsporangium produces many microspores.

Ovules and Production of Eggs The megasporangium is surrounded by protective **integuments** derived from sporophyte tissue. An **ovule** consists of the integument(s), megasporangium, and megaspore. The female gametophyte develops from the megaspore and produces one or more eggs.

Pollen and Production of Sperm A microspore develops into a male gametophyte contained in a sporopollenin-protected **pollen grain.** The transfer of pollen to the part of a plant containing ovules is called **pollination.** When pollen grains germinate, they release sperm through a pollen tube to the female gametophyte.

INTERACTIVE QUESTION 30.1

Why is the evolution of pollen an important terrestrial adaptation?

The Evolutionary Advantage of Seeds An ovule develops into a seed, consisting of a sporophyte embryo surrounded by a food supply and enclosed within a protective coat derived from the integument(s) of the ovule. Seeds enable seed plants to withstand harsh environmental conditions and to disperse offspring. Unicellular spores perform those functions for bryophytes and seedless vascular plants.

INTERACTIVE QUESTION   30.2

Label the parts in the following generalized diagrams of an unfertilized ovule and a seed of a gymnosperm. Indicate whether structures are diploid or haploid tissues.

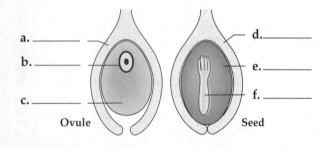

Ovule   Seed

30.2   Gymnosperms bear "naked" seeds, typically on cones

Gymnosperm Evolution   Transitional species of heterosporous, wood-producing, but seedless vascular plants are called **progymnosperms.** The first seed plants appear in the fossil record from about 360 million years ago. The earliest gymnosperm fossils date from 305 million years ago.

During the Carboniferous period, early gymnosperms lived in ecosystems dominated by seedless vascular plants. Drier conditions during the Permian gave gymnosperms a selective advantage. Gymnosperms supported the giant herbaceous dinosaurs of the Mesozoic era. With the environmental changes at the end of the Mesozoic 65 million years ago, most of the dinosaurs became extinct. Angiosperm dominance of ecosystems increased. Many gymnosperms persisted—in particular, the cone-bearing **conifers** found in large forests in northern latitudes.

There are four gymnosperm phyla. Cycadophyta includes cycads, which have large cones and palmlike leaves and thrived during the Mesozoic era. Ginkgophyta has only one extant species: the deciduous, ornamental ginkgo tree. Gnetophyta includes three genera: the bizarre *Welwitschia*, the tropical *Gnetum*, and the desert shrub *Ephedra*. The largest gymnosperm phylum is Coniferophyta. Most conifers are evergreens with needle-shaped leaves. Coniferous trees are some of the largest and oldest living organisms.

The Life Cycle of a Pine: A Closer Look   The pine tree is a heterosporous sporophyte. Pollen cones consist of

many scales (modified leaves or sporophylls) that bear sporangia. Meiosis gives rise to microspores, each of which develops into a pollen grain enclosing a male gametophyte. Scales of an ovulate cone hold two ovules, each of which contains a megasporangium. A megasporocyte undergoes meiosis, and one of the resulting megaspores undergoes repeated divisions to produce a female gametophyte in which a few archegonia develop.

Following pollination, a pollen tube grows and digests its way through the megasporangium. Fertilization may occur more than a year after pollination. The zygote develops into a sporophyte embryo, which is nourished by the remaining female gametophyte tissue and is enclosed within a seed coat derived from the integument of the parent sporophyte.

INTERACTIVE QUESTION   30.3

a. Describe a pollen cone and the formation of a male gametophyte.

b. Describe an ovulate cone and the formation of a female gametophyte.

30.3   The reproductive adaptations of angiosperms include flowers and fruits

The phylum Anthophyta contains more than 90% of all plant species—over 250,000 species.

Characteristics of Angiosperms   The **flower,** the reproductive structure of an angiosperm, has up to four rings of modified leaves (sporophylls) called floral organs. The outer **sepals** are usually green, whereas **petals** are brightly colored in most flowers that are pollinated by insects and birds. **Stamens** produce microspores, which develop into pollen grains containing male gametophytes. A stamen has a stalk, called a **filament,** and a terminal **anther,** in which pollen is produced. **Carpels** produce megaspores, which develop into female gametophytes. The carpel has a sticky **stigma,** which receives pollen, and a **style,** which leads to the **ovary.** The ovary contains ovules, which develop into seeds.

A **fruit** is a mature ovary that functions in the protection and dispersal of seeds. As seeds develop, the ovary enlarges and forms the thickened wall (pericarp) of the fruit. Fruits may be fleshy or dry. The dry fruits of grasses are the major foods for humans.

INTERACTIVE QUESTION 30.4

a. Name the four floral organs that make up a flower.

b. Name the parts of a seed.

c. What is a fruit?

In the angiosperm life cycle, the two highly reduced gametophytes are produced within the flower. Pollen grains contain the male gametophytes, consisting of a *generative cell* (which divides to form two sperm) and a *tube cell.* Ovules contain the female gametophyte, called an **embryo sac,** which consists of a few cells.

Most flowers have some mechanism to ensure **cross-pollination.** A pollen grain germinates on the stigma and extends a pollen tube down the style, through the **micropyle,** and into the ovule, where it releases two sperm cells into the embryo sac. In a process called **double fertilization,** one sperm unites with the egg to form the zygote, and the other sperm fuses with the two polar nuclei in the large central cell of the female gametophyte.

The zygote divides and forms the sporophyte embryo, consisting of a rudimentary root and one or two seed leaves, called **cotyledons.** The **endosperm,** which develops from the triploid central cell, serves as a food reserve for the embryo.

INTERACTIVE QUESTION 30.5

In the following diagram of the life cycle of an angiosperm, label the indicated structures and processes. Which structures represent the male and female gametophyte generations?

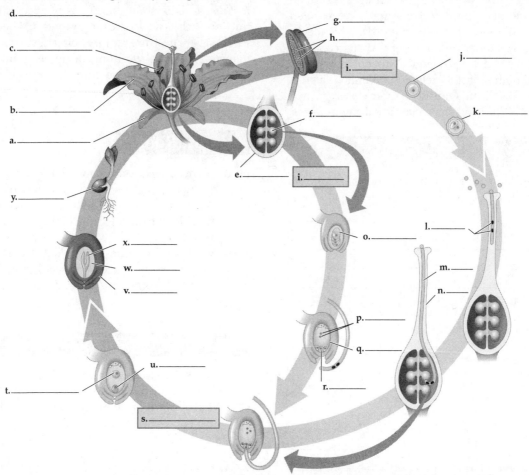

Angiosperm Evolution   Although details of their evolutionary origin remain unclear, angiosperms date from at least 140 million years ago. By 100 million years ago, angiosperms were becoming the dominant plants.

A phylogenetic comparison of living plants and recently discovered fossils of 125-million-year-old angiosperms called *Archaefructus* indicates that this group (which has both derived and primitive traits) may be the earliest-diverging angiosperm group. Unlike the herbaceous, aquatic *Archaefructus*, however, all fossil seed plants proposed to be close relatives of angiosperms were woody, suggesting that the angiosperm common ancestor was woody.

The monophyletic living gymnosperms appear to have diverged from the ancestors of angiosperms 305 million years ago. Studies of plant development and the genes that control flower development may help elucidate the origin of flowering plants.

Angiosperm Diversity   The angiosperms were traditionally divided into two main groups, **monocots** and **dicots**. Recent DNA comparisons reveal that, although the monocots are a monophyletic group, the dicots are not. The clade **eudicots** includes most previously classified dicots, but several other small lineages have been identified. The three early-diverging lineages are called **basal angiosperms**. The oldest lineage is represented today by one species, *Amborella trichopoda*, which lacks the xylem vessels found in more derived angiosperms. The next lineage to diverge includes water lilies, with the lineage of star anise and its relatives diverging later. A lineage known as the **magnoliids** is more closely related to monocots and eudicots than to the basal angiosperms.

INTERACTIVE QUESTION   30.6

Fill in the following table, which compares characteristics of monocots and eudicots.

Characteristic	Monocot	Eudicot
Number of cotyledons		
Leaf venation		
Vascular tissue in stems		
Root system		
Openings in pollen grain		
Floral organs		

Evolutionary Links Between Angiosperms and Animals   Animals influenced the evolution of plants, and vice versa. Both plant defenses against herbivory and herbivores that can overcome plant defenses would be favored by natural selection. Some animals became beneficial as pollinators and seed dispersers.

Pollinators can enter bilateral flowers from only one direction and thus more specifically pick up and disperse pollen to flowers of the same species. A study comparing the number of species in clades with bilaterally symmetrical flowers with the number in sister clades with radially symmetrical flowers, supports the hypothesis that bilateral flower shape promotes speciation, perhaps by reducing gene flow between diverging populations.

30.4   Human welfare depends greatly on seed plants

Products from Seed Plants   Our food crops are angiosperms, and angiosperms are used to feed livestock. The dramatic evolution of domesticated plants is a result of artificial selection in plant breeding. Seed plants also supply wood and medicines.

Threats to Plant Diversity   The growing human population—with its demand for space, food, and natural resources—is pushing hundreds of species toward extinction each year. Tropical rain forests, where plant diversity is greatest, are rapidly being destroyed, resulting in increased global warming, decreased rainfall, and the loss of potential new food crops and medicines.

Word Roots

endo- = inner (*endosperm:* a nutrient-rich tissue, formed by the union of a sperm cell with two polar nuclei during double fertilization, that provides nourishment to the developing embryo in angiosperm seeds)

pro- = before; **gymno-** = naked; **-sperm** = seed (*progymnosperm:* an extinct seedless vascular plant that may be ancestral to seed plants)

Structure Your Knowledge

1. List the characteristics of seed plants that are evolutionary adaptations to a terrestrial habitat.
2. What adaptations helped angiosperms become the most successful and widespread land plants?

3. The angiosperms are grouped into a single phylum, Anthophyta, which traditionally contained the classes monocots and dicots. Morphological and molecular evidence now suggests six angiosperm lineages. Label those lineages on the following tree.

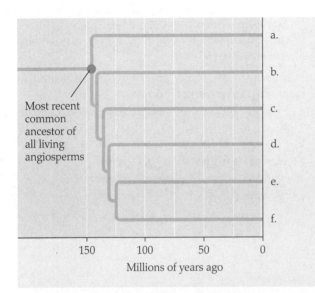

Most recent common ancestor of all living angiosperms

150   100   50   0

Millions of years ago

Test Your Knowledge

TRUE OR FALSE: *Indicate T or F, and then correct the false statements.*

_____ 1. All photoautotrophic, multicellular eukaryotes are plants.

_____ 2. Heterosporous plants produce male and female spores.

_____ 3. The gametophyte generation is most reduced in the gymnosperms.

_____ 4. *Ginkgo*, cycads, and conifers are naked-seed plants.

_____ 5. A sporangium produces spores, no matter what group it is found in.

_____ 6. A fruit consists of an embryo, nutritive material, and a protective coat.

_____ 7. A stamen consists of a filament and an anther that produces microspores, which give rise to pollen grains.

_____ 8. The female gametophyte in angiosperms consists of haploid cells in which a few archegonia develop.

MULTIPLE CHOICE: *Choose the one best answer.*

1. In which of the following groups must sperm no longer swim to reach the female gametophyte?
   a. bryophytes
   b. ferns
   c. most gymnosperms
   d. angiosperms
   e. both c and d

2. Which of the following statements describes a key difference between seedless vascular plants and plants with seeds?
   a. The gametophyte generation is dominant in the seedless plants, whereas the sporophyte is dominant in the seed plants.
   b. Spores are the agents of dispersal in seedless plants; seeds function in dispersal in seed plants.
   c. Seedless plants are heterosporous, whereas seed plants are homosporous.
   d. The embryo is unprotected and totally independent in the seedless plants but retained within the female reproductive structure in the seed plants.
   e. Sporopollenin is not found in the seedless plants.

3. For a pine seed, which of the following choices states the number of generations and correctly describes those generations?
   a. one: the new sporophyte generation
   b. two: seed coat and food supply from female gametophyte and the sporophyte embryo
   c. two: seed coat from integument of the parent sporophyte and new sporophyte embryo
   d. three: seed coat from the parent sporophyte, food supply from the gametophyte, and the sporophyte embryo
   e. three: seed coat from the female gametophyte, food supply from the parent sporophyte, and the sporophyte embryo

4. Gymnosperms rose to dominance during
   a. the Carboniferous period, when they formed important components of the great "coal forests."
   b. the Devonian period, when they successfully competed with the short-statured bryophytes.
   c. the Permian period, when they replaced the seedless vascular plants as the climate became drier.
   d. the Cretaceous period, when cooler climates favored their growth over the ferns and other seedless vascular plants.
   e. the Cretaceous period, when they replaced angiosperms as the largest and most widespread group of plants.

5. In a pine tree, where would you find a micro-sporangium?
   a. within the embryo sac in an ovule
   b. in the pollen sacs in an anther
   c. on a scale in a pollen cone
   d. on a scale in an ovulate cone
   e. forming a seed coat surrounding a pine seed

6. Which of the following correctly describes the path a pollen tube takes to reach the female gametophyte in an angiosperm?
   a. stigma, style, ovary, ovule, embryo sac
   b. anther, stigma, filament, ovule, ovum
   c. stigma, filament, carpel, ovary, ovule
   d. anther, stigma, ovary, ovule, embryo sac
   e. stigma, style, sepal, ovule, ovary

7. Which of the following sentences describes a likely advantage of double fertilization in angiosperms?
   a. Two embryos are produced in a seed, thus increasing the reproductive output of a plant.
   b. Two embryos are produced in a seed, allowing selection for the stronger one to reproduce the plant.

   c. This is a remnant of the two sperm produced by the ancestral angiosperm; there does not appear to be an adaptive advantage to this trait.
   d. The endosperm does not form unless fertilization of the egg occurs, so plants do not waste nutrients on infertile ovules.
   e. The triploid endosperm resulting from the fusion of two sperm with the central cell has a higher nutrient content than the haploid food supply surrounding gymnosperm embryos.

8. Which of the following events may result from the clear-cutting of tropical forests?
   a. a rise in temperature and a decrease in rainfall in the area
   b. a loss of potential medicines
   c. an increase in atmospheric CO_2 levels
   d. extinctions of many plant and animal species
   e. all of the above

Key Concepts

31.1 Fungi are heterotrophs that feed by absorption

31.2 Fungi produce spores through sexual or asexual life cycles

31.3 The ancestor of fungi was an aquatic, single-celled, flagellated protist

31.4 Fungi have radiated into a diverse set of lineages

31.5 Fungi play key roles in nutrient cycling, ecological interactions, and human welfare

Framework

This chapter describes the morphology, life cycles, evolutionary history, diversity, and economic and ecological importance of the kingdom Fungi. Fungi play an essential ecological role, both as decomposers and in their mycorrhizal association with plant roots. A flagellated protist may have been the common ancestor to fungi and animals.

Chapter Review

About 100,000 fungal species have been described; mycologists estimate that as many as 1.5 million fungal species may exist. Fungi are found in nearly every habitat on Earth.

31.1 Fungi are heterotrophs that feed by absorption

Nutrition and Ecology Fungi are heterotrophs that obtain their nutrients by absorption; many secrete digestive enzymes into the surrounding food and absorb the resulting small organic molecules. Fungi may be decomposers that break down nonliving organic material, parasites that absorb nutrients from living cells, or mutualists that feed on, but also benefit, their hosts.

Body Structure Most fungi are composed of multicellular filaments. Some are single-celled **yeasts,** which may inhabit liquid or moist habitats. Some species can grow as both filaments and yeasts.

A typical multicellular fungal body consists of a network of filamentous **hyphae,** which form a mass called a **mycelium.** This body form provides an extensive surface area for absorption of nutrients. The tubular cell walls are composed of **chitin.** Although fungi are nonmotile, the rapid growth of their hyphae enables them to enter new food territory.

The hyphae of most fungi are divided into cells by cross-walls called **septa,** which usually have pores through which nutrients and cell organelles can pass. **Coenocytic fungi** lack septa and consist of a continuous cytoplasmic mass containing many nuclei.

Specialized Hyphae in Mycorrhizal Fungi Symbiotic fungi may penetrate plant cell walls with specialized hyphae called **haustoria,** which can extract or exchange nutrients with the plant hosts. Mutualistic relationships with plant roots are known as **mycorrhizae. Ectomycorrhizal fungi** form hyphal sheaths around a root and also grow into spaces in the root cortex. **Arbuscular mycorrhizal fungi** extend their hyphae through root cell walls, pushing in the plant cell membrane to form tubes within root cells.

INTERACTIVE QUESTION 31.1

In what way(s) do mycorrhizal fungi benefit plants? In what way(s) do plants benefit the fungi?

31.2   Fungi produce spores through sexual or asexual life cycles

Fungi release huge quantities of **spores**, produced either sexually or asexually, which aid in dispersal.

Sexual Reproduction   Nuclei of hyphae and spores of most fungal species are haploid. Sexual reproduction involves the release of sexual signaling **pheromones**, which may cause the hypha of suitable mating types to grow toward and fuse with each other. Such cytoplasmic fusion, called **plasmogamy**, forms a mycelium with genetically different nuclei, or a **heterokaryon**. The haploid nuclei may mingle and exchange genes in a process similar to crossing over. Nuclei from the two parents may pair up in cells and divide in tandem, forming **dikaryotic** cells. After a period of time, the nuclei fuse (**karyogamy**), forming the only diploid stage in most fungi. Meiosis produces haploid spores.

Asexual Reproduction   Many fungi reproduce asexually by spores; some species reproduce only asexually.

Molds are rapidly growing mycelia that reproduce asexually by spores. Many molds can reproduce sexually if they encounter other mating types. Asexual reproduction in single-celled yeasts occurs by cell division or budding.

Molds and yeasts for which no sexual stage is known are traditionally called **deuteromycetes**. When mycologists discover a sexual stage in one of these fungi, they classify the species accordingly. Genetic analysis is also used to classify fungi.

INTERACTIVE QUESTION   31.2

Briefly define each of the following terms that relate to the structure and reproduction of fungi.

a. mycelium

b. septa

c. coenocytic

d. plasmogamy

e. heterokaryon

f. dikaryon

g. karyogamy

31.3   The ancestor of fungi was an aquatic, single-celled, flagellated protist

Fungi and Animalia are more closely related to each other than to plants or to other eukaryotes.

The Origin of Fungi   Although most fungi lack flagella, phylogenetic systematics indicates that they evolved from a flagellated ancestor. The fungi, animals, and their protistan relatives form a clade called **opisthokonts.**

DNA data indicate that fungi are most closely related to **nucleariids,** a group of single-celled protists, suggesting that the ancestor of fungi was unicellular. Because evidence indicates that fungi and animals evolved from different single-celled ancestors, they evolved multicellularity independently. Estimates using the molecular clock suggest that the ancestors of animals and fungi diverged about 1 billion years ago. The oldest fossils of fungi date back 460 million years.

Are Microsporidia Fungi?   Microsporidia are unicellular parasites of animals and protists that do not have conventional mitochondria. Molecular evidence suggests that these highly derived parasites may be an early-diverging lineage of fungi.

The Move to Land   The first vascular plant fossils (from 420 million years ago) include evidence of mycorrhizae. The adaptive radiation of fungi may have begun with this move to land.

INTERACTIVE QUESTION   31.3

Fill in the blanks in the following phylogenetic tree of the opisthokonts.

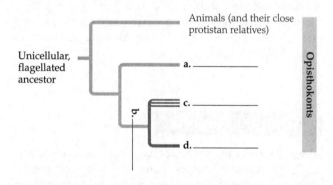

31.4   Fungi have radiated into a diverse set of lineages

Chytrids   The **chytrids** are unique among fungi because of their flagellated **zoospores.** Molecular evidence indicates that this paraphyletic group diverged early in fungal evolution. Chytrids include aquatic decomposers and parasites. Some are mutualists in the digestive tracts of sheep and cattle. One species is implicated in the recent massive die-offs in frog populations.

Zygomycetes **Zygomycetes** are probably a paraphyletic group that includes decomposing molds, parasites, and commensal symbionts.

Rhizopus stolonifer, the black bread mold, is a common zygomycete that spreads horizontal coenocytic hyphae across its food substrate and erects hyphae with bulbous black sporangia containing hundreds of asexual haploid spores. Sexual reproduction involving plasmogamy between mycelia of different mating types produces a heterokaryotic **zygosporangium** with a tough protective coat that remains dormant until conditions are favorable. Karyogamy occurs between paired nuclei, followed by meiosis to produce genetically diverse haploid spores.

Glomeromycetes Analysis of fungal genomes has identified the **glomeromycetes** as a separate clade. Almost all glomeromycetes are arbuscular mycorrhizal fungi. Hyphae that invade plant root cells branch into tiny treelike arbuscles. About 90% of all plants form mycorrhizae with glomeromycetes.

Ascomycetes The **ascomycetes,** or *sac fungi,* are found in a wide variety of habitats. They produce sexual spores in saclike **asci,** usually contained in fruiting bodies called **ascocarps.**

Ascomycetes range in complexity from yeasts to elaborate cup fungi and morels. Some are serious plant pathogens; others are important decomposers. Many ascomycete species form symbiotic associations called lichens with algae or cyanobacteria and some form mycorrhizae with plants.

Asexual reproduction involves spores called **conidia,** which are produced in chains or clusters at the ends of hyphae called conidiophores. Plasmogamy between different mating types produces dikaryotic hyphae. Karyogamy occurs in terminal cells, which develop into asci. Meiosis yields haploid spores called ascospores.

Basidiomycetes The **basidiomycetes,** which include mushrooms and shelf fungi, produce a club-shaped **basidium** in which karyogamy and then meisois occur. These *club fungi* include important decomposers, mycorrhizae-forming mutualists, and the plant parasites rusts and smuts.

In response to environmental stimuli, an elaborate fruiting body called a **basidiocarp** is formed from the long-lived dikaryotic mycelium. Karyogamy and meiosis occur in numerous basidia, producing huge numbers of haploid basidiospores.

Indicate whether the following diagrams (1, 2, 3) are from a zygomycete life cycle, an ascomycete life cycle, or a basidiomycete life cycle. Identify the labeled structures.

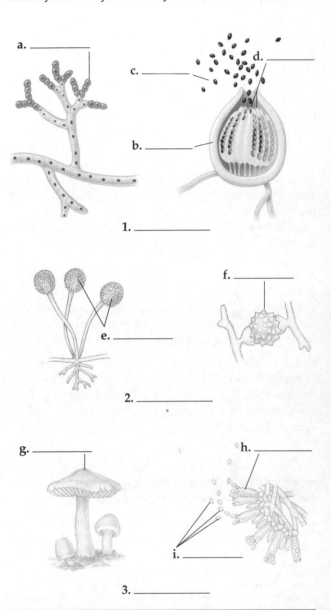

a. _____

c. _____

d. _____

b. _____

1. _____

f. _____

e. _____

2. _____

g. _____

h. _____

i. _____

3. _____

31.5 Fungi play key roles in nutrient cycling, ecological interactions, and human welfare

Fungi as Decomposers Fungi are decomposers of organic matter, facilitating the essential recycling of chemical elements for plant growth.

Fungi as Mutualists Mycorrhizae are very common and important in natural ecosystems and agriculture.

The leaves or other parts of all plant species studied thus far have symbiotic **endophytes,** which benefit plants by either producing toxins that deter herbivores, increasing the plant's tolerance for heat or drought, or playing a role in defending against pathogens.

Fungal inhabitants of the guts of grazing mammals help digest plant material. Farmer leaf-cutter ants raise crops of fungi by gathering leaves and feeding them to the fungi.

Lichens are symbiotic associations of millions of photosynthetic organisms in a lattice of fungal hyphae. The fungus is usually an ascomycete, and the partner is a unicellular or filamentous green alga or a cyanobacterium. The alga provides the fungus with food and, in the case of cyanobacteria, nitrogen. The fungus creates most of the mass of the lichen, provides protection for the alga, and absorbs water and minerals.

Lichens reproduce asexually, either as fragments or as tiny clusters called **soredia.** In addition, it is common for the fungal component to reproduce sexually and for the algal component independently to reproduce asexually.

Lichens are important colonizers of bare rock and soil and may have helped facilitate the colonization of land by plants. Lichens are sensitive to air pollution.

Fungi as Pathogens Fungal parasites of plants are common, killing chestnut and pine trees and spoiling grain and fruit crops. An ascomycete forms ergots on rye that can cause serious symptoms when accidentally milled into flour.

Fungal infections in animals are called **mycoses.** Parasitic fungi cause ringworm, athlete's foot, vaginal yeast infections, and lung infections. Systemic mycoses are very serious. Many mycoses are opportunistic.

Practical Uses of Fungi Commercially cultivated mushrooms are eaten, and fungi are used to ripen some cheeses. *Saccharomyces cerevisiae* is the yeast used in baking, brewing, and molecular research and biotechnology. Some fungi are the source of medical compounds, including antibiotics.

Word Roots

coeno- = common; **-cyto** = cell (*coenocytic fungus:* a fungus that lacks septa, and thus its body consists of a continuous cytoplasmic mass that may contain hundreds or thousands of nuclei)

di- = two; **-karyo** = nucleus (*dikaryotic:* referring to a fungal mycelium with two haploid nuclei per cell, one from each parent)

hetero- = different (*heterokaryon:* a fungal mycelium that contains two or more haploid nuclei per cell)

myco- = fungus; **rhizo-** = root (*mycorrhiza:* a mutualistic association of plant roots and a fungus)

-osis = a condition of (*mycosis:* general term for a fungal infection)

plasmo- = plasm; **-gamy** = marriage (*plasmogamy:* in fungi, the fusion of the cytoplasm of cells from two individuals; occurs as one stage of sexual reproduction, followed later by karyogamy)

Structure Your Knowledge

1. Fill in the following table that summarizes the characteristics of the five major groups of fungi.

2. The kingdom Fungi contains members that are decomposers, parasites, and mutualists. How does each of these lifestyles relate to the ecological and economic importance of fungi?

Phylum	Examples		Key Features and Reproduction
Chytridiomycota	a.	b.	
Zygomycota	c.	d.	
Glomeromycota	e.	f.	
Ascomycota	g.	h.	
Basidiomycota	i.	j.	

Test Your Knowledge

FILL IN THE BLANKS

_____ 1. division between cells in fungal hyphae

_____ 2. hyphae with nuclei from two parents

_____ 3. component of cell walls in most fungi

_____ 4. produced by fungi to absorb nutrients

_____ 5. club-shaped reproductive structure found in mushrooms

_____ 6. fungus living inside a plant leaf that benefits the plant

_____ 7. mutualistic associations between plant roots and fungi

_____ 8. resistant structures produced by zygomycetes

_____ 9. hyphae with many nuclei

_____ 10. fungal group that diverged earliest

_____ 11. clade that includes fungi, animals, and related flagellated protists

_____ 12. distinctive feature of glomeromycetes

MULTIPLE CHOICE: *Choose the one best answer.*

1. The major difference between fungi and plants is that fungi
   a. have an absorptive form of nutrition.
   b. do not have cell walls.
   c. are not eukaryotic.
   d. are multinucleate but not multicellular.
   e. Both a and d are correct.

2. A fungus that is both a parasite and a decomposer
   a. digests the nonliving portions of its host's body.
   b. lives off the sap within its host's body.
   c. first lives as a parasite and then consumes the host after it dies.
   d. lives as a mutualistic symbiont on its host.
   e. causes athlete's foot and vaginal infections.

3. The fact that karyogamy and meiosis do not immediately follow plasmogamy
   a. is necessary to create coenocytic hyphae.
   b. is characteristic of all fungi.
   c. allows fungi to reproduce asexually most of the time.
   d. results in heterokaryotic cells that may benefit from the variation present in two genomes.
   e. is characteristic of yeasts.

4. The traditional group deuteromycetes
   a. represents the most ancient lineage of fungi.
   b. includes the fungal components of lichens.
   c. has abnormal forms of sexual reproduction.
   d. consists of fungi that are predatory.
   e. consists of fungi whose sexual stage is absent or unknown.

5. Fungi and animals appear to be more closely related to each other than either is to plants
   a. because neither of them are photosynthetic.
   b. based on similarities in cell structure.
   c. based on molecular analyses.
   d. because they moved onto land together.
   e. based on homologous ultrastructure of their flagella.

6. How can fungi be classified as opisthokonts, which means "posterior flagella," when most fungi lack flagella?
   a. Most fungal lineages appear to have lost their flagella during their evolution.
   b. The most primitive lineage of fungi, the chytrids, have flagellated zoospores.
   c. Fungi are in the same lineage as all flagellated protists.
   d. Fungi are closely related to animals, and because animals, which do have flagella, are opisthokonts, fungi must be opisthokonts too.
   e. Both a and b are the best reasons given here for their classification as opisthokonts.

7. The names given to three of the fungal phyla are based on
   a. the structures in which karyogamy occurs during sexual reproduction.
   b. the locations of plasmogamy during sexual reproduction.
   c. the locations of the heterokaryotic stage in the life cycle.
   d. the structures that produces asexual spores.
   e. their ancestral origins.

8. Which of the following statements accurately describes ascomycetes?
   a. Sexual reproduction occurs by conjugation.
   b. Spores often line up in a sac in the order in which they were formed by meiosis.
   c. Asexual spores form in sporangia on erect hyphae.
   d. Most hyphae are dikaryotic.
   e. Reproduction is always sexual.

9. Lichens are symbiotic associations that
   a. usually involve an ascomycete and a green alga or cyanobacterium.
   b. can reproduce sexually by forming soredia.
   c. require moist environments to grow.
   d. fix nitrogen for absorption by plant roots.
   e. are unusually resistant to air pollution.

From the following phylogenetic tree showing the proposed relationships among fungi, choose the letter that is associated with each of the lineages named in questions 10 through 14.

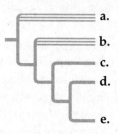

10. Ascomycetes _____

11. Basidiomycetes _____

12. Chytrids _____

13. Glomeromycetes _____

14. Zygomycetes _____

Chapter 32

An Overview of Animal Diversity

Key Concepts

32.1 Animals are multicellular, heterotrophic eukaryotes with tissues that develop from embryonic layers

32.2 The history of animals spans more than half a billion years

32.3 Animals can be characterized by "body plans"

32.4 New views of animal phylogeny are emerging from molecular data

Framework

Animals are multicellular eukaryotic heterotrophs that ingest their food. Fossil evidence for the origin and rapid divergence of animals is limited. A colonial flagellated protist is the probable ancestor, and most animal phyla diversified during the 20 million years of the Cambrian explosion. The traditional phylogenetic tree of animals is based on body plan grades and includes such characteristics as symmetry, tissue layers, body cavities, and development. Molecular systematics splits the protostomes into clades Ecdysozoa and Lophotrochozoa.

Chapter Review

32.1 Animals are multicellular, heterotrophic eukaryotes with tissues that develop from embryonic layers

Nutritional Mode Most animals ingest their food, eating either other organisms or nonliving organic matter.

Cell Structure and Specialization Animals are multicellular, and their cells lack walls. Extracellular proteins, such as collagen, provide support and hold adjacent cells together. Most animals have **tissues**, groups of specialized cells with a common structure and/or function. Muscle and nervous tissues are unique to animals.

Reproduction and Development The diploid stage is usually dominant, and most animals reproduce sexually, with a flagellated sperm fertilizing a larger, nonmotile egg. The zygote undergoes a series of mitotic divisions, called **cleavage,** usually passing through a **blastula** stage during embryonic development. The infolding process of **gastrulation** produces layers of embryonic tissues, resulting in the **gastrula** stage.

All animals have developmental genes, many of which contain DNA sequences called *homeoboxes*. A unique homeobox-containing family of genes, called *Hox* genes, is involved in regulating gene expression in the embryonic development of most animals.

INTERACTIVE QUESTION 32.1

a. What is a **larva**?

b. How does a larva differ from a juvenile form of an animal?

c. By what process does a larva transform into a juvenile?

32.2 The history of animals spans more than half a billion years

The diversity of the animal kingdom includes extinct species, which constitute an estimated 99% of all the species that have ever lived. Some molecular clock calculations indicate that the ancestors of fungi and animals diverged about a billion years ago. The common ancestor of living animals may have lived 675–800 million years ago and probably resembled choanoflagellates. These colonial, flagellated protists are stationary suspension feeders. Evidence that choanoflagellates are closely related to animals includes molecular data and morphological similarities to the collar cells found in sponges and some other animals.

Neoproterozoic Era (1 Billion–542 Million Years Ago)
The earliest generally accepted fossils of animals, known as the **Ediacaran biota,** date from 565–550 million years ago. Some are sponges; others may be related to cnidarians. What appear to be fossilized animal embryos date from 575 million years ago.

Paleozoic Era (542–251 Million Years Ago) Fossils from the **Cambrian explosion** (a burst of evolutionary change about 535–525 million years ago) represent about half of all extant phyla and include the first animals with hard skeletons.

During the Ordovician, Silurian, and Devonian periods, animal diversity continued to increase. Fishes became the top predators of the seas. Arthropods appeared on land by 460 million years ago. Vertebrates began to adapt to land around 365 million years ago. Amphibians and amniotes (reptiles and mammals) are the two terrestrial groups that survive today.

Mesozoic Era (251–65.5 Million Years Ago) The first coral reefs formed; some reptiles returned to water; and flight appeared in pterosaurs and birds. Large predatory and herbivorous dinosaurs emerged, as did the first mammals.

Cenozoic Era (65.5 Million Years Ago to the Present)
Mass extinctions marked the beginning of this period and included the disappearance of the nonflying dinosaurs and marine reptiles. Mammals began to exploit the vacated niches.

INTERACTIVE QUESTION   32.2

Describe the three hypotheses for the rapid diversification of animal phyla during the Cambrian period.

32.3   Animals can be characterized by "body plans"

An animal's set of morphological and developmental traits is often referred to as a **body plan.** Experimental evidence suggests that β-catenin, one of the molecular controls necessary for gastrulation, evolved more than 500 million years ago. Other aspects of body plans may have changed independently in different lineages.

Symmetry Sponges lack body symmetry. An animal with **radial symmetry** has a round or barrel shape, and any plane through its center divides the animal into mirror images. Animals with **bilateral symmetry** have distinct **anterior** (front) and **posterior** (back) ends, and left and right sides. Bilateral animals also have **dorsal** (top) and **ventral** (bottom) sides. Bilateral symmetry is associated with **cephalization,** the concentration of sensory organs and a central nervous system in the head end, which is an adaptation facilitating unidirectional movement.

Tissues True tissues are groups of specialized cells separated from other tissues by membranous layers. Sponges lack true tissues. During gastrulation, an embryo develops concentric layers of cells called *germ layers:* **Ectoderm** develops into the outer body covering and, in some phyla, into the central nervous system; **endoderm** lines the developing digestive tube, or archenteron, and gives rise to the lining of the digestive tract and associated organs. Cnidarians and comb jellies are **diploblastic,** forming only these two germ layers. The bilaterally symmetrical animals are **triploblastic;** they have a third, middle layer, the **mesoderm,** from which arise muscles and most other organs.

Body Cavities A fluid- or air-filled **body cavity** between the digestive tract and the outer body wall is called a **coelom.** A "true" coelom is completely lined by mesodermally derived tissue. Animals with a true coelom are called **coelomates.** Animals known as **pseudocoelomates** have a body cavity, called a pseudocoelom, formed from mesoderm and endoderm. Triploblastic animals that have solid bodies are called **acoelomates.**

A fluid-filled body cavity cushions internal organs, allows organs to grow and move independently of the outer body wall, and also functions as a skeleton in soft-bodied animals.

A *grade* refers to a group of animals that share a similar body plan. During animal evolution, coeloms and pseudocoeloms have evolved and been lost many times. Thus the grade of coelomate or pseudocoelomate cannot be used to distinguish clades (which include an ancestral species and all of its descendants).

Protostome and Deuterostome Development In general, **protostome development** differs from **deuterostome development** in three features: cleavage, coelom formation, and fate of the blastopore.

Protostome development, typical of molluscs and annelids, is characterized by **spiral cleavage,** in which the planes of cell division are diagonal, and newly formed cells fit in the grooves between cells of adjacent tiers. **Determinate cleavage** sets the developmental fate of each embryonic cell very early. Deuterostome development, typical of echinoderms and chordates, involves **radial cleavage,** in which parallel and perpendicular cleavage planes result in aligned tiers of cells. In **indeterminate cleavage,** typical of most animals with deuterostome development, cells from early cleavage divisions retain the capacity to develop into complete embryos.

The **archenteron,** which forms during gastrulation, develops into the digestive tract. In protostome development, the coelom forms from splits within solid masses of mesoderm. In deuterostome development, the coelom forms from mesodermal outpocketings of the archenteron.

The **blastopore** forms during gastrulation and is the opening leading to the developing archenteron. In protostome ("first mouth") development, the blastopore develops into the mouth, and a second opening at the end of the archenteron becomes the anus. In deuterostome development, the blastopore forms the anus, and the second opening develops into the mouth.

INTERACTIVE QUESTION 32.3

Fill in the following table to review some of the differences in protostome and deuterostome development.

	Protostome	Deuterostome
Cleavage	a.	b.
Coelom formation	c.	d.
Blastopore fate	e.	f.

32.4 New views of animal phylogeny are emerging from molecular data

The relationships among the 36 or so recognized animal phyla continue to be debated. Phylogenetic systematics is based on identifying a hierarchy of clades nested within larger clades. Molecular data, new studies of lesser-known phyla, analyses of fossils, and the use of cladistics are contributing to the identification of such clades. A traditional phylogenetic hypothesis is based on morphological and developmental comparisons; a more current view is based primarily on molecular data.

Points of Agreement There are five major points of agreement between the morphological and molecular trees: (1) All animals share a common ancestor and thus represent a clade called Metazoa. (2) Sponges are basal animals, branching from the base of both trees. (3) Eumetazoa is a clade of animals with true tissues. The **eumetazoans** include all animals except sponges and a few other groups. Basal members of the clade include the diploblastic, radially symmetrical phyla Cnidaria and Ctenophora (comb jellies). (4) Most animal phyla belong to the clade Bilateria. The **bilaterians** are defined by the shared derived characters of bilateral symmetry and three germ layers. (5) Chordates and some other phyla belong to the clade Deuterostomia.

Progress in Resolving Bilaterian Relationships The morphology-based tree divides the bilaterians into deuterostomes and protostomes, with the assumption that these developmental differences reflect phylogeny. This tree also groups arthropods with annelids, both of which have segmented bodies. Recent molecular studies distinguish a group of acoelomate flatworms (the Acoela) that appear to be basal bilaterians and three bilaterian clades: the deuterostomes; **ecdysozoans,** which include the arthropods and nematodes; and **lophotrochozoans,** which include the annelids and molluscs. The name Ecdysozoa refers to the shedding of an exoskeleton, called *ecdysis,* a trait shared by some ecdysozoan phyla. The name Lophotrochozoa is based on a feeding apparatus called a **lophophore** (found in animals such as brachiopods) and a larval stage called the **trochophore larva** (which is shared by other members of the lophotrochozoans).

Future Directions in Animal Systematics Large-scale comparisons of multiple genes and morphological traits across many animal phyla are helping systematists to test hypotheses about animal phylogeny.

Word Roots

a- = without; **-koilos** = a hollow (*acoelomate:* a solid-bodied animal lacking a cavity between the gut and outer body wall)

arch- = ancient, beginning (*archenteron:* the endoderm-lined cavity, formed during gastrulation, that develops into the digestive tract of an animal)

bi- = two (*bilaterian:* member of a clade of animals with bilateral symmetry and three germ layers)

blast- = bud, sprout; **-pore** = a passage (*blastopore:* in a gastrula, the opening of the archenteron that typically develops into the mouth in protostomes and the anus in deuterostomes)

cephal- = head (*cephalization:* an evolutionary trend toward the concentration of sensory equipment at the anterior end of the body)

deutero- = second (*deuterostome development:* a developmental mode distinguished by development of the anus from the blastopore, radial cleavage, and body cavity forming as outpockets of mesoderm)

di- = two (*diploblastic:* having two germ layers)

ecdys- = an escape (*ecdysozoan:* member of a group of animal phyla identified as a clade by molecular evidence; many are animals that molt)

ecto- = outside; **-derm** = skin (*ectoderm:* the outermost primary germ layer in animal embryos)

endo- = within (*endoderm:* the innermost primary germ layer in animal embryos)

gastro- = stomach, belly (*gastrulation:* the inward folding of the blastula-stage embryo, producing a three-layered gastrula)

lopho- = a crest, tuft; **-trocho** = a wheel (*lophotrochozoan:* member of a group of animal phyla identified as a clade by molecular evidence; including organisms that have lophophores or trochophore larvae.)

meso- = middle (*mesoderm:* the middle primary germ layer in a triploblastic animal embryo)

meta- = boundary, turning point; **-morph** = form (*metamorphosis:* a developmental transformation that turns an animal larva into either an adult or an adult-like stage that is not yet sexually mature)

proto- = first; **-stoma** = mouth (*protostome development:* a developmental mode distinguished by development of the mouth from the blastopore, spiral cleavage, and formation of the body cavity when solid masses of mesoderm split)

pseudo- = false (*pseudocoelomate:* an animal whose body cavity is lined by tissue derived from mesoderm and endoderm)

radia- = a spoke, ray (*radial symmetry:* symmetry in which the body is shaped like a pie or barrel and can be divided into mirror-imaged halves by any plane through its central axis)

tri- = three (*triploblastic:* possessing three germ layers)

Structure Your Knowledge

1. The following two very simplified trees represent the morphological and molecular hypotheses of animal phylogeny. Fill in the clades for these two trees. In the table that follows, list some of the characteristics that define each group.

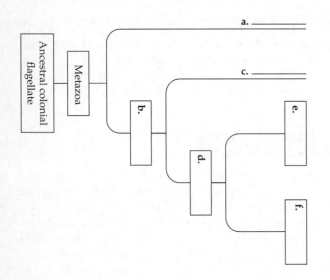

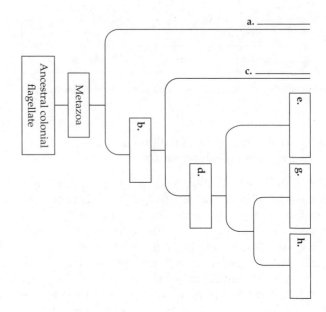

Clade	Characteristics
a.	
b.	
c.	
d.	
e.	
f.	
g.	
h.	

Test Your Knowledge

MULTIPLE CHOICE: *Choose the one best answer.*

1. Which of the following characteristics is found only in animals?
   a. homeobox-containing genes
   b. flagellated sperm
   c. heterotrophic nutrition
   d. *Hox* genes
   e. All of the above are exclusive animal traits.

2. A fly larva
   a. is a miniature version of the adult.
   b. is transformed into an adult by molting.
   c. ensures more genetic variation in the insect life cycle.
   d. is morphologically different from the adult and specialized for eating and growth.
   e. is all of the above.

3. Which of the following statements does *not* support the hypothesis that choanoflagellates are closely related to animals?
   a. Choanoflagellate cells and the collar cells of sponges are morphologically similar.
   b. Similar collar cells have been found in other animals, but never in non-choanoflagellate protists or plants or fungi.
   c. DNA data indicate that choanoflagellates and animals are sister groups.
   d. Some animal signaling and adhesion genes have been discovered in choanoflagellates.
   e. All of the above support the hypothesis.

4. Sponges differ from other animals in that
   a. they are completely sessile.
   b. they have radial symmetry and are suspension feeders.
   c. their simple body structure has no true tissues, and they have no symmetry.
   d. they are not multicellular.
   e. they have no flagellated cells.

5. Cephalization
   a. is associated with motile animals that concentrate sensory organs in a head region.
   b. is the formation of a coelom by budding from the archenteron.
   c. is a diagnostic characteristic of deuterostomes.
   d. is common in radially symmetrical animals.
   e. is the development of bilateral symmetry.

6. A true coelom
   a. has mesoderm-derived tissues that extend from the dorsal and ventral sides and support internal organs.
   b. allows organs to grow and move independently of the outer body wall.
   c. is a fluid-filled cavity completely lined by mesoderm.
   d. may be used as a skeleton by soft-bodied coelomates.
   e. has all of the above characteristics.

7. Which of the following characteristics are descriptive of protostome development?
   a. radial and determinate cleavage, blastopore becomes mouth
   b. spiral and indeterminate cleavage, coelom from a split in a solid mass of mesoderm
   c. spiral and determinate cleavage, blastopore becomes mouth, coelom from a split in mass of mesoderm
   d. spiral and indeterminate cleavage, blastopore becomes mouth, coelom from mesoderm outpockets
   e. radial and determinate cleavage, coelom from mesoderm outpockets, blastopore becomes anus

8. The morphology- and molecular-based phylogenetic trees agree in that each acknowledges
   a. the acoel flatworms as basal bilaterians.
   b. sponges as basal animals, Eumetazoa as animals with true tissues, and Bilateria as the clade that includes most animals.
   c. sponges as the probable ancestor of animals.
   d. the grouping together of the segmented annelids and arthropods.
   e. Deuterostomia as a monophyletic clade that includes chordates, echinoderms, and molluscs.

Chapter 33

An Introduction to Invertebrates

33.1 Sponges are basal animals that lack true tissues

33.2 Cnidarians are an ancient phylum of eumetazoans

33.3 Lophotrochozoans, a clade identified by molecular data, have the widest range of animal body forms

33.4 Ecdysozoans are the most species-rich animal group

33.5 Echinoderms and chordates are deuterostomes

Framework

This chapter surveys the amazing diversity of invertebrate animals. Characteristics and representatives of the major animal phyla are presented. The groups covered are the sponges, cnidarians, clade Lophotrochozoa (flatworms, rotifers, two lophophorate phyla, molluscs, and annelids), clade Ecdysozoans (nematodes and arthropods), and clade Deuterostomia (echinoderms and chordates).

Chapter Summary

More than 95% of known animal species are **invertebrates,** animals that lack a backbone.

33.1 Sponges are basal animals that lack true tissues

Sponges (phylum Porifera) are sessile animals found in fresh and marine waters. Water is drawn through pores in the body wall of these saclike animals into a central cavity, the **spongocoel,** and flows out through the **osculum.** Sponges are **suspension feeders,** collecting food particles by the action of collared, flagellated **choanocytes** lining the spongocoel.

Sponges are basal animals, originating near the base of the animal phylogenetic tree. The resemblance between choanocytes and the cells of choanoflagellates support the molecular evidence that animals originated from a choanoflagellate-like ancestor.

Sponges lack true tissues. In the **mesohyl,** or gelatinous matrix between the two body-wall layers, are **amoebocytes.** These cells take up food, digest it, and carry nutrients to other cells. Amoebocytes also form skeletal fibers, which may be sharp spicules or flexible fibers. Their ability to change into other cell types enables a sponge to reshape its body in response to environmental changes.

Most sponges are **hermaphrodites,** which produce both eggs and sperm, although they typically function sequentially as one sex or the other. Sperm, released into the water, fertilize eggs retained in the mesohyl of neighboring sponges. Flagellated larvae disperse to a suitable substratum and develop into sessile adults. Sponges produce defensive compounds that may have pharmaceutical uses.

INTERACTIVE QUESTION 33.1

Give the locations and functions of each of the following cells:

a. choanocytes

b. amoebocytes

33.2 Cnidarians are an ancient phylum of eumetazoans

One of the oldest lineages of clade Eumetazoa, animals with true tissues, is the phylum Cnidaria. The cnidarians include hydras, jellies, and corals. Their simple anatomy consists of a two-cell-layered sac with a central **gastrovascular cavity** and a single opening serving as both mouth and anus. This radially symmetrical body plan has two forms: **polyp,** which is a sessile, cylindrical form with mouth and tentacles extending upward, and **medusa,** which is a flattened, mouth-down polyp that moves by passive drifting and weak body contractions. Both body forms occur in the life histories of some cnidarians.

Cnidarians use their ring of tentacles, armed with **cnidocytes,** to capture prey. Cnidocytes contain cnidae, which are capsules that can evert and discharge long threads. Specialized stinging cnidae are called **nematocysts.** A gelatinous mesoglea is sandwiched between the epidermis and the gastrodermis. Cells of these two layers have bundles of microfilaments arranged into contractile fibers. A nerve net is associated with simple sensory receptors and coordinates the contraction of these cells against the hydrostatic skeleton of the gastrovascular cavity, producing movement. Phylum Cnidaria includes the following four clades:

Hydrozoans Most hydrozoans alternate between an asexually reproducing polyp and a sexually reproducing medusa form. The common freshwater hydras exist only in polyp form.

Scyphozoans The medusa stage is more prevalent in the scyphozoans. The sessile polyp stage often does not occur in the jellies of the open ocean.

Cubozoans Cubozoans have a box-shaped medusa stage and complex eyes in the fringe of their medusae. Many species, such as the sea wasp, have highly toxic cnidocytes.

Anthozoans Sea anemones and corals occur only as polyps. Corals secrete calcified external skeletons, and the accumulation of such skeletons produces coral.

INTERACTIVE QUESTION 33.2

In the following diagrams, name the two cnidarian body plans and identify the indicated structures.

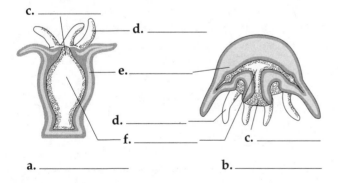

c. _____
d. _____
e. _____
d. _____
f. _____
c. _____
a. _____
b. _____

33.3 Lophotrochozoans, a clade identified by molecular data, have the widest range of animal body forms

Clade Bilateria consists of bilaterally symmetrical animals with triploblastic development. Most have a complete digestive tract and a coelom. Molecular data suggest that there are three major clades of bilaterians: Lophotrochozoa, Ecdysozoa, and Deuterostomia. The clade Lophotrochozoa is named for the *lophophore,* a feeding structure found in some members, and the *trochophore larva* found in others. About 18 animal phyla are included in this morphologically diverse group.

Flatworms Free-living flatworms live in marine, freshwater, and damp terrestrial habitats, and many are parasitic. They are triploblastic acoelomates. Most flatworms have a branching gastrovascular cavity, which functions in both digestion and distribution of food. Gas exchange and diffusion of nitrogenous wastes occur across the body surface. **Protonephridia** are networks of tubules with ciliated *flame bulbs* that pull fluid out of the body and function in osmoregulation.

Early flatworms diverged into two lineages. The Catenulida is a small clade of mostly freshwater species that typically reproduce by posterior asexual buds. The Rhabditophora is a larger, diverse clade of flatworms.

Rhabditophorans include freshwater **planarians** and other, mostly marine, free-living flatworms. Planarians move using cilia to glide on secreted mucus. Eyespots on the head detect light, and lateral head flaps detect chemicals. Their nervous system consists of pairs of anterior ganglia and ventral nerve cords. Planarians reproduce asexually by regeneration or sexually by copulation between hermaphroditic worms.

A majority of rhabditophorans are parasites, living in or on other animals. Many have a tough outer covering, suckers, and extensive reproductive organs. One group, the trematodes, has complex life cycles, usually including asexual and sexual stages and intermediate hosts in which larvae develop. Blood flukes are trematodes that use snails as their intermediate hosts and cause schistosomiasis in humans.

A second group of parasitic rhabditophorans, the tapeworms, have bodies that consist of a scolex with suckers and hooks for attaching to the host's intestinal lining, and a ribbon of proglottids packed with reproductive organs. The life cycles of tapeworms may also include intermediate hosts.

INTERACTIVE QUESTION 33.3

a. Describe the digestive system of a planarian.

b. Explain how tapeworms, which lack a digestive system, obtain nutrients.

Rotifers   Rotifers are smaller than many protists but have an **alimentary canal,** with separate mouth and anus, and other organs. Fluid in the pseudocoelom functions as a hydrostatic skeleton and distributes nutrients. A crown of cilia draws microscopic food particles into the mouth and to the pharynx, with its grinding jaws. Some species reproduce totally asexually by **parthenogenesis,** in which female offspring develop from unfertilized eggs. In other species, two types of eggs develop under certain conditions: one type forms females, and the other develops into males that produce sperm. The resulting zygotes develop into resistant embryos that can survive harsh conditions in a dormant state. Despite the disadvantages of asexual reproduction (such as faster accumulation of harmful mutations), bdelloids are an asexual clade of rotifers that appear to have been reproducing without males for 100 million years.

Lophophorates: Ectoprocts and Brachiopods   These two phyla, known as lophophorates, have a *lophophore,* a crown of ciliated tentacles around the mouth that function in feeding. They have a true coelom.

Ectoprocts, commonly called bryozoans, are tiny, mostly marine animals that live in colonies and are often encased in a hard **exoskeleton** that has pores through which their lophophores extend. Some species are important reef builders. **Brachiopods,** or lamp shells, attach to the seafloor by a stalk and open their hinged shell to allow water to flow through the lophophore.

Molluscs   Molluscs are soft-bodied, mostly marine animals, most of which are protected by a shell. The molluscan body plan has three main parts: a muscular **foot** used for movement, a **visceral mass** containing the internal organs, and a **mantle** that covers the visceral mass and may secrete a shell. In many molluscs, a **mantle cavity**— a water-filled chamber formed by the extension of the mantle—encloses the gills, anus, and excretory pores. A rasping **radula** is used for feeding by many molluscs. The nervous system includes a nerve ring and nerve cords. Most molluscs have an open circulatory system. Nephridia remove wastes from the hemolymph.

Most molluscs have separate sexes. Many marine molluscs have a life cycle that includes a ciliated larva called a trochophore. Four of the seven or eight molluscan clades are discussed in the text.

Chitons are oval marine animals with shells that are divided into eight dorsal plates. Chitons cling to and creep slowly over rocks, feeding on algae.

Most gastropods are marine, although there are many freshwater species, and some snails and slugs are terrestrial. A distinctive feature of this clade is **torsion,** the embryonic rotation of the visceral mass that results in the anus and mantle cavity being above the head. Most gastropods have single coiled shells. Many gastropods have distinct heads with eyes at the tips of tentacles. Moving by the rippling of the foot, most gastropods graze on plant material. Land snails lack gills; the lining of the mantle cavity functions in gas exchange.

Bivalves, such as clams, oysters, mussels, and scallops, have the two halves of their shell hinged at the mid-dorsal line. Most bivalves are suspension feeders; water flows into and out of the mantle cavity through siphons, and food particles are trapped in the mucus that coats the gills and then are swept to the mouth by cilia.

Cephalopods are rapid-moving marine predators. The mouth has beaklike biting jaws and is surrounded by tentacles. The shell is reduced and internal in most species, absent in some octopuses, and external only in the chambered nautilus. The foot has been modified to form parts of the tentacles and the muscular excurrent siphon, which animals may use to jet-propel themselves when water from the mantle cavity is forcibly expelled.

Cephalopods are the only molluscs with a *closed circulatory system.* They have a well-developed nervous system, sense organs, and a complex brain—important features for active predators. Shelled **ammonites** were the dominant invertebrate predators until their extinction at the end of the Cretaceous period.

Many freshwater bivalves and terrestrial gastropods are at risk of extinction due to habitat loss, pollution, introduction of non-native species, and overharvesting.

INTERACTIVE QUESTION   33.4

a. Describe the three parts of the molluscan body plan.

1.

2.

3.

b. Compare the feeding behaviors and activity levels of snails, clams, and squid.

1. Snails:

2. Clams:

3. Squid:

Annelids   Annelids are segmented worms found in marine, freshwater, and damp soil habitats. This coelomate phylum can be divided into two main groups, although their phylogeny is still under debate.

Polychaetes are mostly marine worms with parapodia on each segment that function in locomotion and often in gas exchange. Each parapodium has numerous chaetae (polychaetae), or stiff bristles. Polychaetes may be planktonic, bottom burrowers, or tube dwellers.

The oligochaetes are a diverse clade that includes the earthworms, several aquatic species, and the leeches. Earthworms eat through the soil, and their castings improve soil texture. The earthworm has a closed circulatory system, and respiration occurs across the moist, highly vascularized skin. Septa partition the coelom into segments, in each of which is found a pair of excretory metanephridia. The nervous system consists of cerebral ganglia and fused segmental ganglia along the ventral nerve cord. Earthworms are hermaphrodites; sperm are exchanged between worms during mating.

INTERACTIVE QUESTION 33.5

Identify the structures shown in the following illustration of a body segment of an earthworm.

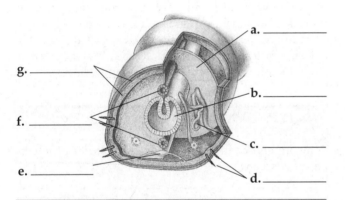

Most leeches inhabit freshwater. Many feed on small invertebrates, whereas others are parasites that temporarily attach to animals, slit or digest a hole through the skin, and suck the blood of their host.

33.4 Ecdysozoans are the most species-rich animal group

The clade Ecdysozoa includes animals that **molt,** shedding their external **cuticle** as they grow. Among the eight ecdysozoan phyla are two of the most abundant and successful of animal groups, the nematodes and the arthropods.

Nematodes Nematodes, or roundworms, inhabit water, soil, and the bodies of plants and animals. These cylindrical worms have a tough cuticle, which is periodically shed as they grow. Fluid in the pseudocoelom circulates nutrients absorbed from the alimentary canal. Reproduction is usually sexual, fertilization is internal, and most zygotes are resistant cells.

Numerous nematode species are ecologically important decomposers. Other nematodes are serious agricultural pests and animal parasites. Parasitic nematodes can enlist the cells of their host either to supply nutrients or to expand to house them.

INTERACTIVE QUESTION 33.6

Compare the locomotion of an earthworm and that of a nematode.

Arthropods In terms of species diversity, distribution, and vast numbers, **arthropods** are the most successful group of animals. Characteristics of arthropods include segmentation, which allows for regional specialization; a hard exoskeleton; and jointed appendages.

The earliest arthropods are found in the fossil record of the Cambrian explosion. The extinct *lobopods*, with identical body segments, may have been the group from which arthropods evolved. Trilobites were early arthropods with fairly uniform appendages. Studies show that onychophorans, a closely related invertebrate group, share all the arthropod *Hox* genes, including the unusual *Ubx* and *abd-A* genes. Thus, changes in the sequence or regulation of these genes (and not the origin of new *Hox* genes) may underlie the increased diversity in arthropod body segments.

Pairs of jointed appendages have become modified for walking, feeding, sensing, mating, or defense. A cuticle of chitin and protein completely covers the body as an exoskeleton, providing protection and points of attachment for muscles. To grow, an arthropod must molt, shedding its exoskeleton and secreting a larger one. The exoskeleton probably first evolved as protection and anchorage for muscles in marine organisms, but as arthropods diversified on land, the exoskeleton's functions came to include protection from desiccation and support.

Arthropods have well-developed sensory organs. A heart pumps hemolymph through an **open circulatory system** consisting of short arteries and a network of sinuses known as the hemocoel. The embryonic coelom becomes reduced during development. Gas exchange in most aquatic species occurs through gills, whereas most insects have tracheal systems of branching internal ducts.

Molecular and morphological data suggest an early divergence into four evolutionary lineages: **chelicerates, myriapods, hexapods,** and **crustaceans.**

Chelicerates (horseshoe crabs, scorpions, spiders, mites, and ticks) have clawlike **chelicerae,** which serve as pincers or fangs used for feeding; an anterior cephalothorax; and an abdomen. The earliest chelicerates were large, predatory **eurypterids.** Horseshoe crabs and sea spiders are the few marine chelicerates that survive today.

Most modern chelicerates are **arachnids.** Most ticks are bloodsucking parasites on reptiles or mammals. Many mites are parasites of plants and animals. In arachnids, the cephalothorax has six pairs of appendages: chelicerae, sensing or feeding appendages called *pedipalps,* and four pairs of walking legs. In most spiders, **book lungs,** consisting of stacked internal plates, function in gas exchange. Many spiders spin characteristic webs of silk from special abdominal glands.

INTERACTIVE QUESTION 33.7

a. What are chelicerae?

b. How do spiders trap, kill, and eat their prey?

Millipedes and centipedes are in the subphylum Myriapoda. These terrestrial groups have a pair of antennae and three pairs of mouthparts, including jaw-like **mandibles.** Millipedes are wormlike vegetarians with two pairs of walking legs per segment. Centipedes are carnivores with one pair of legs per segment and poison claws on their front segment.

Insects and their relatives (subphylum Hexapoda) have more known species than all other forms of life combined. An insect body consists of a head, a thorax, and an abdomen. The head has one pair of antennae, a pair of compound eyes, and several pairs of mouthparts modified for various types of ingestion. The nervous system consists of a cerebral ganglion and paired ventral nerve cords with segmental ganglia. Malpighian tubules, out-pocketings of the digestive tract, function in excretion. Trachael tubes constitute the respiratory system.

The oldest insect fossils are from the Devonian period (about 416 mya), but a major diversification occurred in the Carboniferous and Permian periods with the evolution of flight and the modification of mouthparts for specialized feeding on plants. A diversification of insects appears to have accompanied the radiation of flowering plants about 90 mya.

Flight is a major key to the success of insects. Many insects have one or two pairs of wings that are extensions of the cuticle of the dorsal thorax. Dragonflies were among the first flying insects.

In **incomplete metamorphosis,** the young (nymphs) are smaller versions of the adult and pass through several molts before developing wings and becoming sexually mature. In **complete metamorphosis,** the larva looks entirely different from the adult. Larvae eat and grow; adults primarily reproduce and disperse. Metamorphosis occurs in a pupal stage.

Insect reproduction is usually sexual; fertilization is usually internal.

Insects are classified into more than 30 orders. Insects affect humans as pollinators of crops, as food, as vectors of disease, and as competitors for food.

The mostly aquatic crustaceans have two pairs of antennae; three or more pairs of mouthpart appendages, including the hard mandibles; walking legs on the thorax; and appendages on the abdomen. Larger crustaceans have gills. Nitrogenous wastes pass by diffusion through thin areas of the cuticle, and a pair of glands regulates the salt balance of the hemolymph. Sexes usually are separate. One or more swimming larval stages occur in most aquatic crustaceans.

Isopods are a large group of mostly small aquatic crustaceans but also include terrestrial pill bugs. Lobsters, crayfish, crabs, and shrimp are large crustaceans called **decapods.** Their cuticle is hardened by calcium carbonate, and a carapace covers the dorsal side of their cephalothorax. The small, very numerous **copepods** are important members of marine and freshwater plankton communities. Shrimplike krill are a major food source for baleen whales. Barnacles are sessile crustaceans that strain food from the water with their appendages.

INTERACTIVE QUESTION 33.8

Compare and contrast the following features of insects and crustaceans.

Feature	Insects	Crustaceans
Habitat	a.	b.
Locomotion	c.	d.
Respiration	e.	f.
Excretion	g.	h.
Number of antennae	i.	j.
Appendages	k.	l.

33.5   Echinoderms and chordates are deuterostomes

DNA evidence supports Deuterostomia as a clade that includes echinoderms and chordates; it also indicates that some animals with deuterostome developmental features are not members of that clade.

Echinoderms   Most **echinoderms** are sessile or slow-moving marine animals. They have a thin skin covering an endoskeleton of calcareous plates. A **water vascular system** with a network of hydraulic canals controls extensions, called **tube feet,** that function in locomotion and feeding. Sexual reproduction usually involves separate sexes and external fertilization. Bilateral larvae metamorphose into adults that typically have five parts radiating from the center.

Starfishes (clade Asteroidea) have multiple arms extending from a central disk. Nerve cords radiate from a central nerve ring into the arms. Tube feet lining the undersurfaces of these arms are used to creep slowly and to grasp and open prey. The gripping of tube feet is the result of adhesive chemicals. Starfishes evert their stomach through their mouth and slip it into a slightly opened bivalve shell to eat. Discovered in 1986, the *sea daisies* also belong to clade Asteroidea. They have an armless, disk-shaped body with a five-fold symmetry.

Brittle stars (clade Ophiuroidea) have distinct central disks and move by lashing their long, flexible arms. They may be predators, scavengers, or suspension feeders.

Sea urchins and sand dollars, members of clade Echinoidea, have no arms but are able to move slowly using their five rows of tube feet. Long spines also aid a sea urchin's movement. Its mouth is ringed by complex jaw-like structures used to eat seaweed.

Members of clade Crinoidea, sea lilies live attached to the substrate by stalks; feather stars crawl about. Their long, flexible arms extend upward from around the mouth and are used in suspension feeding.

Sea cucumbers (clade Holothuroidea) are elongated animals that bear little resemblance to other echinoderms other than having five rows of tube feet.

INTERACTIVE QUESTION   33.9

List the key characteristics that distinguish the phylum Echinodermata.

a.

b.

c.

Chordates   Echinoderms and chordates have existed as separate phyla for over 500 million years. Phylum Chordata contains two subphyla of invertebrates in addition to the hagfishes and vertebrates.

Word Roots

arachn- = spider (*arachnid:* a member of the arthropod group chelicerates, which includes spiders, scorpions, ticks, and mites)

arthro- = jointed; **-pod** = foot (*arthropod:* a segmented ecdysozoan with a hard exoskeleton and jointed appendages)

brachio- = the arm (*brachiopod:* a marine lophophorate with a shell divided into dorsal and ventral halves; also called lamp shells)

cheli- = a claw (*chelicerae:* clawlike feeding appendages characteristic of chelicerates)

choano- = a funnel; **-cyte** = cell (*choanocyte:* a flagellated feeding cell found in sponges; also called a collar cell)

cnido- = a nettle (*cnidocyte:* a specialized cell unique to cnidarians that functions in defense and prey capture)

cope- = an oar (*copepod:* any of a group of small crustaceans that are important members of marine and freshwater plankton communities)

cuti- = the skin (*cuticle:* the exoskeleton of an arthropod; also a tough coat that covers the body of a nematode)

deca- = ten (*decapod:* a member of the group of crustaceans that includes lobsters, crayfish, crabs, and shrimps)

echino- = spiny; **-derm** = skin (*echinoderm:* a sessile or slow-moving marine deuterostome with a water vascular system; the group includes sea stars, brittle stars, sea urchins, feather stars, and sea cucumbers)

eury- = broad, wide; **-pter** = a wing, a feather, a fin (*eurypterid:* an extinct carnivorous chelicerate; also called a water scorpion)

exo- = outside (*exoskeleton:* a hard encasement on the surface of an animal)

gastro- = stomach; **-vascula** = a little vessel (*gastrovascular cavity:* a central cavity with a single opening in the body of certain animals that functions in both the digestion and distribution of nutrients)

hermaphrod- = with both male and female organs (*hermaphrodite:* an individual that functions as both male and female in sexual reproduction by producing both sperm and eggs)

in- = without (*invertebrate:* an animal without a backbone)

iso- = equal (*isopod:* a member of one of the largest groups of crustaceans, which includes terrestrial, freshwater, and marine species)

meso- = the middle; **-hyl** = matter (*mesohyl:* a gelatinous region between the two layers of cells in a sponge)

meta- = change; **-morph** = shape (*complete metamorphosis:* the transformation of a larva into an adult that looks very different and often functions in its environment differently than did the larva)

nemato- = a thread; **-cyst** = a bag (*nematocyst:* in a cnidocyte of a cnidarian, a specialized capsule-like organelle containing a coiled thread that when discharged can penetrate the body wall of prey)

oscul- = a little mouth (*osculum:* a large opening in a sponge that connects the spongocoel to the environment)

partheno- = without fertilization; **-genesis** = producing (*parthenogenesis:* a form of asexual reproduction in which females produce offspring from unfertilized eggs)

plan- = flat or wandering (*planarian:* a free-living flatworm found in unpolluted ponds and streams)

proto- = first; **-nephri-** = the kidney (*protonephridia:* an excretory system, such as the flame bulb system of flatworms, consisting of a network of tubules lacking internal openings)

Structure Your Knowledge

1. Use the following phylogenetic tree (which is based on molecular evidence) to help you review the diversity of animal phyla. Label the indicated clades. Then identify each phylum and list some key characteristics and common examples of each. The small sketches should help you identify the phyla. (Note that a sketch of an ectoproct [bryozoan] represents the two lophophorate phyla.)

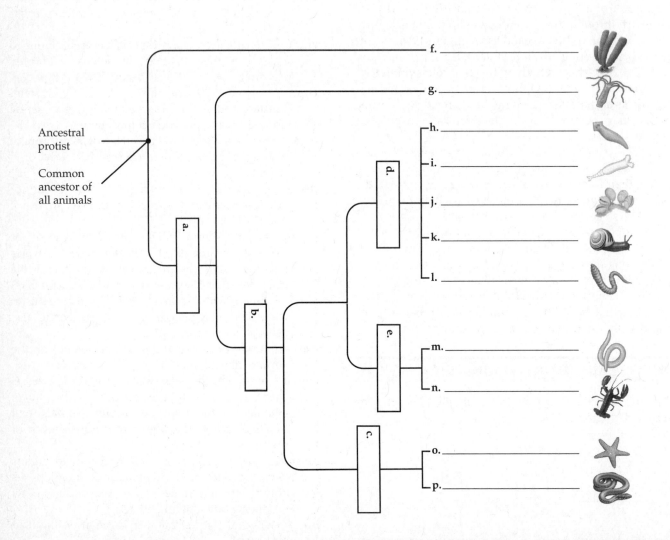

Ancestral protist

Common ancestor of all animals

Test Your Knowledge

MATCHING: *Match the following organisms with their phylum (A–J) and their clade or subphylum (a–m). Answers may be used more than once or not at all.*

Organism	Phylum	Clade or Subphylum
1. jelly (jellyfish)	_____	_____
2. crayfish	_____	_____
3. snail	_____	_____
4. leech	_____	_____
5. tapeworm	_____	_____
6. cricket	_____	_____
7. scallop	_____	_____
8. tick	_____	_____
9. sea urchin	_____	_____
10. hydra	_____	_____
11. planaria	_____	_____
12. octopus	_____	_____

Phyla	Clades or Subphyla
A. Annelida	a. anthozoan
B. Arthropoda	b. arachnid
C. Cnidaria	c. bivalve
D. Echinodermata	d. cephalopod
E. Ectoprocta	e. crustacean
F. Mollusca	f. echinoidean
G. Nematoda	g. gastropod
H. Porifera	h. hydrozoan
I. Platyhelminthes	i. insect
J. Rotifera	j. myriapod
	k. oligochaete
	l. scyphozoan
	m. rhabditophoran

MULTIPLE CHOICE: *Choose the one best answer.*

1. Invertebrates include
   a. all animals except for those in the phylum Vertebrata.
   b. all animals without backbones.
   c. only animals that have hydrostatic skeletons.
   d. members of the protostomes, but not of the deuterostomes.
   e. all animals without an endoskeleton.

2. Which of the following lists provides the best description of the sponges?
   a. no real symmetry, diploblastic, cnidocytes for capturing prey
   b. radial symmetry, triploblastic, nematocysts
   c. no real symmetry, without true tissues, choanocytes for trapping food particles
   d. bilateral symmetry, pseudocoel, flame bulbs for excretion
   e. radial symmetry, osculum and spongocoel for filtering water

3. Which of the following organisms does *not* have a gastrovascular cavity for digestion?
   a. flatworm
   b. hydra
   c. polychaete worm
   d. sea anemone
   e. fluke

4. The basal group in clade Eumetazoa is
   a. Ectoprocta.
   b. Rotifera.
   c. Platyhelminthes (flatworms).
   d. Porifera.
   e. Cnidaria.

5. Hermaphrodites
   a. contain male and female sex organs but usually cross-fertilize.
   b. include sponges, earthworms, and most insects.
   c. include parthenogenic rotifers.
   d. are both a and b.
   e. are a, b, and c.

6. Which of the following statements is *not* true of cnidarians?
   a. An alternation of medusa and polyp stage is common in clade Hydrozoa.
   b. They use a ring of tentacles armed with stinging cells to capture prey.
   c. They include corals, jellies, sponges, and sea anemones.
   d. They have a nerve net that coordinates contraction of microfilaments for movement.
   e. They have a gastrovascular cavity.

7. Which of the following combinations of phylum and characteristics is *incorrect*?
   a. Brachiopoda—lophophore, stalked, marine animals with hinged shells
   b. Rotifera—crown of cilia, microscopic animals, parthenogenesis (in some)
   c. Nematoda—gastrovascular cavity, tough cuticle, ubiquitous
   d. Annelida—segmentation, closed circulatory system, hydrostatic skeleton
   e. Echinodermata—radial anatomy in adults, endoskeleton, water vascular system

8. Which of the following pairings includes an excretory or osmoregulatory structure that is *incorrectly* matched with its clade?
   a. metanephridia—oligochaetes
   b. Malpighian tubules—cephalopods
   c. protonephredia—rhabditophorans
   d. thin region of cuticle—crustaceans
   e. diffusion across cell membranes—hydrozoans

9. Torsion
   a. is embryonic rotation of the visceral mass that results in a U-shaped digestive tract in gastropods.
   b. is part of the process of molting in the incomplete metamorphosis of some insects.
   c. is responsible for the spiral growth of bivalve shells.
   d. describes the thrashing movement of nematodes.
   e. is a stage in the complete metamorphosis of some insects.

10. Bivalves differ from other molluscs in that they
   a. are predaceous.
   b. have no heads and are suspension feeders.
   c. have shells.
   d. have open circulatory systems.
   e. use a radula to feed as they burrow through sand.

11. The exoskeleton of arthropods
   a. functions in protection and anchorage for muscles.
   b. is composed of chitin and cellulose.
   c. is absent in millipedes and centipedes.
   d. expands at the joints when the arthropod grows.
   e. functions in respiration and movement.

12. Which of the following structures does *not* function in suspension feeding?
   a. the lophophore of ectoprocts
   b. the radula of snails
   c. the choanocytes of sponges
   d. the mucus-coated gills of clams
   e. the crown of cilia in rotifers

13. What do nematodes and arthropods have in common?
   a. They are both segmented.
   b. They are both pseudocoelomates.
   c. They include important members of plankton communities.
   d. They both have exoskeletons and undergo ecdysis (molting).
   e. Both a and d are correct.

14. Which of the following structures is *not* associated with capturing prey?
   a. the tube feet of starfish
   b. the mandibles of centipedes
   c. the chaetae of earthworm
   d. the tentacles of squid
   e. the cnidocytes of hydra

15. Which of the following statements is an *incorrect* description of parasitic animals?
   a. Ticks are bloodsucking parasites belonging to clade Arachnida.
   b. Some roundworms (Nematoda) are internal parasites of humans.
   c. Leeches are parasites that secrete hirudin to keep the host's blood from coagulating.
   d. Flukes are flatworms and may have complex life cycles.
   e. Tapeworms are annelids that reproduce by shedding proglottids.

16. Which of the following phenomena is *not* considered a factor in the amazing radiation of insect groups?
   a. the evolution of wings and flight
   b. the acquisition of new *Hox* genes
   c. the evolution of mouthparts specialized for feeding on plants
   d. the radiation of flowering plants, providing new food sources
   e. All of the above contributed to insect diversity.

Chapter 34

The Origin and Evolution of Vertebrates

Key Concepts

Framework

This chapter focuses on the phylogeny of chordates, introducing the derived characters that define the major clades. Fill in the phylogenetic tree of chordates on p. 245 as you work through this chapter. Fossil evidence and hypotheses about human ancestry are also described.

Chapter Review

There are about 52,000 species of **vertebrates**, animals with a vertebral column, or backbone. This group is known for its *disparity*—its widely varying characteristics such as body mass.

34.1 Chordates have a notochord and a dorsal, hollow nerve cord

The deuterostome clade of the bilaterians includes the echinoderms and **chordates**. The chordates include the cephalochordates, urochordates, hagfishes, and vertebrates.

Derived Characters of Chordates Chordates have four structural trademarks, some of which may occur only during the embryonic stage: (1) a flexible rod called a **notochord** that provides skeletal support along the length of chordate embryos and in some adult chordates; (2) a dorsal, hollow nerve cord that develops into the brain and spinal cord; (3) **pharyngeal clefts** or grooves that often develop into **pharyngeal slits**, which open from the pharynx to the outside and function in suspension feeding or as gill slits for gas exchange; and (4) a muscular, post-anal tail containing skeletal elements and muscles.

Lancelets Lancelets (Cephalochordata) are tiny marine animals that retain all four chordate characteristics into the adult stage. The suspension-feeding adults burrow backward into the sand and filter food particles through a mucous net secreted across the pharyngeal slits. A lancelet swims by coordinated contractions of serial muscles that flex the notochord from side to side. These muscle segments develop from blocks of mesoderm called *somites,* which are present in all chordate embryos.

Tunicates Recent molecular studies indicate that **tunicates** (Urochordata) are more closely related to other chordates than are lancelets. In the saclike, sessile adult tunicate, water flows into the incurrent siphon, through pharyngeal slits where food is filtered by a mucous net, and out the excurrent siphon. Tunicates have only nine *Hox* genes, apparently having lost four of these developmental genes after branching off from other chordates.

Early Chordate Evolution   The ancestral chordate may have resembled a lancelet, the basal branch of chordates. Studies have shown that several *Hox* genes that organize the major regions of the vertebrate brain are expressed in the anterior end of the lancelet nerve cord in the same anterior-to-posterior pattern. The sequenced genome of tunicates indicates that whereas genes associated with the vertebrate heart and thyroid gland are present, genes involved with the long-range transmission of nerve impulses are found only in the vertebrate lineage.

INTERACTIVE QUESTION   34.1

Name the animal shown in the following diagram, and indicate the chordate group to which it belongs. Identify the labeled structures and indicate the four chordate characters.

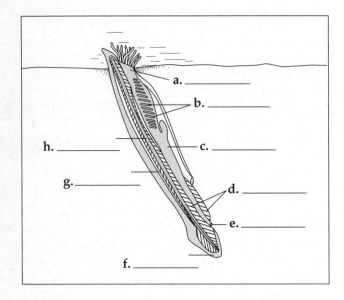

34.2   Craniates are chordates that have a head

The origin of a head with eyes and other sensory organs and a skull enclosing a brain was a major evolutionary transition seen in **craniates.**

Derived Characters of Craniates   Derived characters that distinguish craniates from other chordates include two or more sets of *Hox* genes instead of only one, and duplicated families of genes for signaling molecules and transcription factors. A group of embryonic cells called the **neural crest** is found along the dorsal margin of the embryonic folds that form the nerve cord. These cells migrate in the embryo and contribute to various structures, including teeth, some of the bones and

cartilage of the skull, and several types of neurons. The pharyngeal clefts of aquatic craniates evolved into gill slits with muscles and nerves that facilitate food intake and gas exchange. Craniates also have a higher metabolic rate and more extensive muscles than lancelets and tunicates, a heart with at least two chambers, red blood cells with hemoglobin, and kidneys.

The Origin of Craniates   Newly discovered 530 million-year-old fossils (from the Cambrian explosion) appear to be transitional stages. *Haikouella* resembled lancelets but had craniate characters such as eyes, respiratory gills, and a well-formed brain. Another fossil from this period, *Myllokunmingia,* had a skull with ear and eye capsules and is considered to be a true craniate.

Hagfishes   Hagfishes (clade Myxini) are the basal group of craniates. They have a cartilaginous skull and a notochord. Hagfishes are marine, mostly bottom-dwelling scavengers that produce slime to repulse predators and competing scavengers.

INTERACTIVE QUESTION   34.2

List some of the derived characters of craniates.

34.3   Vertebrates are craniates that have a backbone

Derived Characters of Vertebrates   The gene duplication involving the *Dlx* family of transcription factor genes may be associated with nervous system innovations and the more extensive skull and backbone of vertebrates. In most vertebrates, the vertebrae enclose the spinal cord. The dorsal, ventral, and anal fins supported by bony fin rays of aquatic vertebrates provide thrust and steering control for swimming.

Lampreys   The jawless lampreys of clade Petromyzontida are the basal lineage of extant vertebrates. Lampreys are suspension feeders as larvae, and blood-sucking parasites as adults. The notochord is the main axial skeleton. A flexible sheath around the notochord has pairs of cartilaginous projections that partially enclose the nerve cord.

Fossils of Early Vertebrates   **Conodants** were soft-bodied vertebrates with barbed, mineralized hooks in the mouth, dental elements in the pharynx for processing food, and prominent eyes. Conodonts appeared in the late Cambrian and were abundant for more than 300 million years.

Vertebrates that had paired fins and an inner ear with semicircular canals (also present in lampreys) emerged during the Ordovician, Silurian, and Devonian periods. These animals lacked jaws but had a muscular pharynx and were armored with mineralized bone. They all disappeared by the end of the Devonian.

Origins of Bone and Teeth The origin of mineralization in vertebrates may be associated with the switch from suspension-feeding to scavenging and predation. The earliest known mineralized structures are conodont dental elements. The armor of jawless vertebrates was composed of small, tooth-like structures.

INTERACTIVE QUESTION 34.3

The following phylogenetic hypothesis shows the major clades of chordates with some of the derived characters that define them. As you work through this chapter, fill in the eight large clades along the side and the name of each lineage.

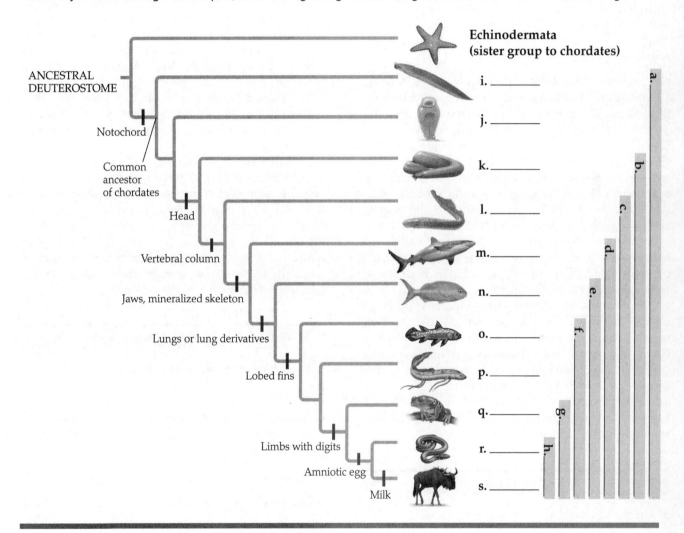

INTERACTIVE QUESTION 34.4

List the derived characters of vertebrates.

34.4 Gnathostomes are vertebrates that have jaws

Derived Characters of Gnathostomes The hinged jaws of **gnathostomes** may have evolved from the skeletal rods of the anterior pharyngeal slits, leaving the remaining gill slits to function in gas exchange.

Other derived characters of gnathostomes include an additional duplication of *Hox* genes (now four sets), an apparent duplication of the whole genome, an enlarged forebrain associated with enhanced senses of smell and vision, and in aquatic gnathostomes, a vibration-sensing **lateral line system.**

Fossil Gnathostomes Beginning about 450 million years ago (mya), gnathostomes became increasingly diverse. Their hinged jaws, paired fins, and tail facilitated a predatory lifestyle. The earliest gnathostomes were the armored **placoderms.** Other groups of jawed vertebrates, the **acanthodians,** also radiated during the Silurian and Devonian periods. Both groups became extinct, but by 420 million years ago three lineages of gnathostomes had diverged: chondrichthyans, ray-finned fishes, and lobe-fins.

Chondrichthyans (Sharks, Rays, and Their Relatives) **Chondrichthyans** have a skeleton made predominantly of cartilage, which appears to be a derived condition.

The largest group of chondrichthyans consists of sharks, rays, and skates. Sharks swim using powerful muscles in their trunk and tail fin. The paired pectoral and pelvic fins enhance maneuverability. Swimming maintains buoyancy and moves water past the gills. Although most sharks are carnivores, the largest sharks and rays are suspension feeders.

Sharks have keen senses: sharp vision, nostrils, and skin regions in the head that can detect electric fields generated by nearby animals. The shark's body transmits sound waves to the inner ear.

Following internal fertilization, **oviparous** species of sharks lay eggs that hatch outside the mother; **ovoviviparous** species retain the fertilized eggs until they hatch; and in the few **viviparous** species, developing young are nourished by a yolk sac placenta, by nutrients in a uterine fluid, or by eating other eggs.

Rays, which propel themselves with their enlarged pectoral fins, are primarily bottom dwellers that feed on molluscs and crustaceans.

INTERACTIVE QUESTION 34.5

a. List the derived characters of gnathostomes.

b. What is the function of the *spiral valve* in a shark's intestine?

c. What is a **cloaca?**

Ray-Finned Fishes and Lobe-Fins Nearly all vertebrates belong to a clade of gnathostomes called Osteichthyes, which historically included only the "bony fishes" but now includes the tetrapods. Almost all living **osteichthyans** have an ossified endoskeleton with a hard calcium-phosphate matrix.

Aquatic osteichthyans (informally called fishes) move water through the mouth and out between the gills by the contraction of muscles and movement of the protective bony flap called the **operculum.** Most fishes have a **swim bladder,** an air sac that controls buoyancy. It appears that lungs arose in early osteichthyans (as a supplement to gas exchange in the gills) and later evolved into swim bladders in some lineages.

The flattened, bony scales that cover the skin of most fishes are coated with mucus to reduce drag. Fishes have a lateral line system. Most species are oviparous, with eggs that are fertilized externally.

Ray-finned fishes (Actinopterygii) originated during the Silurian period and are named for the bony rays that support their fins. They include most of the familiar modern fish.

The **lobe-fins** (Sarcopterygii), which originated during the Silurian period, have muscular pectoral and pelvic fins supported by a series of rod-shaped bones that may have enabled these lobe-fins to swim and "walk" along the underwater substrate. Only three lobe-fin lineages survive: the coelacanths (Actinistia), which were once thought to be extinct; the lungfishes (Dipnoi), found in swamps in the Southern Hemisphere; and the tetrapods, found on land.

INTERACTIVE QUESTION 34.6

a. What three lineages of lobe-fins survive today?

b. Go back to the phylogenetic tree in Interactive Question 34.3 and make sure you have filled in the information up to the letter *p*.

34.5 Tetrapods are gnathostomes that have limbs

About 365 million years ago, the fins of some lobe-fins evolved into the limbs and feet of tetrapods.

Derived Characters of Tetrapods **Tetrapods** have limbs that support their weight on land, and feet with digits. Neck vertebrae enable greater head movement. The pelvic girdle is fused to the backbone, transferring forces generated by the hind legs to the rest of the body.

The adults of almost all living tetrapods lack gills. Pharyngeal clefts formed during development give rise to parts of the ears, glands, and other structures.

The Origin of Tetrapods A wide range of lobe-fins evolved during the Devonian, probably using stout fins to crawl through shallow waters and supplementing gas exchange with lungs. The fossil record documents how fins became progressively more limb-like. The first tetrapods appeared 365 million years ago. A diversity of tetrapods arose during the next 60 million years, most of which probably remained tied to the water.

INTERACTIVE QUESTION 34.7

Describe the 375-million-year-old "fishapod" fossil *Tiktaalik.* Is it considered to be a fish or a tetrapod?

Amphibians **Amphibians** (class Amphibia) include salamanders (order Urodela), frogs (order Anura), and caecilians (order Apoda).

Salamanders, which may be aquatic or terrestrial, move with a lateral bending of the body. Frogs are more specialized for moving on land, using their powerful hind legs for hopping. Tropical caecilians are nearly blind, wormlike (legless) burrowing animals.

Many frogs undergo a metamorphosis from a tadpole—the aquatic, herbivorous larval form with gills, a lateral line system, and a finned tail—to the carnivorous, terrestrial adult form with legs, lungs, and external eardrums. Some amphibians are exclusively aquatic or terrestrial. Amphibians are most abundant in damp habitats, and much of their gas exchange occurs across their moist skin.

Fertilization is generally external. Oviparous species lay their eggs in aquatic or moist environments. Some species show varying types of parental care. During breeding season, many frogs communicate via vocalizations. Some species migrate to specific breeding sites, using various forms of communication and navigation.

34.6 Amniotes are tetrapods that have a terrestrially adapted egg

Amniotes include reptiles (and birds) and mammals.

Derived Characters of Amniotes The **amniotic egg** contains *extraembryonic membranes* that protect and nourish the embryo. Most reptiles and some mammals have shelled amniotic eggs. Most mammals retain the embryo within the mother. Other terrestrial adaptations of the amniotes include a less permeable skin and the movement of the rib cage to ventilate the lungs.

INTERACTIVE QUESTION 34.8

Identify the four extraembryonic membranes in the following sketch of an amniotic egg. List the function of each membrane.

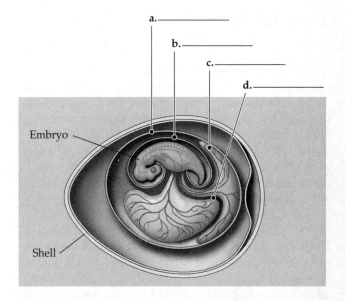

a.

b.

c.

d.

Early Amniotes The common ancestor of living amphibians and amniotes likely lived about 350 million years ago. Over time, early amniotes expanded into drier environments.

Reptiles The **reptile** clade includes a number of extinct groups and tuataras, lizards, snakes, turtles, crocodilians, and birds. The earliest reptiles date from about 310 mya. Their derived characters include waterproof scales, and most have shelled amniotic eggs. Fertilization is internal.

Many reptiles are **ectothermic**—they absorb external heat rather than generating their own—but they may regulate their body temperature through behavioral adaptations. Birds are **endothermic,** warming their bodies with metabolic heat.

The oldest reptilian fossils date from the late Carboniferous period. The first major branch was the **parareptiles,** large herbivores with dermal plates on the skin. The

diapsids diversified next. Their most obvious derived character is a pair of holes, on either side of the skull, through which jaw muscles pass. The two main lineages of diapsids are the **lepidosaurs,** which include tuataras, lizards, and snakes (and a number of now extinct huge marine reptiles), and the **archosaurs,** which produced the crocodilians and the extinct pterosaurs and dinosaurs. The **pterosaurs** had wings formed from a vascularized, muscled membrane stretched between an elongated finger and the body. Two main branches of the highly diverse **dinosaurs** were the herbivorous ornithischians and the saurischians, which gave rise to long-necked giants and the **theropods,** which included the ancestors of birds.

Scientists continue to debate whether dinosaurs were ectothermic or endothermic. The dinosaur ancestor of birds was certainly endothermic. Evidence indicates that many dinosaurs were active, agile, and in some species, social. At the end of the Cretaceous, all dinosaurs (except birds) became extinct.

The lizardlike tuatara, representing one lineage of lepidosaurs, survives only on islands off New Zealand. The other living lineage of lepidosaurs, the squamates, consists of lizards and snakes. Lizards are the most numerous and diverse reptiles (aside from birds). Snakes are limbless, although some species have vestigial pelvic and limb bones. Snakes are carnivorous, with adaptations for locating (heat-detecting and olfactory organs), killing (sharp teeth, which sometimes inject toxins), and swallowing prey (loosely articulated jaws and elastic skin).

The earliest turtle fossil dates from 220 mya. A turtle's protective (usually hard) shell is composed of upper and lower shields fused to the vertebrae, clavicles, and ribs. The phylogeny of turtles and the origin of their shells are still uncertain.

Crocodiles and alligators, collectively called crocodilians, belong to an archosaur lineage. They now live in warm regions and are adapted to aquatic habitats.

INTERACTIVE QUESTION 34.9

The two main lineages of diapsids are archosaurs and lepidosaurs. List several representatives, both living and extinct, of these two lineages.

Birds are archosaurs whose features have become adapted to flight. Many of the derived characters of birds are adaptations that reduce weight. Feathers are extremely light and strong and shape the wing into an airfoil. Large pectoral muscles attached to a keel on the sternum flap the aerodynamic wings. The evolution of strong flight enhanced the abilities to hunt, forage, escape from predators, and migrate to favorable habitats.

Birds are endothermic; feathers help retain metabolic heat. The four-chambered heart and efficient respiratory system facilitate a high metabolic rate.

Birds have excellent vision. Their relatively large brains permit detailed visual processing and motor coordination. Birds display complex behaviors, particularly in courtship rituals and parenting. Eggs are brooded during development by the mother, the father, or both.

Cladistic analysis indicates that birds are theropods, a group of bipedal saurischian dinosaurs. The fossils of feathered but flightless theropods have been found. Functions of early feathers may have been insulation, camouflage, and courtship displays. Flight may have evolved in leaping ground-running dinosaurs, flapping hill-running dinosaurs, or gliding tree-climbing dinosaurs.

Feathered theropods had evolved into birds by 150 million years ago. *Archaeopteryx,* the oldest bird known, had feathered wings with clawed digits, teeth, and a long tail. Fossils from the Cretaceous period show a gradual loss of ancestral dinosaur features and the acquisition of bird innovations.

The clade Neornithes, which includes the 28 orders of living birds, appeared before 65.5 million years ago. The **ratites** include the ostrich, kiwi, and emu. These flightless birds lack a keeled breast-bone and enlarged pectoral muscles. The general body form of many flying birds is similar, with beaks adapted to different foods and variations in foot structure.

INTERACTIVE QUESTION 34.10

List several adaptations for flight found in birds.

34.7 Mammals are amniotes that have hair and produce milk

Derived Characters of Mammals **Mammals** have mammary glands, which produce milk to nourish the young, and hair, which (along with a layer of fat under the skin) helps to insulate these endothermic animals. Active metabolism is provided for by an efficient respiratory system that uses a diaphragm to help ventilate the lungs and a circulatory system with a four-chambered heart.

Mammals generally have large brains. Extended parental care of the young provides them time to observe and learn survival skills. Mammalian teeth come in an assortment of shapes and sizes specialized for eating various foods.

Early Evolution of Mammals Mammals are part of the amniote group known as **synapsids,** distinguished by the temporal fenestra, a single hole behind each eye socket. During the evolution of mammals, the jaw was remodeled and two former jaw-joint bones became incorporated into the middle ear.

Synapsids were the dominant tetrapods during a time in the Permian period. Increasingly mammal-like synapsids emerged over a period of 100 million years. The first true mammals arose and diversified during the Jurassic period (200 to 145 mya). These still small mammals, which were probably nocturnal and insectivorous, coexisted with dinosaurs. Mammals underwent an extensive radiation in the wake of the Cretaceous extinctions.

Monotremes The platypus and echidnas (spiny anteaters) are **monotremes,** egg-laying mammals. Although egg-laying is an ancestral character for amniotes, monotremes have hair and produce milk for their young. Monotremes appear to have diverged from other mammals about 180 million years ago.

Marsupials **Marsupials,** including opossums, kangaroos, and koalas, share derived characters with eutherians, including a high metabolic rate, nipples that provide milk, and giving birth to live young. A **placenta,** formed from the uterine lining and the extraembryonic membranes of the embryo, nourishes the developing embryo. A marsupial is born very early in development and completes its embryonic development attached to a nipple in a maternal pouch called a *marsupium.* Australian marsupials have radiated and filled the niches occupied by eutherian mammals in other parts of the world.

Eutherians (Placental Mammals) **Eutherians** complete development attached to a placenta within the maternal uterus. Eutherians and marsupials may have diverged about 140 million years ago.

Molecular systematics is helping to establish the phylogeny of eutherian orders, although there is still no consensus. There appear to be at least four main clades. One clade evolved in Africa and includes the elephants, manatees, hyraxes, and aardvarks. A second clade radiated in South America and includes the sloths, anteaters, and armadillos. The third and largest clade includes rabbits, rodents, and primates. The fourth clade includes many diverse groups: carnivores, even- and odd-toed hoofed herbivores, bats, some shrews and moles, and cetaceans (dolphins and whales).

Primates include the lemurs, tarsiers, monkeys, and apes (including humans). Derived characters include grasping hands and feet, and flat nails on the digits. Compared to other mammals, primates have a large brain, short jaws, and forward-looking eyes with overlapping visual fields that enhance depth perception. They also exhibit complex social behavior and extensive parental care.

Many primate traits were shaped by natural selection in the tree-dwelling early primates. All primates—except humans—have a wide separation between the big toe and the other toes, allowing them to grasp branches with their feet. Monkeys and apes have an **opposable thumb,** which functions in a grasping "power grip" in monkeys and nonhuman apes but is adapted for precise manipulation in humans.

There are three main groups of living primates: the lemurs, lorises, and pottos; the tarsiers; and the anthropoids. The **anthropoids** include monkeys and apes. Both Old World and New World monkeys probably originated in Africa or Asia. New World monkeys, which colonized South America about 25 million years ago, are strictly arboreal. Most monkeys in both groups are diurnal and live in social bands.

The primates informally called apes include gibbons, orangutans, gorillas, chimpanzees and bonobos, and humans. Apes diverged from Old World monkeys about 20–25 mya. All nonhuman apes live in the Old World tropics. All living apes have relatively long arms, short legs, and no tails. Apes have a larger brain and more flexible behavior than do other primates.

INTERACTIVE QUESTION 34.11

a. List the derived characteristics of mammals.

b. Go back to the phylogenetic tree in Interactive Question 34.3 and make sure you have filled in all the information.

34.8 Humans are mammals that have a large brain and bipedal locomotion

Derived Characters of Humans The species *Homo sapiens* is about 200,000 years old. Characters that distinguish humans from other apes are upright stance and bipedal locomotion; a larger brain with the capacity for language, symbolic thought, and artistic expression; and the ability to manufacture and use complex tools. Humans also have reduced jaws and a shorter digestive tract.

Although the genomes of humans and chimpanzees are 99% identical, recent studies show that they differ in the expression of regulatory genes, which may account for the many differences between the two groups.

The Earliest Hominins **Paleoanthropology** is the study of human origins. Paleoanthropologists have identified approximately 20 extinct species of **hominins**—species that are more closely related to humans than to chimpanzees. The oldest of these, *Sahelanthropus,* lived about 6 to 7 million years ago. These early hominins shared some derived characters of humans, such as reduced canine teeth, relatively flat faces, and a more upright stance, as indicated by the location of the foramen magnum underneath the skull. The increasing bipedalism of early hominins is suggested by fossil pelvis, leg, and foot bones of 4.4-million-year-old *Ardipithecus ramidus.* Different human features have evolved at different rates.

Australopiths   Hominin diversity increased between 4 and 2 million years ago, and many of the hominins from this period are called australopiths, although their phylogeny is unresolved. Two lineages from that time include several "gracile" (slender) *Australopithecus* species and two "robust" *Paranthropus* species that had sturdy skulls with powerful jaws and teeth.

Bipedalism   By about 10 million years ago, the Himalayan range had formed, the climate had became drier, and the forests of Africa and Asia had contracted, presenting a more extensive savannah habitat that may have influenced hominin evolution. But all recently discovered fossils of early hominins show signs of bipedalism, and these hominins lived in mixed forests and open woodlands. Some early hominins appear to have been able to switch between walking upright and climbing trees. About 1.9 million years ago hominins that lived in more arid environments began to walk long distances on two legs.

Tool Use   The first evidence of complex tool use is 2.5-million-year-old cuts on animal bones, suggesting the use of stone tools. *Australopithecus garhi,* whose fossils were found nearby, had a relatively small brain.

Early Homo   Larger-brained fossils dating from 2.4 to 1.6 million years ago are the first to be placed in the genus *Homo.* Sometimes sharp stone tools are found with fossils of *Homo habilis,* or "handy man." Fossils from 1.9 to 1.5 million years ago are recognized by a number of paleoanthropologists as the species *Homo ergaster,* which had a larger brain, long legs and hip joints adapted for long-distance walking, relatively short and straight fingers, smaller teeth, and more sophisticated stone tools.

The sexual dimorphism of size difference between the sexes was also less in *Homo ergaster* than in *Australopithecus.* In living primates, reduced sexual dimorphism is associated with more pair-bonding.

Homo erectus was the first hominin to migrate out of Africa. The oldest fossils of hominins outside of Africa date back 1.8 million years. Fossil evidence indicates that *H. erectus* became extinct 200,000 years ago, although one group may have persisted on Java until around 50,000 years ago.

Neanderthals   *Homo neanderthalensis,* or Neanderthals, lived in Europe from about 350,000 years ago, and later spread to the Near East, Central Asia, and Siberia. They apparently became extinct about 28,000 years ago. They had large brains, buried their dead, and made tools from stone and wood. Comparisons of mtDNA from Neanderthal fossils and living humans suggest that little gene flow occurred between the two species.

Homo sapiens   The oldest known fossils of *Homo sapiens* date from 195,000 years ago. DNA comparisons show that Europeans and Asians share a relatively recent common ancestor, with many African lineages branching off earlier. Comparisons of mitochondrial DNA and the Y chromosomes of various populations support a common African ancestor of all *H. sapiens.*

The oldest fossils outside Africa come from the Middle East and date back about 115,000 years. One or more waves of humans appear to have first spread into Asia, then to Europe and Australia. The date of arrival in the New World may be 15,000 years ago. The 2004 discovery of a small hominin on the island Flores, named *Homo floresiensis,* illustrates how new information contributes to the ongoing inquiry into human origins. Continuing studies of these fossils suggest that this species may have branched off before the origin of the *Homo sapiens,* and its small stature and brain size could relate to its island habitat.

The rapid expansion of our species may be tied to changes in human cognition. The discovery of 77,000-year-old engravings and other works of art document the origin of sophisticated thought in humans.

INTERACTIVE QUESTION   34.12

List some of the derived characters of humans.

Word Roots

arch- = ancient (*archosaur:* member of the reptilian group that includes crocodiles, alligators, dinosaurs, and birds)

crani- = the skull (*craniate:* a chordate with a head)

di- = two (*diapsid:* member of an amniote clade distinguished by a pair of holes on each side of the skull; diapsids include the lepidosaurs and archosaurs)

dino- = terrible; **-saur** = lizard (*dinosaur:* member of an extremely diverse clade of reptiles varying in body shape, size, and habitat)

endo- = inner; **-therm** = heat (*endothermic*: referring to organisms with bodies that are warmed by heat generated by metabolism)

eu- = good (*eutherian*: a placental mammal; a mammal whose young complete their embryonic development within the uterus, joined to the mother by the placenta)

gnantho- = the jaw; **-stoma** = the mouth (*gnathostomes*: members of the vertebrate clade possessing jaws)

homin- = man (*hominin*: a species on the human branch of the evolutionary tree that is more closely related to humans than to chimpanzees)

lepido- = a scale (*lepidosaurs*: member of the reptilian group that includes lizards, snakes, and tuataras)

marsupi- = a bag, pouch (*marsupial*: a mammal, such as a koala, kangaroo, or opossum, whose young complete their embryonic development inside a maternal pouch called the marsupium)

mono- = one (*monotreme*: an egg-laying mammal, represented by the platypus and echidna)

neuro- = nerve (*neural crest cells*: groups of cells along the sides of the neural tube where it pinches off from the ectoderm; the cells migrate to various parts of the embryo and form various structures)

noto- = the back; **-chord** = a string (*notochord*: a flexible rod, made of tightly packed mesodermal cells, that runs along the dorsal anterior-posterior axis of a chordate)

opercul- = a covering, lid (*operculum*: in aquatic osteichthyans, a protective bony flap that covers and protects the gills)

osteo- = bone; **-ichthy** = fish (*osteichthyan*: member of a vertebrate clade with jaws and mostly bony skeletons)

ovi- = an egg; **-parous** = bearing (*oviparous*: referring to a type of development in which young hatch from eggs laid outside the mother's body)

paleo- = ancient; **anthrop-** = man; **-ology** = the science of (*paleoanthropology*: the study of human origins and evolution)

placo- = a plate (*placoderm*: a member of an extinct group of fishlike vertebrates that had jaws and were enclosed in a tough, outer armor)

ptero- = a wing (*pterosaur*: winged reptile that lived during the Mesozoic era)

ratit- = flat-bottomed (*ratite*: member of the group of flightless birds)

syn- = together (*synapsid*: member of an amniote clade distinguished by a single hole on each side of the skull; includes the mammals)

tetra- = four; **-podi** = foot (*tetrapod*: a vertebrate clade whose members have limbs with digits; includes mammals, amphibians, and birds and other reptiles)

tunic- = a covering (*tunicates*: members of the clade Urochordata; sessile marine chordates that lack a backbone)

vivi- = alive (*ovoviviparous*: referring to a type of development in which young hatch from eggs that are retained in the mother's uterus)

Structure Your Knowledge

1. Starting from primates, list the increasingly inclusive clades (from eutherians to chordates) and some of the derived characters that define each of these clades. Reviewing the phylogenetic tree on page 245 may help.
2. Describe several examples from vertebrate evolution that illustrate the common evolutionary theme that new adaptations usually evolve from preexisting structures.
3. How do various tetrapod groups meet the challenges of a terrestrial habitat?

Test Your Knowledge

FILL IN THE BLANKS

_____ 1. invertebrate chordate clade most closely related to craniates

_____ 2. the most basal chordate group

_____ 3. blocks of mesoderm along notochord that develop into muscles

_____ 4. clade of jawed vertebrates

_____ 5. flap over the gills of bony fishes

_____ 6. adaptation that allows reptiles to reproduce on land

_____ 7. egg-laying mammals

_____ 8. structure that helps mammals ventilate their lungs

_____ 9. group of primates that includes monkeys and apes

_____ 10. genus in which the fossil Lucy is placed

MULTIPLE CHOICE: *Choose the one best answer.*

1. Pharyngeal slits appear to have functioned first as
   a. suspension-feeding devices.
   b. gill slits for respiration.
   c. components of the jaw.
   d. portions of the inner ear.
   e. mouth openings.

2. Which of the following is *not* a derived character of craniates?
   a. cranium or skull
   b. neural crest in embryonic development
   c. four sets of *Hox* genes
   d. heart with at least two chambers
   e. cephalization with sensory organs

3. Which of the following groups represents the oldest lineage of vertebrates?
   a. caecilians
   b. sharks and rays
   c. lancelet
   d. hagfishes
   e. lampreys

4. Which of the following organisms is in the lobe-fin clade?
   a. lampreys
   b. sharks and rays
   c. ray-finned fishes
   d. hagfishes
   e. tetrapods

5. Which of the following groups is *not* in the same lineage as the others?
   a. lizards
   b. birds
   c. dinosaurs
   d. crocodilians
   e. pterosaurs

6. Which of the following groups is *incorrectly* paired with its mechanism for gas exchange?
   a. amphibians—skin and lungs
   b. lungfishes—gills and lungs
   c. reptiles—lungs
   d. ray-finned fishes—swim bladder
   e. mammals—lungs with diaphragm to help ventilate

7. Non-bird reptiles have lower caloric needs than do birds of comparable size because they
   a. are ectothermic, whereas birds are endothermic.
   b. have waterproof scales instead of feathers.
   c. have a longer digestive tract and obtain more nutrients from their food.
   d. provide less parental care than do birds.
   e. have a less efficient respiratory system and thus a lower metabolic rate.

8. Which of the following best describes the earliest mammals?
   a. large, herbivorous
   b. large, carnivorous
   c. small, insectivorous
   d. small, herbivorous
   e. small, carnivorous

9. Oviparity is a reproductive strategy that
   a. allows placental mammals to bear well-developed young.
   b. is common in reptiles and some sharks.
   c. is necessary for vertebrates to reproduce on land.
   d. is characteristic of all flying vertebrates.
   e. protects the embryo inside the mother but uses the food resources of the egg.

10. In Australia, marsupials fill the niches that eutherians fill in other parts of the world because
    a. they are better adapted and have outcompeted eutherians.
    b. their offspring complete their development attached to a nipple in a marsupium.
    c. they originated in Australia.
    d. they evolved from monotremes that migrated to Australia about 65 million years ago.
    e. after Pangaea broke up, they diversified in isolation from eutherians.

11. Which of the following groups is *incorrectly* matched to its clade?
    a. mammals—synapsids
    b. snakes—lepidosaurs
    c. birds—theropods
    d. crocodiles—saurischians
    e. dinosaurs—diapsids

12. Which of the following statements is supported by DNA studies?
    a. *Homo sapiens* originated in Africa.
    b. Erect posture preceded the enlargement of the brain in human evolution.
    c. The key differences between chimpanzees and humans appear to involve duplications of *Hox* genes.
    d. Neanderthals are more closely related to Asians than to Europeans.
    e. *Sahelanthropus* was the earliest hominid.

Key Concepts

35.1 Plants have a hierarchical organization consisting of organs, tissues, and cells

35.2 Meristems generate cells for primary and secondary growth

35.3 Primary growth lengthens roots and shoots

35.4 Secondary growth increases the diameter of stems and roots in woody plants

35.5 Growth, morphogenesis, and cell differentiation produce the plant body

Framework

A rather large new vocabulary is needed to name the specialized cells and structures in a study of plant structure and growth. Focus your attention on how a plant's roots, stems, and leaves are specialized to function in absorption, support, transport, protection, and photosynthesis. The plant body is composed of dermal, vascular, and ground tissue systems. Apical meristems at the tips of roots and shoots create primary growth. The lateral meristems—vascular cambium and cork cambium—create secondary growth that adds girth to stems and roots. New techniques and model systems such as *Arabidopsis* are enabling researchers to explore the molecular bases for plant growth, morphogenesis, and cellular differentiation.

Chapter Review

35.1 Plants have a hierarchical organization consisting of organs, tissues, and cells

A plant **organ** performs a specific function and is composed of several types of **tissues,** groups of cells with a common function.

The Three Basic Plant Organs: Roots, Stems, and Leaves As an evolutionary adaptation to the dispersed resources in a terrestrial environment, vascular plants have an underground **root system** for obtaining water and minerals from the soil and an aerial **shoot system** of stems and leaves for absorbing light and CO_2 for photosynthesis. *Photosynthates* from the shoot system nourish the roots. The shoot system includes vegetative shoots and reproductive shoots, which, in angiosperms, bear flowers.

The functions of **roots** include anchorage, absorption, and storage of food. A *taproot system* is found commonly in eudicots and gymnosperms. The one main deep **taproot** often stores nutrients and gives rise to **lateral roots.** Most monocots have a shallower *fibrous root system* in which many small roots grow from the stem. Such roots are called *adventitious,* as they grow in an unusual location.

Most absorption of water and minerals occurs through tiny **root hairs,** extensions of epidermal cells that are clustered near root tips. Modified roots may serve various functions, including support, storage, and oxygen absorption.

A **stem** consists of alternating **nodes,** the points at which leaves are attached, and segments between nodes, called **internodes.** An **axillary bud** is found in the angle (axil) between a leaf and the stem. Located at the tip or apex of a shoot is an **apical bud** (or terminal bud) with developing leaves, as well as compacted internodes. The apical bud may exhibit **apical dominance,** inhibiting the growth of axillary buds and producing a taller plant. Axillary buds may develop into lateral shoots, or branches, each with its own apical bud, leaves, and axillary buds.

Modifications of shoots include stolons (horizontal shoots that enable asexual reproduction), horizontally growing underground rhizomes, and food storage structures (such as tubers and bulbs).

Leaves, the main photosynthetic organs of most plants, usually consist of a flattened **blade** and a **petiole,** or stalk. Many monocots lack petioles. Monocot leaves usually have parallel major **veins** (the vascular tissue of leaves), whereas eudicot leaves have networks of branched veins. Leaves may be simple or compound. Modified leaves may function in support, protection, storage, or reproduction.

INTERACTIVE QUESTION   35.1

Label the parts in the following diagram of a flowering plant.

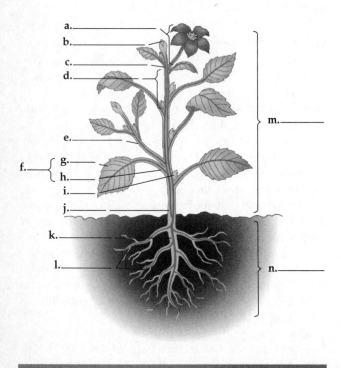

Dermal, Vascular, and Ground Tissues   **Tissue systems** are functional units of tissues that are continuous throughout the plant but have specific characteristics in each plant organ (root, stem, and leaf).

The **dermal tissue system** forms a protective outer layer. The **epidermis** is a single layer of tightly packed cells that covers nonwoody plants. The epidermis of leaves and most stems is covered with a **cuticle,** a waxy coating that prevents excessive water loss. Older stems and roots of woody plants are covered by a protective **periderm.** *Trichomes* are outgrowths of the shoot epidermis that may protect against insects and disease or reduce water loss.

The **vascular tissue system** consists of **xylem** and **phloem** and functions in long-distance transport of water and minerals (xylem) and of sugars (phloem). In angiosperms, the **stele,** or vascular tissue of an organ, takes the form of a solid central *vascular cylinder* in roots but is arranged in *vascular bundles* in stems and leaves.

The **ground tissue system** contains cells that function in photosynthesis, support, and storage. Ground tissue internal to vascular tissue is called **pith;** ground tissue outside the vascular tissue is called **cortex.**

Common Types of Plant Cells   **Parenchyma cells** carry on most of a plant's metabolic functions, such as photosynthesis and food storage. They usually lack secondary walls and have large central vacuoles. Most parenchyma cells retain the ability to divide and differentiate into other types of plant cells.

Collenchyma cells lack secondary walls but have thickened primary walls. Strands or cylinders of collenchyma cells provide flexible support for young parts of the plant and elongate along with the plant.

Sclerenchyma cells have thick secondary walls strengthened with lignin. The protoplasts of these specialized supporting cells often die at maturity. **Fibers** are long, tapered cells that usually occur in threads. **Sclereids** are shorter and irregular in shape, with very thick, lignified cell walls.

The water-conducting cells of xylem also die at functional maturity, leaving behind their secondary walls, which have interspersed thinner pits. **Tracheids,** found in all vascular plants, are long, thin, tapered cells. Water passes through pits from cell to cell. **Vessel elements,** found mainly in angiosperms, are wider, shorter, and thinner walled, with perforations in their end walls. They align to form long tubes known as **vessels.** Both tracheids and vessel elements have lignin-strengthened walls.

In the phloem of angiosperms, sugars flow through chains of cells called **sieve-tube elements,** or sieve-tube members. These cells remain alive at functional maturity but lack nuclei, ribosomes, and vacuoles. Fluid flows through pores in the **sieve plates** in the end walls between cells. The nucleus and ribosomes of an adjacent companion cell, which is connected to a sieve-tube element by numerous plasmodesmata, serve both cells.

INTERACTIVE QUESTION 35.2

a. Which types of plant cells are dead at functional maturity?

b. Which types of plant cells lack nuclei at functional maturity?

35.2 Meristems generate cells for primary and secondary growth

Plants have tissues called **meristems,** which remain undifferentiated and can perpetually divide to form new cells. **Apical meristems,** located at the tips of roots and shoots and in the axillary buds of shoots, produce **primary growth,** resulting in elongation. Herbaceous (nonwoody) plants usually undergo primary growth only. **Secondary growth** is an increase in diameter as new cells are produced by **lateral meristems. Vascular cambium** produces secondary xylem and phloem; **cork cambium** produces the periderm. Cells that remain to divide in a meristem are called *initials* (or *stem cells*), whereas cells that are displaced from the meristem and become specialized in developing tissues are called *derivatives.*

Flowering plants may be *annuals,* which complete their life cycle in a year or less; *biennials,* which have a life cycle spanning two years; or *perennials,* which live many years.

INTERACTIVE QUESTION 35.3

What is the difference between **indeterminate growth** and **determinate growth?**

35.3 Primary growth lengthens roots and shoots

An entire herbaceous plant and the youngest parts of a woody plant consist of primary growth produced by apical meristems.

Primary Growth of Roots The apical meristem of the root tip is protected by a **root cap,** which secretes a lubricating polysaccharide slime. The *zone of cell division*

includes the apical meristem and its derivatives. In the *zone of elongation,* cells lengthen to many times their original size, which pushes the root tip through the soil. As the zone of elongation grades into the *zone of differentiation* (or maturation), cells become specialized in structure and function.

In the vascular cylinder of most eudicot roots, xylem cells radiate from the center in spokes, with phloem located between them. The vascular tissue of a monocot may have pith, a central core of undifferentiated parenchyma cells, inside a ring of xylem and phloem.

The ground tissue consists of parenchyma cells in the cortex. The innermost layer of cortex is the one-cell-thick **endodermis,** which regulates the passage of materials into the vascular cylinder.

Lateral roots develop from the **pericycle,** the outer layer of cells in the vascular cylinder, and push through the cortex and epidermis.

INTERACTIVE QUESTION 35.4

Label the tissues in the following cross sections of a eudicot root and its vascular cylinder. Identify the functions of each of these tissues.

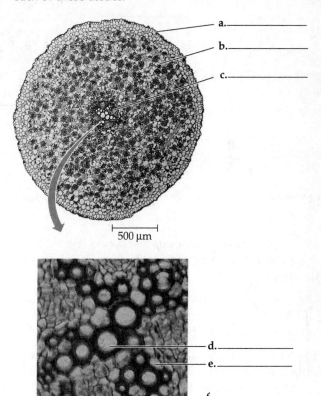

a. _____
b. _____
c. _____

500 μm

d. _____
e. _____
f. _____
g. _____

50 μm

Primary Growth of Shoots   The dome-shaped mass at the tip of the apical bud is the shoot apical meristem. **Leaf primordia** form on the sides of the apical meristem, and axillary bud meristems develop at the bases of the leaf primordia.

Elongation of the shoot occurs by cell elongation within young internodes. In grasses and some other plants, the bases of leaves and stems have *intercalary meristems,* which enable damaged leaves to regrow.

The epidermis of the dermal tissue system covers stems. In most eudicots, vascular bundles are arranged in a ring, with the ground tissues pith inside and cortex outside the ring. Xylem is located internal to the phloem in the vascular bundles. In most monocot stems, the vascular bundles are scattered throughout the ground tissue. A layer of collenchyma just beneath the epidermis and sclerenchyma fiber cells may strengthen the stem.

Stomata, tiny pores flanked by **guard cells** in the cuticle-covered epidermis, permit both gas exchange and evaporation of water from the leaf.

Mesophyll consists of parenchyma cells containing chloroplasts. In many eudicot leaves, columnar *palisade mesophyll* is located above *spongy mesophyll,* which has loosely packed cells surrounding many air spaces.

Vascular tissue in the stem branches into the leaf and divides repeatedly, providing support and vascular tissue to the photosynthetic mesophyll. Each vein is enclosed in a ring of protective cells called a *bundle sheath.*

INTERACTIVE QUESTION   35.5

Name the indicated structures in the following diagram of a leaf.

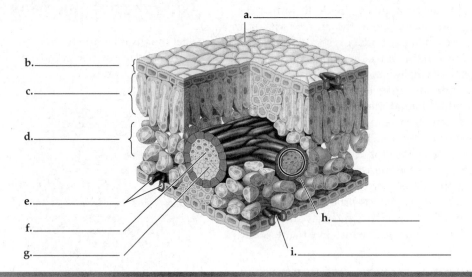

a._____

b._____

c._____

d._____

e._____

f._____

g._____

h._____

i._____

35.4   Secondary growth increases the diameter of stems and roots in woody plants

The vascular cambium and cork cambium produce secondary growth, which occurs in all gymnosperms and in many eudicots but is rare in monocots.

The Vascular Cambium and Secondary Vascular Tissue
In a woody stem, vascular cambium forms a continuous cylinder of meristematic cells that produce secondary xylem to the inside of the cylinder and secondary phloem to the outside. Some cambial initials produce radial lines of parenchyma cells called *vascular rays,* which function in lateral transport of water and nutrients and in storage.

Wood is the accumulation of secondary xylem cells with thick, lignified walls. Annual growth rings in temperate regions result from the seasonal cycle of cambium dormancy, early wood production (with larger, thinner cells in the spring), and late wood production (in the rest of the growing season). Scientists analyzing tree rings (*dendrochronology*) of Mongolian conifers from over the past five centuries have found evidence of warming temperatures during the twentieth century.

In older trees, a central column of *heartwood* consists of older xylem with resin-filled cell cavities; the *sapwood* consists of secondary xylem that still functions in transport.

Older secondary phloem is sloughed off as bark; only that closest to the vascular cambium functions in sugar transport.

The Cork Cambium and the Production of Periderm
The epidermis splits off during secondary growth and is replaced by new protective tissues produced by the cork cambium, a meristematic cylinder that first forms in the outer cortex in stems and in the outer pericycle in roots. The cork cambium produces a thin-layered *phelloderm* to the inside and cork cells to the exterior. Cork cells develop suberin-impregnated walls. The protective coat formed by a cork cambium and its tissues is a layer of periderm. As secondary growth continually splits the outer layers of periderm, new cork cambia develop, eventually forming from parenchyma cells in the secondary phloem. **Lenticels** are spaces between cork cells through which gas exchange occurs. **Bark** refers to phloem and periderm.

INTERACTIVE QUESTION 35.6

Place the letters of the following tissues in the blanks so that they are in the order in which they are located in a tree trunk going from the outside in.

A. primary phloem   E. pith

B. secondary phloem   F. cork cambium

C. primary xylem   G. vascular cambium

D. secondary xylem   H. cork

____ ____ ____ ____ ____ ____ ____ ____

Evolution of Secondary Growth   Studying the molecular basis of secondary growth is complicated by the time and space needed for trees to develop. Studies with *Arabidopsis* have suggested that the accumulation of weight by the stem may cue wood formation. The genetic regulation of both apical meristem and vascular cambium involves common genes, suggesting that these genes have been conserved by evolution.

35.5 Growth, morphogenesis, and cell differentiation produce the plant body

Development involves the growth and differentiation of cells into the specialized form of an organism. Plant development and structure are controlled by both genetic and environmental factors, resulting in *developmental plasticity* in response to local environments.

Growth is an increase in size. The organization of cells to create body form is called **morphogenesis.** **Differentiation** is the process by which cells become specialized for different functions.

Model Organisms: Revolutionizing the Study of Plants
Arabidopsis thaliana has many traits of a model organism: It is a small, easily cultivated plant with a short generation time and prolific seed production. Its genome is small for a plant and was the first plant genome to be sequenced.

INTERACTIVE QUESTION 35.7

Plant molecular biologists use the bacterium *Agrobacterium tumefaciens* to insert its transforming DNA randomly into *Arabidopsis* DNA. How are the resulting "knock-out mutants" being used?

Growth: Cell Division and Cell Expansion   The plane of cell division, which is usually perpendicular to the long axis of a dividing cell, is determined by the *pre-prophase band,* a ring of cytoskeletal microtubules that forms during late interphase.

Studies with the *tangled-1* mutant of maize, in which transverse cell divisions are normal but longitudinal divisions are aberrant, indicate that the generation of leaf shape may not depend solely on the orientation of cell division.

In *asymmetrical cell division,* one daughter cell receives more cytoplasm than the other, often leading to a key developmental event, such as the formation of guard cells. The first asymmetric division of a plant zygote establishes the **polarity** of plants, which have a root end and a shoot end. In the *gnom* mutant of *Arabidopsis*, the first division is not asymmetrical, and the resulting ball-shaped plant lacks leaves and roots.

About 90% of plant cell expansion is due to the uptake of water into vacuoles. This economical means of cell elongation produces the rapid growth of shoots and roots. When enzymes break the cross-links in the cell wall, the restraint on the turgid cell is reduced and water enters the cell by osmosis. Cells expand in a direction perpendicular to the orientation of the cellulose microfibrils in the inner layers of the cell wall.

INTERACTIVE QUESTION 35.8

Review the role of microtubules in the orientation of plant cell division and expansion.

Morphogenesis and Pattern Formation **Pattern formation** is the characteristic development of structures in specific locations. The fate of plant cells does not appear to be determined by *lineage-based mechanisms,* but rather by *position-based mechanisms.* Experiments demonstrate that a cell's fate is determined late in development and depends on signaling from neighboring cells.

Animal cell fate, which depends on lineage-dependent mechanisms, involves transcription factors encoded by homeotic (*Hox*) genes. In many plants, a *Hox*-like gene called *KNOTTED-1* influences leaf morphology. Its overexpression in tomato plants results in "super-compound" leaves.

Gene Expression and Control of Cellular Differentiation Differential gene expression within genetically identical cells leads to their differentiation into the various plant cell types. Cell-to-cell communication provides the positional information that largely determines a plant cell's fate. In the root epidermis of *Arabidopsis,* the homeotic gene *GLABRA-2* is expressed in epidermal cells that are in contact with only one underlying cortical cell, and these cells do not develop root hairs. The gene is not expressed in those cells in contact with two cortical cells, and they differentiate into root hair cells.

Shifts in Development: Phase Changes In a process known as **phase change,** the apical meristem can switch from one developmental stage, or *phase,* to another, such as from producing juvenile vegetative nodes and internodes to laying down adult nodes. A node's developmental phase is fixed when it is produced, and thus both juvenile and mature leaves can coexist along the shoots of a plant.

Genetic Control of Flowering In the transition from vegetative growth to reproductive growth (production of a flower), **meristem identity genes** code for transcription factors that activate the genes involved in the conversion of indeterminate vegetative meristems to determinate floral meristems.

The four floral organs develop from floral primordia in four concentric whorls: sepals, petals, stamens, and carpels. Scientists have identified several **organ identity genes.** These genes, belonging to *the MADS-box* family, code for transcription factors that regulate floral pattern, and their expression is influenced by positional information. Mutations of organ-identity genes result in the placement of one type of floral organ where another type would normally develop. The **ABC hypothesis** of flower formation identifies three classes of organ identity genes, each of which is expressed in two adjacent whorls.

INTERACTIVE QUESTION   35.9

The following diagram depicting the ABC hypothesis shows the active genes in each whorl and the resulting anatomy of a wild-type flower.

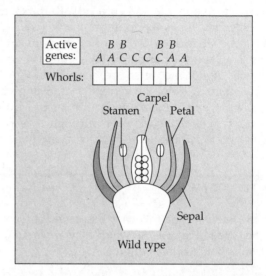

Wild type

a. Fill in the following table to indicate which organs are produced in the whorls in a normal flower. In a mutant that lacks a functional gene *A,* what gene expression pattern and resulting flower organ arrangement would be produced? (Remember that the lack of *A* activity removes the inhibition of gene *C.*)

Whorl	Genes Active	Organs in Normal Flower	Genes Active in Mutant A	Organs in Mutant A Flower
1	A			
2	AB			
3	BC			
4	C			

b. If you had a double-mutant plant that had no gene activity for *B* or *C,* what would the resulting flower look like?

Word Roots

apic- = the tip; **meristo-** = divided (*apical meristem:* embryonic plant tissue, located in the tips of roots and the buds of shoots, whose dividing cells enable the plant to grow in length)

coll- = glue; **-enchyma** = an infusion (*collenchyma cell:* a flexible plant cell type that occurs in strands or cylinders that support young parts of the plant without restraining growth)

endo- = inner; **derm-** = skin (*endodermis:* in plant roots, the innermost layer of the cortex, which surrounds the vascular cylinder)

epi- = over (*epidermis:* the dermal tissue system of non-woody plants, usually consisting of a single layer of tightly packed cells)

inter- = between (*internode:* a segment of a plant stem between the points where leaves are attached)

meso- = middle; **-phyll** = a leaf (*mesophyll:* the ground tissue of a leaf, located between the upper and lower epidermis and specialized for photosynthesis)

morpho- = form; **-genesis** = origin (*morphogenesis:* the development of body shape and organization)

peri- = around; **-cycle** = a circle (*pericycle:* the outermost layer in the vascular cylinder from which lateral roots arise)

phloe- = the bark of a tree (*phloem:* vascular plant tissue consisting of living cells arranged into elongated tubes that transport sugar and other organic nutrients throughout the plant)

sclero- = hard (*sclerenchyma cell:* a rigid, supportive plant cell type usually lacking a protoplast and having thick secondary walls strengthened by lignin)

trachei- = the windpipe (*tracheid:* a long, tapered water-conducting cell found in the xylem of nearly all vascular plants; functioning tracheids are no longer living.)

vascula- = a little vessel (*vascular tissue:* plant tissue consisting of cells joined into tubes that transport water and nutrients throughout the plant body)

xyl- = wood (*xylem:* vascular plant tissue consisting mainly of dead tubular cells that conduct water and minerals from the roots to the rest of the plant)

Structure Your Knowledge

1. How does the indeterminate growth pattern of plants lead to developmental plasticity? What is the adaptive advantage of such plasticity?

2. In the following cross section of a young stem, label the indicated structures. Is this a stem of a monocot or a eudicot? How can you tell? Does this stem illustrate primary growth or secondary growth? How can you tell?

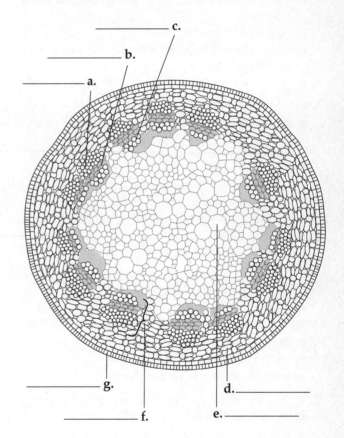

Test Your Knowledge

MULTIPLE CHOICE: *Choose the one best answer.*

1. Which of the following is *not* a reason that *Arabidopsis* is used as a model organism?
   a. It has a short generation time.
   b. It takes little space and is easy to grow in the laboratory.
   c. Its relatively small genome has been sequenced.
   d. Its ability to reproduce only by self fertilization has facilitated genetic studies.
   e. The production of "knockout mutants" allows researchers to study the function of individual genes.

2. Which of the following is an *incorrect* completion to this sentence? Monocots typically have
   a. a taproot rather than a fibrous root system.
   b. leaves with parallel veins rather than branching venation.
   c. no secondary growth.
   d. scattered vascular bundles in the stem rather than bundles in a ring.
   e. pith in the center of the vascular cylinder in the root.

3. Sieve-tube elements
   a. are responsible for lateral transport through a woody stem.
   b. control the activities of phloem cells that lack nuclei and ribosomes.
   c. have spiral thickenings that allow the cell to elongate along with a young shoot.
   d. are transport cells with sieve plates in the end walls between cells.
   e. transport photosynthate from leaves through companion cells.

4. Axillary buds
   a. may exhibit apical dominance over the apical bud.
   b. form at nodes in the angle where leaves join the stem.
   c. grow out from the pericycle layer.
   d. are formed from intercalary meristems.
   e. only develop into vegetative shoots.

5. Which of the following structures provide structural support to a leaf?
   a. petioles
   b. vascular bundles
   c. suberin-impregnated cork cells
   d. epidermal cells with lignin-strengthened cell walls
   e. columnar cells of the palisade mesophyll

6. Which of the following plant structures show determinate growth?
   a. roots
   b. vegetative shoots
   c. adventitious roots
   d. leaves
   e. No parts of a plant show determinate growth.

7. Which of the following is *incorrectly* paired with the length of its life cycle?
   a. pine tree—perennial
   b. rose bush—perennial
   c. carrot—biennial
   d. marigold—annual
   e. corn—biennial

8. Which of the following is *incorrectly* paired with its function?
   a. perpendicularly oriented initials—form radial vascular rays
   b. lenticels—facilitate gas exchange in woody stems
   c. root hairs—absorb water and dissolved minerals
   d. root cap—protects the root as it pushes through soil
   e. periderm—forms protective layer of cork

9. Secondary xylem and phloem are produced in a root by the
   a. pericycle.
   b. endodermis.
   c. vascular cambium.
   d. apical meristem.
   e. cork cambium.

10. Bark consists of
    a. secondary phloem.
    b. layers of periderm.
    c. cork cells.
    d. cork cambium.
    e. all of the above.

11. In what direction does a plant cell enlarge?
    a. toward its basal end as a result of positional information
    b. parallel to the orientation of the preprophase band of microtubules
    c. perpendicular to the orientation of cellulose microfibrils in the cell wall
    d. in the direction from which water flows into the cell
    e. perpendicular to the internodes

12. What do organ-identity genes code for?
   a. transcription factors that control development of floral organs
   b. signals that change a vegetative meristem into a floral meristem
   c. the root and shoot meristems
   d. the three tissue systems
   e. pollen and egg cells

13. Which of the following is essential to establishing the axial polarity of a plant?
   a. expression of different homeotic genes in the shoot and root meristems
   b. orientation of cortical microtubules perpendicular to the ground
   c. the expression of the *GLABRA-2* gene in the root but not in the shoot
   d. the asymmetric first division of the zygote
   e. the proper orientation of microtubules, which then orient cellulose microfibrils

14. The results from genetic studies of which of the following support the idea that plant form is not determined merely by the orientation of cell divisions?
   a. the *gnom* mutant of *Arabidopsis*, in which the first division of the zygote is symmetrical
   b. the *tangled-1* mutant of maize, with its aberrant longitudinal division and normal leaves
   c. the *KNOTTED-1* mutant in tomato plants, with its "super-compound" leaves
   d. mutations in organ identity genes that produce abnormal flowers
   e. the expression of the *GLABRA-2* gene in root epidermal cells that do not develop root hairs

15. Morphogenesis and pattern formation in plants results from
   a. cells moving from the zone of elongation to the zone of differentiation.
   b. the differentiation of cells within the root and stem apical meristem.
   c. differences in the direction of cell expansion as well as positional information from neighboring cells.
   d. the expression of floral meristem identity genes.
   e. the ABC hypothesis of gene expression in an apical meristem, which causes phase changes.

MATCHING: *Match each of the plant tissues (1–10) with its description (A–J).*

_____	1. cambial initials	A.	layer from which lateral roots originate
_____	2. collenchyma	B.	tapered xylem cells with lignin in cell walls
_____	3. endodermis	C.	parenchyma cells with chloroplasts in leaves
_____	4. fibers	D.	protective coat made of cork and cork cambium
_____	5. mesophyll	E.	bundles of long sclerenchyma cells
_____	6. pericycle	F.	supporting cells with thickened primary walls
_____	7. periderm	G.	parenchyma cells inside vascular ring in eudicot stem
_____	8. pith	H.	supporting cells with thickened secondary walls
_____	9. sclerenchyma	I.	cells that produce secondary xylem and phloem
_____	10. tracheids	J.	cell layer in root regulating movement into vascular cylinder

Chapter 36

Resource Acquisition and Transport in Vascular Plants

Framework

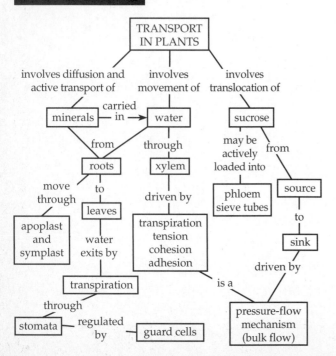

Chapter Review

36.1 Adaptations for acquiring resources were key steps in the evolution of vascular plants

Evolutionary adaptations to life on land involved the specialization of roots to absorb water and minerals, shoots to absorb CO_2 and light, and vascular tissue to transport materials between roots and shoots.

Shoot Architecture and Light Capture Leaf size is often correlated with water availability. **Phyllotaxy,** the arrangement of leaves on the stem of a plant, may be one leaf per node (alternate, or spiral), two leaves per node (opposite), or multiple (whorled). Alternate phyllotaxy, with leaves emerging at a light-maximizing 137.5° angle, is most common in angiosperms.

The *leaf area index* is the ratio of the total area of the top surfaces of the leaves to the area of ground covered by the plant or crop. When the leaf area index value is above 7, the shaded, nonproductive lower leaves may undergo programmed cell death and be shed through *self-pruning.* The orientation of leaves and stems affects light capture. The extent of branching and stem elongation represents a compromise between energy spent growing tall to avoid shading versus energy spent branching and increasing photosynthetic surface area. Increased stem thickness usually accompanies growth in height.

Root Architecture and Acquisition of Water and Minerals The taproot system of gymnosperms and eudicots supports tall growth. Evidence indicates that physiological mechanisms prevent a plant's roots from competing with each other for the same limited resources. **Mycorrhizae,** symbiotic associations between roots and fungi, greatly increase the surface area for water and mineral absorption.

a. What leaf orientation works best in low light levels? In bright sunny conditions?

b. Which results in greater productivity: a high leaf area index or a low leaf area index?

c. How might root branching and physiology change in response to patches of nutrients such as nitrate?

36.2 Different mechanisms transport substances over short or long distances

The Apoplast and Symplast: Transport Continuums

The **apoplast** consists of plant cell walls, intercellular spaces, and interiors of dead xylem vessels. The protoplasts of plant cells are connected by plasmodesmata, forming a cytoplasmic continuum called the **symplast.**

Transport of water and solutes within plant tissues can occur by three routes: the *transmembrane route,* by crossing plasma membranes and cell walls as materials move between cells; the *symplastic route,* moving through plasmodesmata; and the *apoplastic route,* along cell walls and extracellular spaces. Substances may change routes during transit.

Short-Distance Transport of Solutes Across Plasma Membranes

Substances may move across the selectively permeable plasma membrane by passive or active transport, using protein carriers, ion channels, and cotransporters. Gated ion channels open or close in response to stimuli such as voltage, pressure, or chemicals.

A plant proton pump moves H^+ out of the cell, generating an energy-storing proton gradient and a membrane potential. Cotransport with H^+ can move solutes such as sugars and ions such as NO_3^- across the plasma membrane.

Short-Distance Transport of Water Across Plasma Membranes

The diffusion of free water across a membrane is called **osmosis. Water potential,** designated Ψ *(psi),* is a useful measurement for predicting the direction that water will move. It takes into account both solute concentration and physical pressure. Free water (not bound to solutes or surfaces) will flow from a region of higher water potential to one of lower water potential. Water potential is measured in **megapascals** (MPa); 1 MPa is equal to about 10 atmospheres of pressure. The water potential of pure water in an open container under standard conditions is assigned the value of 0 MPa.

The combined effect of solute concentration and pressure is shown by the *water potential equation:* $\Psi = \Psi_S + \Psi_P$. This equation states that the water potential is the sum of the solute potential and the pressure potential.

Solute potential (Ψ_S), also called *osmotic potential,* is proportional to the molarity of a solution. Solutes reduce the number of free water molecules and the capacity of water to do work. Solutes lower the water potential, and a solution's Ψ_S is always negative.

Pressure potential (Ψ_P) measures the pressure on a solution and can have a positive or negative value. Negative pressure is tension. A plant **protoplast** that is expanding due to the osmotic uptake of water pushes against the cell wall, and the rigid cell wall exerts pressure against the protoplast. This positive pressure is called **turgor pressure.**

A **flaccid** plant cell ($\Psi_P = 0$ MPA) bathed in a solution more concentrated than the cell will lose water by osmosis because the solution has a lower (more negative) Ψ than the Ψ_S of the cell. The cell's protoplast shrinks and pulls away from the cell wall (**plasmolysis**). A flaccid cell bathed in pure water has a Ψ that is lower than the Ψ of the water. Water will then enter the cell until enough turgor pressure builds up such that Ψ_P and Ψ_S are equal and opposite in magnitude, and $\Psi = 0$ both inside and outside the cell. Net movement of water will then stop.

Plant cells are usually **turgid;** they have a greater solute concentration than their extracellular environment and turgor pressure keeps them firm. Loss of turgor causes **wilting** of a plant's leaves and stems.

Water-specific transport proteins called **aquaporins** increase the rate of water diffusion across a membrane.

a. A flaccid plant cell has a water potential of -0.6 MPa. Fill in the water potential equation for this cell.

$$\Psi_P =$$
$$+ \Psi_S =$$
$$\Psi =$$

b. The cell is then placed in a beaker of distilled water ($\Psi = 0$ MPa). Fill in the equation for the cell after it reaches equilibrium in pure water. Explain what happens to the water potential of this cell.

$$\Psi_P =$$
$$+ \Psi_S =$$
$$\Psi =$$

c. Explain what would happen to the same cell if it is placed in a solution that has a water potential of -0.8 MPa. Fill in the equation for the cell after it reaches equilibrium in the solution.

$$\Psi_P =$$
$$+ \Psi_S =$$
$$\Psi =$$

Long-Distance Transport: The Role of Bulk Flow Long-distance transport in plants occurs by **bulk flow,** the movement of fluid driven by a pressure gradient. Sieve-tube elements lack most cellular organelles, and vessel elements and tracheids are dead (and empty) at maturity; both factors facilitate more efficient bulk flow.

36.3 Transpiration drives the transport of water and minerals from roots to shoots via the xylem

Absorption of Water and Minerals by Root Cells Much of the absorption of water and mineral ions occurs along root tips, where root hairs are located. The soil solution flows into the hydrophilic walls of epidermal cells and moves along the apoplast into the root cortex, exposing a large surface area of plasma membrane for the uptake of water and minerals. Active transport allows cells to accumulate essential minerals.

Transport of Water and Minerals into the Xylem Water and minerals that entered the symplast through a cortex or epidermal cell pass through plasmodesmata of cells of the **endodermis** and into the stele, or vascular cylinder. A ring of suberin around each endodermal cell, called the **Casparian strip,** prevents water and minerals from the apoplast from entering the stele without passing through a selectively permeable plasma membrane.

By a combination of diffusion and active transport, minerals move from endodermal cells to the apoplast and enter the nonliving xylem tracheids and vessel elements along with water.

INTERACTIVE QUESTION   36.3

Label the following diagram of a section of a root. Letters a and b refer to transport routes of water and minerals; letters c–i refer to cell layers or structures.

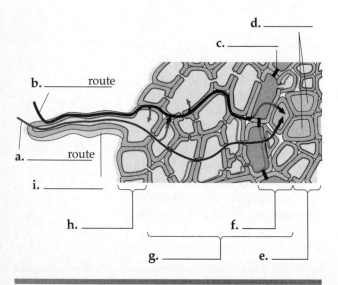

Bulk Flow Transport via the Xylem The water and dissolved minerals of **xylem sap** move by bulk flow from the vascular cylinder of a root to the branching veins of leaves. Through the evaporative loss of water vapor from leaves, called **transpiration,** the plant loses a tremendous amount of water that must be replaced by water transported up from the roots.

At night, transpiration rate is low, but minerals are still actively pumped into the vascular cylinder and prevented from leaking out by the endodermis. Water flowing from the cortex into this area of lower water potential produces **root pressure,** which pushes xylem sap upward. In some plants, root pressure may cause **guttation,** the exudation of water droplets from leaves when more water is forced up the xylem than is transpired by the plant.

According to the **cohesion-tension hypothesis,** formulated in 1894 by J. Joly and H. Dixon, transpiration creates a negative-pressure pull on xylem sap, which is transmitted from shoots to roots by the cohesion of water molecules within the xylem.

Water vapor from saturated air spaces within a leaf exits through stomata. The thin layer of water that coats the mesophyll cells lining the air spaces begins to evaporate. The adhesion of the remaining water to the hydrophilic walls causes a curvature in the air-water interface. Because of the high surface tension of water, this curvature produces tension in the water layer. This negative pressure draws water from the cells, which is replaced by water from the xylem. Water moves along a gradient toward more negative water potentials, from xylem to neighboring cells to air spaces to the drier air outside the leaf.

The transpirational pull on xylem sap is transmitted from the leaves to the root tips by the cohesiveness of water, which results from hydrogen bonding between water molecules. Adhesion of water molecules to the hydrophilic walls of the narrow xylem cells also contributes to overcoming the downward pull of gravity.

The upward transpirational pull on the cohesive sap creates tension within the xylem, lowering the water potential so that water flows passively from the soil, across the cortex, and into the vascular cylinder.

A break in the chain of water molecules by the formation of a water vapor pocket in a xylem vessel, called cavitation, breaks the transpirational pull. The vessel cannot function in transport unless the break is repaired or the chain of water molecules detours through adjacent vessels.

Xylem Sap Ascent by Bulk Flow: A Review The cohesion-tension mechanism results in the bulk flow of water and solutes from roots to leaves. Water potential differences caused by solute concentration and pressure contribute to the diffusion of water from cell to cell, but solar-powered tension caused by transpiration creates the pressure gradient that is responsible for long-distance transport of water and minerals.

INTERACTIVE QUESTION   36.4

Explain the contribution of each of the following phenomena to the long-distance transport of water.

a. Transpiration:

b. Tension:

c. Cohesion:

d. Adhesion:

36.4   The rate of transpiration is regulated by stomata

For photosynthesis to occur, leaves must exchange gases through the stomata and provide a large internal surface area for CO_2 uptake, both of which increase evaporative water loss.

Stomata: Major Pathways for Water Loss   A plant loses most of its water through stomata. Guard cells regulate the size of stomatal openings, which controls the rate of transpiration. Stomatal densities in many plant species are both genetically and environmentally influenced; densities have been shown to relate to CO_2 levels, providing a measure of such levels in past climates.

Mechanisms of Stomatal Opening and Closing   When the kidney-shaped guard cells of most angiosperms become turgid and swell, their radially oriented microfibrils cause them to increase in length and bow outward, increasing the size of the gap between them.

Guard cells can actively accumulate potassium ions (K^+), which lowers water potential and leads to the osmotic inflow of water and an increase in turgor pressure. The flow of K^+ through specific membrane channels into the cell is coupled with the generation of a membrane potential by proton pumps that transport H^+ out of the cell. The exodus of K^+ (with water following) leads to a loss of turgor and stomatal closing.

Stimuli for Stomatal Opening and Closing   The opening of stomata at dawn is related to at least three factors. First, light stimulates guard cells to accumulate K^+, triggered by the illumination of blue-light receptors that activate proton pumps. Second, the depletion of CO_2 within air spaces of the leaf as photosynthesis begins in the mesophyll stimulates stomata to open. The third factor is a daily rhythm of opening and closing that is endogenous to guard cells. Cycles that have intervals of approximately 24 hours are called **circadian rhythms.**

Environmental stress can cause stomata to close during the day. Guard cells lose turgor when water is in short supply. Also, a hormone called **abscisic acid (ABA),** produced in roots and leaves in response to a lack of water, signals guard cells to close stomata.

Effects of Transpiration on Wilting and Leaf Temperature   When transpiration exceeds the water available, leaves wilt as cells lose turgor pressure. Transpiration also produces evaporative cooling, maintaining a cooler temperature for critical enzymes in leaf cells.

Adaptations That Reduce Evaporative Water Loss   Many **xerophytes,** plants adapted to arid climates, have leaves that are highly reduced; photosynthesis is carried out in thick, water-storing stems. Leaf cuticles may be thick, and stomata may be sheltered in depressions. Succulent plants of the family Crassulaceae take in CO_2 at night, so their stomata can be closed during the day, reducing water loss.

INTERACTIVE QUESTION   36.5

What is meant by saying that plants face a photosynthesis–transpiration compromise? Explain how a hot sunny day with a dry wind can affect this compromise in a plant.

36.5   Sugars are transported from sources to sinks via the phloem

Movement from Sugar Sources to Sugar Sinks   **Translocation,** the transport of photosynthetic products throughout the plant, occurs in the sieve-tube elements of phloem. **Phloem sap** may have a sucrose concentration as high as 30% and may also contain minerals, amino acids, and hormones.

Phloem sap flows from a **sugar source,** where it is produced by photosynthesis or the breakdown of starch, to a **sugar sink,** an organ that consumes or stores sugar. The direction of transport in any one sieve tube depends on the location of the source and sink connected by that tube, and the direction may change with the season or the needs of the plant.

In some species, sugar in the leaf moves through the symplast of the mesophyll cells to sieve-tube elements. In other species, sugar first moves through the symplast and then into the apoplast in the vicinity of sieve-tube elements and companion cells, which actively accumulate sugar. In some plants, companion (transfer) cells have specialized ingrowths of the cell wall that increase surface area for movement of solutes.

Phloem loading requires active transport in plants that accumulate high sugar concentrations in the sieve tubes. Proton pumps and the cotransport of sucrose through cotransport proteins along with the returning protons is the mechanism used for active transport. In sink tissues, the concentration gradient favors the diffusion of sugar out of sieve tubes because sugar is either used or converted into starch within sink cells.

INTERACTIVE QUESTION   36.6

Label the components of the following diagram of the mechanism involved in actively transporting sucrose into companion cells or sieve-tube elements.

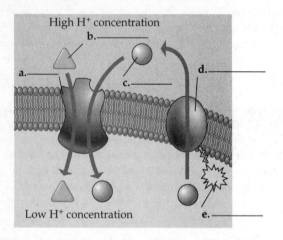

Bulk Flow by Positive Pressure: The Mechanism of Translocation in Angiosperms   The rapid movement of phloem sap from source to sink is driven by *pressure flow*. High solute concentration at the source lowers the water potential, and the resulting movement of water into the sieve tube produces positive pressure. At the sink end, the osmotic loss of water following the exodus of sucrose into the surrounding tissue results in a lower pressure. The difference in these pressures causes sap to move by bulk flow from source to sink. Innovative tests of this hypothesis using aphid stylets support it as the explanation for the flow of sap in the phloem of angiosperms.

In a phenomenon known as *self-thinning*, plants abort flowers, seeds, or fruits when they have more sinks than can be supported by sources.

36.6   The symplast is highly dynamic

Changes in Plasmodesmata   Recent studies have shown that plasmodesmata are dynamic structures that can open and close rapidly, increase in number or size,

and be eliminated as developmental needs change. Plant viruses produce *viral movement proteins* that mimic the cell's regulatory machinery that dilates plasmodesmata. Communication during plant development may occur within interconnected groups of cells called *symplastic domains*.

Phloem: An Information Superhighway   The systemic transport via phloem of certain macromolecules, such as flower-inducing signals or defense signals, helps integrate the functions of the whole plant.

Electrical Signaling in the Phloem   Some studies indicate that phloem can transmit electrical signals that affect gene transcription or other processes in other locations of the plant.

Word Roots

apo- = off, away; **-plast** formed, molded (*apoplast*: everything external to the plasma membrane of a plant cell, including cell walls, intercellular spaces, and the lumen of xylem vessels and tracheids)

aqua- = water; **-pori** a pore, small opening (*aquaporin*: a membrane channel protein that specifically facilitates osmosis, the diffusion of water across the membrane)

circa- = a circle (*circadian rhythm*: a physiological cycle of about 24 hours that persists even in the absence of external cues)

endo- = within, inner; **-derm** = skin (*endodermis*: the innermost layer of the cortex in plant roots)

gutt- = a drop (*guttation*: the exudation of water droplets from leaves, caused by root pressure in certain plants)

mega- = large, great (*megapascal*: a unit of pressure equivalent to about 10 atmospheres of pressure)

myco- = a fungus; **-rhizo** = a root (*mycorrhiza*: a mutualistic association of plant roots and a fungus)

osmo- = pushing (*osmosis*: the diffusion of free water across a selectively permeable membrane)

sym- = with, together (*symplast*: in plants, the continuum of cytoplasm connected by plasmodesmata between cells)

turg- = swollen (*turgor pressure*: the force directed against a plant cell wall after the osmotic influx of water causes the cell to swell)

xero- = dry; **-phyto** = a plant (*xerophyte*: a plant adapted to an arid climate)

Structure Your Knowledge

1. Describe the ways in which solutes may move across the plasma membrane in plants.
2. Both xylem sap and phloem sap move by bulk flow in an angiosperm. Compare and contrast the mechanisms for the movement of each.

Test Your Knowledge

MULTIPLE CHOICE: *Choose the one best answer.*

1. If a plant has a phyllotaxy of alternate leaves that emerge at a 30° angle from each previous leaf, one can surmise that
   a. the leaves are most likely very broad, indicating that the plant grows where there is sufficient water.
   b. the leaves are most likely very broad, indicating that the plant grows in low light levels.
   c. the leaves are most likely very thin, because emerging at such a close angle they would have to be thin to avoid shading each other.
   d. the leaves are most likely very thin, because the plant grows in a sunny, wet area.
   e. the leaves are most likely intermediate in size, but the leaf area index of the plant must be very high with leaves spaced so close together.

2. Which of the following structures are *not* components of the symplast?
   a. sieve-tube elements
   b. xylem tracheids
   c. endodermal cells
   d. root hairs
   e. both a and b

3. Proton pumps in the plasma membranes of plant cells may
   a. generate a membrane potential that helps drive cations into the cell through specific channels.
   b. be coupled to the movement of K^+ into guard cells.
   c. drive the accumulation of sucrose in sieve-tube elements.
   d. contribute to the movement of ions through a cotransport mechanism.
   e. be involved in all of the above.

4. If a turgid plant cell placed in a solution in an open beaker becomes flaccid, which of the following statements is true?
   a. The water potential of the cell was initially higher than that of the solution.
   b. The water potential of the cell was initially equal to that of the solution.

   c. The pressure (Ψ_P) of the cell was initially lower than that of the solution.
   d. The Ψ_S of the cell was initially more negative than that of the solution.
   e. Turgor pressure disappeared because the cell no longer needed support.

5. Which of the following mechanisms facilitates the movement of NO_3^- into epidermal cells of the root?
   a. bulk flow of water into the root
   b. cotransport through a membrane protein
   c. passage through selective channels, aided by the membrane potential created by proton pumps
   d. active transport through a nitrate pump
   e. simple diffusion across the cell membrane down its concentration gradient

6. Aquaporins are
   a. cytoplasmic connections between cortical cells.
   b. pores through the ends of sieve-tube elements through which phloem sap flows.
   c. openings in the lower epidermis of leaves through which water vapor escapes.
   d. openings into root hairs through which water enters.
   e. water-specific channels in membranes that speed the rate of osmosis.

7. The water potential of a plant cell
   a. is equal to 0 MPa when the cell is in pure water and is turgid.
   b. is equal to that of air.
   c. is equal to -0.23 MPa.
   d. becomes greater when K^+ are actively moved into the cell.
   e. becomes 0 MPa due to loss of turgor pressure in a concentrated sugar solution.

8. The Casparian strip prevents water and minerals from entering the vascular cylinder through the
   a. plasmodesmata.
   b. endodermal cells.
   c. symplast.
   d. apoplast.
   e. xylem vessels.

9. Which of the following does *not* increase the surface area of a root available for absorbing water and minerals?
   a. mycorrhizae
   b. numerous branch roots
   c. root hairs
   d. cytoplasmic extensions of tracheids
   e. the large surface area of cortical cells

10. Guttation results from
    a. the pressure flow of sap through phloem.
    b. a water vapor break in a column of xylem sap.
    c. root pressure causing water to flow up through xylem faster than it can be lost through transpiration.
    d. a higher water potential in the leaves than in the roots.
    e. specialized transport cell structures that accumulate sucrose.

11. The formation of a curvature in the air-water interface along the cell walls of cells lining the air space of a leaf contributes to water transport by
    a. creating a more positive water potential than in the surrounding mesophyll cells.
    b. creating tension, thereby lowering pressure and the water potential of the leaf.
    c. raising the water potential of the surrounding saturated air.
    d. increasing the adhesion of water molecules to the cell walls.
    e. increasing the rate of transpiration from the leaf.

12. Adhesion is a result of
    a. hydrogen bonding between water molecules.
    b. the pull on the water column as water evaporates from the surfaces of mesophyll cells.
    c. tension within the xylem caused by a negative pressure.
    d. attraction of water molecules to hydrophilic walls of narrow xylem tubes.
    e. the high surface tension of water.

13. Which of the following factors does *not* stimulate the opening of stomata?
    a. daylight in a CAM plant
    b. depletion of CO_2 in the air spaces of the leaf
    c. stimulation of proton pumps that result in the movement of K^+ into the guard cells
    d. circadian rhythm of guard cell opening
    e. an increase in the turgor of guard cells

14. Your favorite houseplant is wilting. Which of the following statements describes the most likely cause and remedy for its declining condition?
    a. Water potential is too low; apply sugar water.
    b. The plant's stomata won't open; no remedy is available.
    c. The plant's cells are undergoing plasmolysis; water it.
    d. Cavitation has occurred; perform a xylem vessel bypass.
    e. Circadian rhythm has caused the stomata to open; place the plant in the dark.

15. Given the current increase in atmospheric CO_2 levels, which of the following changes would you predict?
    a. an increase in the production of abscisic acid to signal stomatal closing in response to excess CO_2
    b. an increase in transpiration due to increased levels of photosynthesis
    c. a decrease in stomatal density due to increased CO_2 levels
    d. an increase in stomatal density to provide for increased levels of photosynthesis
    e. an increase in distribution of CAM plants, which more effectively take in CO_2 at night

16. Which of the following mechanisms explains the movement of sucrose from source to sink?
    a. evaporation of water and active transport of sucrose from the sink
    b. osmotic movement of water into the sucrose-loaded sieve-tube elements, creating a higher pressure in the source than in the sink
    c. tension created by pressure differences in the source and the sink
    d. active transport of sucrose through the sieve-tube cells driven by proton pumps
    e. the hydrolysis of starch to sucrose in mesophyll cells, which raises their water potential and drives the bulk flow of sap to the sink

17. Which of the following would most likely result in self-pruning or self-thinning?
    a. the transport of viral particles facilitated by viral movement proteins through plasmodesmata
    b. the conduction of electrical signals in phloem
    c. the shading of leaves to the extent that respiration exceeds photosynthesis
    d. a lack of sufficient products of photosynthesis in sources to supply all sinks
    e. both c and d

18. Which of the following statements is *not* correct concerning the dynamic nature of the symplast?
    a. Environmental stress may activate signal transduction pathways that alter membrane transport proteins.
    b. As a leaf matures from a sink to a source, certain of its plasmodesmata may close, thereby stopping phloem unloading.
    c. Electrical signals traveling through the phloem can affect rates of respiration and photosynthesis in other parts of the plant.
    d. Transpiration rates increase on warm, sunny days and decrease on cloudy, damp days.
    e. Proteins and RNAs may travel through symplastic domains to coordinate development within tissues.

Soil and Plant Nutrition

37.1 Soil contains a living, complex ecosystem

37.2 Plants require essential elements to complete their life cycle

37.3 Plant nutrition often involves relationships with other organisms

▮ Framework

The fertility of soil is influenced by its texture and composition. Good agricultural practices incorporate soil and water conservation. The nutritional requirements of plants include essential macronutrients and micronutrients. The rhizosphere includes bacteria and fungi that are supported by plant roots and may enhance plant growth. Plants assimilate nitrogen made available by the decomposition of humus and from nitrogen fixation by bacteria. Mycorrhizae are important associations between fungi and plant roots that increase mineral and water absorption.

▮ Chapter Review

37.1 Soil contains a living, complex ecosystem

Soil Texture The texture of soil depends on particle size, which varies from coarse sand to silt to fine clay. The formation of soil begins with the weathering of rock as frozen water in crevices mechanically fractures rock and soil acids break down rocks chemically. Plant roots accelerate these processes. **Topsoil** is a mixture of mineral particles from rock, **humus** (decomposing organic matter), and living organisms. Two other distinct soil layers, or **soil horizons,** are found under the topsoil layer.

Loams, containing almost equal parts of sand, silt, and clay, are often the most fertile soils, having enough fine particles to provide a large surface area for retaining water and minerals but enough coarse particles to provide air spaces for respiring roots.

Topsoil Composition Positively charged ions, such as K^+, Ca^{2+}, and Mg^{2+}, adhere to the negatively charged surfaces of soil particles and are not easily lost by *leaching*. In **cation exchange,** hydrogen ions displace positively charged mineral ions from soil particles, making the ions available for absorption. Negatively charged ions, such as nitrate (NO_3^-), phosphate ($H_2PO_4^-$), and sulfate (SO_4^{2-}), tend to leach away more quickly.

Humus builds a crumbly soil that retains water, provides good aeration of roots, and supplies mineral nutrients.

The activities of the numerous soil inhabitants—bacteria, fungi, algae, other protists, insects, worms, nematodes, and plant roots—affect the physical and chemical properties of soil.

INTERACTIVE QUESTION 37.1

Describe the characteristics of a fertile soil.

Soil Conservation and Sustainable Agriculture As ancient farmers learned to fertilize the soil, humans were able to establish permanent residences. Soil mismanagement, however, leads to low productivity.

Irrigation can make farming possible in arid regions, but it places a huge drain on surface and underground water resources and leads to soil *salinization*. Excess water removal from *aquifers* results in *land subsidence*, the sinking of Earth's surface. *Drip irrigation*, which slowly releases water to the root zone, reduces some of these problems.

Agriculture removes mineral nutrients from the soil; **fertilization** replaces them. In industrialized nations, commercially produced fertilizers (usually containing nitrogen, phosphorus, and potassium) are used. Such fertilizers may rapidly leach from the soil, polluting streams and lakes. Manure, fishmeal, and compost (so-called "organic" fertilizers) contain organic material that slowly decomposes into inorganic nutrients.

The acidity of the soil affects cation exchange. It can also alter the chemical form of minerals and thus their ability to be absorbed by the plant. Managing soil pH is an important aspect of maintaining fertility.

A large amount of topsoil is lost to water and wind erosion each year. Agricultural use of cover crops, windbreaks, and terracing can minimize erosion, as can **no-till agriculture,** which minimally disturbs the soil.

The use of plants to extract heavy metals and other pollutants from contaminated soils is an emerging technology known as **phytoremediation.**

INTERACTIVE QUESTION   37.2

a. What is **sustainable agriculture?**

b. What is the N-P-K ratio?

37.2   Plants require essential elements to complete their life cycle

Early scientists speculated on whether soil, water, or air provides the substance for plant growth. Inorganic nutrients absorbed from the soil account for only 4% of the dry mass of a plant. Water makes up 80–90% of the weight of a fresh plant and supplies most of the hydrogen incorporated into organic compounds. CO_2 from the air is the source of most of the organic material of a plant. Carbon, oxygen, and hydrogen—the components of carbohydrates such as cellulose—are the most abundant elements making up the dry weight of a plant. Nitrogen, sulfur, and phosphorus are also found in macromolecules and are relatively abundant in plants.

Macronutrients and Micronutrients   **Essential elements** are those required for a plant to complete its life cycle from a seed to an adult that produces more seeds. **Hydroponic culture** has been used to help identify the 17 elements that are essential in all plants.

Nine **macronutrients** are required by plants in relatively large amounts and include the six major elements of organic compounds as well as calcium, potassium, and magnesium.

Eight **micronutrients** are needed by plants in very small amounts, mainly as cofactors of enzymatic reactions.

INTERACTIVE QUESTION   37.3

List some functions of each of the following elements, and indicate whether each is a macronutrient or a micronutrient.

a. magnesium

b. phosphorus

c. iron

Symptoms of Mineral Deficiency   Mobile nutrients move to young, growing tissues, so deficiencies of those nutrients first appear in older parts of the plant. Symptoms of a given mineral deficiency may be distinctive enough to allow diagnosis by a plant physiologist or farmer. Soil and plant analysis can confirm a specific deficiency. Deficiencies of potassium, phosphorus, and especially nitrogen are most common.

INTERACTIVE QUESTION   37.4

a. What is *chlorosis?*

b. Where in a plant would you first expect to see a deficiency of a relatively immobile element?

Improving Plant Nutrition by Genetic Modification: Some Examples   Resistance to aluminum toxicity has been engineered into tobacco and papaya plants by introducing a citrate synthase gene from a bacterium. Citric acid secreted by roots binds to free aluminum ions, lowering the levels of toxic aluminum in the soil.

The gene *Submergence 1A-1* (*Sub1A-1*) codes for a protein that regulates genes normally activated under anaerobic conditions, such as occur during flooding. Increased expression of *Sub1A-1* in flooding-intolerant varieties of rice or other crops may enhance flood tolerance.

When a promoter that responds to a decrease in a nutrient is linked to a "reporter" gene in plants, a grower can tell when a deficiency of that nutrient is imminent and can then fertilize only in response to need.

37.3 Plant nutrition often involves relationships with other organisms

Plants form *mutualistic* relationships with many bacteria and fungi.

Soil Bacteria and Plant Nutrition The **rhizosphere** is the layer of soil surrounding plant roots that contains high populations of soil bacteria called **rhizobacteria,** which are supported by plant secretions.

Some rhizobacteria, called *plant-growth-promoting rhizobacteria,* enhance plant growth by producing growth-stimulating hormones or antibiotics, by absorbing toxins, or by making nutrients more available.

The **nitrogen cycle** describes the movement of nitrogen in an ecosystem, which involves soil microbes, plants, decomposers, and the atmosphere. *Ammonifying bacteria* decompose humus and form NH_3; *nitrogen-fixing bacteria* convert atmospheric N_2 to NH_3.

In the soil solution, ammonia forms ammonium (NH_4^+). Different types of *nitrifying bacteria* perform the two-step oxidization of ammonium to nitrate (NO_3^-), the form of nitrogen most readily absorbed by roots. Most plants incorporate nitrogen into amino acids or other organic compounds in the roots and then transport these compounds of NO_3^- through the xylem to shoots. Some nitrate is lost to the atmosphere by the action of denitrifying bacteria.

Some free-living soil bacteria and bacteria of the genus *Rhizobium,* which are closely associated with the roots of legumes, are able to convert atmospheric nitrogen into ammonia through the process of **nitrogen fixation.** *Nitrogenase* is an enzyme complex that reduces N_2 by adding H^+ and electrons to form NH_3. This process is energetically expensive, and nitrogen-fixing bacteria rely on decaying material or plant roots for carbohydrates.

INTERACTIVE QUESTION 37.5

Fill in the types of bacteria (a–d) that participate in the nitrogen nutrition of plants. Indicate the form (e) in which nitrogen is transported in xylem to the shoot system.

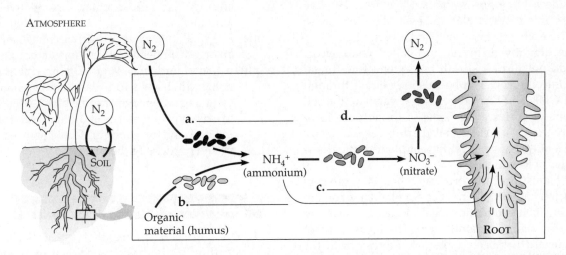

Plants of the legume family have root swellings, called **nodules,** composed of plant cells with vesicles containing bacteria in a form called **bacteroids.** Nitrogen fixation requires an anaerobic environment. Lignified external layers of nodule cells limit gas exchange. In some nodules, the presence of oxygen-binding leghemoglobin keeps the concentration of free O_2 low and regulates the oxygen supply for the bacterial respiration required to provide ATP for nitrogen fixation.

Mutual signaling between roots and bacteria leads to root hair elongation and formation of an infection thread within a root hair. Vesicles containing bacteroids bud from the infection thread into dividing cortical and pericycle cells. These dividing cells fuse to form the nodule, which grows and develops vascular tissue connecting with the vascular cylinder. Bacteria in the nodules are supplied with carbohydrates, and nodule cells use the fixed nitrogen to make amino acids, which are then transported throughout the plant.

Legume seeds are often coated with their specific *Rhizobium* before planting. Rice farmers culture a water fern that has symbiotic cyanobacteria that fix nitrogen, improving the fertility of rice paddies.

INTERACTIVE QUESTION   37.6

a. What is **crop rotation**?

b. What is "green manure"?

Fungi and Plant Nutrition   Most plants have symbiotic associations between their roots and fungi called **mycorrhizae.** The fungus receives sugar from the plant while providing a large surface area for the absorption of water and minerals (especially phosphate) for the plant. The fungus secretes growth hormones that stimulate root growth and branching and may help protect the plant from some soil pathogens.

Especially common in woody plants, **ectomycorrhizae** form a dense sheath or mantle of mycelium over the root surface. Hyphae extending from the mantle provide a huge absorptive surface area. Hyphae also grow into extracellular spaces in the root cortex, facilitating exchange between plant and fungus.

Arbuscular mycorrhizae, also called endomycorrhizae, do not form a sheath but extend fine fungal hyphae into the soil. Hyphae also penetrate through root cell walls and form tubes that invaginate the cell membrane of cortex cells. Some of these tubes form dense branched structures called arbuscles that facilitate nutrient exchange. Over 85% of plant species have arbuscular mycorrhizae.

Most plants form mycorrhizae when they grow in their natural habitat. When seeds are planted in foreign soil lacking their symbiotic fungal species, the resulting plants show signs of malnutrition. The invasive plant garlic mustard has been shown to interfere with the growth of arbuscular mycorrhizae on native trees.

INTERACTIVE QUESTION   37.7

Why is it thought that mycorrhizae were important in the colonization of land by the first plants?

Epiphytes, Parasitic Plants, and Carnivorous Plants
Epiphytes are plants that grow on the surface of another plant but do not take nourishment from it. Many parasitic plants produce roots that function as haustoria that siphon nutrients from their host plant.

Living in acid bogs or other nutrient-poor soils, carnivorous plants obtain nitrogen and minerals by killing and digesting insects and other small animals that are caught in various types of traps formed from modified leaves.

Word Roots

arbo- = tree; **-myco-** = a fungus; **-rhizo** = a root (*arbuscular mycorrhiza:* association of a fungus with a plant root system in which the fungus causes the invagination of the plant cells' plasma membranes; some branch densely, forming arbuscules or "little trees")

ecto- = outside (*ectomycorrhiza:* association of a fungus with a plant root system in which the fungus surrounds the roots but does not cause invagination of the plant cells' plasma membranes)

macro- = large (*macronutrient:* an essential element that an organism must obtain in relatively large amounts)

micro- = small (*micronutrient:* an essential element that an organism needs in very small amounts)

-phyto = a plant (*phytoremediation:* an emerging technology that seeks to reclaim contaminated areas by taking advantage of some plant species' ability to extract heavy metals and other pollutants from the soil and to concentrate them in easily harvested portions of the plant)

Structure Your Knowledge

1. Develop a concept map that organizes your understanding of the basic nutritional requirements of plants.
2. What are the differences between root nodules and mycorrhizae? How is each beneficial to plants?
3. What are the similarities between root nodules and mycorrhizae?

Test Your Knowledge

MULTIPLE CHOICE: *Choose the one best answer.*

1. The most fertile type of soil is usually
   a. sand, because its large particles allow room for air spaces.
   b. loam, which has a mixture of fine and coarse particles.
   c. clay, because the fine particles provide much surface area to which minerals and water adhere.
   d. humus, which is decomposing organic material.
   e. moist and alkaline.

2. An advantage of organic fertilizers over commercially produced fertilizers is that they
   a. are more natural.
   b. release their nutrients over a longer period of time and are less likely to be lost to runoff.
   c. provide nutrients in the forms most readily absorbed by plants.
   d. are easier to mass produce and transport.
   e. are all of the above.

3. Which of the following is an example of phytoremediation?
   a. dusting legume seeds with specific *Rhizobium* strains to increase nodule formation
   b. inoculating seeds with fungal spores to ensure mycorrhizal formation
   c. using plants to remove toxic heavy metals from contaminated soils
   d. genetic engineering of plants to increase protein content
   e. seeding the ocean with iron to create algal blooms that reduce atmospheric CO_2 levels

4. What created the Dust Bowl in the Great Plains in the 1930s?
   a. several years of drought
   b. building dams that removed the region's normal supply of water
   c. removal of prairie grasses to grow wheat and raise cattle
   d. the lack of crop rotation, which destroyed soil fertility
   e. Both a and c were important factors.

5. Negatively charged minerals
   a. are released from clay particles by cation exchange.
   b. are reduced by cation exchange before they can be absorbed.
   c. are converted into amino acids before they are transported through the plant.
   d. are bound when roots release acids into the soil.
   e. are leached away more easily than positively charged minerals.

6. The inorganic compound that contributes most of the mass to a plant's organic matter is
   a. H_2O.
   b. CO_2.
   c. NO_3^-.
   d. $H_2PO_4^-$.
   e. carbohydrate.

7. The effects of mineral deficiencies involving fairly mobile nutrients will first be observed in
   a. older portions of the plant.
   b. new leaves and shoots.
   c. the root system.
   d. the color of the leaves.
   e. the flowers.

8. Micronutrients
   a. may be cofactors in enzymes.
   b. are required in very minute quantities.
   c. may be components of cytochromes.
   d. can be identified using hydroponic culture techniques.
   e. All of the above are true.

9. Chlorosis is
   a. yellowing leaves due to decreased chlorophyll synthesis in response to a mineral deficiency.
   b. a plant's uptake of the micronutrient chlorine, facilitated by symbiotic bacteria.
   c. the production of chlorophyll within the thylakoid membranes of a plant.
   d. a contamination of glassware in hydroponic culture.
   e. a root mold caused by wet soil conditions.

10. The insertion of a *Submergence 1A-1* gene in rice plants
    a. is a form of phytoremediation that enables these genetically modified plants to remove toxic zinc from contaminated soils.
    b. increases the availability of nitrogen by promoting the growth of nitrogen-fixing cyanobacteria.
    c. enables these genetically modified plants to overproduce citric acid, which lowers the level of toxic free aluminum in the soil.
    d. increases the flood tolerance of these genetically modified rice varieties by increasing expression of genes activated under anaerobic conditions.
    e. allows more precise fertilizer use, as this gene serves as a reporter gene that signals a nutrient deficiency before damage has occurred.

11. Nitrogenase
    a. is an enzyme complex that reduces atmospheric nitrogen to ammonia.
    b. is found in *Rhizobium* and other nitrogen-fixing bacteria.
    c. catalyzes the energy-expensive fixation of nitrogen.
    d. requires an anaerobic environment.
    e. is or does all of the above.

12. Epiphytes
    a. have haustoria for anchoring to their host plants and obtaining water and nutrients.
    b. are symbiotic relationships between specialized leaves and specific fungi.
    c. live in poor soil and capture and digest insects to obtain nitrogen.
    d. grow on other plants but do not obtain nutrients from their hosts.
    e. house nitrogen-fixing bacteria within vesicles of their root cells.

13. The nitrogen content of some agricultural soils may be improved by
    a. the synthesis of leghemoglobin by ammonifying bacteria.
    b. mycorrhizae on legumes.
    c. crop rotation with legumes that house nitrogen-fixing bacteria.
    d. cation exchange.
    e. the action of denitrifying bacteria.

14. Which of the following structures takes the form of a mycelial sheath over plant roots, from which hyphae extend and increase the surface area for absorption?
    a. root nodules
    b. haustoria
    c. arbuscular mycorrhizae
    d. ectomycorrhizae
    e. infection threads

15. What result would you predict if a gardener sterilizes all of her soil in an effort to reduce plant diseases?
    a. decreased plant growth due to a decrease in cation exchange
    b. increased plant growth due to lack of soil microbes draining away the photosynthetic output of the plants
    c. decreased growth due to lack of water-retaining bacteria
    d. increased growth due to the destruction of disease-causing microbes
    e. overall decreased plant growth due to destruction of mutualistic associations with microbes in the rhizosphere

Angiosperm Reproduction and Biotechnology

Framework

This chapter describes the sexual and asexual reproduction of flowering plants. The flower produces spores that grow into the haploid gametophyte stages of the life cycle: Microspores in the anther develop into male gametophytes in pollen grains, and a megaspore in the ovule produces an embryo sac. Pollination and the double fertilization of egg and polar nuclei are followed by the development of a seed with an embryo and endosperm, protected in a seed coat and housed within a fruit. Seed dormancy is broken following proper environmental cues and the imbibition of water.

Vegetative propagation allows successful plants to clone themselves. Agriculture makes extensive use of this type of plant reproduction by using cuttings, grafts, and test-tube cloning.

Plant biotechnologists are creating genetically modified (GM or transgenic) plants that have such traits as insect or disease resistance, herbicide tolerance, and improved nutritional value. Opposition to the development of GM organisms focuses on human health concerns and the unknown dangers of introducing transgenic plants into the environment.

Chapter Review

Humans and their crop plants are an example of the mutualistic relationships common between plants and animals.

38.1 Flowers, double fertilization, and fruits are unique features of the angiosperm life cycle

Plants exhibit an alternation of generations between haploid (n) and diploid ($2n$) generations. The diploid sporophyte produces haploid spores by meiosis. Spores develop into multicellular haploid male and female gametophytes, which produce gametes by mitosis. **Fertilization** yields zygotes that grow into new sporophyte plants. In angiosperms, the male and female gametophytes have become reduced to only a few cells.

Flower Structure and Function Flowers are determinant, reproductive shoots that usually contain four whorls of modified leaves—**sepals, petals, stamens,** and **carpels**—which attach to a part of a stem called the **receptacle.** Sepals enclose and protect the unopened floral bud. Petals are generally more brightly colored and may attract pollinators. Stamens consist of a *filament* and an **anther,** which contains pollen sacs (microsporangia). A carpel consists of a sticky **stigma** at the top of a slender **style,** which leads to an **ovary.** The ovary encloses one or more **ovules.** A flower may have a single carpel or multiple fused carpels; either may be referred to as a **pistil.**

A **complete flower** has sepals, petals, stamens, and carpels. **Incomplete flowers** lack one or more of these floral organs. Some incomplete flowers may be sterile or *unisexual.* Floral variations include individual flowers or clusters of flowers called **inflorescences,** as well as diverse shapes, colors, and odors adapted to attracting different pollinators.

INTERACTIVE QUESTION   38.1

Identify the flower parts in the following diagram.

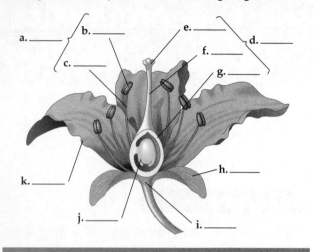

Within each microsporangium (pollen sac), diploid cells called *microsporocytes* undergo meiosis to form four haploid **microspores.** A microspore divides once by mitosis to produce a *tube cell* and a *generative cell,* which moves into the tube cell. The spore wall surrounding the cells thickens into the sculptured coat of the **pollen grain.** After the pollen grain lands on a receptive stigma, the tube cell begins to form the **pollen tube.** The generative cell divides to form two sperm cells. The pollen tube releases the sperm cells near the female gametophyte.

There are many variations in the development of the female gametophyte, also called an **embryo sac.** Two *integuments* surround each megasporangium except at the *micropyle.* The *megasporocyte* in the megasporangium of an ovule undergoes meiosis to form four haploid **megaspores,** only one of which survives. This megaspore grows and divides by mitosis three times, forming the female gametophyte, which typically consists of eight nuclei contained in seven cells. At the micropylar end of the embryo sac, an egg cell is lodged between two cells called synergids, which help attract the pollen tube. Three antipodal cells are at the other end; and two nuclei, called polar nuclei, are in a large central cell.

The transfer of pollen from an anther to a stigma, called **pollination,** may be accomplished by wind, water, or animals, including bees, moths and butterflies, flies, birds, or bats. Many species of flowering plants and their specific pollinators illustrate **coevolution,** the evolution of reciprocal adaptations in interacting species. For instance, natural selection favors traits that enhance flower-pollinator mutualism, in which flowers provide nectar to specific insects while those insects efficiently pick up pollen and transfer it to other flowers.

INTERACTIVE QUESTION   38.2

a. Describe the male gametophyte.

b. Describe the female gametophyte.

Double Fertilization   A pollen grain that lands on a receptive stigma absorbs moisture and germinates. The pollen tube grows through the style, and the generative cell divides to form two sperm. The pollen tube reaches the micropyle and releases its two sperm into the embryo sac. In **double fertilization,** one sperm fertilizes the egg to form the zygote, and the other combines with the polar nuclei to form a triploid nucleus, which will develop into a food-storing tissue called the **endosperm.**

As shown by *in vitro* studies, gamete fusion is immediately followed by an increase in cytoplasmic Ca^{2+} levels in the egg and the establishment of a block to *polyspermy.*

INTERACTIVE QUESTION   38.3

What function does double fertilization serve?

Seed Development, Form, and Function   Each ovule develops into a seed enclosed in the ovary, which develops into a fruit. The triploid nucleus formed by double fertilization divides to form the endosperm, a multicellular mass rich in nutrients that are stored for later use by the seedling. In many eudicots, the food reserves of the endosperm are transferred to the cotyledons before the seed matures.

In the zygote, the first mitotic division creates a basal cell and a terminal cell. The basal cell divides to produce a thread of cells, called the *suspensor,* which anchors the embryo and transfers nutrients to it. The terminal cell divides to form a spherical proembryo, on which the cotyledons (two in eudicots, one in monocots) begin to form. The embryo elongates, and apical meristems develop at the apexes of the embryonic shoot and root.

As it matures, the seed dehydrates and the embryo enters **dormancy.** The embryo and its food supply, the endosperm and/or enlarged cotyledons, are enclosed in a **seed coat** formed from the ovule integuments.

In a eudicot seed such as a bean, the embryo is an elongated embryonic axis attached to fleshy cotyledons. The axis below the cotyledonary attachment is called the **hypocotyl;** it terminates in the **radicle,** or embryonic root. The upper axis is the **epicotyl;** it leads to a shoot tip with a pair of leaves. Together, the epicotyl, young leaves, and shoot apical meristem are called the *plumule.*

The monocot seed found in grasses has a single thin cotyledon, called a *scutellum,* which absorbs nutrients from the endosperm during germination. A sheath called a **coleorhiza** covers the root, and a **coleoptile** encloses the young shoot.

INTERACTIVE QUESTION 38.4

Label the parts in the following diagrams of a bean seed and a corn seed.

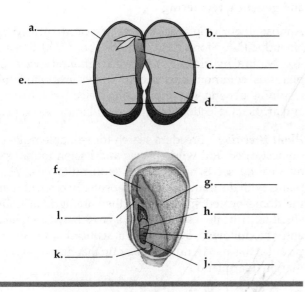

Dormancy increases the chances that the seed will germinate when and where the seedling has a good chance of surviving. The specific cues for breaking dormancy vary with the environment and may include heavy rain, intense heat from fires, cold, light, or chemical breakdown of the seed coat within an animal's digestive tract. The viability of a dormant seed may vary from a few days to decades or longer.

Imbibition, the absorption of water by the dry seed, causes the seed to expand, rupture its coat, and begin a series of metabolic changes. Enzymes digest stored compounds, and nutrients are sent to growing regions.

The radicle emerges from the seed first, followed by the shoot tip. In many eudicots, a hook that forms in the hypocotyl is pushed up through the ground, pulling the delicate shoot and cotyledons behind it. Light stimulates the straightening of the hook, and the first foliage leaves begin to perform photosynthesis.

INTERACTIVE QUESTION 38.5

How does the shoot tip break through the soil in maize and other grasses?

Fruit Form and Function The ovary of the flower develops into a **fruit,** which both protects and helps to disperse the seeds. Hormonal changes following fertilization cause the ovary to enlarge; its wall becomes the pericarp, or thickened wall of the fruit. Fruit usually does not set if a flower has not been pollinated.

A **simple fruit** is derived from a single carpel or several fused carpels; an **aggregate fruit** results from a flower with more than one separate carpel; a **multiple fruit** develops from an inflorescence whose many ovaries fuse together to become one fruit. Other floral parts (such as the receptacle) may develop into a fruit. These fruits—apples, for example—are called **accessory fruits.**

Fruits usually ripen as the seeds are completing their development.

INTERACTIVE QUESTION 38.6

a. List some ways in which fruits aid in seed dispersal.

b. What changes usually occur when a fleshy fruit ripens?

38.2 Flowering plants reproduce sexually, asexually, or both

Mechanisms of Asexual Reproduction Many plant species produce genetically identical copies (clones) of themselves through **asexual reproduction.** Asexual reproduction is an extension of the indeterminate growth of plants in which meristematic tissues can grow indefinitely and parenchyma cells can divide and differentiate into specialized cells. A common type of vegetative reproduction is **fragmentation,** the formation of whole plants from parts of a parent plant. In some species, the root system gives rise to many adventitious shoots that develop into a clone with separate shoot systems. Some plants, such as dandelions, can produce seeds asexually, a process called **apomixis.**

Advantages and Disadvantages of Asexual Versus Sexual Reproduction   Advantages of asexual or **vegetative reproduction** include the production of clones of plants that are well suited to a certain environment and of progeny that are usually stronger as seedlings. Sexual reproduction generates variation in a population, an advantage when the environment changes. Seeds, which are almost always produced sexually, provide a means of dispersal to new locations and dormancy during harsh conditions.

INTERACTIVE QUESTION   38.7

a. What is an advantage of apomixis?

b. Why is "selfing" a desirable trait in plants such as garden peas?

Mechanisms That Prevent Self-Fertilization   Self-fertilization may be prevented by temporal or structural adaptations in flowers. In **dioecious** species, flowers are either staminate or carpellate and cannot self-fertilize because they are on separate plants.

The ability of flowers to reject their own pollen or that of closely related individuals, called **self-incompatibility,** depends on genes called *S*-genes. A plant population may have dozens of alleles of the *S*-gene. Gametophytic self-incompatibility depends on the *S*-allele of the pollen grain. RNA-hydrolyzing enzymes produced by the style destroy the RNA of developing pollen tubes that have a matching allele. In sporophytic self-incompatibility, self-recognition activates a signal-transduction pathway in stigma cells (with the sporophyte *S*-alleles) that blocks pollen germination.

Further research on the molecular basis of self-incompatibility may allow plant breeders to manipulate crop species to assure hybridization.

Vegetative Propagation and Agriculture   New plants may develop from stem cuttings when adventitious roots develop from a mass of dividing cells, called a **callus,** that forms at the cut end of the shoot. Adventitious roots can also form from a node in the shoot fragment.

Twigs or buds of one plant can be grafted onto a plant of a different variety or closely related species. The plant that provides the root system is called the **stock,** and the twig is called the **scion.** Grafting can combine the best qualities of different plants.

In test-tube cloning, whole plants can develop from pieces of tissue from the parent plant. A single plant can be cloned into thousands of plants by subdividing the undifferentiated calluses as they grow in tissue culture. Stimulated by proper hormone balances, calluses sprout shoots and roots and develop into plantlets, which can be transferred to soil to develop.

Foreign DNA may be inserted into individual plant cells, which then grow into **transgenic** or genetically modified (GM) plants using *in vitro* culture.

A technique called **protoplast fusion** is used to create new plant varieties. Protoplasts are cells whose cell walls have been enzymatically removed. Protoplasts from different species can be fused and cultured to form hybrid plantlets.

38.3   Humans modify crops by breeding and genetic engineering

Almost all our crop species were first domesticated by Neolithic (late Stone Age) humans about 10,000 years ago. Natural hybridization between different species of plants is common, and humans have exploited the resulting genetic variation using selective breeding and artificial selection to develop and improve crops.

Plant Breeding   Breeders search for valuable traits in domesticated and wild species and hasten mutations by treating seeds or seedlings with mutagens. Wild plants with desirable traits are repeatedly crossed with the domesticated species until they are agriculturally suitable. Triticale resulted from a cross between wheat and rye, followed by extensive artificial selection to produce this hardy, nutritious grain that grows well on marginal farmland.

Plant Biotechnology and Genetic Engineering   Plant biotechnology refers both to the age-old use of plants to make products for human use and to the use of GM organisms in agriculture. Genetic engineering allows plant biotechnologists to transfer genes between species.

The serious malnutrition of millions of people may be due to inequities in distribution of food resources or may signal that the world is already overpopulated. Plant biotechnology can help increase crop yields.

The use of transgenic crops has increased dramatically. Cotton, maize, and potatoes have been engineered to contain genes from *Bacillus thuringiensis* that code for *Bt* toxin, reducing the need for spraying chemical insecticides. Researchers have developed other transgenic crops that are resistant to some herbicides, allowing farmers to "weed" crops without heavy tillage. Some transgenic plants are more resistant to disease or have improved nutritional quality.

Biomass from fast-growing plants, such as switchgrass and poplars, may help meet the world's energy demands. Enzymatic reactions break down plant cell walls into sugars, which are fermented and distilled to yield **biofuels.**

INTERACTIVE QUESTION 38.8

Biofuels are considered carbon neutral, even though the burning of biofuels releases CO₂, just as the burning of fossil fuels does. Explain.

The Debate over Plant Biotechnology GM organisms (GMOs) may present an unknown risk to human health or the environment. One concern is the transfer of allergens to a food source. Some GM crops may be safer than regular crops; *Bt* maize contains lower levels of a harmful toxin called fumonisin, which is produced by a fungus that infects insect-damaged crops.

Many ecologists are concerned about the effects of GM crops on nontarget organisms. All GM crops need to be field tested for nontarget effects.

A serious concern is the escape of herbicide- or disease-resistance genes through crop-to-weed hybridization that may create "superweeds." Various techniques are being developed to reduce the ability of transgenic crops to hybridize, such as developing male sterility so that transgenic pollen is not produced; engineering apomixis into the crop; introducing genes into chloroplast DNA so that those genes are not present in pollen; and developing flowers that self-pollinate and do not open so that pollen cannot escape.

Word Roots

a- = without; **-pomo** = fruit (*apomixis:* the ability to reproduce asexually through seeds without fertilization)

anth- = a flower (*anther:* the terminal pollen sac of a stamen, where pollen grains containing sperm-producing male gametes form in an angiosperm)

carp- = a fruit (*carpel:* the ovule-producing reproductive organ of a flower, consisting of the stigma, style, and ovary)

coleo- = a sheath; **-rhiza** = a root (*coleorhiza:* the covering of the young root of the embryo of a grass seed)

di- = two (*dioecious:* having male and female reproductive parts on different plants of the same species)

dorm- = sleep (*dormancy:* a condition typified by extremely low metabolic rate and a suspension of growth and development)

endo- = within (*endosperm:* a nutrient-rich tissue, formed by the union of a sperm with two polar nuclei during double fertilization, that provides nourishment to the developing embryo in angiosperm seeds)

epi- = on, over (*epicotyl:* the embryonic axis above the point of attachment of the cotyledon(s) and below the first pair of miniature leaves)

hypo- = under (*hypocotyl:* the embryonic axis below the point of attachment of the cotyledon(s) and above the radicle)

mega- = large (*megaspore:* a spore that develops into a female gametophyte)

micro- = small (*microspore:* a spore that develops into a male gametophyte)

proto- = first; **-plast** = formed, molded (*protoplast:* the contents of a plant cell exclusive of the cell wall)

stam- = standing upright (*stamen:* the pollen-producing reproductive organ of a flower, consisting of an anther and filament)

Structure Your Knowledge

1. Draw a diagram of the major events in the life cycle of an angiosperm.
2. List the advantages and disadvantages of sexual and asexual reproduction in plants.
3. List some of the potential benefits and dangers of plant biotechnology.

Test Your Knowledge

FILL IN THE BLANKS

_____ 1. structure from which fruit typically develops

_____ 2. generation that produces spores by meiosis

_____ 3. fuel produced from organic matter

_____ 4. female gametophyte of angiosperms

_____ 5. embryonic root

_____ **6.** embryonic axis above attachment point of cotyledon

_____ **7.** protects a grass shoot as it breaks through the soil

_____ **8.** twig or stem portion of a graft

_____ **9.** plant cell from which the cell wall is removed

_____ **10.** mass of dividing cells at the cut end of a shoot

MULTIPLE CHOICE: *Choose the one best answer.*

1. A flower on a dioecious plant is
   a. complete.
   b. biennial.
   c. incomplete.
   d. staminate or carpellate.
   e. both c and d.

2. Which of the following structures is haploid?
   a. embryo sac
   b. anther
   c. endosperm
   d. microsporocyte
   e. both a and b

3. The terminal cell of an early plant embryo
   a. develops into the shoot apex of the embryo.
   b. forms the suspensor that anchors the embryo and transfers nutrients.
   c. develops into the endosperm when fertilized by a sperm nucleus.
   d. divides to form the proembryo.
   e. develops into the cotyledons.

4. In angiosperms, sperm are formed by
   a. meiosis in the anther.
   b. meiosis in the pollen grain.
   c. mitosis in the anther.
   d. mitosis in the pollen tube.
   e. double fertilization in the embryo sac.

5. The endosperm
   a. may have its nutrients absorbed by the cotyledons in the seeds of eudicots.
   b. is usually a triploid tissue.
   c. is digested by enzymes in monocot seeds following hydration.
   d. develops in concert with the embryo as a result of double fertilization.
   e. is or does all of the above.

6. A seed consists of
   a. an embryo, a seed coat, and a nutrient supply.
   b. an embryo sac.
   c. a gametophyte and a nutrient supply.
   d. an enlarged ovary.
   e. an immature ovule.

7. Which of the following structures protects a bean shoot as it breaks through the soil?
   a. hypocotyl hook
   b. radicle
   c. coleoptile
   d. coleorhiza
   e. seed coat

8. Which of the following processes is a form of a sexual or vegetative reproduction?
   a. apomixis
   b. grafting
   c. test-tube cloning
   d. fragmentation
   e. All of the above are forms of asexual reproduction.

9. Protoplast fusion
   a. is used to study the fertilization of a plant egg by a sperm.
   b. is the method used to produce test-tube plantlets.
   c. can be used to form new plant species.
   d. occurs within a callus.
   e. occurs following fertilization.

10. In the plant embryo, the suspensor
   a. is produced by vertical mitotic divisions within the proembryo.
   b. connects the early root and shoot apexes.
   c. develops into the endosperm.
   d. is analogous to the umbilical cord in mammals.
   e. is the point of attachment of the cotyledons.

11. Which of the following statements is *false*?
   a. Fly-pollinated flowers may smell like rotten meat.
   b. Bird-pollinated flowers are usually light-colored, strong smelling, and produce copious amounts of nectar.
   c. Insects pollinate approximately 65% of all flowering plants, and bees are the most important of these pollinators.
   d. The flowers of many temperate, wind-pollinated trees appear before the production of leaves, which could interfere with pollen movement.
   e. Moths and bats, which are often active at night, are attracted to light-colored, fragrant flowers.

12. Which of the following is an example of co-evolution?
    a. the dispersal of the sharp fruits of the puncture vine by animals
    b. the dispersal of seeds of edible fruits in animal feces and the germination of those seeds following the chemical breakdown of the seed coat in the animal's digestive tract
    c. the pollination of a yucca plant by a single species of moth, which has appendages that pack pollen onto the stigma of the plant and then deposits its eggs in the flower's ovaries, where the larvae eat some developing seeds
    d. the pollination of carrion flowers by blowflies that mistake the flower for rotting meat and deposit their eggs on it while becoming dusted with pollen that they transport to other flowers
    e. "nectar guides" that help bees (which can see ultraviolet light) locate the nectarines of flowers

13. Self-incompatibility provides a plant
    a. a means of transferring pollen to another plant.
    b. a means of coordinating the fertilization of an egg with the development of stored nutrients.
    c. a means of destroying foreign pollen before it fertilizes the egg cell.
    d. a biochemical block to self-fertilization so that cross-fertilization is assured.
    e. a means of producing seeds without the need for fertilization.

14. Which of the following statements about fruit is *not* true?
    a. The primary functions of fruits include protection of dormant seeds and their dispersal away from the parent plant.
    b. An accessory fruit develops from the endosperm of a seed and is composed of triploid cells.
    c. Wind-dispersed fruits may have wings or "parachutes."
    d. Animals may help disperse seeds by "planting" fruits in underground caches.
    e. Most fruits develop from the ovary of a flower.

15. The formation of clones when multiple shoots emerge from a single root system benefits plant by
    a. providing a strong start for new plants.
    b. dispersing offspring to new habitats.
    c. providing a period of dormancy until specific environmental cues signal regrowth.
    d. providing a strong root system for genetically diverse shoot systems.
    e. allowing the production of seeds without fertilization.

16. Which of the following techniques is being developed to reduce the likelihood that introduced genes for herbicide or insect resistance escape to closely related weed species?
    a. engineering flowers to self-fertilize and remain closed so that pollen is not released
    b. breeding male sterility into transgenic plants so that they have no pollen to be transferred to nearby weeds
    c. engineering the gene of interest into chloroplast DNA, which is inherited from the maternal plant and is not transferred by pollen
    d. engineering crops, such as soybeans, that have no weedy relatives
    e. All of the above would reduce the risk of crop-to-weed transgene escape.

Chapter 39

Plant Responses to Internal and External Signals

Framework

Environmental stimuli and internal signals are linked by signal transduction pathways to cellular responses such as changes in gene expression and activation of enzymes. Plant hormones—auxin, cytokinins, gibberellins, brassinosteroids, abscisic acid, strigolactones, and ethylene—control growth, development, flowering, and senescence, as plants respond and adapt to their environments. Plant movements in response to environmental stimuli include phototropism, gravitropism, and thigmotropism. The biological clock of plants controls circadian rhythms, such as stomatal opening and sleep movements. Blue-light photoreceptors are involved in phototropism and stomatal opening. Phytochromes absorb red and far-red light and are involved in the photoperiodic control of flowering. Plants have various physiological responses to environmental stresses and pathogens.

Chapter Review

Plants are able to sense and adaptively respond to their environments, generally by altering their patterns of growth and development.

39.1 Signal transduction pathways link signal reception to response

The growth pattern of a sprouting potato shoot growing in darkness, called **etiolation,** facilitates the shoot breaking ground. When the shoot reaches sunlight, stem elongation slows, leaves expand, the root system elongates, and chlorophyll production begins—all part of a process known as **de-etiolation** or greening.

Reception Signals are detected by receptors, proteins that change shape in response to a specific stimulus. A *phytochrome* is the photoreceptor involved in de-etiolation, and, unlike most receptors built into the plasma membrane, it is located in the cytosol. Researchers have studied the role of phytochrome in de-etiolation using the tomato mutant *aurea*, which has lower than normal levels of phytochrome.

Transduction **Second messengers** are small molecules that amplify the signal from the receptor and transfer it to proteins that produce the specific response. Phytochrome signal transduction opens calcium channels. Increases in the second messenger Ca^{2+} activate certain protein kinases. The second messenger cyclic GMP activates other protein kinases. Both these pathways affect gene expression and are necessary for the de-etiolation response.

Response The response to a signal transduction pathway usually involves the activation of specific enzymes, either by modification of existing enzymes or by transcriptional regulation.

Post-translational modification of existing proteins usually involves phosphorylations, catalyzed by protein kinases. Cascades of protein kinase activations may lead to the phosphorylation of transcription factors, and thus, ultimately, to changes in gene expression. Protein phosphatases are enzymes that dephosphorylate specific proteins, allowing signal pathways to be turned off when a signal is no longer present.

Several *specific transcription factors*, which bind to DNA and control the transcription of specific genes, are activated during phytochrome-induced de-etiolation. Their activation results from their phosphorylation by protein kinases activated by cGMP or Ca^{2+}. Either positive or negative transcription factors may be activated, leading to changes in gene expression.

INTERACTIVE QUESTION 39.1

Answer the following questions concerning the process known as de-etiolation.

a. What is the signal, and what is the receptor?

b. What are the important steps in the transduction of this signal?

c. What is the plant's growth response? What types of proteins are involved?

39.2 Plant hormones help coordinate growth, development, and responses to stimuli

Hormones, chemical signals transported throughout the plant body, trigger responses in target cells and tissues. Some signal molecules act only locally, and others occur at much higher concentrations than a typical hormone. Thus, many plant biologists use the term *plant growth regulators* to describe compounds that affect specific physiological processes in plants.

The Discovery of Plant Hormones   A **tropism** is a growth response of plant organs toward or away from a stimulus. The growth of a shoot toward light is called positive **phototropism.** A shoot bends toward the light when illuminated from one side because of the elongation of cells on the darker side.

Darwin and his son observed that a grass seedling enclosed in its coleoptile would not bend toward light if its tip were removed or covered by an opaque cap. They postulated that a signal must be transmitted from the tip to the elongating region of the coleoptile. P. Boysen-Jensen demonstrated that the signal was a mobile substance, capable of being transmitted through

a block of gelatin separating the tip from the rest of the coleoptile. In 1926, F. Went concluded that the chemical produced in the tip, which he called auxin, promoted growth and that it was in higher concentration on the side away from the light.

Researchers have not found a light-induced asymmetrical distribution of auxin in eudicots, but certain substances that may act as growth inhibitors have been shown to be more concentrated on the lighted sides of such stems.

A Survey of Plant Hormones   Several major classes of plant hormones have been identified. Very low concentrations of hormones, acting through signal transduction pathways, may control plant growth by affecting cell division, elongation, and differentiation. The effect of a hormone varies depending on its relative concentration, the site of action, and the developmental stage of the plant.

Auxin refers to any substance that stimulates elongation of coleoptiles. The major auxin extracted from plants is indoleacetic acid (IAA). Auxin is transported from the shoot tip apical meristem (where it is synthesized) down the shoot. This *polar transport* involves auxin transporters located only at the basal ends of cells.

INTERACTIVE QUESTION 39.2

Describe the *acid growth hypothesis* for the elongation of cells in response to auxin.

Auxin stimulates a plasma membrane's **a.** _____, resulting in an increased membrane potential and a **b.** _____ pH in the cell wall. This activates enzymes called **c.** _____, which break the cross-links between cellulose microfibrils. The increased membrane potential enhances ion uptake, resulting in the **d.** _____. Thus, the two factors that cause the cell to elongate are **e.** _____ and **f.** _____

For continued growth after this relatively fast initial elongation due to the osmotic uptake of water, the cell must produce more cytoplasm and wall material, processes that rely on changes in gene expression that are also stimulated by auxin.

Auxin is important in the *pattern formation* of a developing plant. The polar transport of auxin, produced by shoot tips, controls branching patterns. A reduced flow from a branch releases lateral buds from dormancy. Polar auxin transport also helps establish *phyllotaxy,* the arrangement of leaves. Peaks in auxin establish the sites of leaf primordials. Auxin flowing from the leaf margin also affects leaf venation patterns. Auxin stimulates the vascular cambium in secondary growth.

Recent evidence indicates that an auxin gradient also regulates the organization of the microscopic female gametophyte.

The auxin IBA affects lateral root formation and is used commercially to enhance formation of adventitious roots at the cut base of stems. Synthetic auxins, such as 2,4-D, are used as herbicides that kill eudicot (broadleaf) weeds with a hormonal overdose. Auxin produced by developing seeds promotes fruit growth; synthetic auxins can induce fruit development in greenhouses.

Cytokinins—so named because they stimulate cytokinesis—are modified forms of adenine. Zeatin is the most common naturally occurring cytokinin. Cytokinins are produced in actively growing roots, embryos, and fruits. Acting with auxin, they stimulate cell division and affect differentiation.

According to the direct inhibition hypothesis, the control of apical dominance involves the interaction between auxin, which is transported down from the terminal bud and restrains axillary bud development, and cytokinins, which are transported up from the roots and stimulate bud growth. It now appears, however, that the polar flow of auxin stimulates the production of strigolactone, a hormone that represses bud growth.

Cytokinins can retard aging of some plant organs, because they stimulate RNA and protein synthesis, mobilize nutrients, and inhibit protein breakdown and **apoptosis,** a type of programmed cell death.

In the 1930s Japanese scientists determined that the fungus *Gibberella* secretes a chemical that causes the hyperelongation of rice stems, or "foolish seedling disease." More than 100 different naturally occurring **gibberellins** (GAs) have now been identified.

Gibberellins, which are produced by roots and young leaves, stimulate growth by both cell elongation and division. Gibberellins may promote elongation by stimulating enzymes that loosen cell walls, facilitating the entry of expansin proteins into the cell wall.

Bolting, the growth of an elongated floral stalk, is caused by a surge of gibberellins. In many plants, both auxin and gibberellins contribute to fruit set. Gibberellins are applied in the production of Thompson seedless grapes. The release of gibberellins from the embryo signals seeds of many plants to break dormancy.

Brassinosteroids are steroids that have effects very similar to those of auxin: They promote cell elongation and division, retard leaf abscission, and promote xylem differentiation. Identification of a brassinosteroid-deficient mutant of *Arabidopsis* helped to establish these compounds as nonauxin plant hormones.

The hormone **abscisic acid (ABA)** generally slows growth. The high concentration of ABA in maturing seeds inhibits germination and stimulates production of proteins that protect the seeds during dehydration.

For dormancy to be broken in some seeds, ABA must be removed or inactivated, or the ratio of gibberellins to ABA must increase.

ABA also reduces drought stress. In a wilting plant, ABA causes stomata in the leaves to close. ABA may be produced in the roots in response to water shortage and then be transported to the leaves.

Strigolactones stimulate seed germination, enhance mycorrhizal associations, and repress the growth of lateral buds in apical dominance. They were first identified as signals, which are released by roots, that stimulate germination of seeds of *Striga,* rootless parasitic plants that are serious agricultural pests in Africa.

Plants produce the gas **ethylene** in response to stress and during fruit ripening and programmed cell death. Ethylene production may be induced by a high concentration of auxin.

The mechanical stress in a seedling pushing against an obstacle as it grows upward through the soil induces the production of ethylene. Ethylene then initiates a growth pattern called the **triple response,** consisting of a slowing of stem elongation, a thickening of the stem, and initiation of horizontal growth. As the effects of ethylene lessen, normal upward growth resumes. Researchers have identified *Arabidopsis* mutants that are ethylene insensitive *(ein),* ethylene overproducing *(eto),* and that undergo the triple response in the absence of ethylene. In these latter constitutive triple-response *(ctr)* mutants, the ethylene signal transduction pathway is permanently turned on, even though ethylene is not present. This mutant gene codes for a protein kinase, suggesting that the normal kinase product is a negative regulator of ethylene signal transduction. Binding of ethylene to the ethylene receptor may lead to the inactivation of the negative kinase, which allows the synthesis of the proteins involved in the triple response.

Senescence is the programmed death of plant organs or the whole plant. Apoptosis requires the synthesis of new enzymes that break down many cellular components, which the plant may salvage. Ethylene is almost always associated with this programmed cell death.

Deciduous leaf loss protects against winter desiccation. Before leaves abscise in the autumn, many of the leaves' compounds are stored in the stem. A change in the balance between auxin and ethylene initiates changes in the abscission layer located near the base of the petiole, including the production of enzymes that hydrolyze polysaccharides in cell walls.

Ethylene initiates the breakdown of cell walls and the conversion of starches to sugars associated with fruit ripening. In an example of positive feedback, ethylene triggers ripening, and ripening triggers more ethylene production. Many commercial fruits are ripened in huge containers perfused with ethylene gas.

Fill in the name of the hormone that is responsible for each of the following functions:

a. _____ promotes fruit ripening; initiates the triple response; is involved in apoptosis

b. _____ stimulate cell division, growth, and germination; are anti-aging compounds

c. _____ inhibits growth; maintains dormancy; closes stomata during water stress

d. _____ stimulates stem elongation, root branching, and fruit development; is involved in apical dominance

e. _____ promote cell elongation and division, and xylem differentiation; also retard leaf abscission

f. _____ stimulate seed germination; help establish mycorrhizal associations; repress lateral bud growth

g. _____ promote stem elongation and seed germination; contribute to fruit set

Systems Biology and Hormone Interactions Systems biology studies properties that emerge from the interactions of many system elements. Now that many plant genomes have been sequenced and all the genes in a plant can be identified, scientists can use microarray and proteomic techniques to determine which genes are activated or inactivated in response to environmental stimuli or during development. A systems-based approach may enable biologists to determine the timing and sites of gene activations that produce plant responses, and to predict the results of genetic manipulations.

39.3 Responses to light are critical for plant success

The effect of light on plant morphology is called **photomorphogenesis.** An **action spectrum** is a graph of a physiological response across different wavelengths of light. An absorption spectrum shows the wavelengths of light a pigment absorbs. Close correlation between an action spectrum for a plant response and the absorption spectrum of a pigment may indicate that the pigment is the photoreceptor

involved in the response. **Blue-light photoreceptors** and **phytochromes,** which absorb mostly red light, are the two major classes of light receptors.

Blue-Light Photoreceptors Molecular biologists have determined that plants use as many as three different types of pigments to detect blue light: *cryptochrome* (in inhibition of hypocotyl elongation when a seedling breaks ground), *phototropin* (in phototropism), and *zeaxanthin* or perhaps phototropin (in stomatal opening).

Phytochromes as Photoreceptors Studies of lettuce seed germination in the 1930s determined that red light (wavelength: 660 nm) increased germination the most, and that far-red light (730 nm) inhibited germination. The effects of red and far-red light are reversible, and a seed's response is determined by the last flash of light it receives.

A phytochrome is a photoreceptor that consists of a protein with two subunits, each of which has two domains: one that has kinase activity, and one that is bonded to a nonprotein light-absorbing *chromophore*. The chromophore alternates between two isomers; one absorbs red light, and the other, far-red light. These two forms of phytochrome are photoreversible; the P_r to P_{fr} interconversion acts as a switch that controls various events in the life of the plant.

The signal that tells the plant that sunlight is present is the conversion of P_r (the phytochrome form the plant synthesizes) to P_{fr}. Subsequently P_{fr} triggers many plant responses to light, such as breaking seed dormancy.

Explain how phytochrome can also indicate the *quality* of light available to a plant. What effect might this have on a shaded tree?

Biological Clocks and Circadian Rhythms A **circadian rhythm** is a physiological cycle with about a 24-hour frequency. These rhythms persist, even when the organism is sheltered from environmental cues. Research indicates that the circadian clock is internal, although it is set (entrained) to a 24-hour period by daily environmental signals.

INTERACTIVE QUESTION   39.5

What are free-running periods?

The molecular mechanism of the biological clock may be cyclical changes in the concentration of transcription factors that inhibit, after a time delay, the expression of their own genes.

Researchers have identified clock mutants of *Arabidopsis* by splicing the gene for luciferase to the promoter for genes for cyclically produced photosynthesis-related proteins. When the biological clock turned on the promoter for these genes, the plants glowed. Plants that glowed for an abnormal amount of time were clock mutants. Some of these mutants had defects in proteins that normally bind photoreceptors, suggesting that a light-dependent mechanism sets the biological clock.

The Effect of Light on the Biological Clock   Both blue-light photoreceptors and phytochrome can entrain the biological clock in plants. In darkness, the phytochrome ratio shifts toward P_r, in part because P_{fr} converts slowly to P_r in some plants, and in part because P_{fr} is degraded and new pigment is synthesized as P_r. When the sun rises, P_r is rapidly converted to P_{fr}, resetting the biological clock each day at dawn.

Photoperiodism and Responses to Seasons   Seasonal events in the life cycle of plants usually are cued by photoperiod, the relative length of night and day. A physiological response to photoperiod is called **photoperiodism.**

Researchers discovered that a variety of tobacco plant flowered only when day length was 14 hours or shorter. They termed it a **short-day plant. Long-day plants** flower when days are longer than a certain number of hours. The flowering of **day-neutral plants** is unaffected by photoperiod.

Researchers have found that night length controls flowering and other photoperiod responses. If the dark period is interrupted by even a few minutes of light, a short-day (long-night) plant will not flower. Photoperiodic responses thus depend on a critical night length.

Red light was found to be the most effective in interrupting a plant's perception of night length. A flash of red light breaks a dark period of sufficient length and prevents short-day plants from flowering, whereas a flash of red light during a dark period that is longer than the critical length will induce flowering in a long-day plant.

Some plants bloom after a single exposure to the required photoperiod. Others respond to photoperiod only after exposure to another environmental stimulus. **Vernalization** is the need for pretreatment with cold to induce flowering.

Leaves detect the photoperiod. The signal for flowering, called **florigen,** travels from leaves to buds. Recent research indicates that the gene *FLOWERING LOCUS T* (*FT*) produces the FT protein in the leaf; the protein then travels to the shoot meristem and initiates flowering.

In order for flowering to occur, the bud meristem must transition from a vegetative state to a flowering state, a change that requires first the activation of meristem-identity genes followed by activation of organ-identity genes.

INTERACTIVE QUESTION   39.6

Indicate whether a short-day plant (a–e) and a long-day plant (f–j) would flower or would not flower under the indicated light conditions.

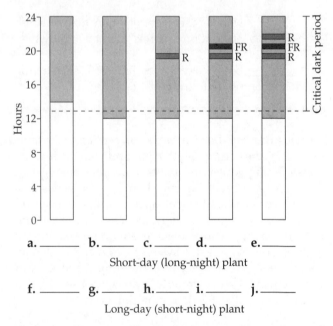

a. ____ b. ____ c. ____ d. ____ e. ____

Short-day (long-night) plant

f. ____ g. ____ h. ____ i. ____ j. ____

Long-day (short-night) plant

How do these results demonstrate the red/far-red photoreversibility of phytochrome?

39.4   Plants respond to a wide variety of stimuli other than light

Gravity   Roots exhibit positive **gravitropism,** whereas shoots show negative gravitropism. According to one hypothesis, the settling of **statoliths**—plastids containing dense starch grains—in cells of the root cap triggers movement of calcium, which causes the lateral transport of auxin. Accumulating on the lower side of the growing root, auxin inhibits cell elongation, causing the root to curve downward. The settling of the protoplast and large organelles may distort the cytoskeleton and also signal gravitational direction.

Mechanical Stimuli   The stunting of growth in height and increase in girth of plants that are exposed to wind is an example of **thigmomorphogenesis,** changes in plant form in response to mechanical stimulation.

Most climbing plants have tendrils that coil rapidly around supports, exhibiting **thigmotropism** or directional growth in response to touch.

The sensitive plant *Mimosa* folds its leaves after being touched due to the rapid loss of turgor by cells in specialized motor organs called pulvini, located at the joints of the leaf. The message travels through the plant from the point of stimulation, perhaps as the result of electrical impulses called **action potentials.** These electrical messages may be used in plants as a form of internal communication.

INTERACTIVE QUESTION   39.7

a. Name the types of tropisms that a stem exhibits.

b. Name and describe the growth mechanism that produces the coiling of a tendril.

Environmental Stresses   Plants respond to both **abiotic** environmental stresses and to **biotic** stresses such as pathogens and herbivores.

Mechanisms that reduce transpiration help a plant respond to water deficit. Guard cells lose turgor and stomata close. Abscisic acid, produced by leaves in response to water deficit, acts on guard cell membranes to keep stomata closed. Wilted leaves may roll up, further reducing transpiration. These responses, however, reduce photosynthesis. During a drought, root growth in shallow, dry soil decreases while deeper roots in moist soil continue to grow.

Plants adapted to wet habitats may have aerial roots that provide oxygen to their submerged roots. When roots of other plants are in waterlogged soils, oxygen deprivation may stimulate ethylene production, causing some root cortical cells to undergo apoptosis, opening up air tubes within the roots.

Excess salts in the soil may lower the water potential of the soil solution, reducing water uptake by roots. Plants may respond to moderate soil salinity by producing organic solutes that lower the water potential of root cells, thus increasing water uptake. Special adaptations for dealing with high soil salinity have evolved in halophytes.

Transpiration creates evaporative cooling for a plant, but this effect may be lost on hot, dry days when stomata close. Above critical temperatures, plant cells produce **heat-shock proteins** that may provide temporary support to reduce protein denaturation.

Plants respond to cold stress by increasing the proportion of unsaturated fatty acids in membrane lipids, which maintain the fluidity of cell membranes. Subfreezing temperatures cause ice to form in cell walls, lowering the extracellular water potential and causing cells to dehydrate. Plants adapted to cold winters have adaptations to deal with freezing stress, such as changing the solute composition of the cytosol. Plants may also produce *antifreeze proteins* that reduce the formation of ice crystals.

INTERACTIVE QUESTION   39.8

a. How might a plant exposed to a spell of very hot and dry weather survive this stress?

b. How might a plant respond to an unusually cold and wet fall?

39.5   Plants respond to attacks by herbivores and pathogens

Defenses Against Herbivores   Physical defenses, such as thorns, and chemical defenses, such as distasteful or toxic compounds, help plants cope with herbivory. Some plants produce *canavanine*, which resembles arginine and, when incorporated into an insect's proteins, alters the shape of its proteins and causes death.

Some plants recruit predators of their herbivores. A combination of eating damage by caterpillars and a compound in caterpillar saliva stimulates leaves to release volatile compounds that attract parasitoid wasps. The wasps lay their eggs inside the caterpillar, and then the developing wasp larvae eat their host. These volatile compounds may also signal neighboring plants, making them less susceptible to herbivory.

Defenses Against Pathogens   The physical barrier of the plant's epidermis and/or periderm is the first line of defense against pathogenic viruses, bacteria, and fungi. Should pathogens enter the plant due to injuries or through openings such as stomata, the plant mounts a chemical defense.

Virulent pathogens are those to which the plant has no specific defense. Plants are generally resistant to most pathogens. These **avirulent** pathogens do not extensively harm or kill the plant.

Specific resistance is based on **gene-for-gene recognition** between the protein products of a plant's disease resistance (*R*) genes and a pathogen molecule coded for by an avirulence (*Avr*) gene. Plants have many different *R* genes, corresponding to many different potential pathogens. The binding of a pathogen molecule called an *effector* to an R protein triggers a signal transduction pathway that produces local and systemic responses.

In the **hypersensitive response,** plant cells produce antimicrobial phytoalexins and *PR proteins* (pathogenesis-related proteins), which may attack the pathogen's cell wall. The hypersensitive response also stimulates strengthening of cell walls to slow the spread of the pathogen, after which the plant cells destroy themselves.

A hypersensitive response also involves the release of alarm signals, which stimulate PR protein production throughout the plant, creating a **systemic acquired resistance.** The signaling molecule is probably *methylsalicylic acid,* which is converted to **salicylic acid** in cells distant from the infection and activates nonspecific resistance to a diversity of pathogens for several days.

INTERACTIVE QUESTION   39.9

a. In what ways may insects harm plants?

b. In what ways may insects benefit plants?

Word Roots

aux- = grow, enlarge (*auxin:* a term usually referring to indoleacetic acid, a natural plant hormone that has a variety of effects, including cell elongation, root formation, secondary growth, and fruit growth)

circ- = a circle (*circadian rhythm:* a physiological cycle of about 24 hours in duration that persists even in the absence of external cues)

cyto- = cell; **-kine** = moving (*cytokinin:* a class of related plant hormones that retard aging and act in concert with auxin to stimulate cell division, influence cell differentiation, and control apical dominance)

gibb- = humped (*gibberellin:* a class of related plant hormones that stimulate growth in the stem and leaves, trigger the germination of seeds and breaking of bud dormancy, and (with auxin) stimulate fruit development)

hyper- = excessive (*hypersensitive response:* a plant's localized defense response to a pathogen, involving the death of cells around the site of infection)

photo- = light; **-trop** = turn, change (*phototropism:* growth of a plant shoot toward or away from light)

phyto- = a plant; **-chromo** = color (*phytochrome:* a type of light receptor that absorbs red light and regulates many plant responses)

stato- = standing, placed; **-lith** = a stone (*statolith:* a specialized plastid that contains dense starch grains and may play a role in detecting gravity)

thigmo- = a touch; **morpho-** = form; **-genesis** = origin (*thigmomorphogenesis:* a response in plants to chronic mechanical stimulation, resulting from increased ethylene production; an example is thickening stems in response to strong winds)

Structure Your Knowledge

1. List some agricultural uses of plant hormones.

2. Develop a concept map to illustrate your understanding of photoperiodism and the control of flowering. Remember to include the role of the biological clock.

3. Briefly describe the steps in the hypersensitive response and in systemic acquired resistance that result from a plant's encounter with an avirulent pathogen.

Test Your Knowledge

TRUE OR FALSE: *Indicate T or F and then correct the false statements.*

_____ 1. Abscisic acid is necessary for a seed to break dormancy.

_____ 2. Cytokinins, synthesized in the root, seem to counteract apical dominance.

_____ 3. Thigmomorphogenesis is a growth response of a seedling that encounters an obstacle while pushing upward through the soil.

_____ 4. Action potentials are electrical impulses that may be used as internal communication in plants.

_____ 5. A physiological response to day or night length is called a circadian rhythm.

_____ 6. Vernalization is the pre-treatment with cold before a plant flowers.

_____ 7. A virulent pathogen is one to which a plant has a specific resistance based on gene-for-gene recognition.

_____ 8. A cryptochrome is the light-absorbing portion of a phytochrome that reverts between two isomeric forms.

MULTIPLE CHOICE: *Choose the one best answer.*

1. The body form within a species of plant may vary more than that within a species of animal because
   a. growth in animals is indeterminate.
   b. plants respond adaptively to their environments by altering their patterns of growth and development.
   c. plant growth and development are governed by many hormones.
   d. plants can respond to environmental stress.
   e. all of the above are true.

2. Polar transport of auxin involves
   a. the accumulation of higher concentrations on the side of a shoot away from light.
   b. the movement of auxin from roots to shoots.
   c. the movement of auxin through transport proteins located only at the basal end of cells.
   d. the unidirectional active transport of auxin into and out of parenchyma cells.
   e. the settling of statoliths.

3. According to the acid growth hypothesis,
   a. auxin stimulates membrane proton pumps.
   b. a lowered pH outside the cell activates expansins, which break cross-links between cellulose microfibrils.
   c. the membrane potential created by the proton pumps enhances ion uptake, increasing the osmotic movement of water into cells.
   d. cells elongate when they take up water by osmosis.
   e. all of the above are involved in cell elongation.

4. Which of the following situations would most likely stimulate the development of axillary buds?
   a. a large quantity of auxin traveling down from the shoot and a small amount of cytokinin produced by the roots
   b. a reduction of auxin traveling from the shoot and thus a reduced production of strigolactone
   c. a 1:1 ratio of gibberellins to auxin

   d. the absence of cytokinins caused by the removal of the terminal bud
   e. an increase in the concentration of brassinosteroids produced by leaves

5. The growth inhibitor in seeds is usually
   a. abscisic acid.
   b. ethylene.
   c. gibberellin.
   d. a small amount of ABA combined with a larger concentration of gibberellins.
   e. a high cytokinin-to-auxin ratio.

6. A circadian rhythm
   a. is controlled by an external oscillator.
   b. is seen in the oscillation of stomatal opening and production of photosynthetic enzymes in a 24-hour period.
   c. involves an internal biological clock that is not set by daily environmental signals.
   d. provides the signal for flowering.
   e. involves all of the above.

7. Which of the following is *not* a component of the signal transduction pathway involved in the de-etiolation process when a shoot breaks ground?
   a. reception of light by a phytochrome located in the cytosol
   b. inhibition of the triple response and initiation of phototropism as the shoot bends toward light
   c. production of second messengers such as cGMP and Ca^{2+}
   d. cascades of protein kinase activations that phosphorylate transcription factors
   e. activation of genes that code for photosynthesis-related enzymes

8. Injecting phytochrome into cells of the tomato mutant *aurea*
   a. causes the plant to undergo the triple response above ground when exposed to ethylene.
   b. causes the plant to undergo the triple response above ground without exposure to ethylene.
   c. helps researchers identify clock mutants.
   d. initiates flowering even when nights are longer than some critical period.
   e. causes the plant to develop a normal de-etiolation response to light.

9. A flash of far-red light during a critical-length dark period will
   a. induce flowering in a long-day plant.
   b. induce flowering in a short-day plant.
   c. not influence flowering.
   d. increase the P_{fr} level suddenly.
   e. be negated by a second flash of far-red light.

10. The conversion of P_r to P_{fr}
    a. occurs slowly at night.
    b. may be the way in which plants sense daybreak and serves to entrain the biological clock.
    c. is the molecular mechanism responsible for the biological clock.
    d. occurs when P_r absorbs far-red light.
    e. occurs in flower buds and induces flowering.

11. A plant may withstand salt stress by
    a. releasing abscisic acid, which closes stomata to prevent salt accumulation.
    b. producing organic solutes, which lower the water potential of root cells.
    c. wilting, which reduces water and salt uptake by reducing transpiration.
    d. producing canavanine, which reduces the toxic effect of sodium ions.
    e. releasing ethylene, which leads to apoptosis of damaged cells.

12. Which of the following hormones would be sprayed on barley seeds to speed germination in the production of malt for making beer?
    a. abscisic acid
    b. auxin
    c. cytokinin
    d. strigolactone
    e. gibberellin

13. Scientists have identified an *Arabidopsis eto* mutant that greatly overproduces ethylene. Which of the following statements accurately describes the growth of such a mutant?
    a. It undergoes the triple response only when it encounters an obstacle, but the response is greater than that of a wild-type plant.
    b. It does not undergo the triple response, because its ethylene receptors are blocked.
    c. *Eto* mutants undergo the triple response even out of the soil.
    d. When experimentally exposed to ethylene, the triple response of such mutants is inhibited.
    e. Treating *eto* mutants with inhibitors of ethylene synthesis would not reduce their triple response when they do not encounter an obstacle.

14. Many plants will flower in response to a specific
    a. photoperiod, which seems to be measured by the length of darkness to which the leaves of the plant are exposed.
    b. effector, which is produced in leaves and travels to a floral meristem.
    c. minimal temperature that appears to signal a seasonal change.
    d. flowering hormone that is produced in the apical bud.
    e. combination of high levels of auxin and FT protein and low levels of cytokinin and strigolactone.

15. Which of the following is *not* a plant defense against herbivory?
    a. production of distasteful compounds
    b. production of toxic compounds such as canavanine
    c. physical defenses such as thorns
    d. initiation of a hypersensitive response with production of phytoalexins and PR proteins
    e. release of volatile compounds that attract parasitoid wasps

16. Which of the following is *not* true of the hypersensitive response?
    a. It relies on *R-Avr* recognition between a specific plant protein and a pathogen molecule.
    b. It increases the production of PR proteins that may have antimicrobial functions.
    c. It enhances the production of ethylene, which serves as a signal molecule that is transported throughout the plant and activates systemic acquired resistance.
    d. It contains an infection by stimulating the cross-linking of cell wall molecules and the production of lignin.
    e. The plant cells involved in the defense undergo apoptosis, leaving lesions that indicate the site of the contained infection.

Chapter 40

Basic Principles of Animal Form and Function

Key Concepts

40.1 **Animal form and function are correlated at all levels of organization**

40.2 **Feedback control maintains the internal environment in many animals**

40.3 **Homeostatic processes for thermoregulation involve form, function, and behavior**

40.4 **Energy requirements are related to animal size, activity, and environment**

Framework

The body structures of an animal are built from four basic tissues: epithelial, connective, muscle, and nervous. Organs function together in organ systems. Structure correlates with function in each of the various hierarchical levels of organization.

The internal environment is regulated by homeostatic feedback mechanisms. Thermoregulatory mechanisms include insulation, circulatory adaptations, evaporative cooling, behavior, and regulation of heat production. Endotherms use metabolic heat to maintain body temperature; ectotherms gain their heat mostly from the environment.

Metabolic rate, the amount of energy used in a unit of time, is higher for endothermic animals and inversely related to body size.

Chapter Review

Anatomy is the study of an organism's structure; **physiology** is the study of biological function.

40.1 Animal form and function are correlated at all levels of organization

Evolution of Animal Size and Shape Physical laws drove natural selection for a fusiform body shape in fast-swimming fishes, birds, and mammals, an example of convergent evolution. Physical laws also influence body size, as larger bodies require stronger skeletons and relatively larger muscle masses for locomotion.

Exchange with the Environment   Every cell must be in an aqueous medium to allow exchange across its plasma membrane. Animals with two-layered, saclike bodies or thin, flat bodies can maintain sufficient cellular contact with the aqueous environment.

Animals with compact bodies, however, must provide extensively branched or folded internal membranes for exchanging materials with the environment. A circulatory system often connects these exchange surfaces with the **interstitial fluid** bathing the body's cells.

INTERACTIVE QUESTION   40.1

List some advantages of a compact, complex body form in which most body cells are not in contact with the external environment.

Hierarchical Organization of Body Plans   **Tissues** are groups of similar cells with a common function. **Organs,** which perform specific functions, often consist of a layered arrangement of tissues. Groups of organs are integrated into **organ systems,** which carry out the main functions required for life.

Tissue Structure and Function   **Epithelial tissues,** or **epithelia,** line the outer and inner surfaces of the body in sheets of tightly packed cells. A simple epithelium has one layer of cells; a stratified epithelium has multiple layers of cells; and a pseudostratified epithelium has a single layer of cells that vary in height. The shape of cells at the *apical* surface facing the lumen or the outside of an organ may be squamous (flat), cuboidal (boxlike), or columnar (pillarlike). Cells at the *basal* surface of an epithelium are attached to a *basal lamina,* a dense layer of extracellular matrix.

INTERACTIVE QUESTION   40.2

Name the two types of epithelia in the following illustration. One of these epithelia lines the mouth, whereas the other lines the intestines. Explain why each type of epithelium is adapted to its location.

a. _____   b. _____

Connective tissue binds and supports other tissues. It is composed of relatively few cells suspended in an extracellular matrix, which consists of protein fibers embedded in a liquid, jellylike, or solid material. The cells within the matrix are **fibroblasts,** which secrete the protein fibers, and **macrophages,** which engulf bacteria and cellular debris.

Connective tissue fibers are of three types. *Collagenous fibers* provide strength and flexibility. *Elastic fibers* can stretch and provide resilience. *Reticular fibers* attach connective tissue to adjacent tissues.

The widespread *loose connective tissue* attaches epithelia to underlying tissues and holds organs in place.

Fibrous connective tissue, with its dense arrangement of parallel collagenous fibers, is found in **tendons,** which attach muscles to bones, and in **ligaments,** which join bones together at joints.

Bone is formed by *osteoblasts,* which deposit a matrix containing collagen. Calcium, magnesium, and phosphate ions form a hard mineral within the matrix. *Osteons* consist of concentric layers of matrix around a central canal containing blood vessels and nerves.

Cartilage is composed of collagenous fibers embedded in a rubbery substance; both are secreted by *chondrocytes.* Cartilage is a strong but somewhat flexible support material.

Adipose tissue pads and insulates the body and stores fat. Each adipose cell contains a large fat droplet.

Blood has a liquid extracellular matrix called plasma. Erythrocytes (red blood cells) carry oxygen; leukocytes (white blood cells) function in defense; and cell fragments called platelets are involved in the clotting of blood.

INTERACTIVE QUESTION 40.3

Identify the types of connective tissue and the indicated structures or cells in the following three micrographs.

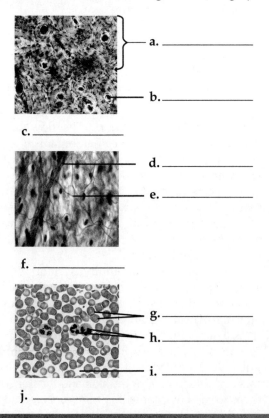

a. _____

b. _____

c. _____

d. _____

e. _____

f. _____

g. _____

h. _____

i. _____

j. _____

The cells of **muscle tissue** contain contracting filaments of actin and myosin. **Skeletal muscle**—also called *striated muscle*—is responsible for voluntary body movements. **Cardiac muscle,** forming the wall of the heart, is also striated, but its cells are branched and joined by intercalated discs. **Smooth muscle** is composed of spindle-shaped cells lacking striations. It is found in the walls of the digestive tract, arteries, and other internal organs, and produces involuntary movements.

Nervous tissue senses stimuli and transmits nerve impulses. The **neuron,** or nerve cell, consists of a cell body and two types of processes: dendrites, which conduct impulses toward the cell body, and axons, which conduct impulses away from the cell body. **Glial cells (glia)** insulate and nourish neurons.

INTERACTIVE QUESTION 40.4

Identify the types of vertebrate muscle tissue. What is the dark band at the end of the line in micrograph **a**, and what is its function? What are the thin vertical stripes in micrograph **b**?

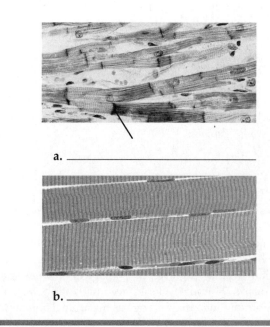

a. _____

b. _____

Coordination and Control The endocrine and nervous systems coordinate the body's activities. Endocrine cells secrete **hormones** into the bloodstream. Body cells with matching receptors respond to these signaling molecules. Neurons deliver signals directly to target cells.

INTERACTIVE QUESTION 40.5

Compare the speed of signaling and the duration of the response in the endocrine and nervous systems. How do the functions of these two communication systems differ?

40.2 Feedback control maintains the internal environment in many animals

Regulating and Conforming For a given environmental variable, an animal may be a **regulator,** using mechanisms to control internal fluctuations, or a **conformer,** allowing internal conditions to vary with external changes.

Homeostasis The maintenance of a "steady state" or relative internal balance is called **homeostasis.** Homeostatic control mechanisms involve a receptor or **sensor,** which detects a **stimulus** (a fluctuation above or below a particular **set point** for a variable), and a *control center,* which triggers a **response** that returns the variable to the set point.

Most homeostatic mechanisms operate by **negative feedback,** in which the response reduces the stimulus. Variables may have a *normal range* within which they fluctuate instead of a single set point.

What is **positive feedback**? Give an example. Are positive feedback mechanisms usually involved in maintaining homeostasis? Explain.

Homeostatic mechanisms allow for *regulated change* in the body's internal environment when necessary. A **circadian rhythm** involves physiological changes that occur over a 24-hour cycle. In **acclimatization,** an animal's normal range may change as the animal adjusts to environmental changes.

40.3   Homeostatic processes for thermoregulation involve form, function, and behavior

Changes in temperature affect most biochemical and physiological processes. Animals maintain internal temperatures within an optimal range through the process of **thermoregulation.**

Endothermy and Ectothermy   Animals that use metabolic heat in generating body warmth are **endothermic** (birds, mammals, some fishes, a few other reptiles, and some insects). Animals that gain their heat mostly from the environment are **ectothermic** (most invertebrates, fishes, amphibians, lizards, snakes, and turtles). Ectotherms require less food energy and tolerate wider internal temperature variations than do endotherms.

Variation in Body Temperature   Poikilotherm refers to animals whose body temperatures vary with the environment; *homeotherm* refers to animals with relatively stable body temperatures.

Explain the following statement: Some ectotherms are homeotherms, and some even have higher body temperatures than those of endotherms.

Balancing Heat Loss and Gain   The flow of heat within an organism and between an organism and its environment is affected by the processes of **conduction,** the direct transfer of thermal motion between objects in contact; **convection,** the transfer of heat by the flow of air or water past a surface; **radiation,** the emission of electromagnetic waves by all objects; and **evaporation,** the loss of heat due to the conversion of the surface molecules of a liquid to a gas.

The mammalian **integumentary system** serves a thermoregulatory function. Insulation, in the form of hair, feathers, and fat layers in mammals and birds, reduces the rate of heat exchange with the environment. A very thick layer of blubber insulates marine mammals.

How can circulatory adaptations contribute to thermoregulation? *Vasodilation* of superficial blood vessels increases both blood flow in the skin and the transfer of body heat to the environment. *Vasoconstriction* reduces blood flow and heat transfer. In **countercurrent exchange,** common in marine mammals and birds, the close proximity of blood vessels servicing the extremities allows heat in arterial blood leaving the body core to be transferred to venous blood returning to the body core. Some sharks and fish that are powerful swimmers, as well as endothermic insects, use countercurrent heat exchangers to maintain high temperatures for muscles involved in swimming or flight.

Mammals and birds may use the cooling mechanism of evaporative heat loss across the skin and respiratory surfaces. Panting, sweating, and bathing may enhance evaporative cooling.

Behavioral responses contribute to thermoregulation in both endotherms and ectotherms. Amphibians, reptiles, and many terrestrial invertebrates may thermoregulate by orienting the body to the sun or moving to cool areas. The social organization of honeybees enables them to retain heat by huddling together and to cool the hive by bringing in water and fanning their wings.

Mammals and birds are able to increase the rate of metabolic heat produced by muscles through moving and shivering. Some mammals generate heat through *nonshivering thermogenesis,* a rise in metabolic rate and the production of heat instead of ATP. Some mammals have *brown fat* stores that are specialized for rapid heat production.

A few large reptiles use shivering to generate heat and become endothermic while incubating eggs. Endothermic flying insects may warm up before taking off by contracting their powerful flight muscles.

Acclimatization in Thermoregulation Birds and mammals often adjust the thickness of their insulating coats to meet seasonal changes. Ectotherms may adjust their cellular physiology (changes in enzymes, membranes, or production of "antifreeze" compounds) to acclimatize to changes in environmental (and thus body) temperature.

Physiological Thermostats and Fever In mammals, a group of neurons in the **hypothalamus** functions as a thermostat that triggers heat loss or gain mechanisms in response to a body temperature outside the normal range. Mammals and birds develop a fever in response to certain infections, apparently due to the resetting of the hypothalamic set point.

INTERACTIVE QUESTION 40.8

a. What cooling mechanisms are activated when the hypothalamus senses a body temperature above the set point?

b. What mechanisms are activated when body temperature falls below the set point?

40.4 Energy requirements are related to animal size, activity, and environment

Energy Allocation and Use An animal's **bioenergetics** involves the overall flow and transformation of energy—the input of energy in the form of food, its use in the body, and its return to the environment as heat or in waste products. Fuel molecules are used to generate ATP for cellular work and biosynthesis.

Quantifying Energy Use The total energy an animal uses in a unit of time is its **metabolic rate.** Metabolic rate can be determined by measuring heat loss, the rate of oxygen consumption, the rate of CO_2 production, or the energy content of food and the energy lost in waste products.

Minimum Metabolic Rate and Thermoregulation The minimal metabolic rate for a nongrowing endotherm

that is at rest, fasting, and nonstressed is called the **basal metabolic rate (BMR).** The **standard metabolic rate (SMR)** is the metabolic rate of a resting, fasting, nonstressed ectotherm determined at a specific temperature.

A human adult male's BMR is about 1,600–1,800 kcal/day; a female's is about 1,300–1,500 kcal/day. The SMR of a comparably sized ectotherm would be much, much less.

Influences on Metabolic Rate BMR increases with increasing body size. However, the energy required to maintain each gram of body weight is inversely related to body size. Smaller mammals have higher metabolic, respiration, and heart rates; have larger blood volumes; and require more food per unit of body mass. With a greater surface-to-volume ratio, small endotherms may have a higher energy cost to maintain a stable body temperature. However, the factors that contribute to this inverse relationship in ectotherms as well as endotherms are not fully understood.

Maximal metabolic rates occur during intense activity. Most terrestrial animals have an average daily rate of energy consumption that is 2 to 4 times BMR or SMR.

Energy Budgets Species vary in the food energy they allocate for BMR or SMR, activity, growth, reproduction, and temperature regulation.

Torpor and Energy Conservation **Torpor** is a physiological state characterized by decreases in metabolism and activity that allow an animal to save energy while avoiding unfavorable environmental conditions.

INTERACTIVE QUESTION 40.9

Describe the following types of torpor. What challenges does each type help an animal meet?

a. Hibernation

b. *Estivation*

c. Daily torpor

Word Roots

con- = with; **-vect** = carried (*convection:* the mass movement of warmed air or liquid to or from the surface of a body or object)

counter- = opposite (*countercurrent exchange:* the exchange of a substance or of heat between two fluids flowing in opposite directions)

-dilat = expanded (*vasodilation:* an increase in the diameter of blood vessels triggered by nerve signals that relax the muscles of the vessel walls)

ecto- = outside; **-therm** = heat (*ectothermic:* referring to organisms for which external sources provide most of the heat for temperature regulation)

endo- = inner (*endothermic:* referring to organisms with bodies that are warmed by heat generated by metabolism, This heat is usually used to maintain a relatively stable body temperature that is higher than that of the external environment)

fibro- = a fiber (*fibroblast:* a type of cell in loose connective tissue that secretes extracellular protein fibers)

homeo- = same; **-stasis** = standing, posture (*homeostasis:* the steady-state physiological condition of the body)

inter- = between (*interstitial fluid:* the internal environment of vertebrates, consisting of the fluid filling the spaces between cells)

macro- = large (*macrophage:* a phagocytic cell present in many tissues that functions in innate immunity by destroying microbes and in acquired immunity)

Structure Your Knowledge

1. Fill in the following table on the structure and function of the four types of animal tissues.

2. Organs are composed of layers of several different tissues. Which of the four animal tissues would you expect to find in all organs? In what types of organs would you predict the remaining tissues occur?

3. Explain the following statement: Endotherms can tolerate wider external temperature fluctuations; ectotherms may be able to tolerate wider internal temperature fluctuations.

4. An animal's energy budget may include basal (or standard) metabolism, reproduction, temperature regulation, growth, and activity. Describe some factors that lead to different energy allocations for different animals.

Tissue	Structural Characteristics	General Functions	Specific Examples

Test Your Knowledge

MULTIPLE CHOICE: *Choose the one best answer.*

1. Tunas, dolphins, penguins, and sharks all have a fusiform body shape. What is the best explanation for this similarity?
   a. This shape is an example of divergent evolution.
   b. They are all vertebrates and evolved from a common ancestor.
   c. The physical laws of hydrodynamics drove the natural selection for this shape.
   d. All aquatic animals have this shape, an example of convergent evolution.
   e. This shape is determined by the thermodynamics of maintaining body temperature in cold water.

2. Which of the following best describes smooth muscle?
   a. spindle-shaped cells; involuntary control
   b. striated, branching cells; involuntary control
   c. spindle-shaped cells connected by intercalated disks; voluntary control
   d. striated cells containing overlapping filaments; involuntary control
   e. spindle-shaped striated cells; voluntary control

3. Which of the following statements is *not* true of connective tissue?
   a. It consists of few cells surrounded by fibers in a matrix.
   b. It includes such diverse tissues as bone, cartilage, adipose, and loose connective tissue.
   c. It connects and supports other tissues.
   d. It forms the internal and external lining of many organs.
   e. It can have a matrix that is a liquid, a gel, or a solid.

4. Which of the following are *incorrectly* paired?
   a. blood—erythrocytes, leukocytes, and platelets in plasma
   b. bone—osteoblasts embedded in a mineral matrix
   c. loose connective tissue—collagenous, elastic, and reticular fibers
   d. adipose tissue—loose connective tissue with fat-storing cells
   e. fibrous connective tissue—chondrocytes embedded in chondroitin sulfate

5. The interstitial fluid of vertebrates
   a. is the internal environment within cells.
   b. surrounds cells and provides for the exchange of nutrients and wastes.
   c. makes up blood plasma.
   d. is not necessary in flat, thin vertebrates.
   e. is less abundant in ectotherms than in endotherms.

6. Which of the following would be *least* likely to be true of an animal that is a regulator?
   a. It can live in a variable climate because of its homeostatic mechanisms.
   b. It may have a larger geographic range than a conformer.
   c. It may acclimatize to winter by increasing the thickness of its insulating coat.
   d. Much of its energy budget can be allocated to reproduction.
   e. It has behavioral as well as physiological mechanisms for responding to changing conditions.

7. Negative feedback circuits are
   a. mechanisms that most commonly maintain homeostasis.
   b. activated only when a physiological variable rises above a set point.
   c. analogous to a radiator that heats a room.
   d. involved in maintaining contractions during childbirth.
   e. found in endotherms but not in ectotherms.

8. Which of the following mechanisms dissipates heat?
   a. countercurrent exchange between vessels that service the extremities
   b. trapping air by raising fur or feathers
   c. vasoconstriction of surface vessels
   d. vasodilation of surface vessels
   e. both a and c

9. Which of the following animals would be likely to produce antifreeze compounds in their cells?
   a. a polar bear or an arctic wolf
   b. a hibernating mammal
   c. an estivating amphibian
   d. an ectotherm in a subzero environment
   e. a chickadee in very cold winter

10. The rate of metabolic heat production can be increased by
    a. nonshivering thermogenesis.
    b. adding to the layer of brown fat.
    c. vasoconstriction.
    d. thick layers of blubber and countercurrent heat exchangers.
    e. all of the above.

11. Which of the following statements is *true*?
    a. BMR is always higher than SMR.
    b. BMR per kg of body mass is inversely related to body size.
    c. SMR per kg of body mass is positively correlated to body size.
    d. Overall metabolic rate is positively correlated with body size.
    e. both b and d

12. Which of the following animals would most likely have the highest metabolic rate?
    a. a horse
    b. a lizard
    c. a giant squid
    d. a whale
    e. a bat

13. Which of the following animals would have the largest percentage of its energy budget available for growth?
    a. a deer mouse from a temperate forest
    b. a python from tropical India
    c. a human adult from a temperate climate
    d. a penguin from Antarctica
    e. They would all have the same percentage, because most of their energy is used to maintain basal or standard metabolic rate.

14. Which of the following animals would be likely to experience a daily torpor?
    a. a bear
    b. a frog
    c. a lizard
    d. a hamster
    e. a hummingbird

15. For most terrestrial animals, the average daily rate of energy consumption is 2–4 times BMR or SMR. Which of the following animals has an unusually low average daily metabolic rate?
    a. a human in a developed country
    b. a wild turkey
    c. a domesticated turkey
    d. a forest-dwelling salamander
    e. a desert snake

Chapter 41

Animal Nutrition

Key Concepts

41.1 An animal's diet must supply chemical energy, organic molecules, and essential nutrients

41.2 The main stages of food processing are ingestion, digestion, absorption, and elimination

41.3 Organs specialized for sequential stages of food processing form the mammalian digestive system

41.4 Evolutionary adaptations of vertebrate digestive systems correlate with diet

41.5 Feedback circuits regulate digestion, energy storage, and appetite

Framework

Chapter Review

Nutrition is the process by which organisms take in and use food. **Herbivores** eat plants or algae; **carnivores** eat animals; and **omnivores** consume both plants and animals. Most animals are opportunistic and eat foods from different categories when they are available.

41.1 An animal's diet must supply chemical energy, organic molecules, and essential nutrients

An animal's diet provides fuel for ATP production, organic carbon and nitrogen molecules for biosynthesis, and **essential nutrients,** which are molecules that an animal requires but cannot synthesize itself.

Essential Nutrients Eight of the 20 amino acids that are needed to make proteins are **essential amino acids** required in the diet of most animals (including adult humans). Protein deficiency develops from a lack of one or more essential amino acids in the diet and may impair development.

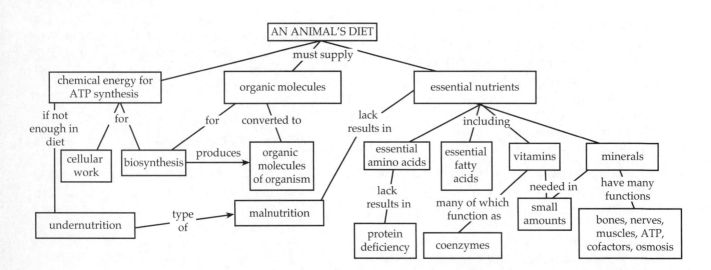

Meat, eggs, and cheese contain "complete" proteins with all essential amino acids. Most plant proteins are "incomplete," deficient in one or more essential amino acids.

INTERACTIVE QUESTION   41.1

How can vegetarians avoid protein deficiency?

Animals produce the enzymes to make most of the fatty acids they need. Linoleic acid is required in the human diet to make some membrane phospholipids. Deficiencies of **essential fatty acids** are rare.

Vitamins are organic molecules required in small amounts in the diet. Thirteen vitamins essential to humans have been identified. Water-soluble vitamins include the B complex, most of which function as coenzymes, and vitamin C, which is required for the production of connective tissue. The fat-soluble vitamins are A, incorporated into visual pigments; D, which aids in calcium absorption and bone formation; E, an antioxidant; and K, required for blood clotting.

Minerals are inorganic nutrients that are usually needed in small amounts. Many minerals function as cofactors of enzymes. Iodine is needed by vertebrates to make the metabolism-regulating thyroid hormones. Sodium, potassium, and chloride are important in nerve function and osmotic balance.

Dietary Deficiencies   *Malnutrition* results from a failure to obtain one or more essential nutrients or sufficient chemical energy. *Undernutrition* refers to a diet deficient in calories. With severe deficiency, the body breaks down its own proteins for energy, eventually causing irreversible damage or death.

Assessing Nutritional Needs   Research on human dietary requirements involves many challenges. Techniques include the study of genetic defects that affect nutrient uptake or use, and *epidemiology*, the study of disease among various populations (in this case, those with different dietary practices).

INTERACTIVE QUESTION   41.2

a. How does undernutrition differ from other types of malnutrition?

b. Which two minerals do vertebrates require in relatively large amounts? Why?

41.2   The main stages of food processing are ingestion, digestion, absorption, and elimination

Ingestion is the act of consuming food. Many aquatic animals are **suspension feeders** or **filter feeders,** eating or sifting small food particles from the water. **Substrate feeders** live in or on their food, eating their way through it. **Fluid feeders** suck fluids from a living plant or animal host. Most animals are **bulk feeders,** which eat relatively large pieces of food.

Digestion splits food into small molecules. Mechanical digestion often precedes chemical digestion, which is the *enzymatic hydrolysis* of bonds between components of large molecules.

In the last two stages of food processing, **absorption** and **elimination,** small molecules are taken into the cells of an animal, and the undigested remainder of the food leaves the digestive system.

INTERACTIVE QUESTION   41.3

Why must large molecules in food be digested into monomers or smaller components?

Digestive Compartments   Following phagocytosis or pinocytosis, food vacuoles fuse with enzyme-containing lysosomes, and *intracellular digestion* occurs safely within a membrane-enclosed compartment.

Most animals break down their food, at least initially, by *extracellular digestion* within a separate compartment that connects to the external environment.

A **gastrovascular cavity** functions in both digestion and the transport of nutrients throughout the body. In the small cnidarian hydra, digestive enzymes secreted into the gastrovascular cavity begin food breakdown. Gastrodermal cells then engulf food particles for hydrolysis within food vacuoles. Undigested materials are expelled through the single opening of the gastrovascular cavity.

Most animals have *complete digestive tracts* or **alimentary canals** with two openings, a mouth and an anus. Food, ingested through the mouth and pharynx, passes through an esophagus that may lead to a crop, stomach, or gizzard—organs specialized for storing or grinding food. In the intestine, digestive enzymes hydrolyze large molecules, and nutrients are absorbed across the intestinal lining. Undigested material exits through the anus.

INTERACTIVE QUESTION   41.4

What are the advantages of an alimentary canal compared to a gastrovascular cavity?

41.3   Organs specialized for sequential stages of food processing form the mammalian digestive system

The mammalian digestive system consists of the alimentary canal and accessory glands that secrete digestive juices through ducts into the canal. Rhythmic waves of muscular contraction called **peristalsis** push food along the tract. Muscular ringlike valves called **sphincters** regulate the passage of material between some segments.

The Oral Cavity, Pharynx, and Esophagus   Physical and chemical digestion begins in the mouth, or **oral cavity,** where teeth grind food to expose a greater surface area to enzyme action. Saliva, released from **salivary glands,** has several components: **mucus,** which contains slippery glycoproteins called mucins that protect the mouth lining from abrasion and lubricate food for swallowing; buffers to neutralize acidity; antibacterial agents; and **amylase,** an enzyme that begins the hydrolysis of starch and glycogen. The tongue tastes and manipulates food, and pushes the food ball, or **bolus,** into the pharynx for swallowing.

The **pharynx** is the intersection leading to both the esophagus and the trachea. During swallowing, the *larynx,* the upper part of the respiratory tract, moves up so that the cartilaginous *epiglottis* covers and blocks the *glottis,* the vocal cords and the opening between them.

Food moves down through the **esophagus** to the stomach, squeezed along by a wave of peristalsis.

Digestion in the Stomach   In the expandable **stomach,** food is combined with **gastric juice,** producing a mixture called **chyme.**

Parietal cells within gastric glands in the stomach secrete hydrogen ions and chloride ions. The resulting hydrochloric acid (HCl) breaks down food tissues, kills bacteria, and denatures proteins. A **protease** called **pepsin,** which hydrolyzes specific peptide bonds in proteins, is synthesized and secreted by *chief cells* in an inactive form called **pepsinogen.** It is activated by HCl and by pepsin itself, which produces more active pepsin through positive feedback.

Mucus helps protect the rapidly dividing epithelium from digestion. Gastric ulcers, damaged areas of the stomach lining, have been linked to infection by the bacterium *Helicobacter pylori.*

Smooth muscles churn the contents of the stomach. Except when a bolus arrives, the opening from the esophagus is closed to prevent backflow of the acidic chyme. A sphincter regulates passage of chyme into the intestine.

INTERACTIVE QUESTION   41.5

Two types of macromolecules have been partially digested by the time chyme moves into the intestine. What are these molecules, where does this digestion take place, and what enzymes are involved?

a.

b.

Digestion in the Small Intestine   Digestive juices are mixed with chyme in the **duodenum,** the first section of the **small intestine.** The **pancreas** produces digestive enzymes and a bicarbonate-rich alkaline solution that offsets the acidity of the chyme. The **liver** produces **bile,** which is stored in the **gallbladder** until needed. Bile salts emulsify fats for easier digestion. Bile also contains pigments that are by-products of the breakdown of red blood cells in the liver. The lining of the duodenum secretes some enzymes; others are bound to the surfaces of epithelial cells.

Most digestion occurs in the duodenum. The *jejunum* and *ileum* function in nutrient and water absorption.

INTERACTIVE QUESTION   41.6

For each of the following molecules, describe their enzymatic digestion in the small intestine.

a. polysaccharides

b. polypeptides

c. nucleic acids

d. fats

Absorption in the Small Intestine Large folds of the small intestine lining are covered with fingerlike projections called **villi.** Each epithelial cell of a villus has microscopic extensions called **microvilli.** This huge surface area for absorption is called the *brush border.* Nutrients may be actively or passively transported across the membranes.

The core of each villus has a net of capillaries and a lymph vessel called a **lacteal.** Monoglycerides and fatty acids absorbed by epithelial cells are recombined to form fats. These fats are coated with proteins, phospholipids, and cholesterol to make tiny globules called **chylomicrons,** which move out of epithelial cells by exocytosis and into a lacteal. They are then transported by the lymphatic system to large veins.

Absorbed amino acids and sugars enter capillaries. The nutrient-laden blood from the small intestine is carried directly to the liver by the **hepatic portal vein.** The liver regulates the nutrient content of the blood and detoxifies foreign molecules.

Absorption in the Large Intestine The small intestine leads into the **large intestine** at a T-shaped junction. One arm of this junction is a pouch called the **cecum,** which in humans has a fingerlike extension called the **appendix.** The other arm of the junction, the **colon,** finishes the reabsorption of the large quantity of water secreted with digestive enzymes into the digestive tract.

Escherichia coli and other mostly harmless bacteria live on unabsorbed organic matter in the large intestine. Some of these bacteria produce B vitamins and vitamin K. The **feces** contain cellulose, other undigested materials, and a large proportion of intestinal bacteria. Feces are stored in the **rectum,** the final section of the large intestine.

41.4 Evolutionary adaptations of vertebrate digestive systems correlate with diet

Dental Adaptations Dentition, the type and arrangement of teeth, correlates with diet.

Stomach and Intestinal Adaptations Vegetation is more difficult to digest than meat. The longer alimentary canal of herbivores facilitates the digestion of plant material.

Mutualistic Adaptations Many herbivorous mammals have fermentation chambers filled with symbiotic bacteria and protists. These microorganisms, often housed in the cecum, digest cellulose to simple sugars and produce a variety of essential nutrients. Rabbits and rodents ingest their feces (*coprophagy*) to capture the nutrients that mutualistic bacteria had released in the large intestine.

Ruminants have an elaborate system involving several stomach chambers, regurgitation and rechewing of the cud, and digestion of symbiotic bacteria, which maximizes the nutrient yield of their grass diet.

INTERACTIVE QUESTION 41.7

Compare and contrast the dentition and alimentary canals of carnivores and herbivores.

41.5 Feedback circuits regulate digestion, energy storage, and appetite

Regulation of Digestion Hormones, such as gastrin, released by the stomach, and secretin and cholecystokinin (CCK), released by the duodenum, stimulate the release of digestive secretions from the stomach, pancreas, and gallbladder. The *enteric division* of the nervous system regulates peristalsis and hormone release.

Regulation of Energy Storage When an animal consumes more calories than are needed to meet its energy requirements, the excess can be stored in the liver and muscles as glycogen, or in adipose tissue as fat (when glycogen stores are full). Glucose homeostasis is maintained by two pancreatic hormones: insulin, which promotes glycogen synthesis in the liver and thus lowers blood glucose, and glucagon, which stimulates the liver to break down glycogen and thus raises blood glucose.

Regulation of Appetite and Consumption Overnourishment, consuming more calories than the body needs, can lead to obesity. This excessive accumulation of fat is associated with a number of health problems.

Appetite-regulating hormones act on the brain's "satiety center," which generates impulses that make one feel hungry or satiated. These hormones include ghrelin, which is secreted by the stomach and triggers hunger, and three hormones that suppress appetite: insulin from the pancreas, PYY from the small intestine, and **leptin** produced by adipose tissue.

Research with mutant mice led to the identification of the *ob* gene, which produces leptin, and of the *db* gene, which codes for the leptin receptor.

Obesity and Evolution Sometimes obesity is adaptive, as in the obese chicks of some petrel species. Gorging on the lipid-rich food supplied by parent birds enables them to obtain sufficient protein for growth as well as survival when food is scarce.

INTERACTIVE QUESTION   41.8

How might the human craving for fatty foods, which is helping to fuel the current rise in obesity, have evolved through natural selection?

Word Roots

chylo- = juice; **micro-** = small (*chylomicron:* a lipid transport globule composed of fats mixed with cholesterol and coated with proteins)

chymo- = juice (*chyme:* the mixture of partially digested food and digestive juices formed in the stomach)

gastro- = stomach; **-vascula** = a little vessel (*gastrovascular cavity:* a central cavity with a single opening that functions in both digestion and nutrient distribution)

herb- = grass; **-vora** = eat (*herbivore:* an animal that mainly eats plants or algae)

hydro- = water; **-lysis** = to loosen (*hydrolysis:* a chemical process that splits molecules by the addition of water)

micro- = small; **-villi** = shaggy hair (*microvilli:* the many fine, fingerlike projections of the epithelial cells in the lumen of the small intestine that increase its surface area)

omni- = all (*omnivore:* an animal that regularly eats animals as well as plants or algae)

peri- = around; **-stalsis** = a constriction (*peristalsis:* rhythmic waves of contraction of smooth muscle that push food along the alimentary canal)

Structure Your Knowledge

1. Label the indicated structures in the following diagram of the human digestive system. Review the functions of these structures.

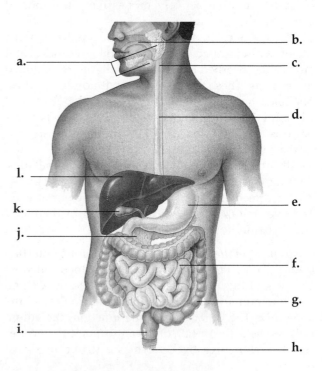

Test Your Knowledge

MULTIPLE CHOICE: *Choose the one best answer.*

1. Why do many vegetarians combine different protein sources in their diet or eat some animal products such as eggs or milk products?
    a. to make sure they obtain sufficient calories
    b. to provide sufficient vitamins
    c. to make sure they ingest all the essential fatty acids
    d. to make their diet more interesting
    e. to provide all the essential amino acids

2. Which of the following is a *true* statement about vitamins?
    a. They may be produced by intestinal microorganisms.
    b. Requirements for them are the same from one species to the next.
    c. They are stored in large quantities in the liver.
    d. They are inorganic nutrients, needed in small amounts, that often function as cofactors.
    e. Because they are water soluble, they must be replaced every day.

3. An amino acid that is not an essential amino acid would be one that an animal
   a. can obtain from its own proteins.
   b. has the needed enzymes for synthesizing.
   c. gets in sufficient quantities in its diet.
   d. obtains from its microbial symbionts.
   e. does not use in protein synthesis.

4. How do molting penguins obtain enough protein to produce new feathers?
   a. They become obese right before molting in order to store protein.
   b. They combine different protein sources to make up for the incomplete proteins in their food.
   c. They become malnourished during the period of new feather production.
   d. They feed on fish, an excellent protein source, while molting.
   e. They stockpile extra muscle proteins prior to molting, which then supply amino acids for protein synthesis.

5. Which of the following substances is mismatched with its function?
   a. most B vitamins—coenzymes
   b. vitamin E—antioxidant
   c. vitamin K—blood clotting
   d. iron—component of thyroid hormones
   e. phosphorus—bone formation, nucleotide synthesis

6. An organism that exclusively uses extracellular digestion
   a. is a sponge.
   b. is a hydra with a gastrovascular cavity.
   c. needs no circulatory system.
   d. must be very thin and elongated.
   e. has a specialized body compartment containing digestive enzymes.

7. Substrate feeders such as earthworms
   a. eat mostly mineral substrates.
   b. filter small organisms from water.
   c. eat autotrophs.
   d. move and eat through their food.
   e. are bulk feeders.

8. In animals, a distinct advantage of extracellular digestion over intracellular digestion is that
   a. polymers are hydrolyzed to monomers by digestive enzymes.
   b. a greater surface area is available for the absorption of digested nutrients.
   c. larger pieces of food can be ingested and then digested.
   d. all four types of macromolecules can be digested instead of just carbohydrates.
   e. the products of extracellular digestion can be absorbed into all body cells, without the need for a transport system.

9. Why does salivary amylase not hydrolyze starch in the duodenum?
   a. Starch is completely hydrolyzed into maltose in the oral cavity.
   b. The acid pH of the stomach denatures salivary amylase, and pepsin begins hydrolyzing it.
   c. Salivary amylase is produced by salivary glands and never leaves the oral cavity.
   d. Pancreatic amylase is a more effective enzyme in the pH of the duodenum.
   e. Salivary amylase can hydrolyze glycogen but not starch.

10. What moves a bolus of food down the esophagus to the stomach?
    a. the closing of the epiglottis over the glottis
    b. a lower pressure in the abdominal cavity compared to that in the oral cavity
    c. the contraction of the diaphragm and the opening of the sphincter to the stomach
    d. peristalsis caused by waves of smooth muscle contraction and relaxation
    e. the action of cilia coated by mucus

11. After a meal of greasy french fried potatoes, which enzymes would you expect to be most active?
    a. salivary and pancreatic amylase, disaccharidases, lipase
    b. lipase, lactase, maltase
    c. pepsin, trypsin, chymotrypsin, dipeptidases
    d. gastric juice, bile, bicarbonate
    e. sucrase, lipase, bile

12. Villi are
    a. folds in the stomach that allow the stomach to expand.
    b. extensions of the lymphatic system that pick up digested fats for transport to the circulatory system.
    c. fingerlike projections of the small intestine lining that increase the surface area for absorption.
    d. microscopic extensions of epithelial cells lining the small intestine that provide more surface area for digestion.
    e. extensions of the intestinal capillaries that join to form the hepatic portal vein.

13. Which of the following is an example of positive feedback?
    a. the release of insulin in response to high blood glucose level, which stimulates the production of glycogen in the liver
    b. the formation of HCl within the stomach lumen after the release of H^+ and Cl^- from gastric glands
    c. the rechewing of a ruminant's cud and subsequent digestion of cellulose-digesting microbes
    d. the conversion of pepsinogen to active pepsin by pepsin
    e. the increased production of leptin from increasing stores of adipose tissue

14. The hepatic portal vein
    a. supplies the capillaries of the intestines.
    b. carries absorbed nutrients to the liver for processing.
    c. carries blood from the liver to the heart.
    d. drains the lacteals of the villi.
    e. supplies oxygenated blood to the liver.

15. The most likely action of antidiarrhea medicine is to
    a. speed up peristalsis in the small intestine.
    b. speed up peristalsis in the large intestine.
    c. kill *E. coli* in the intestine.
    d. increase water reabsorption in the large intestine.
    e. increase salt secretion into the feces.

16. Which of the following is *not* a common component of feces?
    a. intestinal bacteria
    b. cellulose
    c. saturated fats
    d. bile pigments
    e. undigested materials

17. Ruminants
    a. eat antlers or bones in order to obtain phosphorus.
    b. house microorganisms that digest cellulose.
    c. eat their feces to obtain nutrients digested from cellulose by microorganisms.
    d. house symbiotic bacteria and protists in a large cecum.
    e. get all of their nutrition from digested plant material.

18. Which of the following statements is *false*?
    a. The average human has enough stored fat to supply calories for several weeks.
    b. An increase in leptin levels leads to an increase in appetite and weight gain.
    c. Conversion of glucose and glycogen takes place in the liver.
    d. After glycogen stores are filled, excessive calories are stored as fat, regardless of their original food source.
    e. Carbohydrates and fats are preferentially used as fuel before proteins are used.

19. You and two friends are trying a cereal diet to lose weight. You each eat 30 g of cereal, six times a day. You choose an organic cereal with nuts, grains, and no added sugar that has 3 g of fat, 20 g of carbohydrates (4 g of which are nondigestible insoluble fiber), and 7 g of protein. Friend 1 chooses a high-fiber cereal with 1 g of fat, 27 g of carbohydrates (9 g of insoluble fiber), and 2 g of protein. Friend 2 chooses a sugary cereal with 0 g of fat, 28 g of carbohydrates (3 g of insoluble fiber), and 2 g of protein. Predict the order, from fastest to slowest, in which the three of you will lose weight.
    a. you (because you ate the fewest carbs), then Friend 1 (because 9 g of the carbohydrates eaten are not digestible), then Friend 2

    b. Friend 1 (who ate 9 g of insoluble fiber), then you, then your sugary Friend 2
    c. Friend 1, then Friend 2, then you (because each of your fat grams contains approximately two times the energy content of carbohydrate or protein)
    d. Friend 2 (sugar burns the fastest), then Friend 1, then you
    e. you (because proteins are not used as fuel, those 7 g of protein do not count), then Friend 2, then Friend 1

20. Studies with mice have identified the ob^+ gene for the production of leptin (the satiety factor) and the db^+ gene for the leptin receptor. An obese mutant mouse has a genotype of ob^+, db. The amount of leptin found in this mouse's blood would be
    a. a low level because of the mouse's mutant ob gene.
    b. a high level because of the mouse's large amount of adipose tissue.
    c. a high level because the mouse produces an excess of leptin receptor.
    d. a low level because this obese mouse must not be producing the satiety factor and thus continues to overeat.
    e. impossible to predict from this obese mouse's genotype.

MATCHING: *Match the description with the correct enzyme, hormone, or other substance.*

_____ 1. enzyme that hydrolyzes peptide bonds in the stomach

_____ 2. hormone that stimulates secretion of gastric juice

_____ 3. hormone that stimulates release of bile and pancreatic enzymes; also inhibits peristalsis

_____ 4. lipid transport globule

_____ 5. hormone secreted by adipose cells that suppresses appetite

_____ 6. substances that emulsify fats

_____ 7. enzyme that hydrolyzes fats

_____ 8. hormone that is secreted by the stomach and triggers hunger

A. aminopeptidase
B. bile salts
C. cholecystokinin
D. chylomicron
E. gastrin
F. ghrelin
G. insulin
H. leptin
I. lipase
J. pepsin

Circulation and Gas Exchange

Framework

This chapter surveys the circulatory and gas exchange systems found in animals, with special attention given to the human systems and the problems of cardiovas- cular disease. The following concept map organizes some of the chapter's key ideas.

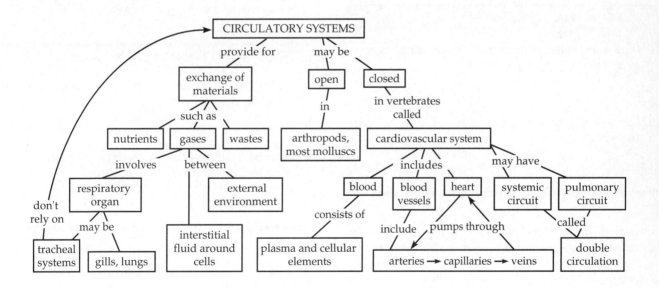

Chapter Review

42.1 Circulatory systems link exchange surfaces with cells throughout the body

A cell's aqueous environment must provide O_2 and nutrients and permit disposal of CO_2 and metabolic wastes. Because diffusion is too slow a process across distances more than a few millimeters, most animals have organs that exchange materials with the environment and a circulatory system to service body cells.

Gastrovascular Cavities   The central gastrovascular cavity inside the two-cell-thick body wall of cnidarians serves for both the digestion and transport of materials. The internal fluid exchanges directly with the aqueous environment through the single opening. Flatworms also have gastrovascular cavities that branch throughout their thin, flat bodies.

Evolutionary Variation in Circulatory Systems   The circulatory systems of more complex animals consist of a pumping **heart,** a circulatory fluid, and transport vessels.

Arthropods and most molluscs have an **open circulatory system.** The *interstitial fluid* in sinuses (spaces) between organs, called **hemolymph,** bathes the internal tissues and provides for chemical exchange.

Cephalopods, annelids, and vertebrates have **closed circulatory systems,** in which **blood** remains in vessels and materials are exchanged between the blood and interstitial fluid bathing the cells.

INTERACTIVE QUESTION   42.1

a. How is the hemolymph of an open circulatory system moved through the body?

b. Compare the advantages of open and closed circulatory systems.

Organization of Vertebrate Circulatory Systems   In the **cardiovascular system** of vertebrates, **arteries** carrying blood away from the heart branch into tiny **arterioles** within organs. Arterioles then divide into microscopic **capillaries,** which infiltrate tissues in networks called **capillary beds.** Capillaries converge to form **venules,** which meet to form the **veins** that return blood to the heart. In a few instances, portal veins connect two capillary beds without the blood first returning to the heart.

The vertebrate heart has one or two **atria,** which receive blood, and one or more **ventricles,** which pump blood out of the heart.

The ventricle of the two-chambered heart of a fish pumps blood to the gills, from which the oxygen-rich blood flows through a vessel to capillary beds in the other organs. Veins return the oxygen-poor blood to the atrium. Passage through two capillary beds slows the flow of blood, but body movements help to maintain circulation.

In the three-chambered heart of amphibians, the single ventricle pumps blood into both the **pulmocutaneous circuit,** which leads to the lungs and skin and then back to the left atrium, and the **systemic circuit,** which carries blood to the rest of the body and back to the right atrium. This double circulation repumps blood after it returns from the capillary beds of the lungs or skin, ensuring a strong flow of oxygen-rich blood to the brain, muscles, and body organs.

The three-chambered heart of lizards, snakes, and turtles has a partially divided ventricle that helps to separate blood flow through the **pulmonary circuit** to the gas exchange tissues in the lungs and through the systematic circuit to the body tissues.

Delivery of oxygen for cellular respiration is most efficient in birds and mammals, which, as endotherms, have high oxygen and energy demands. The left side of the large and powerful four-chambered heart receives and pumps oxygen-rich blood only, whereas the right side receives and pumps oxygen-poor blood.

INTERACTIVE QUESTION   42.2

a. Explain the difference between a **single circulation** and a **double circulation.**

b. Describe the difference between the circulatory system of amphibians and that of mammals.

42.2 Coordinated cycles of heart contraction drive double circulation in mammals

Mammalian Circulation   Review the circulation of blood through a mammalian cardiovascular system by completing Interactive Question 42.3.

INTERACTIVE QUESTION 42.3

In the following diagram of a mammalian circulatory system, label the indicated parts, color the vessels that carry oxygen-rich blood red, and then trace the flow of blood by numbering the circles from 1–11. Start numbering by assigning 1 to the right ventricle.

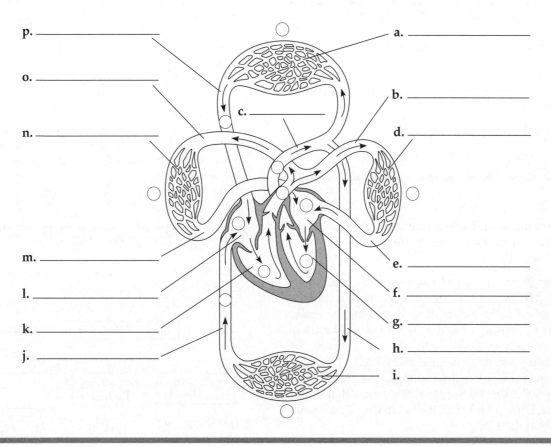

p. _____ a. _____

o. _____

b. _____

c. _____ d. _____

n. _____

m. _____ e. _____

l. _____ f. _____

k. _____ g. _____

h. _____

j. _____ i. _____

The Mammalian Heart: A Closer Look The human heart, located just beneath the sternum, is composed mostly of cardiac muscle. The atria have relatively thin muscular walls, whereas the ventricles have thicker walls. The **cardiac cycle** consists of **systole,** during which the cardiac muscle contracts and the chambers pump blood, and **diastole,** when the heart chambers are relaxed and filling with blood.

The **cardiac output,** or volume of blood pumped by each ventricle per minute, depends on the **heart rate** and **stroke volume** (the quantity of blood pumped by each contraction).

Atrioventricular (AV) valves between each atrium and ventricle are snapped shut when blood is forced against them as the ventricles contract. The **semilunar valves** where the aorta and pulmonary artery exit the heart are forced open by ventricular contraction and close when the ventricles relax.

The heart sounds are caused by the recoil of blood against the closed heart valves. A **heart murmur** is the detectable hissing sound of blood leaking back through a defective valve.

Maintaining the Heart's Rhythmic Beat The rhythm of heart contractions is coordinated by the **sinoatrial (SA) node,** or *pacemaker,* a region of specialized autorhythmic muscle cells located in the wall of the right atrium. The SA node initiates an electrical impulse that spreads via the intercalated disks of the cardiac muscle cells, and the two atria contract. After a 0.1-second delay, the **atrioventricular (AV) node,** located between the left and right atria, relays the impulse through specialized fibers to the ventricles. The electrical currents produced during the heart cycle can be detected by electrodes placed on the skin and recorded in an **electrocardiogram (ECG or EKG).**

The SA node is controlled by sympathetic and parasympathetic nerves (which speed or slow the heart rate) and is influenced by hormones and body temperature.

INTERACTIVE QUESTION   42.4

a. Name the valve between an atrium and ventricle. _____

b. Name the valve between a ventricle and the aorta or pulmonary artery. _____

c. Which node initiates atrial systole? _____

d. Which node initiates ventricular systole? _____

42.3   Patterns of blood pressure and flow reflect the structure and arrangement of blood vessels

Blood Vessel Structure and Function   The wall of an artery or vein consists of three layers: an outer connective tissue layer with elastic fibers that allow the vessel to stretch and recoil; a middle layer of smooth muscle and more elastic fibers; and an inner lining of **endothelium,** a single layer of flattened cells. The outer two layers are thicker in arteries than they are in veins. One-way valves in the larger veins assure that blood cannot backflow away from the heart. Capillaries have only an endothelium.

Blood Flow Velocity   The enormous number of capillaries creates a large total cross-sectional area, resulting in a very slow flow of blood that provides opportunity for exchange of substances with interstitial fluid. Blood flow speeds up within venules and veins due to the decrease in total cross-sectional area.

Blood Pressure   Blood pressure, the force that drives blood from the heart to the capillary beds, is highest during ventricular systole **(systolic pressure).** Resistance caused by the narrow openings of the arterioles impeding the exit of blood from arteries causes the swelling of the arteries during systole, which can be felt as the **pulse.** The recoiling of the stretched elastic arteries creates **diastolic pressure** and maintains a continuous flow of blood into arterioles and capillaries.

Vasoconstriction, a reduction in vessel diameter due to contraction of smooth muscles in arteriole walls, increases resistance and thus increases blood pressure. **Vasodilation** of arterioles as smooth muscles relax increases blood flow into the arterioles and thus lowers

blood pressure. Nervous and hormonal signals control the production of the gas nitric oxide (NO) and the peptide endothelin, which induce vasodilation and vasoconstriction, respectively. Changes in cardiac output also affect blood pressure.

Arterial blood pressure may be measured with a sphygmomanometer. The first number is the pressure during systole (when blood first spurts through the artery that was closed off by the cuff), and the second number is the pressure during diastole.

Blood pressure is negligible by the time blood exits the capillary beds. Contraction of the smooth muscle walls of venules and veins and the contraction of skeletal muscles between which veins are embedded force blood to flow back to the heart. Pressure changes during breathing also draw blood into the large veins in the thoracic cavity.

INTERACTIVE QUESTION   42.5

Complete the following concept map to help you organize your understanding of blood pressure.

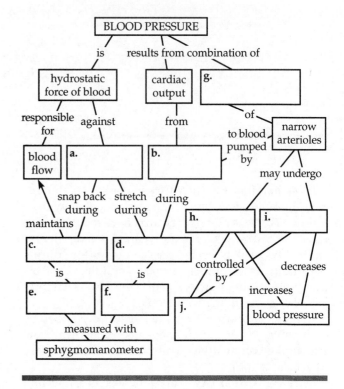

Capillary Function   Only about 5–10% of the body's capillaries have blood flowing through them at any one time. Capillaries branch off thoroughfare channels that are direct connections between arterioles and venules.

The distribution of blood to capillary beds varies with need and is regulated by nerve impulses, hormones, and local chemicals that control contraction of smooth muscles in arteriole walls and in *precapillary sphincters* at the entrance to capillary beds.

The exchange of substances between blood and interstitial fluid may occur by endocytosis and exocytosis by endothelial cells, by passive diffusion of small molecules across endothelial cells, and by movement through microscopic pores in the capillary wall. Water, sugars, salts, and urea move through these openings, whereas blood cells and proteins are too large to fit through.

INTERACTIVE QUESTION 42.6

Explain how edema, the accumulation of fluid in body tissues, can result from a severe dietary protein deficiency that decreases the concentration of blood plasma proteins.

Fluid Return by the Lymphatic System The fluid lost from capillaries and any proteins that may have leaked out are returned to the blood through the **lymphatic system.** Fluid diffuses into tiny lymph vessels intermingled with blood capillaries. The fluid, called **lymph,** moves through lymph vessels containing one-way valves, mainly as a result of the movement of skeletal muscles, and is returned to the circulatory system near the junction of the venae cavae with the right atrium. In **lymph nodes,** the lymph is filtered, and white blood cells attack viruses and bacteria.

42.4 Blood components function in exchange, transport, and defense

Blood Composition and Function Cells and cell fragments make up about 45% of the volume of blood; the rest is a liquid matrix called **plasma.**

The plasma consists of a large variety of solutes dissolved in water. The collective and individual concentrations of the dissolved ions of inorganic salts (electrolytes) are important to osmotic balance, pH levels, and the functioning of muscles and nerves.

Plasma proteins function in the osmotic balance of blood and as buffers, antibodies, escorts for lipids, and clotting factors. Blood plasma from which clotting factors have been removed is called *serum.* Nutrients, metabolic wastes, gases, and hormones are transported in the plasma.

Erythrocytes (red blood cells) transport O_2. The red blood cells of mammals lack nuclei. Their small size and biconcave shape create a large surface area of plasma membrane across which O_2 can diffuse. Erythrocytes are packed with **hemoglobin,** an iron-containing protein that binds O_2.

In **sickle-cell disease,** abnormal hemoglobin molecules cause erythrocytes to distort into a sickle shape, resulting in blocked vessels and other serious effects. Ruptured sickled erythrocytes can lead to anemia.

There are five major types of **leukocytes,** or white blood cells, all of which fight infections. Some white cells are phagocytes that engulf bacteria and cellular debris; lymphocytes give rise to cells that produce the immune response. Leukocytes are also found in interstitial fluid and lymph nodes.

Platelets, pinched-off fragments of specialized bone marrow cells, are involved in blood clotting. The clotting process usually begins when platelets, clumped together along a damaged endothelium, release clotting factors. By a series of steps, the enzyme *thrombin* converts the plasma protein *fibrinogen* to its active form, *fibrin.* The threads of fibrin form a patch. Hemophilia is caused by an inherited defect in any step of the complex clotting process.

Anticlotting factors normally prevent the clotting of blood in the absence of injury. A **thrombus** is a clot that occurs within a blood vessel and blocks the flow of blood.

Red and white blood cells and platelets are produced from multipotent **stem cells** in red bone marrow. Erythrocytes circulate for about 4 months before they are phagocytosed by cells in the liver and spleen and their components recycled. The production of red blood cells is controlled by a negative feedback mechanism involving the hormone **erythropoietin (EPO),** which is produced by the kidney in response to low O_2 supply in tissues.

INTERACTIVE QUESTION   42.7

Complete the following concept map of the components of vertebrate blood and their functions.

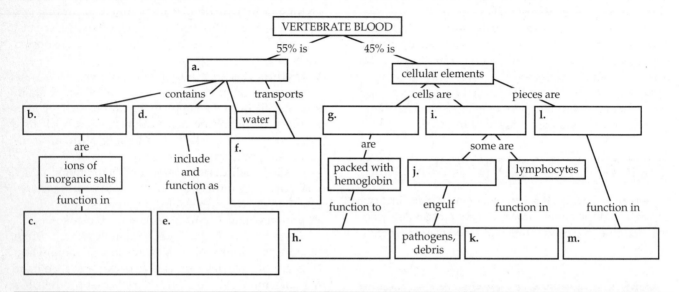

Cardiovascular Disease   Cholesterol metabolism and *inflammation* are factors in the development of cardiovascular diseases, disorders of the heart and blood vessels. Cholesterol is carried to body cells for inclusion in cell membranes in particles called **low-density lipoproteins (LDLs)**. **High-density lipoproteins (HDLs)** return excess cholesterol to the liver. A high LDL/HDL ratio is a risk factor for heart disease.

In the chronic cardiovascular disease known as **atherosclerosis,** fatty deposits called plaques narrow the arteries. Injury to or infection of the lining of an artery can lead to *inflammation*. Leukocytes are attracted and take up accumulating lipoproteins. The plaque grows, becomes covered with a fibrous cap, and may obstruct the artery or rupture.

A **heart attack,** or *myocardial infarction,* is the death of cardiac muscle due to blockage of coronary arteries; a **stroke,** the death of nervous tissue in the brain, results from blockage or rupture of arteries in the head.

Statins are drugs used to lower LDL levels in individuals at risk of cardiovascular disease. High levels of C-reactive protein (CRP), indicative of acute inflammation, also indicate disease risk.

Hypertension, or high blood pressure, is thought to damage the endothelium and initiate plaque formation, promoting atherosclerosis and increasing the risk of heart attack and stroke. This condition can be easily diagnosed and controlled by drugs, diet, and exercise.

INTERACTIVE QUESTION   42.8

What lifestyle choices have been correlated with increased risk of cardiovascular disease?

42.5   Gas exchange occurs across specialized respiratory surfaces

Gas exchange, or respiration, is the uptake of O_2 from and the discharge of CO_2 to the environment.

Partial Pressure Gradients in Gas Exchange   The concentration of a gas in air or dissolved in water is that gas's **partial pressure.** At sea level, the partial pressure of O_2, which makes up 21% of the atmosphere, is 160 mm Hg (0.21 × 760 mm, which is atmospheric pressure at sea level). A gas diffuses from a region of higher partial pressure to a region of lower partial pressure.

Respiratory Media   Air is a respiratory medium with plentiful O_2. Water has much less O_2 and greater density and viscosity. Thus gas exchange in water is often more energetically demanding.

Respiratory Surfaces   The respiratory surface, where gas exchange with the respiratory medium occurs, must be moist, thin, and have a large enough area to supply the metabolic needs of the animal.

In animals with simple body forms, gases can easily diffuse in and out of body cells. Animals with denser bodies usually have a branched or folded respiratory surface associated with a rich blood supply.

Gills in Aquatic Animals Gills are outfoldings of the body surface, ranging from simple bumps on echinoderms to the more complex gills of molluscs, crustaceans, and fishes. To maintain partial pressure gradients of O_2 and CO_2, gills usually require **ventilation,** or movement of the respiratory medium across the respiratory surface.

In the gills of a fish, blood flows through capillaries in a direction opposite that of the flow of water. This arrangement sets up a **countercurrent exchange,** in which the O_2 diffusion gradient favors the movement of O_2 into the blood along the length of the capillary.

Tracheal Systems in Insects **Tracheal systems** are tiny air tubes that branch throughout the body and come into contact with nearly every cell. Thus, an insect's open circulatory system does not function in O_2 and CO_2 transport. In large insects, rhythmic muscle contractions ventilate tracheal systems.

Lungs **Lungs** are invaginated respiratory surfaces restricted to one location. A circulatory system transports gases between the lungs and the body.

The lungs of mammals are located in the thoracic cavity. Air entering through the nostrils is filtered, warmed, humidified, and smelled in the nasal cavity. Air passes through the pharynx and enters the respiratory tract through the glottis. The **larynx** moves up and tips the epiglottis over the glottis when food is swallowed. The larynx functions as a voice box in most mammals; exhaled air vibrates a pair of *vocal folds* (called *vocal cords* in humans). The **trachea,** or windpipe, branches into two **bronchi,** which then branch repeatedly into **bronchioles** within each lung. Ciliated, mucus-coated epithelium lines the major branches of the respiratory tree.

Alveoli are multilobed air sacs encased in a web of capillaries at the tips of the tiniest bronchioles. Gas exchange takes place across the thin, moist epithelium of each alveolus. These tiny sacs are coated with **surfactants,** which decrease surface tension and help keep the sacs from sticking shut.

INTERACTIVE QUESTION 42.9

What causes *respiratory distress syndrome (RDS)* in infants?

42.6 Breathing ventilates the lungs

Vertebrate lungs are ventilated by **breathing,** the alternate inhalation and exhalation of air.

How an Amphibian Breathes A frog uses **positive pressure breathing.** It lowers the floor of the oral cavity, drawing air into the expanding cavity; raising the floor of the oral cavity then pushes air into the lungs.

How a Bird Breathes Birds have air sacs that act as bellows to maintain air flow through the lungs. The air sacs and lungs are ventilated through a circuit that includes a one-way passage through tiny, gas-exchange channels called *parabronchi.*

How a Mammal Breathes Mammals ventilate their lungs by **negative pressure breathing.** Contraction of the rib muscles and the **diaphragm** expands the chest cavity and increases the volume of the lungs. The reduction in air pressure in this increased volume draws air into the lungs. Relaxation of the rib muscles and diaphragm compresses the lungs, increasing the pressure and forcing air out.

Tidal volume is the volume of air inhaled and exhaled by an animal during normal breathing. The maximum volume during forced breathing is called **vital capacity.** The **residual volume** is the air that remains in the alveoli and lungs after forceful exhaling.

Control of Breathing in Humans Breathing is under automatic regulation by neural circuits that form a *breathing control center* in the medulla oblongata. (The pons, near the medulla, also plays a role in regulating breathing.) Nerve impulses from the medulla instruct the rib muscles and diaphragm to contract. When the medulla senses a drop in the pH of the *cerebrospinal fluid*, which is caused by an increase in blood CO_2 concentration, its control circuits increase the depth and rate of breathing. Although breathing centers respond primarily to CO_2 levels, they are alerted by oxygen sensors in the aorta and carotid arteries that react to severe deficiencies of O_2.

INTERACTIVE QUESTION   42.10

For the following animals, indicate the respiratory medium, respiratory surface, and means of ventilation.

Animal	Respiratory Medium	Respiratory Surface	Means of Ventilation
Fish	a.	b.	c.
Grasshopper	d.	e.	f.
Frog	g.	h.	i.
Human	j.	k.	l.

42.7   Adaptations for exchange include pigments that bind and transport gases

Coordination of Circulation and Gas Exchange   Blood entering the capillaries of the lungs has a lower P_{O_2} and a higher P_{CO_2} than does the air in the alveoli, so O_2 diffuses into the capillaries and CO_2 diffuses out. In the tissue capillaries, pressure differences favor the diffusion of O_2 out of the blood into the interstitial fluid and of CO_2 into the blood.

Respiratory Pigments   In most animals, O_2 is carried by **respiratory pigments** in the blood. Copper is the oxygen-binding component in the blue respiratory protein *hemocyanin,* common in arthropods and many molluscs.

Hemoglobin, contained in erythrocytes, is the respiratory pigment of almost all vertebrates. Hemoglobin is composed of four subunits, each of which has a cofactor called a heme group with iron at its center. The binding of O_2 to the iron atom of one subunit induces a change in shape of the other subunits, and their affinity for O_2 increases. Likewise, the unloading of the first O_2 lowers the other subunits' affinity for O_2.

The dissociation curve for hemoglobin shows the relative amounts of O_2 bound to hemoglobin under varying O_2 partial pressures. In the steep part of this S-shaped curve, a slight change in partial pressure will cause hemoglobin to load or unload a substantial amount of O_2. Because of the **Bohr shift** in response to the lower pH of active tissues, hemoglobin releases more O_2 in such tissues.

About 70% of CO_2 is transported in blood as bicarbonate ions. CO_2 enters red blood cells, where it first combines with H_2O to form carbonic acid (catalyzed by carbonic anhydrase) and then dissociates into H^+ and HCO_3^-. Bicarbonate moves into the plasma for transport. The H^+ binds to hemoglobin and other proteins, which minimizes changes in pH during the transport of CO_2. In the lungs, the diffusion of CO_2 out of the blood shifts the equilibrium in favor of the conversion of HCO_3^- back to CO_2, which is unloaded from the blood.

INTERACTIVE QUESTION   42.11

The following graph shows dissociation curves for hemoglobin at two different pH values. Explain the significance of these two curves.

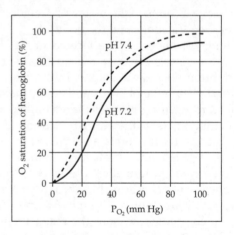

Respiratory Adaptations of Diving Mammals   Special adaptations have enabled some air-breathing animals to make sustained underwater dives. Weddell seals store twice the amount of O_2 per kilogram of body weight as do humans, mostly by having a larger volume of blood and a higher concentration of **myoglobin,** an oxygen-storing muscle protein. Oxygen-conserving adaptations during a dive include decreasing the heart rate, rerouting blood to the brain and essential organs, and restricting blood supply to the muscles, which use fermentation to produce ATP during long dives.

Word Roots

alveol- = a cavity (*alveolus:* one of the dead-end, multi-lobed air sacs where gas exchange occurs in a mammalian lung)

atrio- = a vestibule; **-ventriculo** = ventricle (*atrioventricular node:* a region of specialized heart muscle tissue between the left and right atria where electrical impulses are delayed before spreading to both ventricles and causing them to contract)

cardi- = heart; **-vascula** = a little vessel (*cardiovascular system:* the closed circulatory system characteristic of vertebrates)

counter- = opposite (*countercurrent exchange:* exchange of a substance or heat between two fluids flowing in opposite directions)

endo- = inner (*endothelium:* the simple squamous layer of cells lining the lumen of blood vessels)

erythro- = red; **-poiet** = produce (*erythropoietin:* a hormone that stimulates the production of erythrocytes. It is secreted by the kidney when body tissues do not receive enough oxygen)

hemo- = blood; **-lymph** = clear fluid (*hemolymph:* in invertebrates with an open circulatory system, the body fluid that bathes tissues)

leuko- = white; **-cyte** = cell (*leukocyte:* a white blood cell; functions in fighting infections)

multi- = many; **-potent** = powerful (*multipotent stem cell:* a bone marrow cell that is a progenitor for any kind of blood cell)

myo- = muscle (*myoglobin:* an oxygen-storing, pigmented protein in muscle cells)

pulmo- = a lung; **-cutane** = skin (*pulmocutaneous circuit:* a branch of the circulatory system in many amphibians that supplies the lungs and skin)

semi- = half; **-luna** = moon (*semilunar valve:* a valve located where the aorta leaves the left ventricle and the pulmonary artery leaves the right ventricle)

thrombo- = a clot (*thrombus:* a fibrin-containing clot that forms in a blood vessel and blocks the flow of blood)

Structure Your Knowledge

1. Identify the labeled structures in the following diagram of a human heart. Draw arrows to trace the flow of blood.

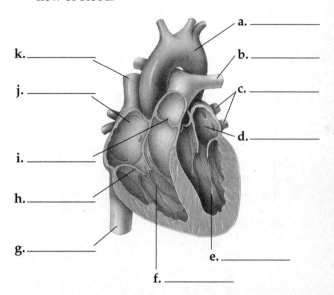

2. Trace the path of a molecule of O_2 from its inhalation into the nasal cavity to its delivery to the kidney.

Test Your Knowledge

MULTIPLE CHOICE: *Choose the one best answer.*

1. A gastrovascular cavity
   a. is found in cnidarians and annelids.
   b. pumps fluid through short vessels from which the fluid spreads throughout the body.
   c. functions in both digestion and distribution of nutrients.
   d. has a single opening for ingestion and elimination, but a separate opening for gas exchange.
   e. involves all of the above.

2. Which of the following statements does *not* describe a similarity between open and closed circulatory systems?
   a. Some sort of pumping device helps to move blood through the body.
   b. Some of the circulation of blood results from body movements.
   c. The blood and interstitial fluid are distinguishable from each other.
   d. All tissues come into close contact with the circulating body fluid so that the exchange of nutrients and wastes can take place.
   e. All of these apply to both open and closed circulatory systems.

3. In a system with double circulation,
   a. blood is usually pumped at two separate locations as it circulates through the body.
   b. there is a countercurrent exchange in the gills.
   c. hemolymph circulates both through a pumping blood vessel and through body sinuses.
   d. outgoing blood is pumped to the gas-exchange organ, and returning blood is pumped to a systemic circuit.
   e. there are always two ventricles and often two atria.

4. Which of the following statements is the best explanation for the presence of four-chambered hearts in both birds and mammals?
   a. They shared a common ancestor that had a four-chambered heart.
   b. They are the only vertebrates with double circulation, which requires four heart chambers.
   c. They are both endotherms, and the evolution of efficient circulatory systems supported the high metabolic rate of endotherms.
   d. This is an example of convergent evolution, because animals that obtain their O_2 from air require both a pulmonary circuit and a systemic circuit.
   e. The more inefficient single atrium of amphibians and most reptiles could not supply the higher O_2 needs of these endotherms.

5. During diastole,
   a. the atria fill with blood.
   b. blood flows passively into the ventricles.
   c. the elastic recoil of the arteries maintains hydrostatic pressure on the blood.
   d. semilunar valves are closed, but atrioventricular valves are open.
   e. all of the above are occurring.

6. An atrioventricular valve prevents the backflow or leakage of blood from
   a. the right ventricle into the right atrium.
   b. the left atrium into the left ventricle.
   c. the aorta into the left ventricle.
   d. the pulmonary vein into the right atrium.
   e. the right atrium into the vena cava.

7. As a general rule, blood leaving the right ventricle of a mammal's heart will pass through how many capillary beds before it returns to the right ventricle?
   a. one
   b. two
   c. three
   d. one or two, depending on the circuit it takes
   e. at least two, but almost always three

8. The nurse tells you that your blood pressure is 112/70. The "70" refers to
   a. your heart rate.
   b. the velocity of blood during diastole.
   c. the systolic pressure from ventricular contraction.
   d. the diastolic pressure from the recoil of the arteries.
   e. the venous pressure caused by the compression of the blood pressure cuff.

9. Blood flows more slowly in arterioles than in arteries because arterioles
   a. have thoroughfare channels to venules that are often closed off.
   b. collectively have a larger cross-sectional area than do arteries.
   c. must provide opportunity for exchange with the interstitial fluid.
   d. have sphincters that restrict blood flow to capillary beds.
   e. are narrower than arteries.

Use the following to answer questions 10–13.

1. vena cava	4. right atrium
2. left ventricle	5. aorta
3. pulmonary vein	6. pulmonary capillaries

10. Which of the following sequences represents the flow of blood through the human body? (Obviously, many structures are not included.)
   a. 1-5-3-6-2-4   d. 1-2-3-6-4-5
   b. 1-2-6-3-4-5   e. 1-4-6-3-2-5
   c. 1-4-3-6-2-5

11. In which vessel is blood pressure the highest?
   a. 1   d. 5
   b. 3   e. 6
   c. 4

12. In which vessel is the velocity of blood flow the lowest?
   a. 1   d. 5
   b. 3   e. 6
   c. 4

13. Of these structures, which has the thickest muscle layer?
   a. 1   d. 4
   b. 2   e. 5
   c. 3

14. Which of the following processes is *not* a factor in the exchange of substances in capillary beds?
   a. endocytosis and exocytosis
   b. passive diffusion
   c. blood pressure at the arterial end of a capillary
   d. bulk flow through pores in the capillary wall
   e. active transport by white blood cells

15. Which of the following processes would tend to reduce the blood supply to a capillary bed?
    a. release of EPO (erythropoietin)
    b. relaxation of precapillary sphincters
    c. vasodilation
    d. closing of thoroughfare channel
    e. release of endothelin

16. Which of the following changes would you expect following an increase in the concentration of blood proteins?
    a. the swelling of body tissues (edema)
    b. the swelling of lymph glands
    c. an increase in the outward movement of fluid at the arterial end of capillaries
    d. a decrease in lymph flow through the lymphatic system
    e. an increase in the delivery of nutrients to body tissues

17. The functions of the kidney include
    a. production of the hormone erythropoietin to increase production of red blood cells.
    b. maintenance of electrolyte balance.
    c. destruction and recycling of red blood cells.
    d. both a and b.
    e. a, b, and c.

18. Fibrinogen is
    a. a blood protein that escorts lipids through the circulatory system.
    b. a cell fragment involved in the blood-clotting mechanism.
    c. a blood protein that is converted to fibrin to form a blood clot.
    d. a leukocyte involved in trapping bacteria.
    e. a lymph protein that regulates osmotic balance in the tissues.

19. A thrombus
    a. may form at a site of plaque formation in an artery.
    b. is a typical complication of hemophilia.
    c. may cause hypertension.
    d. narrows the diameter of arteries due to the buildup of plaque.
    e. is often associated with high levels of HDLs in the blood.

20. Which of the following functions is *not* associated with plasma proteins in a mammal?
    a. pH buffer
    b. osmotic balance
    c. clotting agent
    d. oxygen transport
    e. antibody

21. In countercurrent exchange,
    a. the flow of fluids in opposite directions maintains a favorable diffusion gradient along the length of an exchange surface.
    b. O_2 is exchanged for CO_2.
    c. double circulation keeps oxygenated and deoxygenated blood separate.
    d. O_2 moves from a region of high partial pressure to one of low partial pressure, but CO_2 moves in the opposite direction.
    e. the capillaries of the lung pick up more O_2 than do tissue capillaries.

22. The tracheae of insects
    a. are stiffened with chitinous rings and lead into the lungs.
    b. are filled by positive pressure breathing.
    c. ramify along capillaries of the circulatory system for gas exchange.
    d. are highly branched and come into contact with almost every cell to effect gas exchange.
    e. provide an extensive, fluid-filled system of tubes for gas exchange.

23. What do the alveoli of mammalian lungs, the gill filaments of fish, and the tracheoles of insects all have in common?
    a. They use a circulatory system to transport absorbed O_2.
    b. Their respiratory surfaces are invaginated.
    c. They function using countercurrent exchange.
    d. They have a large, thin surface area for gas exchange.
    e. They have all of the above in common.

24. The cilia in the trachea and bronchi
    a. move air into and out of the lungs.
    b. increase the surface area for gas exchange.
    c. vibrate when air rushes past them to produce sounds.
    d. filter the air that rushes through them.
    e. sweep mucus containing trapped particles up and out of the respiratory tract.

25. When you hold your breath, which of the following leads to the urge to breathe?
    a. falling O_2
    b. falling CO_2
    c. rising CO_2
    d. falling blood pH
    e. both c and d

26. Which of the following mechanisms is *not* involved in increasing the respiration rate?
    a. nervous and chemical signals
    b. stretch receptors in the lungs
    c. impulses from a breathing center in the medulla
    d. severe deficiencies of oxygen
    e. a drop in the pH of cerebrospinal fluid

27. Which of the following statements is the best explanation for why birds can fly over the Himalayas but most humans require bottled oxygen to climb these mountains?

    a. Birds are much smaller and require less O_2.

    b. Birds use positive pressure breathing, whereas humans use negative pressure breathing.

    c. With a one-way flow of air and efficient ventilation, the lungs of birds extract more O_2 from the air they breathe.

    d. The circulatory system of birds is much more efficient in delivering O_2 to body tissues.

    e. The vital capacity of bird lungs, relative to their size, is much greater than that of humans.

28. Gas exchange between maternal blood and fetal blood takes place in the placenta. Fetal hemoglobin and adult hemoglobin differ slightly in composition, and their dissociation curves are different such that O_2 can be transferred from maternal blood to fetal blood. If curve *c* in the following graph represents the dissociation curve of adult hemoglobin, which curve would represent the dissociation curve of fetal hemoglobin? (*Hint:* Must fetal hemoglobin have a higher or lower affinity for O_2 than maternal hemoglobin has?)

29. Why is more O_2 unloaded from hemoglobin in an actively metabolizing tissue than in a resting tissue, even at the same P_{O_2}.

    a. The pH of an active tissue is higher as more CO_2 combines with water to produce H^+ and HCO_3^-.

    b. At the lower pH of an actively metabolizing tissue, hemoglobin has a lower affinity for O_2.

    c. Hemoglobin binds four O_2 due to a shape change caused by the lower pH of active tissues.

    d. More O_2 is bound to myoglobin at lower O_2 concentrations.

    e. More CO_2 is drawn out of tissue cells as it is converted to bicarbonate in red blood cells.

30. Which of the following does *not* contribute to the respiratory adaptations of deep-diving animals?

    a. an unusually large blood volume

    b. a large quantity of myoglobin in the muscles

    c. a reduction in blood flow to noncritical body areas

    d. exceptionally large lungs

    e. movement upward or downward by changes in buoyancy instead of by active swimming

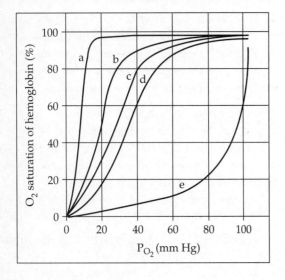

Chapter 43

The Immune System

Key Concepts

43.1 In innate immunity, recognition and response rely on traits common to groups of pathogens

43.2 In adaptive immunity, receptors provide pathogen-specific recognition

43.3 Adaptive immunity defends against infection of body fluids and body cells

43.4 Disruptions in immune system function can elicit or exacerbate disease

Framework

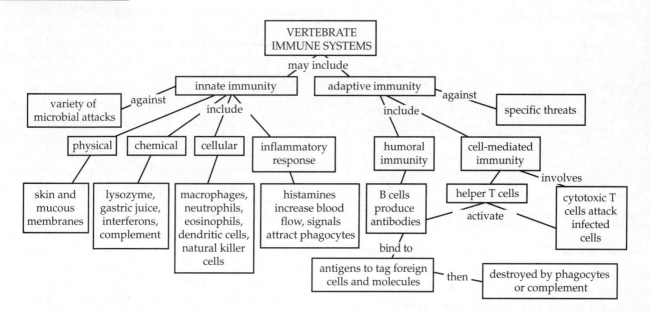

Chapter Review

The **immune system** defends against pathogens—infectious agents such as bacteria, viruses, protists, and fungi—that cause disease. **Innate immunity** includes external physical barriers and internal defenses that are active immediately upon exposure to a pathogen and are the same whether or not the pathogen was previously encountered. The immune system relies on *molecular recognition* to identify internal invaders that are nonself. In innate immunity, immune cells have a small group of receptor proteins that recognize a broad range of pathogens and can trigger a defensive response. **Adaptive immunity,** also called acquired immunity, is a line of defense in vertebrates in which immune cells react specifically to pathogens. A vast array of adaptive immune receptors enables recognition and response to specific pathogens, and the response is enhanced upon re-exposure to a pathogen.

43.1   In innate immunity, recognition and response rely on traits common to groups of pathogens

Innate Immunity of Invertebrates   Insect defenses begin with their protective exoskeleton. Together, the enzyme **lysozyme,** which attacks microbial cell walls, and chitin lining the intestine protect the digestive system. *Hemocytes* are circulating cells that can destroy bacteria by **phagocytosis,** trigger the production of chemicals that entrap multicellular parasites, and secrete *antimicrobial peptides* that kill fungi and bacteria. Different classes of pathogens bind with recognition proteins secreted by immune cells. The complex then activates distinct Toll receptors that initiate pathways for the production of antimicrobial peptides effective against that group of pathogens.

INTERACTIVE QUESTION   43.1

Explain why the data presented in the following graph indicate that defensin and not drosomycin is the antimicrobial peptide in fruit flies that is effective against infection by the bacterium *M. luteus*.

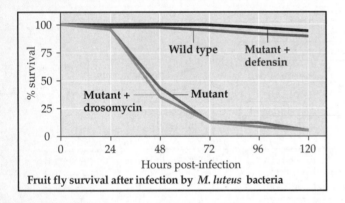

Fruit fly survival after infection by *M. luteus* bacteria

Innate Immunity of Vertebrates   In mammals, skin and the mucous membranes lining the digestive, respiratory, urinary, and reproductive tracts are physical barriers to microbes. The ciliated epithelial lining of the respiratory tract sweeps *mucus* containing trapped microbes upward and out. Lysozyme present in tears, saliva, and mucus kills bacteria. Skin secretions maintain a low pH, which discourages colonization by microbes. The acidity of gastric juice kills most microorganisms that reach the stomach.

Cellular innate defenses involve specific **Toll-like receptors (TLRs)** that recognize molecules that are common to a set of pathogens. Phagocytic white blood cells engulf a recognized microbe, and the resulting vacuole fuses with lysosomes containing toxic gases (such as nitric oxide) and digestive enzymes.

Neutrophils are phagocytic cells that circulate in the blood. **Macrophages** are larger phagocytic cells that migrate through the body or remain permanently in various organs, such as the spleen and other organs of the lymphatic system. Vessels of this system return interstitial fluid (lymph) to the blood after it is filtered through lymph nodes packed with defensive cells.

The primary role of **dendritic cells,** located in tissues in contact with the environment, is to stimulate adaptive immunity. *Eosinophils* are leukocytes that attack multicellular parasitic invaders with destructive enzymes. **Natural killer cells** recognize abnormal surface proteins on some virus-infected or cancerous cells and release chemicals that trigger target cell death.

Various antimicrobial peptides and proteins are released in response to pathogen recognition. Virus-infected cells produce **interferons,** which stimulate neighboring cells to produce substances that inhibit viral reproduction in those cells. Some white blood cells produce an interferon that activates macrophages.

The **complement system** is a group of about 30 proteins in the blood plasma that, when activated by contact with microbes, may lyse cells, trigger inflammation, or assist in adaptive defenses.

Signaling molecules released in response to physical injury or pathogen entry can trigger an **inflammatory response,** characterized by redness, swelling, heat, and pain. Damaged **mast cells** in connective tissue release **histamine,** a chemical that triggers dilation and leakiness of blood vessels; activated macrophages and neutrophils release **cytokines,** signaling molecules that promote blood flow to the damaged area and enhance an immune response. Vasodilation and signaling proteins result in the congregation of phagocytes and antimicrobial compounds. *Pus* is an accumulation of fluid, white blood cells, and dead microbes.

A systemic response to an infection may include an increase in the number of circulating white blood cells and a fever. Fevers, triggered by substances released by activated macrophages, may stimulate phagocytosis and speed tissue repair. *Septic shock* is a dangerous condition resulting from an overwhelming systemic inflammatory response.

Evasion of Innate Immunity by Pathogens   Certain pathogens may evade the innate immune system with an outer capsule that covers their surface molecules or by resisting breakdown and growing within the host's cells.

INTERACTIVE QUESTION 43.2

Complete the following table that summarizes the functions of the cells and chemicals of the innate defense mechanisms.

Cells or Compounds	Functions
Neutrophils	a.
Macrophages	b.
Eosinophils	c.
Dendritic cells	d.
Natural killer (NK) cells	e.
Mast cells	f.
Histamine	g.
Lysozyme	h.
Interferons	i.
Complement system	j.
Cytokines	k.

43.2 In adaptive immunity, receptors provide pathogen-specific recognition

Lymphocytes are the key cells of adaptive immunity. **T cells** migrate to the **thymus** to mature, whereas **B cells** remain in the bone marrow and mature there.

B cells and T cells become activated upon binding with an **antigen,** often a foreign protein or polysaccharide protruding from the surface of a pathogen or a toxin secreted by bacteria. B cells and T cells have thousands of identical membrane-bound **antigen receptors** that enable them to recognize a specific antigen. The small region of an antigen to which an antigen receptor binds is called an **epitope,** or *antigenic determinant.*

Antigen Recognition by B Cells and Antibodies   Each Y-shaped antigen receptor on a B cell consists of four polypeptide chains: two identical **light chains** and two identical **heavy chains,** linked together by disulfide bridges. Both heavy and light chains have *variable (V) regions* at the ends of the two arms of the Y, which form two identical antigen-binding sites. The *constant (C) regions* of the molecule vary little from cell to cell. The constant region of the heavy chains includes an anchoring transmembrane region and a cytoplasmic tail.

Activated B cells give rise to cells that secrete **antibodies,** or **immunoglobulins (Ig),** which are soluble antigen receptors.

Antigen Recognition by T Cells   A T cell antigen receptor, also anchored by a transmembrane region, consists of one α *chain* and one β *chain,* linked by a disulfide bridge. The variable regions at the tip of the molecule form a single antigen-binding site. T cells recognize small fragments of antigens displayed on the surface of cells. The displaying protein is called an **MHC (major histocompatibility complex) molecule.**

Host cells that are infected by or take in a pathogen cleave the antigen into small peptides. In a process called **antigen presentation,** MHC molecules bind with *antigen fragments* within the cell and then display those fragments on the cell's surface, where a T cell receptor can recognize the antigen–MHC molecule complex.

INTERACTIVE QUESTION 43.3

Describe the differences between the antigens that B cell antigen receptors recognize and the antigens that T cell antigen receptors recognize.

B Cell and T Cell Development   The four major characteristics of adaptive immunity include the immense diversity of antigen receptors, the self-tolerance of the system, the huge increase in B cells and T cells specific for an antigen following activation, and *immunological memory,* an enhanced response to a previously encountered foreign molecule.

Each person may have more than 1 million different B cell antigen receptors and 10 million different T cell antigen receptors. The genes coding for this remarkable diversity have numerous coding segments that are randomly and permanently rearranged. For example, in the Ig (immunoglobulin) light-chain gene, there are 40 variable (V) gene segments and five joining (J) gene segments. The J segments are followed by a single C segment, which codes for the constant region. Early in B cell development, one V gene segment is randomly linked to one J segment by an enzyme complex called *recombinase*, producing one of 200 possible light chains. These light chains combine with heavy chains that were produced the same way, yielding a huge diversity of antigen-binding sites.

In the bone marrow and thymus, respectively, the antigen receptors of maturing B and T cells are tested for self-reactivity. Lymphocytes with receptors specific for the body's own molecules are either inactivated or destroyed by *apoptosis*. This critical *self-tolerance* means that normally the body contains no mature lymphocytes that react against self components.

When an antigen interacts with antigen receptors on B cells or T cells that are specific for epitopes of that antigen, those particular lymphocytes are activated to divide repeatedly into a clone of identical cells. Some daughter cells develop into short-lived **effector cells,** which combat the antigen. Others develop into long-lived **memory cells.** By this process of **clonal selection,** a small number of cells is selected by their interaction with a specific antigen to produce thousands of identical cells keyed to that particular antigen.

The body mounts a **primary immune response** upon a first exposure to an antigen. About 10 to 17 days are required for selected lymphocytes to proliferate and yield the maximum response produced by effector T cells and the antibody-producing effector B cells, called plasma cells. Should the body reencounter the same antigen, the resulting **secondary immune response** is more rapid, effective, and prolonged. The long-lived T memory cells and B memory cells are responsible for this immunological memory. The secondary immune response provides long-term protection against a previously encountered pathogen.

INTERACTIVE QUESTION   43.4

Answer the following questions concerning the four major characteristics of adaptive immunity.

a. How is the great diversity of B and T cells produced?

b. What prevents B cells and T cells from reacting against the body's own molecules?

c. Describe clonal selection.

d. What is immunological memory?

43.3   Adaptive immunity defends against infection of body fluids and body cells

The **humoral immune response** involves the production of antibodies that help destroy toxins and pathogens in the blood and lymph. The **cell-mediated immune response** involves specialized T cells that identify and destroy infected host cells. Helper T cells aid both types of immune responses.

Helper T Cells: A Response to Nearly All Antigens
Both humoral and cell-mediated immune responses are triggered by **helper T cells.** Helper T cells recognize specific antigens displayed by **antigen-presenting cells.**

Most body cells have only class I MHC molecules, but dendritic cells, macrophages, and B cells have both class I and class II MHC molecules. These antigen-presenting cells engulf and fragment microbes, then display the antigen fragments in class II MHC molecules. A helper T cell antigen receptor binds to the antigen fragment, and a surface protein called CD4 enhances the binding by attaching to the class II MHC molecule. Signaling between the two cells via cytokines results in the proliferation of activated helper T cells. These helper T cells secrete cytokines that activate both B cells involved in humoral immunity and cytotoxic T cells, which are part of cell-mediated immunity.

The interactions of dendritic cells and macrophages with helper T cells result in a clone of activated helper T cells. A B cell presents antigens to an already activated helper T cell, which in turn activates the B cell.

INTERACTIVE QUESTION 43.5

Label the components in the following diagram that shows a helper T cell being activated by interaction with an antigen-presenting cell and the helper T cell's central role in activating both humoral and cell-mediated immunity.

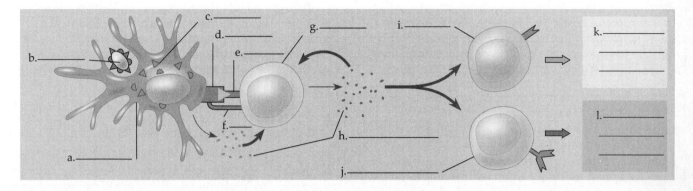

Cytotoxic T Cells: A Response to Infected Cells **Cytotoxic T cells** recognize nonself proteins that are synthesized in infected host cells and displayed with class I MHC molecules. An accessory protein (CD8) helps hold the cells together, and cytokines from nearby helper T cells activate the cytotoxic T cell. The activated cell secretes proteins that destroy the infected cell.

INTERACTIVE QUESTION 43.6

a. What does an activated helper T cell release?

b. What does a cytotoxic T cell attached to an infected body cell release?

B Cells and Antibodies: A Response to Extracellular Pathogens Upon first binding antigen, a B cell takes in a few foreign molecules by receptor-mediated endocytosis and presents antigen fragments in its class II MHC molecules to an activated helper T cell. The helper T cell had been activated by binding with a macrophage or dendritic cell displaying the same antigen. Aided by both direct contact with a helper T cell and the cytokines it secretes, an activated B cell proliferates into **plasma cells,** which produce and secrete antibodies.

When antibodies bind to viral proteins that appear on the surface of an infected cell, natural killer cells may be recruited.

INTERACTIVE QUESTION 43.7

Describe each of the following mechanisms by which antibodies mediate the disposal of antigens.

a. *neutralization*

b. *opsonization*

c. complement activation and formation of *membrane attack complex*

There are five types of constant regions, creating five *classes* of antibodies. IgD is the membrane-bound B cell antigen receptor. IgM is the first type of soluble antibody produced in response to a foreign antigen, followed by IgG, which is the most abundant antibody in blood.

Summary of the Humoral and Cell-Mediated Immune Responses Adaptive immunity defends against pathogens in the blood and lymph with antibodies secreted by B cells (humoral immune response) and against intracellular pathogens and cancer cells with cytotoxic T cells (cell-mediated immune response). Both responses depend on antigen-presenting cells activating helper T cells, which secrete cytokines that activate cytotoxic T cells and B cells.

Active and Passive Immunization **Active immunity** results when the body responds to an infection and develops immunological memory specific to that pathogen. In **passive immunity,** temporary immunity is provided by antibodies supplied through the placenta to a fetus (IgG antibodies) or through milk to a nursing infant. IgA antibodies are present in breast milk, tears, saliva, and mucus.

Active immunity can also develop from **immunization,** also called **vaccination.** A vaccine may be an inactivated toxin, a killed or weakened microbe, a portion of a microbe, or even genes for microbial proteins. In artificial passive immunization, antibodies from an immune animal are injected into a nonimmune animal to provide temporary protection.

Antibodies as Tools The specificity of antigen-antibody binding can provide tools for research, diagnosis, and treatment. Antibodies an animal produces are *polyclonal* because they were formed by several different B cell clones—each specific for a different epitope of an antigen. A technique for making **monoclonal antibodies** can supply quantities of identical antibodies.

Immune Rejection The immune response to the chemical markers that determine ABO blood groups must be considered in blood transfusions. Antibodies to blood group antigens other than the individual's own antigens arise in response to normal bacterial flora. These antigens circulate in the blood plasma and will induce a devastating reaction to transfused blood cells that possess matching antigens.

INTERACTIVE QUESTION 43.8

Fill in the following table to review your understanding of the antigens and antibodies of the ABO blood groups. Remember to compare the antigens on the donor cells with the antibodies in the recipient's plasma. (Assume that these are packed cell transfusions in which the antibodies in the donor's plasma are not a risk factor.)

Blood Type	Antigens on RBCs	Antibodies in Plasma	Can Receive Blood from	Can Donate Blood to

Transplanted tissues and organs are rejected because the foreign MHC molecules are antigenic and trigger immune responses. The use of closely related donors, as well as drugs that suppress immune responses, helps to minimize rejection.

In bone marrow transplants, used to treat leukemia and blood cell diseases, the graft itself may be the source of immune rejection. The lymphocytes in the bone marrow transplant may produce a *graft versus host reaction* if the MHC molecules of the donor and recipient are not well matched.

43.4   Disruptions in immune system function can elicit or exacerbate disease

Exaggerated, Self-Directed, and Diminished Immune Responses Allergies are hypersensitivities to certain antigens called **allergens.** IgE antibodies created in response to initial exposure to an allergen may bind to mast cells. When allergens then bind and cross-link to these cell surface antibodies, the mast cells release histamines. The resulting inflammatory response may include sneezing, a runny nose, and difficulty in breathing due to smooth muscle contractions. Antihistamines are drugs that reduce symptoms by blocking receptors for histamine. *Anaphylactic shock* is a severe allergic response in which the rapid release of mast cell contents causes abrupt dilation of peripheral blood vessels and a drop in blood pressure, as well as constriction of bronchioles.

Sometimes the immune system turns against self, leading to **autoimmune diseases,** such as *lupus, rheumatoid arthritis, Type 1 diabetes mellitus*, and *multiple sclerosis*. These diseases result from a loss of self-tolerance and are influenced by gender, genetics, and environment.

Exercising to exhaustion and psychological stress have been shown to impair immune system function.

An **immunodeficiency** is the inability of the immune system to protect against pathogens. A genetic or developmental defect in the immune system is called an *inborn immunodeficiency;* a defect that arises later in response to chemical or biological agents is called an *acquired immunodeficiency.* An inborn immunodeficiency may occur in any of the components of the immune system. In the rare congenital disease known as severe combined immunodeficiency (SCID), lymphocytes (and therefore adaptive immune responses) are lacking. Acquired immunodeficiency may be the result of drugs used to treat autoimmune diseases or to suppress transplant rejection. **Acquired immunodeficiency syndrome,** or **AIDS,** and certain cancers such as Hodgkin's disease damage the immune system.

Evolutionary Adaptations of Pathogens That Underlie Immune System Avoidance Some viruses and parasites evade the body's defenses through *antigenic variation*, changing their surface epitopes so that the immune system cannot develop an immunological memory of the pathogen. Examples include the parasite that causes sleeping sickness and the influenza virus. The 2009 H1N1 virus caused a *pandemic*, a worldwide outbreak of flu.

Some viruses enter an inactive state called *latency*, in which viral DNA remains in a host cell's nucleus. Should conditions change, the virus can replicate and spread to neighboring cells or to new hosts.

The infectious agent responsible for AIDS is HIV (human immunodeficiency virus). HIV infects helper T cells (by binding with surface CD4 accessory protein), macrophages, and brain cells. HIV RNA is reverse-transcribed into DNA, which is integrated into the host cell genome, from which it directs production of new viral particles. HIV escapes the immune system through both antigenic variation (when it mutates rapidly during replication) and latency (when it integrates into host cell DNA). As HIV infection kills helper T cells, both humoral and cell-mediated immune responses are impaired. Individuals with AIDS are highly susceptible to opportunistic infections and cancers that take advantage of a suppressed immune system.

Although new drug combinations are unable to cure HIV infection, they are slowing the progression to AIDS. HIV is transmitted by the transfer of body fluids, such as blood, semen, and breast milk, containing viral particles or infected cells.

INTERACTIVE QUESTION 43.9

a. Why is AIDS such a deadly disease?

b. Why has it proved so difficult to prevent and cure this disease?

Cancer and Immunity The incidence of certain cancers increases when the immune system is impaired. A healthy immune system may be able to recognize and defend against the 15–20% of human cancers that involve viruses. Vaccines are now available against hepatitis B virus, which is linked to liver cancer, and human papillomavirus (HPV), which is associated with cervical cancer.

Word Roots

anti- = against; **-gen** = produce (*antigen:* a macromolecule that elicits an immune response by binding to receptors of B cells or T cells)

cyto- = cell (*cytokines:* a group of small proteins, secreted by a number of cell types, including macrophages and helper T cells, that regulate the function of other cells)

epi- = over; **-topo** = place (*epitope:* a small, accessible region of an antigen to which an antigen receptor or antibody binds)

immuno- = safe, free; **-glob** = globe, sphere (*immunoglobulin:* one of the classes of proteins that function as antibodies)

macro- = large; **-phage** = eat (*macrophage:* a phagocytic cell present in many tissues that functions in innate immunity by destroying microbes, and in adaptive immunity as an antigen-presenting cell)

neutro- = neutral; **-phil** = loving (*neutrophil:* the most abundant type of white blood cell; neutrophils are phagocytic and tend to self-destruct as they destroy foreign invaders)

Structure Your Knowledge

This chapter contains a wealth of information that is probably fairly new to you. If you take a little time and pull out the key players of the immune system and organize them first into very basic concept clusters and then develop more interrelated concept maps, you will find that this information is both understandable and fascinating.

1. Create a concept map outlining adaptive immunity; include the cells involved in the humoral and cell-mediated responses and their functions.

2. Describe the structure of an antibody molecule, and relate this structure to its function.

3. Although innate and adaptive immunity were presented separately, these defense mechanisms interact in several ways. Describe a few of the chemical and cellular components they share.

Test Your Knowledge

MULTIPLE CHOICE: *Choose the one best answer.*

1. Which of the following statements best describes an insect's immune system?
   a. Insects rely on the barrier defense of an exoskeleton.
   b. Lysozyme and a chitin-lined intestine with a low pH protect an insect's digestive system.
   c. Hemocytes can carry out phagocytosis of bacteria and foreign substances.
   d. Insects produce different antimicrobial peptides in response to the binding of molecules from a particular type of pathogen to Toll receptors.
   e. All of the above are part of an insect's innate immunity.

2. Which of the following defense mechanisms is *incorrectly* paired with its function?
   a. gastric juice—kills bacteria in the stomach
   b. fever—may stimulate phagocytosis
   c. histamine—causes blood vessels to dilate
   d. cytokines—attract phagocytes in the inflammatory response
   e. lysozyme—attacks the cell wall of viruses

3. Which of the following cells would release interferon?
   a. a macrophage that has become an antigen-presenting cell
   b. an injured endothelial cell of a blood vessel
   c. a cell infected by a virus
   d. a mast cell that has bound an antigen
   e. a helper T cell bound to an antigen-presenting cell

4. Which of the following statements is *not* true of Toll-like receptors?
   a. They are located on phagocytic white blood cells.
   b. They recognize molecules specific to individual pathogens.
   c. They resemble the Toll receptor on the immune response cells of insects.
   d. TLR proteins may be located on the plasma membrane or inside vesicles of immune cells.
   e. They trigger innate immune responses.

5. Which of the following statements correctly describes the main difference between innate immunity and adaptive immunity?
   a. Innate immunity responds only to free pathogens in a localized area; adaptive immunity responds only to pathogens that have entered body cells.
   b. Innate immunity involves only leukocytes, whereas adaptive immunity involves only lymphocytes.
   c. Innate immunity relies on phagocytes to destroy pathogens, whereas adaptive immunity does not involve phagocytes.
   d. Innate immunity recognizes molecules common to a set of pathogens, whereas adaptive immunity reacts to specific microbes on the basis of their unique antigens.
   e. Complement proteins participate in adaptive immunity but not in innate immunity.

6. Which of the following destroys a target cell by phagocytosis?
   a. a neutrophil
   b. a cytotoxic T cell
   c. a natural killer cell
   d. complement proteins
   e. a helper T cell

7. Antibodies are
   a. proteins or polysaccharides usually found on the surface of invading bacteria.
   b. proteins embedded in T cell membranes.
   c. proteins circulating in the blood that tag foreign cells for complement destruction.
   d. proteins that consist of two light and two heavy polypeptide chains.
   e. Both c and d are correct.

8. What accounts for the huge diversity of antigens to which B cells can respond?
   a. The antibody genes have millions of alleles.
   b. The recombination of a light and a heavy chain gene during development results in millions of possible antigen receptors.
   c. The antigen-binding sites at the arms of the molecule can assume a huge diversity of shapes in response to the specific antigen encountered.
   d. B cells have thousands of copies of antibodies bound to their plasma membranes.
   e. B cells can be antigen-presenting cells when they take in antigens and display fragments in their class II MHC molecules.

9. A secondary immune response is more rapid and effective than a primary immune response because
   a. histamines cause rapid vasodilation.
   b. the second response is an active immunity, whereas the primary one was a passive immunity.
   c. helper T cells are available to activate other white blood cells.
   d. chemical signals cause the rapid accumulation of phagocytic cells.
   e. memory cells respond to the pathogen and rapidly proliferate into effector cells.

10. Clonal selection is responsible for the
    a. proliferation of effector cells and memory cells specific for an encountered antigen.
    b. recognition of class I MHC molecules by cytotoxic T cells.

c. rearrangement of antibody genes for the light and heavy chains.

d. formation of cell cultures in the commercial production of monoclonal antibodies.

e. transformation of a clone of helper T cells into cytotoxic T cells keyed to a specific antigen.

11. Major histocompatibility complex molecules

a. are involved in the ability to distinguish self from nonself.

b. are a collection of cell surface proteins.

c. may trigger T cell responses after transplant operations.

d. present antigen fragments on infected cells.

e. are or do all of the above.

12. In opsonization,

a. antibodies coat microorganisms and help phagocytes bind to and engulf the foreign cell.

b. a set of complement proteins lyses a hole in a foreign cell's membrane.

c. antibodies coat binding sites of viruses.

d. a flood of histamines is released and may result in anaphylactic shock.

e. *V* gene segments and *J* gene segments are joined by recombinase.

13. A transfusion of type B blood in a person who has type A blood would result in

a. the recipient's anti-B antibodies reacting with the donated red blood cells.

b. the recipient's B antigens reacting with the donated anti-B antibodies.

c. the recipient forming both anti-A and anti-B antibodies.

d. no reaction; B is a universal donor blood type.

e. the introduced blood cells being destroyed by innate defense mechanisms.

14. Which of the following pairs is *incorrectly* matched?

a. variable region—determines antibody specificity for an epitope

b. constant region—determines class and function of antibody

c. IgA—antibody molecules attached to mast cells

d. IgG—most abundant circulating antibodies, confer passive immunity to fetus

e. IgD—membrane-bound B cell receptor

15. What do IgE antibodies, T cell receptors, and MHC molecules have in common?

a. They are found exclusively in cells of the immune system.

b. They are all part of the complement system.

c. They are antigen-presenting molecules.

d. They are or can be membrane-bound proteins.

e. They are involved in the cell-mediated portion of the immune system.

16. How are antibodies and complement related?

a. They are both coded for by genes that have hundreds of alleles.

b. They are both involved in innate defenses.

c. They are both produced by plasma cells.

d. Antibodies bound to antigens on a pathogen's membrane may activate complement proteins to form a membrane attack complex.

e. Complement proteins tag foreign cells for destruction; antibodies destroy cells by opsonization.

17. In an adaptive immune response, a dendritic cell

a. activates complement proteins to form a membrane attack complex.

b. binds to an accessory protein on cytotoxic T cells to activate their production of perforin.

c. releases cytokines to activate B cells to produce clones of plasma cells.

d. activates humoral and cell-mediated immunity by releasing interferons after engulfing a virus.

e. presents peptide antigens of an engulfed pathogen in its class II MHC molecules to helper T cells, and releases cytokines.

18. Which of the lists in choices **a–e** places the following steps in the helper T cell activation of cell-mediated and humoral immunity in the correct order?

1. Macrophage and helper T cell secrete cytokines.

2. Macrophage engulfs pathogen and presents antigen in class II MHC.

3. Plasma cells secrete antibodies, and cytotoxic T cells attack cells with class I MHC molecule-antigen complex.

4. T cell receptor recognizes class II MHC molecule-antigen complex.

5. Activated B cells form plasma cells and memory cells, and activated T cells form cytotoxic T cells and memory cells.

a. 1, 3, 5, 2, 4

b. 5, 1, 2, 4, 3

c. 2, 4, 1, 5, 3

d. 5, 2, 4, 1, 3

e. 2, 1, 4, 5, 3

19. Which of the following statements correctly describes positive feedback during an immune response?
    a. Substances released by macrophages during an inflammatory response increase body temperature, which increases the rate of phagocytosis.
    b. The combined actions of neutrophils, macrophages, eosinophils, dendritic cells, and natural killer cells produce an enhanced innate immune response; the combined actions of B cells, helper T cells, and cytotoxic T cells produce an enhanced adaptive immune response.
    c. Antibodies produced by plasma cells enhance phagocytosis by macrophages, which present antigens to and stimulate helper T cells, which then further stimulate B cells.
    d. Complement proteins and cytokines play a role in both innate and adaptive immunity, thereby magnifying their effectiveness in fighting pathogens.
    e. Helper T cells activate both the humoral and cell-mediated immunities by releasing cytokines after recognizing class II MHC molecule-antigen complexes on an antigen-presenting cell.

20. Which of the following statements about humoral immunity is *correct*?
    a. It is a form of passive immunity produced by vaccination.
    b. It defends against free pathogens with effector mechanisms such as neutralization, opsonization, or complement activation.
    c. It protects against pathogens that have invaded body cells as well as against cancer cells.
    d. It is mounted by lymphocytes that have matured in the thymus.
    e. It requires recognition of class I MHC molecule—antigen complexes to activate its effector mechanism.

21. The accessory protein CD4 is
    a. a surface molecule on a cytotoxic T cell that enhances its binding to a class I MHC molecule displaying a foreign antigen.
    b. a membrane protein on an antigen-presenting cell that helps a helper T cell recognize the MHC molecule-antigen complex.
    c. a receptor that normally functions for cytokines, but which HIV uses as its receptor.
    d. a surface molecule on a helper T cell that enhances its binding to a class II MHC molecule displaying an antigen fragment.
    e. a portion of the class I MHC molecule that is found on all cells and identifies those cells as "self."

22. In which of the following circumstances would B cells display antigens to T cells?
    a. B cells take in a few antigen molecules and display them in class II MHC molecules to activated helper T cells.
    b. B cells phagocytose bacteria and display bacterial peptide antigens in class II MHC molecules to helper T cells.
    c. After being infected by a virus, B cells display viral peptides they have synthesized in class I MHC molecules to cytotoxic T cells.
    d. B cells bind free antigens and display them to helper T cells while they are still attached to their B cell antigen receptors.
    e. Both a and c are possible ways that B cells can display antigens to T cells.

23. Which of the following conditions is *not* considered a disease or malfunction of the immune system?
    a. MHC-induced transplant rejection
    b. SCID (severe combined immunodeficiency)
    c. lupus, multiple sclerosis, and Type I diabetes
    d. AIDS
    e. allergic anaphylactic shock

24. Which of the following procedures may induce a graft-versus-host reaction?
    a. an organ transplant
    b. a blood transfusion
    c. a bone marrow transplant
    d. a skin transplant
    e. gene therapy

25. From which of the following conditions would an AIDS patient be *least* likely to suffer?
    a. Kaposi's sarcoma or other cancers
    b. tuberculosis
    c. pneumonia
    d. rheumatoid arthritis
    e. yeast infections of mucous membranes

Chapter 44

Osmoregulation and Excretion

Key Concepts

44.1 Osmoregulation balances the uptake and loss of water and solutes

44.2 An animal's nitrogenous wastes reflect its phylogeny and habitat

44.3 Diverse excretory systems are variations on a tubular theme

44.4 The nephron is organized for stepwise processing of blood filtrate

44.5 Hormonal circuits link kidney function, water balance, and blood pressure

Framework

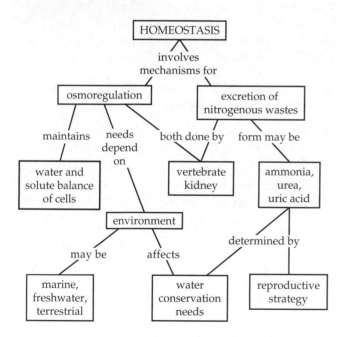

Chapter Review

Animals are able to tolerate the osmoregulatory challenges in their external environments by regulating solute concentrations and water balance (**osmoregulation**) in their internal environments. Animals must also dispose of *nitrogenous* (nitrogen-containing) wastes through the process called **excretion.**

44.1 Osmoregulation balances the uptake and loss of water and solutes

Osmosis and Osmolarity Whatever an animal's habitat or nitrogenous waste product, its water gain must balance water loss.

Osmosis is the diffusion of water across a selectively permeable membrane separating two solutions that differ in **osmolarity** (moles of solute per liter). Osmolarity is expressed in units of milliosmoles per liter (mOsm/L). *Isoosmotic* solutions are equal in osmolarity, and no net osmosis occurs between them. There is a net flow of water from a *hypoosmotic* (more dilute) to a *hyperosmotic* (more concentrated) solution.

Osmotic Challenges **Osmoconformers** are isoosmotic with their surroundings. **Osmoregulators** regulate their internal osmolarity. Osmoregulators must get rid of excess water if they live in a hypoosmotic medium, or they must take in water to offset osmotic loss if they inhabit a hyperosmotic environment.

Most animals are *stenohaline,* able to tolerate only small changes in external osmolarity. Animals that are *euryhaline* can survive large differences in their osmotic environments.

Most marine invertebrates are osmoconformers, whereas many marine vertebrates are osmoregulators. Marine fishes are hypoosmotic to seawater. They gain salt from food and from the large quantities of seawater they drink to replace the water they lose by osmosis. *Chloride cells* in the gills pump out excess Cl^- ions (and Na^+ ions then follow), and other ions are excreted in the scanty urine.

A shark's osmolarity is slightly higher than that of seawater because it maintains in its tissues a high concentration of urea as well as trimethylamine oxide (TMAO), which protects proteins from the damaging effects of urea. Water that enters a shark's body by osmosis or in food, and salt that diffuses in through its gills, are disposed of in urine.

Freshwater animals constantly take in water by osmosis and lose salts by diffusion. Freshwater fishes excrete large quantities of dilute urine. Salt supplies are replaced from their food or by active uptake of ions across the gills.

Some animals are capable of **anhydrobiosis,** surviving dehydration, or *desiccation,* in a dormant state. Anhydrobiotic roundworms produce large quantities of trehalose, a sugar that replaces water around membranes and proteins during dehydration. Many insects survive freezing by producing trehalose.

Adaptations to prevent dehydration in terrestrial animals include water-impervious coverings and behavioral adaptations such as nocturnal lifestyles. Water is lost in urine and feces and through evaporation, and is gained by drinking and eating moist food and by metabolic production during cellular respiration.

Energetics of Osmoregulation Osmoregulation is energetically costly because animals must actively transport solutes to maintain osmotic gradients needed to gain or lose water. The energy cost is related to the difference in osmolarity between an animal and its environment, and the body fluid composition of many animals is adapted to the salinity of their habitats.

INTERACTIVE QUESTION 44.1

Indicate whether the following animals are isoosmotic, hyperosmotic, or hypoosmotic to their environment. Then briefly list their mechanisms of osmoregulation.

Animal	Osmotic Relation to Environment	Osmoregulatory Mechanisms
Marine invertebrate	a.	b.
Shark	c.	d.
Marine bony fish	e.	f.
Freshwater fish	g.	h.
Terrestrial animal	i.	j.

Transport Epithelia in Osmoregulation Animals regulate the solute composition of the fluid that bathes body cells, whether it is hemolymph or interstitial fluid. The movement of specific solutes across a **transport epithelium** is involved in both osmotic regulation and the disposal of metabolic wastes. Transport epithelia are usually arranged in tubular networks that provide large surface areas for exchange. Nasal glands in marine birds utilize countercurrent exchange in removing salt from the blood and producing a concentrated salt secretion.

44.2 An animal's nitrogenous wastes reflect its phylogeny and habitat

Ammonia, a small and toxic molecule, is produced when proteins and nucleic acids are metabolized.

Forms of Nitrogenous Waste Aquatic animals can excrete nitrogenous wastes as ammonia because it easily diffuses through membranes into the surrounding water. In fishes, most ammonia passes across the epithelia of the gills as NH_4^+.

Mammals, most adult amphibians, sharks, and some marine bony fishes and turtles produce **urea,** a much less toxic compound than ammonia. Urea can be tolerated in a more concentrated form and excreted with less loss of water. Ammonia and carbon dioxide are combined in the liver to produce urea, which is then carried by the circulatory system to the kidneys. The production of urea requires energy.

Insects, land snails, and many reptiles, including birds, produce **uric acid,** a compound of low solubility in water that can be excreted as a semisolid with very little water loss. Its synthesis, however, is energetically expensive.

The Influence of Evolution and Environment on Nitrogenous Wastes The form of an animal's nitrogenous wastes relates to its habitat. Terrestrial turtles excrete mainly uric acid, whereas aquatic turtles excrete both urea and ammonia. The mode of reproduction influenced which form of nitrogenous waste product evolved in various animal groups.

INTERACTIVE QUESTION 44.2

a. Of two animals of the same size, which would produce more nitrogenous waste:

an endotherm or an ectotherm?

an herbivore or a carnivore?

Explain your answers.

b. Vertebrates that produce shelled eggs excrete _____.

Mammals produce _____.

What adaptive advantage do these types of nitrogenous wastes provide?

44.3 Diverse excretory systems are variations on a tubular theme

Excretory Processes During **filtration,** water and small solutes are forced out of the blood or body fluids across a transport epithelium into an excretory tubule. Two mechanisms transform this **filtrate** to urine: Valuable solutes are returned from the filtrate by selective **reabsorption,** and excess ions, toxins, and other solutes are added to the filtrate by selective **secretion.** The osmotic movement of water into or out of the filtrate follows the pumping of solutes. The processed filtrate is released from the body by excretion.

Survey of Excretory Systems The **protonephridia** of flatworms are branched systems of closed tubules. Water and solutes from the interstitial fluid pass into flame bulbs at the ends of the tubules, propelled by cilia. Urine is excreted into the external environment. The protonephridia of freshwater flatworms are primarily osmoregulatory. In parasitic flatworms, however, protonephridia function mainly in the excretion of nitrogenous wastes.

Metanephridia are found in most annelids. They occur in pairs in each segment of an earthworm. An open ciliated funnel collects coelomic fluid, which then moves through a folded tubule encased in capillaries. The transport epithelium of the tubule reabsorbs most solutes, which then reenter the blood. Hypoosmotic urine, consisting of nitrogenous wastes and excess water absorbed by osmosis, exits to the outside.

Malpighian tubules are excretory organs in insects and other terrestrial arthropods. These blind sacs, which open into the digestive tract, are lined by transport epithelia that secrete solutes from the hemolymph into the tubule. The fluid passes into the rectum, where most of the solutes are pumped back into the hemolymph, and water follows by osmosis. In this water-conserving system, uric acid is eliminated as dry matter along with the feces.

The numerous excretory tubules of most vertebrates are associated with a network of capillaries and arranged into compact kidneys. The kidneys and the structures that carry urine out of the body constitute the vertebrate excretory system.

INTERACTIVE QUESTION 44.3

Indicate whether each of the following tubular systems functions in osmoregulation, excretion, or both. If they function in osmoregulation, do they help conserve water or remove excess water that had entered by osmosis?

a. Protonephridia of freshwater planaria

b. Metanephridia of earthworms

c. Malpighian tubules of insects

d. Kidneys of terrestrial mammals

Exploring the Mammalian Excretory System Blood enters each of the pair of bean-shaped **kidneys** through a renal artery and leaves by way of a renal vein. Urine exits the kidney through a **ureter** and is temporarily stored in the **urinary bladder.** Urine exits the body through the **urethra.**

The outer **renal cortex** and inner **renal medulla** of the kidney are packed with excretory tubules and blood vessels. Urine collects in the inner **renal pelvis.**

A **nephron** consists of a long tubule, at the blind end of which is a cuplike **Bowman's capsule** enclosing a ball of capillaries called the **glomerulus.** Filtration of the blood occurs as blood pressure forces water, urea, salts, glucose, amino acids, and other small molecules through the porous capillary walls and specialized capsule cells into Bowman's capsule, forming the filtrate. The filtrate, in turn, passes through the **proximal tubule;** the **loop of Henle,** consisting of a descending limb and an ascending limb; and the **distal tubule. Collecting ducts** receive processed filtrate from many nephrons and pass the urine into the renal pelvis.

In the human kidney, 85% of the nephrons are **cortical nephrons** located almost entirely in the renal cortex. **Juxtamedullary nephrons** have long loops of Henle that extend into the renal medulla and allow the formation hyperosmotic urine.

An *afferent arteriole* supplying each nephron subdivides to form the capillaries of the glomerulus, which then converge to form an *efferent arteriole.* This arteriole divides into a second capillary network—the **peritubular capillaries**—that surrounds the proximal and distal tubules. The **vasa recta** includes descending and ascending capillaries that parallel the loop of Henle.

INTERACTIVE QUESTION   44.4

a. What substances in the filtrate move from Bowman's capsule into the tubule?

b. What substances remain in the capillaries?

44.4 The nephron is organized for stepwise processing of blood filtrate

From Blood Filtrate to Urine: A Closer Look The numbered steps in the following discussion correspond to the figure in Interactive Question 44.5.

1. Reabsorption and secretion both take place in the proximal tubule. Salt (NaCl) diffuses from the filtrate into the cells of the proximal tubule, which then actively transport Na^+ across the membrane into the interstitial fluid; Cl^- is transported passively out to balance the positive charge; and water follows by osmosis. Salt and water diffuse from the interstitial fluid into the peritubular capillaries. Glucose, amino acids, K^+, and the buffer bicarbonate (HCO_3^-) are reabsorbed either actively or passively. The transport epithelium secretes H^+, ammonia, and drugs and poisons (processed by the liver and delivered via peritubular capillaries) into the filtrate.

2. The descending limb of the loop of Henle has numerous **aquaporins** and is thus freely permeable to water. It is not permeable to NaCl or other small solutes. The interstitial fluid is increasingly hyperosmotic toward the inner medulla, so water exits by osmosis, leaving behind a filtrate with a high solute concentration.

3. The ascending limb of the loop of Henle is not permeable to water but is permeable to salt, which diffuses out of the thin lower segment of the loop, increasing the osmolarity of the medulla. NaCl is actively transported out of the thick upper portion of the ascending limb. As salt leaves, the filtrate becomes less concentrated.

4. The distal tubule regulates K^+ secretion and NaCl reabsorption, and it helps to regulate pH by secreting H^+ and reabsorbing HCO_3^-.

5. The transport epithelium lining the nephrons and collecting ducts processes about 180 L of filtrate each day to produce about 1.5 L of urine. As the filtrate moves along the collecting duct through the osmotic gradient of the medulla, more and more water exits by osmosis through aquaporin channels. Urea diffuses out of the lower portion of the duct and helps to maintain the osmotic gradient in the medulla. When producing dilute urine, the transport epithelium actively reabsorbs NaCl, and water is not allowed to follow by osmosis.

Review these by steps by completing Interactive Question 44.5.

INTERACTIVE QUESTION 44.5

In the following diagram of a nephron and collecting tubule, label the parts on the indicated lines. Label the arrows to indicate the movement of NaCl, water, nutrients, K^+, HCO_3^-, H^+, and urea into or out of the tubule.

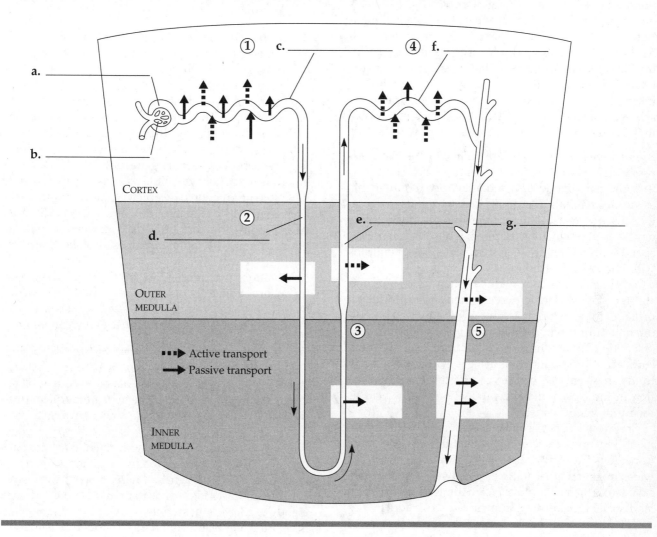

Solute Gradients and Water Conservation The osmolarity gradient in the renal medulla involves energy-requiring active transport by the loops of Henle and collecting ducts and enables the human kidney to produce urine up to four times as concentrated as blood and normal interstitial fluid.

Filtrate leaving Bowman's capsule has an osmolarity of about 300 mOsm/L, the same as blood. Both water and salt are reabsorbed in the proximal tubule; thus, the volume decreases but the osmolarity remains the same. During the filtrate's movement down the descending limb of the loop of Henle, water exits by osmosis, and the filtrate becomes more concentrated.

Salt, which is now in high concentration in the filtrate, diffuses out as the filtrate moves up the salt-permeable but water-impermeable ascending limb—helping to create the osmolarity gradient.

Through a **countercurrent multiplier system,** the thick segment of the ascending limb expends energy to actively transport NaCl from the filtrate and generate the osmotic gradient in the medulla.

The vasa recta also consists of descending and ascending vessels. As blood flows down through the osmotic gradient in the inner medulla, water leaves the descending vessel by osmosis and salt enters; as the ascending vessel carries blood back up toward the cortex, water reenters and salt diffuses out. Thus the capillaries can carry nutrients and other supplies to the medulla without disrupting the osmolarity gradient.

The filtrate is hypoosmotic to body fluids after salt is pumped out of the ascending limb. The filtrate makes one final pass through the medulla, this time in the collecting duct, which is permeable to water and urea but not to salt. Water flows out by osmosis, and as the filtrate becomes more concentrated, urea diffuses into the interstitial fluid, adding to the high osmolarity of the inner medulla. The resulting concentrated urine may achieve an osmolarity as high as 1,200 mOsm/L, isoosmotic to the interstitial fluid of the inner medulla.

Adaptations of the Vertebrate Kidney to Diverse Environments Modifications of nephron structure and function correspond to osmoregulatory requirements in various habitats. Mammals in dry habitats, needing to conserve water and excrete hyperosmotic urine, have exceptionally long loops of Henle. Mammals that spend much of their time in fresh water have nephrons with relatively short loops.

Birds also have juxtamedullary nephrons, but their shorter loops of Henle prevent them from producing urine as hyperosmotic as that of mammals. The production of uric acid is their main means of water conservation. The kidneys of other reptiles have only cortical nephrons, and their urine is isoosmotic to body fluids; these animals conserve water through reabsorption in the cloaca and the production of uric acid as their nitrogenous waste.

Freshwater fishes are hyperosmotic to their environment and must excrete large quantities of very dilute urine. Salts are reabsorbed from the filtrate. When in fresh water, frogs and other amphibians actively absorb certain salts across their skin and excrete a dilute urine. When on land, water is reabsorbed from the urinary bladder.

The kidneys of marine bony fishes have fewer and smaller nephrons and excrete very little urine. They function mainly to get rid of Ca^{2+}, Mg^{2+}, and sulfate ions taken in by drinking seawater. Excess Na^+ and Cl^- are secreted by the gills.

INTERACTIVE QUESTION   44.6

a. What type of nephron enables the production of hyperosmotic urine?

b. Which animal groups have this type of nephron?

44.5   Hormonal circuits link kidney function, water balance, and blood pressure

Volume and osmolarity of urine in mammals can vary depending on the animal's water and salt balance and urea production.

Antidiuretic Hormone   Antidiuretic hormone (ADH), also called *vasopressin*, is produced in the hypothalamus and stored in the pituitary gland. When blood osmolarity rises above a set point of 300 mOsm/L, osmoreceptor cells in the hypothalamus trigger the release of ADH. This hormone increases water permeability of the collecting ducts, increasing water reabsorption from the urine. As blood osmolarity declines, negative feedback decreases the release of ADH. When blood osmolarity is low, little ADH is released, resulting in the production of a large volume of dilute urine; such increased production of urine is called diuresis. ADH binding to receptors leads to an increase in aquaporin proteins in the membranes of collecting duct cells.

Mutations that affect ADH production, ADH receptor action, or aquaporin formation can cause severe dehydration in the condition *diabetes insipidus*. Alcohol inhibits ADH release and can cause dehydration.

The Renin-Angiotensin-Aldosterone System   The renin-angiotensin-aldosterone system (RAAS) is a homeostatic feedback circuit that maintains adequate blood pressure and blood volume. The juxtaglomerular apparatus (JGA), located around the afferent arteriole leading to the glomerulus, responds to a drop in blood pressure or volume by releasing renin. This enzyme converts the plasma protein angiotensinogen to angiotensin II. Angiotensin II raises blood pressure (by constricting arterioles) and stimulates the adrenal glands to release aldosterone. This hormone stimulates Na^+ and water reabsorption in the distal tubules. Some drugs that treat hypertension target the enzyme ACE, which functions in the activation of angiotensin II.

Homeostatic Regulation of the Kidney   Both ADH and the RAAS increase water absorption, but in response to different problems. ADH counters dehydration by responding to an increase in blood osmolarity. The RAAS responds to a loss of both fluid and salts by sensing a drop in blood volume or pressure.

Atrial natriuretic peptide (ANP) counters the RAAS. ANP, released by the atria of the heart in response to increased blood volume and pressure, inhibits the release of renin from the JGA, inhibits NaCl reabsorption by the collecting ducts, and reduces the release of aldosterone from the adrenal glands. To review this material, complete Interactive Question 44.7.

INTERACTIVE QUESTION 44.7

The following concept map may look overwhelming, but working through it should help organize your understanding of the homeostatic regulation of the kidney.

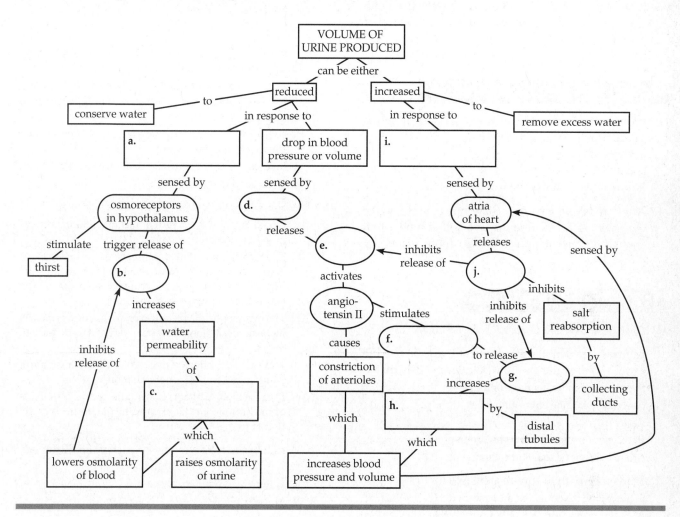

Word Roots

an- = without, **hydro-** = water; **-bios** = life (*anhydrobiosis:* a dormant state involving loss of almost all body water)

anti- = against; **-diure** = urinate (*antidiuretic hormone:* a hormone that promotes water retention by the kidneys)

glomer- = a ball (*glomerulus:* a ball of capillaries surrounded by Bowman's capsule in the nephron; is the site of filtration in the kidney)

juxta- = near to (*juxtaglomerular apparatus:* a specialized tissue in nephrons that releases the enzyme renin in response to a drop in blood pressure or volume)

meta- = with; **-nephri** = kidney (*metanephridium:* an excretory organ, found in many invertebrates, that consists of tubules connecting ciliated internal openings to external openings)

osmo- = pushing; **-regula** = regular (*osmoregulation:* regulation of solute concentrations and water balance by a cell or an organism)

peri- = around (*peritubular capillaries:* the network of tiny blood vessels surrounding the proximal and distal tubules in the kidney)

proto- = first (*protonephridium:* an excretory system, such as the flame bulb system of flatworms, consisting of a network of tubules lacking internal openings)

reni- = a kidney; **-angio** = a vessel; **-tens** = stretched (*renin-angiotensin-aldosterone system:* a hormone cascade pathway that helps regulate blood pressure and volume)

vasa- = a vessel; **-recta** = straight (*vasa recta:* the capillary system in the kidney that serves the loop of Henle)

Structure Your Knowledge

1. Provide the following information concerning the production of urine in mammals.
   a. Substances filtered:
   b. Substances secreted:
   c. Substances reabsorbed:

2. After referring to the diagram in Interactive Question 44.5, describe the two-solute model that enables the production of hyperosmotic urine.

Test Your Knowledge

MULTIPLE CHOICE: *Choose the one best answer.*

1. Contractile vacuoles would most likely be used to expel excess water from unicellular protists that
   a. live in a freshwater environment.
   b. live in a marine environment.
   c. are internal parasites.
   d. are hypoosmotic to their environment.
   e. are isoosmotic to their environment.

2. Transport epithelia are responsible for
   a. pumping water across a membrane.
   b. forming an impermeable boundary at an interface with the environment.
   c. the movement of solutes in osmoregulation or excretion.
   d. transporting urine in the ureter and urethra.
   e. the passive transport of H^+ and HCO_3^- for the regulation of pH.

3. Which of the following organisms would be most likely to produce the sugar trehalose?
   a. a desert mammal
   b. a euryhaline fish
   c. a shark
   d. an anhydrobiotic roundworm
   e. a vampire bat

4. A freshwater fish would be expected to
   a. pump salt out through salt glands in the gills.
   b. produce copious quantities of dilute urine.
   c. excrete urea across the epithelium of the gills.
   d. have scales that reduce water loss.
   e. do all of the above.

5. The production of uric acid is advantageous to birds because uric acid
   a. has low toxicity and can be safely stored in the egg.
   b. is very soluble in water and takes very little energy to produce.
   c. contributes to the production of the egg shell and can thus serve two purposes.
   d. takes less energy to produce than urea and is nontoxic.
   e. Both a and d are correct.

6. Which of the following statements is *incorrect*?
   a. The kidneys of most vertebrates function in both osmoregulation and excretion.
   b. The malpighian tubules of insects secrete nitrogenous wastes and solutes from the hemolymph into the tubules. Most solutes are reabsorbed in the rectum and water follows by osmosis.
   c. The metanephridia of earthworms function primarily in osmoregulation; ammonia diffuses out through the skin.
   d. The main function of the protonephridia of freshwater flatworms is osmoregulation.
   e. Because their habitat is isoosmotic, the main function of the protonephridia of parasitic flatworms is excretion of nitrogenous wastes.

7. Which of the following is *not* part of the filtrate entering Bowman's capsule?
   a. water, salt, and electrolytes
   b. glucose
   c. urea
   d. plasma proteins
   e. amino acids

8. Which of the following correctly traces the pathway of urine flow in vertebrates?
   a. collecting tubule → ureter → bladder → urethra
   b. renal vein → renal ureter → bladder → urethra
   c. nephron → urethra → bladder → ureter
   d. cortex → medulla → bladder → ureter
   e. renal pelvis → medulla → bladder → urethra

9. Which of the following statements is *incorrect*?
   a. Long loops of Henle are associated with steep osmotic gradients and the production of hyperosmotic urine.
   b. Ammonia is a toxic nitrogenous waste molecule that passively diffuses out of the bodies of aquatic invertebrates.
   c. Uric acid is the nitrogenous waste that requires the least amount of water to excrete.
   d. In the mammalian kidney, urea diffuses out of the collecting duct and contributes to the osmotic gradient within the medulla.
   e. Ammonia is produced by a mammalian fetus, removed through the placenta, transported to the mother's liver, and then converted to urea.

10. The mechanism for the filtration of blood within the nephron involves
    a. the active transport of Na^+ and glucose, followed by osmosis.
    b. both active and passive secretion of ions, toxins, and $NH3^+$ into the tubule.
    c. high hydrostatic pressure of blood forcing water and small molecules out of the capillary.
    d. the high osmolarity of the medulla that was created by active and passive transport of salt from the tubule, and passive diffusion of urea from the collecting duct.
    e. a lower osmotic pressure in Bowman's capsule compared to that in the glomerulus.

11. The process of reabsorption in the formation of urine ensures that
    a. excess H^+ is removed from the blood.
    b. drugs and toxins are removed from the blood.
    c. urine is always hyperosmotic to the interstitial fluid.
    d. glucose, salts, and water are returned to the blood.
    e. pH is maintained by balancing H^+ and HCO_3^-.

12. The peritubular capillaries
    a. form the ball of capillaries inside Bowman's capsule.
    b. intertwine with the proximal and distal tubules and exchange solutes with the interstitial fluid.
    c. produce a countercurrent flow of blood through the medulla to supply nutrients without interfering with the osmolarity gradient.
    d. surround the collecting ducts and reabsorb water, helping to create hyperosmotic urine.
    e. rejoin to form the efferent arteriole.

13. Which of the following sections of the mammalian nephron is *incorrectly* paired with its function?
    a. Bowman's capsule and glomerulus—filtration of blood
    b. proximal tubule—secretion of ammonia and H^+ into the filtrate and transport of glucose and amino acids out of tubule
    c. descending limb of loop of Henle—diffusion of urea out of the filtrate
    d. ascending limb of loop of Henle—diffusion and pumping of NaCl out of the filtrate
    e. distal tubule—regulation of pH and K^+

14. Which of the following animals exhibits the most versatile osmoregulatory and excretory systems?
    a. a marine shark
    b. a migratory salmon
    c. a desert kangaroo rat
    d. a vampire bat
    e. both b and d

15. ADH and the RAAS both increase water reabsorption, but they respond to different osmoregulatory problems. Which *two* of the following statements are true?
    1. ADH will be released in response to high alcohol consumption.
    2. ADH is released when osmoregulatory cells in the hypothalamus detect an increase in blood osmolarity.
    3. The RAAS will increase the osmolarity of urine due to the cooperative actions of renin, angiotensin II, and aldosterone.
    4. The RAAS is a response to a rise in blood pressure or volume.
    5. The RAAS is most likely to respond following an accident or a severe case of diarrhea.
       a. 1 and 4
       b. 1 and 5
       c. 2 and 3
       d. 2 and 4
       e. 2 and 5

16. Which of the following stimuli causes the atria of the heart to release ANP (atrial natriuretic peptide)?
    a. a rise in blood pressure
    b. a drop in blood pressure
    c. a drop in blood pH
    d. a rise in blood osmolarity
    e. both a and c

17. Which of the following stimuli causes the juxta-glomerular apparatus to release renin?
    a. a drop in blood pH
    b. a rise in blood pH
    c. a drop in blood pressure
    d. a rise in blood osmolarity
    e. the release of ANP by the atria of the heart

18. Which of the following mechanisms is an effective way for a drug to treat hypertension?
    a. increase the production of ADH
    b. block the production of ANP
    c. produce vasoconstriction of renal arteries
    d. inhibit an enzyme involved in producing angiotensin II
    e. stimulate the production of aquaporins in the collecting tubules

19. The hormone aldosterone
    a. stimulates thirst.
    b. is secreted by the adrenal glands in response to high blood osmolarity.
    c. stimulates the active reabsorption of Na^+ in the distal tubules.
    d. causes diuresis.
    e. is converted from a blood protein by renin.

20. The production of urine that is hyperosmotic to blood and interstitial fluids requires
    a. juxtamedullary nephrons.
    b. an osmotic gradient in the medulla.
    c. the inhibition of ADH release.
    d. a and b only.
    e. a, b, and c.

Hormones and the Endocrine System

Key Concepts

45.1 Hormones and other signaling molecules bind to target receptors, triggering specific response pathways

45.2 Feedback regulation and antagonistic hormone pairs are common in endocrine systems

45.3 The hypothalamus and pituitary are central to endocrine regulation

45.4 Endocrine glands respond to diverse stimuli in regulating homeostasis, development, and behavior

Framework

This chapter introduces the intricate system of chemical control and communication within animals. Endocrine cells (or neurosecretory cells) produce hormones that regulate the activity of other cells and organs. Steroid hormones bind with receptors within their target cells and influence gene expression. The signal transduction pathways of most polypeptide and amine hormones involve binding with cell surface receptors and triggering metabolic reactions in target cells.

The hypothalamus and pituitary gland play coordinating roles by integrating the nervous and endocrine systems and producing many tropic hormones that control the synthesis and secretion of hormones in other endocrine glands and organs.

Chapter Review

An animal **hormone** is a chemical that is released into the extracellular fluid, circulates through the hemolymph or blood, and elicits a specific response from *target cells* (that is, cells with matching receptors). The steroid hormone **ecdysteroid** stimulates the *metamorphosis* of a larval caterpillar into a butterfly. Coordination and communication among the specialized parts of complex animals are achieved by the **endocrine system,** which produces hormones that regulate many biological processes, and the **nervous system,** which conveys messages along neurons that regulate other neurons, muscle cells, and endocrine cells.

45.1 Hormones and other signaling molecules bind to target receptors, triggering specific response pathways

Intercellular Communication In endocrine signaling, hormones travel via the circulatory system throughout the body. Hormones maintain homeostasis and regulate growth, development, and reproduction.

Local regulators are molecules that reach their target cells by diffusion. In **paracrine** signaling, secreted molecules act on neighboring cells; in **autocrine** signaling, the molecules act on the secreting cell itself.

In *synaptic signaling,* **neurotransmitters** are secreted by neurons at synapses with other neurons or muscle cells. In *neuroendocrine signaling,* specialized neurons called neurosecretory cells secrete **neurohormones,** which travel through the bloodstream to target cells.

Pheromones are signaling molecules that are released into the environment and communicate between individuals of the same species.

Endocrine Tissues and Organs Some endocrine cells are located in organs that are part of other organ systems. **Endocrine glands** are ductless secretory organs composed of groups of endocrine cells.

Chemical Classes of Hormones Hormone classes include polypeptides (peptides and proteins), steroids, and amines. Polypeptides and most amines are water-soluble. Because these hydrophilic hormones cannot pass through the plasma membranes, they bind to receptors on the cell surface. Steroids and other nonpolar hormones are lipid-soluble; they can pass through cell membranes and bind to intracellular receptors.

Cellular Response Pathways Water-soluble hormones are released by exocytosis, travel through the bloodstream, and bind to cell-surface receptors, triggering a cytoplasmic response or a change in gene expression. Lipid-soluble hormones travel in the bloodstream bound to transport proteins, enter target cells, and bind to a receptor in the cytoplasm or nucleus, triggering changes in gene transcription.

Binding of water-soluble hormones to plasma membrane receptors initiates **signal transduction,** which converts extracellular signals to specific intracellular responses. For example, after **epinephrine** is released by the adrenal glands in response to stress, it binds to G protein-coupled receptors on liver cells, initiating a signal transduction pathway that involves cAMP as a *second messenger* and activates a protein kinase; the result is glycogen breakdown and the release of glucose into the bloodstream.

Steroid hormones bind to cytoplasmic receptors, and the hormone-receptor complex then moves into the nucleus, where it stimulates transcription of specific genes. Thyroxine, vitamin D, and other non-steroid, lipid-soluble hormones bind with receptors inside the nucleus that then stimulate transcription.

INTERACTIVE QUESTION   45.1

Label the parts of the following diagrams, indicating which parts represent a water-soluble hormone and which a lipid-soluble hormone.

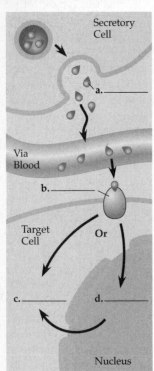

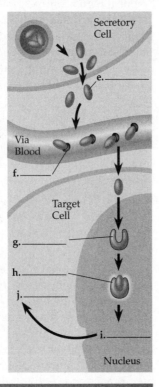

Multiple Effects of Hormones   A given signal can have different effects on different target cells as a result of the types of receptors (for example, the a and b epinephrine receptors), the specific signal transduction pathways

present, or the particular transcription factors that are activated.

Signaling by Local Regulators   Local regulators elicit fast responses from target cells. Cytokines are polypeptide local regulators involved in immune responses. The presence of **growth factors** in the extracellular environment is required for many types of cells to divide and develop.

Nitric oxide (NO) is a gas that can function as a neurotransmitter or as a local regulator that produces vasodilation, increasing blood flow.

Prostaglandins are modified fatty acids that have a wide range of effects on nearby target cells. Prostaglandins help induce labor. Aspirin and ibuprofen inhibit the synthesis of prostaglandins and thus reduce their fever- and inflammation-inducing and pain-intensifying effects.

Coordination of Neuroendocrine and Endocrine Signaling   Reproduction and development are usually controlled through an integration of the endocrine and nervous systems. In insects, *prothoracicotropic hormone* (PTTH), produced by neurosecretory cells in the brain, stimulates the prothoracic glands to secrete ecdysteroid, which triggers molting and metamorphosis. **Juvenile hormone** promotes retention of larval characteristics during molting. When its level is high, ecdysteroid-induced molting produces larger larval stages; only after the level of juvenile hormone has decreased does molting result in a pupa.

45.2   Feedback regulation and antagonistic hormone pairs are common in endocrine systems

Simple Hormone Pathways   In a *simple endocrine pathway,* endocrine cells secrete a hormone in response to a stimulus; the hormone travels in the bloodstream to target cells, where signal transduction produces a physiological response. Endocrine cells of the duodenum secrete the hormone *secretin* in response to a low pH caused by the entry of acidic stomach contents. Target cells in the **pancreas** release bicarbonate, which raises the pH in the duodenum.

In a *simple neurohormone pathway,* a neurosecretory cell secretes a neurohormone in response to a stimulus received by a sensory neuron. The circulating neurohormone triggers a specific response in target cells. For example, in response to nerve signals from the nipples, the hypothalamus triggers the release of the neurohormone **oxytocin,** which stimulates the mammary glands to secrete milk.

Feedback Regulation   In a regulatory loop called **negative feedback,** the response to a stimulus reduces the stimulus, and the pathway shuts off. In **positive feedback,** the response to a stimulus reinforces the stimulus, leading to a further increase in response. The stimulation of milk secretion and oxytocin's role in inducing contraction of uterine muscles during birth are two examples of positive feedback regulation. Negative feedback is typical of pathways involved in maintaining homeostasis. In many instances, a pair of pathways, each counteracting the other, constitutes the control system.

INTERACTIVE QUESTION 45.2

Explain the difference between negative feedback and positive feedback in the regulation of hormonal pathways.

Insulin and Glucagon: Control of Blood Glucose   Two antagonistic hormones regulate glucose concentration in the blood, and negative feedback controls both of these simple endocrine pathways. **Insulin** is released in response to a rise in blood glucose above the set point and triggers the uptake of glucose from the blood. **Glucagon,** released in response to a drop in blood glucose concentration, stimulates the release of glucose into the blood. Scattered within the exocrine tissue of the pancreas are clusters of endocrine cells called

pancreatic islets. Within each islet are *alpha cells* that secrete glucagon, and *beta cells* that secrete insulin.

Insulin lowers blood glucose levels by promoting the uptake of glucose from the blood into body cells, by slowing the breakdown of glycogen in the liver, and by inhibiting the conversion of amino acids and glycerol to glucose. Glucagon raises glucose concentrations by stimulating target cells in the liver to increase glycogen hydrolysis, convert amino acids and glycerol to glucose, and release glucose into the blood.

In **diabetes mellitus,** the absence of insulin in the bloodstream or the loss of response to insulin in target tissues reduces glucose uptake by cells. Glucose accumulates in the blood and is excreted in the urine, with an accompanying loss of water. The body must use fats for fuel, and acids from the breakdown of fats may lower blood pH.

Type 1 diabetes, also known as insulin-dependent diabetes, is an autoimmune disorder in which pancreatic beta cells are destroyed. This type of diabetes is treated by regular injections of genetically engineered human insulin. More than 90% of diabetics have *Type 2 diabetes*, or non-insulin-dependent diabetes, characterized by reduced responsiveness of target cells to insulin. Exercise and dietary control are often sufficient to manage this disease, although it is becoming a growing public health problem.

INTERACTIVE QUESTION 45.3

Complete the following concept map on the regulation of blood glucose levels by hormones of the pancreas.

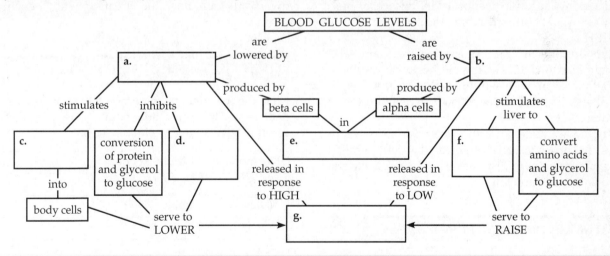

45.3   The hypothalamus and pituitary are central to endocrine regulation

Coordination of Endocrine and Nervous Systems in Vertebrates   The **hypothalamus,** situated in the lower brain, plays a key role in integrating the endocrine and nervous systems. In response to nerve signals it receives from throughout the body, the hypothalamus sends endocrine signals to the pituitary gland located at its base.

The **pituitary gland** has two discrete parts that develop separately and have different functions. The **posterior pituitary** stores and secretes oxytocin and **antidiuretic hormone (ADH).** These two neurohormones are produced by and delivered from neurosecretory cells of the hypothalamus. In addition to regulating milk secretion and uterine contractions, oxytocin influences maternal care, pair bonding, and sexual activity. ADH, or *vasopressin*, increases water retention in the kidneys and plays a role in social behavior.

Most of hormones secreted by the **anterior pituitary** are **tropic hormones,** which regulate other endocrine cells or glands. The anterior pituitary itself is regulated by *releasing hormones* and *inhibiting hormones* secreted by neurosecretory cells of the hypothalamus. These hypothalmic hormones are released into capillaries at the base of the hypothalamus and travel via a short portal vessel to capillary beds in the anterior pituitary. **Prolactin** is an anterior pituitary hormone that stimulates milk production; it is released in response to a *prolactin-releasing hormone.*

Thyroid Regulation: A Hormone Cascade Pathway

In a *hormone cascade pathway,* the hypothalamus secretes a hormone that stimulates or inhibits release of an anterior pituitary hormone, which then acts on a target endocrine tissue, stimulating the release of a hormone that produces a response. For example, thyrotropin-releasing hormone (TRH) stimulates the anterior pituitary to secrete thyrotropin, or thyroid-stimulating hormone (TSH). TSH stimulates the **thyroid gland** to release thyroid hormone.

Thyroid hormone regulates metabolic rate in mammals and helps to maintain normal blood pressure, heart rate, muscle tone, digestion, and reproductive functions. Hormone cascade pathways are typically regulated by negative feedback. In this case, high levels of thyroid hormone inhibit the secretion of TRH and TSH.

Hypothyroidism can cause weight gain and lethargy in adults. Excess thyroid hormone results in hyperthyroidism, with symptoms that include weight loss, irritability, and elevated body temperature and blood pressure.

Thyroid hormones are critical to vertebrate development. In humans, an inherited deficiency called congenital hypothyroidism results in retarded skeletal and mental development.

INTERACTIVE QUESTION   45.4

The thyroid gland produces two hormones, **triiodothyronine (T_3)** and **thyroxine (T_4).** Explain how a lack of iodine in the diet may result in goiter, an enlarged thyroid gland.

Evolution of Hormone Function

The hormone prolactin has effects ranging from milk production and secretion in mammals, to delay of metamorphosis in amphibians, to osmoregulation in fishes. The diversity of actions in different vertebrate species suggests that prolactin is an ancient hormone whose functions diversified during the evolution of vertebrate groups.

Melanocyte-stimulating hormone (MSH) regulates the activity of pigment-containing cells in the skin of most vertebrates, and it also functions in hunger and metabolism in mammals. Activation of a brain receptor for MSH increases fat metabolism and severely depresses appetite. Drugs that inhibit MSH receptors in the brain may be a potential treatment for the wasting condition called cachexia, which is associated with AIDS, TB, and late-stage cancer.

Tropic and Nontropic Hormones

In addition to TSH, the anterior pituitary produces three other tropic hormones. **Follicle-stimulating hormone (FSH)** and **luteinizing hormone (LH),** also called *gonadotropins,* stimulate activity of the testes and ovaries. Their release is regulated by *gonadotropin releasing hormone (GnRH).* **Adrenocorticotropic hormone (ACTH)** stimulates the adrenal cortex to produce and secrete steroid hormones.

Growth hormone (GH) has diverse metabolic nontropic effects. Its main tropic action is signaling the release of *insulin-like growth factors (IGFs),* which are produced by the liver and stimulate bone and cartilage growth. Gigantism and acromegaly are human growth disorders caused by hypersecretion of GH. Pituitary dwarfism can now be treated with genetically engineered GH.

INTERACTIVE QUESTION   45.5

Fill in the following table to review the hormones stored and released by the posterior pituitary (a and b) and those secreted by the anterior pituitary (c–h).

Hormone	Main Actions
Oxytocin	**a.**
ADH	**b.**
Prolactin	**c.**
MSH	**d.**
TSH	**e.**
FSH and LH	**f.**
ACTH	**g.**
GH	**h.**

45.4 Endocrine glands respond to diverse stimuli in regulating homeostasis, development, and behavior

Parathyroid Hormone and Vitamin D: Control of Blood Calcium The four **parathyroid glands** secrete **parathyroid hormone (PTH),** which stimulates the elevation of blood calcium levels through the reabsorption of Ca^{2+} in the kidney and its release from bone. PTH also activates vitamin D, which then increases Ca^{2+} uptake from food in the intestines. If blood calcium rises above the set point, the thyroid gland secretes **calcitonin,** a hormone that lowers blood calcium levels. In humans, calcitonin appears to be required only during the bone growth of childhood.

Adrenal Hormones: Response to Stress In mammals, the **adrenal glands** consist of two different glands: the outer *adrenal cortex* composed of endocrine cells, and the central *adrenal medulla* derived from neural tissue.

The adrenal medulla produces **norepinephrine** (noradrenaline) and **epinephrine** (adrenaline)—both of which are **catecholamines,** a class of compounds synthesized from the amino acid tyrosine. Epinephrine and norepinephrine, released in response to positive or negative stress, increase the availability of energy sources by stimulating glycogen hydrolysis in skeletal muscle and the liver, glucose release from liver cells, and fatty acid release from fat cells. These hormones increase the rate and volume of the heartbeat, dilate bronchioles in the lungs, and influence the contraction or relaxation of smooth muscles to increase blood supply to the heart, brain, and skeletal muscles while reducing the supply to other organs. Nerve signals from the hypothalamus stimulate the release of epinephrine and norepinephrine from the adrenal medulla.

The adrenal cortex responds to endocrine signals released in response to stress. A releasing hormone from the hypothalamus causes the anterior pituitary to release ACTH. This tropic hormone stimulates the adrenal cortex to secrete **corticosteroids.**

Glucocorticoids promote the synthesis of glucose from noncarbohydrates (such as muscle proteins) and thus increase energy supplies during stress. Cortisol has been used to treat inflammatory diseases such as arthritis, but its side effects can be dangerous.

Mineralocorticoids affect salt and water balance. *Aldosterone,* released in response to low blood volume or pressure, stimulates kidney cells to reabsorb sodium ions and water from the filtrate. ACTH secreted by the anterior pituitary in response to severe stress increases aldosterone secretion.

A third group of corticosteroids produced by the adrenal cortex are sex hormones—in particular androgens, which appear to influence female sex drive.

Gonadal Sex Hormones The testes of males and the ovaries of females produce steroids that affect growth, development, and reproductive cycles and behaviors. The three major categories of gonadal steroids—androgens, estrogens, and progestins—are present in different proportions in males and females.

The testes primarily synthesize **androgens,** such as **testosterone,** which determine the sex of the developing embryo and stimulate development of the male reproductive system and male secondary sex characteristics. **Estrogens,** in particular **estradiol,** regulate the maintenance of the female reproductive system and the development of female secondary sex characteristics. In mammals, **progestins,** including **progesterone,** help prepare and maintain the uterus for the growth of an embryo.

In a hormone cascade pathway, a hypothalamic releasing hormone, GnRH, controls the anterior pituitary secretion of the gonadotropins FSH and LH, which then control the synthesis of estrogens and androgens.

An *endocrine disruptor* is a foreign molecule that disrupts a hormone pathway. The synthetic estrogen DES, once prescribed for problem pregnancies, was shown to affect the reproductive system of female offspring. Other potential endocrine disruptors, such as bisphenol A, are under study.

Melatonin and Biorhythms The **pineal gland,** located near the center of the mammalian brain, secretes **melatonin,** which regulates functions related to light and seasonal changes in day length. The secretion of melatonin at night is controlled by the suprachiasmatic nuclei (SCN) in the hypothalamus, a biological clock that receives input from the retina. Nightly increases in melatonin may play an important role in promoting sleep.

Word Roots

andro- = male; **-gen** = produce (*androgens:* any steroid hormone, such as testosterone, that stimulates the development and maintenance of the male reproductive system and male secondary sex characteristics)

anti- = against; **-diure** = urinate (*antidiuretic hormone:* a hormone that promotes water retention by the kidneys)

cata- = down; **-chol** = anger (*catecholamines:* a class of neurotransmitters and hormones, including epinephrine and norepinephrine, synthesized from the amino acid tyrosine)

-cortico = the shell; **-tropic** = to turn or change (*adrenocorticotropic hormone:* a tropic hormone produced by the anterior pituitary that stimulates the production of steroid hormones by the adrenal cortex)

ecdys- = an escape (*ecdysteroid:* a steroid hormone that triggers molting in arthropods)

epi- = above, over (*epinephrine:* a catecholamine that, when secreted as a hormone by the adrenal medulla, mediates "fight-or-flight" responses to short-term stresses; also called adrenaline)

gluco- = sweet (*glucagon:* a hormone, secreted by pancreatic alpha cells, that raises blood glucose levels by promoting glycogen breakdown and the release of glucose by the liver)

lut- = yellow (*luteinizing hormone:* a gonadotropin secreted by the anterior pituitary gland)

melan- = black (*melatonin:* a hormone secreted by the pineal gland involved in regulation of biological rhythms and sleep)

neuro- = nerve (*neurohormone:* a molecule that is secreted by a neuron, travels in body fluids, and acts on specific target cells)

oxy- = sharp, acid (*oxytocin:* a hormone that induces contractions of the uterine muscles and causes the mammary glands to eject milk during nursing)

para- = beside, near (*parathyroid glands:* four endocrine glands, embedded in the surface of the thyroid gland, that secrete parathyroid hormone, which raises blood calcium levels)

pro- = before; **-lact** = milk (*prolactin:* a hormone produced by the anterior pituitary gland with diverse effects in different vertebrates; it stimulates milk production in mammals)

tri- = three; **-iodo** = violet (*triiodothyronine:* one of two iodine-containing hormones, secreted by the thyroid gland, that help regulate metabolism, development, and maturation)

Structure Your Knowledge

1. Briefly describe the two general mechanisms (depending on the location of the receptor) by which hormones trigger responses in target cells.
2. Explain how secretin, which is released in response to a low pH in the intestines and stimulates the release of bicarbonate from the pancreas, illustrates a simple hormone pathway.
3. What is a hormone cascade pathway? Describe how thyroid regulation illustrates such a pathway.

Test Your Knowledge

MATCHING: *Write the letter of the hormone (A–L) and the letter of the gland or organ that produces it (a–k) in the blank next to the description of each hormone's action. Choices may be used more than once, and not all choices must be used.*

Hormones	Gland or Organ
A. ACTH	**a.** adrenal cortex
B. androgen	**b.** adrenal medulla
C. ADH	**c.** anterior pituitary
D. calcitonin	**d.** hypothalamus
E. epinephrine	**e.** pancreas
F. glucagon	**f.** parathyroid gland
G. glucocorticoid	**g.** pineal
H. insulin	**h.** posterior pituitary
I. melatonin	**i.** testis
J. oxytocin	**j.** thymus
K. PTH	**k.** thyroid
L. thyroxine	

Hormone	Gland or Organ	Hormone Action
_____	_____	1. functions in biological clock and seasonal activities
_____	_____	2. breaks down muscle protein for conversion to glucose
_____	_____	3. increases blood sugar, glycogen breakdown in liver
_____	_____	4. stimulates development of male reproductive system
_____	_____	5. stimulates adrenal cortex to synthesize corticosteroids
_____	_____	6. increases available energy, heart rate, metabolism

_____ _____ 7. regulates metabolism, growth, and development

_____ _____ 8. raises blood calcium levels

_____ _____ 9. increases reabsorption of water by kidney

_____ _____ 10. stimulates contraction of uterus, milk secretion

MULTIPLE CHOICE: *Choose the one best answer.*

1. Which of the following statements is *not* accurate?
   a. Not all hormones are secreted by endocrine glands.
   b. Hormones are transported through the circulatory system to their destinations.
   c. Target cells have specific hormone receptors.
   d. Hormones are essential to homeostasis.
   e. Steroid hormones often function as neurotransmitters.

2. The best description of the difference between pheromones and hormones is that
   a. pheromones are small, volatile molecules; hormones are large, hydrophilic molecules.
   b. pheromones are involved in reproduction; hormones are not.
   c. pheromones are a form of neural communication; hormones are a form of chemical communication.
   d. pheromones are signaling molecules that function between organisms; hormones communicate among parts within an organism.
   e. pheromones are local regulators; hormones circulate in blood or hemolymph.

3. Ecdysteroid is
   a. a steroid hormone produced in insects that promotes retention of larval characteristics.
   b. secreted by prothoracic glands in insects and triggers molting and the development of adult form.
   c. a neurohormone that triggers the formation of a pupa.
   d. responsible for color changes in amphibians.
   e. involved in metamorphosis in amphibians.

4. The activation of helper T cells during an immune response by cytokines released by antigen-presenting cells and helper T cells themselves is an example of
   a. endocrine signaling.
   b. synaptic signaling.
   c. autocrine signaling.
   d. paracrine signaling.
   e. both c and d.

5. The anti-inflammatory and pain-relieving effects of ibuprofen are due to
   a. inhibition of the synthesis of the local regulator prostaglandin.
   b. inhibition of the release of NO by pain-sensing neurons and blood vessels.
   c. stimulation of the release of glucocorticoids by the adrenal cortex.
   d. activation of pain-reducing centers of the hypothalamus.
   e. inhibition of MSH receptors in the brain.

6. The finding that MSH (melanocyte-stimulating hormone) causes frog skin cells to darken when added to the interstitial fluid but has no effect on color when injected into the cells supports which of the following ideas?
   a. MSH is a tropic hormone.
   b. MSH is a neurohormone.
   c. MSH is a steroid hormone.
   d. MSH binds to cell surface receptors.
   e. Both c and d are reasonable conclusions.

7. Which one of the following hormones is *incorrectly* paired with its site of production?
   a. releasing hormones—hypothalamus
   b. growth hormone—anterior pituitary
   c. progestins—ovary
   d. TSH—thyroid
   e. mineralocorticoids—adrenal cortex

8. Steroid hormones
   a. are transported by neurosecretory cells directly to target tissues.
   b. form a hormone-receptor complex inside the cell that regulates gene expression.
   c. function via an amplified response involving second messengers.
   d. bind to membrane-bound receptors and initiate a signal transduction pathway.
   e. Both c and d are correct.

9. Which of the following situations is an example of positive feedback?
   a. In response to growth hormone, the liver produces insulin-like growth factors, which promote skeletal growth.
   b. The neurotransmitter acetylcholine causes skeletal muscle to contract, heart muscle to relax, and cells of the adrenal medulla to secrete epinephrine.
   c. Prostaglandins released from placental cells promote muscle contraction during childbirth, and those muscle contractions stimulate further prostaglandin release.
   d. Secretin acts on the pancreas, stimulating the release of bicarbonate.
   e. Elevated levels of stress result in neural stimulation of the adrenal medulla and hormonal stimulation of the adrenal cortex.

10. A tropic hormone is a hormone
    a. that is produced by the hypothalamus but stored and released from the anterior pituitary.
    b. whose target tissue is another endocrine gland.
    c. that acts by negative feedback to regulate its own levels in the body.
    d. from the hypothalamus that regulates hormone secretion of the posterior pituitary.
    e. that is released in response to nervous stimulation.

11. A lack of iodine in the diet can lead to formation of a goiter because
    a. with limited iodine available, triiodothyronine (T3) rather than thyroxine (T_4) is produced.
    b. excess thyroxine production causes the thyroid gland to enlarge.
    c. iodine is a key ingredient of the growth hormones that control growth of the thyroid.
    d. little thyroxine is produced, TRH and TSH production is not inhibited, and thyroid stimulation continues.
    e. the hypothalamus is stimulated to increase its production of TRH.

12. The anterior pituitary
    a. stores oxytocin and ADH produced by the hypothalamus.
    b. receives releasing and inhibiting hormones from the hypothalamus through portal vessels connecting capillary beds.
    c. produces releasing and inhibiting hormones.
    d. is responsible for nervous and hormonal stimulation of the adrenal glands.
    e. produces only tropic hormones.

13. Which of the following statements concerning melanocyte-stimulating hormone is *not* accurate?
    a. MSH controls pigment distribution in skin cells in amphibians, fish, and reptiles.
    b. MSH is a non-tropic hormone produced by the anterior pituitary gland.
    c. In mammals, MSH functions in suppression of hunger and an increase in fat metabolism.
    d. Drugs that block the brain MSH receptor may help in the treatment of the wasting condition called cachexia.
    e. All of the above are accurate statements about MSH.

14. Which of the following hormones is *not* involved in increasing the blood glucose concentration?
    a. glucagon
    b. epinephrine
    c. glucocorticoids
    d. ACTH (adrenocorticotropic hormone)
    e. insulin

15. Which of the following statements about epinephrine is *not* true?
    a. It is secreted by the adrenal medulla.
    b. It dilates bronchioles in the lungs.
    c. Its release is stimulated by ACTH.
    d. It can function as a neurohormone or a neurotransmitter.
    e. It is part of the fight-or-flight response to stress.

Choose from the following hormones to answer questions 16–20.
    a. ACTH
    b. prolactin
    c. norepinephrine
    d. estrogen
    e. insulin

16. Which hormone is a tropic hormone?

17. Which hormone is produced as the result of a hormone cascade pathway?

18. Which hormone is part of a simple neurohormone pathway?

19. Which hormone has diverse effects in different vertebrate species?

20. Which hormone is part of an antagonistic pair of hormones?

Chapter 46

Animal Reproduction

Key Concepts

46.1 Both asexual and sexual reproduction occur in the animal kingdom

46.2 Fertilization depends on mechanisms that bring together sperm and eggs of the same species

46.3 Reproductive organs produce and transport gametes

46.4 The interplay of tropic and sex hormones regulates mammalian reproduction

46.5 In placental mammals, an embryo develops fully within the mother's uterus

Framework

This chapter covers the patterns and mechanisms of animal reproduction. In sexual reproduction, fertilization may occur externally or internally. Development of the zygote may take place internally within the female; externally in a moist environment; or in a protective, resistant egg. Placental mammals provide nourishment as well as shelter for the embryo, and they nurse their young.

The human reproductive system is described, including its organs, glands, hormones, gamete formation, and sexual response. The chapter also covers pregnancy and birth, contraception, and recently developed reproductive technologies.

Chapter Review

46.1 Both asexual and sexual reproduction occur in the animal kingdom

In **sexual reproduction,** two meiotically formed haploid gametes fuse to form a diploid **zygote,** which develops into an offspring. Gametes are usually a relatively large, nonmotile **egg** and a small, motile **sperm. In asexual reproduction,** a single individual produces offspring, with no fusion of egg and sperm.

Mechanisms of Asexual Reproduction Many invertebrates can reproduce asexually by **fission,** in which a parent is separated into two individuals of equal size; by **budding,** in which a new individual grows out from the parent's body; or by *fragmentation*, in which the body is broken into several pieces, each of which develops into a complete animal. *Regeneration*, the regrowth of body parts, is necessary for a fragment to develop into a new organism. In **parthenogenesis,** eggs develop without being fertilized.

Sexual Reproduction: An Evolutionary Enigma Asexual reproduction potentially produces many more offspring, yet most eukaryotic species reproduce sexually. The shuffling of genes in sexual reproduction may allow more rapid adaptation to changing conditions or allow harmful genes to be eliminated more easily.

Reproductive Cycles Periodic reproductive cycles may be linked to favorable conditions or energy supplies. A combination of environmental and hormonal cues controls the timing of **ovulation,** the release of eggs.

Climate change is causing a mismatch between some animal migrations, which are cued by seasonal changes in daylight, and the new plant growth that provides nutrition for successful rearing of offspring.

Variation in Patterns of Sexual Reproductive Animals may reproduce sexually, asexually, or both. The freshwater crustacean *Daphnia*, like aphids and rotifers, produces eggs that develop by parthenogenesis, as well as eggs that are fertilized. In a few parthenogenetic fishes, amphibians, and lizards, doubling of chromosomes after meiosis creates diploid offspring.

In **hermaphroditism,** found in some sessile, burrowing, and parasitic animals, each individual has functioning male and female reproductive systems. Mating between two individuals results in fertilization of both. In some fishes and oysters, individuals reverse their sex during their lifetime.

46.2 Fertilization depends on mechanisms that bring together sperm and eggs of the same species

Fertilization is the union of sperm and egg. In **external fertilization,** eggs and sperm are shed, and fertilization occurs in the environment. External fertilization occurs almost exclusively in moist habitats, where the gametes and developing zygote are not in danger of desiccation.

Internal fertilization involves the placement of sperm in or near the female reproductive tract so that the egg and sperm unite internally. Behavioral cooperation, as well as copulatory organs and sperm receptacles, are required for internal fertilization.

Pheromones are small, volatile or water-soluble chemicals that may function as mate attractants.

Ensuring the Survival of Offspring   Species with internal fertilization usually produce fewer gametes, but a larger fraction of zygotes survive.

The shells of the eggs of birds, other reptiles, and monotremes protect the embryo from water loss or injury. Embryos of marsupial mammals complete development in the mother's pouch, attached to a mammary gland.

Embryos of eutherian (placental) mammals develop in the uterus, nourished by the mother's blood through the placenta. Parental care of young is widespread among vertebrates and even among many invertebrates.

Gamete Production and Delivery   Precursor cells for eggs or sperm are often established early in embryogenesis and later *amplified* in number. The simplest reproductive systems lack **gonads** to produce gametes. In polychaete annelids, eggs or sperm develop from cells lining the coelom.

Most insects have separate sexes and complex reproductive systems. Sperm develop in the testes, are stored in the seminal vesicles, and are ejaculated into the female. Eggs are produced in the ovaries, fertilized in the uterus, and then expelled from the body. Females may have a **spermatheca,** or sperm-storing sac.

With the exception of mammals, many vertebrates have a common opening—a **cloaca**—for the digestive, excretory, and reproductive systems. Most mammals have a separate opening for the digestive tract. The vagina and excretory opening are separate in most female mammals.

Mechanisms that influence which sperm are successful in sequential matings have evolved in some male and female insects.

46.3 Reproductive organs produce and transport gametes

Female Reproductive Anatomy   The external reproductive structures in human females include the clitoris and two sets of labia. The ovaries contain many **follicles,** each of which contains an **oocyte** surrounded by cells that nourish and protect the developing egg during its development. During each menstrual cycle, a maturing follicle produces estradiol, the primary female sex hormone. Following ovulation, the follicle forms a mass called the **corpus luteum,** which secretes progesterone and estradiol.

The egg cell is expelled into the abdominal cavity and swept by cilia into the **oviduct,** or fallopian tube, through which it travels to the **uterus.** The uterine lining, the **endometrium,** is richly supplied with blood vessels. The neck of the uterus, called the **cervix,** opens into the **vagina,** the elastic-walled birth canal and repository for sperm during copulation.

The external genitals are collectively called the **vulva.** Two pairs of skin folds, the thick outer **labia majora** and the thin inner **labia minora,** enclose the separate openings of the vagina and urethra. The vaginal opening is initially covered by a membrane called the **hymen.** The **clitoris** is composed of erectile tissue supporting a **glans,** or head, covered by the **prepuce.**

A **mammary gland,** located within a breast, is composed of fatty tissue and small milk-secreting sacs that drain into ducts that open at the nipple. The low level of estradiol in males prevents the mammary glands from enlarging.

Male Reproductive Anatomy   The external male reproductive organs include the scrotum and penis. Sperm are produced in the highly coiled **seminiferous tubules** of the **testes. Leydig cells** produce testosterone and other androgens. Sperm production requires a cooler temperature than the internal body temperature of most mammals, so the **scrotum** suspends the testes below the abdominal cavity.

Sperm pass from a testis into the coiled **epididymis,** in which they mature and gain motility. During **ejaculation,** sperm are propelled through the **vas deferens,** into a short **ejaculatory duct,** and out through the **urethra,** which runs through the penis.

Three sets of accessory glands add secretions to the **semen.** The **seminal vesicles** contribute an alkaline fluid containing mucus, fructose (an energy source for the sperm), a coagulating enzyme, ascorbic acid, and prostaglandins. The **prostate gland** produces a secretion that contains anticoagulant enzymes and citrate. Benign enlargement of the prostate and prostate cancer are common medical problems in older men. The *bulbourethral glands* produce a mucus that neutralizes urine remaining in the urethra.

The **penis** is composed of spongy tissue that engorges with blood during sexual arousal, producing an erection that facilitates insertion of the penis into the vagina. Some mammals have a baculum, a bone that helps stiffen the penis. A fold of skin called the prepuce, or foreskin, covers the head, or glans, of the penis.

INTERACTIVE QUESTION   46.3

Label the indicated structures in the following diagrams of the human male and female reproductive systems.

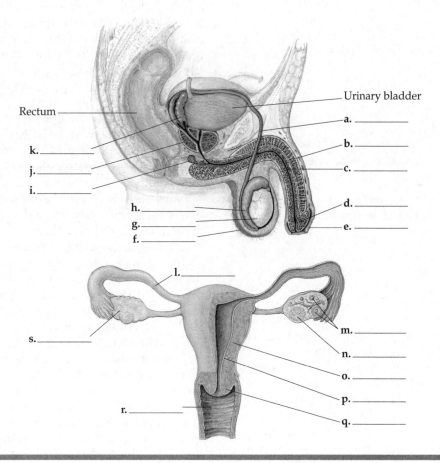

Gametogenesis   The production of gametes, or **gametogenesis,** differs in females and males. **Spermatogenesis** occurs continuously in the seminiferous tubules of the testes. *Primordial* germ cells in male embryos differentiate into spermatogonial stem cells. These cells divide mitotically to form **spermatogonia,** which divide to form spermatocytes. Spermatocytes undergo meiosis to produce four spermatids. The haploid nucleus of a sperm is contained in a head, tipped with an **acrosome,** which contains enzymes that help the sperm penetrate the egg. Mitochondria provide ATP for movement of the flagellum, or tail.

Oogenesis begins in the female embryo. **Oogonia,** produced from primordial germ cells, divide and differentiate into **primary oocytes,** which are arrested in prophase I of meiosis before birth. Beginning at puberty, FSH periodically stimulates a follicle to resume growth. Its enclosed primary oocyte finishes meiosis I and develops into a **secondary oocyte,** which is arrested in metaphase II. In humans, meiosis is completed if a sperm penetrates the oocyte. Meiotic cytokinesis is unequal, producing one large egg and up to three small haploid polar bodies that disintegrate.

INTERACTIVE QUESTION   46.4

List three important ways in which oogenesis differs from spermatogenesis.

a.

b.

c.

46.4   The interplay of tropic and sex hormones regulates mammalian reproduction

Gonadotropin-releasing hormone (GnRH) from the hypothalamus directs the anterior pituitary to produce follicle-stimulating hormone (FSH) and luteinizing hormone (LH), gonadotropins that regulate gametogenesis and sex hormone production in both males and females. The steroid sex hormones, androgens (primarily testosterone) in males and estrogens (chiefly estradiol) and progesterone in females, also regulate gametogenesis.

Androgens are responsible for the embryonic development of male primary sex characteristics. At puberty, androgens direct formation of secondary sex characteristics, such as voice deepening, pubic and facial hair, and muscle growth. In females, estrogens stimulate breast and pubic hair development at puberty. Sex hormones influence sexual behaviors in both sexes.

Hormonal Control of Female Reproductive Cycles In females, the endometrium thickens to prepare for the implantation of the embryo and then is shed in a process called **menstruation** if fertilization does not occur. The human female reproductive cycle involves the integration of the **uterine cycle** (**menstrual cycle**), which involves these cyclic changes in the uterus, and the **ovarian cycle,** which involves follicle growth and ovulation.

The ovarian cycle begins with the release from the hypothalamus of GnRH, which stimulates the anterior pituitary to secrete small amounts of FSH and LH. In this **follicular phase,** FSH stimulates follicular growth, and the cells of the follicle secrete estradiol. The low level of estradiol inhibits the release of FSH and LH.

When the secretion of estradiol by the growing follicle rises sharply, the hypothalamus is stimulated to increase GnRH output, which results in a rise in LH and FSH release. The increase in LH induces maturation of the follicle, which enlarges and forms a bulge near the ovary surface. Ovulation occurs about a day after the LH surge, with the rupture of the follicle and adjacent ovary wall.

In the **luteal phase,** LH stimulates the transformation of the ruptured follicle into the corpus luteum, which secretes estradiol and progesterone. The rising level of these hormones exerts negative feedback on the hypothalamus and pituitary, inhibiting secretion of LH and FSH and thereby preventing another egg from maturing. Without LH to maintain it, the corpus luteum disintegrates, which reduces the levels of estradiol and progesterone. This drop releases the inhibition of the hypothalamus and pituitary, and FSH and LH secretion begins again, stimulating growth of new follicles and the start of the next ovarian cycle.

The uterine (menstrual) cycle is controlled by hormones secreted by the ovaries. Estradiol secreted by the growing follicles causes the endometrium to begin to thicken in the **proliferative phase,** which correlates with the follicular phase of the ovarian cycle. After ovulation, estradiol and progesterone stimulate continued development of the endometrium and of glands that secrete a nutrient fluid. The luteal phase of the ovarian cycle corresponds with the **secretory phase** of the uterine cycle.

The rapid drop in ovarian hormones caused by the disintegration of the corpus luteum reduces blood supply to the endometrium and begins its disintegration, leading to the **menstrual flow phase** of the uterine cycle. By convention, the uterine and ovarian cycles begin with the first day of menstruation.

Endometriosis is a condition in which some endometrial tissue becomes **ectopic**—that is, located outside the uterus. The cyclic swelling and breakdown of this tissue causes pain and bleeding into the abdomen.

Menopause, the cessation of ovulation and menstruation, occurs in older women as the ovaries become less responsive to FSH and LH.

Humans and some other primates have menstrual cycles. Other mammals have **estrous cycles,** during which the endometrium thickens but is reabsorbed (not shed) if fertilization does not occur. Females are receptive to sexual activity only during estrus, or "heat," the period surrounding ovulation.

INTERACTIVE QUESTION 46.5

In the following diagram of the human female reproductive cycle, label the lines indicating the levels of the gonadotropic and the ovarian hormones in the blood and the phases of the ovarian and uterine cycles.

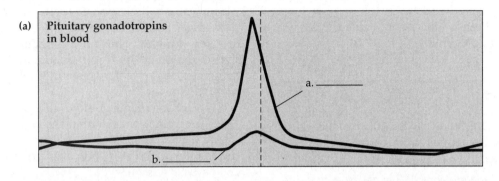

(a)   **Pituitary gonadotropins in blood**

a. _____

b. _____

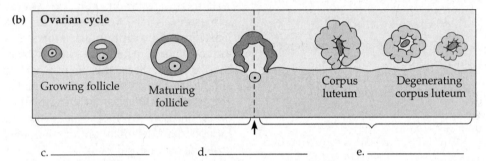

(b)   **Ovarian cycle**

Growing follicle   Maturing follicle   Corpus luteum   Degenerating corpus luteum

c. _____   d. _____   e. _____

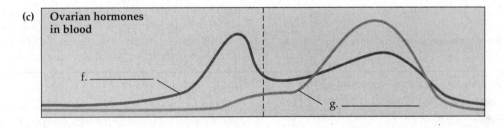

(c)   **Ovarian hormones in blood**

f. _____

g. _____

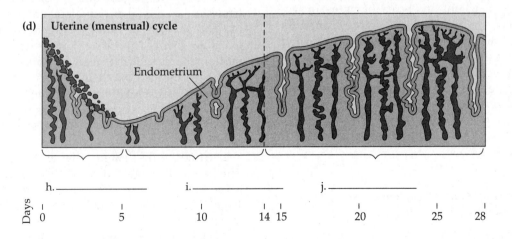

(d)   **Uterine (menstrual) cycle**

Endometrium

h. _____   i. _____   j. _____

Days   0   5   10   14   15   20   25   28

Hormonal Control of the Male Reproductive System

FSH acts on Sertoli cells, which nourish developing sperm. LH acts on Leydig cells, which secrete testosterone and other androgens, promoting spermatogenesis. Testosterone inhibits production of GnRH, FSH, and LH; **inhibin,** produced by Sertoli cells, reduces FSH secretion. These feedback mechanisms maintain proper levels of androgen production.

INTERACTIVE QUESTION 46.6

Fill in the blanks in the following description of the control of male reproductive hormones.

a. _____, produced by the hypothalamus, regulates the release of gonadotropic hormones from the anterior pituitary.

b. _____ stimulates Sertoli cells of the seminiferous tubules, which nourish sperm.

c. _____ stimulates the production of **d.** _____ by Leydig cells.

e. The production of GnRH, LH, and FSH is regulated by _____.

46.5 In placental mammals, an embryo develops fully within the mother's uterus

Conception, Embryonic Development, and Birth The 2–5 mL of ejaculated semen contains 70–130 million sperm/mL. The slightly alkaline semen neutralizes the acidity of the vagina. Semen initially coagulates and then liquifies as the sperm start swimming through the female reproductive tract.

Fertilization, or **conception,** occurs in the oviduct. **Cleavage** produces a ball of 16 cells by the time the embryo reaches the uterus in about 3 to 4 days. The **blastocyst,** a hollow sphere of cells, implants in the endometrium about 7 days after conception. The embryo secretes **human chorionic gonadotropin (hCG),** which maintains the corpus luteum's secretion of progesterone and estrogens through the first few months.

The carrying of one or more embryos in the uterus is called **pregnancy** or **gestation.** The gestation period correlates with body size and the degree of development of the young at birth. In humans, pregnancy averages 266 days (38 weeks).

Human gestation can be divided into three **trimesters.** After implantation, the outer layer of the blastocyst, the **trophoblast,** grows and mingles with the endometrium, helping to form the **placenta,** a disk-shaped organ that is the site of waste removal and gas and nutrient exchange between maternal blood and embryonic capillaries. Splitting of the embryo in the first month can produce *monozygotic* or identical twins. Independent fertilization of two eggs can lead to *dizygotic* or fraternal twins.

Development proceeds to **organogenesis,** and by the eighth week the embryo has all the rudimentary structures of the adult and is called a **fetus.** In the mother, high progesterone levels stimulate growth of the uterus and the maternal part of the placenta, the cessation of menstrual cycling, and breast enlargement.

During the second trimester, hCG declines, the corpus luteum degenerates, and the placenta secretes its own progesterone, which maintains the pregnancy. The fetus grows rapidly and is quite active. The third trimester is a period of rapid fetal growth.

Labor, a series of strong contractions of the uterus, leads to birth. An inflammatory response in the mother in response to fetal hormones may induce labor. Hormones (oxytocin and estradiol) and local regulators (prostaglandins) regulate labor. Through a positive feedback system, uterine contractions stimulate the release of oxytocin, which stimulates further contractions.

During the first stage of labor, the cervix dilates. The second stage consists of contractions that force the fetus out of the uterus and through the vagina. The placenta is delivered in the final stage of labor.

Lactation is unique to mammals. Prolactin secretion by the anterior pituitary stimulates milk production, and oxytocin triggers the release of milk during nursing.

Human Sexual Response The human sexual response cycle includes two types of physiological reactions: **vasocongestion,** increased blood flow to a tissue, and **myotonia,** increased muscle tension. The excitement phase involves vasocongestion and vaginal lubrication, preparing the vagina and penis for **coitus,** or sexual intercourse. The plateau phase continues vasocongestion and myotonia, and breathing rate and heart rate increase. In both sexes, **orgasm** is characterized by rhythmic, involuntary contractions of reproductive structures. In males, emission moves semen into the urethra, and ejaculation occurs when the urethra contracts and semen is expelled. In the resolution phase, vasocongested organs return to normal size, and muscles relax.

INTERACTIVE QUESTION 46.7

List the functions of each of the following hormones and the structures that secrete them:

a. human chorionic gonadotropin

b. progesterone

c. oxytocin

Maternal Immune Tolerance of the Embryo and Fetus
Current research is attempting to explain why a mother does not reject an embryo, which has paternal as well as maternal chemical markers. Changes in immune system regulation appear to occur during pregnancy.

Contraception and Abortion **Contraception,** the deliberate prevention of pregnancy, can be accomplished by several methods: preventing release of egg or sperm, preventing fertilization, or preventing implantation. Complete abstinence from sexual intercourse is the most effective means of birth control. The **rhythm method,** or **natural family planning,** is based on refraining from intercourse during the period in which conception is most likely: several days before and after ovulation.

Barrier methods of contraception include **condoms** and **diaphragms,** which, when used in conjunction with spermicidal foam or jelly, present a physical and chemical barrier to fertilization. Condoms are the only form of birth control that offers some protection against sexually transmitted diseases.

IUDs (intrauterine devices) interfere with fertilization and implantation. The release of gametes may be prevented by chemical contraception, as in **birth control pills,** and by sterilization. Most birth control pills are combinations of a synthetic estrogen and progestin, which act by negative feedback to prevent the release of GnRH by the hypothalamus and thus of FSH and LH by the pituitary, resulting in a cessation of ovulation and follicle development. High doses of combination birth control pills can be taken as morning-after pills (MAPS) following unprotected intercourse. Progestin-only injections or minipills alter the cervical mucus such that it blocks sperm entry to the uterus. Cardiovascular problems are a potential effect of combination birth control pills.

Tubal ligation in women and **vasectomy** in men are sterilization procedures that permanently prevent gamete release.

Abortion is the termination of a pregnancy. Spontaneous abortion, or *miscarriage,* occurs in as many as one-third of all pregnancies. The drug mifepristone (RU486) blocks progesterone receptors in the uterus and can be used to terminate a pregnancy within the first 7 weeks.

INTERACTIVE QUESTION 46.8

List the three general types of birth control methods, and cite examples of each. Indicate which examples are most likely to prevent pregnancy, and which are least likely to do so.

a.

b.

c.

Modern Reproductive Technologies Some genetic diseases and developmental defects can be detected while the fetus is in the uterus, using such techniques as amniocentesis, chorionic villus sampling, ultrasound imaging, and a new technique that identifies fetal cells in the mother's blood, which can then be tested.

Reproductive technology can help couples unable to conceive. Fertilization procedures called **assisted reproductive technology** include *in vitro* **fertilization (IVF),** which involves the removal of eggs from a woman, fertilization within a culture dish, and implantation of the developing embryo in the uterus. In **intracytoplasmic sperm injection (ICSI),** a spermatid or sperm is directly injected into an oocyte.

Word Roots

a- = not, without (*asexual reproduction:* the generation of offspring from a single parent, without the fusion of gametes; offspring are usually genetically identical to the parent)

acro- = tip; **-soma** = body (*acrosome:* a vesicle in the tip of a sperm containing enzymes that help the sperm reach the egg)

blasto- = produce; **-cyst** = sac, bladder (*blastocyst:* the blastula stage of mammalian embryonic development, consisting of an inner cell mass, a cavity, and an outer layer, the trophoblast)

coit- = a coming together (*coitus:* the insertion of a penis into a vagina; also called sexual intercourse)

contra- = against (*contraception:* the deliberate prevention of pregnancy)

-ectomy = cut out (*vasectomy:* the cutting and sealing of each vas deferens to prevent sperm from entering the urethra)

endo- = inside (*endometrium:* the inner lining of the uterus, which is richly supplied with blood vessels)

epi- = above, over (*epididymis:* a coiled tubule adjacent to the testes, where sperm are stored)

labi- = lip; **major-** = larger (*labia majora:* a pair of thick, fatty ridges that encloses and protects the rest of the vulva)

lact- = milk (*lactation:* the continued production of milk by the mammary glands)

menstru- = month (*menstruation:* the shedding of portions of the endometrium during a uterine (menstrual) cycle)

minor- = smaller (*labia minora:* a pair of slender skin folds that surrounds the openings of the urethra and vagina)

myo- = muscle (*myotonia:* increased muscle tension, characteristic of sexual arousal in certain human tissues)

oo- = egg; **-genesis** = producing (*oogenesis:* the process in the ovary that results in the production of female gametes)

partheno- = a virgin (*parthenogenesis:* asexual reproduction in which females produce offspring from unfertilized eggs)

-theca = a cup, case (*spermatheca:* a sac in which sperm are stored in the female reproductive system of many insects)

tri- = three (*trimester:* in human development, one of the three 3-month periods of pregnancy)

vasa- = a vessel (*vasocongestion:* the filling of a tissue with blood, caused by increased blood flow through the arteries of that tissue)

Structure Your Knowledge

1. Use a simple flow diagram to trace the path of a human sperm from the point of production to the point of fertilization; then briefly describe the functions of both the structures the sperm passes through and the associated glands.

2. Answer the following questions concerning the human menstrual cycle.
   a. What does GnRH do?
   b. What does FSH stimulate?
   c. What causes the spike in LH level?
   d. What does this LH surge induce?
   e. What does LH maintain during the luteal phase?
   f. What inhibits the secretion of LH and FSH?
   g. What allows LH and FSH secretion to begin again?

3. Describe how birth control pills work. How does the drug RU486 (mifepristone) function?

Test Your Knowledge

FILL IN THE BLANKS

_____ 1. type of asexual reproduction in which a new individual grows while attached to the parent's body

_____ 2. small, volatile chemicals that may act as mate attractants

_____ 3. development of egg without fertilization

_____ 4. individual with functioning male and female reproductive systems

_____ 5. common opening of digestive, excretory, and reproductive systems in non-mammalian vertebrates

_____ 6. reproductive cycle in which thickened endometrium is reabsorbed

_____ 7. assisted reproductive technology in which oocytes are fertilized in a culture dish and then implanted in the uterus

_____ 8. common duct for urine and semen in mammalian males

_____ 9. filling of a tissue with blood due to increased blood flow

_____ 10. period when ovulation and menstruation cease in human females

MULTIPLE CHOICE: *Choose the one best answer.*

1. An advantage of sexual reproduction is that
   a. it is easier than asexual reproduction.
   b. offspring have a better start when produced from a union of egg and sperm than when simply budded off a parent.
   c. it produces diploid offspring, whereas asexually produced offspring are haploid.
   d. it produces variation in the offspring, and new genetic combinations may be better adapted to a changing environment.
   e. it requires both male and female members of a species to be present in a population and have behavioral interactions.

2. Which of the following animals is *least* likely to be hermaphroditic?
   a. an earthworm
   b. a barnacle (sessile crustacean)
   c. a tapeworm
   d. a grasshopper
   e. a liver fluke

3. External fertilization is *least* likely to be associated with
   a. pheromones.
   b. spermatheca.
   c. large numbers of gametes and zygotes.
   d. behavioral cooperation.
   e. moist environments.

4. Which of the following structures is *incorrectly* paired with its function?
   a. seminiferous tubule—adds fluid containing mucus, fructose, and prostaglandins to semen
   b. scrotum—encases testes and suspends them below the abdominal cavity
   c. epididymis—tubules in which sperm mature
   d. prostate gland—adds fluid to semen
   e. vas deferens—transports sperm from epididymis to ejaculatory duct

5. In which location does fertilization usually take place in a human female?
   a. ovary
   b. oviduct
   c. uterus
   d. cervix
   e. vagina

6. Which of the following statements *correctly* describes how the production of human sperm and eggs differs?
   a. The meiotic production of gametes occurs before a female is born, but does not begin until puberty in males.
   b. Each meiotic division produces four sperm but only two eggs.
   c. Meiosis occurs in the testes of males but in the oviducts of females.
   d. Primary oocytes stop dividing by mitosis before birth, whereas male stem cells continue to divide throughout life.
   e. Meiosis is an uninterrupted process in males, whereas in females it resumes when a follicle matures and is only completed when a sperm penetrates the egg.

7. The function of the corpus luteum is to
   a. nourish and protect the egg cell.
   b. produce prolactin, which stimulates the milk sacs of the mammary gland.
   c. produce progesterone and estradiol during the luteal phase.
   d. eject the egg and then disintegrate following ovulation.
   e. maintain pregnancy by producing human chorionic gonadotropin.

8. The secretory phase of the uterine cycle
   a. begins with falling levels of estradiol and progesterone.
   b. corresponds with the luteal phase of the ovarian cycle.
   c. involves the initial proliferation of the endometrium.
   d. corresponds with the follicular phase of the ovarian cycle.
   e. occurs when the endometrium begins to degenerate and menstrual flow occurs.

Choose from the following hormones to answer questions 9–13.
   a. estradiol
   b. progesterone
   c. LH (luteinizing hormone)
   d. FSH (follicle-stimulating hormone)
   e. hCG (human chorionic gonadotropin)

9. Which hormone stimulates ovulation and the development of the corpus luteum?

10. Which hormone is produced by the developing follicle and initiates thickening of the endometrium?

11. Which hormone is produced by the embryo during the first trimester and is necessary for maintaining a pregnancy?

12. Which hormone stimulates Leydig cells to make testosterone, which in turn stimulates sperm production?

13. Which hormone is produced by the corpus luteum and later by the placenta and is responsible for maintaining a pregnancy?

14. Which of the following hormones is *incorrectly* paired with its function?

    a. androgens—responsible for primary and secondary male sex characteristics

    b. oxytocin—stimulates uterine contractions during labor

    c. estradiol—responsible for secondary female sex characteristics

    d. FSH—acts on Sertoli cells that nourish sperm, promoting spermatogenesis

    e. prolactin—stimulates breast development at puberty

15. Examples of birth control methods that prevent the production or release of gametes are

    a. chemical contraception and sterilization.

    b. birth control pills and IUDs.

    c. condoms and diaphragms.

    d. abstinence and chemical contraception.

    e. the progestin minipill and RU486.

16. Certain maternal diseases, drugs, alcohol, and radiation are most dangerous to embryonic development

    a. during the first 2 weeks, when the embryo has not yet implanted and spontaneous abortion may occur.

    b. during the first 3 months, when organogenesis is occurring.

    c. during the first and second trimesters, when the embryonic liver is not yet filtering toxins.

    d. during the second trimester, when the corpus luteum no longer secretes progesterone.

    e. during the third trimester, when the most rapid growth is occurring.

Chapter 47

Animal Development

Key Concepts

47.1 Fertilization and cleavage initiate embryonic development

47.2 Morphogenesis in animals involves specific changes in cell shape, position, and survival

47.3 Cytoplasmic determinants and inductive signals contribute to cell fate specification

Framework

Fertilization initiates physical and molecular changes in the egg cell. Early embryonic development includes cleavage to form a blastula, gastrulation, and organogenesis. The amount of yolk in the egg and the phylogeny of the animal determine how these processes occur in any particular animal group.

Development and the differentiation of cells result from the control of gene expression by both localized determinants and cell–cell induction. Pattern formation depends on the positional information a cell receives within a developing structure. Developmental biologists are gradually unraveling some of the mechanisms underlying the complex processes of animal development.

Chapter Review

A common sequence of developmental stages, a common set of regulatory genes, and many basic mechanisms of development are shared across many animal species. Biologists use molecular genetics, classical embryology, and **model organisms,** which are easy to study in the lab, to study development.

47.1 Fertilization and cleavage initiate embryonic development

Fertilization **Fertilization** is the union of haploid egg and sperm to form a diploid zygote. Successful fertilization requires that (1) sperm penetrate any protective layers around the egg, (2) recognition molecules on the sperm bind with matching egg receptors, and (3) changes at the egg surface prevent *polyspermy*.

In the external fertilization of sea urchins, the jelly coat surrounding the egg releases molecules that attract the sperm. When a sperm comes in contact with the jelly coat, the **acrosome** at the tip of the sperm discharges hydrolytic enzymes. This **acrosomal reaction** allows the *acrosomal process* to elongate through the jelly coat. Fertilization within the same species is assured when proteins on the tip of the acrosomal process bind to specific receptor proteins extending from the egg's plasma membrane.

In response to the fusion of sperm and egg plasma membranes, ion channels open and sodium ions flow into the egg. The resulting *depolarization* of the membrane prevents other sperm from fusing with the egg, providing a **fast block to polyspermy.** The sperm nucleus then enters the egg cytoplasm.

Membrane fusion also initiates the cortical reaction. In response to a high concentration of Ca^{2+} released from the endoplasmic reticulum, cortical granules located in the outer *cortex* release their contents by exocytosis. The secreted enzymes and other macromolecules clip off sperm-binding receptors and cause the *vitelline layer* (formed from the egg's extracellular matrix) to elevate and harden into the fertilization envelope, which functions as a **slow block to polyspermy.**

The rise in Ca^{2+} concentration also increases cellular respiration and protein synthesis, known as egg activation. Injecting calcium into an egg that lacks a nucleus artificially activates the egg, indicating that cytoplasmic proteins and mRNAs are sufficient for egg activation.

After the sperm nucleus fuses with the egg nucleus, DNA replication begins in preparation for the cell division that begins the development of the embryo.

INTERACTIVE QUESTION   47.1

a. What assures that only a sperm of the correct species fertilizes a sea urchin egg?

b. What assures that only one sperm will fertilize an egg?

Following internal fertilization in mammals, secretions of the female reproductive tract affect the motility and structure of sperm, leading to their *capacitation*. A sperm migrates through the follicle cells released with the egg to the **zona pellucida,** the extracellular matrix of the egg. Complementary binding of a sperm to a receptor in the zona pellucida induces an acrosomal reaction. The hydrolytic enzymes released from the acrosome enable the sperm cell to penetrate the zona pellucida.

Binding of a sperm membrane protein with the egg membrane leads to the release of enzymes from cortical granules. This cortical reaction causes alterations of the zona pellucida and functions as a slow block to polyspermy.

The entire sperm enters the egg. Nuclear envelopes of egg and sperm disperse, and the chromosomes align on the mitotic spindle for the first cell division.

Cleavage   The rapid cell divisions of **cleavage** parcel the egg cytoplasm into smaller cells called **blastomeres.** These first divisions usually produce a **blastula**—a cluster of cells surrounding a fluid-filled cavity called the **blastocoel.**

In frogs and many other animals, cleavage is affected by the distribution of **yolk,** which is much more concentrated toward the **vegetal pole** of the egg than toward the **animal pole.** A *cleavage furrow* marks each cell division. The first division of a frog egg bisects the gray crescent, a lighter region opposite the site of sperm entry. The first two cleavage divisions are vertical, extending from the animal pole to the vegetal pole; the third division is equatorial or perpendicular to the polar axis. Yolk impedes cell division, and smaller cells are produced in the animal hemisphere. In the frog blastula, the blastocoel forms in the animal hemisphere due to unequal cell division.

INTERACTIVE QUESTION   47.2

In the following diagrams of frog embryos, identify the stages and label the indicated regions or structures.

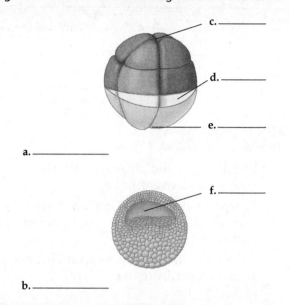

c. _____

d. _____

e. _____

a. _____

f. _____

b. _____

In species in which eggs have little yolk, blastomeres are of equal size and the blastocoel forms centrally. The complete cleavage of eggs with small or moderate amounts of yolk is said to be **holoblastic.**

The large amount of yolk in the eggs of birds, other reptiles, many fishes, and insects results in a type of incomplete cleavage called **meroblastic.**

Divisions of a fertilized bird egg are restricted to the small disk of cytoplasm on top of the yolk. The blastomeres of the resulting cap of cells separate into upper and lower layers, forming a cavity between them comparable to a blastocoel.

In the yolk-rich eggs of most insects, the zygote nucleus undergoes repeated divisions, and the resulting nuclei migrate to the egg surface. Plasma membranes eventually form around the nuclei, creating a blastula made of cells surrounding a mass of yolk.

The ratio of material in the nucleus to that in the cytoplasm appears to regulate the number of cleavage divisions. The nuclei in the resulting smaller blastomeres are now able to produce enough RNA to support the cell's metabolism and program further development.

47.2 Morphogenesis in animals involves specific changes in cell shape, position, and survival

Morphogenesis, the development of body shape, occurs through **gastrulation** and **organogenesis.**

Gastrulation The movement of cells into the blastula produces a two- or three-layered embryo called a **gastrula.** The embryonic **germ layers** produced by gastrulation are the outer **ectoderm,** the middle **mesoderm,** and the inner **endoderm** that lines the embryonic digestive tract. (Diploblastic animals have only ectoderm and endoderm.) Adult tissues and organs derive from one or more of these cell layers.

In a sea urchin, gastrulation occurs as cells from the vegetal pole detach and move into the blastocoel as migratory *mesenchyme cells.* The remaining cells at the vegetal pole flatten and then buckle inward. This *invagination* deepens to form a narrow pouch called the **archenteron,** which will become the digestive tube. The opening to the archenteron, the **blastopore,** develops into the anus. Extensions of mesenchyme cells pull the tip of the archenteron across the blastocoel, where it fuses with the ectoderm to form the mouth. Gastrulation produces a ciliated larva that eventually metamorphoses into an adult sea urchin.

Gastrulation in frog development is more complicated because of the multilayered blastula wall and the large, yolk-filled cells of the vegetal hemisphere. A group of invaginating cells begins gastrulation in the region of the gray crescent, forming a crease. The upper area of the crease is called the **dorsal lip** of the blastopore. In a process called *involution,* surface cells roll over the dorsal lip and move inside, becoming organized into the endoderm and mesoderm. The blastopore eventually forms a circle surrounding a plug of yolk-filled cells. Gastrulation produces an endoderm-lined archenteron surrounded by mesoderm; the outer layer of the gastrula is composed of ectoderm.

In the chick embryo, a linear thickening called the **primitive streak** is the site of gastrulation. Cells of the upper layer, the *epiblast,* move inward through the primitive streak, forming the mesoderm and the endoderm. Cells remaining in the epiblast become ectoderm. The *hypoblast* cells of the original lower layer form part of the yolk sac and connecting stalk.

Because mammalian eggs contain little yolk, cleavage is holoblastic. The **blastocyst** consists of an outer epithelium, called the **trophoblast,** surrounding a cavity into which protrudes a cluster of cells, called the **inner cell mass.** These inner cells develop into the embryo and some of the extraembryonic membranes. The

Label the parts in the following diagrams of gastrulation in a sea urchin and a frog.

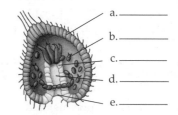

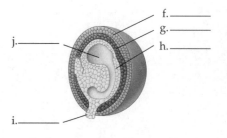

a. _____
b. _____
c. _____
d. _____
e. _____

f. _____
g. _____
h. _____
j. _____
i. _____

trophoblast secretes enzymes that enable the blastocyst to embed in the uterine lining (endometrium) and then extends projections that will contribute to the fetal portion of the placenta.

The inner cell mass forms a flat disk with an upper *epiblast* and lower *hypoblast.* In gastrulation, cells from the epiblast move through a primitive streak to form the mesoderm and endoderm. Four **extraembryonic membranes** develop from and support the embryo.

Label the indicated cell layers in the following diagram of a human blastocyst implanting in the endometrium.

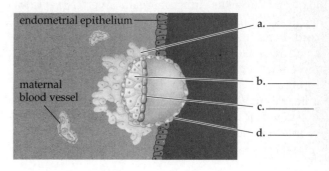

endometrial epithelium

maternal blood vessel

a. _____
b. _____
c. _____
d. _____

Developmental Adaptations of Amniotes Reptiles (including birds) and mammals are **amniotes,** meaning that they provide the aqueous environment necessary for embryonic development within a fluid-filled sac surrounded by a membrane called the amnion. Extraembryonic membranes are essential to development within the egg of reptiles (and monotremes) and in the uterus of mammals.

INTERACTIVE QUESTION 47.5

List the four extraembryonic membranes found in birds and other reptiles, and briefly describe the functions of each.

a.

b.

c.

d.

The extraembryonic membranes of mammals are homologous to those of reptiles. The chorion surrounds the embryo and the other membranes and functions in gas exchange. The amnion encloses the embryo in a fluid-filled amniotic cavity. The yolk sac, enclosing a small fluid-filled cavity, is the site of the early formation of blood cells. The allantois forms blood vessels that connect the embryo with the placenta through the umbilical cord. The exchange of nutrients and gases with maternal blood and the elimination of wastes occur in the placenta.

Organogenesis During organogenesis, organs begin to develop from the embryonic germ layers. *Neurulation* begins the formation of the vertebrate brain and spinal cord. In a frog embryo, the **notochord**—the dorsal rod characteristic of chordate embryos—forms from dorsal mesoderm. Ectoderm above the developing notochord forms a *neural plate,* which curves inward and rolls into a **neural tube,** from which the central nervous system will develop.

In vertebrate embryos, a band of cells called the **neural crest** separates from the side of the neural tube.

These neural crest cells migrate to form peripheral nerves as well as teeth and parts of the bones of the skull.

Mesoderm along the sides of the notochord separates into serial blocks, called **somites.** Mesenchyme (migratory) cells from the somites give rise to vertebrae and to the muscles associated with the vertebral column and ribs, arranged in segmented fashion. Lateral to the somites, the mesoderm splits to form the lining of the coelom.

Organogenesis in the chick proceeds in a fashion similar to that in the frog embryo. The borders of the embryo fold down and join, forming a three-layered tube attached by a stalk to the yolk. In human development, a failure of the neural tube to close properly results in the disabling birth defect called *spina bifada.*

In invertebrates, many of the same mechanisms drive organogenesis: cell migrations, cell signaling between different tissues, and cell shape changes. Although the neural tube forms on the ventral side of an insect embryo, the signaling pathways involved are similar to those of vertebrates.

INTERACTIVE QUESTION 47.6

Label the indicated structures in the following diagram of organogenesis in a frog embryo.

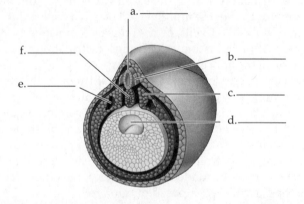

Mechanisms of Morphogenesis Changes in cell shape usually involve reorganization of the cytoskeleton. Extension of microtubules elongates a cell, and then contraction of actin filaments at one end of the cell creates a wedge shape, initiating invaginations and evaginations.

By **convergent extension,** cells of a tissue layer crawl between each other, causing the sheet of cells to extend and narrow. Embryonic cells use their cytoskeleton to extend and retract cellular protrusions in amoeboid movement. Transmembrane glycoproteins called *cell adhesion molecules* help regulate movement by enabling cell interactions. Glycoproteins in the *extracellular matrix (ECM)* help direct migrating cells.

Apoptosis is a common developmental feature in which cells or whole tissues die and are engulfed by surrounding cells. This type of *programmed cell death* may occur because tissues are no longer needed (a tadpole's tail), because more cells are produced than needed (neurons survive only if they make functional connections with each other), or because structures that are part of an ancestral developmental program are later eliminated (webbing between mammalian digits).

47.3 Cytoplasmic determinants and inductive signals contribute to cell fate specification

As a result of differences in gene expression during development, cells and groups of cells become committed to particular fates (**determination**) and then become specialized in structure and function (**differentiation**).

Fate Mapping   In the 1920s, W. Vogt developed a **fate map** for amphibian embryos by labeling regions of the blastula with dye to determine where specific cells showed up in later developmental stages. One type of fate mapping is cell lineage analysis, in which individual blastomeres are marked and their mitotic descendants followed.

The developmental history of every cell has been determined for the transparent nematode *Caenorhabditis elegans*. *Germ cells* give rise to eggs or sperm, and their fate is determined early in development. The first cleavage division in *C. elegans* partitions *P granules* in the posterior cell. Further divisions result in P granules ending up in germ cells in the adult worm's gonad.

Studies of programmed cell death during *C. elegans* development helped identify cell death pathways present in many animals.

In bilateral animals, the three body axes (anterior–posterior, dorsal–ventral, and left–right) are established early in development. In the frog egg, the animal hemisphere has melanin granules in the cortex, whereas the vegetal hemisphere contains the yellow yolk. The animal-vegetal axis determines the anterior-posterior (head-tail) axis of the embryo. Fertilization results in a rotation of the animal pole cortex toward the point of sperm entry, exposing the gray crescent that marks the future dorsal side. This *cortical rotation* leads to inductive interactions that influence the development of dorsal and ventral structures.

In chicks, gravity may be involved in establishing the anterior-posterior axis while the egg moves down the oviduct, and pH differences may determine the dorsal-ventral axis of the blastoderm. In mammals, polarity becomes obvious only after cleavage, but the location of sperm and egg nuclei before fusion may play a role. In insects, morphogen gradients establish the body axes.

The left-right axis is fixed once the other two axes are determined. In vertebrates, cilia direct a flow of fluid, which establishes the left-right asymmetry of several internal organs.

Even though cytoplasmic determinants are asymmetrically distributed in the frog egg, the first cleavage division bisects the gray crescent and equally separates these determinants. H. Spemann found that the *developmental potential* of the first two blastomeres is not yet restricted: If separated, they are **totipotent** and can give rise to all the tissues of a normal embryo.

Mammalian blastomeres remain totipotent through the eight-cell stage, although this may be partly due to a separated cell's ability to respond to its environment. The progressive restriction of a cell's developmental potency is characteristic of animal development. Experimental manipulation shows that the tissue-specific fates of cells of a late gastrula are usually fixed.

INTERACTIVE QUESTION   47.7

If a researcher divides an eight-cell stage of a sea urchin embryo vertically into two four-cell groups, both develop into normal larvae. If the division is made horizontally, abnormal larvae develop. Explain this experimental result.

Cell Fate Determination and Pattern Formation by Inductive Signals   Cells influence the development of other cells through signals that usually switch on sets of genes.

Using transplantation experiments with amphibian embryos, H. Mangold and H. Spemann established that the dorsal lip acts as an "organizer" of the embryo because of its influence in early organogenesis. The molecular basis of induction by *Spemann's organizer* appears to be tied to the inactivation of bone morphogenetic protein 4 (BMP-4), an event that allows cells to make dorsal structures on the dorsal side of the embryo. Proteins similar to BMP-4 and its inhibitors appear to play roles in development in many different organisms.

Pattern formation is the ordering of cells and tissues into characteristic structures in the proper locations in the animal. Cells and their offspring differentiate based on **positional information,** molecular cues that indicate the cells' positions with respect to body axes.

A limb bud of a chick consists of a core of mesoderm beneath a layer of ectoderm. Researchers have identified two organizer regions that secrete proteins that provide positional information to cells.

Cells of the **apical ectodermal ridge (AER)** at the tip of the limb bud secrete proteins of the fibroblast growth factor (FGF) family. These growth signals appear to promote limb bud outgrowth and patterning along the proximal-distal axis. An FGF-secreting AER has been identified in the median fin in a shark, suggesting a long evolutionary history of the AER.

The **zone of polarizing activity (ZPA),** mesodermal tissue located at the posterior attachment of the bud, appears to communicate position along the anterior-posterior axis in a developing limb. Cells of the ZPA secrete a protein growth factor called Sonic hedgehog. Hedgehog proteins have been found to influence development in many different organisms.

A cell's developmental history shapes that cell's response to positional information relating to its location in a developing organ. Thus, patterns of *Hox* gene expression first determine whether cells will become part of a wing or part of a leg; then positional cues indicate what part of the wing or leg the cells will help form.

INTERACTIVE QUESTION   47.8

What would be the effect of implanting cells genetically engineered to produce and secrete Sonic hedgehog to the anterior attachment of a limb bud?

Recent research indicates that the primary cilia, or *monocilia,* found on eukaryotic cells play important developmental roles through their reception of signaling proteins, including Sonic hedgehog.

Word Roots

acro- = the tip (*acrosomal reaction:* the discharge of hydrolytic enzymes from the acrosome when a sperm approaches or contacts an egg)

arch- = ancient, beginning (*archenteron:* the endoderm-lined cavity, formed during gastrulation, that develops into the digestive tract)

blast- = bud, sprout; **-pore** = a passage (*blastopore:* in a gastrula, the opening of the archenteron that typically develops into the mouth in protostomes, and the anus in deuterostomes)

blasto- = produce; **-mere** = a part (*blastomere:* an early embryonic cell arising during the cleavage stage of an early embryo)

ecto- = outside; **-derm** = skin (*ectoderm:* the outermost of the three primary germ layers in animal embryos)

endo- = within (*endoderm:* the innermost of the three primary germ layers in animal embryos)

extra- = beyond (*extraembryonic membrane:* one of four membranes located outside the embryo that support the developing embryo in reptiles and mammals)

fertil- = fruitful (*fertilization:* the union of haploid gametes to produce a diploid zygote)

gastro- = stomach, belly (*gastrulation:* a series of cell and tissue movements in which the blastula-stage embryo folds inward, producing a three-layered embryo, the gastrula)

holo- = whole (*holoblastic:* referring to a type of cleavage in which there is complete division of the egg)

mero- = a part (*meroblastic:* referring to a type of cleavage in which there is incomplete division of a yolk-rich egg, characteristic of avian development)

meso- = middle (*mesoderm:* the middle primary germ layer in an animal embryo)

noto- = the back; **-chord** = a string (*notochord:* a dorsal, flexible rod made of tightly packed mesodermal cells that runs along the anterior-posterior axis of a chordate)

poly- = many (*polyspermy:* fertilization of an egg by more than one sperm)

soma- = a body (*somite:* one of a series of paired blocks of mesoderm just lateral to the notochord in a vertebrate embryo)

tropho- = nourish (*trophoblast:* the outer epithelium of a mammalian blastocyst; forms the fetal part of the placenta)

zona = a belt; **pellucid-** = transparent (*zona pellucida:* the extracellular matrix surrounding a mammalian egg)

Structure Your Knowledge

1. Create a flowchart that shows the sequence of key events in the fertilization of a sea urchin egg, and indicate the functions of each of these events.

2. Fill in the following table, briefly describing these stages of development in sea urchin, frog, bird, and mammalian embryos.

Animal	Cleavage	Blastula	Gastrulation
Sea urchin			
Frog			
Bird			
Mammal			

Test Your Knowledge

MULTIPLE CHOICE: *Choose the one best answer.*

1. The fact that an eunucleated sea urchin egg can be activated by the injection of Ca^{2+} is evidence that
   a. development can proceed parthenogenetically.
   b. the fast block to polyspermy may not always work and two sperm may fertilize an egg.
   c. inactive mRNA and other proteins had been stockpiled during development of the egg.
   d. the cortical reaction is not necessary for development to proceed.
   e. development is a cytoplasmic process that does not require genetic regulation.

2. The slow block to polyspermy
   a. is essential to prevent sperm from other species from fertilizing the egg.
   b. is directly produced by the depolarization of the membrane.
   c. is a result of the formation of the fertilization envelope.
   d. is caused by the expulsion of calcium ions.
   e. involves all of the above.

3. The blastocoel
   a. develops into the archenteron or embryonic gut.
   b. is a fluid-filled cavity in the blastula.
   c. opens to the exterior through a blastopore.
   d. is lined with mesoderm.
   e. Both b and c are correct.

4. In a frog embryo, gastrulation
   a. is slowed by the large amount of yolk.
   b. proceeds by involution as cells roll over the dorsal lip of the blastopore.
   c. produces a blastocoel that is displaced into the animal hemisphere.
   d. occurs along the primitive streak.
   e. involves the formation of the notochord and neural tube.

5. Which of the following body structures is *incorrectly* paired with the embryonic germ layer from which it arose?
   a. muscles—mesoderm
   b. central nervous system—ectoderm
   c. lining of gut—endoderm
   d. heart—endoderm
   e. notochord—mesoderm

6. The primitive streak of mammalian embryos is analogous to
   a. the dorsal lip of frog embryos.
   b. the blastoderm of bird embryos.
   c. the multinucleated stage of fruit fly embryos.
   d. the archenteron of sea urchin embryos.
   e. none of the above.

7. Which of the following phenomena is a result of apoptosis during development?
   a. the birth defect spina bifida
   b. the left/right asymmetry of some vertebrate organs
   c. the concentration of P granules in the formation of germ cells in *C. elegans*
   d. the removal of webbing between mammalian digits
   e. the formation of extraembryonic membranes from the hypoblast

8. Somites are
   a. blocks of mesoderm circling the archenteron.
   b. clumps of cells from which the notochord arises.
   c. serially arranged mesoderm blocks lateral to the notochord in a vertebrate embryo.
   d. structures arising from neural crest cells.
   e. produced during organogenesis in sea urchins, frogs, birds, and mammals, but not in fruit flies.

9. Which of the following structures forms the fetal portion of the placenta?
   a. the epiblast
   b. the trophoblast and some cells from the epiblast
   c. the allantois and yolk sac
   d. the endometrium
   e. the amnion

Use the following diagram of a chick embryo for questions 10–12.

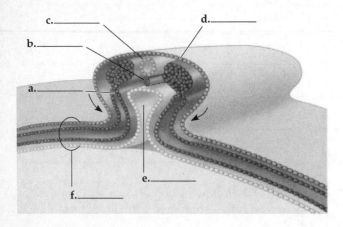

10. Which line points to the notochord?

11. Which line points to the archenteron?

12. To what structure does line f point?
   a. the primitive streak through which gastrulation occurred
   b. the hypoblast
   c. the mesodermal lining of the coelom
   d. the blastocoel enclosed by the epiblast and the hypoblast
   e. the developing extraembryonic membranes

13. The early cleavage divisions of a human zygote most closely resemble
   a. those of a sea urchin because neither species has eggs with a large store of yolk and cleavage in both species is holoblastic.
   b. those of a chick because both birds and mammals share a common reptilian ancestor.
   c. those of a frog because the cells of the animal hemisphere are smaller than those of the vegetal hemisphere in both types of embryos.
   d. those of a *Drosophila* because zygotes of both species pass through an early multinucleated stage.
   e. those of all of the above animals because as deuterostomes, they have similar embryonic patterns.

14. Gastrulation and organogenesis in mammals most closely resemble
   a. those processes in sea urchins because the eggs of both groups have little yolk and cleavage is holoblastic.
   b. those processes in chicks because both birds and mammals share a common reptilian ancestor.

c. those processes in frogs because gastrulation takes place through the dorsal lip of the blastopore in both groups.
   d. those processes in neither chicks nor frogs because mammals have some embryonic cells that develop into extraembryonic membranes.
   e. those processes in *Drosophila* because homeotic genes control pattern development in both groups.

15. Localized determinants
   a. are unevenly distributed materials in the egg that influence the developmental fates of cells.
   b. are involved in regulation of gene expression.
   c. usually include maternal mRNA.
   d. are often separated in the first few cleavage divisions.
   e. may be all of the above.

16. In her transplantation experiments with frog embryos, Mangold found that moving a part of the dorsal lip to another location would result in a second gastrulation, and even in the development of a doubled, face-to-face tadpole. The best explanation for these results is
   a. morphogenetic movements.
   b. separation of localized determinants.
   c. totipotency of cells.
   d. fate map determination.
   e. induction by a dorsal lip "organizer."

17. Pattern formation is most directly influenced by
   a. positional information a cell receives from gradients of molecules called morphogens.
   b. differentiation of cells that then migrate into developing organs.
   c. the movement of cells along fibronectin fibers.
   d. the induction of cells by mesoderm cells found in the center of organs.
   e. gastrulation and the formation of the three tissue layers.

18. Cells of the apical ectodermal ridge secrete
   a. Sonic hedgehog, which communicates location along the anterior-posterior axis.
   b. cell adhesion molecules that hold migrating cells together as they form tissues of the leg.
   c. ZPA, which forms fibronectin fibers that guide migrating cells.
   d. fibroblast growth factors that promote limb bud outgrowth.
   e. morphogens that cause the limb bud to elongate by convergent extension.

Chapter 48

Neurons, Synapses, and Signaling

Key Concepts

48.1 Neuron organization and structure reflect function in information transfer

48.2 Ion pumps and ion channels establish the resting potential of a neuron

48.3 Action potentials are the signals conducted by axons

48.4 Neurons communicate with other cells at synapses

Framework

This chapter describes the transmission of nervous impulses along and between neurons. Ion channels establish the membrane potential of a neuron, and the rapid flow of ions through voltage-gated channels creates the rapid depolarization of an action potential. The release of neurotransmitters at synapses converts electrical signals to chemical signals that pass information to receiving cells.

Chapter Review

Neurons transmit long-distance electrical signals throughout the body and usually communicate between cells using short-distance chemical signals. The higher-order processing of nervous signals may involve clusters of neurons called **ganglia** or more structured groups of neurons organized into a **brain.**

48.1 Neuron organization and structure reflect function in information transfer

Introduction to Information Processing The three stages of information processing are sensory input, integration, and motor output. A **central nervous system (CNS),** found in many animals, includes the brain and longitudinal nerve cord and is involved in integration. The **peripheral nervous system (PNS)** consists of neurons, often bundled into **nerves,** that carry information to and from the CNS.

Sensors detect external stimuli or internal conditions. **Sensory neurons** transmit this information to the brain or ganglia, where **interneurons** integrate it and send output that triggers muscle or gland activity. **Motor neurons** carry impulses to muscle cells.

Neuron Structure and Function A neuron consists of a **cell body,** which contains the nucleus and organelles, and numerous extensions. The highly branched, short **dendrites** (along with the cell body) *receive* signals from other neurons; the single, longer **axon** *transmits* signals to other cells. Signals that travel down an axon originate from a region of the cell body called the *axon hillock.*

Information is transmitted to another cell at a junction called a **synapse.** Terminal branches of axons end in *synaptic terminals,* which usually release **neurotransmitters** that relay signals to another neuron, a muscle, or a gland cell. The transmitting cell is called the *presynaptic cell;* the receiving cell is the *postsynaptic cell.*

The very numerous supporting cells called **glial cells,** or **glia,** give structural integrity and physiological support to neurons.

INTERACTIVE QUESTION 48.1

Label the indicated structures on the following diagram of a neuron. Indicate the direction of impulse transmission. What happens at part e?

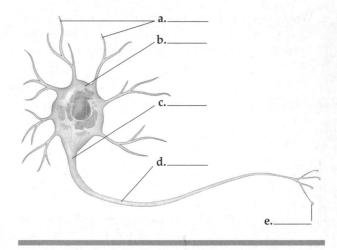

a. _____
b. _____
c. _____
d. _____
e. _____

48.2 Ion pumps and ion channels establish the resting potential of a neuron

Due to the separation of charges, a voltage, or **membrane potential,** exists across the plasma membrane of all cells. In a typical neuron at rest, the membrane potential—called the **resting potential**—is between −60 and −80 mV.

Formation of the Resting Potential   In a mammalian neuron, the concentration of K^+ is much higher inside the cell than outside, and the Na^+ concentration is much greater outside than inside. *Sodium-potassium pumps* maintain these gradients, moving three Na^+ ions out of the cell for every two K^+ ions it pumps in.

Ion channels that are *selectively permeable* allow specific ions to diffuse across the membrane. The diffusion of K^+ out of a cell through open potassium channels results in a buildup of negative charge within the neuron, creating the membrane potential.

Modeling the Resting Potential   Ions diffuse through channels down their concentration gradients until their chemical gradient is balanced by the electrical gradient across the membrane. The membrane voltage at this equilibrium, or **equilibrium potential (E_{ion}),** is determined for a single ion with a 1+ charge at 37°C by the Nernst equation: $E_{ion} = 62$ mV (log ($[ion]_{outside}/[ion]_{inside}$). The equilibrium potential for K^+ (E_K) across a membrane is −90 mV. Using the concentration gradient for Na^+, E_{Na} is +62 mV (the inside of the membrane is more positive than outside).

Neurons at rest have more open K^+ channels than open Na^+ channels, and the resting potential is closer to E_K than to E_{Na}.

INTERACTIVE QUESTION   48.2

a. What is the principal cation inside the cell? outside the cell?

b. Which side of the membrane has a negative charge?

c. What change in the permeability of the cell's membrane to K^+ and/or Na^+ would cause the cell's membrane potential to shift from −70 mV to −90 mV?

48.3 Action potentials are the signals conducted by axons

In addition to the ungated potassium and sodium ion channels that create the resting potential, neurons also have **gated ion channels** that open or close in response to

stimuli and increase or decrease the membrane potential. Electrophysiologists measure membrane potential by placing microelectrodes connected to a voltage recorder inside and outside a cell.

Hyperpolarization and Depolarization   A stimulus that opens gated potassium channels will result in **hyperpolarization,** as K^+ flows out and the membrane potential moves toward E_K (−90 mV). When gated sodium channels open and Na^+ flows in, a **depolarization** occurs as the inside of the cell becomes less negative.

Graded Potentials and Action Potentials   A shift in membrane potential can result in a **graded potential,** whose magnitude is proportional to the strength of the stimulus. An **action potential** is a massive change in membrane voltage that has a constant magnitude and can spread to adjacent regions of the membrane. **Voltage-gated ion channels** open or close in response to a change in membrane potential. A depolarization opens voltage-gated sodium channels; Na^+ inflow increases depolarization and causes more sodium channels to open. Once this *positive feedback* results in depolarization of a neuron to a certain membrane voltage called the **threshold,** an action potential is triggered. The action potential is an *all-or-none* event, always creating the same voltage spike once the threshold is reached.

Generation of Action Potentials: A Closer Look   Both Na^+ and K^+ voltage-gated channels are involved in an action potential. Sodium channels open rapidly in response to depolarization but become inactivated when a loop of the channel protein blocks the opening. Potassium channels open slowly in response to depolarization, but remain open throughout the action potential.

As a stimulus depolarizes the membrane, Na^+ channels open; the influx of Na^+ causes further depolarization, which opens more channels. If the depolarization reaches the threshold, an action potential is triggered. The continuing positive-feedback cycle of Na^+ influx during the *rising phase* brings the membrane potential close to E_{Na}. The inactivation of voltage-gated Na^+ channels and the opening of voltage-gated K^+ channels rapidly bring the membrane potential back toward E_K during the *falling phase*.

During the *undershoot*, the membrane's permeability to K^+ is higher than at rest, and the continued outflow of K^+ temporarily hyperpolarizes the membrane. During the **refractory period,** which occurs while the Na^+ channels remain inactivated, the neuron cannot respond to another stimulus.

Action potentials last only about 1–2 milliseconds. The frequency of action potentials increases with the intensity of the stimulus.

INTERACTIVE QUESTION 48.3

The following diagram shows the changes in voltage-gated channels during an action potential. Label the channels and inactivation gate, the ions, and the components of the graph. Name and describe the five phases of the action potential. Place numbers on the graph to show where each phase is occurring.

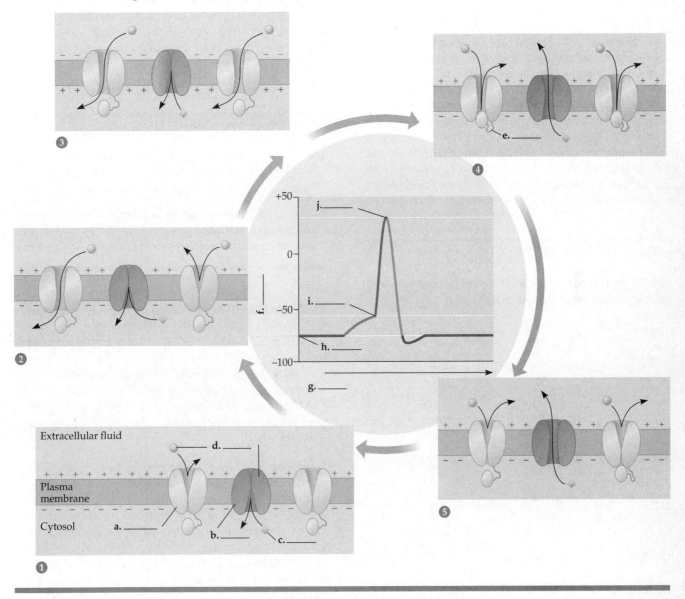

Conduction of Action Potentials   Na⁺ influx in the rising phase depolarizes adjacent sections of the membrane, bringing them to threshold. Local depolarizations and action potentials across the membrane result in the propagation of serial action potentials along the length of the neuron. Because of the brief refractory period, the action potential is propagated in one direction only.

Resistance to current flow is inversely proportional to the cross-sectional area of the conducting "wire."

The greater the axon diameter, the faster action potentials are propagated. Some invertebrates, such as squid, have giant axons that conduct impulses very rapidly.

In vertebrates, **oligodendrocytes** (in the CNS) and **Schwann cells** (in the PNS) insulate axons in multiple layers of membranes called a **myelin sheath.** Voltage-gated ion channels are concentrated in the **nodes of Ranvier,** small gaps between successive Schwann cells. Action potentials can be generated only at these nodes, and a nerve impulse "jumps" from node to node, resulting in a faster mode of transmission known as **saltatory conduction.**

48.4 Neurons communicate with other cells at synapses

At *electrical synapses,* an electrical current flows directly from cell to cell via gap junctions. Electrical synapses are found in the giant axons of squid and lobsters and in many neurons of the vertebrate brain.

A *chemical synapse* involves the release of neurotransmitters. A synaptic terminal contains *synaptic vesicles,* in which thousands of molecules of neurotransmitter are stored. The depolarization of the presynaptic membrane opens voltage-gated calcium channels in the membrane. The influx of Ca^{2+} causes the synaptic vesicles to fuse with the presynaptic membrane and release neurotransmitter into the *synaptic cleft.*

INTERACTIVE QUESTION   48.4

Identify the components of this chemical synapse following the depolarization of the synaptic terminal.

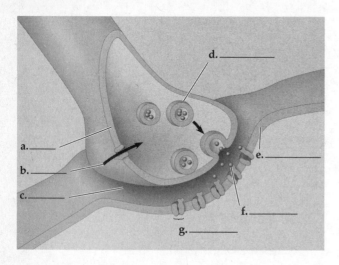

d. _____

a. _____
b. _____
c. _____

e. _____

f. _____

g. _____

Generation of Postsynaptic Potentials Binding of neurotransmitter to a receptor on a **ligand-gated ion channel** (often called an *ionotropic receptor*) in the postsynaptic cell membrane allows ions to cross the membrane, creating a graded *postsynaptic potential.* If both Na^+ and K^+ are able to diffuse through the channel, the net inflow of positive charge depolarizes the membrane, creating an **excitatory postsynaptic potential (EPSP)** that brings the membrane potential closer to threshold. If binding of neurotransmitter opens K^+ or Cl^- channels, the membrane hyperpolarizes, producing an **inhibitory postsynaptic potential (IPSP).**

The effect of neurotransmitters is brief because the molecules diffuse away, are taken up and repackaged into synaptic vesicles, or are broken down by enzymes.

Summation of Postsynaptic Potentials Postsynaptic potentials are graded potentials. Their magnitude depends on the number of neurotransmitter molecules that bind to receptors and the distance a synapse is from the axon hillock. The membrane potential of the axon hillock at any given time is determined by the sum of all EPSPs and IPSPs.

INTERACTIVE QUESTION   48.5

a. _____ summation occurs with repeated release of neurotransmitter from a synaptic terminal before the postsynaptic potential returns to its resting potential.

b. _____ summation occurs when several different presynaptic terminals, usually from different neurons, release neurotransmitter simultaneously.

Modulated Signaling at Synapses Some neurotransmitters bind to *metabotropic receptors* that trigger a signal transduction pathway in the postsynaptic cells. This modulated synaptic transmission begins more slowly but lasts longer. Binding of neurotransmitter to a metabotropic receptor often leads to the production of cAMP as a second messenger and to the amplified phosphorylation of channel proteins, opening or closing many ion channels.

Neurotransmitters Neurotransmitters may have many different ionotropic and metabotropic receptors and may produce very different effects in postsynaptic cells. The more than 100 identified neurotransmitters can be divided into five groups: acetylcholine, amino acids, biogenic amines, neuropeptides, and gases.

Acetylcholine is a common neurotransmitter in invertebrates and vertebrates. Depending on the type of receptor, it can be inhibitory or excitatory. In *neuromuscular junctions,* acetylcholine released from a motor axon produces an EPSP in a muscle cell. The enzyme acetylcholinesterase hydrolyzes the neurotransmitter. The same ionotropic receptor binds nicotine, producing its physiological and psychological stimulant effect.

The amino acids **glutamate** and **gamma aminobutyric acid (GABA)** function as neurotransmitters in the CNS. Glutamate, the most common CNS neurotransmitter, is excitatory and plays a key role in long-term memory formation. GABA is the most common inhibitory transmitter in the brain. Glycine is inhibitory in parts of the CNS outside the brain.

Biogenic amines are neurotransmitters derived from amino acids. **Norepinephrine,** derived from tyrosine, is an excitatory neurotransmitter in the autonomic nervous system. It also acts as a hormone, as does the related biogenic amine *epinephrine.* **Dopamine** is also derived from tyrosine. **Serotonin** (synthesized from tryptophan) and dopamine affect sleep, mood, attention, and learning. Imbalances of these neurotransmitters have been associated with several nervous system disorders.

Neuropeptides are short chains of amino acids that function via metabotropic receptors. *Substance P* is an excitatory neurotransmitter that functions in pain perception. **Endorphins** are neuropeptides that decrease pain perception. Endorphins are produced in the brain during physical or emotional stress, and their effects include pain reduction, euphoria, and other physiological effects. Opiates bind to endorphin receptors in the brain.

Some neurons use nitric oxide (NO) and carbon monoxide as local regulators or neurotransmitters. The release of NO triggers relaxation of smooth muscle cells and vessel dilation. CO produced by neurons in the brain affects the release of hypothalamic hormones.

INTERACTIVE QUESTION   48.6

Acetylcholine stimulates skeletal muscle contraction but inhibits or slows cardiac muscle contraction. How can this neurotransmitter have such opposite effects?

Word Roots

bio- = life; **-genic** = producing (*biogenic amine:* a neurotransmitter derived from an amino acid)

dendro- = tree (*dendrite:* one of usually numerous, short, highly branched extensions of a neuron that receive signals from other neurons)

de- = down, out (*depolarization:* a change in a cell's membrane potential such that the inside of the membrane is made less negative relative to the outside)

endo- = within (*endorphin:* a hormone produced in the brain and anterior pituitary that inhibits pain perception)

glia = glue (*glia:* cells of the nervous system that support, regulate, and augment the functions of neurons)

hyper- = over, above, excessive (*hyperpolarization:* a change in a cell's membrane potential such that the inside of the membrane becomes more negative relative to the outside)

inter- = between (*interneurons:* an association neuron; a nerve cell within the central nervous system that forms synapses with sensory and/or motor neurons and integrates sensory input and motor output)

neuro- = nerve; **trans-** = across (*neurotransmitter:* a molecule that is released from the synaptic terminal of a neuron at a chemical synapse, diffuses across the synaptic cleft, and binds to the postsynaptic cell, triggering a response)

oligo- = few, small (*oligodendrocyte:* a type of glial cell that forms insulating myelin sheaths around the axons of neurons in the central nervous system)

salta- = leap (*saltatory conduction:* rapid transmission of a nerve impulse along an axon resulting from the action potential jumping from one node of Ranvier to another, skipping the myelin-sheathed regions of membrane)

syn- = together (*synapse:* the junction where a neuron communicates with another cell across a narrow gap)

Structure Your Knowledge

1. Develop a flowchart or diagram or write a description of the sequence of events in the (1) creation and propagation of an action potential and (2) in the transmission of this action potential across a chemical synapse.

Test Your Knowledge

MULTIPLE CHOICE: *Choose the one best answer.*

1. Which of the following nervous system components is *incorrectly* paired with its function?
   a. axon hillock—region of neuron where an action potential originates
   b. Schwann cells—produce a myelin sheath around axons in the PNS
   c. synapse—space between presynaptic and postsynaptic cells into which neurotransmitter is released
   d. synaptic terminal—receptor that is part of an ion channel that is keyed to a specific neurotransmitter
   e. dendrite—receives signals from other neurons

2. Interneurons
    a. may connect sensory and motor neurons.
    b. are more common in the PNS than the CNS.
    c. are involved in the integration of sensory information.
    d. typically have more axons than dendrites.
    e. Both a and c are correct.

3. Which of the following statements describes the role of the sodium-potassium pump in establishing a cell's membrane potential?
    a. By pumping 3 Na^+ out of a cell for every 2 K^+ it pumps in, it creates the negative charge inside a cell that establishes a membrane potential of about -70 mV across the membrane.
    b. It is responsible for maintaining the steep gradients of higher K^+ concentration inside the cell and higher Na^+ concentration outside the cell.
    c. Phosphate groups attached to the sodium-potassium pump produce a high-energy membrane.
    d. It is a voltage-gated channel that opens when a cell's membrane potential deviates from its resting value, returning the ion concentrations to their proper concentration gradients.
    e. The equilibrium potential of potassium and sodium establish a cell's membrane potential; the sodium-potassium pump is only involved in reestablishing the resting potential following the refractory phase of an action potential.

4. Which of the following statements concerning the resting potential of a typical neuron is *not* true?
    a. The inside of the cell is more negative than is the outside.
    b. Concentration gradients exist in which more sodium is outside the cell and a higher potassium concentration is inside the cell.
    c. The resting potential is about -70 mV and can be measured by using microelectrodes placed inside and outside the cell.
    d. The resting potential is formed by the sequential opening of sodium and then potassium voltage-gated channels.
    e. The diffusion of K^+ through numerous open potassium channels (coupled with the very few open sodium channels) is the main source of the separation of charges across the membrane.

5. The threshold of a membrane
    a. is exactly midway between E_K and E_{Na}.
    b. opens voltage-gated channels and permits the rapid outflow of sodium ions.
    c. is the depolarization that is needed to generate an action potential.
    d. is a graded potential that is proportional to the strength of a stimulus.
    e. is an all-or-none event.

6. How is an increase in the strength of a stimulus communicated by a neuron?
    a. The spike of the action potential reaches a higher voltage.
    b. The frequency of action potentials generated along the neuron increases.
    c. The length of an action potential (the duration of the rising phase) increases.
    d. The action potential travels along the neuron faster.
    e. All action potentials are the same, so the nervous system cannot discriminate between stimuli of different strengths.

7. After the rapid depolarization of an action potential, the fall in the membrane potential occurs due to the
    a. closing of voltage-gated sodium channels.
    b. closing of potassium and sodium channels.
    c. refractory period, during which the membrane is hyperpolarized.
    d. delay in the action of the sodium-potassium pump.
    e. opening of voltage-gated potassium channels and the closing of sodium inactivation gates.

8. Nodes of Ranvier are
    a. gaps where Schwann cells abut and action potentials are generated.
    b. neurotransmitter-containing vesicles located in the synaptic terminals.
    c. the parts of neurons where action potentials are initiated.
    d. clusters of receptor proteins located on the postsynaptic membrane.
    e. ganglia adjacent to the spinal cord.

9. Movement of an action potential in only one direction along a neuron is a function of
    a. saltatory conduction.
    b. the pathway from dendrite to axon.
    c. the refractory period, when sodium inactivation gates are still closed.
    d. the localized depolarization of the surrounding membrane.
    e. the reaching of threshold, which creates an all-or-none firing.

10. Signal transmission is faster in myelinated axons because
    a. these axons are thinner and thus present less resistance to voltage flow.
    b. these axons use electrical synapses rather than chemical synapses.
    c. the action potential can jump from node to node along the insulating myelin sheath.
    d. these axons are thicker and thus present less resistance to voltage flow.
    e. these axons have higher depolarization values than do unmyelinated axons.

11. Which of the following statements concerning chemical synapses is *not* true?

    a. Synaptic terminals at the ends of branching axons contain synaptic vesicles, which enclose the neurotransmitter.

    b. The influx of sodium when an action potential reaches the presynaptic membrane causes synaptic vesicles to release their neurotransmitter into the cleft.

    c. The binding of neurotransmitter to receptors on the postsynaptic membrane usually opens or closes ligand-gated ion channels.

    d. An excitatory postsynaptic potential forms when sodium channels open and the membrane potential moves closer to threshold.

    e. Neurotransmitter is often rapidly degraded in the synaptic cleft.

12. An inhibitory postsynaptic potential occurs when

    a. sodium flows into the postsynaptic cell.

    b. enzymes do not break down the neurotransmitter in the synaptic cleft.

    c. binding of neurotransmitter opens ion gates, resulting in the membrane becoming hyperpolarized.

    d. acetylcholine is the neurotransmitter.

    e. norepinephrine is the neurotransmitter.

13. Which of the following substances does *not* function as a neurotransmitter or a local regulator released by neurons?

    a. an amino acid such as glutamate

    b. a neuropeptide such as endorphin

    c. a biogenic amine such as dopamine

    d. a steroid

    e. NO

14. The main effect of the neurotransmitter GABA in the CNS is to

    a. increase pain.

    b. create inhibitory postsynaptic potentials.

    c. create excitatory postsynaptic potentials.

    d. induce sleep.

    e. decrease pain and induce euphoria.

15. If the binding of a neurotransmitter to its receptor opens Cl^- channels, what would be the effect on the postsynaptic cell? (Cl^- is in higher concentration outside the cell.)

    a. It would hyperpolarize, producing an IPSP.

    b. It would reach threshold, and an action potential would be generated.

    c. It would depolarize and form an EPSP but probably not generate an action potential.

    d. It would initiate a signal transduction pathway.

    e. Its membrane potential would not change because neither sodium nor potassium channels opened.

Chapter 49

Nervous Systems

Key Concepts

49.1 Nervous systems consist of circuits of neurons and supporting cells

49.2 The vertebrate brain is regionally specialized

49.3 The cerebral cortex controls voluntary movement and cognitive functions

49.4 Changes in synaptic connections underlie memory and learning

49.5 Many nervous system disorders can be explained in molecular terms

Framework

This chapter describes the organization of the vertebrate nervous system and the functional specialization of its components. The branches of the peripheral nervous system include the sensory afferent neurons and the efferent motor system and autonomic nervous system. The central nervous system includes the spinal cord, which mediates reflexes and transmits information, and the regionally specialized brain, which regulates homeostatic functions, integrates sensory information, and controls voluntary movement and cognitive functions.

Chapter Review

The estimated 10^{11} neurons of the human brain communicate in complex circuits. New techniques of brainbow staining of neurons and powerful imaging techniques of brain activity are enhancing the study of the structure and function of the brain.

49.1 Nervous systems consist of circuits of neurons and supporting cells

Mechanisms to sense and respond to changes in the environment evolved in prokaryotes. Modification of this ability allowed for the development of communication between the cells of multicellular organisms.

A simple diffuse **nerve net** controls the radially symmetrical body and gastrovascular cavity in cnidarians. Sea stars have a central nerve ring with radial **nerves** (bundles of axons of neurons) connected to a nerve net in each arm.

Cephalization, the concentration of neurons in the head, evolved in bilaterally symmetrical animals. Flatworms have the simplest *central nervous system (CNS)* consisting of a small brain and two longitudinal nerve cords. Annelids and arthropods have a more complicated brain and a ventral nerve cord with segmentally arranged clusters of neurons called ganglia. Vertebrates have a brain and a dorsal spinal cord; nerves and ganglia make up the *peripheral nervous system (PNS)*.

INTERACTIVE QUESTION 49.1

The organization of an organism's nervous system correlates with how the animal interacts with its environment. How do the different nervous systems of clams and squid illustrate this generalization?

Organization of the Vertebrate Nervous System The complex behavior of vertebrates is integrated by the brain. The spinal cord carries information to and from the brain and controls **reflexes,** which are automatic responses to stimuli.

The CNS develops from the embryonic dorsal hollow nerve cord. Spaces in the brain called **ventricles** are continuous with the narrow **central canal** of the spinal cord, and all are filled with **cerebrospinal fluid.** This fluid, formed by filtration of the blood, carries out circulatory functions and cushions the brain.

Neuron cell bodies, dendrites, and unmyelinated axons make up the **gray matter** of the brain and spinal cord. **White matter** is named for the white color of the myelin sheaths of bundles of axons.

Glia   Glia in the CNS include ciliated ependymal cells, which line the ventricles; microglia, which protect against pathogens; and oligodendrocytes, which function in axon myelination. Schwann cells myelinate axons in the PNS. **Astrocytes** are supporting glia with various functions. Some facilitate information transfer at synapses and increase blood flow to active neurons. During development, astrocytes induce the formation of the *blood–brain barrier*, which restricts the passage of most substances from the blood into the brain. *Radial glia* guide the embryonic growth of neurons. Radial glia and astrocytes can act as stem cells.

The Peripheral Nervous System   Sensory information is carried to the CNS along *afferent* PNS neurons. *Efferent* neurons of the PNS carry instructions to muscles, glands, and endocrine cells. Most nerves contain both afferent and efferent neurons.

Functionally, the PNS has two efferent components. The **motor system** carries signals to skeletal muscles, whose movement is largely under conscious control. The **autonomic nervous system** maintains involuntary control over smooth and cardiac muscles.

The autonomic nervous system is separated into three divisions. The **sympathetic division** accelerates heart rate and metabolic rate, generating energy and arousing an organism for action. The **parasympathetic division** carries signals that enhance self-maintenance activities that conserve energy, such as digestion and slowing the heart rate. The **enteric division** includes complex networks of neurons that control the secretions of the digestive tract, pancreas, and gallbladder, and the contractions of smooth muscles that produce peristalsis.

INTERACTIVE QUESTION   49.2

Complete the following concept map to help you review the organization of the peripheral nervous system.

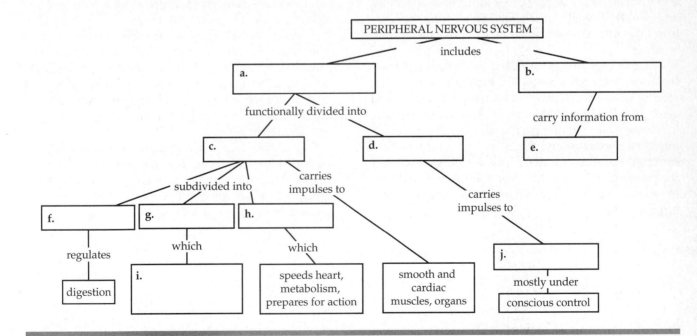

49.2   The vertebrate brain is regionally specialized

Exploring the Organization of the Human Brain   As a human embryo develops, the **forebrain, midbrain,** and **hindbrain** become evident as three anterior bulges of the neural tube. The midbrain and part of the hindbrain give rise to the **brainstem,** which joins the base of the brain to the spinal cord. The rest of the hindbrain gives rise to the **cerebellum.** The forebrain develops into the diencephalon and the telencephalon, which becomes the **cerebrum.** During the second and third months of development, the outer portion of the cerebrum, called the cortex, rapidly expands over much of the brain.

The brainstem consists of the midbrain, the **pons,** and the **medulla oblongata,** or *medulla.* The midbrain receives and integrates sensory information, such as that involved in hearing and visual reflexes, and relays it to specific regions of the forebrain. The pons and medulla transfer information between the PNS and the rest of the brain. They help coordinate large-scale movements, and most axons of motor neurons cross in the medulla, so that the right side of the brain controls much of the movement of the left side of the body, and vice versa.

The medulla contains control centers for such homeostatic functions as breathing, swallowing, heart and blood vessel actions, and digestion. The pons functions with the medulla in some of these activities.

The cerebellum is involved in learning and remembering motor skills, as well as in coordination and error checking during motor and perceptual functions. The cerebellum integrates information from the auditory and visual systems with sensory input from the joints and muscles as well as motor commands from the cerebrum to provide coordination of movements and balance.

The diencephalon gives rise to the thalamus, hypothalamus, and epithalamus. The **thalamus** is the major sorting center for sensory information going to the cerebrum. The **hypothalamus** is the major brain region for homeostatic regulation. It produces the posterior pituitary hormones and the releasing hormones that control the anterior pituitary. The *epithalamus* includes the pineal gland, which produces melatonin, and one of several clusters of capillaries that produce cerebrospinal fluid.

The cerebrum is divided into right and left **cerebral hemispheres.** The **cerebral cortex** is involved in perception, voluntary movement, and learning. The right side of the cortex receives information from and controls the movement of the left side of the body, and vice versa. Communication between the two hemispheres travels through the **corpus callosum,** a thick band of axons. The *basal nuclei,* embedded in the inner white matter, are important in planning and learning movements.

Arousal and Sleep   Arousal is a state in which an individual is aware of the external world. While asleep, an individual does not consciously perceive external stimuli, although EEGs record active brain waves. Sleep and dreaming may be involved in consolidating learning and memory. Several brainstem centers control sleep and arousal. The **reticular formation** is a diffuse network of neurons that filters sensory information going to the cerebral cortex, affecting the level of arousal. Sleep-producing centers are located in the pons and medulla. A center that causes arousal is found in the midbrain. Melatonin, produced by the pineal gland, appears to play a role in sleep/wake cycles.

Biological Clock Regulation   Most organisms appear to have daily cycles of biological activity. Studies show that an internal timekeeper, called the **biological clock,** is important for maintaining these circadian rhythms. In mammals, the **suprachiasmatic nucleus (SCN),** located in the hypothalamus, functions as the biological clock. Sensory neurons provide visual information to the SCN to keep the clock synchronized with natural cycles of light and dark.

Emotions   The amygdala, hippocampus, and sections of the thalamus form a ring around the brainstem called the *limbic system.* This system interacts with sensory and other areas of the cerebrum in generating emotions, both basic emotions such as laughing and crying, and emotional feelings associated with the survival-related functions of the brainstem, such as aggression, feeding, and sexuality. The **amygdala** functions in creating emotional memories.

Functional magnetic resonance imaging (fMRI) is used to identify active parts of the brain in the study of emotion, cognition, and consciousness. Consciousness appears to be an emergent property of the brain that involves many areas of the cortex.

Identify the structures (a–g) in the following illustration of the human brain. Then match the functions (1–7) to these structures.

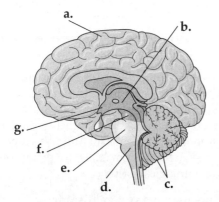

Structure   Function

a. _____ _____
b. _____ _____
c. _____ _____
d. _____ _____
e. _____ _____
f. _____ _____
g. _____ _____

Functions

1. Coordinates balance and movement

2. Aids the medulla in some functions, conducts information between the brain and spinal cord

3. Sorts and relays information to the cerebrum

4. Regulates breathing, heart rate, and digestion

5. Integrates sensory and motor information, learning, emotion, memory, and perception

6. Produces hormones, functions in homeostatic regulation

7. Sends sensory information to the forebrain, is involved in hearing and visual reflexes

49.3 The cerebral cortex controls voluntary movement and cognitive functions

Each side of the cerebral cortex is divided into four lobes—frontal, temporal, occipital, and parietal—each with specialized functions.

Language and Speech Studies of brain injuries and functional imaging of brain activity show that several areas in different lobes are involved in understanding and generating language. *Broca's area* in the left frontal lobe is involved in speech generation; *Wernicke's area* in the left temporal lobe is involved in speech comprehension. Other areas are involved in reading, speaking, and attaching meaning to words.

Lateralization of Cortical Function Lateralization results in two hemispheres that differ in function. The left hemisphere usually becomes specialized for language, math, and logic, whereas the right hemisphere is involved in pattern recognition, spatial perception, and nonverbal thinking. Some of this lateralization relates to handedness.

Information Processing Sensory information (visual, auditory, and *somatosensory*) is directed to primary sensory areas located in different lobes of the cortex. Association areas near the primary sensory areas integrate ("associate") the different aspects of sensory inputs. The prefrontal cortex generates motor response plans, which the primary motor cortex directs to the brainstem and spinal cord, from which action potentials travel in motor neurons to skeletal muscles.

Frontal Lobe Function Injury to the frontal lobe often results in flawed decision making and diminished emotional responses.

Evolution of Cognition in Vertebrates The *neocortex,* the outer six layers of neurons running along the cerebrum, is highly convoluted, with greatly increased surface area in primates and cetaceans. These convolutions were once thought to be required for *cognition,* the perception and reasoning associated with knowledge. The now-documented cognitive ability of birds is linked to the *pallium,* the outer portion of the avian brain, which contains clusters of nuclei.

What are the functions of the primary somatosensory cortex, located along the front of the parietal lobe, and the primary motor cortex, located along the rear of the frontal lobe? Describe the spatial distribution of neurons in these two areas.

49.4   Changes in synaptic connections underlie memory and learning

During embryonic development, neurons that reach the proper target receive growth supporting factors; those that do not undergo programmed cell death. A developing neuron forms numerous synapses; although some of these synapses are stabilized and retained into adulthood, more than half of them are lost.

Neural Plasticity   The nervous system can be remodeled in response to its activity, a capacity called **neural plasticity.** A high level of activity at a synapse may lead to the recruitment of additional synaptic terminals from the presynaptic neuron. A lack of activity may lead to the loss of connections with a neuron. Signals at a synapse can also be strengthened or weakened when several synapses are active at the same time on a postsynaptic cell.

The developmental disorder *autism* may involve a disruption of activity-dependent synapse remodeling. Soon after birth the visual cortex is reorganized in response to visual stimuli.

Memory and Learning   Human memory consists of **short-term memory,** information held briefly, and **long-term memory,** the retention of this information for later recall. Transferring information from short-term memory, which involves links in the hippocampus, to long-term memory requires the making of more permanent connections within the cerebral cortex. Long-term memories become integrated with existing knowledge.

Skill memories are formed by repetition of motor activities and do not require conscious recall of specific steps. Learning and remembering skills and procedures seem to involve nerve cells making new connections. Remembering facts and numbers may involve changes in the strength of existing nerve connections.

Long-Term Potentiation   In **long-term potentiation (LTP),** the strength of synaptic transmission is increased. AMPA receptors are inserted in a postsynaptic membrane following a burst of action potentials and glutamate release from a presynaptic cell that occur when the post-synaptic membrane has been depolarized by activity at nearby synapses. After LTP is established, glutamate released by the presynaptic neuron binds to and activates both NMDA and AMPA receptors, triggering strong post-synaptic potentials and initiating action potentials. LTP may be involved in learning and memory storage.

INTERACTIVE QUESTION   49.5

Action potentials are initiated more readily in synapses exhibiting long-term potentiation. Explain why it initially takes at least two signaling neurons to establish LTP in a synapse.

Stem Cells in the Brain   In 1998, F. Gage and P. Eriksson reported the presence of neural stem cells in the adult human brain. Research has identified mouse neural stem cells that produce new cells that become incorporated into the hippocampus. A goal of future research is to find ways to use neural stem cells to repair or replace damaged neurological tissue.

49.5   Many nervous system disorders can be explained in molecular terms

Chemical or anatomical changes in the brain can lead to nervous system disorders. Research efforts seek to identify genetic and environmental components of these diseases for both prevention and development of better drug treatments.

Schizophrenia   **Schizophrenia** is characterized by psychotic episodes involving hallucinations and delusions. The disease has both environmental and genetic components. Relatives of people with schizophrenia have higher risks of developing the disease. Treatments have focused on drugs that block dopamine receptors, although evidence indicates that glutamate signaling is also involved. Newer drugs can reduce the major symptoms while producing fewer negative side effects.

Depression   Depressive illnesses include **major depressive disorder,** which involves a persistent low mood with changes in sleep, appetite, and energy; and **bipolar disorder,** which involves swings of mood from very high to low. Drug treatments for depression often increase the activity of biogenic amines.

Drug Addiction and the Brain's Reward System   Drug addiction is the compulsive consumption of drugs that increase the activity of the brain's reward system. The reward pathways involve neurons of the *ventral tegmental area (VTA),* located near the base of the brain, that release dopamine at synapses in regions of the cerebrum, including the *nucleus accumbens.* Addictive drugs such as cocaine, heroin, nicotine, amphetamines, and alcohol affect the reward system in different ways—by stimulating dopamine-releasing VTA neurons, blocking the removal of dopamine from synapses, or decreasing the activity of inhibitory neurons in the pathway. Long-lasting changes in the reward circuitry result in increased craving for the drug.

Alzheimer's Disease   **Alzheimer's disease** is dementia characterized by confusion and memory loss. This age-related, progressive disease involves the death of neurons in large areas of the brain. Postmortem examination reveals neurofibrillary tangles (bundles of the protein tau, which normally support nutrient movement along axons) and amyloid plaques (aggregates of β-amyloid, an insoluble peptide that accumulates outside neurons and appears to trigger death of neighboring neurons).

Parkinson's Disease   **Parkinson's disease** is a progressive motor disorder characterized by difficulty in movement, rigidity, and muscle tremors. The death of neurons in the midbrain that normally release dopamine in the basal nuclei leads to the symptoms of this disease. Brain surgery, deep-brain stimulation, and drugs such as L-dopa (a dopamine precursor) are used to manage the symptoms. A potential cure may be to implant dopamine-secreting neurons in the midbrain or basal ganglia.

Word Roots

astro- = a star; **-cyte** = cell (*astrocyte*: a glial cell with diverse functions, including providing structural support for neurons, regulating the interstitial environment, facilitating synaptic transmission, and assisting in regulating the blood supply to the brain)

auto- = self (*autonomic nervous system*: an efferent branch of the vertebrate peripheral nervous system that regulates the internal environment; consists of the sympathetic, parasympathetic, and enteric divisions)

hypo- = below (*hypothalamus*: the ventral part of the vertebrate forebrain; functions in maintaining homeostasis, especially in coordinating the endocrine and nervous systems)

para- = near (*parasympathetic division*: one of three divisions of the autonomic nervous system; generally enhances body activities that gain and conserve energy)

supra- = above, over (*suprachiasmatic nuclei*: a group of neurons in the hypothalamus of mammals that functions as a biological clock)

Structure Your Knowledge

1. List the location and functions of the following important brain nuclei or functional systems.
   a. reticular formation
   b. suprachiasmatic nucleus
   c. basal nuclei
   d. limbic system
   e. amygdala and hippocampus

Test Your Knowledge

MULTIPLE CHOICE: *Choose the one best answer.*

1. Which of the following animals is *incorrectly* paired with the description of its nervous system?
   a. sea star (echinoderm)—modified nerve net, central nerve ring with radial nerves
   b. hydra (cnidarian)—ring of ganglia leading to multiple nerve cords
   c. annelid worm—brain, ventral nerve cord with segmental ganglia
   d. vertebrate—central nervous system of brain and spinal cord and peripheral nervous system
   e. squid (mollusc)—large complex brain, ventral nerve cord, some giant axons

2. Which of the following statements concerning the autonomic nervous system is *not* true?
   a. It is a subdivision of both the central and peripheral nervous systems.
   b. It consists of the sympathetic, parasympathetic, and enteric divisions.
   c. Two of its divisions have generally antagonistic effects.
   d. It controls smooth and cardiac muscles.
   e. Control is generally involuntary.

3. If you needed to obtain nuclei from the cell bodies of neurons for an experiment, you would want to make preparations of
   a. the corpus callosum.
   b. motor nerves.
   c. the neocortex.
   d. sympathetic ganglia near the spinal cord.
   e. either c or d.

4. The amygdala and hippocampus
   a. control biorhythms and are found in the thalamus.
   b. are the parts of the reticulum system found in the midbrain.
   c. are located in the parietal and frontal lobes, respectively, and are involved in language and speech.
   d. are nuclei in the midbrain involved in hearing and vision.
   e. are found in the inner cortex and are involved in memory storage.

5. Which of the following structures is *incorrectly* paired with its function?
   a. pons—conducts information between the spinal cord and the brain
   b. cerebellum—contains the tracts by which motor neurons cross from one side of the brain to the other side of the body
   c. thalamus—sorts and relays incoming impulses to the cerebrum
   d. corpus callosum—band of axons connecting the left and right hemispheres
   e. hypothalamus—homeostatic regulation, pleasure centers

6. Which of the following statements provides evidence indicating that schizophrenia, bipolar disorder, and major depression have a genetic as well as an environmental component?
   a. All three conditions seem to involve an excess of dopamine.
   b. All three conditions can be treated with the same drugs.
   c. If one identical twin has one of these disorders, the other twin has a much higher risk, indicating that genes and the environment both contribute to these disorders.
   d. The genes for these disorders have been identified, and they share a common sequence.
   e. Identical twins raised separately do not share these disorders; those raised in the same home are both likely to be affected by the disorder.

7. Long-term potentiation seems to be involved in
   a. memory storage and learning.
   b. enhanced response of a presynaptic neuron resulting from strong depolarizations due to bursts of action potentials.
   c. the attachment of emotion to events.
   d. the release of the excitatory neurotransmitter glutamate, which opens voltage-gated channels.
   e. all of the above.

8. The white matter of the spinal cord is composed of
   a. the ventricles.
   b. myelinated sheaths of axons.
   c. cerebrospinal fluid.
   d. motor and interneuron cell bodies.
   e. sympathetic ganglia.

9. Lateralization of the cerebrum refers to the fact that
   a. visceral functions are controlled by the hindbrain, whereas thinking and language are centered in association areas of the cortex.
   b. the right side of the brain controls the left side of the body, and vice versa.
   c. the primary somatosensory cortex receives sensory information that is mapped onto one region of the cerebral cortex, whereas the primary motor cortex localized in another region controls motor output to the body.
   d. the left hemisphere usually controls verbal and analytic ability, whereas the right hemisphere is specialized for nonverbal thinking and pattern recognition.
   e. memory formation involves circuits that include the limbic system, hippocampus, and association areas of the cortex.

10. A function of the reticular formation is to
    a. keep you awake during a lecture.
    b. coordinate movement and balance.
    c. produce emotions.
    d. control visceral functions such as breathing and heart rate.
    e. create consciousness.

11. Neural stem cells
    a. are responsible for making new synapses during the process of learning.
    b. retain the ability to divide and form new neurons in the adult brain.
    c. are similar to embryonic stem cells in that they can differentiate into all kinds of body cells.
    d. have been transplanted into individuals with spinal cord injuries and are able to grow and make connections with the proper target cells.
    e. are types of glia that form tracks along which newly formed neurons migrate during development.

12. Which of the following nervous system components does *not* contain efferent neurons?
    a. a spinal nerve
    b. the motor system
    c. the enteric division
    d. sensory neurons
    e. All of the above contain efferent neurons.

13. The pallium of birds is involved in _____ and is homologous to the _____ of mammals.
    a. circadian rhythms . . . suprachiasmatic nuclei
    b. flight . . . cerebellum
    c. cognition . . . neocortex
    d. parenting behavior . . . the reward system
    e. vision . . . optic lobes

14. Which of the following statements correctly describes neural plasticity?
    a. Excesses or deficiencies of neurotransmitters such as dopamine or serotonin are implicated in most nervous system disorders.
    b. Large body movements are controlled by the brainstem, cerebellum, and motor cortex.
    c. The visual cortex is reorganized soon after birth in response to visual stimulation.
    d. The number and activity of synapses can change in the process of learning motor skills and developing long-term memories.
    e. Both c and d correctly describe neural plasticity.

Chapter 50

Sensory and Motor Mechanisms

Key Concepts

50.1 Sensory receptors transduce stimulus energy and transmit signals to the central nervous system

50.2 The mechanoreceptors responsible for hearing and equilibrium detect moving fluid or settling particles

50.3 Visual receptors in diverse animals depend on light-absorbing pigments

50.4 The senses of taste and smell rely on similar sets of sensory receptors

50.5 The physical interaction of protein filaments is required for muscle function

50.6 Skeletal systems transform muscle contraction into locomotion

Framework

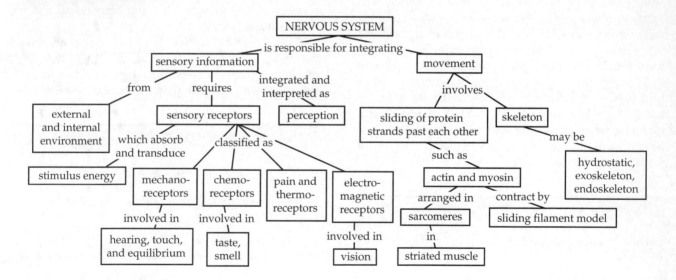

Chapter Review

Animals are constantly exploring and sensing their environment and processing that information in the brain, which then directs their next movements.

50.1 Sensory receptors transduce stimulus energy and transmit signals to the central nervous system

Stimuli are forms of energy that a sensory receptor converts to a change in membrane potential, which can result in action potentials sent to the central nervous system. Integration of sensory input by the brain leads to motor responses.

Sensory Pathways   **Sensory reception** is the detection of energy from a particular stimulus by sensory cells, which are usually modified neurons or epithelial cells that occur singly or in groups within sensory organs. **Sensory receptors** include organs, cells, or structures within cells that detect specific stimuli from the external or internal environment.

Sensory transduction is the conversion of a physical or chemical stimulus into a **receptor potential,** a graded change in the membrane potential of a receptor cell that is proportional to the strength of the stimulus. Receptor potentials result from the opening or closing of ion channels in response to a stimulus.

Transmission of sensory information to the CNS occurs as action potentials. The magnitude of a receptor potential correlates with either the frequency of action potentials generated (when the sensory receptor is a neuron) or the quantity of neurotransmitter released (when a nonneuronal sensory receptor synapses with a sensory neuron). Stronger stimuli may also affect the number of receptors that are activated. Many sensory neurons generate action potentials spontaneously at a low rate, and reception of a stimulus changes the frequency of action potentials.

Integration of sensory information begins with the summation of graded potentials from receptors. Sensory information is further integrated within complex receptors and the CNS. Impulses arriving from sensory neurons are routed to different parts of the brain that interpret them, producing **perceptions** of various stimuli.

The strengthening of the energy of a stimulus, either by accessory structures of sense organs or as part of signal transduction pathways, is called **amplification.** Ongoing stimulation may lead to **sensory adaptation** in which continuous stimulation results in a decline in sensitivity of the receptor cell.

INTERACTIVE QUESTION   50.1

List and describe the four functions common to all sensory pathways.

a.

b.

c.

d.

Types of Sensory Receptors   Sensory receptors can be grouped into five categories.

Mechanoreceptors respond to the mechanical energy of pressure, touch, stretch, motion, and sound. Bending or stretching of external cell structures linked to ion channels increases their permeability to ions, leading to depolarization or hyperpolarization.

Vertebrate stretch receptors monitor the length of skeletal muscles. Dendrites of sensory neurons wound around certain muscle fibers are stimulated by stretching of the muscles. In humans, dendrites of sensory neurons function as touch receptors; the type of stimulus to which they respond best depends on their location and the structure of their connective tissue covering.

Chemoreceptors include both general receptors that monitor total solute concentration and specific receptors that respond only to a single kind of important molecule. Membrane permeability of the sensory cell changes in response to binding of a stimulus molecule.

Electromagnetic receptors respond to various forms of electromagnetic energy, such as visible light, electricity, and magnetic fields.

Thermoreceptors in the skin and hypothalamus respond to heat or cold and send information to the body's thermostat. Researchers found that receptor proteins for capsaicin, a component of hot peppers, also respond to hot temperatures. Mammals have a number of thermoreceptors that respond to specific temperature ranges.

Pain receptors (nociceptors) in humans are naked dendrites. These receptors respond to excess heat, pressure, or chemicals. Prostaglandins released from injured tissues increase pain by sensitizing receptors.

INTERACTIVE QUESTION   50.2

a. Where are pain receptors located in humans?

b. How do aspirin and ibuprofen reduce pain?

50.2   The mechanoreceptors responsible for hearing and equilibrium detect moving fluid or settling particles

Sensing of Gravity and Sound in Invertebrates   Most invertebrates have **statocysts,** which often consist of a layer of ciliated receptor cells lining a chamber containing **statoliths,** dense granules or grains of sand. The stimulation of mechanoreceptors beneath the statoliths provides positional information to the animal.

Most insects sense sounds with body hairs that vibrate in response to sound waves of specific frequencies. Localized "ears" are found on many insects, consisting of a tympanic membrane with attached receptor cells stretched over an internal air-filled chamber.

Hearing and Equilibrium in Mammals In mammals and most terrestrial vertebrates, the sensory organs for hearing and balance are in the ear. In *hearing*, pressure waves in the air are transduced into nerve impulses that the brain perceives as sound.

The ear consists of three regions: The external pinna and the auditory canal make up the **outer ear.** The **tympanic membrane** (eardrum), which separates the outer ear from the **middle ear,** transmits sound pressure waves to three small bones—the malleus (hammer), incus (anvil), and stapes (stirrup)—which conduct the waves to the inner ear by way of a membrane called the **oval window.** The **Eustachian tube,** connecting the pharynx and the middle ear, equalizes pressure within the middle ear. The **inner ear** contains fluid-filled channels. The **semicircular canals** function in equilibrium, and the **cochlea** functions in hearing.

The coiled cochlea has two large, connected, fluid-filled canals—an upper vestibular canal and a lower tympanic canal—separated by the smaller cochlear duct. The **organ of Corti,** located on the basilar membrane that forms the floor of the cochlear duct, contains sensory receptors called **hair cells.** Their hairs extend into the cochlear duct, and some attach to the tectorial membrane, which overhangs the organ of Corti.

Pressure waves in air—transmitted by the tympanic membrane, the three bones of the middle ear, and the oval window—produce pressure waves in the perilymph, which travel from the vestibular canal through the tympanic canal and dissipate when they strike the **round window.** These pressure waves vibrate the basilar membrane, bending the hairs of its receptor cells against the tectorial membrane. When bent in one direction, the hair cells depolarize, increasing neurotransmitter release and the frequency of action potentials in sensory neurons that travel in the auditory nerve to the brain. When bent in the other direction, hair cells hyperpolarize, which reduces neurotransmitter release and the generation of action potentials.

Volume is a result of the amplitude, or height, of the sound wave; a stronger wave bends the hairs more and results in more action potentials. *Pitch* is related to the frequency of sound waves, usually expressed in hertz (Hz). Different regions of the basilar membrane vibrate in response to different frequencies, and neurons associated with the vibrating region transmit action potentials to those auditory regions of the cortex where a particular pitch is perceived.

INTERACTIVE QUESTION   50.3

Label the parts in the following diagram of the human ear.

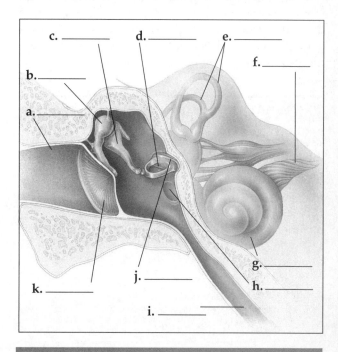

Within the inner ear are two chambers, the **utricle** and **saccule,** and three semicircular canals responsible for detection of body position and movement. Hair cells in the utricle and saccule project into a gelatinous material containing calcium carbonate particles called otoliths. These heavy particles exert a pull on the hairs in the direction of gravity; different head angles stimulate different hair cells, thereby altering the output of neurotransmitters and the resulting action potentials of sensory neurons of the vestibular nerve. Clusters of hair cells in the swellings at the base of the semicircular canals project into a gelatinous mass called the cupula and sense angular motion in any direction by detecting the pressure of the moving semicircular canal fluid against the cupula.

Hearing and Equilibrium in Other Vertebrates In fish, soundwave vibrations pass through the skeleton of the head to the inner ears, within which hair cells are stimulated by the movement of otoliths. Some fishes have a series of bones that transmit vibrations from the swim bladder to the inner ear.

Clusters of hair cells with sensory hairs embedded in a gelatinous cap or cupula are contained in the **lateral line system** of fishes and aquatic amphibians. Water moving through a tube along both sides of the body bends the cupula and stimulates the hair cells, enabling the fish to perceive its own movement, water currents, and the pressure waves generated by other moving objects.

The inner ear is the organ of hearing and equilibrium in terrestrial vertebrates. In amphibians, sound vibrations are conducted by a tympanic membrane on the body surface to a single middle ear bone to the inner ear. A cochlea has evolved in birds and other reptiles, as well as in mammals.

INTERACTIVE QUESTION   50.4

List the organs used to sense movement or equilibrium in the following groups of animals. Then briefly discuss how those organs are similar.

a. Mammals:

b. Fishes:

c. Invertebrates:

d. Similarities:

50.3   Visual receptors in diverse animals depend on light-absorbing pigments

Evolution of Visual Perception   All light detectors in the animal kingdom involve **photoreceptors** containing light-absorbing pigment molecules. All photoreceptors share common developmental genes, indicating that photoreceptors were present in the earliest bilaterian animals.

Most invertebrates have light-detecting organs. The ocelli or eyespots of planarians, which detect light intensity and direction, consist of a layer of pigmented cells that allow light to strike photoreceptors from one direction only. The brain compares impulses from the left and right ocelli to help the animal navigate a direct path away from a light source.

The **compound eye** of insects, crustaceans, and some polychaete worms contains up to thousands of light detectors called **ommatidia,** each with its own lens. The differing intensities of light entering the many ommatidia produce a visual image. The compound eyes of insects are adept at detecting movement and color, and some detect ultraviolet radiation.

Some jellies, polychaetes, spiders, and many molluscs have a **single-lens eye,** in which light is focused through a movable single lens onto a layer of photoreceptors. The **iris** changes the diameter of the **pupil** to adjust the amount of light entering the eye.

The Vertebrate Visual System   The eyeball consists of a tough, outer connective tissue layer called the sclera, and a thin, pigmented inner layer called the choroid. At the front of the eye, the sclera becomes the transparent *cornea,* and the choroid forms the colored *iris,* which regulates the amount of light entering through the pupil. The **retina,** the innermost layer of the eyeball, contains neurons and photoreceptors. The optic nerve attaches to the eye at the optic disk, forming a "blind spot" on the retina.

The transparent **lens** focuses an image onto the retina. The *aqueous humor* fills the anterior eye cavity; jellylike *vitreous humor* fills the posterior cavity.

Rods and **cones** are the photoreceptors in the retina. Several rods or cones relay information to each *bipolar cell,* several of which synapse with *ganglion cells,* whose axons form the optic nerve. *Horizontal* and *amacrine cells* in the retina help to integrate visual information.

Rods are more light sensitive and enable night vision, whereas cones distinguish colors.

INTERACTIVE QUESTION   50.5

Label the parts of the vertebrate eye in the following illustration.

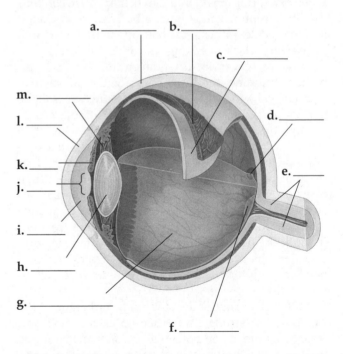

a._____   b._____

c._____

m._____

l._____   d._____

k._____   e._____

j._____

i._____

h._____

g._____

f._____

Both rods and cones have visual pigments embedded in a stack of disks in the outer segment of each cell. The light-absorbing molecule **retinal** is bonded to a membrane protein called an **opsin,** forming **rhodopsin,** the visual pigment in rod cells. When retinal absorbs light, it changes from the *cis* isomer to the *trans* isomer, activating rhodopsin.

Light-activated rhodopsin activates a membrane-bound G protein called transducin, which then activates an enzyme that detaches cGMP bound to Na^+ channels in the plasma membrane. Sodium channels close, hyperpolarizing the cell. Except in bright light when rhodopsin remains active, enzymes convert retinal to its original shape, and the signal transduction pathway shuts off.

Visual processing begins in the retina, where rods and cones synapse with bipolar cells. In the dark, rods and cones are depolarized and continually release glutamate, which, depending on the postsynaptic receptor, either depolarizes or hyperpolarizes bipolar cells. When rods and cones hyperpolarize after absorbing light, glutamate release ceases. In response, the bipolar cells reverse their polarization, either becoming hyperpolarized or depolarized, depending on the receptor.

Signals from rods and cones may follow a pathway directly from photoreceptors to bipolar cells to ganglion cells. Other pathways involve lateral integration by horizontal cells, which carry signals from one receptor to others and to several bipolar cells. **Lateral inhibition** occurs when a horizontal cell inhibits more distant cells and is a form of integration that enhances contrast. Lateral inhibition also occurs as amacrine cells relay information from one bipolar cell to several ganglion cells, as well as at other levels of visual processing in the brain.

The left and right optic nerves meet at the **optic chiasma** at the base of the cerebral cortex. Information from the left visual field of both eyes travels to the right side of the brain, whereas what is sensed in the right field of view goes to the left side.

Most axons of the ganglion cells go to the **lateral geniculate nuclei,** where neurons lead to the **primary visual cortex** in the cerebrum. Neurons also carry information to other visual processing and integrating centers in the cortex. At least 30% of the cerebral cortex may be devoted to creating the three-dimensional perception of what we see.

You focus on close objects by accommodation, in which ciliary muscles contract, causing suspensory ligaments to slacken and the elastic lens to become rounder. In the human eye, rods are most concentrated toward the edge of the retina, whereas the center of the visual field, the **fovea,** is filled with cones.

The relative proportion of rods and cones correlates with the activity pattern of a mammal. Red, green, and blue cones, each with its own type of opsin that binds with retinal to form a *photopsin,* are named for the color of light they optimally absorb. Squirrel monkeys have only one X-linked opsin gene which can have either an

allele for red or an allele for green (as in humans, their blue opsin gene is on an autosome). Researchers used gene therapy to inject the missing allele into the retina of male monkeys, enabling them to distinguish red from green and demonstrating that visual neural circuits can form or be activated even in adults.

INTERACTIVE QUESTION 50.6

The *receptive field* of a ganglion cell includes the rods and/or cones that supply information to it.

a. Would a large or a small receptive field produce the sharpest image?

b. Where on the retina are the smallest receptive fields found?

50.4 The senses of taste and smell rely on similar sets of sensory receptors

For terrestrial animals, the sense of **gustation** (taste) detects chemicals called **tastants** present in a solution, whereas **olfaction** (smell) detects airborne **odorants.** Both rely on chemoreceptors.

Sensory hairs containing taste receptors, are found on the feet and mouthparts of insects. Insects also have olfactory hairs, usually located on their antennae.

Taste in Mammals **Taste buds,** which contain groups of modified epithelial cells, are scattered on the tongue and in the mouth. The receptor cells recognize five types of tastants, producing the taste perceptions—sweet, sour, salty, bitter, and umami. Taste buds contain receptor cells for all five tastes. Individual taste cells express a single receptor type. Experiments indicate that taste perception depends on which sensory neurons are activated: A sweet taste cell reprogrammed to express a receptor for a bitter molecule caused mice to be attracted to that bitter chemical.

A G-protein coupled receptor (GPCR) is involved with sweet, umami, and bitter tastes. Humans have more than 30 different bitter receptors, but only one type of sweet and umami receptor. The sour tastant receptor is a transient receptor protein (TRP), similar to capsaicin and the other thermoreceptor proteins. In either type of taste receptor, depolarization leads to activation of a sensory neuron, which transmits action potentials to the brain.

Smell in Humans Olfactory receptor cells are neurons that line the upper part of the nasal cavity. Binding of an odorant to a GPCR protein, called an odorant receptor (OR), on an olfactory cilium triggers a signal transduction pathway, which opens ion channels. The membrane depolarizes and sends action potentials to the olfactory bulb of the brain. Approximately 3% of the genes in the human genome are OR genes.

INTERACTIVE QUESTION   50.7

For what purposes do animals use their senses of gustation and olfaction?

50.5   The physical interaction of protein filaments is required for muscle function

Vertebrate Skeletal Muscle   Vertebrate **skeletal muscle** consists of a bundle of multinucleated muscle cells, called fibers, running parallel to the length of the muscle. Each fiber contains a bundle of **myofibrils,** each composed of (1) **thin filaments,** which consist of two strands of actin coiled with two strands of regulatory protein, and (2) **thick filaments,** which are arrays of myosin molecules.

The regular arrangement of filaments produces repeating light and dark bands; thus, skeletal muscle is also called **striated muscle.** The repeating units are called **sarcomeres,** which are the contractile units of a muscle. Thin filaments attach to the Z line and project toward the center of the sarcomere. Thick filaments lie in the center, attached at the M lines.

According to the **sliding-filament model** of muscle contraction, the thick and thin filaments do not change length but simply slide past each other, increasing their area of overlap.

The mechanism for the sliding of filaments is the hydrolyzing of ATP by the globular head of a myosin molecule, which then changes to a high-energy form that binds to actin and forms a cross-bridge. The myosin head then bends and pulls the attached thin filament toward the center of the sarcomere. When a new molecule of ATP binds to the myosin head, it breaks its bond to actin, and the cycle repeats with the myosin head attaching to an actin molecule farther along on the thin filament. Energy for muscle contraction comes from creatine phosphate, which regenerates ATP, and from glycogen, which is broken down to glucose. Glucose can generate ATP by glycolysis or aerobic respiration.

INTERACTIVE QUESTION   50.8

Identify the components in the following diagram of a section of a skeletal muscle fiber. Explain how and why this diagram will look different when this muscle fiber is fully contracted.

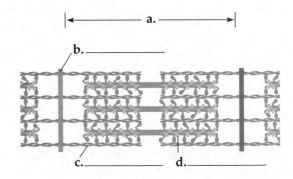

When the muscle is at rest, the myosin-binding sites of the actin molecules are blocked by the regulatory protein **tropomyosin,** which is associated with another set of regulatory proteins, the **troponin complex.** When troponin binds to Ca^{2+}, the regulatory proteins change shape, and myosin-binding sites on the thin filament are exposed.

Calcium ions are actively transported into the **sarcoplasmic reticulum (SR)** of a muscle cell. When an action potential of a motor neuron causes the release of acetylcholine into the synapse with a muscle fiber, the fiber depolarizes and an action potential spreads into the **transverse (T) tubules,** infoldings of the muscle cell's plasma membrane. The action potential opens Ca^{2+} channels in the sarcoplasmic reticulum, releasing Ca^{2+} into the cytosol. Calcium binding with troponin exposes myosin-binding sites, and contraction begins. Contraction ends when motor neuron input stops, the sarcoplasmic reticulum pumps Ca^{2+} back out of the cytosol, and the tropomyosin–troponin complex again blocks the binding sites.

The response of a single muscle fiber to an action potential of a motor neuron is an all-or-none contraction, producing a twitch. Graded contraction of whole muscles may result from involving additional fibers. A single motor neuron and all the muscle fibers it innervates make up a **motor unit.** The strength of a muscle contraction depends on the size and number of motor units involved. Muscle force (tension) can be increased by the activation of additional motor units, called recruitment. Prolonged contraction can cause muscle fatigue due to depletion of ATP and the loss of ion gradients required for depolarization.

Repeated, rapid-fire action potentials produce summation and a state of smooth and sustained contraction called **tetanus.**

Oxidative fibers, which rely mostly on aerobic respiration, have many mitochondria, a good blood supply, and **myoglobin,** a pigment that extracts oxygen from the blood and stores it. Glycolytic fibers primarily use glycolysis to produce ATP. **Fast-twitch fibers** produce rapid and powerful contractions. **Slow-twitch fibers,** common in muscles that must sustain long contractions, have less sarcoplasmic reticulum and pump calcium more slowly; thus, calcium remains in the cytosol longer and twitches last longer.

INTERACTIVE QUESTION 50.9

a. Identify the three distinct types of skeletal muscle fibers.

b. Which type of fiber has few mitochondria, has little myoglobin, and produces brief, powerful contractions?

c. Which type of fiber would you expect to be present in the highest proportion in postural muscles?

Other Types of Muscle Vertebrate **cardiac muscle** fibers are connected by **intercalated disks** through which action potentials spread to all the cells of the heart. Ion channels in the plasma membrane of a cardiac muscle cell produce rhythmic depolarizations that generate action potentials. The action potentials of cardiac muscle cells last quite long.

The scattered arrangement of actin and myosin filaments in **smooth muscle** accounts for its nonstriated appearance. Smooth muscle fibers lack T tubules, a troponin complex, and a well-developed sarcoplasmic reticulum. Calcium ions enter the cell in response to an action potential and bind to calmodulin, which activates an enzyme that phosphorylates myosin. Contractions are slower and may be triggered by the autonomic nervous system or by the muscle cells themselves.

Invertebrates have muscle cells similar to the skeletal and smooth muscle cells of vertebrates. The flight muscles of insects are capable of independent and rapid contraction.

50.6 Skeletal systems transform muscle contraction into locomotion

Skeletons function in support, protection, and movement. Because muscles can only contract, antagonistic muscle pairs are required to move body parts in opposite directions.

Types of Skeleton Systems Fluid under pressure in a closed body compartment creates a **hydrostatic skeleton.** As muscles change the shape of the fluid-filled compartment, the animal elongates or moves. Earthworms and other annelids move by **peristalsis,** using rhythmic waves of contractions of circular and longitudinal muscles.

Typical of molluscs and arthropods, **exoskeletons** are hard coverings deposited on the surface of animals. The cuticle of an arthropod contains fibrils of the polysaccharide **chitin** embedded in a protein matrix. Where protection is most needed, this flexible support is hardened by cross-linking proteins and the addition of calcium salts.

The supporting elements of an **endoskeleton** are embedded in the soft tissues of the animal. Sponges have endoskeletons composed of spicules or fibers, and echinoderms have hard plates beneath the skin. The endoskeletons of chordates are composed of cartilage and/or bone. The vertebrate skeleton consists of a skull, vertebral column, rib cage, the pectoral and pelvic girdles, and the limb bones. Joints allow for flexibility in body movement.

INTERACTIVE QUESTION 50.10

List an advantage and a disadvantage of each of the following types of skeletons, and indicate in which animal groups each type is found.

a. hydrostatic skeleton

b. exoskeleton

c. endoskeleton

Body proportions of small and large animals vary, partly as a result of the physical relationship between the diameter of a support (strength increases with the square of its diameter) and the strain produced by increasing weight (weight increases with the cube of its dimensions). Posture is an important structural factor in supporting body weight. Most of the weight of large land mammals is supported by muscles and tendons, which hold the legs relatively straight and directly under the body.

Types of Locomotion   Locomotion, the active movement from place to place, requires energy to overcome the forces of friction and gravity. Animals move to obtain food, escape from danger, and find mates.

Walking, running, hopping, or crawling on land requires the animal to support itself and move against gravity; however, air offers little resistance to movement. Powerful leg muscles and strong skeletal support are important for supporting an animal and propelling it forward. A running or hopping animal may store some energy in its tendons to facilitate the next step or hop. Maintaining balance is another requirement for upright movement. A crawling animal must overcome the friction produced by its contact with the ground.

Overcoming gravity is less of a problem for swimming animals due to the buoyancy of water. Most fast swimmers have a fusiform shape that reduces friction or drag caused by the density of water. Animals may swim by using their legs as oars, using water to jet-propel themselves, or by undulating their bodies and tails from side to side or up and down.

A flying animal uses its wings to provide enough lift to overcome the force of gravity. Wings shaped as airfoils attach to lightweight, fusiform bodies.

Energy Costs of Locomotion   The cost of locomotion depends on the mode of locomotion and the environment. Adaptations that maximize efficiency of locomotion increase fitness in that more of an animal's energy is available for growth and reproduction.

INTERACTIVE QUESTION   50.11

a. Which mode of locomotion is the most energy efficient per distance traveled?

b. Which mode is the most energetically expensive per distance traveled?

c. Which mode consumes the most energy per unit of time?

d. In animals specialized for a given mode of transport, which animal expends more energy per kilogram of body mass in traveling the same distance: a larger animal or a smaller animal?

Word Roots

chemo- = chemical (*chemoreceptor:* a sensory receptor that responds to a chemical stimulus)

coch- = a snail (*cochlea:* the complex, coiled organ of hearing that contains the organ of Corti)

electro- = electricity (*electromagnetic receptor:* a receptor of electromagnetic energy, such as visible light, electricity, or magnetism)

endo- = within (*endoskeleton:* a hard skeleton buried within the soft tissues of an animal)

exo- = outside (*exoskeleton:* a hard outer structure of an animal's body that provides protection and points of attachment for muscles)

fovea- = a pit (*fovea:* the place on the retina at the eye's center of focus, where cones are highly concentrated)

gusta- = taste (*gustation:* the sense of taste)

hydro- = water (*hydrostatic skeleton:* a skeletal system composed of fluid held under pressure in a closed body compartment; the main skeleton of most cnidarians, flatworms, nematodes, and annelids)

inter- = between; **-cala** = insert (*intercalated disk:* a specialized junction between cardiac muscle cells that provides direct electrical coupling between the cells)

mechano- = an instrument (*mechanoreceptor:* a sensory receptor that detects physical deformation in the body's environment associated with pressure, touch, stretch, motion, and sound)

myo- = muscle; **-fibro** = fiber (*myofibril:* a longitudinal bundle in a muscle cell (fiber) that contains thin filaments of actin and regulatory proteins and thick filaments of myosin)

noci- = harm (*nociceptor:* a sensory receptor that responds to noxious or painful stimuli)

olfact- = smell (*olfaction:* the sense of smell)

omma- = the eye (*ommatidium:* one of the facets of the compound eye of arthropods and some polychaete worms)

peri- = around; **-stalsis** = a constriction (*peristalsis:* a type of movement on land produced by rhythmic waves of muscle contraction passing from front to back, as in many annelids)

photo- = light (*photoreceptor:* an electromagnetic receptor that detects the radiation known as visible light)

rhodo- = red (*rhodopsin:* a visual pigment consisting of retinal and opsin)

sacc- = a sack (*saccule:* a chamber in the vestibule behind the oval window that participates in the sense of balance)

sarco- = flesh; **-mere** = a part (*sarcomere:* the fundamental, repeating unit of striated muscle, delimited by the Z lines)

semi- = half (*semicircular canals:* a three-part chamber of the inner ear that functions in maintaining equilibrium)

stato- = standing; **-lith** = a stone (*statolith:* in invertebrates, a particle that settles in response to gravity and is found in sensory organs that function in equilibrium)

tetan- = rigid, tense (*tetanus:* the maximal, sustained contraction of a skeletal muscle, caused by a very high frequency of action potentials elicited by continual stimulation)

thermo- = heat (*thermoreceptor:* a receptor stimulated by either heat or cold)

trans- = across; **-missi** = send (*transmission:* the passage of a nerve impulse along an axon)

tropo- = turn, change (*tropomyosin:* the regulatory protein that blocks the myosin-binding sites on actin molecules)

tympan- = a drum (*tympanic membrane:* another name for the eardrum, the membrane between the outer ear and the middle ear)

utric- = a leather bag (*utricle:* a chamber in the vestibule behind the oval window that opens into the three semicircular canals)

Structure Your Knowledge

1. Describe the path of a light stimulus from where it enters the vertebrate eye to its transmission as an action potential through the optic nerve.
2. Arrange the following structures in an order that enables you to describe the passage of sound waves from the external environment to the perception of sound in the brain.

round window	basilar membrane
oval window	cerebral cortex
pinna	vestibular and tympanic canals
auditory nerve	tectorial membrane
cochlea	tympanic membrane
auditory canal	malleus, incus, stapes
cochlear duct	organ of Corti

3. Trace the sequence of events in muscle contraction from an action potential in a motor neuron to the relaxation of the muscle.

Test Your Knowledge

MULTIPLE CHOICE: *Choose the one best answer.*

1. Sensory perception is
   a. the reception of a stimulus by a sensory cell.
   b. the conversion of stimulus energy into a receptor potential by a sensory receptor or sensory neuron.
   c. the interpretation of a stimulus according to the region of the brain that receives the nerve impulse.
   d. the amplification and transmission of sensory data to the CNS.
   e. the emotional response to a stimulus.

2. What of the following sensory receptors are *not* present in human skin?
   a. thermoreceptors
   b. mechanoreceptors
   c. pain receptors
   d. electromagnetic receptors
   e. neither b nor b

3. The function of the three bones of the middle ear is to
   a. respond to different frequencies of sound waves.
   b. transmit pressure waves to the oval window.
   c. equalize pressure within the inner ear by communicating through the Eustachian tube.
   d. sense rotation in three different planes (*x, y,* and *z* axes).
   e. support the tympanic membrane (eardrum).

4. The volume of a sound is determined by
   a. the area in the brain that receives the signal.
   b. the section of the basilar membrane that vibrates.
   c. the number of hair cells that are stimulated.
   d. the frequency of the sound wave.
   e. the degree to which hair cells are bent and the resulting increase in action potentials in the sensory neuron.

5. Rotation of the head of a mammal is sensed by
   a. the organ of Corti located in the utricle and saccule.
   b. changes in the action potentials of hair cells in response to bending against the tectorial membrane.
   c. the bending of hair cells embedded in a cupula in a semicircular canal in response to the movement of internal fluid.
   d. hair cells in response to movement of fluid in the coiled cochlea.
   e. statocysts when statoliths stimulate hair cells.

6. Fish can perceive pressure waves from moving objects with
   a. a lateral line system containing mechanoreceptor units.
   b. semicircular canals in which sensory hairs are embedded in a gelatinous cap.
   c. inner ears, within which sensory hairs are stimulated by the movement of otoliths.
   d. a series of bones that transmit vibrations from the swim bladder to the inner ear.
   e. clusters of sensory hairs in the lateral line system.

7. What do the lateral line of fish, statocysts in invertebrates, and the cochlea of your ear all have in common?
   a. They are used to sense sound or pressure waves.
   b. They are organs of equilibrium.
   c. They use hair cells as mechanoreceptors.
   d. They use a second-messenger pathway of signal transduction.
   e. They use granules to stimulate their receptor cells.

8. The compound eye of insects and crustaceans
   a. is composed of thousands of prisms that focus light onto a few photoreceptor cells.
   b. cannot sense colors.
   c. contains many ommatidia, each of which contains a lens that focuses light from a tiny portion of the field of view.
   d. may detect infrared and ultraviolet radiation.
   e. All of the above are true.

9. Which of the following statements is *not* true?
   a. The iris regulates the amount of light entering the pupil.
   b. The choroid is the thin, pigmented inner layer of the eye that contains the photoreceptor cells.
   c. Rounding of the lens enables focus on nearby objects.
   d. Thin aqueous humor fills the anterior eye cavity.
   e. The sclera is the tough outer connective tissue layer of the eye.

10. Which of the following structures is *incorrectly* paired with its function?
    a. cones—respond to different light wavelengths, producing color vision
    b. horizontal cells—produce lateral inhibition of receptor cells and bipolar cells
    c. bipolar cells—relay impulses between rods or cones and ganglion cells
    d. ganglion cells—synapse with the optic nerve at the optic disk or blind spot
    e. fovea—center of the visual field, containing only cones in humans

11. The molecule rhodopsin
    a. dissociates when struck by light and is functional only in the dark.
    b. is formed from retinal and opsin by enzymes when struck by light.
    c. is more sensitive to light than are photopsins and is involved in black-and-white vision in dim light.
    d. activates a signal transduction pathway that increases a rod cell's permeability to sodium.
    e. is or does all of the above.

12. The absorption of light by rhodopsin in a rod cell leads to
    a. an increase in the release of neurotransmitter.
    b. the opening of sodium channels caused by the binding of cGMP.
    c. a hyperpolarized membrane and a decrease in the release of neurotransmitter.
    d. the bleaching of rhodopsin, which increases the sensitivity of rod cells to light.
    e. the closing of sodium channels and the release of neurotransmitter.

13. A receptive field in the retina involves a *single*
    a. bipolar cell.
    b. photoreceptor (rod or cone) cell.
    c. amacrine cell.
    d. horizontal cell.
    e. ganglion cell.

14. Which of the following statements regarding the senses of gustation and olfaction is *not* true?
    a. There is no distinction between these senses in aquatic animals.
    b. There are as many as 1,000 more genes for odorant receptors than for tastant receptors.
    c. Olfactory sensory cells are neurons; gustatory receptor cells are modified epithelial cells that synapse with sensory neurons.
    d. Each taste cell has a single type of receptor; each olfactory cell can have receptors that respond to several odorant molecules.
    e. Taste receptors of insects are located in sensory hairs on the feet and mouthparts; olfactory hairs are located on the antennae.

15. When striated muscle fibers contract,
    a. the Z lines are pulled closer together.
    b. the sarcomere expands.
    c. the thin filaments become shorter.
    d. the thick filaments become longer.
    e. a and c occur.

16. Tetanus
    a. is muscle fatigue resulting from a lack of ATP and loss of ion gradients.
    b. is the all-or-none contraction of a single muscle fiber.
    c. is the result of stimulation of additional motor neurons.
    d. is the result of a volley of action potentials, whose summation produces an increased and sustained contraction.
    e. is the rigidity of muscles following death.

17. The role of ATP in muscle contraction is
    a. to form cross-bridges between thick filaments and thin filaments.
    b. to release the myosin head from actin when it binds to myosin, and to provide energy when hydrolyzed to form myosin's high-energy form.
    c. to remove the tropomyosin–troponin complex from blocking the binding sites on actin.
    d. to bend the cross-bridge and pull the thick filaments toward the center of the sarcomere.
    e. to replace the supply of creatine phosphate required for the movement of myosin past actin.

18. How does calcium affect muscle contraction?
    a. It is released from the T tubules in response to an action potential and initiates contraction.
    b. The binding of acetylcholine opens calcium channels in the plasma membrane, creating an action potential that travels down the T tubules.
    c. It binds to tropomyosin and helps to stabilize cross-bridge formation.
    d. Its binding to troponin causes tropomyosin to move away from the myosin-binding sites on the actin filament.
    e. Its release from the sarcoplasmic reticulum changes the membrane potential of the muscle cell so that contraction can occur.

19. A motor unit is
    a. a sarcomere that extends from one Z line to the next Z line.
    b. a motor neuron and the muscle fibers it innervates.
    c. the myofibrils of a fiber and the transverse tubules that connect them.
    d. the muscle fibers that make up a muscle.
    e. the antagonistic set of muscles that flex and extend a body part.

20. Which of the following characteristics does *not* pertain to cardiac muscle?
    a. scattered arrangement of actin and myosin filaments
    b. intercalated disks that spread action potentials between cells
    c. action potentials that last a long time
    d. ability to generate action potentials without nervous input
    e. striations

21. Smooth muscle contracts relatively slowly because
    a. the only ATP available is supplied by fermentation.
    b. its contraction is stimulated by hormones, not motor neurons.
    c. it does not have a well-developed sarcoplasmic reticulum, and Ca^{2+} enters the cell through the plasma membrane during an action potential.
    d. it is not striated.
    e. it is composed exclusively of slow-twitch muscle fibers.

22. Hydrostatic skeletons are used for movement by all of the following animals *except*
    a. cnidarians.
    b. arthropods.
    c. nematodes.
    d. annelids.
    e. flatworms.

23. When you pull up your lower arm and "make a muscle" in your biceps, you are
    a. contracting a flexor.
    b. relaxing a flexor.
    c. contracting an extensor.
    d. contracting a tendon.
    e. producing a simple muscle twitch.

24. Which of the following modes of locomotion is the most energy efficient for an animal that is specialized for moving in that fashion?
    a. crawling
    b. running
    c. flying
    d. hopping
    e. swimming

Chapter 51

Animal Behavior

Key Concepts

51.1 Discrete sensory inputs can stimulate both simple and complex behaviors

51.2 Learning establishes specific links between experience and behavior

51.3 Selection for individual survival and reproductive success can explain most behaviors

51.4 Inclusive fitness can account for the evolution of behavior, including altruism

Framework

This chapter introduces the complex and fascinating subject of animal behavior.

Behaviors can range from simple fixed-action patterns in response to specific stimuli to problem-solving in novel situations. Behaviors result from interactions among environmental stimuli, experience, and individual genetic makeup. The parameters of behavior are controlled by genetics and thus are acted upon by natural selection. Behavioral ecology focuses on the ultimate cause of reproductive fitness, which can be used to interpret foraging behavior, mating patterns, and altruistic behavior. Sociobiology extends evolutionary interpretations to human social behavior.

Chapter Review

51.1 Discrete sensory inputs can stimulate both simple and complex behaviors

Behavior is an action performed in response to a stimulus. Niko Tinbergen's set of questions to guide behavioral studies emphasizes the importance of both *proximate causation*, the immediate cause of a behavior in terms of the stimuli that trigger it, the mechanisms that produce it, and how experience modifies a response; and *ultimate causation*, which concerns the benefit to survival and reproduction and the evolutionary basis of the behavior. **Behavioral ecology** is the study of the ecological and evolutionary basis of animal behavior.

Fixed Action Patterns A **fixed action pattern** is a sequence of unlearned behaviors that, once begun, is usually carried through to completion. It is triggered by a simple external stimulus called a **sign stimulus**.

Migration Many animals use environmental cues to guide regular, long-distance changes in location called **migrations.** Some migratory animals alter their orientation to the sun during the day with the aid of their *circadian clock.* Some animals navigate using Earth's magnetic field, perhaps as a result of its effect on magnetite-containing brain structures or on photoreceptors.

Behavioral Rhythms Many behaviors follow a circadian rhythm mediated by the biological clock. *Circannual rhythms* are linked to the yearly cycle of seasons and are influenced by changing lengths of daylight and darkness. The mating behavior of fiddler crabs is linked to the lunar cycle, which corresponds to times of greatest tidal movement and the subsequent dispersal of larvae.

Animal Signals and Communication Communication between animals involves the transmission and reception of special stimuli called **signals.** Courtship often involves a *stimulus response chain*, in which each response is the stimulus for the next behavior. Communication may be *visual, auditory, chemical,* or *tactile,* depending on the lifestyle and sensory specializations of a species.

Pheromones are chemical signals commonly used by mammals and insects in reproductive behavior to attract mates and to trigger specific courtship behaviors. Pheromones can also function as alarm signals.

Why is most communication among mammals olfactory and auditory, whereas communication among birds is visual and auditory?

51.2 Learning establishes specific links between experience and behavior

Behavior that is performed virtually the same by all individuals in a population, regardless of developmental or environmental differences, is developmentally fixed and called **innate behavior.**

Experience and Behavior In a **cross-fostering study,** young of one species are raised by adults of another species. Such studies examine how an animal's social and physical surroundings influence behavior. Male California mice provide extensive parental care and are always highly aggressive toward other mice. When cross-fostered in nests of white-footed mice, California mice showed reduced aggression and reduced parental behavior. Early experiences that alter parental behavior extend this environmental influence to the next generation.

Human **twin studies** involving identical twins raised apart or in the same household indicate that both genetics and the environment contribute to various behavioral disorders in humans.

Learning Learning is the modification of behavior as a result of experience.

Imprinting is characterized by a limited **sensitive period** during which learning may occur and involves a behavioral response to an individual or object that is usually long lasting. The ability to respond is innate; the environment provides the *imprinting stimulus.*

The establishment of a memory of the spatial structure of the environment is called **spatial learning.** Animals may learn and use a particular set of landmarks, or location indicators, to find their way within their area.

More complicated than a set of learned landmarks, **cognitive maps** are neural representations of the spatial relationships of objects in an animal's surroundings. Evidence for cognitive maps comes from research with nutcrackers, birds that are able to retrieve stored food from thousands of caches, perhaps by employing abstract geometric rules.

In **associative learning,** animals learn to associate one stimulus with another. In *classical conditioning,* an arbitrary stimulus is associated with a particular outcome. *Operant conditioning* refers to trial-and-error learning through which an animal associates a behavior with a reward or punishment. An animal's nervous system enables the association of behaviors with those stimuli that are most likely to occur in the animal's natural habitat.

Cognition refers to an animal's ability to learn using awareness, reasoning, recollection, and judgment. Research with bees indicates that they are capable of categorizing environmental objects as "same" or "different."

Problem-solving, the ability to achieve an end in the face of obstacles, is most often observed in mammals, especially in primates and dolphins. Such behavior has also been documented in some bird species.

Development of learned behaviors may occur over a short period of time or in distinct stages. For example, white-crowned sparrows appear to have a 50-day sensitive period in which they memorize the song of their species. If raised in isolation during this period, a bird does not develop a typical adult song. In a second learning phase, a juvenile bird sings a subsong, which gradually improves, apparently as the bird compares its own singing with the memorized song. The sparrow then sings this "crystallized" song for the rest of its life.

Many animals use the behavior of others as information in problem solving. Vervet monkeys have an innate ability to give alarm calls in response to threatening objects. They learn to discriminate in their calls by observing other members of the group and by receiving social confirmation of their calls. **Social learning,** learning through the observation of others, forms the basis of **culture**—a system of information transfer that involves teaching and/or social learning and influences behavior in a population.

Indicate the type of learning illustrated by the following examples:

a. Ewes will adopt and nurse a lamb shortly after they give birth to their own lamb but will butt and reject a lamb introduced a day or two later.

b. A dog whose early "accidents" were cleaned up with paper towels accompanied with harsh discipline hides any time a paper towel is used in the household.

c. In experiments, honey bees have the ability to distinguish between patterns that are the same or different.

d. Tinberger showed that female digger wasps use landmarks to locate their hive.

e. Young chimpanzees learn to crack oil palm nuts with two stones by observing others.

51.3 Selection for individual survival and reproductive success can explain most behaviors

Foraging Behavior **Foraging** is behavior involved with searching for, recognizing, obtaining, and consuming food. Natural selection is expected to refine behaviors that improve foraging. Laboratory studies of *Drosophila* have documented an evolutionary change in foraging path length in populations of high or low density. The frequency of the *for^R* (rover) allele increased in high-density populations, whereas the frequency of the *for^S* (sitter) allele increased in low-density populations.

According to the **optimal foraging model,** feeding behaviors maximize energy intake over energy expenditure and the risk of being eaten while foraging. Mule deer appear to forage more in open areas, where food may be slightly less available but the deer are less likely to fall prey to mountain lions.

INTERACTIVE QUESTION   51.3

Explain how the study of whelk-eating crows supports the optimal foraging model.

Mating Behavior and Mate Choice Many species have **promiscuous** mating, in which no strong pair bonds form. Longer-lasting relationships may be **monogamous** or **polygamous.** Polygamous relationships are most often *polygynous* (one male and many females), although a few are *polyandrous.* Monogamous species are less likely to exhibit *sexual dimorphism.*

The needs of offspring are an important factor in the evolution of reproductive patterns. If young require more food than one parent can supply, a male may increase his reproductive fitness by helping to care for offspring rather than going off in search of more mates. With mammals, the female often provides all the food, and males are often polygynous.

Certainty of paternity also influences mating behavior and parental care. With internal fertilization, the acts of mating and egg laying or birth are separated, and paternity is less certain than when eggs are fertilized externally.

INTERACTIVE QUESTION   51.4

Exclusive male parental care is observed much more frequently in species with external fertilization. Explain how such behavior could evolve.

Sexual selection may be *intrasexual,* involving competition among members of one sex for mates, or *intersexual,* in which mates are chosen by one sex on the basis of particular characteristics.

Female choice in stalk-eyed fruit flies has been a selection factor in the evolution of long eyestalks, which correlate with male quality. Experiments involving feather ornaments added to zebra finch parents suggest that females imprint on their fathers, and that mate choice may play a role in the evolution of ornamentation in male zebra finches.

In **mate choice copying,** individuals may copy the mate choice of others in a population. Experiments with guppies have shown that females will preferentially mate with males whom they have observed engaged in courtship with other females. Such mate choice copying was shown to mask a genetically controlled preference for a particular male coloration.

As with female choice, male competition for females can also reduce variation among males. *Agonistic behavior* involves a contest to determine which competitor gains access to a resource, such as food or a mate. The encounter may include a test of strength or, more commonly, symbolic behavior or ritual.

In studying mating behavior, behavioral ecologists apply **game theory,** which evaluates strategies in situations in which the outcome depends on both an individual's strategy and the strategies of others. Researchers studying the coexistence of three male phenotypes of the side-blotched lizard found that the reproductive success of each phenotype depended on the frequency of the other phenotypes.

INTERACTIVE QUESTION   51.5

In populations of the side-blotched lizard, aggressive orange throat males defend large territories with many females; blue throat males defend smaller territories and fewer females; and yellow throat males mimic females and use "sneaky" tactics to obtain matings. Starting with a high abundance of orange throats in the population, which type of male will tend to increase its mating success and increase in frequency next, and which type of male will then replace this second type after it increases in numbers? Explain your answer.

51.4 Inclusive fitness can account for the evolution of behavior, including altruism

Genetic Basis of Behavior The gene called *fru* (for *fruitless*) in fruit flies is a master regulatory gene that controls the sex-specific development of the nervous system and thus the male courtship ritual. Males lacking a functional *fru* gene fail to court, and females expressing a male *fru* gene court other females.

Studies have shown that the courtship song of the green lacewing is genetically controlled by several genes. Hybrid offspring produce songs that have components of the songs of both parent species.

Male prairie voles form *pair bonds,* help care for the young, and are aggressive to intruders. Male meadow voles do not show these behaviors. The gene for receptors for the neurotransmitter vasopressin is highly expressed in the brains of prairie voles but not in those of meadow voles. Male meadow voles with inserted prairie vole receptor genes developed more vasopressin receptors in their brains and showed mating behaviors similar to those of male prairie voles.

Genetic Variation and the Evolution of Behavior Closely related species often exhibit behavioral differences. Variations in behavior between populations within a species may correlate with variations in the environment.

Most laboratory-born garter snakes from coastal areas (where banana slugs are an abundant source of prey) ate slugs when they were offered, whereas few laboratory-born snakes from inland populations ate the slugs. The researcher proposed that snakes in coastal areas with the ability to recognize slugs by chemoreception had higher fitness, leading to the evolution of this difference in prey selection behavior.

Field observations and laboratory studies have documented recent changes in migratory behavior in European blackcaps. Instead of migrating southwest and then south to Africa for the winter, some German blackcaps are migrating west to Great Britain. Offspring of these new migrants had similar migratory orientations, as documented in funnel cage studies.

INTERACTIVE QUESTION 51.6

Why do experiments examining behavioral variations among natural populations often raise and test animals in the laboratory?

Altruism Many social behaviors are selfish, benefiting one individual's reproductive success at the expense of others. Selflessness, or **altruism,** is behavior that reduces an individual's fitness while increasing the fitness of other individuals.

Inclusive Fitness Natural selection favors traits that increase reproductive success, thereby propagating the

genes for those traits. W. Hamilton explained altruistic behavior in terms of **inclusive fitness,** the ability of an individual to pass on its genes by producing its own offspring and by helping close relatives produce their offspring.

A quantitative measure, called **Hamilton's rule,** predicts that natural selection would favor altruistic acts among related individuals if $rB > C$, where B and C are the benefit to the recipient and the cost to the altruist, respectively, as measured by the change in the average number of offspring produced as a result of the altruistic act, and r is the **coefficient of relatedness,** the fraction of genes that are shared by the two individuals. **Kin selection** is the term for the natural selection of altruistic behavior that enhances the reproductive success of related individuals.

Studies show that most cases of altruistic behavior involve close relatives, such as females in Belding's squirrel populations and worker bees in a hive, and thus improve the individual's inclusive fitness.

When altruistic behavior involves nonrelated animals, the explanation offered is **reciprocal altruism;** there is no immediate benefit for the altruistic individual, but some future benefit may occur if the helped animal "returns the favor." Although "cheaters" could obtain a large benefit, behavioral ecologists, using game theory, propose a *tit for tat* behavioral strategy. If cheating is immediately retaliated against, reciprocal altruism may evolve and persist in a population.

INTERACTIVE QUESTION 51.7

a. According to kin selection, would an individual be more likely to exhibit altruistic behavior toward a parent, a sibling, or a first (full) cousin?

b. Explain your answer in terms of the coefficient of relatedness and Hamilton's rule.

Evolution and Human Culture **Sociobiology** relates evolutionary theory to social behavior and to human culture. In his 1975 book, *Sociobiology,* E. O. Wilson speculated on the evolutionary basis of certain social behaviors of humans.

The parameters of human social behavior may be set by genetics, but the environment undoubtedly shapes behavioral traits just as it influences the expression of physical traits. Due to our capacity for learning, human behavior appears to be quite plastic. Our structured societies—with their codification of acceptable behaviors and the exclusion of some behaviors that might otherwise enhance an individual's fitness—may be a unique characteristic separating humans and other animals.

Word Roots

agon- = a contest (*agonistic behavior:* a type of behavior involving a contest of some kind that determines which competitor gains access to some resource, such as food or mates)

mono- = one; **-gamy** = reproduction (*monogamous:* a type of relationship in which one male mates with just one female)

poly- = many (*polygamous:* a type of relationship in which an individual of one sex mates with several individuals of the other sex)

socio- = a companion (*sociobiology:* the study of social behavior based on evolutionary theory)

Structure Your Knowledge

1. How does the nature-versus-nurture controversy apply to the study of behavior?
2. How does the concept of evolutionary fitness apply to all aspects of behavior?

Test Your Knowledge

MULTIPLE CHOICE: *Choose the one best answer.*

1. Behavioral ecology is the
   a. study of the behavior of animals, focusing on stimulus and response.
   b. application of human emotions and thoughts to other animals.
   c. study of animal cognition.
   d. study of animal behavior from an evolutionary perspective of fitness.
   e. behavioral study of ecology.

2. Proximate causes
   a. explain the evolutionary significance of a behavior.
   b. are immediate causes of behavior such as environmental stimuli.
   c. are environmental, whereas ultimate causes are genetic.
   d. are endogenous, although they may be set by exogenous cues.
   e. show that nature is more important than nurture.

3. The behavior that maximizes an animal's energy intake-to-expenditure ratio is called
   a. optimal foraging.
   b. Hamilton's rule.
   c. a fixed-action pattern.
   d. cognition.
   e. learning.

4. Which of the following descriptions is an example of a fixed-action pattern?
   a. a crane in a captive-breeding program imprinting on its human caregiver
   b. a male stickleback chasing a red-bellied object from its territory
   c. a blackcap migrating to its winter territory
   d. a songbird learning its song after listening to a tape of its species' song
   e. a digger wasp returning to its nest with the aid of landmarks

5. Which of the following descriptions is *not* an example of social learning?
   a. chimpanzees using stones to crack nuts
   b. mate choice copying in guppies
   c. alarm calls of vervet monkeys
   d. garter snakes from coastal areas eating slugs
   e. human culture

6. A sensitive period
   a. is the time right after birth.
   b. usually follows the reception of a sign stimulus.
   c. is a limited time in which imprinting can occur.
   d. is the period during which birds can learn to fly.
   e. is the time during which mate selection occurs.

7. In operant conditioning,
   a. an animal improves its performance of a fixed-action pattern.
   b. an animal learns as a result of associating a benefit or harm with an action.
   c. an animal learns a behavior by watching others.
   d. a bird can learn the song of a related species if it hears only that song.
   e. an irrelevant stimulus can elicit a response because of its association with a normal stimulus.

8. Which of the following types of intraspecies communication signal is best suited to a fruitfly's stimulus response chain during mating behavior?
   a. an auditory signal
   b. a visual signal
   c. a chemical signal
   d. a tactile signal
   e. all of the above

9. A female bird would most likely increase her fitness by
   a. mating with as many males as possible.
   b. being polygynous.
   c. reproducing only once in her lifetime.
   d. choosing a mate based on evidence that he has "good genes."
   e. always foraging in a large flock.

10. Which of the following examples of behavior provides evidence of animal cognition?
    a. a chimpanzee stacking up boxes to reach a banana overhead
    b. ravens pulling up string to obtain a food item attached to it
    c. trained honeybees that can discriminate between the concepts of "same" and "different" by matching colors or patterns
    d. a biology student using various resources to study for an exam
    e. All of the above show evidence of information processing and animal cognition.

11. In a species in which females provide all the needed food and protection for the young,
    a. males are likely to be promiscuous.
    b. mating systems are likely to be monogamous.
    c. mating systems are likely to be polyandrous.
    d. females will bond in social groups.
    e. females will have higher fitness than males.

12. A crow that aids its parents in raising its siblings is increasing its
    a. survival success.
    b. altruistic behavior.
    c. inclusive fitness.
    d. coefficient of relatedness.
    e. certainty of paternity.

13. Sociobiology
    a. explores the evolutionary basis of behavioral characteristics within animal societies.
    b. applies evolutionary explanations to human social behaviors.
    c. studies the roles of culture and genetics in human social behavior.
    d. considers communication, mating systems, and altruism from the viewpoint of fitness.
    e. does all of the above.

14. According to the concept of kin selection,
    a. an animal would be more likely to aid a stranger if the "kindness" could be reciprocated.
    b. an animal would aid its parent before it would help its sibling.
    c. animals are more likely to choose close relatives as mates.
    d. examples of altruism usually involve close relatives and increase an animal's inclusive fitness.
    e. evolution is the ultimate cause of animal behavior.

15. According to Hamilton's rule, natural selection would favor altruistic acts when
    a. the probability that the altruist will lose its life is less than 0.5 and the coefficient of relatedness is greater than 0.25 ($rC = B$).
    b. the cost to the altruist times the coefficient of relatedness is less than the benefit to the receiver ($rC < B$).
    c. the benefit to the receiver times the coefficient of relatedness is greater than the cost to the altruist ($rB > C$).
    d. the cost to the receiver times the coefficient of relatedness is greater than the cost to the altruist ($rC > B$).
    e. the benefit to the altruist times the coefficient of relatedness is less than the cost to the receiver ($rB < C$).

16. The cross-fostering of California mice in white-footed mice nests provides evidence for
    a. the genetic control of aggression and parenting behavior.
    b. the relationship between the distribution of vasopressin receptors and parenting behavior.
    c. an imprinting period during which behaviors related to aggression and parenting are set.
    d. the influence of the early social environment on the expression of aggressive and parental behaviors.
    e. the cognitive ability of mice to change their behavior in different environments.

Chapter 52

An Introduction to Ecology and the Biosphere

Key Concepts

52.1 Earth's climate varies by latitude and season and is changing rapidly

52.2 The structure and distribution of terrestrial biomes are controlled by climate and disturbance

52.3 Aquatic biomes are diverse and dynamic systems that cover most of Earth

52.4 Interactions between organisms and the environment limit the distribution of species

Framework

This chapter describes the scope of ecology, global climate patterns, the major terrestrial and aquatic biomes, and the factors that help determine the distribution of species.

Chapter Review

Ecology is the study of interactions of organisms with each other and with their environment. Although ecology had its foundation in the observation of nature, rigorous experimental designs are now commonly used to investigate complex ecological questions.

The **biosphere** includes all of Earth's ecosystems and landscapes. **Global ecology** is the study of the effect of regional energy and material exchanges on the distribution and functioning of organisms across the biosphere.

A **landscape** or seascape consists of several ecosystems in a region; **landscape ecology** is the study of the flow of energy, materials, and organisms across connecting ecosystems.

An **ecosystem** includes all the organisms in an area and the abiotic factors with which they interact, and **ecosystem ecology** addresses such topics as the flow of energy and chemical cycling.

A **community** includes the populations of different species in an area; **community ecology** looks at such interactions as predation and competition and how those interactions between species affect community structure.

Population ecology is concerned with the factors that affect the size of **populations,** which are groups of individuals of the same species occupying a particular area.

Organismal ecology, which may include the disciplines of behavioral, physiological, and evolutionary ecology, considers the responses and adaptations of an organism to its environment.

52.1   Earth's climate varies by latitude and season and is changing rapidly

The **climate,** or prevailing weather conditions of an area, is influenced by temperature, precipitation, sunlight, and wind. **Macroclimate** is the climatic pattern over an entire region; **microclimate** is the small-scale variations within a particular habitat.

Global Climate Patterns   The absorption of solar radiation heats the atmosphere, land, and water, setting patterns for temperature variations, air circulation, and water evaporation that cause latitudinal variations in climate. The shape of Earth and the tilting of its axis create seasonal variations in day length and temperature, which increase with latitude. The **tropics,** located between latitudes 23.5° north and 23.5° south, receive the greatest amount of and least variation in solar radiation.

The global circulation of air begins as intense solar radiation near the equator causes warm, moist air to rise and release its water, producing the characteristic wet tropical climate, the arid conditions around 30° north and south as dry air descends, the fairly wet though cool climate at about 60° latitude as air rises again, and the cold and rainless climates of the polar regions. Air flowing along Earth's surface produces global wind patterns, such as the cooling trade winds blowing from east to west in the tropics and the prevailing westerlies in temperate zones.

Regional and Local Effects on Climate   The changing angle of the sun over the year produces wet and dry seasons around 20° north and 20° south latitude. Seasonal wind pattern changes affect ocean currents, sometimes causing upwellings of cold, nutrient-rich water.

Coastal areas are generally wetter than inland areas, and large bodies of water moderate the climate. Water warmed at the equator flows toward the North Atlantic and warms the coast of western Europe as the Gulf Stream. A *Mediterranean climate* is created when cool, dry ocean breezes warm as they cross land, absorbing moisture and creating a hot, rainless climate.

INTERACTIVE QUESTION   52.1

Describe how mountains affect local climate with respect to the following three factors:

a. sunlight

b. temperature

c. rainfall

Microclimate   Climate also varies on a very small scale. Differences in **abiotic** features created by trees, open spaces, or even fallen logs or large stones affect the local distributions of organisms. **Biotic** factors, or living organisms, also influence the distribution and abundance of organisms.

Global Climate Change   Scientists predict that increasing concentrations of CO_2 and other greenhouse gases will warm the Earth 1–6°C by the end of the twenty-first century. Such climate change will have great effects on the distributions of plants and animals. Fossil pollen deposits have documented the rates of northward expansion of the ranges of various tree species following the last continental glaciers. As their geographic limits change with climatic warming, seed dispersal may not be rapid enough for some species of plants to move into new suitable habitat as their former ranges become inhospitable. Changes in distribution have already been documented for many plant and animal groups.

52.2   The structure and distribution of terrestrial biomes are controlled by climate and disturbance

Biomes are major types of ecosystems characterized by the predominant vegetation or (in aquatic biomes) by the physical environment.

Climate and Terrestrial Biomes   A **climograph** plots annual mean temperature and rainfall for a region; generally these values correlate with the distribution of various biomes. Overlaps of biomes on a climograph indicate the importance of variation in the seasonal patterns of rainfall and temperatures.

General Features of Terrestrial Biomes   Biomes are usually named for their predominant vegetation and major climatic features. Each biome also has characteristic microorganisms, fungi, and animals. The area of transition between biomes is called an **ecotone.**

Terrestrial biomes have vertical stratification, such as the layers in a forest from upper **canopy,** low-tree layer, shrub understory, herbaceous plant ground layer, forest floor (litter), to root layer. Vertical stratification of vegetation provides diverse habitats for animals.

Species composition of any one biome varies locally. Often, convergent evolution has produced a superficial resemblance of unrelated species.

Disturbance and Terrestrial Biomes   A **disturbance** is an event such as a fire, a storm, or human activity that can modify a community and alter resource availability. The patchiness of most biomes results from disturbances. Grasslands, savannas, chaparral, and many coniferous forests are maintained by the periodic disturbance of fire. Urban and agricultural communities now cover a large portion of Earth's land mass.

Exploring Terrestrial Biomes   Tropical forests occur in equatorial and subequatorial regions. Variations in rainfall result in **tropical dry forests,** where rainfall is seasonal, and **tropical rain forests,** where rainfall is more constant and abundant. Temperatures are uniformly high. The tropical rain forest has pronounced vertical stratification, including emergent trees above a closed canopy and abundant epiphytes. Animal diversity is higher than in any other terrestrial biome. Agriculture and development are destroying many tropical forests.

Characterized by low and variable precipitation, **deserts** occur around 30° north and south latitude and in the interior of continents. Deserts may be hot or cold, and temperature varies seasonally and daily. Plants may use C_4 and CAM photosynthesis and have reduced leaf surface area, feature different water storage adaptations, and produce protective spines and toxins. Desert animals have physiological and behavioral adaptations to dry conditions. Irrigated agriculture and urbanization are now common in deserts, reducing natural biodiversity.

Savannas are equatorial and subequatorial grasslands with scattered trees and long dry seasons. Fires are common, and the dominant grasses and forbs (small nonwoody plants) are fire-adapted, drought-tolerant, and able to withstand grazing. Large grazing mammals and their predators are common, although insects are the dominant herbivores. Cattle ranching and over-hunting have reduced the large-mammal populations of savannas.

Chaparral, common along coastlines in midlatitude continents, has cool, rainy winters and hot, dry summers. The dominant vegetation—evergreen shrubs and small trees—is maintained by and adapted to periodic fires. The high plant diversity present also includes many grasses and herbs. Browsing animals and small mammals are common. Urbanization and agriculture have reduced areas of chaparral.

Temperate grasslands are maintained by fire, seasonal drought, and grazing by large mammals. Winters are generally cold and dry; summers are hot and wet. Soils are deep and fertile, and most North American grasslands have been converted to farmland.

Characterized by broad-leaved deciduous trees, **temperate broadleaf forests** grow in midlatitude regions in the Northern Hemisphere, with smaller areas on other continents. Temperate broadleaf forests have distinct vertical layers. Winters are cold and summers hot and humid. During the cold winters, many mammals hibernate and birds migrate. Humans have heavily logged broadleaf forests, clearing land for agriculture and development.

The largest terrestrial biome, the **northern coniferous forest,** or *taiga,* is characterized by harsh winters with heavy snowfall and periodic droughts. Summers may be hot. Birds and large mammals are common animals. Old-growth stands of coniferous trees are rapidly being logged.

Tundra, covering large areas of the Arctic, is characterized by long, cold winters; short, cool summers; and dwarfed or matlike vegetation. A layer of frozen soil called permafrost restricts root growth. Migratory large mammals and birds are common. The *alpine tundra,* found at all latitudes on very high mountains, has similar plant communities. Mineral and oil extraction are becoming common in areas of arctic tundra.

INTERACTIVE QUESTION 52.2

Temperature and precipitation are two of the key factors that influence the vegetation found in a biome. On the following climograph, label the North American biomes (arctic and alpine tundra, northern coniferous forest, desert, temperate grassland, temperate broadleaf forest, and tropical forest) represented by each plotted area of temperature and precipitation.

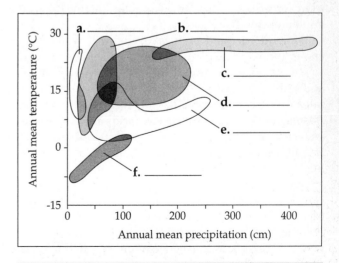

52.3   Aquatic biomes are diverse and dynamic systems that cover most of Earth

One of the chemical differences in aquatic biomes is salt concentration—less than 0.1% for freshwater biomes versus an average of 3% for marine biomes. Three-fourths of Earth is covered by oceans, which influence global rainfall, climate, and wind patterns. Marine algae and photosynthetic bacteria produce a large portion of the world's O_2 and consume enormous amounts of CO_2. Freshwater biomes are influenced by the surrounding terrestrial biome.

Zonation in Aquatic Biomes   Many aquatic biomes are layered both vertically and horizontally. The **photic zone** receives sufficient light for photosynthesis, whereas little light penetrates into the lower **aphotic zone.** The **pelagic zone** includes the waters of the photic and aphotic zone. The very deep ocean region is called the **abyssal zone.** The bottom substrate, called the **benthic zone,** is home to organisms collectively called **benthos.** Settling **detritus** (dead organic material) provides food for many benthic species.

In the ocean and in most lakes, a narrow layer of abrupt temperature change called a **thermocline** separates warmer surface waters from the cold bottom layer. In temperate lakes, seasonal temperature changes produce the **turnover** of water that sends oxygenated water to the bottom and brings nutrient-rich water to the surface.

Exploring Aquatic Biomes   **Oligotrophic lakes** are often deep, nutrient poor, and generally oxygen rich. The nutrient-rich waters of **eutrophic lakes** may be seasonally depleted of oxygen in deeper regions due to decomposition. Rooted and floating plants are found in the **littoral zone** close to shore; phytoplankton and cyanobacteria inhabit the **limnetic zone.** Heterotrophs include floating zooplankton, benthic invertebrates, and fishes. Dumping of municipal wastes and runoff from fertilized lands can cause algal blooms, oxygen depletion, and fish kills.

Wetlands, defined as areas covered with water often enough to support plants adapted to water-saturated soil, are among the most productive of biomes. Water and soils may be periodically oxygen-poor due to high rates of decomposition. Topography creates *basin, riverine,* and *fringe wetlands.* Wetlands filter nutrients and chemical pollutants and reduce flooding. Up to 90% of these richly diverse areas has been lost to draining and filling.

Streams and rivers are flowing habitats whose physical and chemical characteristics vary from the headwaters to the mouth. Oxygen levels are high in turbulently flowing water and low in murky, warm waters. Overhanging vegetation contributes to nutrient content. Diverse fishes and invertebrates inhabit the vertical zones of unpolluted rivers and streams. Human impact on streams and rivers includes pollution and damming.

Where a freshwater river meets the ocean, an **estuary** is formed. Salinity varies both spatially and daily with the rise and fall of tides. Salt marsh grasses and algae are the major producers. Estuaries serve as feeding and breeding areas for marine invertebrates, fish, and waterfowl. Pollution, filling, and dredging have extensively disrupted these highly productive biomes.

INTERACTIVE QUESTION   52.3

In the following table, indicate with a + or a − whether the listed biomes are relatively high or low in oxygen level, nutrient content, and productivity.

Biome	Oxygen Level	Nutrient Content	Productivity
Oligotrophic lake	a.		
Eutrophic lake	b.		
Headwater stream	c.		
Turbid river	d.		
Wetlands	e.		
Estuary	f.		

In **intertidal zones** the daily cycle of tides exposes the shoreline to variations in water, nutrients, and temperature, and to the mechanical force of wave action. Rocky intertidal communities are vertically stratified, with diverse marine algae and animals adapted to attaching firmly to the hard substrate. Sandy intertidal zones are home to burrowing worms, clams, and crustaceans. Oil pollutants have damaged many intertidal areas.

The water of the **oceanic pelagic biome** is typically nutrient poor but oxygen rich. Phytoplankton in the photic region are grazed on by numerous types of zooplankton. Free-swimming animals include squid, fishes, and marine mammals. Overfishing and waste dumping have damaged the Earth's oceans.

Found in the photic zone of clear tropical waters, **coral reefs** are highly diverse and productive biomes. The structure of the reef is produced by the calcium carbonate skeletons of the coral (various cnidarians) and serves as a substrate for red and green algae. The coral animals are nourished by symbiotic unicellular algae. Overfishing, coral collecting, pollution, and global warming are destroying coral reefs.

The **marine benthic zone** consists of the ocean floor. The shallow benthic areas of the **neritic zone** have diverse seaweeds and algae, as well as numerous invertebrates and fishes. The rest of the marine benthic zone receives no sunlight, and nutrients fall as detritus from the waters above. Inhabitants of the abyssal zone are adapted to cold and to high water pressure. Chemoautotrophic prokaryotes form the basis of a collection of organisms adapted to the hot, low-oxygen environment surrounding **deep-sea hydrothermal vents.** Overfishing has eliminated many benthic fish populations.

INTERACTIVE QUESTION 52.4

Different marine environments can be classified on the basis of light penetration, distance from shore, and open water vs. bottom. Match the following zones to their corresponding numbers on the diagram:

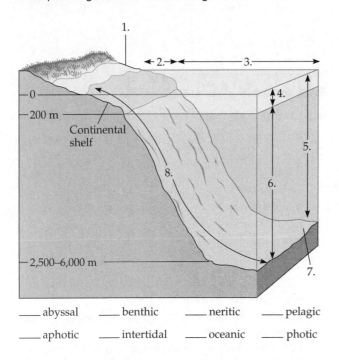

_____ abyssal   _____ benthic   _____ neritic   _____ pelagic

_____ aphotic   _____ intertidal   _____ oceanic   _____ photic

---

52.4 Interactions between organisms and the environment limit the distribution of species

Interactions between organisms and their environments occur within *ecological time.* The cumulative effects of these interactions are seen in the adaptation of organisms to their environment over an *evolutionary time* frame. The distribution of species reflects these ecological and evolutionary interactions with both biotic and abiotic factors. Ecologists often work through a series of four logical questions to determine what factors limits the geographic distribution of a species.

Dispersal and Distribution   **Dispersal** is the movement of individuals away from their original area. The cattle egret is an example of a species that has naturally extended its range through dispersal. Occasionally, such long-distance dispersal can lead to the adaptive radiation of many new species.

Transplants of a species can indicate whether dispersal limits its distribution. A successful transplant shows that the *potential* range of a species is larger than its *actual* range. Species that are introduced to new areas, either purposely or accidentally, often disrupt their new ecosystem.

Behavior and Habitat Selection   Sometimes habitat selection behaviors keep organisms from occupying all of their potential range. Habitat selection by ovipositing insects, which often choose only certain plants, may limit the insects' distribution to the locations of their host plants.

Biotic Factors   The distribution of a species may be limited by predation, disease, parasitism, competition, or lack of mutual symbiosis. "Removal and addition" experiments test whether predators or herbivores limit the distribution of prey species or plants. Sea urchins were shown to limit the abundance and distribution of seaweeds.

Abiotic Factors   A species' geographic distribution may be influenced by regional differences in abiotic factors.

Temperature is an important environmental factor because of its effects on metabolism. Most organisms function best within a narrow range of environmental temperatures.

The availability of water in different habitats can vary greatly. The distribution of organisms reflects their ability to obtain sufficient water and avoid desiccation in terrestrial habitats. Oxygen concentrations can be low in deep waters, in sediments rich in organic matter, and in flooded soils.

The ability of organisms to osmoregulate determines their distribution in freshwater, saltwater, or other high-salinity habitats.

Light energy drives almost all ecosystems. The quantity of sunlight is a limiting factor in aquatic environments and on forest floors. Excessive light intensity at high elevations and in deserts also limits the survival of plants and animals.

Soils, which vary in their physical structure, pH, and mineral composition, affect the distribution of plants and in turn, the distribution of animals. Substrate composition in aquatic environments influences water chemistry and the types of organisms that can inhabit those areas.

INTERACTIVE QUESTION   52.5

List and give examples of the four factors that ecologists examine to understand the geographic distribution of a species.

a.

b.

c.

d.

Word Roots

a- = without; **bio-** = life (*abiotic:* nonliving; referring to physical and chemical properties of an environment)

abyss- = deep, bottomless (*abyssal zone:* the part of the ocean's benthic zone between 2,000 m and 6,000 m deep)

bentho- = the depths of the sea (*benthic zone:* the bottom surface of an aquatic environment)

estuar- = the sea (*estuary:* the area where a freshwater stream or river merges with the ocean)

eu- = good, well; **troph-** = food, nourishment (*eutrophic lake:* a lake with a high rate of productivity and nutrient cycling)

hydro- = water; **therm-** = heat (*deep-sea hydrothermal vent:* a dark, hot, oxygen-deficient environment associated with volcanic activity near the seafloor; the producer organisms are chemoautotrophic prokaryotes)

inter- = between (*intertidal zone:* the shallow zone of ocean adjacent to land and between the high- and low-tide lines)

limn- = a lake (*limnetic zone:* the well-lit, open surface waters of a lake far from shore)

littor- = the seashore (*littoral zone:* the shallow, well-lit waters of a lake close to shore)

micro- = small (*microclimate:* very fine-scale patterns of climate, such as the specific climatic conditions underneath a log)

oligo- = small, scant (*oligotrophic lake:* a nutrient-poor, clear lake with few phytoplankton)

pelag- = the sea (*pelagic zone:* the open-water component of aquatic biomes)

-photo = light (*aphotic zone:* the part of an ocean or lake beneath the photic zone, where light does not penetrate sufficiently for photosynthesis to occur)

thermo- = heat; **-clin** = slope (*thermocline:* a narrow stratum of abrupt temperature change in the ocean and in many temperate-zone lakes)

Structure Your Knowledge

1. a. Define ecology.
   b. How does ecology relate to evolutionary biology?
2. a. What are biomes?
   b. What accounts for the similarities in life forms found in the same type of biome in geographically separated areas?

Test Your Knowledge

MULTIPLE CHOICE: *Choose the one best answer.*

1. Which level of ecology considers the effects of predation, parasitism, and competition on species distribution?
   a. landscape
   b. community
   c. ecosystem
   d. organismal
   e. population

2. Ecologists often use mathematical models and computer simulations because
   a. ecological experiments are always too broad in scope to be performed.
   b. most ecologists are mathematicians.
   c. ecology is becoming a more descriptive science.

d. these approaches allow ecologists to study the interactions of multiple variables and to simulate large-scale experiments.

e. variables can be manipulated with computers but cannot be manipulated in field experiments.

3. The ample rainfall of the tropics and the arid areas around 30° north and south latitudes are caused by

a. ocean currents that flow clockwise in the northern hemisphere and counterclockwise in the southern hemisphere.

b. the global circulation of air initiated by intense solar radiation near the equator that produces wet and warm air.

c. the tilting of Earth on its axis and the resulting seasonal changes in climate.

d. the heavier rain on the windward side of mountain ranges and the drier climate on the leeward side.

e. the location of tropical rain forests and deserts.

4. Which of the following statements reflects a concern about the effects of global climate change on tree species?

a. The increased ozone levels may damage leaf cells, reducing photosynthetic rates.

b. Trees may not be able to disperse fast enough to reach new habitats that meet their climatic requirements.

c. Warmer temperatures may speed tree growth, producing trees that are too tall and spindly.

d. The additional CO_2 in the atmosphere may actually increase photosynthetic rates and prove beneficial to tree growth.

e. All of the above are concerns.

5. Why do the tropics and the windward side of mountains receive more rainfall than areas around 30° latitude or the leeward side of mountains?

a. Rising air expands, cools, and drops its moisture.

b. Descending air condenses and drops its moisture.

c. The tropics and the windward side of mountains are closer to the ocean.

d. Solar radiation is greater in the tropics and on the windward side of mountains.

e. The rotation of Earth determines global wind patterns.

6. The permafrost of the arctic tundra

a. prevents plants from getting established and growing.

b. protects small animals during the long winters.

c. anchors plant roots in the frozen soil, helping them withstand the area's high winds.

d. prevents plant roots from penetrating deep into the soil.

e. Both b and c are correct.

7. Many plant species have adaptations for dealing with the periodic fires typical of a

a. savanna.

b. chaparral.

c. temperate grassland.

d. temperate broadleaf forest.

e. a, b, and c.

8. The best way to explain how two communities can have the same annual mean temperature and rainfall but very different biota and characteristics is that the communities

a. are found at different altitudes.

b. are composed of species that have very low dispersal rates.

c. are found on different continents.

d. receive different amounts of sunlight.

e. have different seasonal temperatures and patterns of rainfall throughout the year.

9. In which of the following biomes is sunlight most likely to be a limiting factor?

a. desert

b. estuary

c. ocean pelagic zone

d. grassland

e. coral reef

10. Which of the following aquatic zones is *incorrectly* paired with its description?

a. neritic zone—shallow region of ocean over the continental shelf

b. abyssal zone—very deep benthic region of the ocean

c. limnetic zone—shallow waters close to shore in a lake

d. intertidal zone—shallow area of the ocean adjacent to land

e. pelagic zone—area of open water

11. In which of the following biomes would you expect to find organisms with adaptations for tolerating changes in salinity?

a. desert

b. wetland

c. deep-sea hydrothermal vent

d. estuary

e. intertidal zone

12. Phytoplankton are the basis of the food chain in

a. the headwaters of streams.

b. wetlands.

c. the oceanic photic zone.

d. rocky intertidal zones.

e. deep-sea hydrothermal vents.

13. Upwellings in oceans
    a. support reef communities.
    b. occur over deep-sea hydrothermal vents.
    c. are responsible for ocean currents.
    d. bring nutrient-rich water to the surface.
    e. are most common in tropical waters, where they bring oxygen-rich water to the surface.

14. Which of the following factors affect the distribution of a species?
    a. dispersal ability
    b. interactions with mutualistic symbionts
    c. climate and physical factors of the environment
    d. behavior, such as habitat selection
    e. All of the above factors influence where species are found.

15. An experiment that removed a species from an area would enable an ecologist to test
    a. whether moving the species to a new, varied habitat could lead to adaptive radiation.
    b. whether competition or predation from that species was limiting the local distribution of other species.
    c. whether dispersal ability was limiting the range of the species.
    d. whether abiotic factors were limiting the dispersal of the species.
    e. whether behavior or habitat selection had limited the distribution of the species.

MATCHING: *Match the biotic description with its biome.*

Biome	Biotic Description
_____ 1. chaparral	A. broad-leaved deciduous trees
_____ 2. desert	B. lush growth, vertical layers
_____ 3. savanna	C. evergreen shrubs, fire-adapted vegetation
_____ 4. coniferous forest	D. scattered trees, grasses, and forbs
_____ 5. temperate broadleaf forest	E. tall stands of conebearing trees
_____ 6. temperate grassland	F. low shrubby or matlike vegetation, lichens
_____ 7. tropical rain forest	G. grasses adapted to fire and drought
_____ 8. tundra	H. widely scattered shrubs, cacti, and euphorbs

Chapter 53

Population Ecology

53.1 Dynamic biological processes influence population density, dispersion, and demographics

53.2 The exponential model describes population growth in an idealized, unlimited environment

53.3 The logistic model describes how a population grows more slowly as it nears its carrying capacity

53.4 Life history traits are products of natural selection

53.5 Many factors that regulate population growth are density dependent

53.6 The human population is no longer growing exponentially but is still increasing rapidly

Chapter Review

Population ecology is the study of the influence of the abiotic and biotic environment on population size, density, and distribution.

53.1 Dynamic biological processes influence population density, dispersion, and demographics

A **population** is a localized group of individuals of the same species. Members of a population use the same resources and have a high probability of interacting and breeding with each other. Populations may be described by their size and geographic boundaries; ecologists define such boundaries based on the type of organism and the research question being asked.

Density and Dispersion The number of individuals per unit area or volume is a population's **density;** the pattern of spacing of those individuals is referred to as **dispersion.**

Scientists often use one of a variety of sampling techniques to estimate population density. Indirect

Framework

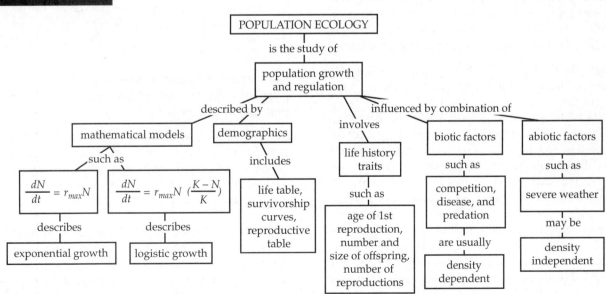

POPULATION ECOLOGY
is the study of
population growth and regulation

described by — mathematical models
such as

$$\frac{dN}{dt} = r_{max}N$$
describes
exponential growth

$$\frac{dN}{dt} = r_{max}N \left(\frac{K-N}{K}\right)$$
describes
logistic growth

demographics
includes
life table, survivorship curves, reproductive table

involves — life history traits
such as
age of 1st reproduction, number and size of offspring, number of reproductions

influenced by combination of

biotic factors
such as
competition, disease, and predation
are usually
density dependent

abiotic factors
such as
severe weather
may be
density independent

indicators, such as burrows or nests, also may be used. In the **mark-recapture method,** a population is sampled twice. First, animals are trapped and marked (or identified), and released. Animals are then captured or sampled again. The proportion of the total second trapping that is marked indicates the proportion of the total population that was initially marked, The calculation uses the equation $x/n = s/N$ (where x is recaptured marked animals; n is number in second sampling; s is the number of animals initially marked and released; and N is estimated population size). Solving for $N = sn/x$.

INTERACTIVE QUESTION   53.1

In a mark-recapture study, an ecologist traps, marks, and releases 25 voles in a small wooded area. A week later she resets her traps and captures 30 voles, 10 of which are marked. What is her estimate of the vole population in that area?

Changes in population density reflect both additions of members through birth (including all forms of reproduction) and **immigration,** and removal of members through death and **emigration.**

Individuals may be dispersed in the population's geographic range in several patterns. *Clumping* may indicate a heterogeneous environment or social interactions between individuals.

Uniform distribution may be related to competition for resources and result from interactions between individuals. **Territoriality,** the defense of a bounded physical space, can lead to uniform dispersion. *Random* spacing, indicating the absence of strong interactions among individuals or a fairly uniform habitat, is less common.

Demographics   The study of the vital statistics of a population, such as birth and death rates, is called **demography.**

A **life table** presents age-specific survival data for a population. It can be constructed by following a **cohort** of organisms from birth to death and calculating the proportion of the cohort surviving in each age group.

A **survivorship curve** shows the number or proportion of members of a cohort still alive at each age. Survivorship curves are often based on a convenient beginning cohort of 1,000 individuals and typically use a logarithmic scale on the y-axis and a relative scale on the x-axis, so that species with different life spans can be compared on the same graph. There are

three general types of survivorship curves. Type I has low mortality during early and middle age and a rapid increase with old age. In a Type II curve, death rate is relatively constant throughout the life span. A Type III curve has high initial mortality, with the few offspring that survive likely to reach adulthood. Many species show intermediate or more complex survivorship patterns.

INTERACTIVE QUESTION   53.2

Identify the types of survivorship curves in the following graph. For each type of curve, describe the characteristics of representative species and then cite some examples of those species.

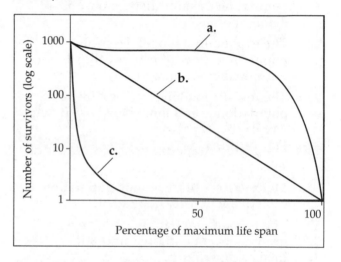

In sexually reproducing species, demographers usually follow only the reproduction of females through the generations. A **reproductive table,** or fertility schedule, gives the age-specific reproductive rates in a population.

53.2   The exponential model describes population growth in an idealized, unlimited environment

Per Capita Rate of Increase   A small population in a very favorable environment may exhibit unrestricted growth. Ignoring immigration and emigration, the change in population size during a specific time period is equal to the number of births minus the number of deaths.

Births and deaths can be expressed in terms of the average number per individual during a time period—that is, as a *per capita birth rate (b)* and a *per capita death rate (m,* for mortality). These rates can be calculated from estimates of population size (N) and from data in life tables and reproductive tables. The population growth equation using per capita birth and death rates becomes $\Delta N/\Delta t = bN - mN$. The *per capita rate of increase (r)* is the difference between the per capita birth and death rates: $r = b - m$. **Zero population growth (ZPG)** occurs when $r = 0$. The equation describing the change in the population at a particular instant in time uses differential calculus and is written as $dN/dt = r_{inst}N$.

Exponential Growth Under ideal conditions, a population may exhibit **exponential population growth,** or geometric population growth. Expressed as $dN/dt = r_{max}N$, exponential population growth produces a J-shaped growth curve when graphed. The larger the population (N) becomes, the faster the population grows. Periods of exponential growth may occur in some populations that exploit an unfilled environment or rebound from a catastrophic event.

53.3   The logistic model describes how a population grows more slowly as it nears its carrying capacity

A population may grow exponentially for only a short time before its increased density limits the resources available to its members. The **carrying capacity** (K) is the maximum population size that a particular environment can support. Crowding and resource limitations may lead to decreased per capita birth rates and increased per capita death rates.

The Logistic Growth Model The per capita rate of increase decreases from its maximum at low population size to zero as carrying capacity is reached. The mathematical model of **logistic population growth** is $dN/dt = r_{max}N(K-N)/K$. The equation includes the expression $(K-N)/K$ to reflect the impact of increasing N on the per capita rate of increase as the population approaches the carrying capacity.

When N is small, $(K-N)/K$ is close to 1, and growth is approximately exponential ($r_{max}N$). As population size approaches the carrying capacity, the $(K-N)/K$ term becomes a smaller fraction, and per capita rate of increase is small. When N reaches K, the term $(K-N)/K$ is 0, and the population stops growing. The logistic

model produces an S-shaped growth curve, and maximum increase in population numbers occurs when N is intermediate in size.

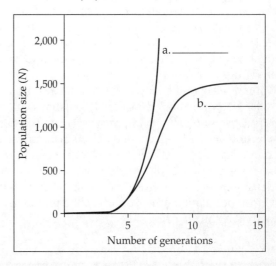

The Logistic Model and Real Populations Some laboratory populations of small animals and microorganisms show logistic growth. Natural populations often fluctuate around carrying capacity.

The logistic model makes the assumption that any increase in population size will have a negative effect on population growth. The *Allee effect* is seen in some populations, however, when individuals have a harder time reproducing if the population size is too small.

53.4   Life history traits are products of natural selection

The **life history** of an organism from birth through reproduction to death reflects evolutionary tradeoffs between survival and reproduction. Life history traits include the age at first reproduction, how often an organism reproduces, and the number of offspring produced in each reproductive episode.

Evolution and Life History Diversity Some species put all their reproductive resources into a single effort, called **semelparity,** or big-bang reproduction. Other species follow the strategy of **iteroparity,** making repeated reproductive efforts over a span of time. Two factors thought to be important in the selection for and evolution of these two reproductive strategies are the survival rate of young offspring and the probability that an adult will survive to reproduce more than once.

"Tradeoffs" and Life Histories   Because organisms have a finite energy budget, they cannot maximize all life history traits simultaneously. The costs of reproduction often include a reduction in survival. The production of large numbers of offspring is related to the selective pressures of high mortality rates of offspring in uncertain environments or from intense predation. Parental investments in the physical size of offspring (or seeds) and extended care increase the survival chances of offspring.

Natural selection will favor different life history traits at different population densities. Populations at high density, close to their carrying capacity, may experience *K*-selection or density-dependent selection for traits such as competitive ability and efficient resource use. In environments in which population density is low, *r*-selection (or density-independent selection) would favor traits that maximize population growth, such as the production of numerous, small offspring.

INTERACTIVE QUESTION   53.5

a. Explain why the life history of an organism cannot be to reproduce early and often, to have large numbers of offspring at a time, and to live a long life.

b. Indicate whether the following life history traits would be considered to be *r*-selected or *K*-selected:

early age at first reproduction; many small offspring produced

few, relatively large offspring produced every year

53.5   Many factors that regulate population growth are density dependent

Population Change and Population Density   A birth rate or a death rate that does not change as population density changes is said to be **density independent.** The death rate is **density dependent** if it rises with increasing population density; the birth rate is density dependent if it falls with increasing population density. An equilibrium density may be reached in a population as long as birth rate or death rate or both are density dependent.

Mechanisms of Density-Dependent Population Regulation   Density-dependent decreases in birth rate and increases in death rate may regulate populations through negative feedback.

Competition for resources such as nutrients or food may limit reproductive output. The availability of territorial space may be the limiting resource for some animals. Increased population densities may affect the transmission rate of diseases in both plant and animal populations. Predation may be a density-dependent factor when a predator feeds preferentially on a prey population that has reached a high density. The accumulation of toxic metabolic wastes may also be a limiting factor for some organisms.

Intrinsic factors may also regulate population size. Studies of mice have shown that even when food or shelter is not limiting, aggressive interactions and hormonal changes inhibit reproduction and increase mortality.

Population Dynamics   All populations show fluctuations in numbers. **Population dynamics** is the study of these variations in population size and the biotic and abiotic factors that cause them. Fluctuations in populations of some large mammals, such as Soay sheep and moose on Isle Royale, may be linked to the severity of winter, the availability of food, or predation.

The 10-year cycles in the densities of snowshoe hares and lynx in northern forests have been studied to determine whether winter food shortages, predator overexploitation, or sunspot activity causes the cyclic collapses in hare populations. The decrease in ozone and increase in UV radiation associated with low sunspot activity result in plants producing more sunscreen chemicals and fewer herbivore-discouraging chemicals. Experimentally increasing the food supply raised the carrying capacity for hares, but the density cycles continued to occur in both experimental and control areas. Field ecologists determined that 90% of hare deaths are due to predation. Experiments that excluded predators from one area, and both excluded predators and added food to another area, support the hypothesis that it is mainly excessive predation that drives hare cycles, but available winter food supply is a contributing influence. Population peaks followed low sunspot activity, suggesting that sunspot activity also regulates hare populations.

Predator cycles most likely follow the population cycles of their prey, and these cycles may be accentuated when predators begin preying on one another when their usual prey becomes scarce.

Emigration may increase with population density. The aggregation of single-celled amoeba into a multicellular slug improved the ability of these protists to reach new food sources.

A group of populations occupying separated suitable habitats may be linked through immigration and emigration to form a **metapopulation.** Emigrants may serve to repopulate habitats in which local populations become extinct.

INTERACTIVE QUESTION 53.6

a. List some density-dependent factors that may limit population growth.

b. List some abiotic factors that may cause population fluctuations.

53.6   The human population is no longer growing exponentially but is still increasing rapidly

The Global Human Population   After 1650, it took 200 years for the global population to double to 1 billion. In the next 80 years, it doubled to 2 billion; it doubled again to 4 billion in the next 45 years; and now, some 35 years later, it is more than 6.8 billion. The rate of growth has begun to slow, partly due to diseases such as AIDS and voluntary population control, although population ecologists project a population of 7.8–10.8 billion by 2050.

Population stability can be reached in one of two ways: ZPG = high birth rates − high death rates, or low birth rates − low death rates. The movement from the first configuration to the second is called the **demographic transition.** Death rates declined rapidly in most developing countries after 1950, but birth rate decline has been more uneven—from rapid in China to slower in sub-Saharan Africa and India. The world's annual population growth rate is regionally variable; it is near equilibrium (0.1%) in industrialized nations, with reproductive rate below the replacement level of 2.1 children per female in some countries. Less industrialized countries, where 80% of the world's population lives, contribute the most to the global growth rate of 1.2%.

Human population growth is unique in that it can be consciously controlled by voluntary contraception and family planning. The key to the demographic transition is reduced family size. In many cultures, women are receiving more education and delaying marriage and reproduction, thus slowing population growth.

The **age structure** of a population affects present and future growth and can be predictive of future social conditions and needs.

Infant mortality and *life expectancy at birth* vary among human populations, with mortality being much higher and life expectancy lower in less industrialized countries.

INTERACTIVE QUESTION 53.7

What do the age structure pyramids in Figure 53.24 predict for the future population growth of Afghanistan and Italy?

Global Carrying Capacity   Estimates of Earth's carrying capacity have varied greatly and have averaged about 10–15 billion. These estimates involve different assumptions, such as the logistic equation, the amount of habitable land, or food as the limiting factor.

The concept of an **ecological footprint** takes into account multiple human needs and the land and water area required to meet those needs and absorb wastes. The amount of ecologically productive land area per person on Earth is estimated to be about 2 hectares (ha). Land for parks and conservation reduces this estimate to 1.7 ha per person. In the United States, a typical person, with an ecological footprint of about 10 ha, is using an unsustainable share of Earth's resources.

Ecological footprints may also be based on the amount of energy used. A typical person in the United States consumes roughly 30 times the energy consumed by an individual in central Africa.

The ultimate carrying capacity of Earth may be determined by food and water supplies, the availability of space or nonrenewable resources, degradation of the environment, or several interacting factors. When and how we reach zero population growth is an issue of great social and ecological consequence.

Word Roots

co- = together (*cohort*: a group of individuals of the same age in a population)

demo- = people; **-graphy** = writing (*demography*: the study of statistics relating to births and deaths in populations)

itero- = to repeat; **-parity** = to beget (*iteroparity*: type of reproduction in which adults produce offspring over many years; also known as repeated reproduction)

semel- = once (*semelparity*: type of reproduction in which an organism produces all of its offspring in a single event; also known as big-bang reproduction)

Structure Your Knowledge

1. Create a concept map to organize your understanding of the exponential and logistic equations—the mathematical models of population growth.

2. What is the best collection of life history traits that would maximize reproductive success?

Test Your Knowledge

MULTIPLE CHOICE: *Choose the one best answer.*

1. In an area with a heterogeneous distribution of suitable habitats, the dispersion pattern of a population is probably
   a. clumped.
   b. uniform.
   c. random.
   d. unpredictable.
   e. dense.

2. Which of the following statements about life tables is *not* true?
   a. They were first used by life insurance companies to estimate survival patterns.
   b. They show the age-specific death rate for a population.
   c. They are used to predict logistic growth.
   d. They can be used to construct survivorship curves.
   e. They are often constructed by following a cohort from birth to death.

3. A Type I survivorship curve is level at first, with a rapid increase in mortality in old age. This type of curve is
   a. typical of many invertebrates that produce large numbers of offspring.
   b. typical of humans and other large mammals.
   c. found most often in *r*-selected populations.
   d. almost never found in nature.
   e. typical of most species of birds.

4. The middle of the S-shaped growth curve in the logistic growth model
   a. shows that at middle densities, individuals of a population do not affect each other.
   b. is best described by the term rN.
   c. is the point where population growth begins to slow due to the Allee effect.
   d. is the period when competition for resources is the highest.
   e. is the period when the population growth rate is the highest.

5. The term $(K - N)/K$
   a. is the carrying capacity for a population.
   b. is greatest when K is very large.
   c. is zero when population size equals carrying capacity.
   d. increases in value as N approaches K.
   e. accounts for the overshoot of carrying capacity.

6. The carrying capacity for a population is estimated at 500; the population size is currently 400; and r_{max} is 0.1. What is dN/dt?
   a. 0.01
   b. 0.8
   c. 8
   d. 40
   e. 50

7. In order to maintain the largest sustainable fish harvest, fishing efforts should
   a. take only postreproductive fish.
   b. maintain the population close to its carrying capacity.
   c. reduce the population to a very low number to take advantage of exponential growth.
   d. maintain the population density close to $\frac{1}{2} K$.
   e. be prohibited.

8. Immigration and emigration are likely to play a role in population dynamics in
   a. metapopulations.
   b. exponential growth.
   c. demographic transition.
   d. big-bang reproduction.
   e. territoriality.

9. In a population in which offspring survival is quite low and the environment is inconsistent, one might expect
   a. the production of a small number of large offspring.
   b. the production of a large number of large offspring.
   c. iteroparity or repeated reproduction with a small number of offspring.
   d. semelparity.
   e. more *K*-selected traits.

10. Which of the following factors is *not* a density-dependent factor limiting a population's growth?
   a. increased specialization by a predator
   b. a limited number of available nesting sites
   c. a stress syndrome that alters hormone levels
   d. a very early fall frost
   e. intraspecific competition

11. A few members of a population have reached a favorable habitat containing few predators and unlimited resources, but their population growth rate is slower than that of the parent population. A possible explanation for this is that
    a. the genetic makeup of these founders may be less favorable than that of the parent population.
    b. the parent population may still be in the exponential part of its growth curve and not yet limited by density-dependent factors.
    c. the Allee effect may be operating; there are not enough population members present for successful survival and reproduction.
    d. a density-independent factor such as a severe storm has limited their growth.
    e. all of the above may apply.

In questions 12–16, use the following choices to indicate how these factors would be affected by the described changes.
    a. an increase
    b. a decrease
    c. no change
    d. the effect cannot be predicted

12. For a population regulated by density-dependent factors, what change would occur in clutch size or seed crop size with increased population density?

13. In the study of European kestrels described in the textbook (Figure 53.13), what effect did reducing the brood size by transferring chicks have on the survivorship of the parents in the following winter?

14. In a population showing exponential growth, what change would occur in dN/dt with an increase in N?

15. As a population approaches zero population growth, what change would occur in the per capita rate of increase (r)?

16. For an r-selected population regulated primarily by density-independent factors, what change would occur in the population growth rate if the carrying capacity of its environment were increased?

17. Experimental studies of the population cycles of the snowshoe hare and the lynx have shown that
    a. the hare population is regulated by its food resources because adding food increased the carrying capacity of experimental areas.
    b. hares are as likely to die of starvation as of predation.
    c. lynx are the only predators of hares, and that increases in the lynx population cause the cycles in the hare populations.
    d. the stress of overcrowding causes the population cycles in both hare and lynx.
    e. the hare population is regulated by a combination of predators (not just the lynx) and by the effect of sunspot activity on food quality; the lynx population appears to cycle in response to the availability of its prey.

18. The human population is growing at such a fast rate because
    a. the age structure of many countries is highly skewed toward younger ages.
    b. the death rate has greatly decreased since the Industrial Revolution.
    c. technology has increased Earth's carrying capacity.
    d. fertility rates in many developing countries are above the 2.1 children per female replacement level.
    e. all of the above are true.

19. The demographic transition is the gradual shift from
    a. a Type III survivorship curve to a Type I curve.
    b. iteroparity to semelparity.
    c. an age structure skewed toward the younger ages to an even age distribution.
    d. high birth rates and high death rates to low birth rates and low death rates.
    e. exponential growth to logistic growth.

20. An ecological footprint is an estimate of
    a. the carrying capacity of a nation.
    b. the number of offspring an adult produces and the resulting demand on resources.
    c. the land and water area needed per person to meet the current demand on resources.
    d. the size of a population in relationship to the resources it uses.
    e. how much energy is used to produce food for a vegetarian versus a meat eater.

Community Ecology

■ Key Concepts

54.1 Community interactions are classified by whether they help, harm, or have no effect on the species involved

54.2 Diversity and trophic structure characterize biological communities

54.3 Disturbance influences species diversity and composition

54.4 Biogeographic factors affect community diversity

54.5 Pathogens alter community structure locally and globally

■ Framework

Communities are composed of populations of various species that may interact through competition, predation, herbivory, symbiosis, or facilitation. The structure of a community—its species composition and relative abundance—is determined by these interactions between species. Disturbances keep most communities in a state of nonequilibrium. Species diversity relates to community size and geographic location. Identifying the hosts and vectors of pathogens helps to combat zoonotic diseases.

■ Chapter Review

The collection of populations of different species living close enough to interact is called a biological **community.** Community ecologists study the interactions between species and the factors involved in determining a community's structure— the composition and the relative abundance of its species.

54.1 Community interactions are classified by whether they help, harm, or have no effect on the species involved

Interspecific interactions occur between the different species living in a community. The positive or negative effect of these interactions on the survival and reproduction of a population can be signified by + or − signs.

Competition If populations of two species use the same limited resource, **interspecific competition** may negatively affect one or both populations.

G. F. Gause's laboratory experiments showed that two species of *Paramecium* that rely on the same limited resource could not coexist in the same community. The **competitive exclusion** principle predicts that the less efficient competitor will be eliminated locally.

An organism's **ecological niche** is described as its role in an ecosystem—its use of biotic and abiotic resources. The competitive exclusion principle holds that two species with identical niches cannot coexist permanently in a community.

Resource partitioning, slight variations in niche that allow ecologically similar species to coexist, provides indirect evidence that competition was a selection factor in the evolution of a species' niche. Due to competition, a species' *realized niche* might be smaller than its *fundamental niche*. **Character displacement** of some morphological trait or in resource use enables closely related sympatric species to avoid competition. When these species are allopatric (geographically separate), their differences may be much smaller.

Whenever these two mouse species coexist, *Acomys cahirinus* is nocturnal, whereas *A. russatus* is active during the day. When all *A. cahirinus* were removed from a research site, diurnal *A. russatus* mice became nocturnal. How were resources partitioned between these species when they coexisted, and what do these results indicate about the niche of *A. russatus*?

Predation   **Predation** involves a predator killing and eating prey. Predator adaptations include acute senses, speed and agility, camouflage coloration, and physical structures such as claws, fangs, teeth, and stingers.

Animals can defend against predation by hiding, fleeing, or forming herds or schools. Potential prey may use camouflage in the form of **cryptic coloration.** Mechanical and chemical defenses discourage predation. Some animals passively accumulate toxic compounds from the food they eat; others may synthesize their own toxins. Bright, conspicuous, **aposematic coloration** in prey species "warns" predators that the prey possesses chemical defenses.

Prey may use mimicry to exploit the warning coloration of other species. Predators may use mimicry to "bait" their prey.

Name the type of mimicry described in each of the following descriptions:

a. harmless species resembling a poisonous or distasteful species

b. mutual imitation by two or more unpalatable species

Herbivory   In **herbivory,** an herbivore eats parts of a plant or alga. Most herbivores are small invertebrates such as insects, which may have chemical sensors that recognize nontoxic and nutritious plants. Herbivores may have teeth or digestive systems adapted for processing vegetation.

Plants may defend themselves with mechanical devices, such as thorns, or chemical compounds.

Distasteful or toxic chemicals include such well-known compounds as strychnine, nicotine, tannins, and various spices.

Symbiosis   **Symbiosis** may be defined as the relationship between organisms of two species that live in direct contact.

In **parasitism,** a **parasite** obtains its nourishment from its **host.** Parasites that live within a host are called **endoparasites;** those that feed on the surface of a host are called **ectoparasites.** Parasitoid insects lay eggs on or in hosts, on which their larvae then feed.

Parasites may have complex life cycles with a number of hosts. Parasites can have a substantial effect on the density of their host population.

In **mutualism,** interactions between species benefit both participants. In *obligate mutualism,* at least one species is unable to survive on its own; in *facultative mutualism,* both species can survive alone. Mutualistic interactions may involve the coevolution of related adaptations in both species.

In **commensalism,** only one member appears to benefit from the interaction. Examples include "hitchhiking" species and species that feed on food incidentally exposed by another species.

Facilitation   In **facilitation,** a species positively affects the survival and reproduction of other species without the intimate association of symbiosis. This type of interaction is common in plant ecology, in which one plant species improves the soil in ways that benefit other species.

Name and give examples of the interspecific interactions symbolized in the table.

	Interspecific Interaction	Examples
$-/-$	a.	
$+/+$	b.	
$+/0$	c.	
$+/-$	d.	
$+/-$	e.	
$+/-$	f.	
$+/+$ or $0/+$	g.	

54.2   Diversity and trophic structure characterize biological communities

Species Diversity   The **species diversity** of a community is determined both by **species richness**, the number of different species present, and by **relative abundance,** the proportional abundance of the different species.

A widely used quantitative diversity index is **Shannon diversity** (*H*), which is based on species richness and relative abundance:

$$H = - (p_A \ln p_A + p_B \ln p_B + p_C \ln p_C + \cdots)$$

where *p* is the proportion of each species (A, B, C, and so on) in the community, and ln is the natural log. Thus Shannon diversity (*H*) for a community is the negative sum of *p* ln *p* for all species present.

INTERACTIVE QUESTION   54.4

Tide pool 1 has three species of sea urchins with the following numbers: A = 8, B = 6, C = 6; tide pool 2 has four species of sea urchins with the following numbers: A = 14, B = 2, C = 2, D = 2.

a. Compute the Shannon diversity index of the sea urchin communities in the two pools.

b. Which pool has the greater species richness? Which community is the more diverse according to the Shannon diversity index?

Estimating the number and relative abundance of species requires various sampling techniques and may be difficult due to the rarity of most species in a community. The diversity and richness of bacterial communities may be determined using molecular tools such as RFLP analysis.

Diversity and Community Stability   Long-term experiments on plant diversity have shown that increased diversity correlates with increased productivity and stability of a community. The tunicate experiment in Long Island Sound indicated that more diverse communities use more of the available resources, making it more difficult for an **invasive species** to become established outside its native range.

Trophic Structure   The feeding relationships in a community determine its **trophic structure.** A **food chain** shows the transfer of food energy from one trophic level to the next: from producers to herbivores (primary consumers) to carnivores (secondary,

tertiary, or quaternary consumers) and eventually to decomposers.

A **food web** diagrams the complex trophic relationships within a community. The complicated connections of a food web arise because many consumers feed at various trophic levels. Food webs can be simplified by grouping species into functional groups (such as primary consumers) or by isolating partial food webs.

Within a food web, each food chain usually consists of five or fewer links. According to the **energetic hypothesis,** food chains are limited by the inefficiency of energy transfer from one trophic level to the next. Only about 10% of the **biomass** (total mass of all individuals) of one trophic level is converted to the biomass of the next. The **dynamic stability hypothesis** suggests that short food chains are more stable. Any environmental disruption will be magnified at higher trophic levels as food supply is reduced all the way up the chain. The increasing size of animals at successive trophic levels may also limit food chain length.

INTERACTIVE QUESTION   54.5

Experimental data from tree hole communities showed that food chains were longest when food supply (leaf litter) was greatest. Which hypothesis about food chain length do these results support?

Species with a Large Impact   A **dominant species** in a community has the greatest abundance or largest biomass and is a major influence on the presence and distribution of other species. A species may become dominant due to its more competitive use of resources or its success at avoiding predation or disease. Invasive species may reach high biomass levels due to the lack of natural predators and pathogens. The removal of a dominant species from a community may adversely affect any species that relied exclusively on that species, but its role may quickly be filled by other species.

A **keystone species** has a large impact on community structure as a result of its ecological role. Paine's study of a predatory sea star demonstrated its role in maintaining species richness in an intertidal community by reducing the density of mussels, a highly competitive prey species.

Ecosystem engineers or "foundation species" influence community structure by changing the physical environment.

Bottom-Up and Top-Down Controls   Arrows can be used to indicate the effect an increase in the biomass in one trophic level has on another trophic level. $V \rightarrow H$ indicates that an increase in vegetation (V) increases the number of herbivores (H); $V \leftarrow H$ indicates that an increase in herbivores decreases vegetation biomass. $V \leftrightarrow H$ means that each trophic level is affected by changes in the other. According to the **bottom-up model** of community organization, $N \rightarrow V \rightarrow H \rightarrow P$; that is, an increase in mineral nutrients (N) yields an increase in biomass at each succeeding trophic level: vegetation, herbivores, and predators (P). The **top-down model,** $N \leftarrow V \leftarrow H \leftarrow P$, assumes that predation controls community organization, with a series of $+/-$ effects cascading down the trophic levels. According to this *trophic cascade model,* an increase in predator abundance decreases herbivore abundance, which results in increased vegetation and lowered nutrient levels.

INTERACTIVE QUESTION   54.6

Many freshwater communities appear to be organized along the top-down model. What actions might ecologists take if they wanted to use **biomanipulation** to control excessive algal blooms in a lake with four trophic levels (algae, zooplankton, primary predator fish, and top predator fish)?

54.3   Disturbance influences species diversity and composition

Traditionally, biological communities were viewed as existing in a state of equilibrium maintained by interspecific interactions. The ability of a community to reach and maintain this relatively constant species composition is known as *stability*. The **nonequilibrium model,** in contrast, emphasizes that communities are constantly changing as a result of **disturbances.** Disturbances such as fire, drought, storms, overgrazing, or human activities may change resource availability and reduce or eliminate some populations.

Characterizing Disturbance   According to the **intermediate disturbance hypothesis,** moderate intensity or frequency of disturbance may result in greater species diversity than either low or high levels of disturbance. Small-scale disturbances may enhance environmental patchiness, helping to maintain species diversity. Human prevention of some natural disturbances, such as small fires, may lead to large-scale disturbances, such as large, destructive fires to which species may not be adapted.

Ecological Succession   The sequential transitions in species composition in a community, usually following a severe disturbance, are known as **ecological succession.** If no soil was originally present, as on a new volcanic island or on the moraine left by a retreating glacier, the process is called **primary succession** and may take hundreds or thousands of years. A series of colonizers usually begins with autotrophic and heterotrophic prokaryotes and protists and then moves through lichens, mosses, grasses, shrubs, and trees until the community reaches its prevalent form of vegetation.

Secondary succession occurs when a community is disrupted by fire, logging, or farming, but the soil remains intact. Herbaceous species may colonize first, followed by woody shrubs and eventually trees.

Early colonizers may either *facilitate* the arrival of other species by improving the environment or *inhibit* the establishment of later species. Or species may be independent in their colonization and *tolerate* the conditions created by early species.

Ecologists have studied moraine succession during the retreat of glaciers at Glacier Bay in Alaska. The first pioneering plant species include mosses and fireweed; *Dryas,* a mat-forming shrub, dominates after about 30 years. Alder thickets are the dominant vegetation a few decades later, followed by Sitka spruce. The community becomes a spruce-hemlock forest by the third century after glacial retreat, except in flat, poorly drained areas, where sphagnum mosses invade. These mosses make the soil waterlogged and acidic, killing the trees and creating sphagnum bogs.

INTERACTIVE QUESTION   54.7

a. Describe the effects of the alder stage on soil fertility during the succession following the retreat of a glacier.

b. What is the effect of the spruce forest on soil pH?

Human Disturbance   Human activities, such as converting land to agriculture, logging, clearing for urban development, and ocean trawling, have altered the structure of communities all over the world. A common result is a reduction in species diversity.

54.4   Biogeographic factors affect community diversity

Latitudinal Gradients   Surveys of plant and animal species have documented much greater numbers of species in tropical habitats than in temperate and polar regions. Tropical communities are older, partly because of their longer growing season and partly because they have not had to "start over" after glaciation, as has been the case several times for many polar and temperate communities.

Solar energy input and water availability are important climatic explanations for the latitudinal gradient in diversity. **Evapotranspiration,** the amount of water evaporated from soil and transpired by plants, is determined by solar energy, temperature, and water availability. *Potential evapotranspiration* is determined by just solar radiation and temperature. Evapotranspiration rates have been shown to correlate with species richness for trees and vertebrates in North America.

INTERACTIVE QUESTION   54.8

Why would the fact that tropical communities are "older" than temperate or polar communities contribute to greater species diversity?

Area Effects   A **species-area curve** represents the correlation between the geographic size of a community and the number of species in it. In general, the larger the area, the greater is the diversity of habitats, and the greater the species richness. Use of such curves in conservation biology can enable scientists to predict how a loss of habitat may affect diversity.

Island Equilibrium Model   Any habitat surrounded by a significantly different habitat is considered an island and allows ecologists to study factors that affect species diversity. In the 1960s, R. MacArthur and E. O. Wilson developed an island equilibrium model, which states that the size of the island and its closeness to the mainland (or source of dispersing species) are important variables directly correlated with species diversity. The rates of immigration and extinction change as the number of species on the island increases. When the rates become equal, species diversity reaches an equilibrium, although species composition may continue to change.

INTERACTIVE QUESTION   54.9

Many biogeographic studies have found that large islands have greater species richness than small islands. Label the lines for the immigration and extinction rates for a large and small island on the following graph, which shows how these rates vary with the number of species on the island. Indicate the location of the equilibrium number on the *x*-axis for a small island and for a large island.

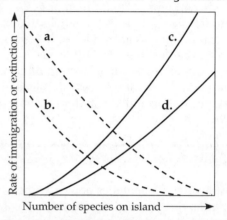

54.5   Pathogens alter community structure locally and globally

Pathogens are disease-causing microorganisms, viruses, viroids, or prions. They can be particularly virulent when introduced into a new habitat, where host species have not yet evolved resistance or where natural controls are lacking.

Pathogens and Community Structure   Pathogens have altered terrestrial ecosystems when dominant tree species are killed and are changing the community structure in coral reef communities. Global travel and trade contributes to the spread of pathogens.

Community Ecology and Zoonotic Diseases   Many human diseases are caused by **zoonotic pathogens,** which are disease-causing agents transferred to humans from other animals, often by means of a **vector** such as ticks or mosquitoes. Understanding species interactions and identifying both hosts and vectors help to combat and control the spread of such diseases. Researchers used genetic and ecological data to identify two inconspicuous shrew species as the primary host of the Lyme pathogen. Infected ticks then transmit Lyme disease to people.

INTERACTIVE QUESTION   54.10

Why are ecologists trapping and testing migrating birds in Alaska?

Word Roots

crypto- = hidden, concealed (*cryptic coloration:* camouflage that makes a potential prey difficult to spot against its background)

ecto- = outer (*ectoparasite:* a parasite that feeds on the external surface of a host)

endo- = inner (*endoparasite:* a parasite that lives within a host)

herb- = grass; **-vora** = eat (*herbivory:* an interaction in which an organism eats parts of a plant or alga)

inter- = between (*interspecific competition:* competition for resources between individuals of two or more species when resources are in short supply)

mutu- = reciprocal (*mutualism:* a symbiotic relationship in which both participants benefit)

Structure Your Knowledge

1. Complete the following concept map to organize your understanding of the important factors that structure a community.

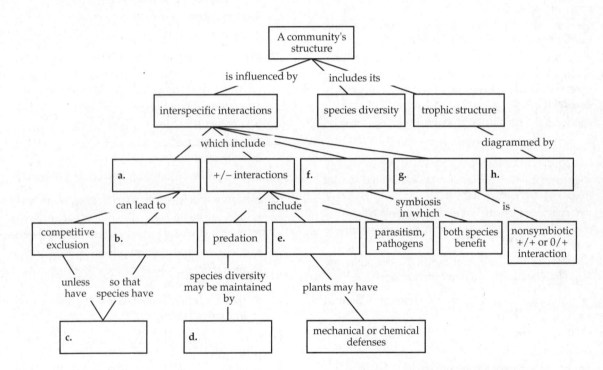

2. Community ecologists develop models or hypotheses to describe community structure and the factors that contribute to such structure. Briefly explain the following models described in this chapter.
   a. Competitive exclusion principle
   b. Energetic hypothesis
   c. Dynamic stability hypothesis
   d. Bottom-up model
   e. Top-down (trophic cascade) model
   f. Nonequilibrium model
   g. Island equilibrium model

Test Your Knowledge

MULTIPLE CHOICE: *Choose the one best answer.*

1. Two allopatric species of Galápagos finches have beaks of similar size. The observation that significant differences in beak size occur when the two species occur on the same island is an example of
   a. competitive exclusion.
   b. species diversity.
   c. commensalism.
   d. character displacement.
   e. a trophic cascade.

2. Aposematic coloring is most commonly found in
   a. prey whose body morphology is cryptic.
   b. predators who are able to sequester toxic plant compounds in their bodies.
   c. prey species that have chemical defenses.
   d. palatable prey that resemble each other.
   e. plants that have toxic secondary compounds.

3. Two species, A and B, occupy adjoining environmental patches that differ in several abiotic factors. When species A is experimentally removed from a portion of its patch, species B colonizes the vacated area and thrives. When species B is experimentally removed from a portion of its patch, species A does not successfully colonize the area. From these results you might conclude that
   a. both species A and species B are limited to their range by abiotic factors.
   b. species A is limited to its range by competition, and species B is limited by abiotic factors.
   c. both species are limited to their range by competition.
   d. species A is limited to its range by abiotic factors, and species B is limited to its range because it cannot compete with species A.
   e. species A is a predator of species B.

4. Through resource partitioning,
   a. two species can compete for the same prey item.
   b. slight variations in niche allow closely related species to coexist in the same habitat.
   c. two species can share identical niches in a habitat.
   d. competitive exclusion results in the success of the superior species.
   e. two species with identical niches do not share the same habitat and thus avoid competition.

5. A palatable (good-tasting) prey species may defend against predation by
   a. Müllerian mimicry.
   b. Batesian mimicry.
   c. producing secondary compounds.
   d. cryptic coloration.
   e. either b or d.

6. The species richness of a community refers to
   a. the relative numbers of individuals in each species.
   b. the number of different species found in the community.
   c. the feeding relationships or trophic structure within the community.
   d. the species diversity of that community.
   e. the community's stability or ability to persist through disturbances.

7. Which of the following organisms is *mismatched* with its trophic level?
   a. algae—producer
   b. phytoplankton—primary consumer
   c. carnivorous fish larvae—secondary consumer
   d. eagle—tertiary or quaternary consumer
   e. Both b and c are mismatched.

8. Ecologists survey the tree species in two forest plots. Plot 1 has six different species, and 95% of all trees belong to just one species. Plot 2 has five different species, each of which is represented by approximately 20% of the trees. Compared with plot 1, plot 2 has
   a. higher species richness.
   b. greater species diversity.
   c. lower relative abundance.
   d. lower species richness.
   e. Both b and d are correct.

9. When one species was removed from a tide pool, the pool's species · diversity was significantly reduced. The removed species was probably
   a. a strong competitor.
   b. a potent parasite.
   c. a resource partitioner.
   d. a keystone species.
   e. the species with the highest relative abundance.

10. Invasive species often reach a large biomass because
    a. they are better competitors than native species.
    b. they are usually producers and not top predators.
    c. they often lack natural predators or pathogens.
    d. their superior ability to disperse enables them to spread to new niches.
    e. they are often protected by the humans who have introduced them.

11. Why do most food chains consist of only three to five links?
    a. There are only five trophic levels: producers; primary, secondary, and tertiary consumers; and decomposers.
    b. Most communities are controlled bottom-up by mineral nutrient supply.
    c. The dominant species in most communities consumes the majority of prey; thus, not enough food is left to support higher predators.
    d. According to the energetic hypothesis, the inefficiency of energy transfer from one trophic level to the next limits the number of links.
    e. According to the trophic cascade model, increasing the biomass of top trophic levels causes a decrease in the biomass of lower levels, so that the top levels are limited.

12. According to the top-down (trophic cascade) model of community control, which trophic level would you *decrease* if you wanted to *increase* the vegetation level in a community?
    a. nutrients
    b. vegetation
    c. secondary consumers (carnivores)
    d. tertiary consumers
    e. omnivores

13. During succession, inhibition by early species
    a. may prevent the achievement of a stable community.
    b. may slow down both the rate of immigration and the rate of extinction, depending on the size of the area and the distance from the source of dispersing species.
    c. results from the frequent disturbances that often eliminate early colonizers.
    d. may slow down the successful colonization by other species.
    e. may involve changes in soil pH or accelerated accumulation of humus.

14. According to the nonequilibrium model,
    a. chance events such as disturbances play major roles in the structure and composition of communities.
    b. species composition in a community is always in flux as a result of human interventions.
    c. food chains are limited to a few links because long chains are more unstable in the face of environmental disturbances.
    d. the communities with the most diversity have the least stability or resistance to change.
    e. early colonizers inhibit other species, whereas later colonizers facilitate the arrival of new species.

15. Which of the following organisms is *mismatched* with its community role?
    a. beaver—ecosystem engineer
    b. a black rush *Juncus* in a salt marsh—facilitator
    c. a sea otter in the North Pacific—keystone predator
    d. trees in a spruce-hemlock forest—dominant species
    e. alder and *Dryas* (a mat-forming shrub)— inhibitors

16. An island that is small and far from the mainland, in contrast to a large island close to the mainland, would be expected to
    a. have lower species diversity.
    b. be in an earlier successional stage.
    c. have higher species diversity but a much lower abundance of organisms.
    d. have a higher rate of colonization but a higher rate of extinction.
    e. have a lower rate of colonization and a lower rate of extinction.

17. A major explanation for the decline in species richness along an equatorial-polar gradient is the correlation of high levels of solar radiation and water availability with diversity. Which of the following factors is also thought to contribute to the high species richness of tropical communities?
    a. the inverse relationship between diversity and evapotranspiration
    b. the greater age of these communities (longer growing season and fewer climatic setbacks), providing more time for speciation events
    c. the larger area of the tropics and corresponding richness predicted by the species-area curve
    d. the lack of disturbances in tropical areas
    e. the greater immigration rate and lower extinction rate found on large tropical islands

18. Which of the following descriptions best describes a zoonotic pathogen?
    a. a pathogen that affects insects
    b. a pathogen that requires a vector to spread from animal to animal
    c. a disease-causing agent that is transmitted to humans from other animals
    d. a pathogen that is found in zoos due to the unnatural habitat provided for animals
    e. an ectoparasite that is transferred from animals to humans

Chapter 55

Ecosystems and Restoration Ecology

Key Concepts

55.1 Physical laws govern energy flow and chemical cycling in ecosystems

55.2 Energy and other limiting factors control primary production in ecosystems

55.3 Energy transfer between trophic levels is typically only 10% efficient

55.4 Biological and geochemical processes cycle nutrients and water in ecosystems

55.5 Restoration ecologists help return degraded ecosystems to a more natural state

Framework

This chapter describes energy flow and chemical cycling through ecosystems. Producers convert solar energy into chemical energy, which passes, with a loss of energy at each trophic level, through the ecosystem and dissipates as heat. Chemical elements are cycled within the ecosystem from inorganic reservoirs through producers, consumers, and detritivores, and back to the reservoirs.

An ecosystem's primary production may be limited by nutrients, temperature, or moisture. The low trophic efficiency in the transfer of energy from one level to the next is reflected in pyramids of production and biomass.

Bioremediation and biological augmentation are both used by restoration ecologists to help restore polluted or damaged ecosystems.

Chapter Review

An **ecosystem** consists of all the organisms in a given area and the abiotic factors with which they interact. Most ecosystems are powered by energy from sunlight, which is transformed to chemical energy by autotrophs, passed to heterotrophs in the organic compounds of food, and continually dissipated in the form of heat. Chemical elements are cycled among the abiotic and biotic components of the ecosystem as autotrophs incorporate them into organic compounds and the processes of metabolism and decomposition return them to the soil, air, and water.

55.1 Physical laws govern energy flow and chemical cycling in ecosystems

Conservation of Energy According to the first law of thermodynamics, energy can neither be created nor destroyed, only transferred or transformed. Energy flows through ecosystems from its input as solar radiation to its conversion into chemical energy to its dissipation as heat. The second law of thermodynamics states that every energy exchange increases entropy. Thus, in each energy conversion, some energy is lost as heat.

Conservation of Mass The **law of conservation of mass** states that matter cannot be created or destroyed. Ecosystem ecologists measure the inputs and outputs of elements as well as chemical cycling within ecosystems.

INTERACTIVE QUESTION 55.1

What may happen if the input of an element into an ecosystem is less than the output of that element from the ecosystem?

Energy, Mass, and Trophic Levels Most **primary producers,** or autotrophs, use light energy to synthesize organic compounds for use as fuel and building materials. Heterotrophs depend on autotrophs for their organic compounds. **Primary consumers** are herbivores; **secondary consumers** are carnivores. **Tertiary consumers** eat other carnivores.

Detritivores, or **decomposers,** consume **detritus,** which includes organic wastes, fallen leaves, and dead organisms. Detritivores often form a link between producers and consumers. Fungi and prokaryotes are important detritivores in most ecosystems, converting organic materials from all trophic levels to inorganic compounds that can be recycled by autotrophs.

INTERACTIVE QUESTION 55.2

Compare the movement of energy and chemical elements in ecosystems.

55.2 Energy and other limiting factors control primary production in ecosystems

Primary production is the amount of light energy converted to chemical energy during a period of time—that is, the photosynthetic output of an ecosystem's autotrophs.

Ecosystem Energy Budgets Only a small portion of incoming solar radiation strikes photosynthetic organisms, and, of that, only about 1% is converted to chemical energy. Nevertheless, worldwide photosynthetic production is about 150 billion metric tons of organic material per year.

Net primary production (NPP), which represents the chemical energy available to consumers in an ecosystem, is equal to an ecosystem's **gross primary production (GPP)** minus the energy used by autotrophs in their cellular respiration (R_a):

$$NPP = GPP - R_a$$

Net primary production can be expressed as energy ($J/m^2 \cdot yr$) or as new biomass produced, expressed as dry weight of vegetation ($g/m^2 \cdot yr$). *Standing crop* is the total biomass of photosynthetic organisms in an ecosystem.

Because photosynthesizing vegetation absorbs more wavelengths of visible light, scientists can estimate primary production using data from satellites that compare the wavelengths of light reflected from Earth's surface. Primary production and the contribution to Earth's total production vary by ecosystem.

Net ecosystem production (NEP), which reflects *total biomass accumulation* in an ecosystem, deducts the energy used in respiration by all autotrophs and heterotrophs (R_T) from gross primary production:

$$NEP = GPP - R_T$$

Scientists estimate NEP by measuring the net flux of CO_2 from terrestrial ecosystems, or the net flux of CO_2 or O_2 in oceans. Researchers using oxygen sensors on profiling floats that rise through the water column have shown greater phytoplankton productivity in nutrient-poor regions of the oceans than previously thought.

Primary Production in Aquatic Ecosystems The depth to which light penetrates and the availability of nutrients affect primary production in oceans and lakes. Nitrogen or phosphorus is often the **limiting nutrient** in the photic zone of the ocean. Nutrient enrichment experiments can identify, for example, whether nitrogen or phosphate pollution is causing algal "blooms," or which nutrient limits phytoplankton growth in the Sargasso Sea.

In freshwater lakes, nutrient limitation also affects production. **Eutrophication,** the rapid growth of cyanobacteria and algae, has been linked to phosphorus pollution from sewage and fertilizer runoff.

Primary Production in Terrestrial Ecosystems There is usually a positive correlation between primary production and precipitation. Another predictor of primary production—*actual evapotranspiration*, the annual amount of water evaporated from a landscape and transpired by plants—reflects both precipitation and temperature.

Nutrients affect primary production, and nitrogen and/or phosphorus are the limiting nutrients in many terrestrial ecosystems. Plant adaptations that increase uptake of nutrients include symbiotic relationships between roots and nitrogen-fixing prokaryotes, and mycorrhizal associations with fungi that provide phosphorus and other elements to plants. Plant root hairs increase absorptive surface area, and roots release substances that increase the availability of nutrients.

INTERACTIVE QUESTION 55.3

a. List some ecosystems with high primary production.

b. List some ecosystems with low primary production.

c. The oceans have low primary production yet contribute as much global net primary production as terrestrial areas. Explain.

d. Areas of the Antarctic Ocean are often more productive than tropical seas, even though they are colder and receive lower light intensity. Explain.

55.3   Energy transfer between trophic levels is typically only 10% efficient

The production of new biomass from the energy contained in consumers' food during a given time is called **secondary production.**

Production Efficiency   Herbivores consume only a fraction of the plant material produced; they cannot digest all they eat; and much of the energy they do absorb is used for cellular respiration. Only the chemical energy stored as growth or in offspring is available as food to secondary consumers. The proportion of assimilated food energy (not including losses in feces) that is used for net secondary production (growth and reproduction) is a measure of the efficiency of energy transformation: **production efficiency** = (net secondary production × 100%)/assimilation of primary production. Production efficiencies vary from 1–3% for endothermic birds and mammals, to 10% for fishes, to around 40% for insects and microorganisms.

Trophic Efficiency and Ecological Pyramids   **Trophic efficiency** is the percentage of the production of one trophic level that makes it to the next level; it typically ranges from 5% to 20%. A *pyramid of net production* shows this multiplicative loss of energy.

A *biomass pyramid* illustrates the standing crop biomass of organisms at each trophic level. This pyramid usually narrows rapidly from producers to the top trophic level. Some aquatic ecosystems have inverted biomass pyramids in which zooplankton (consumers) outlive and outweigh the highly productive, but heavily consumed, phytoplankton. Phytoplankton have a short **turnover time,** the time required to replace a population's standing crop; turnover time is determined by dividing standing crop biomass by production. The production pyramid for such an aquatic ecosystem, however, is normal in shape.

INTERACTIVE QUESTION 55.4

a. Why is production efficiency higher for fishes than for birds and mammals?

b. Assuming a 10% trophic efficiency (transfer of energy to the next trophic level), approximately what proportion of the chemical energy produced in photosynthesis makes it to a tertiary consumer?

55.4   Biological and geochemical processes cycle nutrients and water in ecosystems

Chemical elements are passed between abiotic and biotic components of ecosystems through **biogeochemical cycles.**

Biogeochemical Cycles   Gaseous forms of carbon, oxygen, sulfur, and nitrogen have global cycles. Heavier elements, such as phosphorus, potassium, and calcium, have a more local cycle, especially in terrestrial ecosystems.

Nutrients are found in four types of reservoirs: organic material in living organisms or detritus, which is available to other organisms; unavailable organic material in "fossilized" deposits; available inorganic elements and compounds in water, soil, or air; and unavailable elements in rocks. Nutrients may leave the unavailable reservoirs through the weathering of rock, erosion, or the burning of fossil fuels.

Ecologists study the movement of elements by tracking added radioactive isotopes or by following naturally occurring nonradioactive isotopes through ecosystems.

Exploring Nutrient Cycles   The water cycle involves solar-energy-driven evaporation from oceans and evapotranspiration from land, condensation of water vapor into clouds, and precipitation. Runoff and groundwater flow return water to the oceans.

In the carbon cycle, producers use CO_2 from the atmosphere in photosynthesis, producing organic compounds that are used by producers and consumers. Organisms release CO_2 in cellular respiration. Fossil fuel combustion is increasing atmospheric CO_2. Reservoirs of carbon include fossil fuels, dissolved carbon compounds in oceans, plant and animal biomass, CO_2 in the atmosphere, and sedimentary rocks such as limestone.

Plants require nitrogen in the form of ammonium (NH_4^+) or nitrate (NO_3^-). Animals obtain nitrogen in organic form from plants or other animals. The major reservoir is the atmosphere, and nitrogen enters ecosystems through *nitrogen fixation* when soil bacteria convert N_2 into compounds that can be assimilated by plants. Fertilizers and the planting of legume crops now add more nitrogen to ecosystems than do natural processes. Large quantities of reactive nitrogen gases are released to the atmosphere by human activities.

Weathering of rock adds phosphorus to the soil in the form of PO_4^{3-}, which is absorbed by plants. Organic phosphate is transferred from plants to consumers and returned to the soil through the action of decomposers or in animal excretion. Soil particles bind phosphate, keeping it available locally for recycling. Sedimentary rocks of marine origin are the largest reservoirs.

INTERACTIVE QUESTION 55.5

What is the biological importance of water, carbon, nitrogen, and phosphorus?

Decomposition and Nutrient Cycling Rates Temperature and the availability of water influence decomposition rates; thus, nutrient cycling times vary in different ecosystems. Nutrients cycle rapidly in a tropical rain forest; the soil contains only about 10% of the ecosystem's nutrients. Decomposition is slower in temperate forests, and 50% of the ecosystem's organic material may be found in detritus and soil. Decomposition rates are slow where oxygen is limited, such as in peat lands and aquatic sediments.

Case Study: Nutrient Cycling in the Hubbard Brook Experimental Forest A team of scientists has looked at nutrient cycling in the Hubbard Brook forest ecosystem since 1963. The mineral budget for each of six valleys

was determined by measuring the input of key nutrients in rainfall and their outflow through the creek that drained each watershed. Most minerals were recycled within the forest ecosystem.

The effect of deforestation on nutrient cycling was measured for three years in a valley that was completely logged. Compared with a control area, water runoff from the deforested valley increased 30–40%; net loss of minerals such as Ca^{2+} and K^+ was large. Nitrate increased in concentration in the creek 60-fold, removing this critical soil nutrient and contaminating the water.

INTERACTIVE QUESTION 55.6

a. In which natural ecosystem do nutrients cycle the fastest? Why?

b. In which natural ecosystems do nutrients cycle slowly? Why?

c. What is the effect of loss of vegetation on nutrient cycling?

55.5 Restoration ecologists help return degraded ecosystems to a more natural state

Areas degraded by farming, mining, or environmental pollution are often abandoned. In order to speed the successional processes involved in a community's recovery, restoration ecologists attempt to identify and manipulate the factors that most limit recovery time. In some cases, the physical structure of a site must first be reconstructed.

Bioremediation Bioremediation uses prokaryotes, fungi, or plants to detoxify polluted ecosystems. Some plants are able to extract metals from contaminated soils; harvesting those plants removes the metals from the ecosystem. Prokaryotes can metabolize dangerous elements to less soluble forms.

Biological Augmentation Biological augmentation uses organisms to add essential substances to degraded ecosystems. Restoration ecologists may also reintroduce animals to restored sites.

Restoration Projects Worldwide Restoration ecologists often apply adaptive management, in which they experiment with promising management approaches and learn as they work in each unique and complex disturbed ecosystem.

INTERACTIVE QUESTION 56.7

Give an example of bioremediation and of biological augmentation.

Word Roots

bio- = life; **geo-** = Earth (*biogeochemical cycles:* the various chemical cycles involving both biotic and abiotic components of ecosystems)

detrit- = wear off; **-vora** = eat (*detritivore:* a consumer that derives its energy and nutrients from nonliving organic material)

Structure Your Knowledge

1. Two processes that emerge at the ecosystem level of organization are energy flow and chemical cycling. Develop a concept map that describes, compares and contrasts these two processes.
2. What factors limit primary production in aquatic ecosystems? In terrestrial ecosystems?
3. What processes mediate the interconversion of the nutrients contained in organic materials and inorganic materials?

Test Your Knowledge

MULTIPLE CHOICE: *Choose the one best answer.*

1. Which of the following components are absolutely essential to the functioning of any ecosystem?
   a. producers and primary consumers
   b. producers and secondary consumers
   c. primary, secondary, and tertiary consumers
   d. primary consumers and detritivores
   e. producers and detritivores

2. Which of the following statements about ecosystems is true?
   a. Energy is recycled through the trophic structure.
   b. Energy is usually converted from sunlight by primary producers, passed to secondary producers in the form of organic compounds, and lost to detritivores in the form of heat.
   c. Chemicals are recycled between the biotic and abiotic sectors, whereas energy makes a one-way trip through the food web and is eventually dissipated as heat.
   d. There is a continuous process by which energy is lost as heat, and chemical elements leave the ecosystem through runoff.
   e. A food web shows that all trophic levels may feed off each other.

3. Primary production
   a. is equal to the standing crop of an ecosystem.
   b. is limited by light, nutrients, and moisture in all ecosystems.
   c. is the amount of light energy converted to chemical energy per unit time in an ecosystem.
   d. is inverted in some aquatic ecosystems.
   e. is all of the above.

4. In the experiment in which iron was added to areas of the Pacific Ocean, the growth of eukaryotic phytoplankton was stimulated because
   a. iron was the limiting nutrient for eukaryotic phytoplankton growth.
   b. the iron interacted with bottom sediments, releasing nitrogen into the water.
   c. the iron interacted with phosphorus, making that nutrient available to the phytoplankton.
   d. the iron reached the critical level necessary to promote photosynthesis.
   e. the iron stimulated the growth of nitrogen-fixing cyanobacteria, which then made nitrogen available for phytoplankton growth.

5. The open ocean and tropical rain forest are the two largest contributors to Earth's net primary production because
   a. both have high rates of net primary production.
   b. both cover huge surface areas of Earth.
   c. nutrients cycle rapidly in these two ecosystems.
   d. the ocean covers a huge surface area and the tropical rain forest has a high rate of production.
   e. both a and b are correct.

6. Production in terrestrial ecosystems is affected by
   a. temperature.
   b. light intensity.
   c. availability of nutrients.
   d. availability of water.
   e. all of the above.

7. Which of the following statements concerning net ecosystem production and net primary production is *false?*
   a. NEP will always be less than NPP because it takes the respiration of heterotrophs and decomposers into account.
   b. NEP is a means of measuring the total biomass accumulation in an ecosystem over a period of time.
   c. If measurements indicate that more CO_2 enters an ecosystem than leaves, NEP would be positive.
   d. If measurements indicate that more O_2 leaves an ecosystem than enters, NEP would be negative.
   e. An increase in the rate of decomposition in an ecosystem would lower NEP, but NPP would probably remain the same or perhaps increase.

8. Secondary production
   a. is measured by the standing crop.
   b. is the rate of biomass production in consumers.
   c. is greater than primary production.
   d. is 10% less than primary production.
   e. is the gross primary production minus the energy used for respiration.

9. Which of the following statements concerning a pyramid of net production is *not* true?
   a. Only about 10% of the energy in one trophic level passes into the next level.
   b. Because of the loss of energy at each trophic level, most food chains are limited to three to five links.
   c. The pyramid of production of some aquatic ecosystems is inverted because of the large zooplankton primary consumer level.
   d. Eating grain-fed beef is an inefficient means of obtaining the energy trapped by photosynthesis.
   e. A biomass pyramid is usually the same shape as a pyramid of production.

10. In which of the following groups of organisms would you expect production efficiency to be the greatest?
    a. amphibians
    b. mammals
    c. insects and microorganisms
    d. fishes
    e. birds

11. Biogeochemical cycles are global for elements
    a. that are found in the atmosphere.
    b. that are found mainly in the soil.
    c. such as carbon, nitrogen, and phosphorus.
    d. that are dissolved in water.
    e. Both a and c are correct.

12. Which of these processes is *incorrectly* paired with its description?
    a. nitrification—oxidation of ammonium in the soil to nitrite and nitrate
    b. nitrogen fixation—reduction of atmospheric nitrogen to ammonia
    c. denitrification—return of N_2 to air; occurs when denitrifying bacteria metabolize nitrate
    d. ammonification—decomposition of organic compounds to ammonium
    e. deforestation—increased supply of nitrate in the soil

13. Clear-cutting tropical forests yields agricultural land with limited productivity because
    a. it is too hot in the tropics for most food crops.
    b. the tropical forest regrows rapidly and chokes out agricultural crops.
    c. few of the ecosystem's nutrients are stored in the soil; most are in the forest trees.
    d. phosphorus, not nitrogen, is the limiting nutrient in those soils.
    e. decomposition rates are high but primary production is low in the tropics.

14. Which of the following was *not* shown by the Hubbard Brook Experimental Forest study?
    a. Most minerals recycle within a forest ecosystem.
    b. Deforestation results in a large increase in water runoff.
    c. Mineral losses from a valley were great following deforestation.
    d. Nitrate was the mineral that showed the greatest loss.
    e. Acid rain increased as a result of deforestation.

15. Which of the following actions is an example of bioaugmentation?
    a. adding fertilizer to degraded soils
    b. restoring the natural flow of river channels
    c. planting nitrogen-fixing lupines in soils disturbed by mining
    d. establishing habitat corridors to connect restored sites with undisturbed sites
    e. adding ethanol to stimulate the growth of uranium-reducing bacteria in contaminated groundwater

Chapter 56

Conservation Biology and Global Change

Key Concepts

56.1 Human activities threaten Earth's biodiversity

56.2 Population conservation focuses on population size, genetic diversity, and critical habitat

56.3 Landscape and regional conservation help sustain biodiversity

56.4 Earth is changing rapidly as a result of human actions

56.5 Sustainable development can improve human lives while conserving biodiversity

Framework

Biodiversity at the genetic, species, and ecosystem level is crucial to human welfare. This chapter explores the threats to biodiversity and several of the approaches to preserving the diversity of species on Earth. Conservation biologists focus on the population size, genetic diversity, and habitat needs of endangered species and on establishing and managing nature reserves that are often in human-dominated landscapes. Human activities are causing global change through nutrient enrichment, the release of toxins, and increasing CO_2 levels in the atmosphere. Ecological research and our biophilia may help to achieve the goal of sustainable development—the long-term perpetuation of human societies *and* the ecosystems that support them.

Chapter Review

Conservation biology integrates all areas of biology in the effort to sustain ecosystem processes and biodiversity.

56.1 Human activities threaten Earth's biodiversity

Three Levels of Biodiversity The current high rate of extinction threatens Earth's biological diversity. Loss of the genetic diversity within and between populations lessens a species' adaptive potential.

A second level of biodiversity is species diversity. An **endangered species,** according to the U.S. Endangered Species Act (ESA), is one that is "in danger of extinction throughout all or a significant portion of its range." A **threatened species** is defined as one that is likely to become endangered. There are many well-documented examples of recent extinctions and endangered species in most taxonomic groups.

The third level of biodiversity is ecosystem diversity. A loss of a species can negatively affect other species in the ecosystem. Human activities have altered many ecosystems.

Biodiversity and Human Welfare There are both moral and practical reasons for preserving biodiversity. A loss of biodiversity is a loss of the genetic potential held in the genomes of species, such as genes for traits that could improve crops. Biodiversity is a natural resource that can provide medicines, fibers, and food.

Humans depend on Earth's ecosystems. **Ecosystem services** include such things as purification of air and water, detoxification and decomposition of wastes, nutrient cycling, flood control, pollination of crops, and soil creation and preservation. The monetary value of ecosystem services is huge.

Threats to Biodiversity The greatest threat to biodiversity is habitat destruction, caused by agriculture, urban development, forestry, mining, pollution, and global climate change. According to the International Union for Conservation of Nature and Natural Resources (IUCN), habitat destruction is implicated in 73% of species that have been designated as extinct, endangered, vulnerable, or rare. Fragmentation of natural habitats is a common occurrence and almost always leads to species loss. Both terrestrial ecosystems and aquatic habitats have been damaged.

Introduced species, sometimes called non-native or exotic species, compete with or prey upon native species. Humans have transplanted thousands of species, intentionally and unintentionally, with huge economic costs in terms of damage and control efforts.

Overharvesting involves harvesting wild plants or animals at rates higher than the populations' abilities to reproduce. Species of large animals with low reproductive rates and species on small islands are particularly vulnerable. Overfishing, particularly using new harvesting techniques, has drastically reduced populations of many commercially important fish species. The tools of molecular genetics enable conservation biologists to determine whether tissues have come from threatened or endangered species.

Global change is a threat to biodiversity at regional to global levels through alterations in climate, atmospheric chemistry, and ecological systems. The burning of fossil fuels and wood, for example, releases oxides of sulfur and nitrogen, which form sulfuric and nitric acid in the atmosphere. These acids contribute to *acid precipitation,* defined as rain, snow, sleet, or fog with a pH less than 5.2. Drifting emissions from factories create acid precipitation, which has damaged forests and fish populations in lakes across North America and Europe. New regulations and technologies have reduced sulfur dioxide emissions, and the acidity of precipitation has gradually been reduced.

INTERACTIVE QUESTION 56.1

Give an example of how each of the following threats to biodiversity has reduced population numbers or caused extinctions.

a. habitat destruction

b. introduced species

c. overharvesting

d. global change

56.2 Population conservation focuses on population size, genetic diversity, and critical habitat

Small-Population Approach According to the small-population approach, the inbreeding and genetic drift characteristic of a small population may draw it down an **extinction vortex,** in which the loss of genetic variation leads, by positive feedback loops, to smaller and smaller numbers until the population becomes extinct. Low genetic diversity, however, does not always doom a population.

The story of the greater prairie chicken provides an example of an averted extinction vortex. As agriculture fragmented their habitat, the number of prairie chickens in Illinois declined from millions in the 19th century to 50 in 1993. A comparison with DNA from museum specimens indicated decreased genetic variation in the threatened population. After importing birds from larger populations in other states, researchers noted an increase in egg viability, and the Illinois population rebounded.

Computer models that integrate many factors are used to estimate **minimum viable population (MVP),** the minimum population size necessary to sustain a population.

The **effective population size** (N_e) is based on a population's breeding potential and is determined by a formula that includes the number of individuals that breed and the sex ratio of the population: $N_e = (4N_f N_m)/(N_f + N_m)$. Alternative formulas take into account life history or genetic factors. Conservation efforts should be based on maintaining the minimum number of *reproductively active* individuals needed to prevent extinction.

INTERACTIVE QUESTION 56.2

Is the effective population size usually larger or smaller than the actual number of individuals in the population? Explain.

Population viability analysis uses MVP to predict long-term viability of a population—the probability of survival over a particular period of time. M. Shaffer performed a population viability analysis as part of a long-term study of grizzly bears in Yellowstone National Park. He estimated that the minimum viable population size for threatened grizzly bear populations was 100 bears. Current estimates of the grizzly population in the greater Yellowstone ecosystem indicate a population of about 400. The effective population size, however, is only 25% of the total population size, or 100 bears. Genetic analyses indicate that the Yellowstone grizzly population has less genetic variability than other populations in North America. Migration between isolated populations could increase both the effective size and genetic variation of the Yellowstone grizzly bear population.

INTERACTIVE QUESTION   56.3

Explain the basic premise of the small population approach. What conservation strategy is recommended for preserving small populations?

Declining-Population Approach   The emphasis of the declining-population approach is to identify populations that may be declining, determine the environmental factors that caused the decline, and then recommend corrective measures.

The following logical steps are part of the declining-population approach: assess population trends and distributions to establish that a species is in decline; determine its environmental requirements; list all possible causes of the decline and the predictions that arise from each of these hypotheses; test the most likely hypothesis to see if the population rebounds if this suspected factor is altered; and apply the results to the management of the threatened species.

Several factors have driven the red-cockaded woodpecker into decline. Logging and agriculture have fragmented its mature pine forest habitats, and strict fire control has resulted in excessively thick undergrowth around the pine trees. Recognition of this species' social organization and of the factors that slow its dispersal to new territories has aided in its recovery. Management strategies now include protection of some longleaf pine forests, controlled fires, and the excavation of breeding cavities in unoccupied habitats to encourage establishment of new breeding groups.

INTERACTIVE QUESTION   56.4

Describe the declining-population approach to the conservation of endangered species.

Weighing Conflicting Demands   Preserving habitat for endangered species often conflicts with the economic and recreational desires of humans. Keystone species exert more influence on community structure and ecosystem processes, and prioritizing the species to be saved on the basis of their ecological role may be crucial to the survival of whole communities.

56.3   Landscape and regional conservation help sustain biodiversity

Conservation efforts increasingly are directed at sustaining the biodiversity of whole communities and ecosystems. A goal of landscape ecology is to make biodiversity conservation a part of ecosystem management and landscape use.

Landscape Structure and Biodiversity   Landscape dynamics are important to conservation efforts because species often use more than one ecosystem or live on borders between ecosystems. Landscapes include ecosystems separated by boundaries or *edges*, which have their own sets of physical conditions and communities of organisms. The increase in edge communities due to human fragmentation of habitats may serve to reduce biodiversity as edge-adapted species become predominant. The long-term Biological Dynamics of Forest Fragments Project has shown that species adapted to forest interiors decline the most when habitat fragments are small.

Movement corridors are narrow strips or clumps of habitat that connect isolated patches. Artificial corridors are sometimes constructed when habitat patches have been separated by major human disruptions.

INTERACTIVE QUESTION   56.5

What are some potential benefits of corridors? How might they be harmful?

Establishing Protected Areas   Currently, about 7% of Earth's land area has been set aside as reserves. Conservation biologists consider landscape dynamics in the designation and management of these protected areas. **Biodiversity hot spots**—small areas with very high concentrations of endemic, threatened, and endangered species—are good choices for nature reserves. Global change will complicate the use of the hot-spot approach to conservation.

Even though nature reserves provide islands of protected habitat, the nonequilibrium model of natural disturbances applies to them. Patch dynamics, edges, and corridor effects must be considered in the design and management of reserves.

New information on the requirements for minimum viable population sizes indicates that most national parks and reserves are much too small—the area needed to sustain a population is usually much larger than the legal boundary set in a reserve.

INTERACTIVE QUESTION   56.6

What factors would favor the creation of larger, extensive reserves? What factors favor smaller, unconnected reserves?

Zoned reserves have protected core areas surrounded by buffer zones in which the human social and economic climate is stable and activities are regulated to promote the long-term viability of the protected zones. Costa Rica has established eight zoned reserves, but deforestation has continued in some buffer zones. The Florida Keys National Marine Sanctuary, a zoned marine reserve established in 1990, has increased marine life and generated revenue from recreational divers.

56.4   Earth is changing rapidly as a result of human actions

Nutrient Enrichment   The harvesting of crops removes nutrients that would otherwise be recycled in the soil. Once the organic and inorganic reserves of soils become depleted, crops require the addition of fertilizers. The addition of nitrogen fertilizers, increased legume cultivation, and burning of fossil fuels has more than doubled Earth's supply of fixed nitrogen.

Nitrogen in excess of the **critical load,** the amount of added nutrient that can be absorbed by plants without damaging the ecosystem, can contaminate groundwater, degrade lakes and rivers, and drain into the ocean. Elevated nitrogen levels create phytoplankton blooms. The decomposition of these phytoplankton lowers oxygen levels, creating "dead zones" in coastal waters. Eutrophication of lakes due to nutrient runoff can kill fish and harm other organisms.

Toxins in the Environment   Humans release a huge variety of toxic chemicals into the environment. Organisms absorb these toxins from food and water and may retain them within their tissues. In a process known as **biological magnification,** the concentration of such compounds increases in each successive link of the food chain. Chlorinated hydrocarbons (such as DDT) and polychlorinated biphenyls (or PCBs) have been implicated in endocrine system problems in many animal species. Many toxic chemicals dumped into ecosystems are nonbiodegradable; others, such as mercury, may become more harmful as they react with other environmental factors.

Greenhouse Gases and Global Warming   The concentration of CO_2 in the atmosphere has been increasing since the Industrial Revolution as a result of the combustion of fossil fuel and deforestation. If C_3 plants become able to outcompete C_4 plants with the increase in CO_2, species composition in natural and agricultural communities may be altered.

In the Forest–Atmosphere Carbon Transfer and Storage (FACTS-I) experiment, scientists have monitored the effects of elevated CO_2 levels on sample plots in a forest ecosystem over many years. Trees in experimental plots have produced 15% more wood each year, although this increase is less than expected. Because of nutrient limitation, increased CO_2 levels will increase plant production somewhat less than had been predicted.

Through a phenomenon known as the **greenhouse effect,** CO_2, other greenhouse gases, and water vapor in the atmosphere absorb infrared radiation reflected from Earth and re-reflect it back, thus warming Earth.

Scientists use various models in estimating the extent and consequences of increasing CO_2 levels. A number of studies predict a doubling of CO_2 levels and a temperature rise of 3°C by the end of the twenty-first century. To predict the impact of increasing temperatures, ecologists study the effects on vegetation of previous global warming trends. Because plants cannot disperse rapidly, many may not be able to survive global climate change. One solution, although potentially dangerous, may be **assisted migration,** in which a species is moved to a habitat beyond its native range to protect it from human-caused threats.

Controlling the levels of CO_2 emissions in increasingly industrialized societies is a huge international challenge.

INTERACTIVE QUESTION   56.7

List some of the ways by which we may slow global warming.

Depletion of Atmospheric Ozone   A layer of ozone molecules (O_3) in the stratosphere absorbs damaging ultraviolet radiation. This layer has been gradually thinning since the mid-1970s, largely as a result of the accumulation of breakdown products of chlorofluoro-carbons in the atmosphere. The dangers of ozone depletion may include increased incidence of skin cancer and cataracts and unpredictable effects on phytoplankton, crops, and natural ecosystems.

56.5   Sustainable development can improve human lives while conserving biodiversity

Sustainable Biosphere Initiative   **Sustainable development** is economic development that meets the needs of people today without limiting the ability of future generations to meet their needs. The Ecological Society of America endorses a research agenda, the Sustainable Biosphere Initiative, that encourages studies of global change, biodiversity, and maintenance of the productivity of natural and artificial ecosystems. An important goal is developing the ecological knowledge necessary to make intelligent and responsible decisions concerning Earth's resources.

Case Study: Sustainable Development in Costa Rica   Partnerships between the government, non-government organizations (NGOs), and citizens have contributed to the success of conservation in Costa Rica. Living conditions in the country have improved, as evidenced by a decrease in infant mortality, an increase in life expectancy, and a high literacy rate. A projected increase in population from 4 million to 6 million in the next 50 years, however, will present challenges to the goal of sustainable development.

The Future of the Biosphere   E. O. Wilson calls our attraction to Earth's diversity of life and our affinity for natural environments *biophilia*. Perhaps this connection is innate and will provide the ethical resolve needed to protect species from extinction and ecosystems from destruction. By coming to know and understand nature through the study of biology, we may become more able to appreciate and preserve the processes and diversity of the biosphere.

Word Roots

bio- = life (*biodiversity hot spot:* a relatively small area with numerous endemic species and a large number of endangered and threatened species)

Structure Your Knowledge

1. List the four major threats to biodiversity.
2. How does the loss of biodiversity threaten human welfare?
3. What do edges and movement corridors have to do with habitat fragmentation?

Test Your Knowledge

MULTIPLE CHOICE: *Choose the one best answer.*

1. According to the Endangered Species Act, a threatened species is
   a. an exotic species that cannot successfully compete with native organisms.
   b. an endemic species that is found nowhere else in the world.
   c. a species that is found in disturbed habitats.
   d. a species that is in danger of extinction in all or a large part of its range.
   e. a species that is likely to become endangered.

2. Ecosystem services include all of the following *except*
   a. pollination of crops.
   b. production of antibiotics and drugs.
   c. purification of air and water.
   d. decomposition of wastes.
   e. reduced impact of weather extremes.

3. The most serious threat to biodiversity is
   a. competition from introduced species.
   b. environmental toxins.
   c. habitat destruction.
   d. overharvesting.
   e. disruptions of community dynamics.

4. Some grassland and conifer forest reserves have effective fire prevention programs. The most likely result of such programs is
   a. an increase in species diversity because fires are prevented.
   b. a change in community composition because fires are natural disturbances that maintain the community structure.

c. the preservation of endangered species in the area.

d. no change in the species composition of the community.

e. succession to a deciduous forest.

5. According to the small population approach, the most important remedy for preserving an endangered species is to

   a. establish a large nature reserve around its habitat.

   b. control the populations of its natural predators.

   c. determine the reason for its decline.

   d. encourage dispersal and an increase in genetic diversity.

   e. set up artificial breeding programs.

6. Which of the following characteristics is typical of biodiversity hot spots?

   a. a large number of endemic species

   b. a high rate of habitat degradation

   c. little species diversity

   d. a large land or aquatic area

   e. large populations of migratory birds

7. Which of the following phenomena may occur when a population falls below its minimum viable population size?

   a. genetic drift

   b. a further reduction in population size

   c. inbreeding

   d. a loss of genetic diversity

   e. All of the above are characteristics of an extinction vortex that the population may enter.

8. Movement corridors are

   a. strips or clumps of habitat that connect isolated habitats.

   b. the routes taken by migratory animals.

   c. connections within a landscape that includes several different ecosystems.

   d. the areas forming the boundary or edge between two ecosystems.

   e. buffer zones for human traffic that promote the long-term viability of protected areas.

9. What does it mean if a population's effective population size (N_e) is the same as its actual population size?

   a. The population is not in danger of becoming extinct.

   b. The population has high genetic diversity.

   c. All the members of the population breed.

   d. The population's minimum viable population will not sustain the population.

   e. The population is being drawn into an extinction vortex.

10. The focus of the declining-population approach to conservation is to

    a. predict a species' minimum viable population size.

    b. transplant members from other populations to increase genetic diversity.

    c. perform a population viability analysis to predict the long-term viability of a population in a particular habitat.

    d. determine the cause of a species' decline and take remedial action.

    e. establish zoned reserves that ensure that human landscapes surrounding reserves support the protected habitats.

11. Given their limited resources, conservation biologists working to preserve biodiversity should assign the highest priority to

    a. the northern spotted owl.

    b. declining keystone species in a community.

    c. a commercially important species.

    d. endangered and threatened vertebrate species.

    e. all declining species.

12. The finding of harmful levels of DDT in grebes (fish-eating birds) following years of trying to eliminate bothersome gnat populations in a lakeshore town is an example of

    a. eutrophication.

    b. biological magnification.

    c. nutrient enrichment.

    d. an edge effect of a landscape ecosystem.

    e. increasing resistance to pesticides.

13. Which of the following phenomena is a direct effect of the thinning of the ozone layer?

    a. a reduction in species diversity

    b. global warming

    c. acid precipitation

    d. an increase in harmful UV radiation reaching Earth

    e. dead zones in the Antarctic Ocean

14. Sustainable development

    a. uses nature reserves to save endangered species so that the rest of Earth's ecosystems can be used for human development.

    b. requires the establishment of zoned reserves.

    c. is economic development that meets the needs of people today without limiting the ability of people to meet their needs in the future.

    d. is the research agenda of the Sustainable Biosphere Initiative to study global change, biodiversity, and Earth's ecosystems.

    e. sustains the productivity of agricultural ecosystems through the use of chemical and organic fertilizers and irrigation.

Answer Section

CHAPTER 1: INTRODUCTION: THEMES IN THE STUDY OF LIFE

INTERACTIVE QUESTIONS

1.1 **a.** The biosphere includes the environments inhabited by life—most regions of land and water, the sediments and rocks below Earth's surface, and the atmosphere up to several kilometers.
b. An ecosystem includes all the living organisms in an area and the nonliving parts of the environment with which they interact.
c. A community consists of all the organisms inhabiting a particular area. Each diverse form of life is called a *species*.
d. A population includes all the individuals of a single species in an area.
e. An organism is an individual living entity.
f. Organs, found in more complex organisms, are body parts that perform a particular function. Organ systems are groups of organs that perform a function.
g. Tissues are groups of cells that collectively perform a specialized function. Several tissues make up an organ.
h. Cells are the fundamental units of life.
i. Organelles are the functional components of cells.
j. Molecules are composed of two or more atoms. Molecules are the chemical units that make up living organisms.

1.2 Systems biology could be useful in predicting how a particular drug or treatment may affect various aspects of the body's functions, and in predicting the effects on various parts of the ecosystem as CO_2 concentration climbs.

1.3 Chemical nutrients are recycled as photosynthesizing plants absorb water, minerals, CO_2, and sunlight, and produce sugar and release O_2; consumers eat plants and other organisms; and decomposers return nutrients to the ecosystem. Energy flows through ecosystems, entering as solar energy, being transformed to chemical energy that drives cellular work, and exiting as heat.

1.4 The order of nucleotides in a DNA molecule that makes up a gene "spells" the instructions, which are transcribed into RNA and then translated into a protein with a specific function.

1.5 These organisms are characterized to a large extent by their mode of nutrition. Plants are photosynthetic, fungi absorb their nutrients from decomposing organic material, and animals ingest other organisms.

1.6 Most species tend to produce more offspring than can survive. Organisms with heritable traits best suited to the environment will tend to survive and leave more offspring. Over time, favorable adaptations will accumulate in a population. New species may arise as small populations are exposed to different environments and natural selection favors different traits.

1.7 **a.** Since most encounters with coral snakes are fatal, predators did not "learn" to avoid them. But predators with genes that somehow made them instinctively avoid coral snakes would have been more likely to pass on those genes to offspring, gradually adapting the predator population to the presence of venomous snakes in their environment.
b. This experiment could not control for the number of predators in each area and thus for the total number of attacks on snakes. By presenting data as the percentage of total attacks, the effect of this variable was eliminated.

1.8 **a.** A hypothesis is less broad in scope than a theory; it is a tentative explanation for a smaller set of observations. A theory can generate many testable hypotheses and is supported by a large body of evidence. Both hypotheses and theories are revised when new data fail to support their predictions.

b. Science seeks to understand natural phenomena; technology is the practical use of scientific knowledge.

SUGGESTED ANSWERS TO STRUCTURE YOUR KNOWLEDGE

1. **a.** With every increase in organizational level, the organization and interactions of component parts lead to new, emergent properties of the dynamic system.

    b. Organisms exchange materials with the living and nonliving components of their environment, resulting in the cycling of chemical nutrients in an ecosystem.

    c. All organisms require energy. Energy flows through ecosystems from sunlight to chemical energy in producers and consumers and then escapes as heat.

d. At each level of biological organization, structure and function are correlated.

e. Cells are the basic units of life. They come in two distinct forms: prokaryotic and eukaryotic.

f. DNA is the molecule of inheritance; it codes information for proteins and the functions of a cell and is passed on from one generation to the next.

g. Organisms regulate their internal environment, usually through negative feedback mechanisms that match supply to demand.

h. Evolution explains the unity and diversity of life. All organisms are modified descendants of common ancestors. Darwin's theory of natural selection leading to unequal reproductive success accounts for the adaptation of populations to Earth's varying environments.

ANSWERS TO TEST YOUR KNOWLEDGE

Multiple Choice:

1. d	**3.** b	**5.** a	**7.** e	**9.** e
2. d	**4.** b	**6.** d	**8.** c	

CHAPTER 2: THE CHEMICAL CONTEXT OF LIFE

INTERACTIVE QUESTIONS

2.1   calcium, phosphorus, potassium, sulfur, sodium, chlorine, magnesium

2.2   neutrons; 15; 15; 16; 31 daltons

2.3   absorb; loses

2.4   **a.** Carbon, $_6$C   **c.** Oxygen, $_8$O

b. Nitrogen, $_7$N   **d.** Magnesium, $_{12}$Mg

2.5   Although each orbital can hold a pair of electrons, each electron fills a separate orbital until no empty orbitals remain.

2.6   **a.** protons
         b. atomic number
         c. element
         d. neutrons

e. mass number or atomic mass
f. isotopes
g. electrons
h. electron shells
i. valence shell

2.7   **a.** H = 1, O = 2, N = 3, C = 4

b. H:H   Ö::Ö   :N:::N: H:C:H with H above and H below

2.8   **a.** Nonpolar; even though N has a high electronegativity, the three pairs of electrons are shared equally between the two N atoms because each atom has an equally strong attraction for the electrons.

b. Polar; N is more electronegative than H and pulls the shared electrons in each covalent bond closer to itself.

c. Nonpolar; C and H have similar electronegativities and share electrons fairly equally between them.

d. The bond is polar because O is more electronegative than C; the C—H bonds are relatively nonpolar.

2.9   **a.** $CaCl_2$
         b. Ca^{2+} is the cation.

2.10

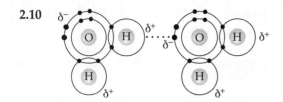

2.11 Water's two covalent bonds with H are spread apart at a 104.5° angle due to the hybridization of the *s* and 3 *p* orbitals.

2.12 6, 6, 6

SUGGESTED ANSWERS TO STRUCTURE YOUR KNOWLEDGE

1.

Particle	Charge	Mass	Location
Proton	+1	1 dalton	nucleus
Neutron	0	1 dalton	nucleus
Electron	−1	negligible	orbitals in electron shells

2. a. The atoms of each element have a characteristic number of protons in their nuclei, referred to as the *atomic number*. In a neutral atom, the atomic number also indicates the number of electrons. The *mass number* is an indication of the approximate mass of an atom and is equal to the number of protons and neutrons in the nucleus.

The *atomic mass* is equal to the mass number and is measured in daltons. Protons and neutrons both have a mass of approximately 1 dalton.

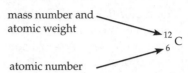

b. The *valence*, an indication of bonding capacity, is the number of unpaired electrons that an atom has in its valence shell.

c. The valence of an atom is most related to the chemical behavior of an atom because it is an indication of the number of bonds the atom will make, or the number of electrons the atom must share in order to reach a filled valence shell.

3. In nonpolar covalent bonds, electrons are equally shared between two atoms. Polar covalent bonds form when a more electronegative atom pulls the shared electrons closer to it, producing a partial negative charge associated with that portion of the molecule and a partial positive charge associated with the atom from which the electrons are pulled. In the formation of ions, an electron is completely pulled away from one atom and transferred to another, creating negatively and positively charged ions (anions and cations). These oppositely charged ions may be attracted to each other in an ionic bond.

ANSWERS TO TEST YOUR KNOWLEDGE

Multiple Choice:

1. b	**7.** c	**13.** e	**19.** b	**25.** c
2. d	**8.** c	**14.** e	**20.** a	**26.** d
3. b	**9.** b	**15.** e	**21.** e	**27.** d
4. d	**10.** e	**16.** d	**22.** c	**28.** a
5. e	**11.** c	**17.** b	**23.** c	**29.** e
6. b	**12.** a	**18.** d	**24.** b	**30.** e

CHAPTER 3: WATER AND LIFE

INTERACTIVE QUESTIONS

3.1

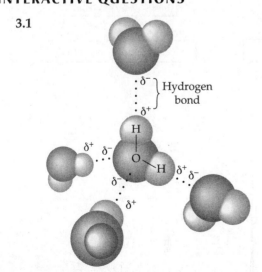

3.2 a. polar water molecules
b. absorbed
c. released
d. specific heat
e. heat of vaporization
f. evaporative cooling
g. solar heat
h. rain
i. ice forms

3.3 a. olive oil: hydrophobic mostly nonpolar
b. sugar: hydrophilic polar
c. salt: hydrophilic ionic
d. candle wax: hydrophobic nonpolar

3.4 a. The molecular mass of $C_3H_6O_3$ is 90 d, the combined atomic masses of its atoms. A mole of lactic acid = 90 g. A 0.5 M solution would require ½ mol or 45 g.
b. 3.01×10^{23}

3.5

[H⁺]	[OH⁻]	pH	Acidic, Basic, or Neutral?
10^{-3}	10^{-11}	3	acidic
10^{-8}	10^{-6}	8	basic
10^{-7}	10^{-7}	7	neutral
10^{-1}	10^{-13}	1	acidic

3.6

$$\underset{\substack{\text{carbonic acid} \\ H^+\text{donor}}}{H_2CO_3} \rightleftharpoons \underset{\substack{\text{bicarbonate} \\ H^+ \text{ acceptor}}}{HCO_3^-} + \underset{\text{hydrogen ion}}{H^+}$$

a. Bicarbonate acts as a base to accept excess H^+ ions when the pH starts to fall; the reaction moves to the left.
b. When the pH rises, H^+ ions are donated by carbonic acid, and the reaction shifts to the right.

3.7 a. $CO_2 + H_2O \rightleftharpoons H_2CO_3 \rightleftharpoons HCO_3^- + H^+$
Increasing $[CO_2]$ will drive these reactions to the right, increasing $[H^+]$.
b. $HCO_3^- \rightleftharpoons CO_3^{2-} + H^+$

Increasing $[H^+]$ will drive this reaction to the left, thus decreasing $[CO_3^{2-}]$.
c. $CO_3^{2-} + Ca^{2+} \rightleftharpoons CaCO_3$. With less CO_3^{2-} available to react with Ca^{2+}, calcification rates would be expected to decrease.

SUGGESTED ANSWERS TO STRUCTURE YOUR KNOWLEDGE

1. a. Cohesion, adhesion
b. A water column is pulled up through plant vessels.
c. Heat is absorbed or released when hydrogen bonds break or form. Water absorbs or releases a large quantity of heat for each degree of temperature change.
d. High heat of vaporization
e. Solar heat is dissipated from tropical seas.
f. Evaporative cooling
g. Evaporation of water cools surfaces of plants and animals.
h. Hydrogen bonds in ice space water molecules farther apart, making ice less dense.
i. Floating ice insulates bodies of water so they don't freeze solid.
j. Versatile solvent
k. Polar water molecules surround and dissolve ionic and polar solutes.

2.

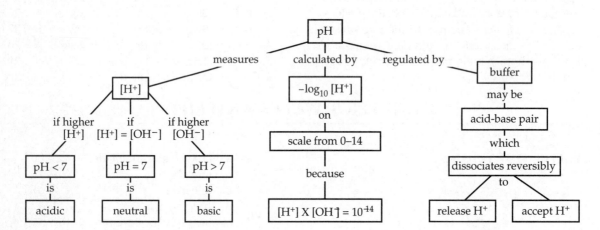

ANSWERS TO TEST YOUR KNOWLEDGE

Multiple Choice:

1. c	**4.** e	**7.** b	**10.** a	**13.** b	**16.** e	**19.** c	**22.** b
2. a	**5.** b	**8.** e	**11.** b	**14.** b	**17.** d	**20.** e	**23.** d
3. e	**6.** a	**9.** e	**12.** d	**15.** c	**18.** e	**21.** b	**24.** d

CHAPTER 4: CARBON AND THE MOLECULAR DIVERSITY OF LIFE

INTERACTIVE QUESTIONS

4.1 The abiotic synthesis of organic compounds from inorganic molecules would be a first step in the origin of life. A variety of organic compounds, including amino acids, were produced in Miller's apparatus, which attempted to simulate the physical and chemical conditions on early Earth.

4.2 Ethanol and dimethyl ether, *structural isomers*, have the same number and kinds of atoms but a different bonding sequence and very different properties. Maleic acid and fumaric acid are *cis-trans isomers* whose C=C bonds fix the spatial arrangement of the molecule. Maleic acid is the *cis* isomer; the same groups are on the same side of the double bond. Although the *enantiomers* L- and D-lactic acid look similar in a flat representation of their structures, they are not superimposable due to the arrangement of four different atoms or groups around an asymmetric carbon.

SUGGESTED ANSWERS TO STRUCTURE YOUR KNOWLEDGE

1.

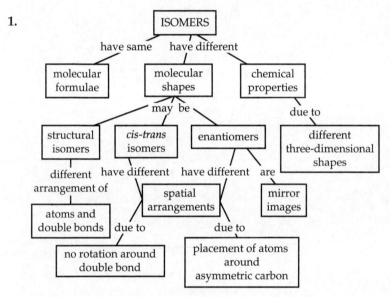

2.

Chemical Group	Molecular Formula	Name and Characteristics of Compounds Containing Group
Hydroxyl	—OH	Alcohols; polar group
Carbonyl	$>C=O$	Aldehyde or ketone; polar group
Carboxyl	—COOH	Carboxylic acid; release H^+
Amino	—NH$_2$	Amines; basic, accept H^+
Sulfhydryl	—SH	Thiols; cross-links stabilize protein structure
Phosphate	—OPO$_3$$^{2-}$	Organic phosphates; involved in energy transfers, adds negative charge
Methyl	—CH$_3$	Methylated compounds; addition may alter expression of genes

ANSWERS TO TEST YOUR KNOWLEDGE

Multiple Choice:

1. e	3. b	5. d	7. c	9. d
2. c	4. e	6. d	8. a	10. a

Matching:

1. a, c	4. e	7. b, f	10. a, d
2. b, f	5. e	8. e	11. e
3. a, d, e	6. a, c, d, e	9. a, c	12. c

CHAPTER 5: THE STRUCTURE AND FUNCTION
OF LARGE BIOLOGICAL MOLECULES

INTERACTIVE QUESTIONS

5.1 dehydration reactions/removal; hydrolysis/addition

5.2
  a. hydroxyl
  b. carbonyl
  c. aldose (carbonyl at end of carbon skeleton)
  d. ketose
  e. rings

5.3

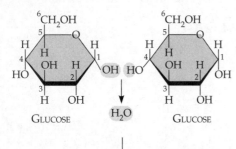

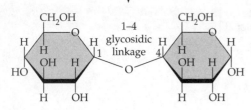

5.4
  a. monosaccharides
  b. (CH$_2$O)
  c. energy compounds
  d. carbon skeletons, monomers
  e. disaccharides
  f. glycosidic linkages
  g. polysaccharides
  h. glycogen
  i. animals
  j. starch
  k. cellulose
  l. chitin

5.5

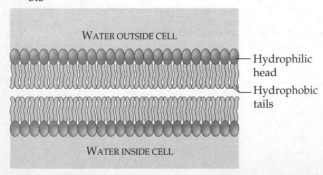

5.6
  a. fats, triacylglycerides
  b. phospholipids
  c. glycerol
  d. fatty acids
  e. unsaturated: has some C═C bonds
  f. saturated: no C═C
  g. phosphate group
  h. cell membranes
  i. steroids
  j. animal cell membrane component (cholesterol), hormones

5.7
  a.

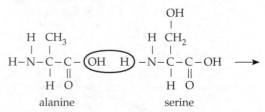

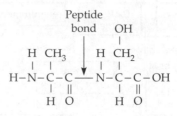

  b. serine's R group is polar; alanine's R group is nonpolar
  c. a polypeptide backbone

5.8
  a. hydrogen bond
  b. hydrophobic interactions and van der Waals interactions
  c. disulfide bridge
  d. ionic bond

These interactions between R groups produce tertiary structure.

5.9

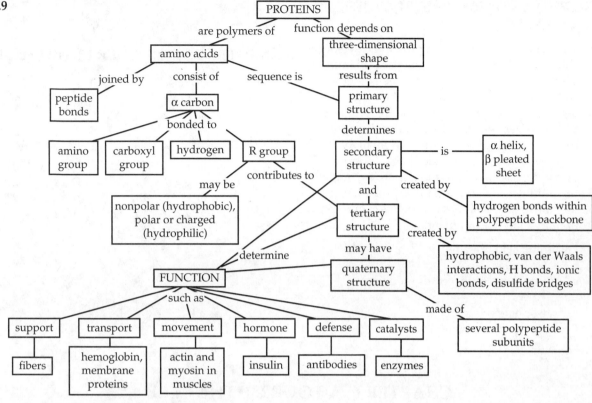

5.10 a.

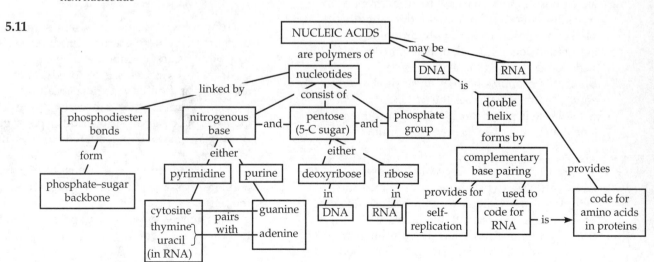

Phosphate
group

Attachment site
for phosphate of
next nucleotide

Deoxyribose

Nitrogenous
base

b. pyrimidine; base has single ring
c. DNA; 2′ carbon in sugar is lacking O

5.11

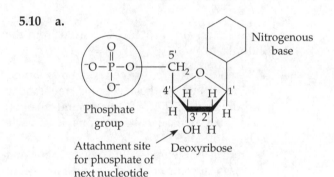

SUGGESTED ANSWERS TO STRUCTURE YOUR KNOWLEDGE

1. The primary structure of a protein is the specific, genetically coded sequence of amino acids in a polypeptide chain. The secondary structure involves the coiling (α helix) or folding (β pleated sheet) of the protein, stabilized by hydrogen bonds along the polypeptide backbone. The tertiary structure involves interactions between the side chains (R groups) of amino acids and produces a characteristic three-dimensional shape for a protein. Quaternary structure occurs in proteins composed of more than one polypeptide.

2. a. amino acid (glycine)
    b. fatty acid
    c. nitrogenous base, purine (adenine)
    d. glycerol
    e. phosphate group
    f. sugar (pentose, ribose)
    g. sugar (triose)

1. b, d	3. c, e, f	5. c	7. e, f
2. a	4. f, g	6. a	8. b

ANSWERS TO TEST YOUR KNOWLEDGE

Matching:

1. A	3. D	5. C	7. B	9. A
2. B	4. C	6. D	8. C	10. A

Multiple Choice:

1. e	6. c	11. d	16. c	21. a
2. c	7. c	12. a	17. c	22. d
3. a	8. d	13. d	18. c	23. c
4. e	9. a	14. e	19. b	24. b
5. c	10. d	15. b	20. b	

CHAPTER 6: A TOUR OF THE CELL

INTERACTIVE QUESTIONS

6.1 a. the study of cell structure
     b. the internal fine structure of cells
     c. the three-dimensional surface topography of a specimen
     d. Light microscopy enables the study of living cells and may introduce fewer artifacts than do TEM and SEM.

6.2 a. a phospholipid bilayer with the hydrophobic tails clustered in the interior and the phosphate heads facing the hydrophilic outside and inside of the cell; proteins are embedded in and attached to the membrane.
     b. area is proportional to a linear dimension squared; 10^2, or 100 times the surface area
     c. volume is proportional to a linear dimension cubed; 10^3, or 1,000 times the volume

6.3 The genetic instructions for specific proteins are transcribed from DNA into messenger RNA (mRNA), which then passes into the cytoplasm to complex with ribosomes, where it is translated into the primary structure of proteins. Ribosomal subunits are synthesized in nucleoli from rRNA (transcribed from DNA) and proteins imported from the cytoplasm.

6.4 a. smooth ER—in different cells may house enzymes that synthesize lipids; metabolize carbohydrates; detoxify drugs and alcohol; store and release calcium ions in muscle cells
     b. nuclear envelope—double membrane that encloses nucleus; pores regulate passage of materials
     c. rough ER—attached ribosomes produce proteins that enter cisternae; produces secretory proteins and membranes
     d. transport vesicle—carries products of ER and Golgi apparatus to various locations
     e. Golgi apparatus—processes products of ER; makes polysaccharides; packages products in vesicles targeted to specific locations
     f. plasma membrane—selective barrier that regulates passage of materials into and out of the cell
     g. lysosome—houses hydrolytic enzymes that digest macromolecules

6.5   **Mitochondrion**

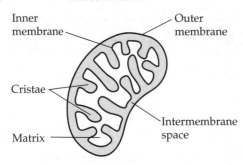

Inner membrane

Outer membrane

Cristae

Matrix

Intermembrane space

Chloroplast

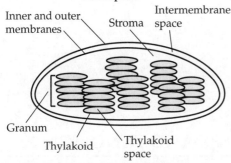

Inner and outer membranes

Stroma

Intermembrane space

Granum

Thylakoid

Thylakoid space

6.6 Peroxisomes do not bud from the endomembrane system, but grow by incorporating proteins and lipids made in cytosol, the ER, or the peroxisome itself; they increase in number by dividing.

6.7 **a.** hollow tube, formed from columns of tubulin dimers; 25-nm diameter
b. cell shape and support (compression resistant), tracks for moving organelles, chromosome movement, beating of cilia and flagella
c. two twisted chains of actin subunits; 7-nm diameter
d. muscle contraction, maintain and change cell shape, pseudopod movement, cytoplasmic streaming, sol-gel transformations
e. supercoiled fibrous proteins of keratin family; 8–12 nm
f. reinforce cell shape (tension bearing); anchor nucleus; form nuclear lamina

6.8   **Two adjacent plant cells**

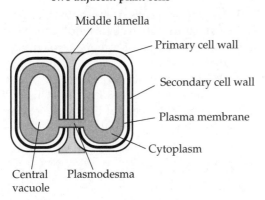

Middle lamella

Primary cell wall

Secondary cell wall

Plasma membrane

Cytoplasm

Central vacuole   Plasmodesma

6.9 **a.** See answer for 6.8.
b. structural support, signaling between environment and cell via fibronectins, integrins, and the cytoskeleton (may affect gene expression), orientation of cell movements

SUGGESTED ANSWERS TO STRUCTURE YOUR KNOWLEDGE

1. **a.** nucleus, chromosomes, centrioles, microtubules (spindle), microfilaments (actin-myosin aggregates pinch apart cells)
  b. nucleus, chromosomes, DNA → mRNA → ribosomes → enzymes and other proteins
  c. mitochondria
  d. ribosomes, rough and smooth ER, Golgi apparatus, transport vesicles
  e. smooth ER (peroxisomes also detoxify substances)
  f. lysosomes, food vacuoles, vesicles enclosing damaged organelles
  g. peroxisomes
  h. cytoskeleton: microtubules, microfilaments, intermediate filaments; extracellular matrix
  i. cilia and flagella (microtubules), microfilaments (actin) in muscles and pseudopodia
  j. plasma membrane, transport vesicles
  k. desmosomes, tight and gap junctions, ECM

2. **a.** structural support, middle lamella glues cells together
  b. storage, waste disposal, protection (toxic compounds), growth
  c. photosynthesis, production of sugars
  d. starch storage
  e. cytoplasmic connections between cells

3. **a.** rough endoplasmic reticulum
  b. smooth endoplasmic reticulum
  c. chromatin
  d. nucleolus
  e. nuclear envelope
  f. nucleus
  g. ribosomes
  h. Golgi apparatus
  i. plasma membrane
  j. mitochondrion
  k. lysosome
  l. cytoskeleton
  m. microtubules
  n. intermediate filaments
  o. microfilaments
  p. microvilli
  q. peroxisome
  r. centrosome (contains a pair of centrioles)
  s. flagellum

4.

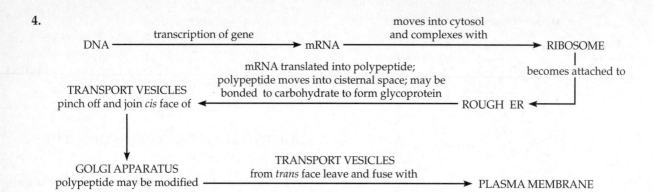

ANSWERS TO TEST YOUR KNOWLEDGE

Multiple Choice:

1. b	4. c	7. e	10. d	13. a	16. e	19. c	21. b
2. a	5. e	8. b	11. d	14. b	17. d	20. e	22. a
3. a	6. b	9. c	12. e	15. c	18. d		

CHAPTER 7: MEMBRANE STRUCTURE AND FUNCTION

INTERACTIVE QUESTIONS

7.1 **a.** hydrophilic phosphate head
b. phospholipid bilayer
c. hydrophobic hydrocarbon tail
d. hydrophobic region of protein
e. hydrophilic region of protein

7.2 **a.** In hybrid human/mouse cells, membrane proteins rapidly intermix (Frye and Edidin, 1970).
b. The cell membranes of the cold lake population would have a higher proportion of unsaturated fatty acids in their phospholipids than that of the population living in warm temperatures.

7.3 transport, enzymatic activity, signal transduction, cell–cell recognition, intercellular joining, attachment to cytoskeleton and ECM (providing support and communication)

7.4 Ions and larger polar molecules, such as glucose, are impeded by the hydrophobic center of the plasma membrane's lipid bilayer. Passage through the center of a lipid bilayer is not fast even for small, polar water molecules.

7.5 Side A initially has fewer free water molecules; side B initially has more. More water molecules are clustered around the glucose in the 1 *M* solution than around the fructose and sucrose, whose combined concentration is 0.9 *M*. Water will move by osmosis from side B to side A.

7.6 The protists will gain water from their hypotonic environment. They have membranes that are less permeable to water and contractile vacuoles that expel excess water.

7.7 **a.** isotonic
b. hypotonic

7.8 Although it speeds diffusion, facilitated diffusion is still passive transport because the solute is moving down its concentration gradient; the process is driven by the concentration gradient and not by energy expended by the cell.

7.9 Three Na^+ are pumped out of the cell for every two K^+ pumped in, resulting in a net movement of positive charge from the cytoplasm to the extracellular fluid.

7.10 **a.** Human cells use receptor-mediated endocytosis to take in cholesterol, which is used for the synthesis of other steroids and membranes.
b. LDL receptor proteins in the plasma membrane are defective, and low-density lipoprotein particles cannot bind and be transported from the blood into the cell.

SUGGESTED ANSWERS TO STRUCTURE YOUR KNOWLEDGE

1.

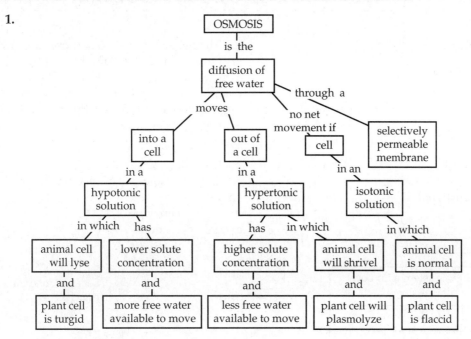

a. II represents facilitated diffusion. The solute is moving through a transport protein and down its concentration gradient. The cell does not expend energy in this transport. Polar molecules and ions may move by facilitated diffusion.

b. III represents active transport because the solute is clearly moving against its concentration gradient and the cell is expending ATP to drive this transport against the gradient.

c. I illustrates diffusion through the lipid bilayer. The solute molecules must be hydrophobic (nonpolar) or very small polar molecules.

d. I and II. Both diffusion and facilitated diffusion are considered passive transport because the solute moves down its concentration gradient and the cell does not expend energy in the process.

ANSWERS TO TEST YOUR KNOWLEDGE

Multiple Choice:

1. b	**3.** c	**5.** b	**7.** c	**9.** a*	**11.** d**	**13.** b	**15.** a	**17.** b	**19.** c
2. e	**4.** e	**6.** d	**8.** e	**10.** c	**12.** a	**14.** d	**16.** e	**18.** d	**20.** a

*Explanation for answer to question 9: This problem involves both osmosis and diffusion. Although the solutions are initially equal in molarity, glucose will diffuse down its concentration gradient until it reaches dynamic equilibrium with a 1.5 M concentration on both sides. The increasing solute concentration on side A will cause water to move into this side, and the water level will rise.

**Explanation for answer to question 11: As the solute in the solution crosses the cell membrane, it increases the concentration of solutes within the cell, reducing the hypertonicity of the solution. As the solute reaches an equal concentration inside and outside the cell, it no longer causes osmotic changes in the cell.

CHAPTER 8: AN INTRODUCTION TO METABOLISM

INTERACTIVE QUESTIONS

8.1
  a. capacity to cause change
  b. kinetic
  c. motion
  d. potential
  e. position
  f. conserved
  g. created nor destroyed
  h. first
  i. transformed or transferred
  j. entropy
  k. second

8.2

	System with High Free Energy	System with Low Free Energy
Stability	low	high
Spontaneous	process will be	process will not be
Equilibrium	moves toward	is at
Work capacity	high	low

8.3

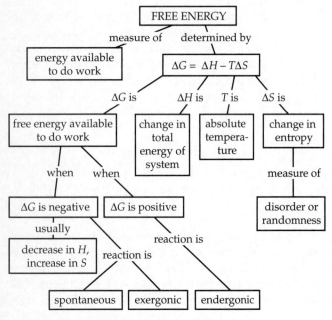

8.4
  a. adenine
  b. ribose
  c. three phosphate groups
  d. A hydrolysis reaction breaks the terminal phosphate bond and releases a molecule of inorganic phosphate: ATP + H_2O → ADP + $\textcircled{P}_i$ + energy

  e. The negatively charged phosphate groups are crowded together, and their mutual repulsion makes this area instable. The chemical change to a more stable state of lower free energy accounts for the relatively high release of energy.

8.5
  a. free energy
  b. transition state
  c. E_A (free energy of activation) without enzyme
  d. E_A with enzyme
  e. ΔG of reaction

8.6

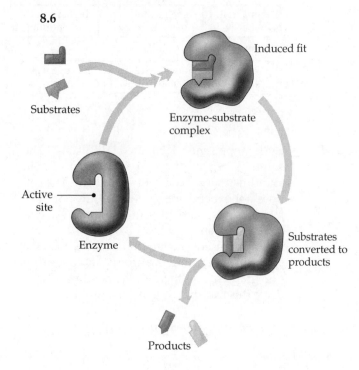

8.7 A competitive inhibitor would mimic the shape of the substrates and compete with them for the active site. A noncompetitive inhibitor would be a shape that could bind to another site on the enzyme molecule and would change the shape of the enzyme such that the active site functions less effectively.

8.8 ATP would act as an inhibitor to catabolic pathways, slowing the breakdown of fuel molecules if the supply of ATP exceeds demand. If ATP supplies drop, ADP would act as an activator of these catabolic enzymes, and more ATP would be produced.

SUGGESTED ANSWERS TO STRUCTURE YOUR KNOWLEDGE

1. Metabolism is the totality of chemical reactions that take place in living organisms. To create and maintain the structural order required for life requires an input of free energy—from sunlight for photosynthetic organisms, and from energy-rich food molecules for other organisms. A cell couples catabolic, exergonic reactions ($-\Delta G$) with anabolic, endergonic reactions ($+\Delta G$), using ATP as the primary energy shuttle between the two.

2. Enzymes are essential for metabolism because they lower the activation energy of the specific reactions they catalyze and allow those reactions to occur extremely rapidly at a temperature conducive to life. By regulating the enzymes it produces, a cell can regulate which of the myriad of possible chemical reactions take place at any given time. Metabolic control also occurs through allosteric regulation and feedback inhibition. The compartmental organization of a cell facilitates a cell's metabolism.

ANSWERS TO TEST YOUR KNOWLEDGE

Multiple Choice:

1. c	6. e	11. c	16. b	21. a
2. c	7. a	12. e	17. d	
3. b	8. e	13. e	18. c	
4. b	9. b	14. e	19. b	
5. d	10. c	15. d	20. c	

Fill in the Blanks:

1. metabolism	6. activation energy
2. anabolic	7. competitive inhibitors
3. potential	8. coenzymes
4. heat	9. feedback inhibition
5. entropy	10. phosphorylated intermediate

CHAPTER 9: CELLULAR RESPIRATION AND FERMENTATION

INTERACTIVE QUESTIONS

9.1 $C_6H_{12}O_6$; 6 CO_2; energy (ATP + heat)

9.2
 a. oxidized
 b. reduced
 c. donates (loses)
 d. oxidizing agent
 e. accepts (gains)

9.3
 a. O_2
 b. glucose
 c. Some is stored in ATP and some is released as heat.

9.4
 a. electron carrier (or acceptor). It is a coenzyme that works with enzymes called dehydrogenases.
 b. NADH

9.5
 a. glycolysis: glucose $\rightarrow$ pyruvate (not technically considered part of cellular respiration)
 b. oxidation of pyruvate and citric acid cycle
 c. oxidative phosphorylation: electron transport and chemiosmosis
 d. substrate-level phosphorylation
 e. substrate-level phosphorylation
 f. oxidative phosphorylation

9.6 The top two arrows show electrons carried by NADH (and $FADH_2$, another electron carrier) to the electron transport chain.
 a. 2 ATP
 b. 2 three-carbon sugars (glyceraldehyde-3-phosphate)
 c. 2 NAD^+
 d. 2 NADH + 2H^+
 e. 4 ATP
 f. 2 pyruvate

9.7
 a. pyruvate
 b. CO_2
 c. NADH + H^+
 d. coenzyme A
 e. acetyl CoA
 f. oxaloacetate
 g. citrate
 h. NADH + H^+
 i. CO_2
 j. CO_2
 k. NADH + H^+
 l. GTP (may make ATP)
 m. $FADH_2$
 n. NADH + H^+

9.8   **a.**   intermembrane space
    b.   inner mitochondrial membrane
    c.   mitochondrial matrix
    d.   electron transport chain
    e.   NADH
    f. NAD^+
    g. $FADH_2$
    h. $2\,H^+ + \frac{1}{2}\,O_2$
    i. H_2O
    j.   chemiosmosis
    k.   ATP synthase
    l.   ADP + Ⓟ$_i$
    m.   ATP

9.9   **a.** -2
    b.   4

c.   citric acid cycle
d.   26 or 28
e.   32
f.   2
g.   6
h.   2
i.   2

9.10   Respiration yields up to 16 times more ATP than does fermentation. By oxidizing pyruvate to CO_2 and passing electrons from NADH (and $FADH_2$) through the electron transport chain, respiration can produce a maximum of 32 ATP compared to the 2 net ATP that are produced by fermentation.

SUGGESTED ANSWERS TO STRUCTURE YOUR KNOWLEDGE

1.   Use Interactive Questions 9.5, 9.6, 9.7, and 9.8 to help you review these pathways.

2.

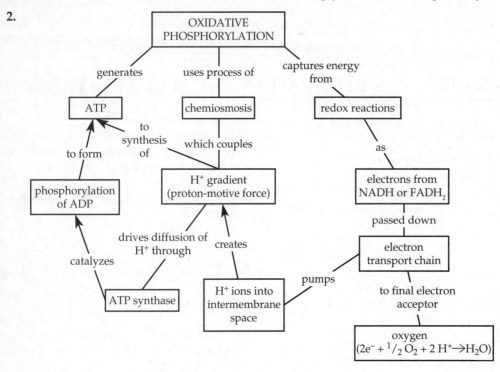

3.

Process	Main Function	Inputs	Output
Glycolysis	Oxidation of glucose to 2 pyruvate, 2 ATP net	glucose 2 ATP 2 NAD$^+$ 4 ADP + P_i	2 pyruvate 4 ATP (2 net) 2 NADH + 2 H$^+$ 2 H$_2$O
Pyruvate to acetyl CoA	Oxidation of pyruvate to acetyl CoA, which then enters citric acid cycle	2 pyruvate 2 CoA 2 NAD$^+$	2 acetyl CoA 2 CO$_2$ 2 NADH + 2 H$^+$
Citric acid cycle	Acetyl CoA is combined with oxaloacetate to produce citrate, which is cycled back to oxaloacetate as redox reactions produce NADH and FADH$_2$; ATP is formed by substrate-level phosphorylation, and CO$_2$ is released.	2 acetyl CoA 2 ADP + P_i 6 NAD$^+$ 2 FAD	4 CO$_2$ 2 ATP 6 NADH + 6 H$^+$ 2 FADH$_2$
Oxidative phosphorylation (Electron transport and chemiosmosis)	NADH (from glycolysis and citric acid) and FADH$_2$ transfer electrons to electron transport chain. In a series of redox reactions, H$^+$ is pumped into intermembrane space, and electrons are delivered to ½ O$_2$. Proton-motive force drives H$^+$ through ATP synthase to make ATP.	10 NADH + 10 H$^+$ 2 FADH$_2$ H$^+$ + O$_2$ 28 ADP + P_i	10 NAD$^+$ 2 FAD H$_2$O 28 ATP (max)
Fermentation	Anaerobic catabolism: glycolysis followed by oxidation of NADH to NAD$^+$ so glycolysis can continue. Pyruvate is either reduced to ethyl alcohol and CO$_2$ or to lactate.	See glycolysis above 2 pyruvate 2 NADH	2 ATP 2 NAD$^+$ 2 ethanol and 2 CO$_2$ or 2 lactate

ANSWERS TO TEST YOUR KNOWLEDGE

Multiple Choice:

1. a	**4.** e	**7.** b	**10.** a	**13.** e	**16.** e	**19.** e	**22.** d
2. b	**5.** c	**8.** b	**11.** a	**14.** e	**17.** b	**20.** c	**23.** c
3. c	**6.** d	**9.** d	**12.** d	**15.** d	**18.** c	**21.** b	

CHAPTER 10: PHOTOSYNTHESIS

INTERACTIVE QUESTIONS

10.1
  a. outer membrane
  b. granum
  c. inner membrane
  d. thylakoid space
  e. thylakoid
  f. stroma

10.2
  a. light
  b. H$_2$O
  c. light reactions in thylakoid membranes
  d. O$_2$
  e. ATP
  f. NADPH
  g. CO$_2$

  h. Calvin cycle in stroma
  i. [CH$_2$O] (sugar)

10.3 The solid line is the absorption spectrum; the dotted line is the action spectrum. Some wavelengths of light, particularly in the blue and the yellow-orange range, result in a higher rate of photosynthesis than would be indicated by the absorption of those wavelengths by chlorophyll *a*. These differences are partially accounted for by accessory pigments, such as chlorophyll *b* and carotenoids, which absorb light energy from different wavelengths and make that energy available to drive photosynthesis.

10.4 A photosystem contains light-harvesting complexes of pigment molecules (chlorophyll *a*, chlorophyll *b*, and carotenoids) bound to particular proteins, and a reaction center complex, which includes two chlorophyll *a* molecules (P680 or P700) and a primary electron acceptor.

10.5
  a. photosystem II
  b. photosystem I
  c. water (H_2O)
  d. oxygen ($\frac{1}{2} O_2$)
  e. P680, reaction-center chlorophyll *a*
  f. primary electron acceptor
  g. electron transport chain
  h. photophosphorylation by chemiosmosis
  i. ATP
  j. P700, reaction-center chlorophyll *a*
  k. primary electron acceptor
  l. $NADP^+$ reductase
  m. NADPH

ATP and NADPH provide the chemical energy and reducing power for the Calvin cycle.

10.6
  a. Ferredoxin (Fd) passes the electrons to the cytochrome complex in the electron transport chain, from which they return to $P700^+$.
  b. Electrons from P680 are not passed to P700. Without the oxidizing agent $P680^+$, water is not split. Fd does not pass electrons to $NADP^+$ reductase to form NADPH.
  c. Electrons do pass down the electron transport chain, and the energy released by their "fall" drives chemiosmosis.

10.7
  a. in the thylakoid space (pH of about 5)
  b. (1) transport of protons into the thylakoid space as Pq transfers electrons to the cytochrome complex; (2) protons from the splitting of water remain in the thylakoid space; (3) removal of H^+ in the stroma during the reduction of $NADP^+$.

10.8
  a. carbon fixation
  b. reduction
  c. regeneration of CO_2 acceptor (RuBP)
  d. $3 CO_2$
  e. ribulose bisphosphate (RuBP)
  f. rubisco
  g. 3-phosphoglycerate
  h. $6 ATP \rightarrow 6 ADP$
  i. 1,3-bisphosphoglycerate
  j. $6 NADPH \rightarrow 6 NADP^+$
  k. $6 \,\textcircled{P}_i$
  l. glyceraldehyde-3-phosphate (G3P)
  m. G3P
  n. glucose and other organic compounds
  o. $3 ATP \rightarrow 3 ADP$

9 ATP and 6 NADPH are required to synthesize one G3P.

10.9 Photorespiration may be an evolutionary relic from the time when there was little O_2 in the atmosphere and the ability of rubisco to distinguish between O_2 and CO_2 was not critical. Photorespiration appears to protect plants from damaging products of the light reactions that build up when the Calvin cycle slows due to a lack of CO_2.

10.10
  a. in the bundle-sheath cells
  b. Carbon is initially fixed into a four-carbon compound in the mesophyll cells by PEP carboxylase. When this compound is broken down in the bundle-sheath cells, CO_2 is maintained at a high enough concentration that rubisco does not accept O_2 and cause photorespiration.
  c. ATP is used to convert pyruvate, returning from the bundle-sheath cells, to PEP in the mesophyll cells. Bundle-sheath cells produce this ATP using PS I and cyclic electron flow. (They lack PS II and thus do not generate O_2, which would compete with CO_2 for binding to rubisco.)

SUGGESTED ANSWERS TO STRUCTURE YOUR KNOWLEDGE

1.

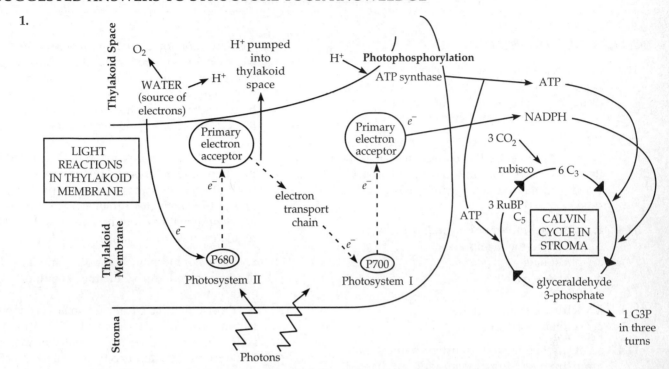

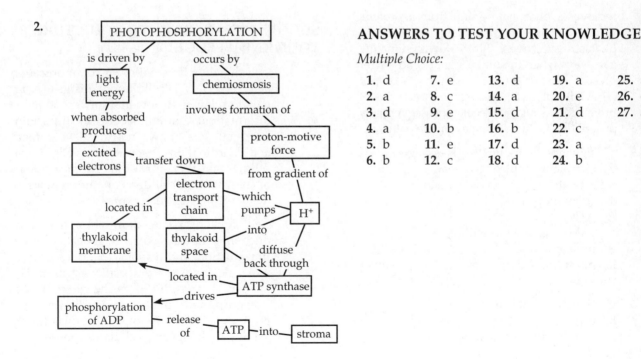

ANSWERS TO TEST YOUR KNOWLEDGE

Multiple Choice:

1. d	**7.** e	**13.** d	**19.** a	**25.** a
2. a	**8.** c	**14.** a	**20.** e	**26.** d
3. d	**9.** e	**15.** d	**21.** d	**27.** a
4. a	**10.** b	**16.** b	**22.** c	
5. b	**11.** e	**17.** d	**23.** a	
6. b	**12.** c	**18.** d	**24.** b	

CHAPTER 11: CELL COMMUNICATION

INTERACTIVE QUESTIONS

11.1 Plant hormones (often called *plant growth regulators*) may reach their target cells by traveling through vessels, between cells via plasmodesmata, or even through the air as a gas.

11.2 The G protein is also a GTPase that hydrolyzes its bound GTP to GDP and inactivates itself. It then dissociates from the enzyme it had activated, and that enzyme returns to its original state.

11.3
a. signaling molecules (ligands)
b. α helix in the membrane
c. phosphates
d. 6 ATP → 6 ADP
e. tyrosines
f. activated relay proteins
g. cellular responses

11.4
a. A protein kinase transfers a phosphate group from ATP to a protein; adding a charged phosphate group causes a shape change that usually activates the protein.
b. A protein phosphatase is an enzyme that removes a phosphate group from a protein, usually inactivating the protein. Protein phosphatases effectively shut down signaling pathways when the initial signal is no longer present.
c. A phosphorylation cascade is a series of protein kinase relay molecules that sequentially phosphorylate the next kinase in the pathway.

11.5
a. signaling molecule (first messenger)
b. G protein-coupled receptor
c. activated G protein (GTP bound)
d. adenylyl cyclase
e. ATP
f. cAMP (second messenger)
g. protein kinase A
h. phosphorylation cascade to cellular response

11.6
a. signaling molecule
b. G protein
c. membrane phospholipid
d. IP_3
e. Ca^{2+}
f. endoplasmic reticulum

11.7
a. Signaling molecules reversibly bind to receptors, and when they leave a receptor, the receptor reverts to its inactive form. The concentration of signaling molecules influences how many bound at any time, with a threshold number of receptors with bound molecules required for the response to occur.
b. Activated G proteins are inactivated when the GTPase portion of the protein converts GTP to GDP.
c. Phosphodiesterase converts cAMP to AMP, thus damping out this second messenger.
d. Protein phosphatases remove phosphate groups from activated proteins. The balance of active protein kinases and active phosphatases regulates the activity of many proteins.

11.8
a. The basic mechanism for programmed cell death evolved early in the evolution of eukaryotes.
b. Programmed cell death in humans is important in the normal development of the nervous system, fingers, and toes; for normal functioning of the immune system; and to prevent the development of cancerous cells (by the internal triggering of apoptosis in cells with DNA damage).

SUGGESTED ANSWERS TO STRUCTURE YOUR KNOWLEDGE

1. Cell signaling is essential for communication between cells. Unicellular organisms use signals to relay environmental or reproductive information. Communication in multicellular organisms allows for the development and coordination of specialized cells. Extracellular signals control the crucial activities of cells, such as cell division, differentiation, metabolism, and gene expression.

2. Cell signaling occurs through signal transduction pathways that include reception, transduction, and response. First a signaling molecule binds to a specific receptor. The message is transduced as the receptor activates a protein that may relay the message through a sequence of activations, finally leading to the specific cellular response.

3. The sequence described in the previous answer results in the activation of cellular proteins. When the signal is transduced to activate a transcription factor, however, the cellular response is a change in gene expression and the production of new proteins.

4. In an enzyme cascade, each step in the pathway activates multiple substrates of the next step, thus amplifying the original message to produce potentially millions of activated proteins and thus a large cellular response to a few signals.

Multiple Choice:

1. b	**4.** b	**7.** d	**10.** a	**13.** b
2. d	**5.** e	**8.** c	**11.** c	**14.** c
3. a	**6.** c	**9.** e	**12.** e	**15.** b

CHAPTER 12: THE CELL CYCLE

INTERACTIVE QUESTIONS

12.1 **a.** 46
 b. 23
 c. 92

12.2 **a.** Growth—most organelles and cell components are produced continuously throughout these subphases.
 b. DNA synthesis and chromosome duplication

12.3 Refer to main text Figure 12.6 for chromosome diagrams (although you were asked to draw only four chromosomes, and the text shows six).
 a. G$_2$ of interphase
 b. prophase
 c. prometaphase
 d. metaphase
 e. anaphase
 f. telophase and cytokinesis
 g. centrosomes (with centrioles)
 h. chromatin (duplicated)
 i. nuclear envelope
 j. nucleolus
 k. early mitotic spindle
 l. aster
 m. nonkinetochore microtubules
 n. metaphase plate
 o. spindle
 p. cleavage furrow
 q. nuclear envelope forming

12.4 **a.** MPF is a complex of cyclin and Cdk (cyclin-dependent kinase) that initiates mitosis by phosphorylating proteins and other kinases.
 b. MPF concentration is high as it triggers the onset of mitosis but is reduced at the end of mitosis because it depends on the concentration of cyclin in the cell. The Cdk level is constant throughout the cell cycle, but the level of cyclin varies because active MPF starts a process that degrades cyclin. Thus, MPF regulates its own level and can only become active when sufficient cyclin accumulates after being synthesized in the S and G$_2$ phases of the next interphase.

SUGGESTED ANSWERS TO STRUCTURE YOUR KNOWLEDGE

1. **a.** anaphase
 b. interphase
 c. late telophase
 d. metaphase

2. Interphase: 90% of cell cycle; growth and DNA replication.
 • G$_1$ phase: The chromosome consists of a long, thin chromatin fiber made of DNA and associated proteins. Growth and metabolic activities occur.
 • S phase—synthesis of DNA: The chromosome is duplicated; two exact copies, called sister chromatids, are produced and held together by cohesins along their length. Growth and metabolic activities continue.
 • G$_2$ phase: Growth and metabolism continue.
 • Mitotic phase: cell division
 • Prophase: The sister chromatids, held together by sister chromatid cohesion and at the centromere, become tightly coiled and condensed.
 • Prometaphase: Kinetochore fibers from opposite ends of the mitotic spindle attach to the kinetochores of the sister chromatids; the chromosome moves toward midline.
 • Metaphase: The centromere of the chromosome is aligned at the metaphase plate along with the centromeres of the other chromosomes.
 • Anaphase: Cohesins are cleaved and the sister chromatids separate (now considered

individual chromosomes) and move to opposite poles of the cell.

• Telophase: Chromatin fiber of chromosome uncoils and is surrounded by reforming nuclear membrane.

3.

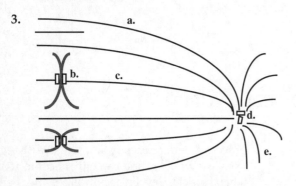

a. **nonkinetochore microtubule:** elongate cell by lengthening and "walking" past microtubules from the opposite pole

b. **kinetochore:** protein and DNA structure in region of centromere where microtubules attach

c. **kinetochore microtubule:** move chromosomes to metaphase plate and separate chromosomes as motor proteins of kinetochores "walk" toward opposite poles and the microtubules disassemble

d. **centrosome and centrioles:** region of mitotic spindle formation; organizes microtubules

e. **aster:** radiating spindle fibers (in animal cells)

ANSWERS TO TEST YOUR KNOWLEDGE

Fill in the Blanks:

1. G_0
2. anaphase
3. prophase
4. cytokinesis
5. S phase
6. metaphase
7. telophase
8. prophase
9. prometaphase
10. G_1 phase

Multiple Choice:

1. d	4. e	7. c	10. a	13. e
2. c	5. b	8. a	11. c	14. d
3. d	6. c	9. b	12. c	

CHAPTER 13: MEIOSIS AND SEXUAL LIFE CYCLES

INTERACTIVE QUESTIONS

13.1
a. 14; 7
b. 28; 1
c. 56. Sister chromatids are produced when a chromosome duplicates before cell division. They are joined at the centromere and attached along their lengths. Nonsister chromatids are found on different chromosomes of a homologous pair (homologs).

13.2
a. meiosis
b. fertilization
c. zygote
d. mitosis
e. gametes
f. fertilization
g. zygote
h. meiosis
i. spores
j. gametes
k. fertilization
l. 2n
m. zygote
n. meiosis

13.3
a. metaphase II
b. prophase I
c. anaphase I
d. metaphase I
e. anaphase II
f. telophase I

Proper sequence: b. d. c. f. a. e.

13.4
a. 2^{23}, approximately 8.4 million
b. about 70 trillion ($2^{23} \times 2^{23}$)
c. This number of combinations does not take into account the additional variation of recombinant chromosomes produced by crossing over.

SUGGESTED ANSWERS TO STRUCTURE YOUR KNOWLEDGE

1. a. sister chromatids
   b. centromere
   c. pair of homologous chromosomes
   d. nonsister chromatids

2. a. Chromosome duplication; sister chromatids attached at centromere and by sister chromatid cohesion along their lengths
   b. Chromosomes condense. Synapsis of homologous pairs (held by synaptonemal complex); crossing over (exchange of corresponding DNA segments) is evident at chiasmata

c. Homologous pairs line up independently at metaphase plate

d. Homologous pairs of chromosomes separate and homologs move toward opposite poles; sister chromatids remain attached at centromere

e. Haploid set of chromosomes, each consisting of two sister chromatids, aligns at metaphase plate; sister chromatids not identical due to crossing over

f. Sister chromatids separate and move to opposite poles as individual chromosomes

3.

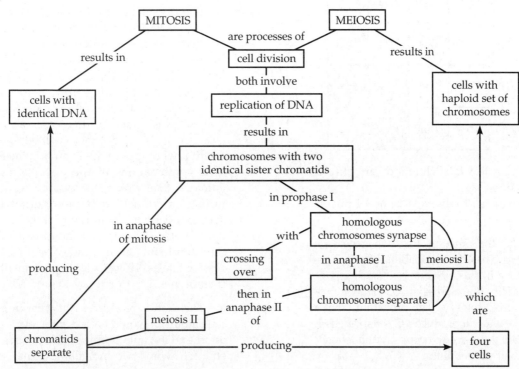

ANSWERS TO TEST YOUR KNOWLEDGE

Multiple Choice:

1. c	**4.** e	**7.** c	**10.** e	**13.** b	**16.** c	**19.** e	**21.** e
2. e	**5.** b	**8.** b	**11.** b	**14.** d	**17.** b	**20.** b	**22.** c
3. b	**6.** a	**9.** d	**12.** a	**15.** b	**18.** d		

CHAPTER 14: MENDEL AND THE GENE IDEA

INTERACTIVE QUESTIONS

14.1
  a. *R*
  b. *r*
  c. F_1 Generation
  d. *Rr*
  e. F_2 Generation
  f. *R*
  g. *r*
  h. *Rr* ◯
  i. *Rr* ◯
  j. *rr* 〰

  k. 3 round:1 wrinkled
  l. 1 *RR*:2*Rr*:1*rr*

14.2
  a. all tall (*Tt*) plants
  b. 1:1 tall (*Tt*) to dwarf (*tt*)

You can use a Punnett square to determine the expected outcome of a test cross, but a shortcut is to use only one column for the recessive individual's gametes, since they produce only one type of gamete. For example, in Figure 14.7 you could use only one column for sperm because all sperm contain the recessive *p* allele.

14.3 **a.** all tall purple plants
b. *TtPp*
c. *TP, Tp, tP, tp*
d.

sperm

e. 9 tall purple:3 tall white:3 dwarf purple: 1 dwarf white
f. 12:4 or 3:1 tall to dwarf; 12:4 or 3:1 purple to white

14.4 **a.** Consider the outcome for each gene as a monohybrid cross. The probability that a cross of *Aa* × *Aa* will produce an *A_* offspring is $\frac{3}{4}$. The probability that a cross of *Bb* × *bb* will produce a *B_* offspring is $\frac{1}{2}$. The probability that a cross of *cc* × *CC* will produce a *C_* offspring is 1. To have all of these events occur simultaneously, multiply their probabilities: $\frac{3}{4} \times \frac{1}{2} \times 1 = \frac{3}{8}$.
b. First determine what genotypes would fill the requirement of at least two dominant traits. Offspring could be *A_bbC_*, *aaB_C_*, or *A_B_C*. The genotype *A_B_cc* is not possible. Can you see why?

Probability of *A_bbC_* = $\frac{3}{4} \times \frac{1}{2} \times 1 = \frac{3}{8}$
Probability of *aaB_C_* = $\frac{1}{4} \times \frac{1}{2} \times 1 = \frac{1}{8}$
Probability of *A_B_C_* = $\frac{3}{4} \times \frac{1}{2} \times 1 = \frac{3}{8}$

Probability of offspring showing at least two dominant traits is the sum of these independent probabilities, or $\frac{7}{8}$.
c. There is only one type of offspring that can show only one dominant trait (*aabbCc*). Can you see why? Its probability is $\frac{1}{4} \times \frac{1}{2} \times 1 = \frac{1}{8}$.

14.5 **a.** A: I^AI^A and I^Ai
b. B: I^BI^B and I^Bi
c. AB: I^AI^B
d. O: *ii*

14.6 The ratio of offspring from this *MmBb* × *MmBb* cross would be 9:3:4, a common ratio when one gene is epistatic to another. All epistatic ratios are modified versions of 9:3:3:1.

Phenotype	Genotype	Ratio
Black	*M_B_*	$\frac{3}{4} \times \frac{3}{4} = \frac{9}{16}$
Gray	*M_bb*	$\frac{3}{4} \times \frac{1}{4} = \frac{3}{16}$
White	*Mm_ _*	$\frac{1}{4} \times 1 = \frac{1}{4}$ or $\frac{4}{16}$

14.7 **a.** The parental cross produced 25-cm tall F_1 plants, all *AaBbCc* plants with 3 units of 5 cm added to the base height of 10 cm.
b. As a general rule, in the polygenic inheritance of a quantitative character the number of phenotypes resulting from a cross of heterozygotes equals the number of alleles involved plus one. In this case, six alleles (*AaBbCc*) + 1 = 7. So, there will be seven different phenotypes in the F_2 among the 64 possible combinations of the eight types of F_1 gametes. (See why you wouldn't want to go through a Punnett square to figure that out!) These phenotypes will range from six dominant alleles (40 cm), five dominant (35 cm), four dominant (30 cm), and so on, to all six recessive alleles (10 cm), yielding a total of seven phenotypes.

14.8 **a.** This trait is recessive. If it were dominant, then albinism would be present in every generation, and it would be impossible to have albino children with two nonalbino (homozygous recessive) parents.
b. father *Aa*; mother *Aa*, because neither parent is albino and they have albino offspring (*aa*)
c. mate 1 *AA* (probably); mate 2 *Aa*; grandson 4 *Aa*
d. The genotype of son 3 could be *AA* or *Aa*. If his wife is *AA*, then he could be *Aa* (since both his parents are carriers), and the recessive allele never would be expressed in his offspring. Even if he and his wife were both carriers (heterozygotes), there would be a $^{243}/_{1024}$ ($\frac{3}{4} \times \frac{3}{4} \times \frac{3}{4} \times \frac{3}{4} \times \frac{3}{4}$) or 24% chance that all five children would be normally pigmented.

14.9 **a.** $\frac{1}{4}$
b. $\frac{2}{3}$. Of offspring with a normal phenotype, $\frac{2}{3}$ would be predicted to be heterozygotes and, thus, carriers of the recessive allele.

14.10 Both sets of prospective grandparents must have been carriers. The prospective parents do not have the disorder, so they are not homozygous recessive. Thus, each has a $\frac{2}{3}$ chance of being a heterozygote carrier. The probability that both parents are carriers is $\frac{2}{3} \times \frac{2}{3} = \frac{4}{9}$; the chance that two heterozygotes will have a recessive homozygous child is $\frac{1}{4}$. The overall chance that a child will inherit the disease is $\frac{4}{9} \times \frac{1}{4} = \frac{1}{9}$. The fact that the first two children are unaffected does not establish the genotype of the parents. Thus the third child would also have a $\frac{1}{9}$ chance of inheriting the disorder. Should the third child have the disease, however, this would establish that both parents are carriers, and the chance that a subsequent child would have the disease is now estimated at $\frac{1}{4}$.

SUGGESTED ANSWERS TO STRUCTURE YOUR KNOWLEDGE

1. Mendel's law of segregation occurs in anaphase I, when alleles segregate as homologs and move to opposite poles of the cell. The two cells formed from this division have one-half the number of chromosomes and one copy of each gene (although sister chromatids still have to separate in anaphase II). Mendel's law of independent assortment relates to the lining up of attached homologous chromosomes at the equatorial plate in a random fashion during metaphase I. Genes on different chromosomes will assort independently into gametes.

2. Aa = 2 different gametes $AaBb$ = 4 gametes
   $AaBbCc$ = 8 gametes $AABbCc$ = 4 gametes
   A general formula is 2^n, where n is the number of gene loci that are heterozygous.

3. Always look to the F_1 heterozygote resulting from a cross of true-breeding parents. If alleles show complete dominance/recessiveness, then the heterozygote will have a phenotype identical to one parent. In incomplete dominance, the F_1 phenotype will be intermediate between that of the parents. If codominant, the phenotype of both alleles will be exhibited in the heterozygote.

ANSWERS TO GENETICS PROBLEMS

1. White alleles are dominant to yellow alleles. If yellow were dominant, then you should be able to get white squash from a cross of two yellow heterozygotes.

2. **a.** $\frac{1}{4}$ [$\frac{1}{2}$ (to get AA) $\times \frac{1}{2}$ (bb)]
   b. $\frac{1}{8}$ [$\frac{1}{4}$ (aa) $\times \frac{1}{2}$ (BB)]
   c. $\frac{1}{2}$ [1 (Aa) $\times \frac{1}{2}$ (Bb) $\times$ 1 (Cc)]
   d. $\frac{1}{32}$ [$\frac{1}{4}$ (aa) $\times \frac{1}{4}$ (bb) $\times \frac{1}{2}$ (cc)]

3. Since flower color shows incomplete dominance, use symbols such as C^R and C^W for those alleles. There will be six phenotypic classes in the F_2 instead of the normal four classes found in a 9:3:3:1 ratio. You could find the answer with a Punnett square, but multiplying the probabilities of the monohybrid crosses is more efficient.

Tall red	$T\_C^RC^R$	$\frac{3}{4} \times \frac{1}{4} = \frac{3}{16}$
Tall pink	$T\_C^RC^W$	$\frac{3}{4} \times \frac{1}{2} = \frac{3}{8}$ or $\frac{6}{16}$
Tall white	$T\_C^WC^W$	$\frac{3}{4} \times \frac{1}{4} = \frac{3}{16}$
Dwarf red	$tt\ C^RC^R$	$\frac{1}{4} \times \frac{1}{4} = \frac{1}{16}$
Dwarf pink	$tt\ C^RC^W$	$\frac{1}{4} \times \frac{1}{2} = \frac{1}{8}$ or $\frac{2}{16}$
Dwarf white	$tt\ C^WC^W$	$\frac{1}{4} \times \frac{1}{4} = \frac{1}{16}$

4. Determine the possible genotypes of the mother and child. Then find the blood groups for the father that could not have resulted in a child with the indicated blood group.
   a. no groups exonerated
   b. A or O
   c. A or O
   d. AB only
   e. B or O

5. The parents are $CcBb$ and $Ccbb$. Right away you know that all cc offspring will die and no BB black offspring are possible because one parent is bb. Only four phenotypic classes are possible. Determine the proportion of each type by applying the law of multiplication.

Lethal ($cc\_\_$)	$\frac{1}{4}$ offspring die
Normal brown ($CCBb$)	$\frac{1}{4} \times \frac{1}{2} = \frac{1}{8}$
Normal white ($CCbb$)	$\frac{1}{4} \times \frac{1}{2} = \frac{1}{8}$
Deformed brown ($CcBb$)	$\frac{1}{2} \times \frac{1}{2} = \frac{1}{4}$
Deformed white ($Ccbb$)	$\frac{1}{2} \times \frac{1}{2} = \frac{1}{4}$
Ratio of viable offspring:	1:1:2:2

6. Father's genotype must be Pp since polydactyly is dominant and he has had one normal child. Mother's genotype is pp. The chance of the next child having normal digits is $\frac{1}{2}$ or 50% because the mother can only donate a p allele and there is a 50% chance that the father will donate a p allele.

7. The genotypes of the puppies were $\frac{3}{8}$ $B\_S\_$, $\frac{3}{8}$ B_ss, $\frac{1}{8}$ $bbS\_$, and $\frac{1}{8}$ $bbss$. Because recessive traits show up in the offspring, both parents had to have had at least one recessive allele for both genes. Black:chestnut occurs in a 6:2 or 3:1 ratio, indicating a heterozygous cross. Solid:spotted occurs in a 4:4 or 1:1 ratio, indicating a cross between a heterozygote and a homozygous recessive. Parental genotypes were $BbSs \times Bbss$.

8. The 1:1 ratio of the first cross indicates a cross between a heterozygote and a homozygote. The second cross indicates that the hairless hamsters cannot have a homozygous genotype, because then only hairless hamsters would be produced. Let us say that hairless hamsters are *Hh* and normal-haired are *HH*. The second cross of heterozygotes should yield a 1:2:1 genotype ratio. The 1:2 ratio indicates that the homozygous recessive genotype may be lethal, with embryos that are *hh* never developing.

9. **a.** Tail length appears to be a quantitative character. A general rule for the number of alleles in a cross involving a quantitative character is one less than the number of phenotypic classes. Here there are five phenotypic classes, thus four alleles or two gene pairs. In this cross, the phenotypic ratio is approximately 1:4:6:4:1. The ratio base of 16 also indicates a dihybrid cross. Each dominant gene appears to add 5 cm in tail length. The 15-cm pigs are double recessives; the 35-cm pigs are double dominants.
b. A cross of a 15-cm pig and a 30-cm pig would be equivalent to *aabb* × *AABb*. Offspring would have either one or two dominant alleles and would have tail lengths of 20 or 25 cm.

10. **a.** The black parent would be *CC^{ch}*. The Himalayan parent could be either *C^{h}C^{h}* or *C^{h}c*. First, list the alleles in order of dominance: *C, C^{ch}, C^{h}, c*. To approach this problem you should write down as much of the genotype as you can be sure about for each phenotype involved. The parents were black (*C_*) and Himalayan (*C^{h}_*). The offspring genotypes would be black (*C_*) and Chinchilla (*C^{ch}_*). Right away you can tell that the black parent must have been *CC^{ch}* because half of the offspring are Chinchilla; *C^{ch}* is dominant over *C^{h}* and *c* and could not have been present in the Himalayan parent. The genotype of this parent could be either *C^{h}C^{h}* or *C^{h}c*. A test-cross with a white rabbit would be necessary to determine this genotype.
b. You cannot definitely determine the genotype of the parents in the second cross. You can, however, eliminate some genotypes. The black parent could not be *CC* or *CC^{ch}*, because both the *C* and *C^{ch}* alleles are dominant to *C^{h}*, and Himalayan offspring were produced. The Chinchilla parent could not be *C^{ch}C^{ch}* for the same reason. One or both parents had to have *C^{h}* as their second allele; one, but not both, might have *c* as their second allele.

11. **a.** First figure out possible genotypes: _ _*E*_ = golden (any *B* combination with at least one *E*), *B_ee* = black, *bbee* = brown. All you know at the start is that both parents are _ _*E*_. Since you see non-golden offspring, you know that both parents had to be heterozygous for *Ee*, and at least one was heterozygous for *Bb* (to get black and brown offspring). Since black and brown are in a 1:1 ratio (like a test-cross ratio), then one hamster was *Bb* and the other was *bb*. You need to consider the results from the second cross to know which hamster was *Bb*.
b. For the second cross, you now know that the black hamster 3 is *B?ee* and the golden hamster 2 is *?bEe*. A ratio of approximately 3:1 black to brown looks like the results of a monohybrid cross, so both parents must be *Bb*. So hamster 3 is *Bbee*, and hamster 2 must be *BbEe*. That means that hamster 1 must be *bbEe*.

12.

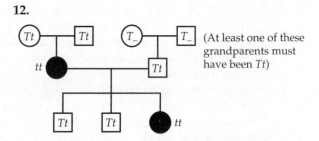

13. Since the parents were true-breeding for two characters, the F₁s would be dihybrids. Since all F₁s had red, terminal flowers, those two traits must be dominant, and their genotypes could be represented as *RrTt*. One would predict an F₂ phenotypic ratio of 9 red, terminal:3 red, axial:3 white, terminal:1 white, axial. If 100 offspring were counted, one would expect approximately 19 ($^3/_{16}$ × 100) plants with red, axial flowers.

14. Because the F₂ generation is a ratio based on 16, one can conclude that two genes are involved in the inheritance of this flower color character. The parent plants were probably double homozygotes (*AABB* for red and *aabb* for white). The all-red F₁s must be *AaBb*, and the red color must require at least one dominant allele of both genes. The 9:6:1 ratio of the F₂s indicates that 9 plants had at least one dominant allele of both genes to produce the red color; 1 plant was doubly homozygous recessive and thus white; and the 6 pale purple were caused by having at least one dominant allele of only one of the two genes involved in flower color (*A_bb* or *aaB_*).

ANSWERS TO TEST YOUR KNOWLEDGE

Multiple Choice:

1. c	**5.** a	**9.** d**	**13.** b
2. b	**6.** c	**10.** d	**14.** a
3. c	**7.** d*	**11.** e	**15.** c
4. c	**8.** c	**12.** b	**16.** a

* There are three different ways to get this outcome: HHT, HTH, THH. Each outcome has a probability of $\frac{1}{8}$.

**The 34-cm plant would be quadruply homozygous dominant, and the F_1 would be quadruply heterozygous. The number of phenotypic classes in the F_2 would equal the number of alleles plus 1.

CHAPTER 15: THE CHROMOSOMAL BASIS OF INHERITANCE

INTERACTIVE QUESTIONS

15.1

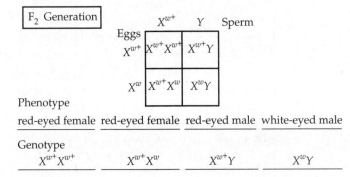

F₂ Generation

Phenotype
red-eyed female red-eyed female red-eyed male white-eyed male

Genotype
$X^{w^+}X^{w^+}$ $X^{w^+}X^w$ $X^{w^+}Y$ X^wY

15.2 The gene is X-linked, so a good notation is X^N, X^n, and Y so that you will remember that the Y does not carry the gene. Capital N indicates normal sight. Genotypes are:

1. X^NY
2. X^NX^n
3. X^NX^N or X^NX^n (probably X^NX^N since four sons are X^NY)
4. X^NX^n
5. X^NY
6. X^nY
7. X^NX^n

15.3 **a.** tall, purple-flowered and dwarf, white-flowered
b. tall, white-flowered and dwarf, purple-flowered

15.4 If linked genes have their loci close together on the same chromosome, they travel together during meiosis and *more* parental offspring are produced. Recombinants are the result of crossing over between nonsister chromatids of homologous chromosomes.

15.5 Solving a linkage problem is often a matter of trial and error. Sometimes it helps to lay out the loci with the greatest distance between

them and fit the other genes between or on either side by adding or subtracting map unit distances.

```
───────── 15 ─────────
m ── 9 ── j ── 6 ── l ── 6 ── k
       ───────── 12 ─────────
```

15.6 **a.** An organism with a trisomy ($2n + 1$) has an extra copy of one chromosome, usually caused by a nondisjunction during meiosis. A triploid organism ($3n$) has an extra set of chromosomes, possibly caused by a total nondisjunction in gamete formation.
b. A trisomy would probably disrupt the genetic balance more than having a complete extra set of chromosomes.

15.7 translocation, deletion, and inversion

15.8 Aneuploidies of sex chromosomes appear to upset genetic balance less, perhaps because relatively few genes are located on the Y chromosome and extra X chromosomes are inactivated as Barr bodies.

SUGGESTED ANSWERS TO STRUCTURE YOUR KNOWLEDGE

1. Genes that are not linked assort independently, and the ratio of offspring from a testcross with a dihybrid heterozygote should be 1:1:1:1 (*AaBb* × *aabb* gives *AaBb*, *Aabb*, *aaBb*, *aabb* offspring and a recombination frequency of 50%—50% of offspring are recombinants). Genes that are linked and do not cross over should produce a 1:1 ratio in such a test-cross (*AaBb* and *aabb* because a heterozygote derived from a parental cross of *AABB* × *aabb* produces only *AB* and *ab* gametes). If crossovers between distant genes almost always occur, the heterozygote will

produce equal quantities of *AB*, *Ab*, *aB*, and *ab* gametes, and the genotype ratio of offspring will be 1:1:1:1, the same as it is for unlinked genes. Because each 1% recombination frequency is equal to 1 map unit, this measurement ceases to be meaningful at relative distances of 50 or more map units. However, crosses with intermediate genes on the chromosome could establish both that the genes *A* and *B* are on the same chromosome and that they are a certain map unit distance apart.

2. A cross between a mutant female fly and a normal male should produce all normal females (who get a wild-type allele from their father) and all mutant male flies (who get the mutant allele on the X chromosome from their mother). $X^m X^m \times X^{m+} Y$ produces $X^{m+} X^m$ and $X^m Y$ offspring (normal females and mutant males).

3. The serious phenotypic effects that are associated with these chromosomal alterations indicate that normal development and functioning are dependent on genetic balance. Most genes appear to be vital to an organism's existence, and extra copies of genes upset genetic balance. Inversions and translocations, which do not disrupt the balance of genes, can alter phenotype because of the effect of neighboring genes on gene functioning.

ANSWERS TO GENETICS PROBLEMS

1. **a.** The trait is recessive and probably X-linked. Two sets of unaffected parents in the second generation have offspring with the trait, indicating that it must be recessive. More males than females display the trait. Females #3 and #5 must be carriers because their father has the trait, and they each pass it on to a son.
    b. Using the symbols X^T for the dominant allele and X^t for the recessive:

    1. $X^t Y$
    2. $X^T X^t$
    3. $X^T X^t$
    4. $X^T Y$
    5. $X^T X^t$
    6. $X^T X^t$ or $X^T X^T$
    7. $X^T X^t$

c. Female #6 has a brother who has the trait and her mother's father had the trait, so her mother must be a carrier of the trait, meaning there is a $\frac{1}{2}$ probability that #6 is a carrier. If she is mated to a phenotypically normal male, none of her daughters will show the trait (0 probability). Her sons have a $\frac{1}{2}$ chance of having the trait if she is a carrier. There is a $\frac{1}{2} \times \frac{1}{2} = \frac{1}{4}$ probability that her sons will have the trait. There is a $\frac{1}{2}$ chance that a child would be male, so the probability of an affected child is $\frac{1}{2} \times \frac{1}{4}$, or $\frac{1}{8}$.

2. c, a, b, d

3. The genes appear to be linked because the parental types appear most frequently in the offspring. Recombinant offspring represent 10 out of 40 total offspring for a recombination frequency of 25%, indicating that the genes are 25 map units apart.

4. One of the mother's X chromosomes carries the recessive lethal allele. One-half of male fetuses would be expected to inherit that chromosome and spontaneously abort. Assuming an equal sex ratio at conception, the ratio of girl to boy children would be 2:1, or six girls and three boys.

5. **a.** For female chicks to be black, they must have received a recessive allele from the male parent. If all female chicks are black, the male parent must have been $Z^b Z^b$. If the male parent was homozygous recessive, then all male offspring will receive a recessive allele, and the female parent would have to be $Z^B W$ to produce all barred males.

    b. For female chicks to be both black and barred, the male parent must have been $Z^B Z^b$. If the female parent were $Z^B W$, only barred male chicks would be produced. To get an equal number of black and barred male chicks, the female parent must have been $Z^b W$.

ANSWERS TO TEST YOUR KNOWLEDGE

Multiple Choice:

1. e	5. d	9. c	13. e	17. d
2. a	6. d	10. e	14. d	18. e
3. b	7. e	11. a	15. a	19. a
4. d	8. b	12. b	16. e	20. c

CHAPTER 16: THE MOLECULAR BASIS OF INHERITANCE

INTERACTIVE QUESTIONS

16.1 a. They grew T2 with *E. coli* with radioactive sulfur to tag phage proteins.
b. T2 was grown with *E. coli* in the presence of radioactive phosphorus to tag phage DNA.
c. Radioactivity was found in the liquid, indicating that the phage protein did not enter the bacterial cells.
d. In the samples with the labeled DNA, most of the radioactivity was found in the bacterial cell pellet.
e. They concluded that viral DNA is injected into the bacterial cells and serves as the hereditary material for viruses.

16.2 a. sugar-phosphate backbone
b. 3' end of chain
c. hydrogen bonds
d. cytosine
e. guanine
f. adenine (purine)
g. thymine (pyrimidine)
h. 5' end of chain
i. nucleotide
j. deoxyribose
k. phosphate group
l. 3.4 nm
m. 0.34 nm
n. 2 nm

16.3

	DNA molecules	Density bands
First generation (grown on light ^{14}N medium)		
Second generation (grown on light ^{14}N medium)		

16.4 The phosphate end of each strand is the 5' end, and the hydroxyl group extending from the 3' carbon of the sugar marks the 3' end. See Figure 16.7 in the text.

16.5 a. single-strand binding protein
b. DNA pol III
c. leading strand
d. 5' end of parental strand
e. 3' end

f. helicase
g. RNA primer
h. primase
i. DNA pol III
j. Okazaki fragment
k. DNA pol I (replacing primer)
l. lagging strand
m. DNA ligase

16.6 No DNA nucleotides can be added after the RNA primer is removed because no 3' end is available for DNA polymerase. Thus, the daughter strand is shorter. Further rounds of replication will continue to produce shorter DNA molecules. See text Figure 16.20.

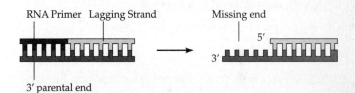

16.7 nucleosomes (10-nm fiber of nucleosomes and linker DNA); 30-nm fiber; looped domains (300-nm fiber); coiling and folding of looped domains into highly condensed metaphase chromosome.

SUGGESTED ANSWERS TO STRUCTURE YOUR KNOWLEDGE

1. Watson and Crick used the X-ray diffraction photo of Franklin to deduce that DNA was a helix, 2 nm wide, with nitrogenous bases stacked 0.34 nm apart, and making a full turn every 3.4 nm. Franklin had concluded that the sugar-phosphate backbones were on the outside of the helix with the bases extending inside. Using molecular models of wire, Watson and Crick experimented with various arrangements and finally paired a purine base with a pyrimidine base, which produced the proper diameter. Specificity of base pairing (A with T and C with G) is ensured by hydrogen bonds.

2. Replication bubbles form where proteins recognize specific base sequences and open up the two strands. *Helicase*, an enzyme that works at the replication fork, unwinds the helix and separates the strands. *Single-strand binding proteins* support the separated strands while replication takes place. *Topoisomerase* eases the twisting ahead of the replication fork by breaking, untwisting, and rejoining DNA strands. *Primase* synthesizes a primer of 5–10 RNA bases to start the new strand. After a proper base pairs up on the exposed template, *DNA polymerase III* joins the nucleotide to the 3' end of the new strand. On the lagging strand, short Okazaki fragments are formed by primase and DNA pol III (again moving 5' → 3'). *DNA polymerase I* replaces the primer with DNA, adding to the 3' end of the leading strand or fragments. *Ligase* joins the 3' end of one fragment to the 5' end of its neighbor. Proofreading enzymes check for mispaired bases, and nucleases, DNA polymerase, and other enzymes repair damage or mismatches.

ANSWERS TO TEST YOUR KNOWLEDGE

Multiple Choice:

1. c	**6.** b	**11.** e	**16.** b	**21.** d
2. a	**7.** a	**12.** d	**17.** c	**22.** e
3. b	**8.** e	**13.** e	**18.** e	
4. d	**9.** c	**14.** a	**19.** c	
5. b	**10.** a	**15.** d	**20.** a	

CHAPTER 17: FROM GENE TO PROTEIN

INTERACTIVE QUESTIONS

17.1 DNA $\xrightarrow{\text{transcription}}$ RNA $\xrightarrow{\text{translation}}$ protein

17.2 Met Pro Asp Phe Lys stop

17.3 **a.** Initiation: Transcription factors bind to promoter and facilitate the binding of RNA polymerase II, forming a transcription initiation complex; RNA polymerase II separates DNA strands at initiation site.
b. Elongation: RNA polymerase II moves along DNA strand, connecting RNA nucleotides that have paired to the DNA template to the 3' end of the growing RNA strand.
c. Termination: After polymerase transcribes past a polyadenylation signal sequence, the pre-mRNA is cut and released.

17.4 A 5' cap consisting of a modified guanine nucleotide is added to the 5' UTR. A poly-A tail consisting of up to 250 adenine nucleotides is attached to the 3' UTR. Spliceosomes have cut out the introns and spliced the exons together.

17.5

DNA Triplet 3' → 5'	mRNA Codon 5' → 3'	Anticodon 3' → 5'	Amino Acid
TAC	AUG	UAC	methionine
GGA	CCU	GGA	proline
TTC	AAG	UUC	Lysine
ATC	UAG	AUC	Stop

17.6 **1.** Codon recognition: An elongation factor (not shown) helps an aminoacyl tRNA into the A site where its anticodon base-pairs to the mRNA codon; hydrolysis of GTP increases accuracy and efficiency.
2. Peptide bond formation: Ribosome catalyzes peptide bond formation between new amino acid and polypeptide held in the P site.
3. Translocation: The empty tRNA in the P site is moved to the E site and released; the tRNA now holding the polypeptide is moved from the A to the P site, taking the mRNA with it; GTP is required.
4. Termination: A release factor binds to stop codon in the A site. Free polypeptide is released from the P site and leaves through the exit tunnel. Ribosomal subunits and other assembly components separate. GTP is required.

a. amino end of growing polypeptide
b. aminoacyl tRNA
c. large subunit
d. A site
e. small subunit
f. 5' end of mRNA
g. peptide bond formation
h. E site
i. release factor
j. stop codon
k. P site
l. free polypeptide

17.7 A ribosome that is translating an mRNA that codes for a secretory or membrane protein will become bound to the ER when an initial signal peptide on the polypeptide is bound by an SRP (signal-recognition particle), which then attaches to an ER receptor protein.

17.8 **a.** Silent: a nucleotide-pair substitution producing a codon that still codes for the same amino acid.

b. Missense: a nucleotide-pair substitution or frameshift mutation that results in a codon for a different amino acid.

c. Nonsense: a nucleotide-pair substitution or frameshift mutation that creates a stop codon and prematurely terminates translation.

d. Frameshift: an insertion or deletion of one, two, or more than three nucleotides that disrupts the reading frame and creates extensive missense and nonsense mutations.

SUGGESTED ANSWERS TO STRUCTURE YOUR KNOWLEDGE

1. Messenger RNA (mRNA) carries the code from DNA that specifies an amino acid sequence to ribosomes, where the RNA's sequence of nucleotides is translated into a polypeptide.
Transfer RNA (tRNA) carries a specific amino acid to its position in a polypeptide based on matching its anticodon to an mRNA codon.
Ribosomal RNA (rRNA) makes up about two-thirds of a ribosome and has specific binding and catalytic functions.
Small nuclear RNA is part of spliceosomes and plays a catalytic role in splicing pre-mRNA.
The versatility of RNA molecules stems from their ability to base-pair with other RNA or DNA molecules, to form specific three-dimensional shapes by base-pairing within itself, and to act as a catalyst.

2.

	Transcription	Translation
Template	DNA	RNA
Location	nucleus (cytoplasm in prokaryotes)	cytoplasm; ribosomes can be free or attached to ER
Molecules involved	RNA nucleotides, DNA template strand, RNA polymerase, transcription factors	amino acids; tRNA; mRNA; ribosomes; ATP; GTP; enzymes; initiation, elongation, and release factors
Enzymes involved	RNA polymerases, spliceosomes (ribozymes)	aminoacyl-tRNA synthetase, ribosomal enzymes (ribozymes)
Control—start and stop	transcription factors locate promoter region with TATA box and start point, polyadenylation signal sequence	initiation factors, initiation sequence (AUG), stop codons, release factor
Product	primary transcript (pre-mRNA)	polypeptide
Product processing	RNA processing: 5' cap and poly-A tail, splicing of pre-mRNA—introns removed by snRNPs in spliceosomes	spontaneous folding, disulfide bridges, signal peptide removed, cleaving, quaternary structure, modification with sugars, etc.
Energy source	ribonucleoside triphosphate	ATP and GTP

3. The genetic code consists of the RNA triplets that code for amino acids. The order of nucleotides in these codons is specified by the sequence of nucleotides in DNA, which is transcribed into the codons found on mRNA and translated into their corresponding sequence of amino acids. There are 64 possible mRNA codons created from the four nucleotides used in the triplet code (4^3).

Redundancy of the code refers to the fact that several triplets may code for the same amino acid. Often these triplets differ only in the third nucleotide. The wobble phenomenon explains the fact that there are only about 45 different tRNA molecules that pair with the 61 possible codons (three codons are stop codons). The third nucleotide of many tRNAs can pair with more than one type of nucleotide. Because of the redundancy of the genetic code, these wobble tRNAs still place the correct amino acid in position.

The genetic code is nearly universal; each codon codes for exactly the same amino acid in almost all organisms. (Some exceptions have been found.) This universality points to an early evolution of the code in the history of life and the evolutionary relationship of all life on Earth.

4.

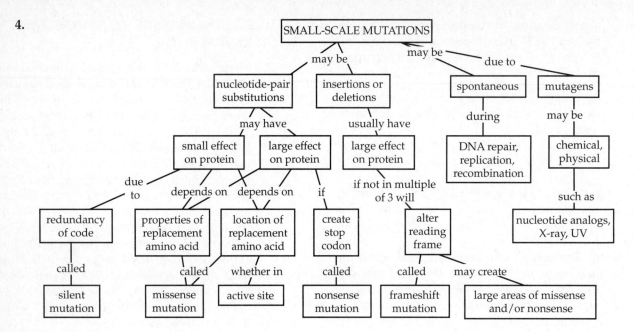

ANSWERS TO TEST YOUR KNOWLEDGE

Multiple Choice:

1. c	**4.** b	**7.** b	**10.** e	**13.** c	**16.** e	**19.** b	**22.** d	**25.** d	**28.** a
2. a	**5.** e	**8.** c	**11.** d	**14.** c	**17.** e	**20.** a	**23.** e	**26.** c	**29.** d
3. b	**6.** c	**9.** c	**12.** d	**15.** b	**18.** b	**21.** a	**24.** e	**27.** e	**30.** c

CHAPTER 18: REGULATION OF GENE EXPRESSION

INTERACTIVE QUESTIONS

18.1 **a.** regulatory gene
  b. promoter
  c. operator
  d. genes coding for enzymes
  e. operon
  f. RNA polymerase
  g. active repressor
  h. inducer (allolactose)
  i. mRNA for enzymes

18.2 **a.** anabolic; corepressor; on; inactive
  b. catabolic; inducers; off; active

18.3 **a.** Barr body—compacted X chromosome in cells of female
  b. Histone tail deacetylation would decrease transcription because it would make genes in the nucleosome less accessible.

18.4 **a.** distal control elements in enhancer
  b. activators
  c. DNA-bending protein
  d. promoter
  e. mediator proteins
  f. general transcription factors
  g. TATA box
  h. RNA polymerase

18.5 **a.** Regulatory proteins may bind to sequences in the 5′ or 3′ UTR and block attachment of ribosomes, thereby decreasing gene expression. Sequences in the 3′ UTR may affect the length of time an mRNA remains intact, thereby either increasing or decreasing gene expression.
  b. A cell marks a protein for destruction by attaching molecules of ubiquitin to the protein. Giant proteasomes encircle and chop up the marked proteins.

18.6 **a.** The primary miRNA transcript folds on itself by hydrogen bonding between complementary bases, forming loops called hairpins. Each hairpin is cut, then trimmed by Dicer, and one strand is degraded. The single-stranded miRNA forms a complex with proteins. The complex binds to mRNA with complementary base sequences, and the mRNA is degraded or translation is blocked.

b. MicroRNAs are coded for by RNA coding genes in a cell and processed as described in (**a**). Small interfering RNAs are longer, double-stranded RNA molecules that may be introduced into a cell (by an experimenter or a virus) or produced by the cell, and are then processed by the cell and function like miRNAs.

c. Evidence indicates that they do both. miRNAs and siRNAs both affect translation by degrading mRNA or blocking translation. Several experiments have shown that siRNAs may silence transcription by changing chromatin structure.

18.7 **a.** A cell is said to be determined when its developmental fate is set. Its series of gene activations and inactivations has set it on the path to express the genes for tissue-specific proteins. When it produces these proteins and develops its characteristic structure, the cell has become differentiated.

b. The action of MyoD must depend on a combination of regulatory proteins, some of which may be lacking in the cells that it is not able to transform into muscle cells.

18.8 Bicoid mRNA was shown to be localized at one end of the unfertilized egg; later in development, Bicoid protein occurred in a gradient that was most concentrated in the anterior cells of the embryo. Also, injection of bicoid mRNA into various regions of early embryos caused anterior structures to form at those sites.

18.9 **a.** Mutations may result in (1) more copies of the gene being present than normal (amplification); (2) translocation, which may bring the gene under the control of a more active promoter or control element; or (3) a change in a nucleotide sequence in either a control element that increases gene expression or in the gene that creates a more active or resilient protein.

b. Tumor-suppressor proteins may function in repair of damaged DNA, control of cell adhesion, or inhibition of the cell cycle.

SUGGESTED ANSWERS TO STRUCTURE YOUR KNOWLEDGE

1. **a.** operons
   b. promoter
   c. operator
   d. negative control
   e. repressor
   f. inactive

   g. active
   h. corepressor
   i. inducer
   j. anabolic
   k. activate
   l. inactivate
   m. catabolic
   n. cAMP
   o. lack of glucose

2. **a.** DNA packing into nucleosomes; histone tail acetylation increases, whereas deacetylation and methylation of tails decreases transcription; methylation of DNA may be involved in long-term inactivation of genes; ncRNAs may promote heterochromatin formation
   b. Specific transcription factors (activators) bind with control elements in enhancers, then interact with mediator proteins and promoter region to form transcription initiation complex; repressors can inhibit transcription
   c. alternative splicing of primary RNA transcript, 5' cap and poly-A tail added
   d. nucleotide sequences in the 3' UTR affect lifespan of mRNA, and miRNAs and siRNAs target mRNA for degradation
   e. Repressor proteins and miRNA or siRNA may prevent translation (or short poly-A tail length can allow mRNA stockpiling in ovum); activation of initiation factors
   f. Protein processing by cleavage or modification; transport to target location; selective degradation by proteosomes of proteins marked with ubiquitin

3. Most cytoplasmic determinants are mRNA for transcription factors that are divided by the first few mitotic divisions. They are present in the cells, and their translated product can enter the nucleus and regulate transcription. Inducers must communicate between cells. They are often proteins that bind to cell surface receptors and initiate a signal transduction pathway involving a cascade of enzyme activations, usually leading to the activation of transcription factors within the target cell.

ANSWERS TO TEST YOUR KNOWLEDGE

Multiple Choice:

1. b	**5.** b	**9.** b	**13.** e	**17.** c	**21.** e
2. c	**6.** b	**10.** b	**14.** d	**18.** d	**22.** e
3. e	**7.** e	**11.** d	**15.** a	**19.** e	**23.** a
4. a	**8.** c	**12.** b	**16.** a	**20.** c	**24.** b

CHAPTER 19: VIRUSES

INTERACTIVE QUESTIONS

19.1 **1.** Phage attaches to host cell and injects DNA.

2. Phage DNA forms a circle. Certain factors determine which cycle is entered.

3. New phage DNA and proteins are synthesized and self-assemble into phages.

4. Bacterium lyses, releasing phages.

5. Phage DNA integrates into bacterial chromosome.

6. Bacterium reproduces, passing prophage to daughter cells.

7. Large population of infected bacteria forms.

8. Occasionally, prophage exits bacterial chromosome and begins lytic cycle.

a. phage DNA

b. bacterial chromosome

c. new phages

d. prophage

e. replicated bacterial chromosome with prophage

19.2 RNA → DNA → RNA; viral reverse transcriptase, host RNA polymerase

19.3 Viral particles spread easily through plasmodesmata, the cytoplasmic connections between plant cells. As there are no cures for plant viral diseases, reducing the spread of infection and breeding resistant varieties are the best approaches.

SUGGESTED ANSWERS TO STRUCTURE YOUR KNOWLEDGE

1.

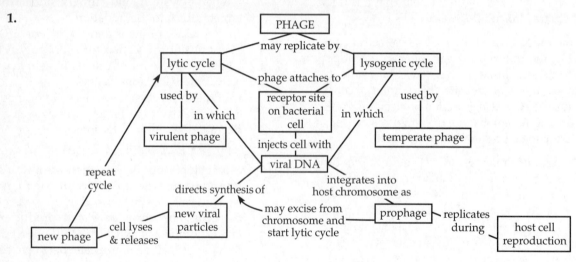

2. **a.** DNA
b. RNA
c. protein capsid
d. host cell
e. bacterium
f. lytic or lysogenic cycle
g. animal
h. membraneous envelope
i. reverse transcriptase
j. plant
k. viroids

l. plasmids
m. transposons

ANSWERS TO TEST YOUR KNOWLEDGE

Multiple Choice:

1. b	**4.** e	**7.** b	**10.** d
2. d	**5.** a	**8.** b	**11.** b
3. c	**6.** d	**9.** c	

CHAPTER 20: BIOTECHNOLOGY

INTERACTIVE QUESTIONS

20.1 The third sequence, because it has the same sequence running in opposite directions. The enzyme would probably cut between G and A, producing AATT and TTAA sticky ends.

20.2 A genomic library contains copies of DNA segments from the entire genome. Thus, all genes should be represented, along with the regulatory sequences and introns. A cDNA library allows you to sequence only the exons of a gene, and also indicates which genes are expressed either in different cell types or at different stages of development in the same cell type.

20.3 **a.** bacterial plasmid
  b. *lacZ* gene
  c. restriction site
  d. *amp*^R (ampicillin resistance) gene
  e. gene of interest
  f. complementary sticky ends
  g. human DNA fragments
  h. recombinant plasmids
  i. recombinant bacteria
  j. plate with ampicillin and X-gal

  1. Plasmids are obtained and human DNA is isolated.
  2. Both plasmid and DNA are cut with the same restriction enzyme. Single cut in plasmid disrupts *lacZ* gene; multiple fragments of human DNA formed.
  3. Fragments are mixed and some foreign fragments base-pair with plasmid. DNA ligase seals ends.
  4. Recombinant plasmids transform bacteria with mutation in their *lacZ* gene.
  5. Plate bacteria on agar containing ampicillin and X-gal.

  Cells containing recombinant plasmids are identified by their ability to grow in the presence of the antibiotic and by their white color. Blue colonies contain plasmids that resealed and thus have a functioning *lacZ* gene. Identify clones carrying the gene of interest with a nucleic acid probe.

20.4 **a.** By amplifying the gene prior to cloning, the later task of identifying clones carrying the desired gene is simplified.
  b. There is a limit to the number of accurate copies that can be made due to the accumulation of relatively rare copying errors. Large quantities of a gene are better prepared by DNA cloning in cells.

20.5 Southern blotting procedure:

  a. restriction enzyme treatment of samples
  b. gel electrophoresis
  c. DNA transfer by blotting onto membrane
  d. hybridization with radioactive probe
  e. autoradiography

  Crime samples contain blood from the victim and, presumably, the perpetrator. After identifying the bands from the victim, the remaining fragments match those of Suspect 2's genetic profile.

20.6 **a.** If similar sequences occur in known genes in other species, the function of a gene may be inferred.
  b. RT-PCR: The mRNA from different tissue samples could be isolated. Reverse transcriptase would be used to make cDNA of all the mRNA, and PCR using specific primers could amplify only the gene of interest. Running the samples on a gel would show bands only in the tissues that were expressing the gene.
  c. DNA microarray assay: As in RT-PCR, mRNA is isolated from different tissues, and cDNA is made and labeled with fluorescent dye. The cDNA is applied to a microarray (single-stranded DNA fragments of the genes of an organism arranged in a grid). The different cDNA will hybridize with the genes that were expressed in the tissue. The intensity with which the hybridized spots fluoresce indicates the relative amount of mRNA that was in the tissue.
  d. Researchers analyzed the phenotypes of the worms that developed with each lacking gene and classified most of the genes into a few functional groups.

20.7 DNA methylation and histone acetylation help to regulate gene expression. An adult cell must have these epigenetic changes in its chromatin reprogrammed in order to support normal gene expression during development. The DNA of many cloned embryos has been found to be improperly methylated.

20.8 The major difficulty is to assure that proper control mechanisms are present so that the gene is expressed at the proper time, in the proper place, and to the proper degree. Insertion of the therapeutic gene must not harm other cell functions. Ethical considerations include whether genetic engineering should be done on any human cells or, in particular, on germ cells, which would influence the genetic makeup of future generations.

SUGGESTED ANSWERS TO STRUCTURE YOUR KNOWLEDGE

1. Agricultural applications: (1) improvement of the genomes of agricultural plants and animals to improve quality and yield, (2) "pharm" animals to produce pharmaceutical proteins, (3) development of plant varieties that have genes for resistance to diseases, herbicides, and insects Medical applications: (1) diagnosis of genetic and infectious diseases, (2) treatment of genetic disorders or other diseases, (3) production of insulin, human growth factor, TPA, and other useful products

2. **a.** Bacterial enzymes that cut DNA at restriction sites, creating "sticky ends" that can base-pair with other fragments. Uses: make recombinant DNA, form restriction fragments used for many other techniques
b. Mixture of molecules applied to gel in electric field; molecules separate, moving at different rates due to charge and size. Uses: separate restriction fragments into pattern of distinct bands; fragments can be removed from gel and retain activity or can be identified with probes
c. mRNA isolated from cell is treated with reverse transcriptase to produce a complementary DNA strand, and then a double-stranded DNA gene, minus introns and control regions. Uses: creates genes that are easier to clone in bacteria; produces library of genes that are expressed in cell
d. Radioactively or fluorescently labeled single-stranded DNA or mRNA used to base-pair with complementary sequence of DNA or RNA. Uses: locate gene in clone of bacteria; identify bands on gels; DNA microarray assay; diagnose infectious diseases
e. DNA fragments separated by gel electrophoresis, transferred by blotting onto paper, labeled probe added, rinsed, autoradiography. Uses: analyze DNA; genetic profile

f. Single-stranded DNA fragments are incubated with four nucleotides, DNA polymerase, and four labeled dideoxy nucleotides that interrupt synthesis; samples separated by size, sequence of nucleotides read from sequence of fluorescent tags. New automated techniques now used. Use: determine nucleotide sequence
g. Polymerase chain reaction: DNA is mixed with heat-resistant DNA polymerase, nucleotides, and primers having complementary sequences for targeted DNA section and repeatedly heated to separate, cooled to pair with primers and replicate. Use: rapidly produce multiple copies of a gene or section of DNA *in vitro*
h. Fluorescently labeled cDNA is made from a cell's mRNA and applied to a DNA microarray of single-stranded DNA from many different genes attached to a glass grid. The intensity and location of fluorescence indicates gene expression in the cell. Uses: test thousands of genes simultaneously to compare gene expression in different tissues or at different developmental stages or conditions
i. mRNA is isolated; cDNA is synthesized by reverse transcriptase and DNA polymerase; cDNA is amplified using primers specific for the gene; gel electrophoresis shows band in samples which had mRNA from that gene. Use: determine whether a particular gene is expressed in a sample
j. Changes introduced in sequence of a cloned gene; mutated gene is inserted into a cell; phenotype of mutant is studied. Use: technique for determining function of a gene
k. Double-stranded RNAs that match a gene sequence are inserted into cell; they trigger breakdown or block translation of that gene's mRNA. Use: silence the expression of genes to study their functions and interactions between genes

ANSWERS TO TEST YOUR KNOWLEDGE

Multiple Choice:

1. c	**5.** d	**9.** a	**13.** d	**17.** b	**21.** d
2. e	**6.** b	**10.** b	**14.** a	**18.** e	**22.** d
3. e	**7.** a	**11.** c	**15.** c	**19.** a	
4. c	**8.** c	**12.** b	**16.** d	**20.** c	

CHAPTER 21: GENOMES AND THEIR EVOLUTION

INTERACTIVE QUESTIONS

21.1 The public consortium followed a hierarchy of three stages: (1) genetic (linkage) mapping that established about 200 markers/chromosome; (2) physical mapping that cloned and ordered smaller and smaller overlapping fragments (using YAC or BAC vectors for cloning fragments); and (3) DNA sequencing of each small fragment, followed by assembly of the overall sequence. The Celera whole-genome shotgun approach omitted the first two stages. Each chromosome was cut by several restriction enzymes into small fragments, which were sequenced, and powerful computers assembled the overlapping fragments to determine the overall sequence.

21.2 Comparing nucleotide sequences may reveal matches with other genes of known function and can identify similarities with closely or distantly related species that help trace a species' evolutionary history.

21.3 **a.** the archaean *Archaeoglobus fulgidus*; humans
b. rice, *Oryza sativa*; the bacterium *Haemophilus influenzae*
c. a plant, *Fritillaria assyriaca*; *H. influenzae*
d. less than 21,000; alternative splicing of exons can allow each gene to code for more than one polypeptide; post-translational processing can also alter polypeptides

21.4 They are first transcribed into RNA; thus, the original retrotransposon remains in place. Reverse transcriptase converts the RNA to DNA, which is inserted into another site in the genome.

21.5 1. D; 1.5
2. I; 20
3. H; 5
4. B; 44
5. E; 10
6. G; 17
7. C; 15
8. F; 5-6
9. A; 3

21.6 The lysozyme gene, which codes for a bacterial infection-fighting enzyme, was present in the last common ancestor of birds and mammals. After their lineages split, the gene underwent a duplication event in the mammalian lineage, and a copy of the lysozyme gene evolved into a gene coding for a protein involved in milk production.

21.7 **a.** Exon shuffling can happen through an error in meiotic recombination, which may occur between two homologous transposable elements, or by the inclusion or tagging along of an exon with a transposable element, which moves a copy of the exon to a new location.
b. Exons often code for domains of a protein. Providing a new domain to a protein may enhance or change its function.

21.8 The homeobox codes for a DNA-binding homodomain, while other domains specific to each regulatory gene interact with transcription factors to recognize particular enhancers or promoters and thus control different batteries of developmental genes.

SUGGESTED ANSWERS TO STRUCTURE YOUR KNOWLEDGE

1. Many noncoding sequences are highly conserved across species, indicating that they perform some important, but as yet unidentified, function. As more data accumulate, the functions of these noncoding DNA sequences may become clear. One example is the newly recognized miRNAs that are involved in regulation of gene expression.
   a. Transposable elements include transposons, which are DNA segments, and retrotransposons, which are produced from RNA transcribed into DNA. These transposable elements are very common and move DNA to new locations in the genome.
   b. *Alu* elements are about 300 nucleotides long and make up about 10% of the human genome. Some are transcribed into RNA, but their function is unknown. They may provide alternate splice sites for RNA processing.
   c. L1 sequences are long retrotransposons that rarely move about. They are found in many introns and may help regulate gene expression.
   d. Simple sequence DNAs are highly repetitive tandem sequences found at centromeres and teleomeres of a chromosome. They appear to have structural functions in organizing chromatin. Sequences with fewer repetitions are called STRs (short tandem repeats) and are used in genetic profiling.
   e. Pseudogenes are remnants of genes that are no longer functional because mutations have altered their regulatory sequences.

2. Chromosome duplication provides additional copies of genes that may undergo mutation and produce new proteins. Chromosomal rearrangements such as duplications, inversions, and translocations may create reproductive barriers between populations that lead to the formation of new species. Errors during meiosis can lead to duplications of genes or exons, or rearrangements of exons, providing genetic material that may take on related or novel functions. Transposable elements can facilitate recombination between different chromosomes, can disrupt genes or control elements, and can carry genes or exons to new locations.

ANSWERS TO TEST YOUR KNOWLEDGE

Multiple Choice:

1. b	**4.** a	**7.** a	**10.** c	**13.** e
2. d	**5.** d	**8.** d	**11.** d	**14.** d
3. c	**6.** b	**9.** e	**12.** b	

CHAPTER 22: DESCENT WITH MODIFICATION: A DARWINIAN VIEW OF LIFE

INTERACTIVE QUESTIONS

22.1 a.
1. E f
2. A b
3. B e
4. C d, g
5. F a
6. D c
7. G g

b. a, f, d, b, e, g, c

22.2 Observation 1: Members of a population vary in their inherited traits.
Observation 2: All species can produce more offspring than the environment can support, and many offspring do not survive.

22.3 a. When a new antibiotic is developed, it is effective at killing most pathogens. A few, however, may have an inherent resistance, such as an enzyme like penicillinase that destroys penicillin. As the continued use of the new antibiotic selects against bacteria without an antibiotic-resistance gene, the resistant bacteria can multiply rapidly and the "beneficial" gene would become more common.

b. Resistance to multiple antibiotics can evolve if bacteria that are already resistant to one antibiotic are exposed to a different one. The new antibiotic, again, will probably kill most bacteria but "select" any that happen to have a variation that provides resistance to it. The exchange of genes between different strains and even different species of bacteria may also produce multi-drug resistant strains.

22.4 a. biogeography
b. fossil record
c. homologies

d. island species and mainland species or neighboring island species
e. ancestral and transitional forms
f. homologous structures
g. functions and form
h. molecular comparisons
i. DNA and proteins
j. descent from a common ancestor

SUGGESTED ANSWERS TO STRUCTURE YOUR KNOWLEDGE

1. The two main components of Darwin's evolutionary theory are that all of life has descended from a common ancestor and that this evolution has been the result of natural selection. The theory of natural selection is based on several key observations and inferences. Members of a population often vary in their inherited traits. The overproduction of offspring in conjunction with limited resources leads to the unequal reproductive success of those organisms best suited to the local environment. The increased survival and reproduction of the best-adapted individuals in a population leads to the gradual accumulation of favorable characteristics in a population.

ANSWERS TO TEST YOUR KNOWLEDGE

Multiple Choice:

1. c	**4.** e	**7.** b	**10.** d	**13.** c
2. e	**5.** d	**8.** a	**11.** c	**14.** b
3. a	**6.** d	**9.** e	**12.** d	

CHAPTER 23: THE EVOLUTION OF POPULATIONS

INTERACTIVE QUESTIONS

23.1 **a.** mutation, with some genetic recombination
b. sexual reproduction
c. Prokaryotes and viruses have very short generation times, and a new beneficial mutation can increase in frequency rapidly in an asexually reproducing population. Although mutations are the source of new alleles, they are so infrequent that their contribution to genetic variation in a large, diploid population is minimal. However, the production and union of gametes produces zygotes with fresh combinations of alleles each generation.

23.2 **a.** The frequencies of genotypes are 0.49 *BB*, 0.42 *Bb*, and 0.09 *bb* ($^{98}/_{200}$, $^{84}/_{200}$, $^{18}/_{200}$).
b. The 98 *BB* mice contribute 196 *B* alleles, and the 84 *Bb* mice contribute 84 *B* alleles to the gene pool. These 84 *Bb* mice also contribute 84 *b* alleles, and the 18 *bb* mice contribute 36 *b* alleles. Of a total of 400 alleles, 280 are *B* and 120 are *b*. Allele frequencies are 0.7 *B* and 0.3 *b*. Another way to determine allele frequencies is from genotype frequencies: Add the frequency of the homozygous dominant genotype and ½ the frequency of the heterozygote to determine the frequency of *p*. For *q*, add the homozygous recessive frequency and ½ the heterozygote frequency. Or determine *q* from $1 - p$.

23.3 0.7; 0.3; 0.49; 0.42; 0.09

23.4 **a.** 0.36; 0.48; 0.16. Plug *p* (0.6) and *q* (0.4) into the expanded binomial: $p^2 + 2pq + q^2 = 1$.
b. 0.6; 0.4. Add the frequencies of *AA* + ½ *Aa* to get *p*. Add the frequencies of *aa* + ½ *Aa* to get *q*. Alternatively, to determine *q*, take the square root of the homozygous recessive frequency if you are sure the population is in Hardy-Weinberg equilibrium. The frequency of *p* is then $1 - q$.
c. These results suggest that the population is not in Hardy-Weinberg equilibrium and thus is evolving.

23.5 **a.** natural selection
b. genetic drift
c. gene flow
d. better reproductive success
e. small population
f. founder effect
g. bottleneck effect
h. genetic variation between populations

23.6 **a.** Diploidy—The sickle-cell allele is hidden from selection in heterozygotes.
b. Heterozygote advantage—Heterozygotes are protected from the most severe effects of malaria and have a selective advantage in areas where malaria is a major cause of death.

SUGGESTED ANSWERS TO STRUCTURE YOUR KNOWLEDGE

1. **a.** The Hardy-Weinberg principle states that allele and genotype frequencies within a population will remain constant from one generation to the next as long as only Mendelian segregation and recombination of alleles are involved. This equilibrium requires five conditions: The population is large, mutation is negligible, migration is negligible, mating is random, and no natural selection occurs. The Hardy-Weinberg equilibrium provides a null hypothesis to enable researchers to test for evolution.
    b. $p^2 + 2pq + q^2 = 1$. In the Hardy-Weinberg equation, *p* and *q* refer to the frequencies of two alleles in the gene pool. The frequency of homozygous offspring is ($p \times p$) or p^2 and ($q \times q$) or q^2. Heterozygous individuals can be formed in two ways, depending on whether the egg or sperm carries the *p* or *q* allele, and their frequency is equal to $2pq$.

2. Genetic variation is retained within a population by diploidy and balancing selection. Diploidy masks recessive alleles from selection when they occur in the heterozygote. Thus, less adaptive or even harmful alleles are maintained in the gene pool and are available should the environment change. Balancing selection maintains two or more phenotypic forms in a population. In heterozygote advantage, two alleles will be retained in stable frequencies within the gene pool. Frequency-dependent selection, in which the more common phenotype is selected against by predators or other factors, is another type of balancing selection.

ANSWERS TO TEST YOUR KNOWLEDGE

Multiple Choice:

1. e	**5.** b	**9.** d	**13.** d	**17.** b	**21.** d
2. b	**6.** a	**10.** e	**14.** b	**18.** c	**22.** d
3. d	**7.** c	**11.** b	**15.** d	**19.** e	
4. b	**8.** c	**12.** d	**16.** a	**20.** e	

CHAPTER 24: THE ORIGIN OF SPECIES

INTERACTIVE QUESTIONS

24.1 **a.** reduced hybrid fertility   **b.** post-
    c. gametic isolation   **d.** pre-
    e. mechanical isolation   **f.** pre-
    g. temporal isolation   **h.** pre-
    i. hybrid breakdown   **j.** post-
    k. behavioral isolation   **l.** pre-
    m. reduced hybrid viability   **n.** post-
    o. habitat isolation   **p.** pre-

24.2 **a.** reproductive isolation
    b. morphological
    c. ecological
    d. smallest group that shares a common ancestor, a single branch on the tree of life

24.3 **a.** 20 chromosomes
    b. 24 chromosomes

24.4 **a.** In allopatric speciation, a new species forms while geographically isolated from the parent population. In sympatric speciation, some reproductive barrier isolates the gene pool of a subgroup of a population within the same geographic range as that of the parent population.
    b. Reproductive barriers may evolve as a byproduct of the genetic change associated with the isolated population's adaptation to a new environment, genetic drift, or sexual selection. Sympatric speciation by polyploidy is common in plants. A change in resource use or sexual selection may reproductively isolate a subset of an animal population.

24.5 **a.** Reinforcement: when hybrid offspring are less viable or reproduce less successfully, the reproductive barriers between the two species may strengthen. Traits that prevent or discourage crossbreeding will increase in frequency while traits that allow for mating with the other species will decrease in frequency as the individuals with those traits produce fewer viable offspring.
    b. Fusion: If hybrids are successful in mating with each other and with the parent species, reproductive barriers may weaken. Increased gene flow between the two species may eventually lead to their fusion into a single species.
    c. Stability: Hybrids continue to be produced between the two species in the area of their overlap, but the gene pools of both parent species remain distinct. Several cases of such stability among hybridizing species have been documented.

24.6 The total time between speciation events has been estimated to range from 4,000 to 40 million years (with an average of 6.5 million years). Thus, even if the process of speciation is relatively rapid once it begins, it may be millions of years before that new species gives rise to another new species. The biodiversity of Earth may take a very long time to recover from the current increased extinction rate.

SUGGESTED ANSWERS TO STRUCTURE YOUR KNOWLEDGE

1. Speciation, which leads to an increase in biological diversity, is the process through which one species splits into two or more species. Microevolution involves changes in the gene pool of a population as a result of either chance events or natural selection. If the makeup of an isolated gene pool changes enough, microevolution may lead to speciation.

2. Reproductive barriers do not develop in order to create new species; they develop as a side consequence of the genetic changes that occur as a species adapts to a new environment or undergoes its own genetic drift. Gene flow between two species tends to break down reproductive barriers between them. Since gene flow is much less likely between allopatric species, one would predict that reproductive barriers are more likely to evolve. In the case of sympatric speciation by polyploidy, however, reproductive barriers arise in one generation.

3. Punctuated equilibria describes the common pattern seen in the fossil record in which species appear rather suddenly and then seem to change little for the rest of their existence. Thus, evolution seems often to occur in spurts of relatively rapid change interspersed with long periods of stasis.

ANSWERS TO TEST YOUR KNOWLEDGE

Multiple Choice:

1. c	**4.** c	**7.** e	**10.** d
2. d	**5.** c	**8.** c	**11.** a
3. e	**6.** b	**9.** b	**12.** e

CHAPTER 25: THE HISTORY OF LIFE ON EARTH

INTERACTIVE QUESTIONS

25.1 Reproduction that faithfully passes genetic information on to offspring is a characteristic of life. But a cell needs metabolic machinery to provide the energy and building blocks needed to replicate its genetic information and reproduce. Thus both self-replicating molecules that encode genetic information for metabolism and the resulting metabolic processes are required for life to have evolved.

25.2 17,190 years. Carbon-14 has a half-life of 5,730 years. In 5,730 years the ratio of C-14 to C-12 would be reduced by $\frac{1}{2}$; in 11,460 years it would be reduced by $\frac{1}{4}$; and in 17,190 years it would be reduced by $\frac{1}{8}$ (three half-lives; $3 \times 5,730$).

25.3 The inner membranes of both mitochondria and chloroplasts have enzymes and transport systems homologous to those in the plasma membranes of modern prokaryotes. They replicate by a splitting process similar to that of some prokaryotes. They contain a single, circular DNA molecule that is not associated with histones. The ribosomes of mitochondria and plastids are more similar to prokaryotic ribosomes than to the cytoplasmic ribosomes of eukaryotic cells.

25.4 The change in nutrient procurement from herbivory, filter-feeding, and scavenging to predation: predators with claws and other prey-capturing adaptations as well as prey with heavy body armor appeared in the fossil record during the Cambrian explosion.

25.5 Marsupials migrated to Australia through South America and Antarctica while the continents were still joined. When Pangaea broke up, marsupials diversified on isolated Australia.

25.6 a. Early mammals were restricted in size and diversity because they were probably outcompeted or eaten by the larger and more diverse dinosaurs. After dinosaurs (except for birds) became extinct, mammals diversified and filled the vacated niches.
b. The rise in large predators in the Cambrian explosion, the diversification of terrestrial plants and vertebrates, the adaptive radiation of insects that ate or pollinated plants.

c. The Hawaiian Archipelago is a series of relatively young, isolated, and physically diverse islands whose thousands of endemic species are examples of adaptive radiation following multiple colonization and speciation events.

25.7 a. This study provided experimental evidence that a specific change in the sequence of a developmental gene could produce a major evolutionary change—the suppression of legs associated with origin of the six-legged insect from a crustacean-like ancestor with many legs.
b. The coding sequence of the developmental gene was identical in the two populations; the gene was not expressed in the ventral spine region of developing lake sticklebacks. Thus, the loss of ventral spines results from a change in gene regulation, not from a change in the gene itself.

25.8 a. The successful species that last the longest before extinction and generate the most new species will determine the direction of an evolutionary trend, just as the individuals that produce the most offspring determine the direction of adaptation in a population.
b. Both microevolution and macroevolution are primarily driven by natural selection resulting from interactions between organisms and their environments. Evolutionary trends are ultimately dictated by environmental conditions; if conditions change, an evolutionary trend may end or change direction.

SUGGESTED ANSWERS TO STRUCTURE YOUR KNOWLEDGE

1. The abiotic synthesis of small organic molecules; the joining of small molecules into macromolecules; the packaging of these molecules into membrane-bound protocells; the origin of self-replicating molecules, probably RNA, that made inheritance and natural selection possible
    a. Archaean
    b. Proterozoic
    c. Phanerozoic
    d. Paleozoic
    e. Mesozoic
    f. Cenozoic
    g. Prokaryotes
    h. Atmospheric oxygen
    i. Single-celled eukaryotes

j. Multicellular eukaryotes
k. Animals
l. Colonization of land

2. *Plate tectonics:* When continents form one supercontinent, many habitats are destroyed or changed. As continents change latitude, their climates either warm or cool. The separation of continents creates major geographic separation events, making possible allopatric speciations. Indeed, continental drift helps explain much of biogeography.

3. *Mass extinctions* result in the loss of many evolutionary lineages. The resulting empty ecological roles, however, may be exploited by species that survived extinction.
    Adaptive radiations are multiple speciation events that fill newly formed or newly emptied niches. Thus, adaptive radiations may follow mass extinctions, the colonization of new regions, or the evolution of novel adaptations. The adaptive radiation of one group may provide new resources (such as food or habitat) that support the radiation of other groups.

4. New morphological forms may evolve as a result of mutations in developmental genes or changes in the regulation of such genes. Such changes may affect the rate or timing of development or the spatial arrangement of body parts (e.g., *Hox* genes).

ANSWERS TO TEST YOUR KNOWLEDGE

Multiple Choice:

1. b	5. c	9. c	13. b	17. b
2. c	6. e	10. a	14. b	18. d
3. c	7. d	11. c	15. e	
4. e	8. e	12. a	16. b	

CHAPTER 26: PHYLOGENY AND THE TREE OF LIFE

INTERACTIVE QUESTIONS

26.1. **a.** #1
       b. A
       c. #4
       d. B and C

26.2. Adaptations to different environments may lead to large morphologic differences within related groups, and convergent evolution can produce similar structures in unrelated organisms in response to similar selective pressures. Examples are the varied Hawaiian silversword plants and the similar Australian and North American burrowing moles.

26.3.

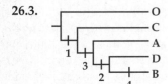

26.4. The four-chambered heart of birds and mammals is analogous, not homologous. It evolved independently in the two groups. An abundance of evidence supports the hypothesis that birds and mammals evolved from different reptilian ancestors.

26.5. **a.** rRNA genes
       b. mtDNA (mitochondrial DNA)

26.6. Assuming that genes have a fairly constant rate of mutation, in a protein that has a crucial function and whose amino acid sequence is central to that function, mutations that are harmful would be quickly removed from the population. Therefore, fewer mutations would be neutral and remain in the genome. If the exact sequence of amino acids in a protein is less critical to survival, more mutations would be neutral and remain. Thus, in the same amount of time, the sequence of the second gene will change more than the sequence of the more crucial gene.

26.7. **a.** Bacteria
       b. Eukarya
       c. Archaea

       #2 represents gene transfer between mitochondrial ancestor and ancestor of eukaryotes

       #3 represents gene transfer between chloroplast ancestor and ancestor of green plants

SUGGESTED ANSWERS TO STRUCTURE YOUR KNOWLEDGE

1.

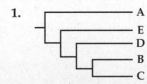

2. In order to reconstruct the evolutionary history of a group of organisms, systematists consider morphological and molecular homologies as well as evidence from the fossil record. Molecular homologies may be identified using computer programs that search for and align common sequences in homologous sections of DNA. Statistical analysis can help sort distant molecular homologies from coincidental homoplasies. Using the techniques of cladistics, the ingroup to be studied is compared with an outgroup and with each other. Shared ancestral characters and shared derived characters are used to determine the sequence of evolutionary divergence. Computer programs can determine the most parsimonious tree (requiring the fewest evolutionary changes). When choosing among several phylogenetic hypotheses, the principle of maximum likelihood states that the most likely tree must take into account rules about how DNA changes over time.

3. A molecular clock is based on the observations that certain regions of DNA evolve (accumulate base changes) at a steady rate. By plotting the number of differences in DNA sequences between groups known to have diverged at specific times, scientists can apply this rate of molecular evolution to infer the timing of other evolutionary divergences.

ANSWERS TO TEST YOUR KNOWLEDGE

Multiple Choice:

1. c	**6.** a	**10.** b	**14.** b	**18.** b
2. d	**7.** e	**11.** b	**15.** d	**19.** b
3. e	**8.** e	**12.** c	**16.** a	**20.** e
4. c	**9.** e	**13.** c	**17.** b	**21.** d
5. d				

CHAPTER 27: BACTERIA AND ARCHAEA

INTERACTIVE QUESTIONS

27.1
  a. spheres (cocci), rods (bacilli), or spirals
  b. 0.5–5 μm in diameter
  c. cell wall made of peptidoglycan; gramnegative bacteria also have outer lipopolysaccharide membrane; archaea lack peptidoglycan; sticky capsule for adherence; fimbriae and pili for attachment or genetic exchange
  d. flagella; may show taxis to stimuli
  e. infoldings of plasma membrane may function in cellular respiration; thylakoid membranes in cyanobacteria
  f. circular DNA molecule with little associated protein found in nucleoid; may have plasmids with other genes
  g. binary fission; rapid population growth; rapid adaptive evolution

27.2
  a. circular chromosome
  b. F plasmid, R plasmid
  c. mutation
  d. transformation
  e. naked DNA
  f. transduction
  g. bacteriophage (phage)
  h. conjugation
  i. F⁺ or HRF cell and F⁻ cell
  j. adaptation to environment/evolution

27.3
  a. endospore; thick-walled, resistant cell that is formed in response to a lack of nutrients and can persist for long periods of time, breaking dormancy when conditions are favorable; chemoheterotroph—it is not photosynthetic and obtains both energy and carbon from organic compounds.
  b. photoautotroph—uses light energy and CO_2 to synthesize organic compounds; the cell is a heterocyte, specialized for nitrogen fixation.

27.4
  a. cyanobacteria
  b. Genetic prospecting is a technique of sampling genetic material directly from the environment. Before its development, researchers could study only those species that could be cultured in the lab. This technique has led to the discovery of new species and even new branches in prokaryotic phylogeny.

27.5 The photosynthesis of cyanobacteria brings carbon from CO_2 into the food chain and releases O_2. Cyanobacteria also fix atmospheric nitrogen, supplying other organisms with the nitrogen they need to make proteins.

SUGGESTED ANSWERS TO STRUCTURE YOUR KNOWLEDGE

1. Prokaryotes are adapted to live in the diverse habitats on (and in) Earth. They have a wide range of metabolic abilities, including the nutritional categories of chemoautotrophy and photoheterotrophy. Their genetic diversity, along with their short generation time, huge

populations, and mechanisms of genetic exchange, enable rapid adaptation to changing environments. The formation of biofilms consisting of multiple species and their many symbiotic associations are further examples of the remarkable abilities of prokaryotes.

2. Decomposers—recycle nutrients; bioremediation: sewage treatment, clean up oil spills.
    Nitrogen fixers—provide nitrogen to the soil by bacteria in nodules on legume roots or by cyanobacteria.
    Mutualism—intestinal bacteria, some digest food and produce vitamins.
    Biotechnology—production of antibiotics, hormones, and many useful products.

ANSWERS TO TEST YOUR KNOWLEDGE

Multiple Choice:

1. b	**5.** c	**9.** d
2. c	**6.** a	**10.** a
3. d	**7.** e	**11.** d
4. e	**8.** b	**12.** e

Fill in the Blanks:

1. cocci	**6.** endospore
2. nucleoid	**7.** biofilms
3. Gram stain	**8.** exotoxins
4. fimbriae	**9.** bioremediation
5. taxis	**10.** anaerobic respiration

CHAPTER 28: PROTISTS

INTERACTIVE QUESTIONS

28.1 The four membranes indicate secondary endosymbiosis of a green alga (whose chloroplasts originated from the primary endosymbiont, a cyanobacterium). The inner two membranes belonged to the green alga's chloroplast. The third membrane is from the engulfed alga's plasma membrane, and the outer membrane is from the heterotrophic eukaryote's food vacuole. The vestigial nucleus indicates that this process occurred relatively recently, and its genetic sequences point to a green algal origin. In more ancient secondary endosymbioses, many components of the engulfed alga have been lost.

28.2 Frequent global changes in the surface proteins of these parasites enable them to evade the host's immune system.

28.3 *Paramecium* is in the subgroup of ciliates in the clade Alveolata in the supergroup Chromalveolata.
   a. cilia
   b. oral groove, leading to cell mouth
   c. food vacuoles, combine with lysosomes
   d. contractile vacuole
   e. release of wastes at region that functions as an anal pore
   f. macronucleus
   g. micronucleus

28.4 These seaweeds may have a holdfast that maintains a strong hold on intertidal rocks.

Biochemical adaptations include cellulose cell walls containing gel-forming polysaccharides that protect the algae from the abrasive action of waves and reduce drying when exposed at low tides.

28.5 **a.** The calcareous tests of forams are long-lasting fossils in marine sediments and sedimentary rocks. Foram fossils are useful in correlating ages of sedimentary rocks in different areas.
   b. Evidence indicates that chromatophores were derived from a different cyanobacterium than the one from which all other plastids were derived—a remarkable second case of primary endosymbiosis.

28.6 **a.** diploid
   b. multinucleate mass, plasmodium
   c. fruiting bodies that function in sexual reproduction
   d. haploid
   e. solitary amoeboid cells
   f. amoeboid cells aggregate and form asexual fruiting body

28.7 Higher sea surface temperatures increase the differences between warm and cold layers of water, reducing the upwelling of cold, nutrient-rich waters from below that contribute to the growth of marine producers. A reduction in this base of marine food webs would affect all consumers in the ecosystem, including important fish species. If diatom populations decrease, less CO_2 would be "pumped" from the atmosphere to the deep ocean floor.

SUGGESTED ANSWERS TO STRUCTURE YOUR KNOWLEDGE

1.
   a. cyanobacterium
   b. primary endosymbiosis
   c. red alga
   d. green alga
   e. secondary endosymbiosis
   f. dinoflagellates
   g. apicomplexans
   h. stramenopiles
   i. euglenids

ANSWERS TO TEST YOUR KNOWLEDGE

Multiple Choice:

1. d	3. d	5. b	7. e	9. a
2. c	4. c	6. d	8. b	10. c

Matching:

1. C	3. B	5. H	7. A
2. F	4. G	6. E	8. D

CHAPTER 29: PLANT DIVERSITY I: HOW PLANTS COLONIZED LAND

INTERACTIVE QUESTIONS

29.1 Benefits: bright, unfiltered sunlight; plentiful CO_2 and soil minerals; few herbivores and pathogens. Challenges: lack of structural support, relative scarcity of water, and problem of dehydration.

29.2
   a. gametes
   b. mitosis
   c. zygote
   d. mitosis
   e. sporophyte
   f. spores
   g. meiosis
   h. gametophyte

29.3 Although lycophytes and pterophytes are both seedless plants, they do not form a monophyletic group. Pterophytes (ferns) share a more recent common ancestor with seed plants than with lycophytes.

29.4
   a. gametophyte
   b. archegonia
   c. antheridia
   d. swim
   e. sporophyte
   f. meiosis
   g. capsule (sporangium)
   h. gametophyte (germinate to form a protonema)

29.5
   a. meiosis
   b. spore
   c. gametophyte
   d. antheridium
   e. sperm
   f. archegonium
   g. egg
   h. fertilization

   i. zygote
   j. gametophyte
   k. young sporophyte
   l. mature sporophyte
   m. sporangia (in sori on undersides of leaves)

The gametophyte (upper portion of the diagram between spore and fertilization) is haploid; the sporophyte (lower portion) is diploid. Lignified xylem provides the mechanical support for aerial growth. Vascular tissues convey water and minerals to photosynthesizing leaves and transport sugars throughout the plant. Roots both absorb water and minerals and anchor the plant. Tall plants could outcompete shorter plants for access to sunlight, and their spores could disperse farther.

SUGGESTED ANSWERS TO STRUCTURE YOUR KNOWLEDGE

1. alternation of generations with protected and nourished embryo, sporopollenin-walled spores produced in sporangia, apical meristems, multicellular gametangia, cuticle, secondary compounds (protection from UV radiation and herbivores), stomata, vascular tissue strengthened with lignin (vascular plants)

2.
   a. land plants
   b. nonvascular plants (bryophytes)
   c. vascular plants
   d. seedless vascular plants
   e. seed plants
   f. liverworts
   g. mosses
   h. hornworts
   i. pterophytes (ferns, horsetails, whisk ferns)
   j. gymnosperms
   k. angiosperms

1. origin of land plants (about 475 mya)
2. origin of vascular plants (about 425 mya)
3. origin of extant seed plants (about 305 mya)

CHAPTER 30: PLANT DIVERSITY II: THE EVOLUTION OF SEED PLANTS

INTERACTIVE QUESTIONS

30.1 Resistant, airborne or animal-carried pollen grains can transport the male gametophyte, even over long distances, to the female gametophyte. A germinating pollen tube delivers sperm directly to an ovule, eliminating the need for a moist environment for sperm to swim to reach eggs.

30.2　a. integument ($2n$)
b. megaspore (n)
c. megasporangium ($2n$)
d. seed coat (from integument) ($2n$)
e. food supply (from female gametophyte tissue) (n)
f. embryo (new sporophyte) ($2n$)

30.3　a. The smaller pollen cones have many scalelike structures holding microsporangia. Microsporocytes (microspore mother cells) undergo meiosis to produce microspores that develop into pollen grains, which enclose the male gametophyte. Two sperm nuclei are eventually discharged from a pollen tube.
b. Ovulate cones have scales that hold two ovules, each containing a megasporangium. A megasporocyte (megaspore mother cell) undergoes meiosis to form four megaspores. The surviving megaspore develops into a multicellular female gametophyte containing a few archegonia, each with an egg.

30.4　a. sepals, petals, stamens, carpels
b. an embryo surrounded by food enclosed in a seed coat
c. mature ovary that may be modified to help disperse seeds, enlisting the aid of wind or animals

30.5　a. sepal
b. petal
c. anther (top of stamen)
d. stigma (top of carpel)
e. ovary
f. ovule with megasporangium
g. microsporangium

h. microsporocytes
i. meiosis
j. microspore (n)
k. pollen grain (containing *male gametophyte* with generative cell and tube cell)
l. two sperm
m. pollen tube
n. style
o. surviving megaspore (n)
p. central cell with two nuclei
q. embryo sac (*female gametophyte*)
r. egg
s. fertilization
t. endosperm nucleus ($3n$)
u. zygote
v. seed coat
w. endosperm
x. embryo
y. germinating seed

30.6

Characteristic	Monocot	Eudicot
Number of cotyledons	one	two
Leaf venation	parallel	netlike
Vascular tissue in stems	scattered	in ring
Root system	fibrous	taproot
Openings in pollen grain	one	three
Floral organs	multiples of three	multiples of four or five

SUGGESTED ANSWERS TO STRUCTURE YOUR KNOWLEDGE

1. reduced male and female gametophytes develop from spores retained within the micro- and megasporangia of the sporophyte; the female gametophyte protected and nourished in ovule retained on sporophyte plant;

nonswimming sperm protected and dispersed in pollen; seed for protection, dispersal, and nourishment of embryo

2. flower as enclosed reproductive structure; animal pollinators; fruit for protection and dispersal of seeds

3. **a.** *Amborella*
   b. water lilies
   c. star anise and relatives
   d. magnoliids
   e. monocots
   f. eudicots

ANSWERS TO TEST YOUR KNOWLEDGE

True or False:

1. False—add *or algae*.
2. True
3. False—change *gymnosperms* to *angiosperms*.

4. True
5. True
6. False—change *fruit* to *seed*.
7. True
8. False—change *angiosperms* to *gymnosperms*, or change *haploid cells . . .* to *an embryo sac with a few haploid cells*.

Multiple Choice:

1. e	**5.** c
2. b	**6.** a
3. d	**7.** d
4. c	**8.** e

CHAPTER 31: FUNGI

INTERACTIVE QUESTIONS

31.1 The fungus provides phosphate ions and other minerals to the plants; the plants supply the fungi with organic nutrients.

31.2 **a.** network of hyphae, typical body form
b. cross-walls between cells, with pores
c. multinucleated hyphae, lack septa
d. cytoplasmic fusion of two parent hyphae
e. nuclei from two different parents exist together in hyphae
f. hyphae with two different nuclei paired up in cells
g. fusion of nuclei from two parents to form diploid state

31.3 **a.** Nucleariids
b. Fungi
c. Chytrids
d. Other fungi (zygomycetes, glomeromycetes, ascomycetes, basidiomycetes)

31.4 **1.** ascomycete
2. zygomycete
3. basidiomycete

a. conidia (asexual spores)
b. ascocarp
c. ascospores

d. asci
e. sporangia (forms asexual spores)
f. zygosporangium
g. basidiocarp
h. basidium
i. basidiospores

SUGGESTED ANSWERS TO STRUCTURE YOUR KNOWLEDGE

1. **a.** chytrids, aquatic and soil decomposers or parasites
   b. flagellated zoospores
   c. zygomycetes, molds and parasites; black bread mold
   d. resistant zygosporangia in sexual stage; asexual spores in sporangium on aerial hyphae
   e. glomeromycetes; almost all form mycorrhizae
   f. 90% of all plants have arbuscular mycorrhizae
   g. ascomycetes; sac fungi, yeasts, molds, morels, mycorrhizae
   h. dikaryotic hyphae in ascocarp, karyogamy in asci, ascospores; asexual spores called conidia at tips of hyphae

i. basidiomycetes; club fungi, mushrooms, shelf fungi, rusts

j. fruiting body called basidiocarp, basidiospores produced in basidia

2. The role of fungi as decomposers of organic material is central to the recycling of chemicals between living organisms and their physical environment. Parasitic fungi have economic and health effects on humans. Plant pathogens cause extensive loss of food crops and trees. Mycorrhizae are found on the roots of almost all vascular plants. These mutualistic associations greatly benefit the plant. Endophytes living inside leaves benefit plants by deterring herbivores and protecting against heat and pathogens. Lichens colonize barren habitats and prepare the way for plant succession.

ANSWERS TO TEST YOUR KNOWLEDGE

Fill in the Blanks:

1. septum	8. zygosporangia
2. heterokaryon	9. coenocytic
3. chitin	10. chytrids
4. enzymes	11. opisthokonts
5. basidium	12. arbuscular
6. endophyte	mycorrhizae
7. mycorrhizae	

Multiple Choice:

1. a	4. e	7. a	10. d	13. c
2. c	5. c	8. b	11. e	14. b
3. d	6. e	9. a	12. a	

CHAPTER 32: AN OVERVIEW OF ANIMAL DIVERSITY

INTERACTIVE QUESTIONS

32.1 a. A larva is a free-living, sexually immature stage in some animal life cycles that may differ from the adult in habitat, nutrition, and body form.

b. A juvenile resembles the adult but is not sexually mature; a larva differs from the adult form.

c. metamorphosis

32.2 (1) The emergence of new predator–prey relationships (increased locomotion and protective shells) may have triggered various evolutionary adaptations. (2) The accumulation of atmospheric oxygen may have supported the active metabolisms of mobile and larger animals. (3) The origin of the *Hox* complex of regulatory genes may have facilitated the evolution of the diverse body plans that appear during the Cambrian explosion. All three of these may have played a role in the Cambrian explosion.

32.3 a. spiral and determinate

b. radial and indeterminate

c. split in solid mass of mesoderm

d. mesodermal buds from archenteron

e. mouth, second opening for anus

f. anus, second opening for mouth

SUGGESTED ANSWERS TO STRUCTURE YOUR KNOWLEDGE

1. **a.** Porifera (sponges): assymetrical; lack true tissues

b. Eumetazoa: "true" animals; have true tissues

c. Cnidaria: radial symmetry; diploblastic

d. Bilateria: bilateral symmetry; triploblastic

e. Deuterostomia: radial, indeterminate cleavage; coelom from outpockets of archenteron; blastopore forms anus

f. Protostomia: spiral, determinate cleavage; coelom from split in blocks of mesoderm; blastopore forms mouth

g. Lophotrochozoa: molecular similarities; named for lophophore feeding structure and trochophore larva of some phyla

h. Ecdysozoa: molecular similarities; named for molting of exoskeleton typical of some phyla

ANSWERS TO TEST YOUR KNOWLEDGE

Multiple Choice:

1. d	3. e	5. a	7. c
2. d	4. c	6. e	8. b

CHAPTER 33: AN INTRODUCTION TO INVERTEBRATES

INTERACTIVE QUESTIONS

33.1 **a.** line body cavity (spongocoel); create water current, trap and engulf food particles, may produce gametes

b. wander through mesohyl; digest food and distribute nutrients, produce skeletal fibers, may produce gametes and other types of sponge cells

33.2 **a.** polyp
b. medusa
c. mouth/anus
d. tentacle
e. mesoglea
f. gastrovascular cavity

33.3 **a.** Planarians have an extensively branched gastrovascular cavity; food is drawn in and undigested wastes are expelled through the mouth at the tip of a muscular pharynx.

b. Tapeworms are bathed in predigested food in their host's intestine.

33.4 **a.** **1.** muscular foot for movement
2. visceral mass containing internal organs
3. mantle that may secrete shell and enclose visceral mass, forming a mantle cavity that houses gills
b. **1.** Snails use their foot to creep slowly and their radula to rasp up algae.
2. Clams are sedentary suspension feeders that use siphons to draw water over the gills, where food is trapped in mucus and swept by cilia to the mouth.
3. Squid are active predators that capture prey with tentacles and crush it with jaws, aided by a poison in their saliva.

33.5 **a.** septum
b. intestine
c. metanephridium
d. chaetae
e. ventral nerve cord
f. dorsal and ventral blood vessels
g. longitudinal and circular muscles

33.6 Earthworms have longitudinal and circular muscle layers and a segmented body cavity that enable them to move by alternately elongating and contracting body sections and anchoring to the substrate with chaetae. Nematodes are not segmented and have only longitudinal muscles. These pull against the hydrostatic skeleton of the roundworm's tapered body, producing thrashing movements.

33.7 **a.** anterior feeding appendages modified as pincers or fangs
b. Many spiders trap their prey in webs they have spun from abdominal gland secretions. Spiders kill prey with poison-equipped, fang-like chelicerae. After secreting digestive juices, they suck up partially digested food.

33.8 **a.** mostly terrestrial, also in fresh water
b. mostly aquatic
c. fly, walk, burrow
d. swim, float (planktonic)
e. tracheal system
f. gills or body surface
g. Malpighian tubules
h. diffusion across cuticle, glands regulate salt balance
i. one pair
j. two pairs
k. mandibles and modified mouthparts for chewing, piercing, or sucking; thorax has three pairs of walking legs and may have wings (extensions of cuticle)
l. mandibles and multiple mouthparts, walking legs on thorax, abdominal appendages

33.9 **a.** radial anatomy as adults, usually five spokes; sessile or slow moving; bilaterally symmetrical larvae
b. endoskeleton of calcareous plates
c. water vascular system with tube feet for locomotion and feeding

SUGGESTED ANSWERS TO STRUCTURE YOUR KNOWLEDGE

1. **a.** Eumetazoa
b. Bilateria
c. Deuterostomia
d. Lophotrochozoa
e. Ecdysozoa
f. Porifera (sponges); lack true tissues; choanocytes used in suspension feeding
g. Cnidaria (hydras, jellies, sea anemones, corals); diploblastic; radial symmetry; gastrovascular cavity; polyp and medusa forms; cnidocytes
h. Platyhelminthes (flatworms: planaria, flukes, tapeworms); acoelomate; gastrovascular cavity
i. Rotifera (rotifers); pseudocoelomate; alimentary canal; crown of cilia around mouth; parthenogenesis (in some)

j. Lophophorates (brachiopods and ecto-procts); coelomate; lophophore feeding structure; exoskeleton or bivalve shell

k. Mollusca (clams, snails, squids); coelomates; muscular foot; visceral mass; mantle; most with hard shell

l. Annelida (segmented worms: polychaetes, earthworms, and leeches); coelomates; segmented body wall and organs

m. Nematoda (roundworms); pseudocoelomate; unsegmented, cylindrical worms; ecdysis

n. Arthropoda (crustaceans, insects, spiders, millipedes); coelomate; segmented body; jointed appendages; exoskeleton

o. Echinodermata (starfishes, sea urchins); coelomate; endoskeleton; water vascular system

p. Chordata (vertebrates, two invertebrate subphyla); coelomates; segmentation; notochord

ANSWERS TO TEST YOUR KNOWLEDGE

Matching:

1. C l	**4.** A k	**7.** F c	**10.** C h
2. B e	**5.** I m	**8.** B b	**11.** I m
3. F g	**6.** B i	**9.** D f	**12.** F d

Multiple Choice:

1. b	**5.** a	**9.** a	**13.** d
2. c	**6.** c	**10.** b	**14.** c
3. c	**7.** c	**11.** a	**15.** e
4. e	**8.** b	**12.** b	**16.** b

CHAPTER 34: THE ORIGIN AND EVOLUTION OF VERTEBRATES

INTERACTIVE QUESTIONS

34.1 lancelet, Cephalochordata
a. mouth
b. *pharyngeal slits
c. digestive tract
d. segmental muscles
e. anus
f. *post-anal tail
g. *dorsal, hollow nerve cord
h. *notochord
* Chordate characters

34.2 head with eyes and other sense organs, skull enclosing the brain, two or more sets of *Hox* genes, other duplicated families of genes, neural crest, pharyngeal clefts that develop into gill slits in aquatic species, higher metabolism, more muscles, chambered heart, red blood cells with hemoglobin, and kidneys that remove wastes

34.3 a. Chordates
b. Craniates
c. Vertebrates
d. Gnathostomes
e. Osteichthyans
f. Lobe-fins
g. Tetrapods
h. Amniotes
i. lancelets (Cephalochordata)
j. tunicates (Urochordata)
k. hagfishes (Myxini)
l. lampreys (Petromyzontida)
m. sharks, rays (Chondrichthyes)
n. ray-finned fishes (Actinopterygii)
o. coelacanths (Actinistia)
p. lungfishes (Dipnoi)
q. frogs, salamanders (Amphibia)
r. turtles, snakes, crocodiles, birds (Reptilia)
s. mammals (Mammalia)

34.4 more extensive skull; vertebrae (backbone) that encloses spinal cord, and in aquatic vertebrates, bony fin rays; duplication of *Dlx* gene family

34.5 a. hinged jaws, lateral line system, enlarged forebrain with better olfaction and vision, four sets of *Hox* genes, and apparent duplication of entire genome
b. increase the area for absorption and slow the passage of food through the short shark intestine
c. common opening for the reproductive tract, excretory system, and digestive tract

34.6 a. Coelacanths, lungfishes, and terrestrial tetrapods
b. See answer to Interactive Question 34.3.

34.7 It has the fish characters of scales, fins, and gills and lungs, but the tetrapod characters of neck, ribs, flat skull, eyes on top of skull, and bones in its front fins with the basic tetrapod pattern of humerus, radius and ulna, and wrist bones. It did not have limbs with digits and thus, is considered to be a lobe-fin, not a tetrapod.

34.8 **a.** chorion—functions in gas exchange
b. amnion—encases embryo in fluid, prevents dehydration, and cushions shocks
c. allantois—stores metabolic wastes, functions with chorion in gas exchange
d. yolk sac—contains yolk; blood vessels transport stored nutrients into embryo

34.9 Archosaurs include the extinct dinosaurs and pterosaurs as well as crocodilians and birds. Lepidosaurs include extinct huge marine reptiles, tuatara, lizards, and snakes.

34.10 lightweight strong skeleton, no teeth and absence of some organs; feathers that shape wings into airfoil; strong keel on sternum to which flight muscles attach; high metabolic rate supported by efficient circulatory system (four-chambered heart), respiratory system, and endothermy; well-developed visual and motor areas of brain

34.11 **a.** hair; milk produced by mammary glands; endothermic with active metabolism; diaphragm to ventilate lungs; four-chambered heart; larger brains, increased learning capacity; differentiated teeth and remodeled jaw
b. See answer to Interactive Question 34.3.

34.12 increased brain size; change in jaw shape and dentition; bipedal posture; reduced size difference between the sexes; capacity for language and symbolic thought; manufacture and use of complex tools

SUGGESTED ANSWERS TO STRUCTURE YOUR KNOWLEDGE

1. Primates: grasping hands and feet, flat nails, moveable thumb, forward-facing eyes, large brain

2. Eutherians: placental mammals; young complete development within uterus
Mammals: mammary glands that produce milk, hair, diaphragm, four-chambered heart
Amniotes: amniotic egg with four extraembryonic membranes, less permeable skin, use of rib cage to breathe
Tetrapods: limbs with digits, neck vertebrae, pelvic girdle fused to backbone

Lobe-fins: muscular fins supported by bones
Osterichthyans: bony skeleton, lungs or lung derivatives
Gnathostomes: hinged jaws, four sets of *Hox* genes, mineralized skeleton, enlarged forebrain with enhanced senses
Vertebrates: vertebral column surrounding nerve cord; *dlx* gene duplication
Craniates: head with eyes; skull enclosing the brain; duplication of *Hox* genes; neural crest cells

3. (1) Pharyngeal slits used for suspension feeding became adapted for gas exchange; (2) skeletal supports for gills became adapted for use as hinged jaws; (3) lungs of bony fish were transformed to air bladders, aiding in buoyancy; (4) fleshy fins supported by skeletal elements developed into limbs for terrestrial animals; (5) feathers, probably originally used for insulation, came to be used for flight; (6) dexterous hands important for an arboreal life, evolved to manipulate tools.

4. Amphibians remain in damp habitats, burrowing in mud during droughts; some secrete foamy protection for eggs laid on land. Nonbird reptiles have scaly, waterproof skin and behavioral adaptations to modulate changing temperatures. The amniote egg provides an aquatic environment for the developing embryo. Endothermy of birds and mammals maintains stable body temperature in the more variable terrestrial climate. Gestation protects a developing embryo. Tetrapods use limbs for locomotion on land.

ANSWERS TO TEST YOUR KNOWLEDGE

Fill in the Blanks:

1. Urochordata (tunicates)
2. Cephalochordata (lancelets)
3. somites
4. gnathostomes
5. operculum
6. amniotic egg
7. monotremes
8. diaphragm
9. anthropoids
10. *Australopithecus*

Multiple Choice:

1. a	**4.** e	**7.** a	**10.** e
2. c	**5.** a	**8.** c	**11.** d
3. e	**6.** d	**9.** b	**12.** a

CHAPTER 35: PLANT STRUCTURE, GROWTH, AND DEVELOPMENT

INTERACTIVE QUESTIONS

35.1 **a.** reproductive shoot (flower)
b. apical bud
c. node
d. internode
e. vegetative shoot
f. leaf
g. blade
h. petiole
i. axillary bud
j. stem
k. taproot
l. lateral roots
m. shoot system
n. root system

35.2 **a.** sclerenchyma (fibers and sclereids), tracheids, and vessel elements
b. sieve-tube elements (and the cells listed in **a.**)

35.3 Most plants exhibit **indeterminate growth,** continuing to grow as long as they live. Animals, as well as some plant organs such as leaves and flowers, have **determinate growth** and stop growing after reaching a certain size.

35.4 **a.** epidermis (dermal)—root hairs for absorption
b. cortex (ground)—uptake of minerals and water, food storage
c. vascular cylinder (vascular)—transport
d. xylem—water and mineral transport
e. phloem—nutrient transport
f. endodermis—regulates water and mineral movement into stele
g. pericycle—origin of lateral roots

35.5 **a.** cuticle
b. upper epidermis
c. palisade mesophyll
d. spongy mesophyll
e. guard cells
f. xylem
g. phloem
h. vein
i. stoma

35.6 H, F, A, B, G, D, C, E

35.7 They are helping to determine each gene's function and eventually to track the complex genetic and biochemical pathways involved in plant development.

35.8 Microtubules of the preprophase band leave behind ordered actin microfilaments that determine the orientation of cell division. Microtubules in the outer cytoplasm orient the cellulose microfibrils in cell walls, which then determine the direction in which cells can expand.

35.9 **a.**

Whorl	Genes Active	Organs in Normal Flower	Genes Active Mutant *A*	Organs in Mutant *A* Flower
1	*A*	sepals	*C*	carpels
2	*AB*	petals	*BC*	stamens
3	*BC*	stamens	*BC*	stamens
4	*C*	carpels	*C*	carpels

b. Gene *A* would be expressed alone in each whorl, producing a flower with four whorls of sepals.

SUGGESTED ANSWERS TO STRUCTURE YOUR KNOWLEDGE

1. Except for leaves and flowers, which stop growing when they reach maturity (determinate growth), plant roots and shoots continue to grow throughout the life of a plant. Although certain characteristic forms have been favored by natural selection in species adapted to particular environments, the specific environment in which a plant grows greatly influences its individual body form. For example, a plant may grow taller or produce different shapes of leaves depending on its light exposure. This developmental plasticity helps compensate for a plant's immobility.

2. This drawing is a eudicot stem, indicated by the ring of vascular bundles and the central pith. It represents primary growth because there is no layer of vascular cambium between the layers of xylem and phloem.
   a. sclerenchyma
   b. phloem
   c. xylem
   d. cortex
   e. pith
   f. vascular bundle
   g. epidermis

ANSWERS TO TEST YOUR KNOWLEDGE

Multiple Choice:

1	d	4	b	7	e	10	e	13	d
2	a	5	b	8	e	11	c	14	b
3	d	6	d	9	c	12	a	15	c

Matching:

1.	I	3.	J	5.	C	7.	D	9.	H
2.	F	4.	E	6.	A	8.	G	10.	B

CHAPTER 36: RESOURCE ACQUISITION AND TRANSPORT IN VASCULAR PLANTS

INTERACTIVE QUESTIONS

36.1   **a.** At low light intensities, horizontal leaf orientation would maximize exposure to sunlight. In bright sun, a vertical orientation would shield leaves from too high a light intensity and allow light to penetrate to lower leaves.

b. Plant productivity would increase with an increasing leaf area index as more leaf area is available to capture sunlight. If this index gets too high, however, then lower branches and leaves are too shaded to photosynthesize and they may die through a process of self-pruning.

c. The branching pattern of roots can increase to take advantage of the local availability of nitrate, as can the production of nutrient transporters within the root cells.

36.2   **a.** $\Psi_p = 0$, $\Psi_s = -0.6$, $\Psi = -0.6$ MPa

b. $\Psi_p = 0.6$, $\Psi_s = -0.6$, $\Psi = 0$ MPa. The turgor pressure of the cell that develops with movement of water into the cell finally offsets the solute potential of the cell. No net osmosis occurs, and the water potentials of both cell and solution are equal and $= 0$ MPa.

c. $\Psi_p = 0$, $\Psi_s = -0.8$, $\Psi = -0.8$ MPa. The cell would plasmolyze, losing water until the solute potential of the cell would equal that of the bathing solution (both -0.8 MPa). There would be no turgor pressure.

36.3   **a.** symplastic
b. apoplastic
c. Casparian strip
d. xylem vessels
e. vascular cylinder (stele)
f. endodermis
g. cortex
h. epidermis
i. root hair

36.4   **a.** The loss of water vapor from the air spaces through the stomata causes evaporation from the water film coating the mesophyll cells, leading to the negative pressure that draws water up from the roots.

b. Evaporation from the water film increases the surface tension of the curving air–water interface. This tension, or negative pressure, produces a gradient of water potentials that extends from the leaf to the root.

c. Water molecules hold together due to hydrogen bonding, and a pull on one water molecule is transmitted throughout the column of water.

d. Water molecules adhere to the hydrophilic walls of narrow xylem cells, helping to support the column of water against gravity.

36.5   In order to provide sufficient CO_2 for photosynthesis, stomata must be open, and large amounts of water will be lost to transpiration. A sunny, windy, dry day will increase the rate of transpiration, perhaps to the point that insufficient water is available in the soil. Due to a lack of water, guard cells may lose turgor (and also be signaled by abscisic acid) and stomata will close, slowing transpiration. In reducing excessive water loss, however, the plant mesophyll cells now receive insufficient CO_2, and the photosynthetic rate will decline.

36.6   **a.** cotransporter
b. sucrose
c. H^+
d. proton pump
e. ATP

SUGGESTED ANSWERS TO STRUCTURE YOUR KNOWLEDGE

1.   Solutes may move across membranes by passive transport when they move down their electrochemical gradient. Transport proteins, which speed this passive transport, may be specific carrier proteins or selective channels. Active transport usually involves a proton pump that creates a membrane potential as H^+ ions are moved out of the cell. Cations such as K^+ may now move through specific

channels down their electrochemical gradient. In cotransport, the inward diffusion of H^+ down its concentration gradient through a cotransporter moves a solute against its concentration gradient.

2. The cohesion–tension hypothesis explains the ascent of xylem sap. Tension created by the evaporation of water and the resulting curvature of the air–water interface lowers the water potential in the leaf and sets up the bulk flow of water. The cohesion of water molecules transmits the pull resulting from transpiration throughout the water column, and the adhesion of water to the hydrophilic walls of the xylem vessels aids the flow.

   A pressure flow mechanism also explains the movement of phloem sap, but the plant must expend energy to create the differences in water potential that drive translocation. The active accumulation of sugars in sieve-tube elements lowers water potential at the source end, resulting in an inflow of water. The removal of sugar from the sink end of a phloem tube is followed by the osmotic loss of water. The resulting difference in pressure between the source and sink end of the phloem tube causes the bulk flow of phloem sap.

ANSWERS TO TEST YOUR KNOWLEDGE

Multiple Choice:

1. c	5. b	9. d	13. a	17. e
2. b	6. e	10. c	14. c	18. d
3. e	7. a	11. b	15. c	
4. a	8. d	12. d	16. b	

CHAPTER 37: SOIL AND PLANT NUTRITION

INTERACTIVE QUESTIONS

37.1 A loam, which is a mixture of sand, silt, and clay, has enough large particles to provide air spaces and enough small particles to retain water and minerals. Cations bind to soil particles and are made available to roots by cation exchange. Humus in a soil helps to retain water and provide a steady supply of mineral nutrients.

37.2 **a.** the variety of conservation-minded, environmentally safe, and profitable farming methods that permit the use of soil as a renewable resource
b. the percentage of nitrogen, phosphorus, and potassium that is in a fertilizer

37.3 **a.** macronutrient; component of chlorophyll, cofactor and activator of enzymes
b. macronutrient; component of nucleic acids, phospholipids, ATP, some coenzymes
c. micronutrient; component of cytochromes, cofactor for chlorophyll synthesis

37.4 **a.** the yellowing of leaves due to a lack of chlorophyll, often indicative of a deficiency of magnesium
b. in the younger, growing parts of a plant

37.5 **a.** nitrogen-fixing bacteria
b. ammonifying bacteria
c. nitrifying bacteria
d. denitrifying bacteria
e. nitrate and nitrogenous organic compounds

37.6 **a.** Legume crops, which increase nitrogen in the soil, are often followed the next growing season with a nonlegume crop.
b. "Green manure" is a legume crop that is plowed under to improve soil fertility by adding both humus and nitrogen.

37.7 The fossil record dates the appearance of mycorrhizae back to 460 million years ago. These symbiotic associations may have been an important adaptation in helping plants to colonize nutrient-poor soils.

SUGGESTED ANSWERS TO STRUCTURE YOUR KNOWLEDGE

1.

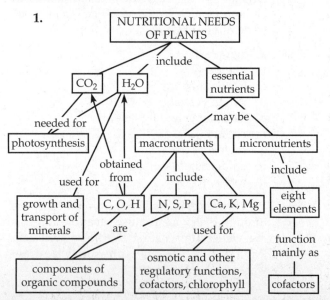

2. Root nodules are mutualistic associations between legumes and nitrogen-fixing bacteria of the genus *Rhizobium*. These bacteria receive nourishment from the plant and provide the plant with fixed nitrogen. Mycorrhizae are mutualistic associations between the roots of most plants and various species of fungi. The fungi stimulate root growth, increase surface area for absorption, and secrete antibiotics that may help protect the plant. In return, the fungi receive nourishment from the plant.

3. Both these types of symbiotic relationships involve chemical recognition between specific plant species and bacterial or fungal species. Both are mutualistic interactions.

ANSWERS TO TEST YOUR KNOWLEDGE

Multiple Choice:

1. b	**4.** e	**7.** a	**10.** d	**13.** c
2. b	**5.** e	**8.** e	**11.** e	**14.** d
3. c	**6.** b	**9.** a	**12.** d	**15.** e

CHAPTER 38: ANGIOSPERM REPRODUCTION AND BIOTECHNOLOGY

INTERACTIVE QUESTIONS

38.1
a. stamen
b. anther
c. filament
d. carpel
e. stigma
f. style
g. ovary
h. sepal
i. receptacle
j. ovule
k. petal

38.2
a. The male gametophyte consists of a tube cell and a generative cell (which moves into the tube cell and will divide to form two sperm). These cells and the spore wall that encloses them constitute a pollen grain.
b. The female gametophyte is the embryo sac, often containing seven cells and eight nuclei.

38.3 Double fertilization conserves resources by allowing for nutrient development only when fertilization has occurred.

38.4
a. seed coat
b. epicotyl
c. hypocotyl
d. cotyledons with endosperm
e. radicle
f. cotyledon (scutellum)
g. endosperm
h. epicotyl

i. hypocotyl
j. radicle
k. coleorhiza
l. coleoptile

38.5 The coleoptile pushes through the soil, and the shoot tip is protected as it grows up through the tubular sheath.

38.6
a. Buoyant fruits may be dispersed by water; winged fruits may "fly" seeds to new destinations; seeds in edible fruits may be dispersed in feces; prickly seeds may stick to animals, and some animals may "plant" seeds in their underground caches.
b. Hormonal interactions induce softening of the pulp, a change in color, and an increase in sugar content.

38.7
a. Apomixis allows for the dispersal value of seeds, which in this case are clones of the parent plant.
b. Selfing is an advantage in certain crop plants because it ensures that each ovule will develop into a seed.

38.8 Burning of fossil fuels releases CO_2 that was removed from the atmosphere millions of years ago. CO_2 that is released by the burning of biofuels is absorbed for photosynthesis as new biofuel crops grow.

SUGGESTED ANSWERS TO STRUCTURE YOUR KNOWLEDGE

1. One version of the life cycle of an angiosperm. See also textbook Figures 38.2 and 38.3.

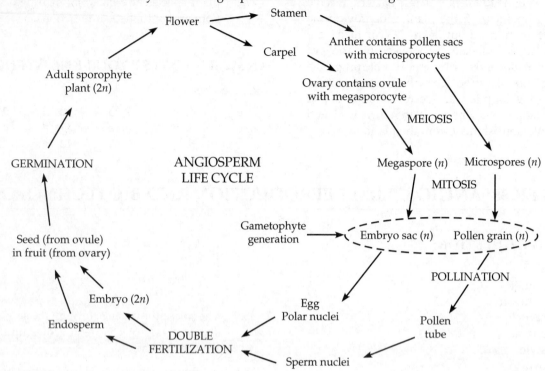

2. *Sexual:* Advantages include increased genetic variability, which provides the potential to adapt to changing conditions, and the dispersal and dormancy capabilities provided by seeds. Disadvantages are that the seedling stage is very vulnerable, sexual reproduction is very energy intensive (since producing flowers, pollen, and fruit takes energy and most seeds and seedlings don't survive), and genetic recombination may separate adaptive traits.

   Asexual: Advantages include the hardiness of vegetative propagation and the maintenance of genetically well-adapted plants in a given environment. Disadvantages relate to the advantages of sexual reproduction: There is no genetic variability from which to "choose" should conditions change, and there are no seeds for dormancy or dispersal.

3. *Benefits:* Reduce use of chemical pesticides; remove weeds with herbicides instead of erosion-producing tillage; increase nutritional value of crops; produce disease-resistant crops.

   Dangers: Introduce allergens into human foods; harm nontarget species; create hybrid "superweeds" due to crop-to-weed transgene escape; produce other unanticipated harmful results that cannot be stopped once GMOs are released into the environment.

ANSWERS TO TEST YOUR KNOWLEDGE

Fill in the Blanks:

1. ovary
2. sporophyte
3. biofuel
4. embryo sac
5. radicle
6. epicotyl
7. coleoptile
8. scion
9. protoplast
10. callus

Multiple Choice:

1. e	5. e	9. c	13. d
2. a	6. a	10. d	14. b
3. d	7. a	11. b	15. a
4. d	8. e	12. c	16. e

CHAPTER 39: PLANT RESPONSES TO INTERNAL AND EXTERNAL SIGNALS

INTERACTIVE QUESTIONS

39.1 **a.** The signal is light; the receptor is a phytochrome located in the cytosol.

b. Light-activated phytochrome activates at least two pathways, one producing the second messenger cGMP and one leading to an increase in cytosolic Ca^{2+} levels. Both activate specific kinases. Some kinase cascades may activate various transcription factors that change gene expression.

c. The plant response involves growth changes (slowing of stem elongation, expansion of leaves, elongation of roots). The de-etiolation process involves changes in levels of growth-regulating hormones and the activation or new production of enzymes involved in producing chlorophyll or in photosynthesis.

39.2 **a.** proton (H^+) pumps
  b. lower
  c. expansins
  d. osmotic uptake of water
  e. increased turgor pressure
  f. loosened walls (increased expandability of walls)

39.3 **a.** ethylene
  b. cytokinins
  c. abscisic acid
  d. auxin
  e. brassinosteroids
  f. strigolactones
  g. gibberellins

39.4 The relative amounts of red and far-red light are communicated to a plant by the ratio of the two forms of phytochrome. Canopy trees absorb red light, and a shaded tree will have a higher ratio of P_r to P_{fr}, which induces the tree to grow taller.

39.5 Free-running periods are circadian rhythms that vary from exactly 24 hours when an organism is kept in a constant environment without environmental cues.

39.6 **a.** not flower
  b. flower
  c. not flower
  d. flower
  e. not flower
  f. flower
  g. not flower
  h. flower
  i. not flower
  j. flower

In the next to last light regimen, a flash of far-red light cancels the effect of the red flash, and the red flash in the last one cancels the effect of the far-red flash.

39.7 **a.** positive phototropism, negative gravitropism
  b. thigmotropism; mechanical stimulation causing unequal growth rates of cells on opposite sides of the tendril

39.8 **a.** reduce transpirational water loss by stomatal closing and wilting; deep roots continue to grow in moister soil; heat-shock proteins produced to stabilize plant proteins and help new proteins fold correctly
  b. change membrane lipid composition by increasing unsaturated fatty acids; air tubes may develop in cortex of root

39.9 **a.** by eating plant tissues and reducing the photosynthetic capacity of plants; by causing wounds through which pathogens can enter a plant
  b. by pollinating plants; by eating other insects that are eating plant tissues (such as parasitoid wasps)

SUGGESTED ANSWERS TO STRUCTURE YOUR KNOWLEDGE

1. cytokinin spray used to keep flowers fresh; ethylene used to ripen stored fruit; synthetic auxins to induce seedless fruit set; gibberellins sprayed on Thompson seedless grapes; 2,4-D (auxin) used as herbicide.

2.

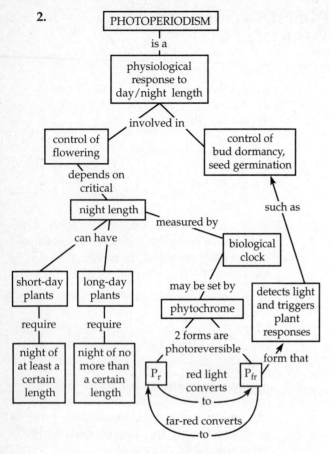

3. Specific resistance begins with an *R-Avr* match. A pathogen molecule (effector) binds to a specific plant receptor (coded for by an *R* gene) and triggers a signal transduction pathway that initiates a hypersensitive response and production of an alarm signal. The hypersensitive response includes production of antimicrobial molecules, sealing off of infected areas, and destruction of infected cells. The distribution of the alarm signal stimulates distant cells to produce salicylic acid, activating a systemic acquired resistance. Molecules are produced that protect the cell against a diversity of pathogens for several days.

ANSWERS TO TEST YOUR KNOWLEDGE

True or False:

False—change *abscisic acid* to *gibberellin*, or use *The removal of abscisic acid*.

True

False—change *thigmomorphogenesis* to *triple response*; or use *Thigmomorphogenesis is a change in growth form in response to mechanical stress*.

True

False—change *circadian rhythm* to *photoperiodism*; or change *day or night length* to *on a 24-hour cycle*.

True

False—change *A virulent* to *An avirulent*

False—change *cryptochrome* to *chromophore*; or use *Cryptochrome is a blue-light photoreceptor*.

Multiple Choice:

1. b	**5.** a	**9.** c	**13.** c
2. c	**6.** b	**10.** b	**14.** a
3. e	**7.** b	**11.** b	**15.** d
4. b	**8.** e	**12.** e	**16.** c

CHAPTER 40: BASIC PRINCIPLES OF ANIMAL FORM AND FUNCTION

INTERACTIVE QUESTIONS

40.1 Complexity allows for more specialization. Specialized organs, which can be protected within the body, can meet the animal's needs and control the composition of the interstitial fluid. A complex body is especially advantageous in highly variable terrestrial environments.

40.2 **a.** Stratified squamous; lining of mouth; thick layer is protective, and new cells produced near basal lamina replace those abraded off.
b. Simple columnar; lining of intestines; cells with large cytoplasmic volumes are specialized for secretion or absorption.

40.3 **a.** osteon
b. central canal
c. bone

d. collagenous fiber
e. elastic fiber
f. loose connective tissue
g. red blood cells
h. white blood cells
i. plasma
j. blood

40.4 **a.** cardiac muscle; dark band is intercalated disk that relays signals from cell to cell during heartbeat.
b. skeletal muscle; the striations are the contractile units (sarcomeres)

40.5 Hormones are released and travel more slowly throughout the body in the bloodstream. Nerve impulses travel rapidly through neurons directly to target cells. Hormone signals last longer and usually mediate responses of longer duration than the shorter responses to nervous signals. The endocrine system coordinates more gradual functions such as growth, reproduction, and metabolism. The nervous system regulates rapid responses such as locomotion and behavior.

40.6 Positive feedback is a mechanism in which a change in a variable serves to amplify rather than reverse the stimulus. An example is the increasing uterine contractions during childbirth. Homeostatic mechanisms tend to return a fluctuating variable to the normal range. Thus, positive feedback does not typically contribute to homeostasis.

40.7 Because the environmental temperature determines its body temperature, an ectotherm that lives where temperatures fluctuate very little will be homeothermic. If environmental temperatures are very high, an ectotherm may have a body temperature higher than that of an endotherm, which may have mechanisms for cooling the body.

40.8 **a.** vasodilation of surface arterioles, activation of sweat glands, panting
b. vasoconstriction of surface vessels, erection of fur, shivering by contraction of skeletal muscles, nonshivering thermogenesis

40.9 **a.** In hibernation, the body temperature is maintained at a lower level, greatly conserving energy and allowing the animal to withstand long periods of cold temperature and decreased food supply.
b. In estivation, animals survive long stretches of elevated temperature and diminished water supply by entering a period of inactivity and lowered metabolism.
c. Many small mammals and birds with very high metabolic rates enter a daily torpor during periods when they are not feeding or when temperatures are low.

SUGGESTED ANSWERS TO STRUCTURE YOUR KNOWLEDGE

1.

Tissue	Structural Characteristics	General Functions	Specific Examples
Epithelial	Tightly packed cells; basal lamina; cuboidal, columnar, or squamous shapes; simple, stratified, or pseudostratified	Protection, absorption, secretion; lines body surfaces	Mucous membrane, may be ciliated; skin; lining of blood vessels, kidney tubules, and intestines
Connective	Few cells that secrete extracellular matrix of protein fibers in liquid, gel, or solid matrix	Connect and support other tissues	Loose connective; adipose; fibrous connective; bone; cartilage; blood
Muscle	Long cells (fibers) with filaments of actin and myosin	Contraction, movement	Skeletal (voluntary); smooth (involuntary); cardiac
Nervous	Neurons with cell bodies, axons, and dendrites; supportive glial cells	Sense stimuli, conduct impulses	Nerves, brain

2. All organs are covered with epithelial tissue, and internal lumens, ducts, and blood vessels are lined with epithelia. Most organs contain some types of connective tissue. Muscle tissue would be present if the organ must contract. Nervous tissue innervates most organs.

3. Endotherms, who warm their bodies with metabolic heat, have thermoregulatory mechanisms that enable them to maintain a constant internal temperature across a wider range of environmental temperatures. Even though ectotherms may use behavioral mechanisms to reduce temperature fluctuations, they may still be exposed to greater internal fluctuations as environmental temperatures change.

4. Due to the higher energy cost of maintaining a warm body temperature, the BMR of an endotherm would always be higher than the SMR of a similarly sized ectotherm. Energy for reproduction would vary by sex, time of year, size and number of offspring produced, and amount of parental care. The cost of temperature regulation for an endotherm varies depending on the environment and the thermoregulation adaptations of the animal. The growth and activity expenditures vary depending on whether the animal is growing and the amount of activity required for obtaining food, building shelter, or escaping predators. Even as the proportions (pieces of the pie)

devoted to each activity change depending on the organism and the environment, the overall energy expenditure per unit mass is almost always higher for smaller animals. This inverse relationship for endotherms may be related to the greater surface area to volume ratio of small animals across which heat may be lost.

ANSWERS TO TEST YOUR KNOWLEDGE

Multiple Choice:

1. c	**4.** e	**7.** a	**10.** a	**13.** b
2. a	**5.** b	**8.** d	**11.** e	**14.** e
3. d	**6.** d	**9.** d	**12.** d	**15.** a

CHAPTER 41: ANIMAL NUTRITION

INTERACTIVE QUESTIONS

41.1 Vegetarians must eat a combination of plant foods that are complementary in amino acids, such as corn and beans.

41.2 **a.** Undernutrition results from a deficiency of chemical energy—a diet that lacks sufficient calories. Other types of malnutrition result from the lack of essential nutrients, such as essential amino acids (protein deficiency), essential fatty acids, vitamins, or minerals.
b. Calcium and phosphorus are required by vertebrates for bone and tooth construction.

41.3 The carbohydrates, proteins, nucleic acids, and fats that make up food are too large to pass through cell membranes, and the molecules in food are not identical to those an animal assembles for its own functions and tissues.

41.4 An alimentary canal can have specialized compartments for the sequential processing of nutrients. Because food moves in one direction, an animal can ingest more food while still digesting a previous meal.

41.5 **a.** Polysaccharides (starch and glycogen) broken down in the mouth by salivary amylase to smaller polysaccharides and maltose; hydrolysis continues until salivary amylase is inactivated by the low pH in the stomach.
b. Proteins are hydrolyzed by pepsin to smaller polypeptides in the stomach.

41.6 **a.** The digestion of starch and glycogen into disaccharides is continued by pancreatic amylase. Disaccharidases, which are enzymes of the brush border, split disaccharides into monosaccharides.
b. Protein digestion is completed in the small intestine by pancreatic trypsin and chymotrypsin, enzymes specific for peptide bonds adjacent to certain amino acids, and by pancreatic carboxypeptidase; as well as enzymes of the brush border: carboxypeptidase and aminopeptidase, which split amino acids off from opposite ends of a polypeptide; and by dipeptidases.
c. Pancreatic nucleases are a group of enzymes that hydrolyze DNA and RNA into their nucleotide monomers. Nucleotidases, nucleosidases, and phosphatases dismantle nucleotides.
d. The digestion of fats is aided by bile salts, which coat or emulsify tiny fat droplets so they do not coalesce, leaving a greater surface area for pancreatic lipase to hydrolyze the fat molecules into glycerol, fatty acids, and monoglycerides.

41.7 Carnivore dentition is characterized by sharp incisors, fanglike canines, and jagged premolars and molars. Herbivores have broad premolars and molars that grind plant material, and incisors and canines modified for biting off vegetation. Carnivorous vertebrates often have large, expandable stomachs. The longer intestine of an herbivore facilitates digestion of plant material, and its cecum may contain cellulose-digesting microorganisms.

41.8 Human craving for fatty foods may have evolved from the feast-and-famine existence of our ancestors. Natural selection may have favored individuals who gorged on and stored high-energy molecules, as they were more likely to survive famines.

SUGGESTED ANSWERS TO STRUCTURE YOUR KNOWLEDGE

1. **a.** salivary glands—produce saliva
    b. oral cavity—teeth and tongue mix food with saliva
    c. pharynx—throat, opening to esophagus and trachea
    d. esophagus—transports bolus to stomach
    e. stomach—churns food with gastric juice

f. small intestine—digestion and absorption
g. large intestine (colon)—absorbs water, compacts feces
h. anus—exit of alimentary canal
i. rectum—stores feces until expelled
j. pancreas—produces enzymes, bicarbonate, and hormones
k. gallbladder—stores bile from liver
l. liver—produces bile; processes nutrients; detoxifies

ANSWERS TO TEST YOUR KNOWLEDGE

Multiple Choice:

1. e	5. d	9. b	13. d	17. b
2. a	6. e	10. d	14. b	18. b
3. b	7. d	11. a	15. d	19. c
4. e	8. c	12. c	16. c	20. b

Matching:

1. J	3. C	5. H	7. I
2. E	4. D	6. B	8. F

CHAPTER 42: CIRCULATION AND GAS EXCHANGE

INTERACTIVE QUESTIONS

42.1 **a.** The heart pumps hemolymph through vessels into sinuses, and body movements squeeze the sinuses, forcing hemolymph through the body and back to the heart. When the heart relaxes, hemolymph is drawn in through pores, which have valves that close when the heart pumps.
b. Open circulatory systems take less energy to operate. Higher blood pressures enable more efficient delivery of nutrients and O_2 in closed circulatory systems, which also allow for greater regulation of blood distribution.

42.2 **a.** In the **single circulation** typical of fish, blood is pumped only once. It flows first through the gill capillaries, through a vessel to the systemic capillaries, and then back to the heart. In a **double circulation**, blood is pumped twice as it circulates through the body: once to the gas exchange organs, from which it returns to the heart, and again to the systemic circuit.
b. The amphibian heart has three chambers: two atria and a ventricle. A ridge in the ventricle helps to direct oxygen-rich blood from the left atrium into the systemic circuit, and oxygen-poor blood from the right atrium into the pulmocutaneous circuit. In the four-chambered heart of mammals, the right side of the heart receives and then pumps oxygen-poor blood to the lungs, and the left side of the heart receives and then pumps oxygen-rich blood to the systemic capillaries.

42.3 **a.** capillaries of head and forelimbs
b. left pulmonary artery
c. aorta
d. capillaries of left lung
e. left pulmonary vein
f. left atrium

g. left ventricle
h. aorta
i. capillaries of abdominal organs and hind limbs
j. inferior vena cava
k. right ventricle
l. right atrium
m. right pulmonary vein
n. capillaries of right lung
o. right pulmonary artery
p. superior vena cava

The pulmonary veins, left side of the heart, and aorta and arteries leading to the systemic capillary beds carry oxygen-rich blood. See text Figure 42.6 for numbers. Note that oxygen-rich blood in the aorta splits and travels either to the head region or lower body, returning to the heart through the superior or inferior vena cava, respectively. Thus, numbers 7–10 do not represent the sequential flow of blood.

42.4 **a.** atrioventricular valve
b. semilunar valve
c. SA (sinoatrial) node
d. AV (atrioventricular) node

42.5 **a.** artery walls
b. contraction of ventricle (systole)
c. diastole
d. systole
e. diastolic pressure (lower number)
f. systolic pressure (higher number)
g. resistance
h. vasoconstriction
i. vasodilation
j. nerves, hormones (NO, endothelin)

42.6 Blood pressure forces fluid out of capillaries by bulk flow. The osmotic pressure due to the proteins that remain within the capillary tends to counter this fluid movement out of the capillary. A reduction in plasma proteins would reduce this counterbalancing force and result in an increased loss of fluid from capillaries and fluid accumulation in body tissues.

42.7 **a.** plasma
b. blood electrolytes
c. osmotic balance, buffering, muscle and nerve functioning
d. proteins
e. buffers, osmotic factors, lipid escorts, antibodies, clotting factors
f. nutrients, wastes, respiratory gases, hormones
g. erythrocytes (red blood cells)
h. transport O_2; help transport CO_2
i. leukocytes (white blood cells)
j. phagocytes
k. defense and immunity
l. platelets
m. blood clotting

42.8 smoking, not exercising, and a diet rich in animal fats or trans fats

42.9 An absence of surfactants, which are not produced by fetal lungs until about 33 weeks of development.

42.10 **a.** water
b. gill
c. gill cover and mouth pump water into mouth, across gills, and out of body
d. air
e. tracheoles of tracheal system
f. body movements compress and expand air tubes
g. air (and water)
h. lungs and moist skin
i. positive pressure breathing: fill mouth, close nostrils, raise floor of mouth, and blow up lungs
j. air
k. alveoli in lungs
l. negative pressure breathing: lower diaphragm and raise ribs, thereby increasing volume and decreasing pressure of lungs; air flows in

42.11 This graph shows the Bohr shift. The dissociation curve for hemoglobin is shifted to the right at a lower pH, meaning that at any given partial pressure of O_2, hemoglobin is less saturated with O_2. A rapidly metabolizing tissue produces more CO_2, which lowers pH, and hemoglobin will unload more of its O_2 to that tissue.

SUGGESTED ANSWERS TO STRUCTURE YOUR KNOWLEDGE

1. **a.** aorta
b. left pulmonary artery
c. left pulmonary veins
d. left atrium
e. left ventricle
f. right ventricle
g. inferior vena cava
h. atrioventricular valve
i. semilunar valve
j. right atrium
k. superior vena cava

   Blood flow from venae cavae → right atrium → right ventricle → pulmonary artery → lung → pulmonary vein → left atrium → left ventricle → aorta

2. Nasal cavity → pharynx → through glottis to larynx → trachea → bronchus → bronchiole → alveoli → interstitial fiuid → alveolar capillary → hemoglobin molecule → venule → pulmonary vein → left atrium → left ventricle → aorta → renal artery → arteriole → capillary in kidney

ANSWERS TO TEST YOUR KNOWLEDGE

Multiple Choice:

1. c	7. b	13. b	19. a	25. e
2. c	8. d	14. e	20. d	26. b
3. d	9. b	15. e	21. a	27. c
4. c	10. e	16. d	22. d	28. b
5. e	11. d	17. d	23. d	29. b
6. a	12. e	18. c	24. e	30. d

CHAPTER 43: THE IMMUNE SYSTEM

INTERACTIVE QUESTIONS

43.1 The mutant flies that were engineered to express defensin survived as well as the wild type flies, whose immune system was intact. The mutant flies that made no antimicrobial peptides and those mutants that were engineered to produce drosomycin survived equally poorly following infection.

43.2
a. phagocytic white blood cells
b. large phagocytes that engulf microbes
c. white blood cells that attack parasitic worms with enzymes
d. white blood cells located in skin and lymphatic tissue that stimulate adaptive immunity
e. white blood cells that attack the body's infected or cancerous cells
f. release histamine to initiate inflammatory response
g. causes vasodilation and increased permeability of blood vessels
h. enzyme that attacks bacterial cell walls; in mammals, found in tears, saliva, and mucus
i. proteins released by virus-infected cells that stimulate neighboring cells to produce substances that inhibit viral replication
j. set of blood proteins that cause lysis of microbes and are involved with innate and adaptive defenses
k. signaling proteins that enhance an immune response and promote blood flow to injury site

43.3 B cell antigen receptors recognize epitopes of intact antigens that are either molecules on the surfaces of infectious agents or molecules free in the body. T cell antigen receptors recognize pieces of antigens that have complexed with an MHC molecule inside a cell and are then presented on the cell surface.

43.4
a. During development of B cells and T cells, the *V* and *J* gene segments of one light-chain allele and one heavy-chain allele are randomly recombined. The joining of these diverse light and heavy chains produces a huge number of different antigen-binding specificities.
b. As lymphocytes mature in the thymus or bone marrow, they are tested for self-reactivity, and those that react against self components are inactivated or destroyed.
c. Once an antigen receptor binds its antigen, the selected cell divides into a clone of effector cells and memory cells specific for that antigen.

d. Immunological memory is an enhanced response to a previously encountered foreign molecule that arises from the presence of memory B and T cells, which rapidly form clones of effector cells specific to that antigen.

43.5
a. antigen-presenting cell
b. pathogen
c. antigen fragment
d. class II MHC molecule
e. T cell antigen receptor
f. accessory protein (CD4)
g. helper T cell
h. cytokines stimulate helper T cells, cytotoxic T cells, and B cells
i. cytotoxic T cell
j. B cell
k. cell-mediated immunity (attack on infected cells)
l. humoral immunity (secretion of antibodies by plasma cells)

43.6
a. cytokines that stimulate cytotoxic T cells, B cells, and themselves
b. perforin molecules that form pores in the target cell and granzymes that initiate apoptosis of the target cell

43.7
a. neutralization—neutralizes the ability of a virus to infect a host cell by blocking binding sites; binds to and neutralizes toxins in body fluids
b. opsonization—the binding of multiple antibodies to bacterium makes it more recognizable to macrophages or neutrophils; the cross-linking of bacteria, virus particles, or antigens into aggregates facilitates phagocytosis
c. complement activation and formation of membrane attack complex—antigen-antibody complexes on microbes activate the complement system. Complement proteins form a membrane attack complex, which produces a pore in the membrane of a foreign cell and causes it to lyse.

43.8

Blood Type	Antigens on RBCs	Antibodies in Plasma	Can Receive Blood from	Can Donate Blood to
A	A	anti-B	A, O	A, AB
B	B	anti-A	B, O	B, AB
AB	AB	none	A, B, AB, O	AB
O	none	anti-A and anti-B	O	A, B, AB, O

43.9 **a.** HIV destroys helper T cells, thereby crippling the humoral and cell-mediated immune systems and leaving the body unable to fight HIV and other opportunistic diseases such as Kaposi's sarcoma and *Pneumocystis* pneumonia.
b. HIV is readily passed through unprotected sex and needle sharing. The frequent mutational changes during replication generate drug-resistant strains of HIV. And frequent mutational changes in surface antigens have made development of an effective vaccine difficult.

SUGGESTED ANSWERS TO STRUCTURE YOUR KNOWLEDGE

1.

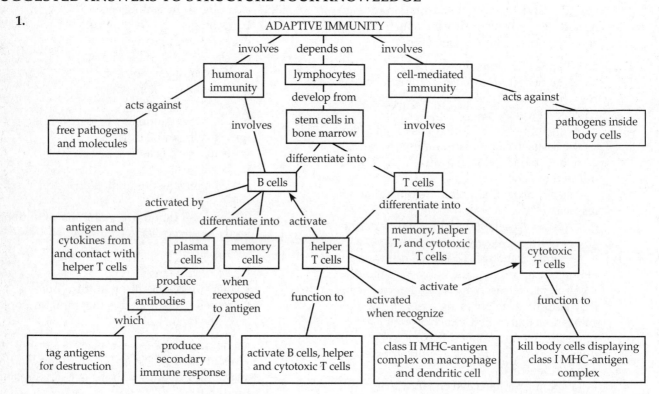

2. An antibody is a Y-shaped protein that consists of two identical light polypeptide chains and two identical heavy chains, held together by disulfide bonds. The amino acid sequences in the variable sections of the light and heavy chains in the arms of the Y account for the specificity in binding between antibodies and epitopes of an antigen. The two arms of an antibody allow it to cross-link antigens, which enhances phagocytosis. The constant region of the antibody tail determines its effector function; there are five classes of antibodies.

3. Macrophages and dendritic cells of innate immunity phagocytose microbes in response to broad categories of molecules, but also present peptide antigens in their class II MHC molecules to helper T cells. This interaction, along with synthesis of cytokines by these phagocytes, helps to activate helper T cells. The complement system functions in both innate and adaptive immunity. Some activated complement proteins promote inflammation or stimulate phagocytosis. The complement system may lyse microbes that are bearing antigen–antibody complexes.

ANSWERS TO TEST YOUR KNOWLEDGE

Multiple Choice:

1. e	**6.** a	**11.** e	**16.** d	**21.** d
2. e	**7.** e	**12.** a	**17.** e	**22.** e
3. c	**8.** b	**13.** a	**18.** c	**23.** a
4. b	**9.** e	**14.** c	**19.** c	**24.** c
5. d	**10.** a	**15.** d	**20.** b	**25.** d

CHAPTER 44: OSMOREGULATION AND EXCRETION

INTERACTIVE QUESTIONS

44.1
a. isoosmotic
b. conformer but may regulate specific solutes
c. slightly hyperosmotic
d. maintain urea and TMAO at high concentrations; rectal gland and kidneys remove salt
e. hypoosmotic
f. drink water to compensate for loss to hyperosmotic seawater; gills pump out salt; little urine excreted
g. hyperosmotic
h. copious dilute urine to compensate for osmotic gain; may pump salts in through gills
i. hyperosmotic
j. body coverings reduce evaporation; behavior adaptations; drinking and eating moist foods

44.2
a. An endotherm—it must eat more food to maintain its higher metabolic rate.
A carnivore—it eats more protein and thus produces more nitrogenous wastes.
b. *Uric acid* can be stored safely within the egg while the embryo develops. *Urea* is soluble and less toxic than ammonia and can be removed by the mother's blood. Urea is less energetically expensive to produce than uric acid.

44.3
a. osmoregulation; remove excess water
b. both; remove excess water
c. both; conserve water
d. both; usually conserve water

44.4
a. water, salts, nitrogenous wastes, glucose, vitamins, and other small molecules
b. blood cells and large molecules such as plasma proteins

44.5
a. Bowman's capsule
b. glomerulus
c. proximal tubule
d. descending limb, loop of Henle
e. ascending limb, loop of Henle
f. distal tubule
g. collecting duct
1. HCO_3^-, NaCl*, H_2O, nutrients*, and K^+ are reabsorbed from the proximal tubule. H^{+*} and NH_3 are secreted into the proximal tubule.
2. Water is reabsorbed from the descending limb of the loop of Henle.
3. NaCl diffuses from the thin segment of the ascending limb and is pumped out* of the thick segment.
4. NaCl*, H_2O, and HCO_3^{-*} are reabsorbed from the distal tubule. K* and H* are secreted into the tubule.
5. NaCl* is reabsorbed; urea and H_2O diffuse out of the collecting tubule.

* indicates active transport.

44.6
a. Juxtamedullary nephrons; the longer the loop of Henle, the more hyperosmotic the urine can become.
b. Mammals and birds; loops of Henle are especially long in those taxa inhabiting dry habitats.

44.7
a. high blood osmolarity
b. ADH, antidiuretic hormone
c. distal tubules and collecting ducts
d. JGA, juxtaglomerular apparatus
e. renin
f. adrenal glands
g. aldosterone
h. Na^+ and water reabsorption
i. increase in blood pressure or volume
j. ANP, atrial natriuretic peptide

SUGGESTED ANSWERS TO STRUCTURE YOUR KNOWLEDGE

1. a. water, salts, glucose, amino acids, vitamins, urea, other small molecules
b. ammonia (to offset acidity), drugs and poisons, H^+, K^+
c. water, glucose, amino acids, vitamins, K^+, NaCl, bicarbonate

2. Two solutes, NaCl and urea, contribute to the high osmolarity of the medulla. As the filtrate moves up the ascending limb, NaCl first diffuses and is then pumped out, contributing to the gradient of NaCl in the medulla. As the filtrate flows through the collecting duct, some urea diffuses out and also contributes to the osmotic gradient. As the filtrate in the collecting duct moves through the high osmolarity of the medulla, water flows out by osmosis, creating a more concentrated urine.

ANSWERS TO TEST YOUR KNOWLEDGE

Multiple Choice:

1. a	**4.** b	**7.** d	**10.** c	**13.** c	**16.** a	**19.** c
2. c	**5.** a	**8.** a	**11.** d	**14.** e	**17.** c	**20.** d
3. d	**6.** c	**9.** e	**12.** b	**15.** e	**18.** d	

CHAPTER 45: HORMONES AND THE ENDOCRINE SYSTEM

INTERACTIVE QUESTIONS

45.1 **a.** water-soluble hormone
b. signal receptor
c. signal transduction produces cytoplasmic response
d. signal transduction leads to gene regulation
e. lipid-soluble hormone
f. transport protein
g. signal receptor
h. hormone-receptor complex acts as transcription factor
i. gene regulation
j. cytoplasmic response

45.2 In negative feedback, the response reduces the initial stimulus, thereby turning off the response. In positive feedback, such as in the release of milk during nursing, the stimulus is reinforced, leading to an even greater response.

45.3 **a.** insulin
b. glucagon
c. uptake of glucose
d. glycogen breakdown in liver
e. islets of Langerhans in pancreas
f. hydrolyze glycogen and release glucose
g. blood glucose level

45.4 Insufficient iodine results in reduced production of T_3 and T_4. Thus, there is no negative feedback to turn off production of TRH by the hypothalamus and of TSH by the anterior pituitary. TSH continues to stimulate the thyroid, which enlarges.

45.5 **a.** stimulates contraction of uterus and milk release; influences reproductive and parenting behaviors
b. increases water reabsorption in kidney; plays role in social behavior
c. various effects, depending on species, including stimulation of milk production and secretion

d. regulates skin pigment cells in many vertebrates; affects hunger and metabolism in mammals
e. stimulates thyroid gland
f. stimulates activity of gonads
g. stimulates adrenal cortex to produce and secrete steroid hormones
h. promotes growth (especially bones) by stimulating release of insulin-like growth factors and affects metabolic functions

45.6 **a.** Nervous stimulation from the hypothalamus to the adrenal medulla causes the secretion of epinephrine and norepinephrine, which increase blood pressure, rate and volume of heartbeat, breathing rate, and metabolic rate; stimulates glycogen hydrolysis and fatty acid release; and changes blood flow patterns.
b. A releasing hormone from the hypothalamus stimulates ACTH release from the pituitary, which stimulates the adrenal cortex to release corticosteroids. Glucocorticoids increase blood glucose through conversion of proteins and fats. Mineralocorticoids increase blood volume and pressure by stimulating the kidney to reabsorb sodium ions and water.

SUGGESTED ANSWERS TO STRUCTURE YOUR KNOWLEDGE

1. Water-soluble signaling molecules that bind to plasma membrane receptors initiate signal transduction pathways that may activate cellular enzymes or affect gene expression. Steroid hormones and other lipid-soluble molecules bind with protein receptors inside a cell, and the hormone-receptor complex acts as a transcription factor to turn on (or off) specific genes.

2. In a simple hormone pathway, a stimulus triggers the release of a hormone, whose effect is to

reduce the stimulus, which, by negative feedback, stops triggering the release of the hormone. Secretin, released into the bloodstream in response to an acid pH, triggers the release of bicarbonate from the pancreas, which raises the pH of the duodenum. The resulting rise in pH removes the stimulus for secretin release.

3. A hormone cascade pathway involves a hormone sequence in which the hypothalamus secretes releasing hormones that stimulate the anterior pituitary to secrete a tropic hormone, which then stimulates target cells to secrete hormones that produce some physiological or developmental effect. For example, in response to a decrease in body temperature, the hypothalamus secretes TRH, which stimulates the anterior pituitary to secrete TSH, which acts on the thyroid gland to stimulate release of thyroid hormone, which increases metabolic rate. Thyroid hormone then acts by negative feedback to block the further release of TSH and TRH.

ANSWERS TO TEST YOUR KNOWLEDGE

Matching:

1. I g	**5.** A c	**9.** C d, released from h
2. G a	**6.** E b	**10.** J d, released from h
3. F e	**7.** L k	
4. B i	**8.** K f	

Multiple Choice:

1. e	**5.** a	**9.** c	**13.** e	**17.** d
2. d	**6.** d	**10.** b	**14.** e	**18.** c
3. b	**7.** d	**11.** d	**15.** c	**19.** b
4. e	**8.** b	**12.** b	**16.** a	**20.** e

CHAPTER 46: ANIMAL REPRODUCTION

INTERACTIVE QUESTIONS

46.1 a. Perpetuation of successful genotypes in stable habitats; elimination of need to locate mate; production of a larger number of offspring over time since all offspring can reproduce.
b. Varying genotypes and phenotypes of offspring may enhance reproductive success and adaptation of a population to fluctuating environments. Harmful alleles may be removed from a population more readily as genes are shuffled during sexual reproduction.
c. Sexual reproduction because it involves union of egg and sperm, and offspring will be genetically varied due to genetic recombination during meiosis and fertilization.

46.2 a. courtship behaviors that trigger the release of gametes and may provide a means for mate selection
b. pheromones that function as mate attractants
c. environmental signals (temperature, day length)

46.3 a. vas deferens
b. erectile tissue of penis
c. urethra
d. glans
e. prepuce (foreskin)
f. scrotum
g. testis
h. epididymis
i. bulbourethral gland
j. prostate gland
k. seminal vesicle
l. oviduct
m. follicles
n. corpus luteum
o. uterine wall (muscular layer)
p. endometrium (lining of uterus)
q. cervix
r. vagina
s. ovary

46.4 a. Spermatogenesis occurs continuously as stem cells and spermatogonia continue to divide from puberty onward. A female's supply of primary oocytes is established before birth.
b. Each meiotic division produces four sperm, but only one egg (and polar bodies that disintegrate).
c. Spermatogenesis is an uninterrupted process. Oogenesis occurs in stages: prophase I before birth; after puberty, meiosis I and II up to metaphase II in a maturing follicle just before ovulation; and completion of meiosis II (in humans) when a sperm cell penetrates the oocyte.

46.5 a. LH
b. FSH
c. follicular phase
d. ovulation

    e. luteal phase
    f. estradiol
    g. progesterone
    h. menstrual flow phase
    i. proliferative phase
    j. secretory phase

46.6 **a.** GnRH (gonadotropin-releasing hormone)
    b. FSH (follicle-stimulating hormone)
    c. LH (luteinizing hormone)
    d. androgens (primarily testosterone)
    e. negative feedback loops involving testosterone and inhibin

46.7 **a.** secreted by embryo, maintains corpus luteum (and thus progesterone) in first trimester
    b. secreted by corpus luteum and later by placenta, regulates growth and maintenance of placenta, formation of mucus plug in cervix, cessation of ovulation, growth of uterus

c. secreted by posterior pituitary; stimulates uterine contractions during birth; triggers release of milk during nursing

46.8 **a.** prevent release of gametes—sterilization, combination birth control pills (or injection, patch, or vaginal ring)
    b. prevent fertilization—abstinence, rhythm method, condom, diaphragm, progestin mini-pill, *coitus interruptus,* cervical cap, spermicides
    c. prevent implantation of embryo—morning after pills (MAP), IUD
The most effective methods are abstinence, sterilization, and chemical contraception. Least effective are the rhythm method and *coitus interruptus.*

SUGGESTED ANSWERS TO STRUCTURE YOUR KNOWLEDGE

   1. Path of sperm from formation to fertilization

seminiferous tubules in testes	site of spermatogenesis, Leydig cells produce testosterone
epididymis	tubules in which sperm mature and become motile
vas deferens	duct running from epididymis
ejaculatory duct	short duct joining vas deferens from both testes
seminal vesicles	contribute alkaline fluid containing mucus, fructose, coagulant, prostaglandins
prostate gland	contributes fluid containing anticoagulants and citrate (a nutrient) to semen
urethra	carries urine or semen through penis
bulbourethral glands	secretes small amount of mucus prior to ejaculation
penis	erection allows for insertion into vagina
vagina	receptacle for sperm
cervix	narrow opening into uterus
uterus	organ in which fetus develops
oviduct	tube leading to ovary, sperm fertilizes egg here

   2. **a.** GnRH is a releasing hormone of the hypothalamus that stimulates the anterior pituitary to secrete FSH and LH.
    b. FSH stimulates growth of follicles.
    c. The LH surge is caused by an increase in GnRH production that resulted from increasing levels of estradiol produced by the developing follicle.
    d. The LH surge induces the maturation of the follicle and ovulation, and it transforms the ruptured follicle to the corpus luteum.
    e. LH maintains the corpus luteum, which secretes estradiol and progesterone.
    f. High levels of estradiol and progesterone act on the hypothalamus and pituitary to inhibit the secretion of LH and FSH.
    g. The lack of LH causes the corpus luteum to disintegrate. Thus, the production of estradiol and progesterone ceases, which allows LH and FSH secretion to begin again.

   3. Birth control pills, which contain a combination of synthetic estrogens and progestin, act by negative feedback to prevent the release of GnRH by the hypothalamus and FSH and LH by the pituitary, thereby preventing the development of follicles and ovulation. RU486 blocks progesterone receptors in the uterus, thereby preventing progesterone from maintaining the endometrium.

ANSWERS TO TEST YOUR KNOWLEDGE

Fill in the Blanks:

1. budding
2. pheromomes
3. parthenogenesis
4. hermaphrodite
5. cloaca
6. estrous cycle
7. *in vitro* fertilization
8. urethra
9. vasocongestion
10. menopause

Multiple Choice:

1. d	5. b	9. c	13. b
2. d	6. e	10. a	14. e
3. b	7. c	11. e	15. a
4. a	8. b	12. c	16. b

CHAPTER 47: ANIMAL DEVELOPMENT

INTERACTIVE QUESTIONS

47.1 **a.** specific binding between receptors on the plasma membrane and proteins on the acrosomal process
b. fast block to polyspermy caused by membrane depolarization during acrosomal reaction and slow block to polyspermy caused by formation of the fertilization envelope during cortical reaction

47.2 **a.** 8-cell stage
b. blastula
c. animal pole
d. gray crescent
e. vegetal pole
f. blastocoel

47.3 **a.** ectoderm
b. mesenchyme cells
c. endoderm
d. archenteron
e. blastopore (becomes anus)
f. ectoderm
g. mesoderm
h. endoderm
i. yolk plug in blastopore
j. archenteron

47.4 **a.** expanding region of trophoblast—forms fetal portion of placenta
b. epiblast—forms three primary germ layers
c. hypoblast
d. trophoblast

47.5 **a.** The yolk sac grows to enclose the yolk and develops blood vessels to carry nutrients to the embryo.
b. The amnion encloses a fluid-filled sac, which provides an aqueous environment for development and acts as a shock absorber.
c. The allantois serves as a receptacle for certain metabolic wastes.
d. The chorion (pressed against the inside of the egg shell) provides gas exchange for the developing embryo.

47.6 **a.** neural tube
b. neural crest cells
c. somite
d. archenteron
e. coelom
f. notochord

47.7 Localized cytoplasmic determinants in the sea urchin egg must have an unequal polar distribution. The developmental fates of cells in the animal and vegetal halves of the embryo are determined by the first horizontal division.

47.8 The developing limb would be a mirror image, with posterior digits facing both forward and backward. This is the same effect as transplanting a donor ZPA to that location.

SUGGESTED ANSWERS TO STRUCTURE YOUR KNOWLEDGE

1.

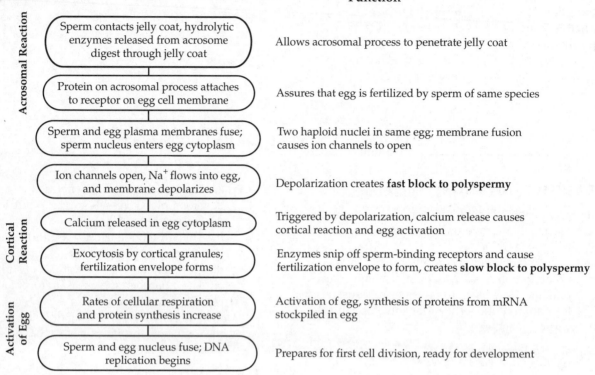

	Event	Function
Acrosomal Reaction	Sperm contacts jelly coat, hydrolytic enzymes released from acrosome digest through jelly coat	Allows acrosomal process to penetrate jelly coat
	Protein on acrosomal process attaches to receptor on egg cell membrane	Assures that egg is fertilized by sperm of same species
	Sperm and egg plasma membranes fuse; sperm nucleus enters egg cytoplasm	Two haploid nuclei in same egg; membrane fusion causes ion channels to open
	Ion channels open, Na⁺ flows into egg, and membrane depolarizes	Depolarization creates **fast block to polyspermy**
Cortical Reaction	Calcium released in egg cytoplasm	Triggered by depolarization, calcium release causes cortical reaction and egg activation
	Exocytosis by cortical granules; fertilization envelope forms	Enzymes snip off sperm-binding receptors and cause fertilization envelope to form, creates **slow block to polyspermy**
Activation of Egg	Rates of cellular respiration and protein synthesis increase	Activation of egg, synthesis of proteins from mRNA stockpiled in egg
	Sperm and egg nucleus fuse; DNA replication begins	Prepares for first cell division, ready for development

2.

Animal	Cleavage	Blastula	Gastrulation
Sea Urchin	Holoblastic; equal-sized blastomeres	Hollow ball of cells, one cell thick, surrounding blastocoel	Invagination through blastopore to form endoderm of archenteron; migrating mesenchyme forms mesoderm
Frog	Holoblastic, but yolk impedes divisions; unequal-sized blastomeres	Blastocoel in animal hemisphere; walls several cells thick	Involution of cells at dorsal lip of blastopore; migrating cells form mesoderm and endoderm around archenteron
Bird	Meroblastic, only in cytoplasmic disk on top of yolk	Blastoderm; blastocoel cavity between epiblast and hypoblast	Migration of epiblast cells through primitive streak forms mesoderm and endoderm; lateral folds join to form archenteron
Mammal	Holoblastic, no apparent polarity to egg, blastomeres equal	Blastocyst with inner cell mass (layers are epiblast and hypoblast); trophoblast surrounds embryo	Migration of epiblast cells through primitive streak forms mesoderm and endoderm

ANSWERS TO TEST YOUR KNOWLEDGE

Multiple Choice:

1. c	**4.** b	**7.** d	**10.** b	**13.** a	**16.** e
2. c	**5.** d	**8.** c	**11.** e	**14.** b	**17.** a
3. b	**6.** a	**9.** b	**12.** e	**15.** e	**18.** d

CHAPTER 48: NEURONS, SYNAPSES, AND SIGNALING

INTERACTIVE QUESTIONS

48.1 **a.** dendrites
b. cell body
c. axon hillock
d. axon
e. synaptic terminal. The impulse moves from the axon hillock along the axon to the synaptic terminals. Neurotransmitter is released into the synapse with another cell at the synaptic terminal.

48.2 **a.** K^+ inside the cell; Na^+ outside the cell
b. The inside of the cell is more negative compared to outside the cell.
c. The membrane potential moved closer to E_K; more potassium channels must have opened, and/or sodium channels must have closed.

48.3 **a.** sodium channel
b. potassium channel
c. K^+
d. Na^+
e. sodium inactivation gate
f. membrane potential (mV)
g. time
h. resting potential
i. threshold
j. action potential

1. Resting state: Voltage-gated Na^+ and K^+ channels are closed.
2. Depolarization: Some Na^+ channels open; depolarization opens more Na^+ channels. If threshold is reached, action potential is triggered.

3. Rising phase: Most Na^+ channels open and Na^+ influx makes inside of cell positive.
4. Falling phase: Na^+ channels become inactivated, K^+ channels open, K^+ leaves cell, and inside of cell becomes negative.
5. Undershoot: Na^+ channels closed (many are still inactivated), K^+ channels still open, and membrane becomes hyperpolarized. As these channels close, membrane returns to resting potential.

See textbook Figure 48.11 for location of stages on the graph.

48.4 **a.** presynaptic membrane
b. Ca^{2+} flowing in through voltage-gated calcium channel
c. synaptic cleft
d. synaptic vesicle containing neurotransmitter
e. postsynaptic membrane
f. receptor with bound neurotransmitter
g. closed ligand-gated ion channel

48.5 **a.** Temporal
b. Spatial

48.6 The type of postsynaptic receptor and its mode of action determine neurotransmitter function. Binding of acetylcholine to metabotropic receptors in heart muscle activates a signal transduction pathway that makes it more difficult to generate an action potential, thereby reducing the strength and rate of contraction. The acetylcholine ionotropic receptor in skeletal muscle cells opens ion channels and depolarizes the muscle cell membrane.

SUGGESTED ANSWERS TO STRUCTURE YOUR KNOWLEDGE

1.

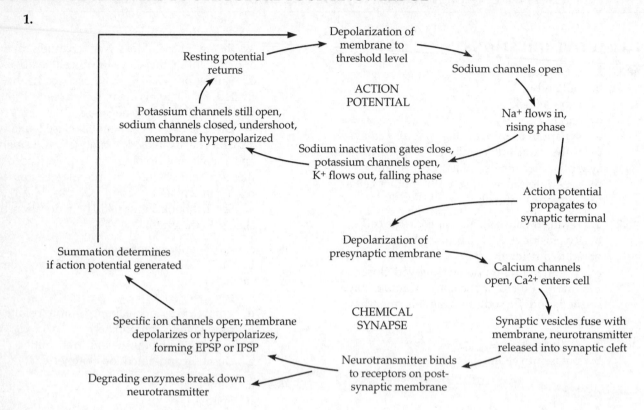

ANSWERS TO TEST YOUR KNOWLEDGE

Multiple Choice:

1. d	**3.** b	**5.** c	**7.** e	**9.** c	**11.** b	**13.** d	**15.** a
2. e	**4.** d	**6.** b	**8.** a	**10.** c	**12.** c	**14.** b	

CHAPTER 49: NERVOUS SYSTEMS

INTERACTIVE QUESTIONS

49.1 Sessile molluscs such as clams have little cephalization and only simple sense organs. Cephalopod molluscs such as squid have large brains, image-forming eyes, and giant axons, all of which contribute to their active predatory life and their ability to learn.

49.2
  a. efferent neurons
  b. afferent (sensory) neurons
  c. autonomic nervous system
  d. motor system
  e. sensory organs and cells
  f. enteric
  g. parasympathetic

  h. sympathetic
  i. slows heart, conserves energy
  j. skeletal muscles

49.3
  a. cerebrum 5
  b. thalamus 3
  c. cerebellum 1
  d. medulla 4
  e. pons 2
  f. midbrain 7
  g. hypothalamus 6

49.4 The primary somatosensory cortex receives information from somatosensory receptors in the body about touch, pain, pressure, temperature, and position of muscles and limbs. The motor cortex generates motor commands that travel to skeletal muscles. The neurons are distributed in an orderly fashion, from those associated with the legs and feet toward the midline of the parietal or frontal lobe, to those associated with the head located along the sides. The proportion of the primary somatosensory cortex and primary motor cortex devoted to each part of the body is correlated with the importance of that area.

49.5 Glutamate released by the presynaptic neuron opens NMDA receptors, but they remain blocked by magnesium ions until the postsynaptic membrane is depolarized by activity at nearby synapses. The flow of Ca^{2+} through the now-open NMDA receptor channels triggers insertion of AMPA glutamate receptors in the membrane. With both types of receptors in the membrane, glutamate release by the presynaptic neuron is now sufficient to initiate action potentials without input from other synapses.

SUGGESTED ANSWERS TO STRUCTURE YOUR KNOWLEDGE

1. **a.** diverse network of neurons extending through the brainstem; regulates state of arousal; filters sensory information going to the cortex
b. in the hypothalamus; functions as biological clock for circadian rhythms
c. clusters of neurons deep in white matter of cerebrum; centers for motor coordination and planning and learning movements
d. functional center in parts of thalamus, hypothalamus, and inner cerebral cortex (olfactory bulb, hippocampus, and amygdala); centers of emotions and memories in humans
e. two structures in inner cerebral cortex; involved in limbic system and memory; amygdala is involved in forming emotional memories; hippocampus functions in short-term memory and transfer to long-term memory

ANSWERS TO TEST YOUR KNOWLEDGE

Multiple Choice:

1. b	4. e	7. a	10. a	13. c
2. a	5. b	8. b	11. b	14. d
3. e	6. c	9. d	12. d	

CHAPTER 50: SENSORY AND MOTOR MECHANISMS

INTERACTIVE QUESTIONS

50.1 **a.** sensory reception: the detection of the energy of a stimulus by sensory cells
b. sensory transduction: the conversion of stimulus energy to a change in membrane potential of a sensory receptor cell
c. transmission: the passage of an action potential along sensory neurons to the brain
d. perception: the interpretation of sensory input by the brain

50.2 **a.** Most are concentrated in the skin; some are associated with other organs.
b. They inhibit prostaglandin synthesis.

50.3 **a.** auditory canal
b. malleus (hammer)
c. incus (anvil)
d. stapes (stirrup)
e. semicircular canals
f. auditory nerve
g. cochlea
h. round window
i. Eustachian tube
j. oval window
k. tympanic membrane (ear drum)

50.4 **a.** utricle and saccule for position and linear movement, semicircular canals for angular movement
b. lateral line system
c. statocysts containing statoliths
d. All of these mechanoreceptors are hair cells. The equilibrium organs of mammals have otoliths, and those of invertebrates contain statoliths. These granules settle by gravity and stimulate hair cells to signal position and movement. Sensory hairs on the hair cells in the lateral line system of fishes are embedded in a gelatinous cap, which is bent by moving water. Hair cells at the base of the semicircular canals also project into a gelatinous cap, which is bent by moving fluid.

50.5   **a.** sclera
b. choroid
c. retina
d. fovea (center of visual field)
e. optic nerve
f. optic disk (blind spot)
g. vitreous humor
h. lens
i. aqueous humor
j. pupil
k. iris
l. cornea
m. suspensory ligament

50.6   **a.** small
b. fovea

50.7   find mates, recognize territory, navigate, communicate, and locate and taste food

50.8   **a.** sarcomere
b. Z line
c. thin filaments (actin)
d. thick filament (myosin)
The Z lines will be closer together as the overlap between thick and thin filaments increases. This sliding of filaments occurs by the following sequence: Binding of ATP releases myosin heads from the actin filament; hydrolysis of ATP converts the head to a high-energy form that binds to actin, forming a cross-bridge, and then pulls the thin filament toward the center of the sarcomere.

50.9   **a.** fast-twitch glycolytic, fast-twitch oxidative, slow-twitch oxidative
b. fast-twitch glycolytic
c. slow-twitch oxidative

50.10   **a.** Hydrostatic skeleton: no extra materials needed, good shock absorber; not much protection or support for lifting animal off the ground; cnidarians, flatworms, nematodes, and annelids
b. Exoskeleton: quite protective; in arthropods, must be molted in order for the animal to grow, restricts size; most molluscs and arthropods
c. Endoskeleton: various types of joints allow for flexible movement in vertebrates, good structural support; may not be very protective; sponges, echinoderms, chordates

50.11   **a.** swimming
b. running
c. flying
d. smaller

SUGGESTED ANSWERS TO STRUCTURE YOUR KNOWLEDGE

1. Light enters the eye through the pupil and is focused by the lens onto the retina. When rhodopsin (visual pigment in rods) absorbs light energy, it sets off a signal transduction pathway that decreases the receptor cell's permeability to sodium and hyperpolarizes the membrane. The reduction in the release of neurotransmitter by a rod cell releases some connected bipolar cells from inhibition, which in turn generate action potentials in ganglion cells—the sensory neurons that form the optic nerve. Horizontal and amacrine cells provide lateral integration of visual information.

2. Sound waves are collected by the *pinna* and travel through the *auditory canal* to the *tympanic membrane*, where they are transmitted by the *malleus, incus,* and *stapes*. Vibration of the stapes against the *oval window* sets up pressure waves in the perilymph in the *vestibular canal* within the *cochlea*, from which they are transmitted to the *tympanic canal* and then dissipated when they strike the *round window*. The pressure waves vibrate the *basilar membrane*, on which the *organ of Corti* is located within the *cochlear duct*. Tips of the hair cells arising from the organ of Corti are embedded in the *tectorial membrane*. When vibrated, they bend, triggering a depolarization, release of neurotransmitter, and initiation of action potentials in the sensory neurons of the *auditory nerve*, which leads to the *cerebral cortex*.

3. A motor neuron releases acetylcholine into the synapse with a muscle fiber, initiating an action potential that spreads into the center of the muscle cell along the transverse tubules. The action potential changes the permeability of the sarcoplasmic reticulum membrane, which releases Ca^{2+} into the cytosol. The calcium binds with troponin, moving tropomyosin and exposing the myosin-binding sites of the actin molecules. Heads of myosin molecules in their high-energy form bind to these sites, forming cross-bridges. The myosin head bends, pulling the thin filament toward the center of the sarcomere. When ATP binds to myosin, the cross-bridge breaks. Hydrolysis of the ATP returns the myosin to its high-energy form, and it binds farther along the actin molecule. This sequence continues as long as there is ATP and until calcium is pumped back into the sarcoplasmic reticulum, when the tropomyosin–troponin complex again blocks the actin sites.

ANSWERS TO TEST YOUR KNOWLEDGE

Multiple Choice:

1. c	4. e	7. c	10. d	13. e	16. d	19. b	22. b
2. d	5. c	8. c	11. c	14. d	17. b	20. a	23. a
3. b	6. a	9. b	12. c	15. a	18. d	21. c	24. e

CHAPTER 51: ANIMAL BEHAVIOR

INTERACTIVE QUESTIONS

51.1 Mammals are mostly nocturnal; birds are diurnal.

51.2
a. imprinting
b. associative learning (operant conditioning)
c. cognition
d. spatial learning
e. social learning

51.3 Researchers measured the average number of drops required to break whelk shells at various heights and calculated the average total effort required by computing the *total fight height* (average number of drops × height per drop). They measured the average fight height for crows in their whelk-breaking behavior (5.23 m) and found it was very close to the 5-meter height predicted based on an optimal compromise between energy gained versus energy expended.

51.4 If a male's parental care resulted in his greater reproductive success, then the genes for his behavior would increase in frequency in a population. If he were parenting offspring that included those of other males, then he would not pass his "parental care" genes on to the next generation in a higher proportion because he was also helping to perpetuate the "non-parental care" genes of other males.

51.5 Orange throats cannot successfully defend all their females from sneaky yellow throats, so yellow throats mate more often, and their frequency increases in the population. But blue throats restrict the access of yellow throats to their small number of females, and blue throat numbers will increase. The aggressive orange throats can take over blue throat territories, and once again, orange throat numbers will increase.

51.6 Raising animals in the lab enables researchers to search for evidence of genetic bases of behavior by controlling for environmental influences. In some cases, crossbreeding experiments provide evidence of genetic control of behavior.

51.7
a. a parent or a sibling
b. The individual shares more genes in common with a parent or a sibling. The coefficient of relatedness is 0.5, whereas it is only 0.125 for a cousin. Thus, with Hamilton's rule $rB > C$, the benefit to the recipient is discounted less by a larger value of r.

SUGGESTED ANSWERS TO STRUCTURE YOUR KNOWLEDGE

1. The scientific study of animal behavior seeks to identify those behaviors that are innate and genetically programmed as well as those that are a product of experience and learning. Fixed-action patterns are clearly developmentally fixed. In other cases, genetics may set the parameters for an organism's behavior; however, experience can modify behavior, and learning is clearly evident.

2. According to the concept of fitness, an animal's behavior should help to increase its chance of survival and production of viable offspring. Survival behaviors that are genetically programmed would be most likely to be passed on, and natural selection would refine innate behaviors for foraging, migrating, mating, and care of offspring. The evolution of cognitive ability would help individuals deal with novel situations. Social behaviors and communication may increase an individual's fitness. Parental investment and certainty of paternity may result in differences in mating systems and parental care that influence an individual's reproductive success. Altruistic behavior may be explained on the basis of kin selection; the inclusive fitness of an animal increases if its altruistic behavior benefits related animals who share many genes.

ANSWERS TO TEST YOUR KNOWLEDGE

Multiple Choice:

1. d	**3.** a	**5.** d	**7.** b	**9.** d	**11.** a	**13.** e	**15.** c
2. b	**4.** b	**6.** c	**8.** e	**10.** e	**12.** c	**14.** d	**16.** d

CHAPTER 52: AN INTRODUCTION TO ECOLOGY AND THE BIOSPHERE

INTERACTIVE QUESTIONS

52.1 a. South-facing slopes in the northern hemisphere receive more sunlight and are warmer and drier than north-facing slopes.
b. Air temperature drops with an increase in elevation, and high-altitude communities may be similar to communities in higher latitudes.
c. The windward side of a mountain range receives much more rainfall than the leeward side. The warm, moist air rising over the mountain releases moisture, and the drier, cooler air absorbs moisture as it descends the other side.

52.2 a. desert
b. temperate grassland
c. tropical forest
d. temperate broadleaf forest
e. northern coniferous forest
f. arctic and alpine tundra

52.3 a. $+--$
b. $-++$
c. $+--$
d. $-+-$
e. $-++$
f. $+++$

52.4 $\underline{7}$ abyssal $\underline{2}$ neritic
$\underline{6}$ aphotic $\underline{3}$ oceanic
$\underline{8}$ benthic $\underline{5}$ pelagic
$\underline{1}$ intertidal $\underline{4}$ photic

52.5 a. dispersal: an area may be beyond the dispersal ability of a species
b. behavior and habitat selection: insect larvae may be able to feed on more plants, but females oviposit on a single type of plant
c. biotic factors: the presence of predators, herbivores, parasites, mutualists, or competitors may restrict a species' range

d. abiotic factors: sunlight, water, oxygen, temperature, salinity, and soil characteristics may determine whether a species can inhabit an area

SUGGESTED ANSWERS TO STRUCTURE YOUR KNOWLEDGE

1. **a.** Ecology is the study of how organisms interact with their abiotic and biotic environments.
    b. The interactions of organisms with their environment can result in changes in the gene pool of a population, or evolution. Interactions occurring within an ecological time frame translate into adaptations to the environment that are evident on the scale of evolutionary time.

2. **a.** Biomes are characteristic ecosystem types—usually identified by the predominant vegetation for terrestrial biomes or the physical environment for aquatic biomes—which range over broad geographic areas.
    b. Convergent evolution, common adaptations of organisms of different evolutionary lineages to similar environments, accounts for similarities in life forms within geographically separated biomes.

ANSWERS TO TEST YOUR KNOWLEDGE

Multiple Choice:

1. b	**4.** b	**7.** e	**10.** c	**13.** d
2. d	**5.** a	**8.** e	**11.** d	**14.** e
3. b	**6.** d	**9.** c	**12.** c	**15.** b

Matching:

1. C	**3.** D	**5.** A	**7.** B
2. H	**4.** E	**6.** G	**8.** F

CHAPTER 53: POPULATION ECOLOGY

INTERACTIVE QUESTIONS

53.1 If 10 of the 30 voles in the second trapping were marked, she would estimate that 1/3 of the vole population is marked. Since 25 voles were initially marked, the total population size is estimated at 75 (25/0.333). Using the textbook formula, $N = sn/x$, or [(number marked and released × total second catch)/number of marked recaptures], also yields 75 [(25 × 30)/10].

53.2 **a.** Type I is typical of populations that produce relatively few offspring and provide parental care, such as humans and many large mammals.
b. Type II has a constant death rate over the organism's life span, such as Belding's ground squirrels and some other rodents, some annual plants, various invertebrates, and some lizards.
c. Type III is typical of populations that produce many offspring, most of which die off rapidly, such as many fishes and marine invertebrates, and long-lived plants.

53.3 The maximum per capita rate of increase possible for a species, when a population is in an unlimited environment and members can reproduce at their physiological capacity

53.4 **a.** exponential growth; $dN/dt = r_{max}N$
b. logistic growth; $dN/dt = r_{max}N(K - N)/K$; K is 1,500

53.5 **a.** An organism has limited resources to divide between growth, survival, and reproduction.
b. r-selected; K-selected

53.6 **a.** competition for resources, availability of territories, disease, accumulation of toxic wastes, predation, intrinsic limiting factors
b. extremes in weather, natural disasters, fires, sunspot activity (which then influences a plant's production of UV-protective or herbivore-deterring compounds)

53.7 A large proportion of individuals of reproductive age or younger, as seen in the bottom-heavy Afghanistan pyramid, may result in more rapid population growth now and in the near future. The small base of the Italy pyramid predicts a population decrease in Italy.

SUGGESTED ANSWERS TO STRUCTURE YOUR KNOWLEDGE

1.

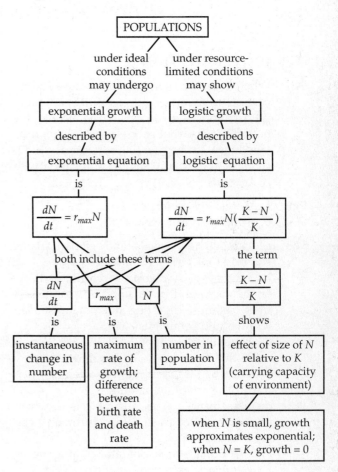

2. Reproductive success is measured in the number of offspring that survive and live to reproduce. Many "choices" are available in life history traits: big-bang versus repeated reproduction, age at first reproduction, number of reproductive episodes, number of offspring, parental investment in size of offspring or care. There are always trade-offs between reproduction and survival due to limited energy budgets. In addition, different environments (with varying degrees of stability and diverse biotic and abiotic factors) and different population densities (both influencing K- or r-selection) create different selection pressures on a population. Thus, there is no one "best" reproductive strategy.

ANSWERS TO TEST YOUR KNOWLEDGE

Multiple Choice:

1. a	**3.** b	**5.** c	**7.** d	**9.** d	**11.** e	**13.** a	**15.** b	**17.** e	**19.** d
2. c	**4.** e	**6.** c	**8.** a	**10.** d	**12.** b	**14.** a	**16.** c	**18.** e	**20.** c

CHAPTER 54: COMMUNITY ECOLOGY

INTERACTIVE QUESTIONS

54.1 Where these two similar spiny mouse species coexist, behavioral changes result in a temporal partitioning of resources. The niche of the naturally nocturnal *A. russatus* (as shown by laboratory studies and this removal study) changes to diurnal when in competition with *A. cahirinus*.

54.2 **a. Batesian mimicry**
b. Müllerian mimicry

54.3 **a.** competition: weeds and garden plants compete for nutrients and water
b. mutualistic symbiosis: mycorrhizae, flowering plants and pollinators, ants on acacia trees, cellulose-digesting microorganisms in termites and ruminants
c. commensal symbiosis: cattle egrets and cattle that flush insects (although cattle may benefit when ectoparasites are eaten or when the birds warn of predators)
d. parasitism: a type of symbiosis in which endo- and ectoparasites feed in or on host; tapeworms, ticks
e. predation: animal predators killing prey
f. herbivory: herbivores eating parts of plants
g. facilitation: black rush (*Juncus gerardi*) prevents salt buildup and oxygen depletion in salt marshes by shading soil and transporting oxygen to its roots

54.4 **a.** Pool 1: $H = -(0.4 \ln 0.4 + 0.3 \ln 0.3 + 0.3 \ln 0.3) = -(-0.37 + -0.36 + -0.36) = 1.09$
Pool 2: $H = -(-0.25 + -0.23 + -0.23 + -0.23) = 0.94$
b. Pool 2 has more species and thus a greater species richness. Pool 1 has the higher diversity index.

54.5 energetic hypothesis

54.6 The $+/-$ cascade that would be needed to end with a decrease in algae would require an increase in zooplankton, which would require a decrease in primary predators caused by an increase in top predators. More top predators could be added to the lake, or primary predators could be removed.

54.7 **a.** Soil nitrogen levels begin quite low but rise due to the symbiotic nitrogen-fixing bacteria associated with alder.
b. The soil pH changes from about 7.0 to 4.0 as the acidic spruce leaves decompose.

54.8 More interspecific interactions would have had time to develop, and this longer span of evolutionary time would allow for more speciation events to have occurred.

54.9 **a.** immigration—large island
b. immigration—small island
c. extinction—small island
d. extinction—large island

The small island equilibrium number of species is projected down from the intersection of lines b and c. The large island equilibrium number is higher and is projected down from the intersection of lines a and d. The larger island is projected to have a larger equilibrium number of species.

54.10 They are testing for the presence of the H5N1 strain of avian flu virus in migrating waterfowl to monitor the disease's possible entry into North America.

SUGGESTED ANSWERS TO STRUCTURE YOUR KNOWLEDGE

1. **a.** competition $(-/-)$
b. resource partitioning and character displacement
c. slightly different niches
d. keystone predator
e. herbivory
f. mutualism $(+/+)$
g. facilitation
h. food chain or food web

2. **a.** No two species with the same niche can coexist permanently in a habitat; the more

competitive species will cause the local elimination of the other.

b. Food chains are limited to a few links because of the inefficiency of energy transfer (only about 10%) from one trophic level to the next.

c. Food chains are more stable with fewer links because the effect of environmental disruptions becomes magnified in higher trophic levels.

d. Each trophic level controls the next higher level; adding nutrients will increase the biomass of all other trophic levels.

e. Community organization is controlled from the top by predation, with a cascade of +/− effects down the trophic levels with changes in predator numbers.

f. Most communities are constantly changing in composition due to the effects of disturbances.

g. The rates of immigration and extinction on "islands" of habitat are affected by the size of the island, the closeness to the "mainland," and the number of species currently on the island. When these rates are equal, an equilibrium number of species is reached.

ANSWERS TO TEST YOUR KNOWLEDGE

Multiple Choice:

1. d	**5.** e	**9.** d	**13.** d	**17.** b
2. c	**6.** b	**10.** c	**14.** a	**18.** c
3. d	**7.** b	**11.** d	**15.** e	
4. b	**8.** e	**12.** d	**16.** a	

CHAPTER 55: ECOSYSTEMS AND RESTORATION ECOLOGY

INTERACTIVE QUESTIONS

55.1 The element may limit production in that ecosystem.

55.2 Whereas energy makes a one-way trip through ecosystems, chemical elements move through the trophic levels and decomposers and are recycled back to producers.

55.3 **a.** tropical rain forest, coral reef, swamp and marsh, estuary
b. desert, tundra, lake and stream, open ocean
c. The open ocean covers 65% of Earth's surface area.
d. Upwellings in these cold seas bring nitrogen and phosphorus to the surface. The lack of these nutrients limits production in many tropical waters.

55.4 **a.** Much of a bird's or a mammal's assimilated energy is used to maintain a warm body temperature and is not available for net secondary production (growth and reproduction).
b. $^1/1000$ (10% of 10% of 10%) or 0.1%

55.5 Water is essential to all organisms, and its availability in terrestrial ecosystems influences primary production and decomposition. Carbon forms the backbone for all organic molecules, which are essential to all organisms. Nitrogen is a component of amino acids and nucleic acids. Phosphorus is found in nucleic acids, phospholipids, and ATP, and as a mineral in bones and teeth.

55.6 **a.** tropical rain forests, because decomposition proceeds rapidly in the warm, wet climate, and nutrients are rapidly assimilated by new growth
b. lakes and oceans; lack of oxygen slows decomposition and, unless there are upwellings, nutrients in sediments are not available to producers. Decomposition is also very slow in cold, wet peatlands.
c. Nutrients are not recycled and may leave the ecosystem through runoff.

55.7 Prokaryotes that can metabolize uranium are being used to treat contaminated groundwater. Plants that can fix nitrogen are often used in biological augmentation to enrich nutrient-poor soils and facilitate the recolonization of native species.

SUGGESTED ANSWERS TO STRUCTURE YOUR KNOWLEDGE

1.

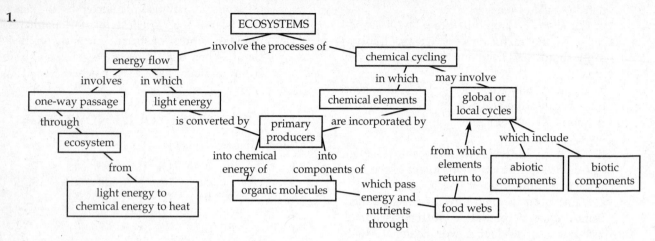

2. Primary production is limited in aquatic ecosystems by the penetration of light and the availability of nutrients, especially nitrogen and phosphorus. In terrestrial ecosystems, primary production is influenced by temperature and precipitation (reflected as actual evapotranspiration), as well as by nutrients such as nitrogen and phosphorus.

3. Photosynthetic and chemosynthetic organisms assimilate inorganic elements and compounds and incorporate them into organic compounds. Animals obtain their organic compounds by eating producers or other consumers. Elements are returned to inorganic form by the processes of respiration, excretion, and decomposition.

ANSWERS TO TEST YOUR KNOWLEDGE

Multiple Choice:

1. e	4. e	7. d	10. c	13. c
2. c	5. d	8. b	11. a	14. e
3. c	6. e	9. c	12. e	15. c

CHAPTER 56: CONSERVATION BIOLOGY AND GLOBAL CHANGE

INTERACTIVE QUESTIONS

56.1 Many examples are provided in the text.
a. About 93% of coral reefs have been damaged by human activities; 40–50% of the reefs could be lost in the next few decades. About a third of all marine fish species utilize these reefs.
b. The introduction of the brown tree snake to Guam resulted in the extinction of multiple species of birds and lizards.
c. Commercial harvest and/or illegal hunting have reduced populations of whales, the African elephant, and many fishes.
d. It may take decades for aquatic ecosystems to recover from the damage done by acid precipitation, and damage continues to forests in central and eastern Europe.

56.2 Smaller, because usually not all individuals in a population successfully breed. Conservation programs attempt to sustain the effective population size above MVP in order to retain genetic diversity in a population.

56.3 The loss of genetic variation within a small population due to inbreeding and genetic drift can force it into an extinction vortex in which the population grows smaller and smaller. Promoting migration between small populations or introducing individuals from other populations to increase genetic variation is proposed as an urgent conservation need.

56.4 This proactive approach relies on early detection of population decline, identification of the species' habitat needs, testing to determine which factor is contributing to the decline, recommending corrective measures, and monitoring results.

56.5 Corridors promote dispersal between populations and may be essential to species that migrate between different habitats. However, they may also contribute to the spread of diseases.

56.6 Large reserves are required for large, far-ranging animals that require extensive habitats. They also have proportionally less border area and thus have fewer edge effects. An advantage of smaller reserves that collectively have the same area as a large one is the slower spread of disease within a population.

56.7 Global warming may be slowed by agreements on an international strategy to reduce CO_2 emissions through increased energy efficiency, use of renewable solar and wind power, use of nuclear power, changes in industrial processes and personal lifestyles, and reductions in deforestation.

SUGGESTED ANSWERS TO STRUCTURE YOUR KNOWLEDGE

1. habitat destruction, introduced species, overharvesting of wild species, and global change

2. Biodiversity is a natural resource from which we obtain medicines, crops, fibers, and other products; many potentially valuable species will become extinct before they are known to scientists. A loss of biodiversity may disrupt ecosystem processes in harmful ways. Humans have evolved within the context of living communities and may be affected in unknown ways by changes in our ecosystem.

3. Fragmentation of habitats produces more interfaces or *edges* between different ecosystems (between forests and cleared areas, between deserts and housing developments). The species that inhabit edges are able to use both types of ecosystems. As edges proliferate, edge-adapted species may become more dominant than the species in the adjoining habitats. Strips of quality habitat that connect fragmented habitat patches may serve as *movement corridors* that promote dispersal between isolated populations and help maintain genetic variation.

ANSWERS TO TEST YOUR KNOWLEDGE

Multiple Choice:

1. e	**4.** b	**7.** e	**10.** d	**13.** d
2. b	**5.** d	**8.** a	**11.** b	**14.** c
3. c	**6.** a	**9.** c	**12.** b